2nd International Conference on Automotive Innovation and Green Vehicle (AiGEV 2016)

MATEC Web of Conferences Volume 90 (2017)

Cyberjaya, Malaysia
2 - 3 August 2016

Editors:

S.A. Che Ghani
W.A. Wan Hamzah
A. Alias

ISBN: 978-1-5108-3507-8

Printed from e-media with permission by:

Curran Associates, Inc.
57 Morehouse Lane
Red Hook, NY 12571

Some format issues inherent in the e-media version may also appear in this print version.

Printed by Curran Associates, Inc. (2017)

For additional information, please contact EDP Sciences – Web of Conferences at the address below.

EDP Sciences – Web of Conferences
17, Avenue du Hoggar
Parc d'Activité de Courtabœuf
BP 112
F-91944 Les Ulis Cedex A
France

Phone: +33 (0) 1 69 18 75 75
Fax: +33 (0) 1 69 28 84 91

contact-edps@webofconferences.org

Additional copies of this publication are available from:

Curran Associates, Inc.
57 Morehouse Lane
Red Hook, NY 12571 USA
Phone: 845-758-0400
Fax: 845-758-2633
Email: curran@proceedings.com
Web: www.proceedings.com

TABLE OF CONTENTS

MATEC Web of Conferences
Open access conference proceedings in Materials science, Engineering and Chemistry
A flexible, fast and efficient publication service, for an open access recording of your event and a wide dissemination of your proceedings
www.matec-conferences.org
edp open
Open Access Publishing from EDP Sciences

ISSN (electronic edition): 2261-236X

THE 2ND NTERNATIONAL CONFERENCE ON AUTOMOTIVE INNOVATION AND GREEN ENERGY VEHICLE (AIGEV 2016)

Automotive Engineering research has been playing a vital role and has become indispensable for the growth of any nation. In this context, research activities are being carried out at the extremes and many problems are yet to be solved. Thus, the pursuit of automotive studies continue to remain the most exciting and challenging areas in furthering the frontiers of knowledge.

As expanding the future research in automotive area, AEC has been promoting and organizing the The 2nd International Conference On Automotive Innovation And Green Energy Vehicle (AiGEV 2016) was held in Malaysia Automotive Institute (MAI) Cyberjaya, Selangor on August 3 – 4, 2016. 1st International Conference On Automotive Innovation And Green Energy Vehicle (AiGEV 2014) was held in Swiss Garden Resort & Spa, Kuantan Pahang, Malaysia on August 26 - 27, 2014. AiGEV 2016 provides a venue for researchers and practitioners in related fields from academia, industry and government to meet in a forum where the latest research results are presented and prospect for future developments are discussed.

The conference was inaugurated by Chief Executive Officer of Malaysia Automotive Institute (MAI), Datuk Madani Sahari and attended by UMP Deputy Vice Chancellor (Academic & International), while the President of Malaysia Automotive Association (MAA) Datuk Aishah Ahmad, Vice Chancellor of DRB-HICOM University Of Automotive Malaysia (DHU), Professor Dr. Masjuki Haji Hassan (University Malaya) and Professor Dr. Abd Rashid Abd Aziz (Universiti Teknologi Petronas) were also there to deliver keynote speeches during the conference,

The AiGEV 2016 paper research included the topics above:

- Advanced Manufacturing in Automotive Industries
- Alternative Energy & Fuel
- Automotive Design
- Automotive Mechatronics
- Computational & Simulation Methods in Automotive Applications
- Fuel & Combustion
- Green Energy Technology
- Hybrid & Electric Vehicle
- Materials Characterisation
- Mathematical Modelling & Optimisation in Automotive Applications
- Noise, Vibration & Harshness
- Safety, Comfort & Ergonomics
- Signal Processing & Automotive Electronics
- Standards, Policies & Regulations
- Traffic & Transport
- Vehicle Dynamics & Control
- Other Related Topics in Automotive Engineering

All accepted papers of AiGEV 2016 will be published in one of the journal such as International Journal of Automotive & Mechanical Engineering (IJAME), Journal of Mechanical Engineering and Sciences (JMES) and MATEC Web Of Conference.This conference successfully attracted 115 local and international participants and also industries presenting their research paper works.

AEC hopes with the success of the first conference, many more conferences will be successfully organised in the future. The organizing committee of the conference thanks all the participants for their fruitful work and personal contribution to the development of these conference proceedings.

Associate Professor Dr. Abdul Adam Abdullah

Chairman of the AiGEV 2016

AiGEV2016 would like to thank the following reviewers for their technical knowledge and time in expertly reviewing the proceeding papers.

1 PROF MADYA DR ABDUL ADAM BIN ABDULLAH

2 DR MOHD SHAHRIR BIN MOHD SANI

3 DR DANDI BACHTIAR

4 DR GIGIH PRIYANDOKO

5 DR FTWI YOHANESS HAGOS

6 DR JANUAR PARLAUNGAN SIREGAR

7 DR WAN AZMI BIN WAN HAMZAH

8 DR SAIFUL ANWAR BIN CHE GHANI

9 DR MOHD ADNIN BIN HAMIDI

10 DR AZRI HIZAMI BIN RASID

11 DR MOHAMAD HEERWAN BIN PEEIE

12 DR AZRI BIN ALIAS

13 DR AHMED NURYE OUMER

14 DR UMMU KULTHUM BINTI JAMALUDDIN

15 DR FATIMAH BINTI DZAHARUDIN

16 DR SITI RABIATUL AISHA BINTI IDRIS

19 DR WAN SHARUZI WAN HARUN

20 DR NURUL AKMAL BINTI CHE LAH

21 DR. IRNIE AZLIN@NUR AQILAH ZAKARIA

Statement of Peer review

In submitting conference proceedings to MATEC Web of Conferences, I certify to the Publisher that I adhere to the **Policy on Publishing Integrity** of the journal in order to safeguard good scientific practice in publishing.

1. All articles have been subjected to peer review administered by the proceedings editors.
2. Reviews have been conducted by expert referees, who have been requested to provide unbiased and constructive comments aimed, whenever possible, at improving the work.
3. Proceedings editors have taken all reasonable steps to ensure the quality of the materials they publish and their decision to accept or reject a paper for publication has been based only on the merits of the work and the relevance to the journal.

The 2nd International Conference on Automotive Innovation and Green Vehicle (AiGEV2016), 2-3 Aug. 2016, MAI Cyberjaya, Malaysia

Title, date and place of the conference

Dr. Saiful Anwar Che Ghani
Dr. Wan Azmi Wan Hamzah
Dr. Azri Alias

Proceedings editor(s):

28.12.2016

Saiful Anwar Che Ghani

Date and editor's signature

MATEC Web of Conferences **90**, 01001 (2017) DOI: 10.1051/matecconf/20179001001
AiGEV 2016

Development of battery management systems (BMS) for electric vehicles (EVs) in Malaysia

P.M.W. Salehen[1], *M.S.* Su'ait[1], *H.* Razali[1,*], and *K.* Sopian[1,2]

[1]Solar Energy Research Institute (SERI), Universiti Kebangsaan Malaysia, 43600 Bangi, Selangor, Malaysia
[2]Department of Mechanical Engineering, Faculty of Engineering and Built Environment, Universiti Kebangsaan Malaysia, 43600 Bangi, Selangor, Malaysia

Abstract. Battery Management Systems (BMS) is an electronic devices component, which is a vital fundamental device connected between the charger and the battery of the hybrid or electric vehicle (EV) systems. Thus, BMS significantly enable for safety protection and reliable battery management by performing of monitoring charge control, state evaluation, reporting the data and functionalities cell balancing. To date, 97.1% of Malaysian CO_2 emissions are mainly caused by transportation activities and the numbers will keep rising as numbers of registered car increase close up to 1 million yearly; double the amounts in the last two decades. The uncertainty of a battery's performance poses a challenge to predict the extended range of EVs, which need BMS implementation of optimization of optimum power management. Hence, using MATLAB/SIMULINK software is one of the potential methods of BMS optimization with power generated by Hybrid Energy Storage system of lithium-ion battery. Therefore, this paper address through reviewing previous literatures initially focuses on the BMS optimization for EVs (car) in Malaysia as prognostic technology model improvement on performance management of EVs.

1 Introduction

Battery Management Systems (BMS) is an electronic devices component, connected between the charger and the battery of the hybrid or electric vehicle (EV) systems. BMS can be consider as the brains behind of battery packs or as vital fundamental components of the battery base electrical vehicle. There are two mains critical functions of BMS which are function as safety protection (the most single important) and energy management. Thus, BMS are able to manage the output, command the charge-discharge input and provide notification on the status of the battery packs. They also provide critical safeguards to protect the battery from damage. While for energy management, BMS be able to monitor the state of battery and evaluation such as State of Charge (SOC) and State of Health (SOH) for the purpose of driving range estimation by shown with BMS indicator [1]. This energy management concept is just about the fuel gauge meter that we have in conventional gasoline car or internal combustion engine (ICE).

* Corresponding author: whitesuchy@gmail.com

The safety and reliability not only subject to the technology of the battery but also the battery management system. For that reason, there is a crucial needs to invent on optimizing vehicle operation and improving battery performance in a safe and reliable manner for a BMS, due to on chemistry history, a lithium-ion battery is possible to explode if overcharged because of instability, entropy changes and flammability. Therefore, BMS functions could be comprehensively determined as for safety protection, data achievement, transport of battery status and authentication to a user interface, ability to control battery charging and discharging, ability to determine and predict the state of the battery, communication with all battery components, thermal management, cell balancing and prolonged battery life. There is an imperative needs to develop a mature and comprehensive BMS for future of overall due to the swift growth global market of the EVs and HEVs [1].

Thus, this present paper is review on BMS, focussing study for optimization of BMS for EVs that will lead to reliability of BMS and optimize power performance of EVs. Analysis from the literature search, there is a need on the improvement of BMS performance in future for EVs. Therefore, Hybrid Energy Storage System (HESS) is the potential outlook of the future BMS which should integrate of ultracapacitor combined with lithium compound battery in order to have prognostic technical model improvement on deploying performance management of EVs. Therefore, there is a very significant for Malaysia to have an expert on the development of BMS due to find a prognostic technology model improvement on performance management of EVs. Since as we know that hybrid car only can deliver less emission compare to EVs can deliver 'zero emission' nor pollution, energy saving and environmental friendly.

2 Status electric vehicles (EV) in Malaysia

Malaysia commencement with green technology initiative on year 2009 when Malaysia Government has committed to reduce greenhouse gases by 40% of year 2005 levels by the year 2020 right after UN Climate Change Summit, Copenhagen (COPs 15) in Denmark. Subsequently, the Malaysian Green Technology Corporation (MGTC) and the Malaysia Automotive Institute (MAI) established as green authority agencies spearheaded with the aims to fast-track the implementation of the Electric Vehicles (EV) policy and regulations for public and private transportation, targeting for 2,000 electric buses and 100,000 electric cars on the road by 2020. On top of that, to emphasis for the development and promotion of green vehicle the Prime Minister of Malaysia was launched the National Automotive Policy (NAP) and the National Green Technology Policy on 24th July 2009. One of the main point stated inside of National Green Technology policy is "Transportation is one of the main emphasized sector that contributed as the 2nd highest for carbon emission. While NAP stated "Promote Hybrid and Electric Vehicles and Development of related infrastructure [2][3].

Malaysia already committed to reduce up to 40% greenhouse gas (GHG) emissions by the year 2020. The study also found that a vast portion of national GHG emissions in Malaysia was coming from the transport sector - in technology journal about Assessment of Greenhouse Gas Emission Reduction Measures in Transportation Sector of Malaysia [4]. To date, 97.1% of Malaysian CO_2 emissions are mainly caused by transportation activities and the numbers will keep rising as numbers of registered car increase closed up to 1 million yearly; double the amounts in the last two decades [5].

PROTON Holding Berhad as the leading car manufacturer in Malaysia already done a pilot project EV called as Fleet Test Vehicle (FTV) with a numbers of 200 REEV Exora with 50 EV Saga around city of Putrajaya and Cyberjaya. The driving range about 140km and 130km respectively with normal mode charging overnight 8 hours. This programme is under phase 2 pilot project of Roadmap for the Deployment of EV for Malaysia, trial for 12 month started September 2011 until end of September 2012 as shown in Figure 1 [6].

Fig. 1. Pilot Project for PROTON FTV location around Putrajaya and Cyberjaya [6].

Proton supposed to meet the projected mass production by 2014 as National roll out (phase 4) but decided to extend the projection by 2017 due for certain reasons. Other project is CMS Consortium private/public partnership between Malaysia Automotive Institute (MAI) and Malaysian Green Technology Corporation (MGTC) roll with first EV car sharing programme, called Cohesive Mobility Solution (COMOS), and was officially launched during 2014 as shown in Figure 2. This initiative, is the first kind in the ASEAN region. COMOS programme only available in hot spot prime city of Peninsula Malaysia, with the total of 30 -40 EV as for public charging stations. Eventually, the aim is have 3,500 EVs in the COMOS loop nationwide by 2020. COMOS also having plans to have EV charging stations at every R&R along the North-South highway by the fifth year of the programme. Average driving range for COMOS vehicle programme is 144km with a single charge 30 minutes to 9 hours. The objective of COMOS programme as a part of awareness campaign for the public about the EVs and know how to operate and get use to EV with seeing is believing. Then eventually build up the interest of the public to own the vehicles in substitution of conventional ICEs [7].

Recently, there is a project undertaken by China's Beijing Auto International Cooperation (BAIC), the country's top and world's second biggest electric vehicle manufacturer, in collaboration with a Malaysian partner, Amber Dual Sdn. Bhd. with targeted EVs production by July 2016 and aspire to lead Malaysia's for EV technology [8]. Even electric cars status in Malaysia is still new and on growing development but it is decent to see that Malaysia is on effort taking measures to ensure that the future of the electric car will be a promising one. As the impact of electric cars are more energy efficient and definitely have cleaner emissions than a HEV and better on conservation environment compare to conventional ICEs, which EVs will be the best option for environmentally-friendly in the future endeavour. Therefore, this BMS's study is very significant that will provides future insight for knowledge contribution and innovation invention for EV industries towards year 2020 as the first time ever in Malaysia. As well as for training in public and private institution to design on the optimization of BMS for EVs in Malaysia.

Fig. 2. COMOS programme [7].

3 Battery and battery management system

The switch from conventional ICEs to EVs is well underway. However, limited mileage of current EVs due to the confined energy storage capability of available battery systems is a major reason why these vehicles are not more common on the road [9]. Research revealed that with fast charging ten-minute period, be able to move the vehicles for one hundred mile and ability to provide enough energy in battery packs of EV. Thus, with the aim of to lengthen the batteries lifespan it is important of the BMS in the EVs playing the vital roles as a component device to control the operation of the battery's [10]. Also very important for BMS to well-maintained the battery reliability and safety, ensure the state monitoring and evaluation, cell balancing and charge control are well functional. A battery chemistry as an electrochemical devices, also will acts differently under different operational and environmental conditions. Due to that uncertainty of a battery's performance poses a challenge to the implementation of these functions [1].

Lithium-ion battery (LIB) has been extensively determined as high technology applied and developed since past a go decade when the BMSs development for EVs arisen globally, due to higher energy density, lower self-discharge, and longer life cycle. Figure 3 is adapted by Rohan which represented Commercial Evolution of Rechargeable Batteries to Higher Density.

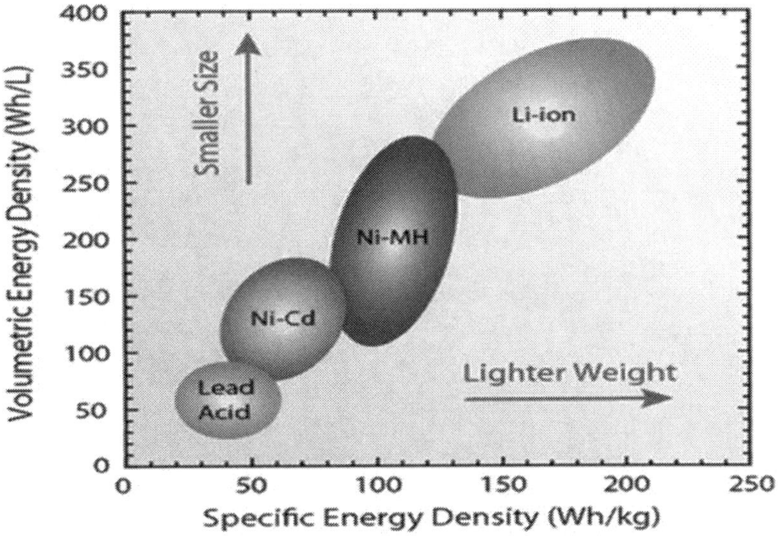

Fig. 3. Commercial evolution of rechargeable batteries to higher density [11].

LIB chemistry also well acknowledged as the technology of choice for energy storage in EVs towards sustainable mobility. Nevertheless, there are still needs of more research points and still open for enhancement in the future endeavours. These included the best choices of the cell materials and the development of electronic circuits and algorithms for a more effective battery utilization for EVs [12]. The battery parameter to consider for BMS optimization were voltage, current, and temperature can be applied to forecast the SOC of battery [13-16]. Another challenges imposed of the performance of vehicle related to BMS is thermal management system would be required to maintain the optimal cell performance and also to achieve a full battery lifespan [14,17].

4 A proposed BMS optimization

The needs of more power will lead to have a large voltage supply that's very difficult to arrange the battery power. The development of battery technology by using the ultra-capacitor (UCs) in automotive solved this problem[18]. This development combined form of battery and ultra-capacitor as hybrid energy or can called as Hybrid Energy Storage Systems (HESS) as shown in Figure 4. UCs have a high power density, but a lower energy density compared to batteries. The combination inherently offers a better performance in comparison to the use of them alone of either one. Figure 5 is sketches which represented proposed HESS configuration. The concept of optimization using the MATLAB/SIMULINK shown as in Figure 6.

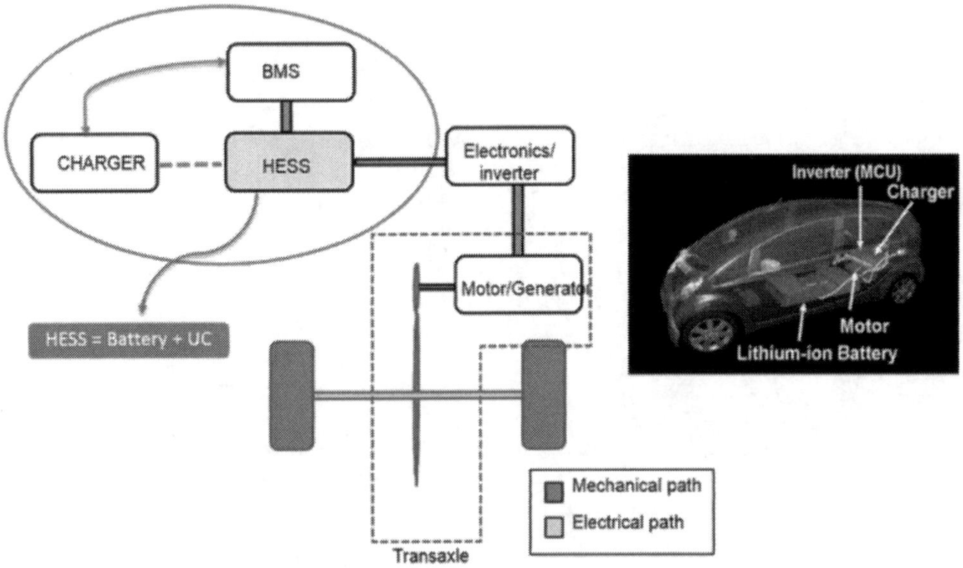

Fig. 4. Proposed EV schematic diagram.

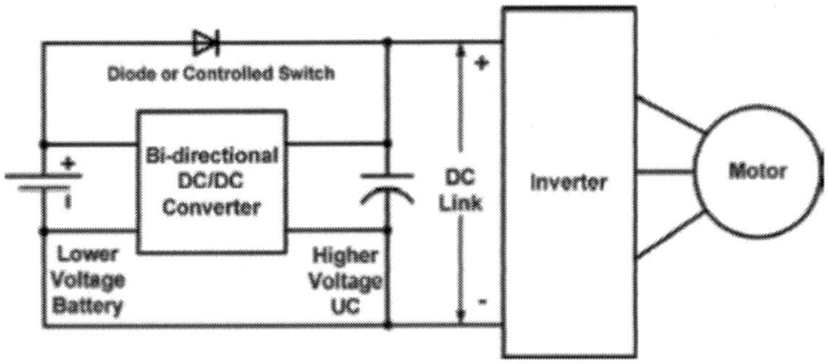

Fig. 5. Proposed HESS configuration [18].

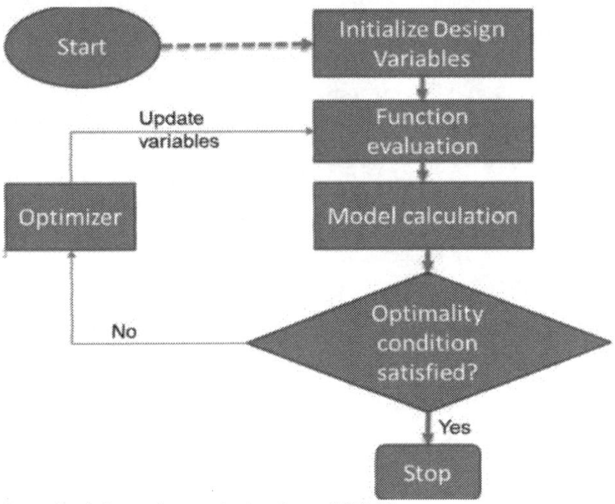

Fig. 6. Experimental methodology for optimization of BMS.

So that in this study, this paper initially focus on the optimization of power generated by lithium-ion cathode with graphene anode using MATLAB/SIMULINK software in order to obtained experimental performance data (charge-discharge, power capacity and life cycles). Therefore, based on the reviewed analysis, this report would open the horizon of new method of BMS optimization to satisfy the objective on the BMS prognostic model in order to have a management performance and driving range extended of the electric car. Thus, increasing the performances of EV will lead to reliability of Battery Management and optimize power performance of EV. This will become essential in the literature especially in Malaysia. Thus, as the first time for the development of the BMS as well as for battery electric vehicles in Malaysia.

5 Conclusions

It is very important for BMS to well-maintained the battery reliability and safety, the state monitoring and evaluation, cell balancing and charge control are well functional. Thus, this present paper is review on BMS, focussing study for optimization of BMS for EVs that will lead to reliability of BMS and optimize power performance of EVs. Analysis from the literature search, there is a need on the improvement of BMS performance in future for EVs. Therefore, Hybrid Energy Storage System (HESS) is the potential outlook of the future BMS which should integrate of ultracapacitor combined with lithium compound battery in order to have prognostic technical model improvement on deploying performance management of EVs. It is expected that to provide future innovation invention to the EV industries towards year 2020 or beyond. Moreover, the target reducing greenhouse gas (GHG) emissions by up to 40% towards year 2020 will be achievable. Nevertheless, due to varying situation, there is no standard solution setting for BMS performance and different strategies still need to be applies to improve the performance of BMS for future HEVs and EVs.

The authors would like to extend their gratitude towards Universiti Kebangsaan Malaysia (UKM) for allowing this research to be carried out. This work is supported by the Research University Grant, TD-2015-08.

References

1. Y. Xing, E. W. M. Ma, K. L. Tsui, M. Pecht, "Battery management systems in electric and hybrid vehicles," *Energies*, **4**, no. 11. pp. 1840–1857, 2011.
2. Malaysia Automotive Institute (MAI), "National Automotive Policy (NAP) 2014," 2014. [Online]. Available: http//www.mai.org.my.
3. Malaysia Automotive Institute (MAI), "Transportation is one of the main emphasized sector as it is the 2nd. Highest contributor in term of carbon emission.," 2014. [Online]. Available: http//www.mai.org.my.
4. S. Shamsudin, & Minhans, A., O. C. Puan, "Assessment of Greenhouse Gas Emission Reduction Measures in Transportation Sector of Malaysia," *J. Teknol.*, **4**, no. 70, pp. 1–8, 2014.
5. N. Yahya, N. Ghazali, F. Ishak, "A review of carbon dioxide in Malaysia : Current status and Challenges," 2013, pp. 8–10.
6. G. T. & W. Ministry of Energy, "Electric Vehicle Roadmap For Malaysia," 2011. [Online]. Available: http://www.kettha.gov.my. [Accessed: 05-May-2015].
7. Anthony Lim, "COMOS EV Car-Sharing programme to launch in August 2014 - rent a Twizy, Zoe of Leaf in KL, by hour," *News, Paul Tan's Automotive*, 2014. [Online]. Available: http://paultan.org/2014/05/07/comos-ev-rental-programme/. [Accessed: 07-Nov-2015].

8. D. Tan, "Malaysia's EV production to start in July 2016," *News, Paul Tan's Automotive*, 2015. [Online]. Available: http://paultan.org/2015/05/18/baic-to-start-making-evs-in-malaysia-from-july-2016/. [Accessed: 03-Jan-2016].
9. C. Botsford, A. Szczepanek, "Fast Charging vs . Slow Charging : Pros and cons for the New Age of Electric Vehicles," in *EVS24*, 2009, no. May, pp. 1–9.
10. R. Adany, D. Aurbach, S. Kraus, "Switching algorithms for extending battery life in Electric Vehicles," *J. Power Sources*, vol. 231, pp. 50–59, 2013.
11. J. F. Rohan, M. Hasan, S. Patil, D. P. Casey, T. Clancy, *Energy Storage : Battery Materials and Architectures at the Nanoscale*. 2014.
12. M. Brandl, H. Gall, M. Wenger, V. Lorentz, M. Giegerich, F. Baronti, G. Fantechi, L. Fanucci, R. Roncella, R. Saletti, S. Saponara, A. Thaler, M. Cifrain, W. Prochazka, "Batteries and battery management systems for electric vehicles," *Des. Autom. Test Eur. Conf. Exhib. (DATE), 2012*, no. September 2015, pp. 971–976, 2012.
13. K. W. E. Cheng, B. P. Divakar, H. Wu, K. Ding, H. F. Ho, "Battery-management system (BMS) and SOC development for electrical vehicles," *IEEE Trans. Veh. Technol.*, vol. 60, no. 1, pp. 76–88, 2011.
14. L. Lu, X. Han, J. Li, J. Hua, M. Ouyang, "A review on the key issues for lithium-ion battery management in electric vehicles," *Journal of Power Sources*, vol. 226. pp. 272–288, 2013.
15. Y. Wang, Y. Liu, "Electronic control system design and test of pure electric vehicle battery management system," *Mechanic Automation and Control Engineering (MACE), 2011 Second International Conference on*. pp. 1289–1292, 2011.
16. A. Vezzini, *Laithium-Ion Battery Mnagement*. Elsevier, 2014.
17. S. Chacko, Y. M. Chung, "Thermal modelling of Li-ion polymer battery for electric vehicle drive cycles," *J. Power Sources*, vol. 213, pp. 296–303, 2012.
18. M. Gopikrishnan, "Battery/ultra capacitor hybrid energy storage system for electric, hybrid and plug-in hybrid electric vehicles," *Middle - East J. Sci. Res.*, vol. 20, no. 9, pp. 1122–1126, 2014.

Effect of annealing, thickness ratio and bend angle on springback of AA6061-T6 with non-uniform thickness section

M.F. Adnan[1], *A.B.* Abdullah[1,*], and *Zahurin* Samad[1]

[1]School of Mechanical Engineering, Universiti Sains Malaysia, 14300 Nibong Tebal, Pulau Pinang, Malaysia.

Abstract. Non-uniform thickness section section is considered one of the most effective approaches to reduce automotive part weight. Reduction in term of mass and size result in less fuel consumption and greenhouse gases. Thickness is the most significant parameter to formability, therefore forming a section with non-uniform thickness becomes a great challenge. Improper process and incorrect decision may lead to severe defect and one of the main concerns is the springback. This study will focus on springback behaviour of non-uniform thickness AA6061 strip with complex profile using Taguchi Method. Profile projector (PC 3000) is used to measure the spring back and two-line technique is applied to measure angles (after loading) between two lines. Three parameters (i.e. annealing temperature, thickness ratio and bend angle) are studied, and results determine that the most significant parameter is bend angle, followed by thickness ratio, and then by the annealing temperature of the specimen during bending process.

1 Introduction

The application of lightweight construction is a one of the central challenge in modern transportation engineering. Optimization of lightweight construction can be achieve if material is used only in the areas where stresses appear and if the material used is charged near yield stress [1]. Lightweight components can be classified into two types. Firstly, lightweight construction deals with the application of light materials. The use of light materials keep remain the workpiece geometry but reducing the weight of the whole component. The other one, lightweight construction which deals with different design strategies including the process of forming the component [2]. Formability of the material depends on many factors including thickness, properties of the material and complexity of the formed part. These factors may initiate defect such as springback lead to higher part rejection.

In the manufacturing sector, there are several important things that need to be considered especially safety of passengers in reducing the vehicle weight. Thus, lightweight and crashworthiness are two crucial aspects of the automobile design.

* Corresponding author: mebaha@usm.my

DOI: 10.1051/matecconf/20179001002

For example, B-pillar as shown in Figure 1 that has been designed with various thickness mainly to reduce the weight. With non-uniform thickness, the springback behavior becomes more complex and difficult to be predicted.

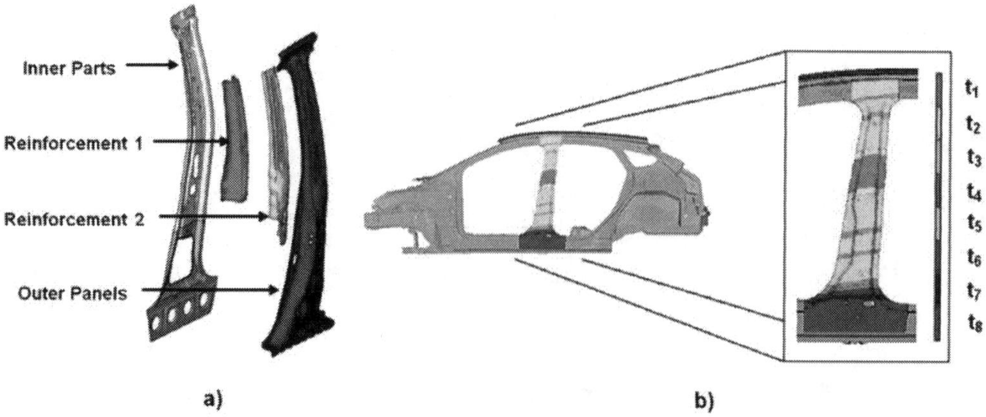

Fig. 1. Illustration of B-Pillar component; a) B-Pillar parts, b) Partition of B-Pillar with various thickness [3].

There are various steps taken to compensate the springback. Some researchers had study the effect of the process parameter to springback behaviour. Nasrollahi & Arezoo [4] found that springback phenomenon can be separated into two categories. First are factors which are more related to material issue such as Young's modulus, yield stress, strength coefficient, strain hardening, Poisson's ratio and anisotropic coefficient. Second factor are factors which are related to the bending process like tooling geometry, clearance, pad force and etc. There are various studies dealing with these parameters and their effect on springback. According to Thipprakmas [5], springback is highly affected by material thickness. While in spring-go, the bending angle has the major influence and closely followed by the material thickness. Thin material provides lighter structure but unable to prevent the presence of the springback. Deng et al. and Vasudevan et al. [6], [7] have investigated the influence of various parameters on springback for sheet metals during air v-bending. They concluded that most parameters affecting springback are identified thickness, orientation, punch travel (stroke), punch radius, die opening and punch velocities. Nandedkar [8] concluded that sheet thickness plays a very important parameter affecting springback and residual stresses. Han et al. [9] found that the springback will reduce as the forming temperatures range between 100°C to 200°C. In other study by Chikalthankar et al. and Ján et al. [10], [11], found that the greater bend depth causes higher bending angle and the higher bending result in higher amount of springback.

Traditional experimental design procedures are complicated, time consuming, and costly. Traditional trial-and-error method fails to provide an overall planning for the parameters. In general, a large number of experimental works must be conducted when the number of process parameters increases. The Taguchi method is used as a tool in designing experiments to avoid the need to conduct numerous experiments [12]. Here, Taguchi method are very useful to study the effect of many factors (variables) on the desired quality characteristic economically.

There have been consider a lot of studies on springback with uniform section, but is still lacking that involved thickness changes (non-uniform section). By knowing the effect of the thickness,variation of the thickness and other process parameter to the springback behaviour, the springback may be predicted and controlled. This paper focus on the optimization of the certain parameters to the springback pattern via v-bending method using Taguchi method.

2 Methodology

2.1 Selection of the springback parameters and their levels

There are 6 sets of specimen with different thickness ratio as listed in Table 1. Here, thickness ratio can be defined as a ratio of minimum thickness to the maximum thickness. The feasible space for the experiment parameters was defined by varying the thickness range in the range of 2.60 to 3.00 mm, the alignment of the puncher in the range 7mm to 9mm, punch stroke in the range 3 m to 7 mm, annealing temperature in the range 24 °C (RT) to 300 °C and rolling direction in the range of 0° to 90°. The other two parameters were selected as shown in Table 1.

Table 1. Springback parameter and their levels.

Factor	Parameters	Level 1	Level 2	Level 3	Level 4	Level 5	Level 6
A	Thickness Ratio	1 (3.00mm)	0.985 (2.96mm)	0.975 (2.93mm)	0.947 (2.84mm)	0.907 (2.72mm)	0.871 (2.61mm)
B	Stroke (Bend Angle)	3 mm	5 mm	7 mm	-	-	-
C	Annealing Temp.	24°C (RT)	150°C	300°C	-	-	-

In this study, an L_{18} ($6^1 \times 3^2$) orthogonal array with three column and eighteen rows was appropriate and used in this study. The experimental layout for springback parameters using L_{18} OA is shown in Table 2. Each row of this table represents an experiment with different combination of parameters and their levels.

2.2 Specimen preparation

For this experiment, the aluminum strip has to be cold formed into non uniform profile (curved shape) and cut into desired dimension as shown in Figure 2, which has various thickness and the data summarized in Table 3. "X mm" indicate the length of curved area and the measured area. For the deformation, the die and puncher were made from D2 steel material and the process was performed using the 100 Tonne mechanical press forging machines. The specimen cut into desired dimension at 50 x 20 mm. There are 6 sets of specimen with different thickness ratio as listed in Table 3. Here, thickness ratio can be defined as a ratio of minimum thickness to the maximum thickness.

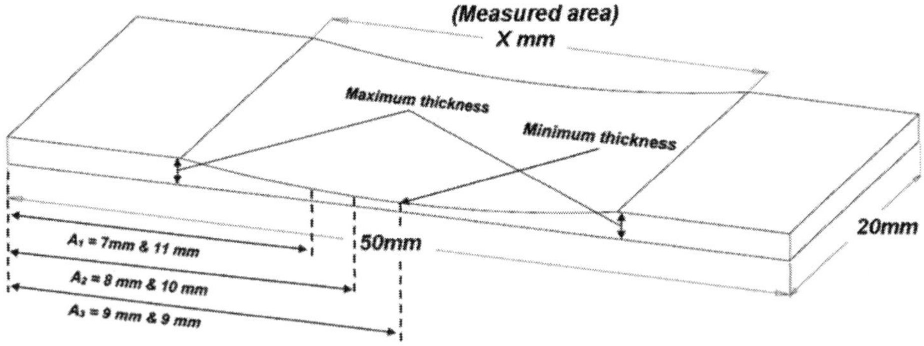

Fig. 2. Aluminium strip for non uniform profile (Curved shaped).

Table 2. Experimental layout based on an L_{18} orthogonal array.

Experiment Number	Parameter			Parameters Setting
	Thickness Ratio (A)	Bend Angle (B)	Annealing Temp (C)	
1	1	3	24 (RT)	A1 B1 C1
2	1	5	150	A1 B2 C2
3	1	7	300	A1 B3 C3
4	0.985	3	24 (RT)	A2 B1 C1
5	0.985	5	150	A2 B2 C2
6	0.985	7	300	A2 B3 C3
7	0.975	3	150	A3 B1 C2
8	0.975	5	300	A3 B2 C3
9	0.975	7	24 (RT)	A3 B3 C1
10	0.947	3	300	A4 B1 C3
11	0.947	5	24 (RT)	A4 B2 C1
12	0.947	7	150	A4 B3 C2
13	0.907	3	150	A5 B1 C2
14	0.907	5	300	A5 B2 C3
15	0.907	7	24 (RT)	A5 B1 C1
16	0.871	3	300	A6 B1 C3
17	0.871	5	24	A6 B2 C1
18	0.871	7	150	A6 B3 C2

Table 3. Aluminium strip data

Set	Length (mm)	Wide (mm)	Minimum Thickness (mm)	Thickness ratio
1.	50	20	3.00	1.000
2.	50	20	2.955	0.985
3.	50	20	2.925	0.975
4.	50	20	2.841	0.947
5.	50	20	2.721	0.907
6.	50	20	2.613	0.871

2.3 Experimental setup

In the measurement of angle after loading, the die valley radius, punch radius and the die opening are constant at 90°, 2 mm and 32 mm respectively. The die radius and punch radius is assume to be equal in this study. The bend radius is 10, 17 and 24 degree at 3, 5 and 7mm stroke respectively. To validate the springback measurement by profile projector, an Alicona Infinite Focus Machine (IFM) is used to as shown in Figure 3. The results in term of average of springback were obtained after conducting the v-bending test for all specimens. The measurement data by both method show less difference in range between -0.0435 to 0.0105.

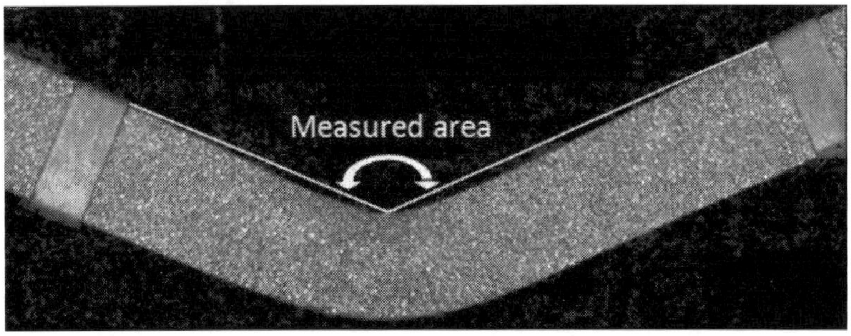

Fig. 3. Springback measurement using Alicona Infinite Focus Machine (IFM).

3 Results and discussion

3.1 V-bending experiment

Measurement of springback can be described as a difference between the angle after loading and after unloading. In the v-bending experiment, observations were made at different stroke. The maximum stress is different for each annealing temperature where specimens with no heat treatment have the highest maximum stress that is 700Mpa approximately, followed by 600 Mpa for 150°C temperature and 300 Mpa for 300°C. The results of this experiment are summarized in Table 4.

Strain is also play significant role in springback result where the increasing number of strain hardening will increase the springback. Figure 4 shows the stress strain behavior of the material for maximum stroke, 7mm. Strain hardening (also called work-hardening or cold-working) is the process of making a metal harder and stronger through plastic deformation. This scenario explains why the flow stress increases with the increase of the axial strain amount. Strain hardening increases the hardness of the section with a minimum thickness. Figure 5 shows the relation between the thickness of a workpiece and Vickers hardness pattern. The average values are obtained from five readings at the same location. Strain also plays a significant role in the occurrence of springback; increased strain hardening increases the springback. Strain hardening constraint, which known as strain hardening index is represented by n. This constant is used to calculate stress-train behavior in work hardening. The work-hardening expression can be determined at higher strain values which in the range plasticity zone. By changing the value of true stress-strain on the curve to logarithmic graph the value of slope, slope of n (strain hardening exponent) and an intercept of log K (K as strength coefficient value) can be obtained. Generally, if n value is 0 it is shown that the material is perfectly plastic solid and if the value is 1 it represents the material is totally elastic. Figure 6 shows the relation between effect of annealing temperature of specimen and Vickers hardness number. Figure 7 show the values known as the strain-hardening index for rolling direction of 0°, 45° and 90° which are 0.30, 0.284 and 0.271 respectively. The results almost coincided with studies made by Dwivedi et al [13].

Table 4. Experimental result for v-bending on an L_{18} orthogonal array.

Experiment	Parameter Setting	% Springback 1	% Springback 2	S/N Ratio
1	A1 B1 C1	0.0041	0.0103	41.016
2	A1 B2 C2	0.0189	0.0194	33.671
3	A1 B3 C3	0.0497	0.0361	26.904
4	A2 B1 C1	0.0159	0.0187	34.501
5	A2 B2 C2	0.0159	0.0108	36.254
6	A2 B3 C3	0.0101	0.0146	37.105
7	A3 B1 C2	-0.0029	0.0008	63.271
8	A3 B2 C3	0.0109	0.0102	38.282
9	A3 B3 C1	0.006	0.0107	40.011
10	A4 B1 C3	-0.0014	0.0002	59.359
11	A4 B2 C1	0.0181	0.0192	33.906
12	A4 B3 C2	0.0296	0.0275	30.408
13	A5 B1 C2	0.0089	0.0037	41.105
14	A5 B2 C3	0.0087	0.0071	40.310
15	A5 B1 C1	0.0304	0.0248	30.624
16	A6 B1 C3	0.0049	0.0053	43.439
17	A6 B2 C1	0.0121	0.0129	37.030
18	A6 B3 C2	0.0353	0.0434	27.754

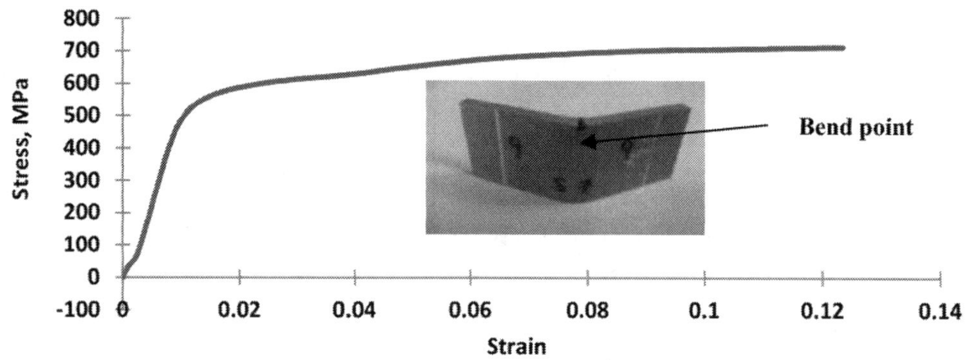

Fig. 4. The stress strain behaviour of the material.

3.2 Regression analysis

The correlation between factors (thickness ratio, bend angle, annealing temperature) and response (springback percentage) were obtained by multiple linear regressions. The statistical software Minitab was used to derive the regression Eq. (1) with $R^2 = 0.7418$.

$$\text{Springback percentage} = 0.01252 + 0.0(A6) - 0.00869(A5) - 0.00493(A4) - 0.01330(A3) - 0.00585(A2) + 0.00137(A1) + 0.0(B1) + 0.00676(B2) + 0.01968(B3) + 0.0(C1) + 0.00150(C2) - 0.00385(C3) \quad (1)$$

In multiple linear regression analysis, R^2 is the regression coefficient ($R^2 > 0.90$) for the models. It indicates that the fit of the experiment is satisfactory. In this case, the parameters numbers influenced the regression coefficient value as shown in Figs. 5-7. Since this model have mixed level of parameters, 0.7418 is possible for this study and coincide with analysis done by Buang et al. [14].

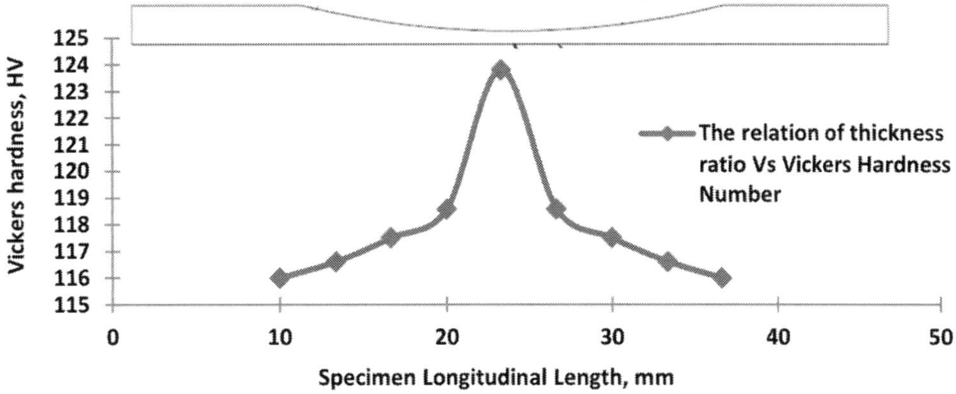

Fig. 5. The relation of Thickness Vs Vickers Hardness Number.

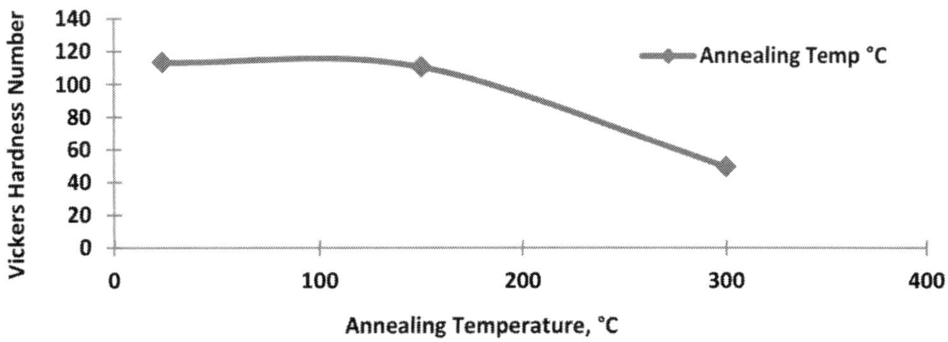

Fig. 6. The relation of Annealing Temperature Vs Vickers Hardness Number.

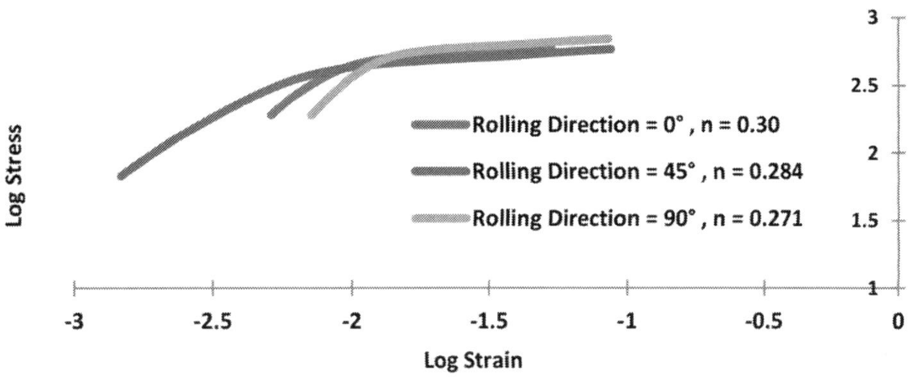

Fig. 7. Stress-strain in various rolling direction in logarithmic diagram taking *n* as parameter.

3.3 Analysis of the signal-to-noise ratio (S/N ratio)

The signal to noise ratio (S/N Ratio) is meant to be used as a measure of the effect of noise factor on target characteristics. The springback values measured from the experiments and their corresponding S/N ratio value are listed in Table 4. The springback response table for thickness ratio, bend angle and annealing temperature was created in the integrated form and the result shown in Table 5. Based on the Figure 8, the optimal performance for the percentage of springback when the parameter setting of specimen is A3, B1, C3, Table 5 show the effect of the process parameters on the springback percentage values. Based on the Table 5, the delta for bend angle (stroke), is the highest which is 14.98. The bend angle (stroke) is the main parameter that affects the springback percentage in this experiment. The results are in agreement with studies that have been done by Abdullah & Samad [15].

Table 5. S/N ratio value for springback percentage by factor level.

Level	Thickness Ratio	Bend Angle	Annealing Temperature
1	36.07	47.12	36.18
2	37.35	36.58	38.74
3	41.22	32.13	40.9
4	47.19		
5	35.95		
6	33.86		
Delta	13.32	14.98	4.72
Rank	2	1	3

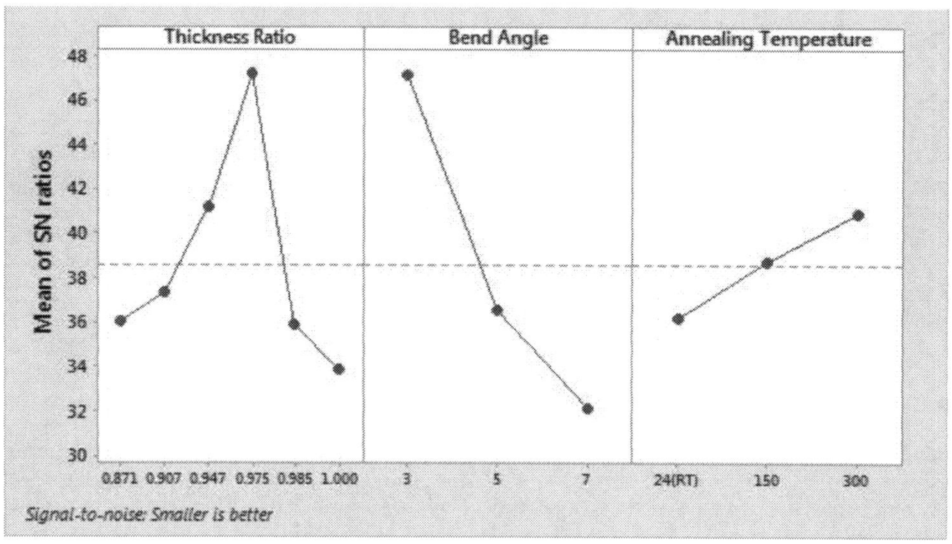

Fig. 8. Effect of process parameters on springback percentage.

3.4 Analysis of variance (ANOVA)

The main goal of ANOVA is to evaluate the experimental error and test its significance. The result of ANOVA and response result for S/N ratio smaller-is-better is represented in Table 6. Based from the result, the F value for the parameter (bend angle) is 7.94 which is the highest. It means that the change of the parameter has significant effect on the percentage springback. The P-value reports the significance level (suitable and unsuitable). The P-value for parameter B is 0.013± which lower than 0.05. So, effect parameter B is significant to the result of the experiment. The percentage number on Table 6 depict that the parameter B, stroke (bend angle) have significant effects on springback percentage. From the table, the contribution of parameter B is 51.25%, parameter A 19.04% and parameter C 3.89%.

3.5 Confirmation test

The final stage in verifying the result drawn based on Taguchi design approach is confirmation test. It has a maximum and minimum value between which the observed value should fall. From Table 7, confidence interval comes out to be ± 0.0005; hence values of spring back for confirmation experimental runs should fall within the range -0.001466± 0.0005 i.e. -0.001966 to 0.000996. Table 8 shows a comparison between the estimated value of springback percentage at optimum condition as shown in figure 8 and the experimental value. A small difference (0.000213%) can be observed between these values. This indicates that the experimental value is close to the estimated value. Therefore, this verifies that the experimental result is highly correlated with the estimated result.

Table 6. ANOVA result.

Symbol	**Parameter**	**Degree of freedom, DOF**	**Sum of Square, SS_T**	**Mean Square, SS_m**	**F-Value**	**P-Value**	**ρ % Contribution**
A	**Thickness Ratio**	5	0.000446	0.000089	1.18	0.397	19.04
B	**Bend Radius**	2	0.0012	0.0006	7.94	0.013	51.25
C	**Annealing Temperature**	2	0.000091	0.000046	0.6	0.57	3.89
Error		8	0.000604	0.000076			25.82
Total		17	0.002341				100

Table 7. Confidence interval.

Response	No. of samples	Mean	Standard Deviation	C.I
Springback Percentage	1	-0.001466	0.0008332	± 0.0005

Table 8. Optimal combination of parameters and their level.

	Estimation	**Experiment**	**Difference**
Level	A3 B1 C3	A3 B1 C3	
Springback percentage, %	-0.001466	-0.001253	0.000213
S/N ratio	57.9865	-	

4 Conclusions

In this study has discussed an application of Taguchi method for investigating the effect of parameters on the springback behaviour of Aluminium 6061-T6 strip with non-uniform thickness. Statistically designed experiments based on Taguchi method were performed using L_{18} orthogonal arrays to analyze the springback behaviour as response variable. Conceptual S/N ratio and ANOVA approaches for data analysis resulting similar conclusion. The regression coefficient, R^2 is equal to 0.7418. Since this an L_{18} $(6^1 \times 3^2)$ orthogonal array is a combination of multi level parameters, possibility to achieve ≥ 0.90 is difficult. The optimum percentage of springback is -0.001466 with S/N ratio of 57.9865. Stroke (bend angle) is the most significant parameter while annealing temperature is the least significant parameter. The contribution of parameter to percentage of springback is Bend radius (51.25%), thickness ratio (19.04%) and annealing temperature (3.89%).

The authors want to acknowledge Ministry of Higher Education for their sponsorship through Fundamental Research Grant Scheme (Acc. No. **203/PMEKANIK/6071308**). For Mr. Fakhrul, who help in conducting the experiments.

References

1. M. Kleiner, M. Geiger, M, Klaus (2003) Manufacturing of Lightweight Components by Metal Forming. *CIRP Ann. - Manufacturing Technology.* **vol. 52**, no. 2, pp. 521–542.
2. R. Jeya Girubha, S. Vinodh (2012) Application of fuzzy VIKOR and environmental impact analysis for material selection of an automotive component. *Material Design.* **vol. 37**, pp. 478–486.
3. Z. Yang, Q. Peng, J. Yang (2012) Lightweight Design of B-pillar with TRB Concept Considering Crashworthiness. *2012 Third International Conference on Digital Manufacturing & Automation Lightweight.* pp. 510–513.
4. V. Nasrollahi, B. Arezoo (2012) Prediction of springback in sheet metal components with holes on the bending area, using experiments, finite element and neural networks. *Material Design.* **vol. 36**, pp. 331–336.
5. S. Thipprakmas, W. Phanitwong (2011) Process parameter design of spring-back and spring-go in V-bending process using Taguchi technique. *Material Design*, **vol. 32**, no. 8–9, pp. 4430–4436.
6. D. Vasudevan, R. Srinivasan, P. Padmanabhan (2011) Effect of process parameters on springback behaviour during air bending of electrogalvanised steel sheet. *Applied Physic Engineering.*, **vol. 12**, no. 3, pp. 183–189.
7. W. Deng, S. Sulaiman, M. Ariffin, S. Y. Lai (2012) Springback Behaviour in Sheet Metal Forming for Automotive Door. *AASRI Procedia*, **vol. 3**, pp. 224–229.
8. V. Nandedkar, S. Chikalthankar, S. Naikwade (2015) Factors Affecting on Residual Stresses & Springback in Sheet Metal Bending : A Review. *International Journal for Scientific Research & Development.* **vol. 3**, no. 08, pp. 258–262.
9. X. Han, Z. Shi Hong, Z. Rong, L. De Hong, (2012) Springback characteristics of AZ31 magnesium alloy as-extruded profile in warm tension-rotation bending process. *Transactions of Nonferrous Metals Society of China (English Ed).* **vol. 22**, pp. 416–421.
10. S. Ján, M. Jurcisin (2012) Springback Prediction in Sheet Metal. *Journal for Technology of Plasticity.* **vol. 37**, no. 1, pp. 93–103.

11. S. Chikalthankar, G. Belurkar, V. Nandedkar (2014) Factors Affecting on Springback in Sheet Metal Bending : A Review. *International Journal of Engineering and Advanced Technology*. **vol. 3**, no. 4, pp. 247–251.
12. S. Kamaruddin, Z. Khan, S. Foong, (2010) Application of Taguchi Method in the Optimization of Injection Moulding Parameters for Manufacturing Products from Plastic Blend. *International Journal of Engineering and Technology*. **vol. 2**, no. 6, pp. 574–580.
13. J. Dwivedi, S. Shah, P. Upadhyay, N. Talukder (2002) Springback analysis of thin rectangular bars of non-linear work-hardening materials under torsional loading. *International Journal of Mechanical Sciences*, **vol. 44**, no. 7, pp. 1505–1519.
14. S. Buang, S. Azam, J. Saedon (2014) Effect of Die Gap and Punch Travel on Springback in Air V-Bending Process Using Taguchi Method. *Applied Mechanics and Materials*. **vol. 660**, no. October, pp. 317–321.
15. A. B. Abdullah, Z. Samad, (2013) An experimental investigation of springback of AA6061 aluminum alloy strip via V- bending process. *IOP Conference Series: Materials Science and Engineering*. **vol. 50**, p. 012069.

MATEC Web of Conferences **90**, 01003 (2017) DOI: 10.1051/matecconf/20179001003
AiGEV 2016

Ergonomics study for workers at food production industry

Hamizatun Mohd Fazi[1,*], *Nik Mohd Zuki* Nik Mohamed[1], *Mohd Fadzil Faisae* Ab Rashid[1] and *Ahmed Nasser* Mohd Rose[1]

[1]Faculty Mechanical Engineering, University Malaysia Pahang, 26600 Pekan, Pahang

Abstract. The health constraint faced by production workers affects the quality of the work. The productivity of the workers is affected by the Work-related Musculoskeletal Disorder (WMSD) which limits the movement of the workers. The comfort workplace condition, known as ergonomic environment is important to prevent the occurrence of the WMSD. Proper ergonomic workplace considers the condition of the workers while doing the assigned work. The objectives of this study are to identify the current problems related to ergonomic in food production process, to analyse the actual production data by using Rapid Upper Limb Assessment (RULA) and Rapid Entire Body Assessment (REBA) and to recommend the ergonomic workplace environment based on the condition of the study. The study was done at a Small and Medium Enterprises (SME) food production company in the Klang Valley of Malaysia. The condition of the workers affects the productivity of the company due to workers' health deficiency. From the findings, the workers are exposed to the awkward postures which leads to the Work-Musculoskeletal Disorders (WMSDs). Besides, the best height of the worker at the study area (critical area) to prevent the worker from WMSDs is within 155 cm to 160 cm. The results show that the workers are exposed to the WMSD in different level of risks which causes high absenteeism among the workers.

1 Introduction

As a developing country, Malaysia faced a rapid industrial growth in various sectors such as foods and services sector. The contributions of every sector are really important to generate a stable economy towards the glorious nation. The food industries in Malaysia had vast growth and contributed to the economy improvement of the country as food is necessity for the human beings [1]. In the food manufacturing sector, the health of the employee may be exposed to the hazards such as the work-related musculoskeletal disorder. Normally, the injuries happen either at the muscles, tendons, nerves, blood vessel or ligaments.

Musculoskeletal disorder (MSDs) may occur because of continuously performing repetitive tasks, working in repeated and sustained or difficult postures, performing strenuous physical work, and using forceful exertion [2]. To improve the productions for

* Corresponding author: hamizatunfazi@gmail.com

higher profit, employee productivity is important [3]. Work productivity as an indicator has been a general subject for examination in several studies on musculoskeletal disorders that influence the workers condition [4]. If the workers facing health problem, it indicates that the losses faced by the company [5]. Heavy lifting and awkward work postures are the physical working condition that relate with sick-leave [6]. As the condition of the workers affects the productivity, hence the worker needs to have a comfortable workplace which is ideally free from hazards.

The comfortable workplace is known as the proper ergonomic working environment. Ergonomics is the information such as the behavior of the human, limitation and capacity of human that applies on the machines, designs of tools, tasks, and environment for secure, comfort and beneficial for human use [7]. A bad worksite design leads to the difficulties for the workers such as fatigue and injuries. Besides, the injuries relate to the low productivity of the workers and increase the cost to the company where the workers need to rest and the company needs to bear the losses [8]. In the study of the occupational risks, there are few ergonomics analysis tools available to determine the risks of the worker at the workplace. There methods identified and classified the risks into several parts which are self-reports, observational methods and direct measurements [9]. In this study, the ergonomic tool used to identify the occupational risks on the workers are Rapid Upper Limb Assessment (RULA) and Rapid Entire Body Assessment (REBA). The selection of the RULA and REBA are based on the tasks doing by the workers which are awkward, repetitive work and involves the entire body parts.

2 Methodology

This study was conducted at one of the food industry in Malaysia. The first step for this research was to study the process flow of the food processing by assessing the overall working areas. Then, by doing interview sessions and observations, the critical workplace was identified. The chosen subjects were from the filling and stamping workstation since it was the critical workplace. The subjects are taken from both gender while the height of the workers are within 150 cm to 180 cm. Next, the subjects were required to conduct their working cycle as usual and the pictures and videos of the process were taken. The postures that were repetitive and awkward were chosen. Later, the RULA and REBA analysis were performed to assess the subjects' posture level of discomfort. Figure 1 and Figure 2 show the RULA and REBA assessments method used in this study. The level of MSD risk for RULA is shown in Table 1 while the level of MSD risk for REBA is shown in Table 2.

The movements of the workers were divided into two different positions which were Posture 1 and Posture 2. Posture 1 was the positions where the workers were taking the products from the filling and the stamping machine and put the products into the basin filling water to prevent the products from defective. The position where the workers were arranging the products in the container is called Posture 2.

The movements of the workers were divided into two different positions which were Posture 1 and Posture 2. Posture 1 was the positions where the workers were taking the products from the filling and the stamping machine and put the products into the basin filling water to prevent the products from defective. The position where the workers were arranging the products in the container is called Posture 2.

 DOI: 10.1051/matecconf/20179001003

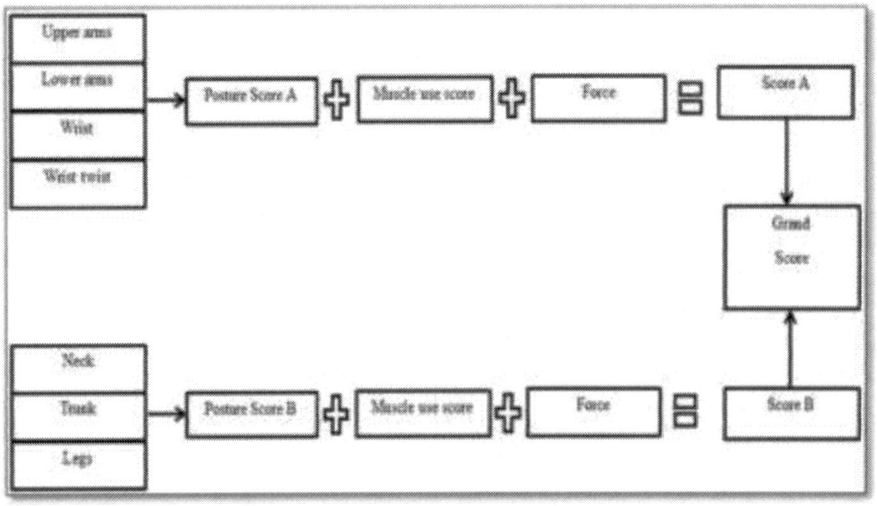

Fig. 1. RULA assessment method.

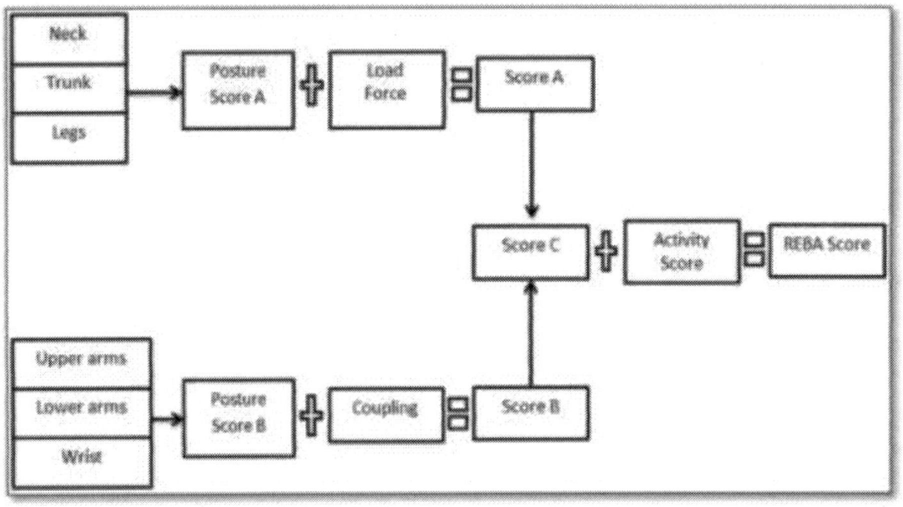

Fig. 2. REBA assessment method.

Table 1. Level of MSD Risk (RULA) [10].

Score	Level of MSD risk
1-2	Negligible risk, no action required
3-4	Low risk, change may be need
5-6	Medium risk, further investigation, change soon
6+	Very high risk, implement change now

Table 2. Level of MSD Risk (REBA) [11].

Score	Level of MSD risk
1	Negligible risk, no action required
2-3	Low risk, change may be need
4-7	Medium risk, further investigation, change soon
8-10	Very high risk, implement change now
11+	Very high risk, implement change

3 Results and discussion

From the interviews and the observations, it was shown clearly that the subjects were exposed to the awkward postures during long working hours. There were two postures chosen involving three subjects for each posture. From the subjects working cycle recorded, the assessments were carried out to identify the problems. Figure 3 shows the RULA grand scores versus postures for all three workers that involved in this study. The results were analysed by referring to the level of MSD risk for RULA in Table 1.

The result for the RULA grand score is shown in Figure 3. For Posture 1, it shows that worker A and C are exposed in the high risks that need to be changed immediately since both scores are 7 while worker B is in medium risks that need further investigations as the score is 6. Based on the results, it indicates that the workplace is too hazardous for workers A and C. For Posture 2, the graph denotes that worker A is highly exposed to MSD as the score is 7. Worker B is still in the low-risk condition while worker C is in medium risk condition. The score for workers B and C are 4 and 6 respectively. The results for RULA assessment reveal that worker A is in the high risks for both postures. The change is needed immediately to prevent the worker A from MSD. Besides, by comparing the results for each posture, all workers are in awkward condition at Posture 1. Therefore, the condition of the workplace for Posture 1 should be improved.

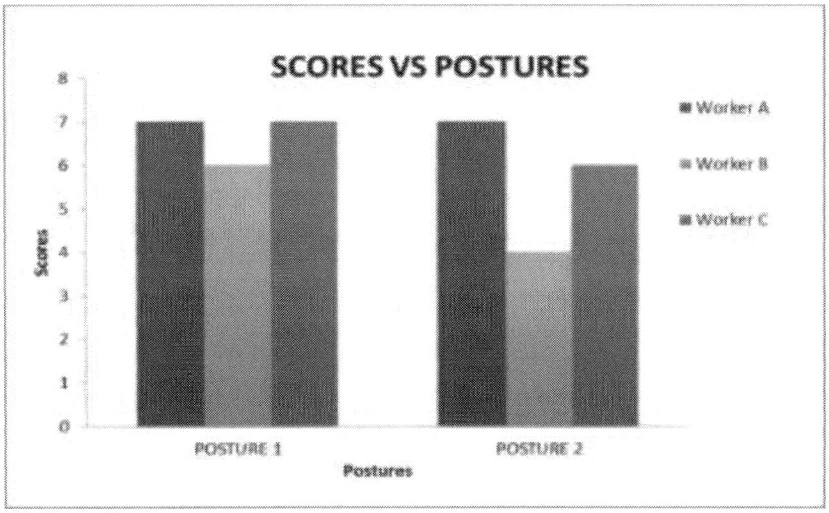

Fig. 3. RULA grand scores versus postures.

The graphical results for REBA scores versus postures are visualized in Figure 4. The results are analyze by referring to the level of MSD risk for REBA as in Table 2. From the graph in Figure 4, it is found that workers B and C are exposed in the medium risks that need implement change in the near future, since both scores are 6 while worker A is in very high risks as the score is 9. The Posture 1 is really dangerous to worker A and needs the implementation of the changes immediately.

DOI: 10.1051/matecconf/20179001003

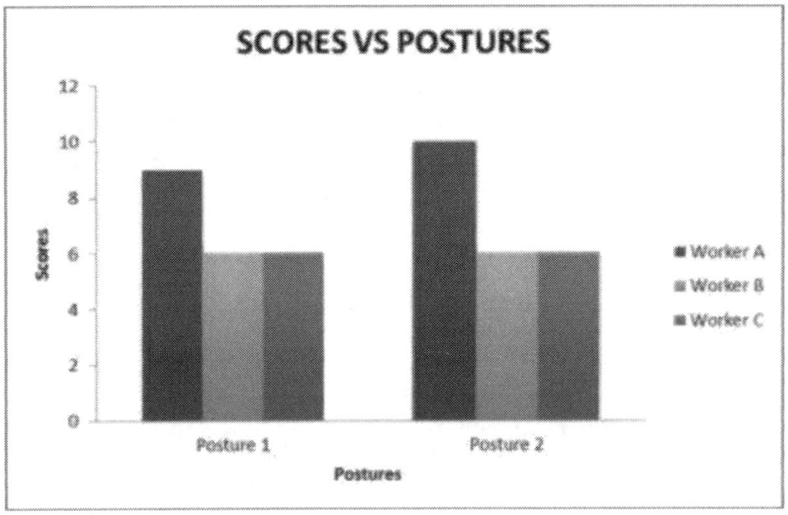

Fig. 4. REBA grand scores versus postures.

Based on the results for Posture 2, it indicates that the position is really at high risk level for worker A as its score is 10. Furthermore, the graph denotes that workers B and C are in medium risk condition. Both scores for worker B and C are 6. The results for REBA assessment divulge that worker A is in the high risks for both postures. The change is needed to be implemented immediately to ensure that the workers are free from MSD.

The comparison between the RULA and REBA are made in Figure 5. The graph indicates that the worker A is always in the awkward condition as compared to the other workers as the grand scores for A are always higher than 6 which represent the medium to high risks states. Worker A is the tallest participants among the others. The results reveal that the worker who has the height of 172 cm is not suitable to work at the study area. Therefore, worker A should be transferred to the other workstations.

Contrarily, the highest scores for worker B and C are 6 and 7 respectively which are still in the medium risk. Therefore, conclusion can be made from the outcomes of the study where the heights of the workers affect the comfortable condition in the workplace. These studies reveal that the height of the worker which suitable to do the job task at the study area and both posture is in the range of 155 cm-160 cm only.

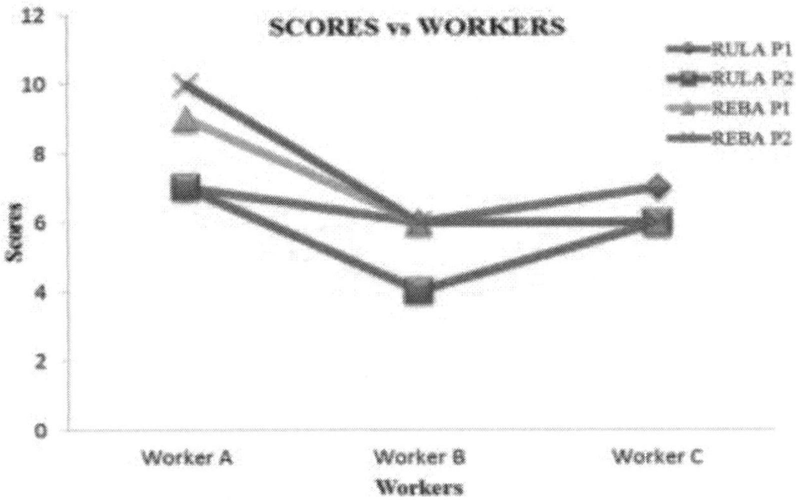

Fig. 5. Grand scores versus workers.

4 Recommendations

Based on the study, below are the items that need to be implemented:

- Worker's Hiring Standard - The suitable height that resulted from the research should be within the range of 155 cm-160 cm, therefore, it can be set as the standard or the compulsory requirement for the company to hire new workers. The anthropometry data of the workers are important to prevent the WMSDs that lead to the absenteeism and poor condition of the workers.
- Conveyor - The conveyor to connect the filling and stamping machine process. The height of the conveyor should be able to be adjusted to the ordinary height of the workers within the range of 75 cm-80 cm based on the samples study. The improved layout helps the workers from the MSD problem where the workers just sit while ensuring the product moves smoothly on the conveyor. Besides, the job task of the workers also decreases since the products directly move to the expiry stamping without the need to be moved manually.
- Using Ergonomic chairs - For this study, the ergonomics chairs are important to prevent the back pain and the lower part of the body while doing the assigned work. The elements of a good ergonomic chair that should be considered during selecting the chair which has lumbar support, foot rest and appropriate height level [12].

5 Conclusions

This paper has presented the approach to find out the occupational risk of the workers at the SME's food production company by using two types of analysis tools. The analysis tools used are RULA and REBA which are able to determine the condition of the workers at the critical areas which are not ergonomically designed. From the findings, the workers are exposed to the awkward posture which leads to the Work-Musculoskeletal Disorders (WMSDs). Besides, the best height of the worker at the study area (critical area) is within 155 cm to 160 cm to prevent the worker from WMSDs. Therefore, the employer should consider the height of the worker while placing the worker at the workstation.

References

1. Dolage, D.A.R., A.B. Sade, E.M. Ahmed, *The influence of Flexible Manufacturing Technology adoption on productivity of Malaysian manufacturing industry.* Economic Modelling, 2010. **27**(1): p. 395-403.
2. Deros, B.M., D.D.I. Daruis, I.M. Basir, *A Study on Ergonomic Awareness among Workers Performing Manual Material Handling Activities.* Procedia - Social and Behavioral Sciences, 2015. **195**: p. 1666-1673.
3. Dollard, M.F., D.Y. Neser, *Worker health is good for the economy: Union density and psychosocial safety climate as determinants of country differences in worker health and productivity in 31 European countries.* Social Science & Medicine, 2013. **92**: p. 114-123.
4. Andersen, L.N., et al., *Efficacy of 'Tailored Physical Activity' on reducing sickness absence among health care workers: A 3-months randomised controlled trial.* Manual Therapy, 2015. **20**(5): p. 666-671.
5. Carrillo-Castrillo, J.A., et al., *Estimation of the relative risks of musculoskeletal injuries in the Andalusian manufacturing sector.* International Journal of Industrial Ergonomics, 2016. **52**: p. 69-77.
6. Laaksonen, M., et al., *Work Arrangements, Physical Working Conditions, and Psychosocial Working Conditions as Risk Factors for Sickness Absence: Bayesian*

Analysis of Prospective Data. Annals of Epidemiology, 2010. **20**(5): p. 332-338.
7. Probst, C.A., et al., *Human factors engineering approaches to patient identification armband design.* Applied Ergonomics, 2016. **52**: p. 1-7.
8. Falck, A.-C., M. Rosenqvist, *A model for calculation of the costs of poor assembly ergonomics (part 1).* International Journal of Industrial Ergonomics, 2014. **44**(1): p. 140-147.
9. Kathy Cheng, H.-Y., C.-Y. Cheng, Y.-Y. Ju, *Work-related musculoskeletal disorders and ergonomic risk factors in early intervention educators.* Applied Ergonomics, 2013. **44**(1): p. 134-141.
10. McAtamney, L., E. Nigel Corlett, *RULA: a survey method for the investigation of work-related upper limb disorders.* Applied Ergonomics, 1993. **24**(2): p. 91-99.
11. Hignett, S., L. McAtamney, *Rapid Entire Body Assessment (REBA).* Applied Ergonomics, 2000. **31**(2): p. 201-205.
12. Amick Iii, B.C., et al., *A field intervention examining the impact of an office ergonomics training and a highly adjustable chair on visual symptoms in a public sector organization.* Applied Ergonomics, 2012. **43**(3): p. 625-631.

The identification of Malaysian driving styles using the multidimensional driving style inventory

Juffrizal Karjanto[1, 2,*], *Nidzamuddin* Md. Yusof[1, 2], *Jacques* Terken[1], *Muhammad Zahir* Hassan[1, 3], *Frank* Delbressine[1], *Hanneke* Hooft van Huysduynen[1], and *Matthias* Rauterberg[1]

[1]Department of Industrial Design, Eindhoven University of Technology, Eindhoven, The Netherlands
[2]Faculty of Mechanical Engineering, Universiti Teknikal Malaysia Melaka, Melaka, Malaysia
[3]Faculty of Engineering Technology, Universiti Teknikal Malaysia Melaka, Melaka, Malaysia

Abstract. A study was conducted to investigate the driving styles within the geographical region of Malaysia. The first part of the study determined Malaysian drivers' driving styles using the Multidimensional Driving Style Inventory approach. Respondents had various backgrounds in terms of age, gender and experience of driving. A statistical factor analysis was done for 338 respondents revealing four driving styles; careful, risky, anxious-dissociative and angry. In addition, a comparison with previous findings from Netherlands and Israel was also done in this study. The next part of the study focused on the associations of these driving styles with two personality traits (sensation seeking and desire for control), sociodemographic factors and experience of driving. The last part of the paper discusses the relevance of the study against the background of future automotive development.

1 Introduction

In Malaysia, the number of road accidents is steadily increasing while deaths remain static every year (Fig. 1). Although motorcycles are the number one contributor in accident fatalities in Malaysia, most of the accidents involve the motorcycle and passenger car [1]. Safety measures, like for example the implementation of Automated Enforcement System (AES) cameras which are able to identify drivers who are driving over the designated speed limit or running over the red light [2], were designed to penalize drivers by producing summons. Campaigns through media such as advocacy programmes are other initiatives to increase road safety awareness educate drivers, especially during festive season [3]. Generally, these kinds of safety measures are believed to create a safer driving environment and also helping in decreasing the accident rates in Malaysia.

On the other hand, researchers reported that sociodemographic factors and personalities are directly influencing the increment in the number of traffic violations and road accidents [4]–[6]. According to Risk Homeostasis Theory (RTH) developed by Wilde [7], drivers

* Corresponding author: j.karjanto@tue.nl

will usually weigh the cost and benefits of alternate actions, thus they will accept certain level of risk which they actively target (target risk). As an example, running late for an interview could be compensated by speeding up as one will gain more time although it would cost the driver of getting a speeding ticket and higher chances of getting into traffic accident. Furthermore, according to the driver models by Näätänen and Summala [8], Summala [9], and Vaa [10], human driver drives a car in a satisficing manner where emotions are the strong driven factors. They elaborated that some drivers like the sensations out of speed, acceleration, driving with very close distance to the preceding vehicle and being competitive on the roads. In addition, some drivers are driven by motivations like social pressure and the need for control.

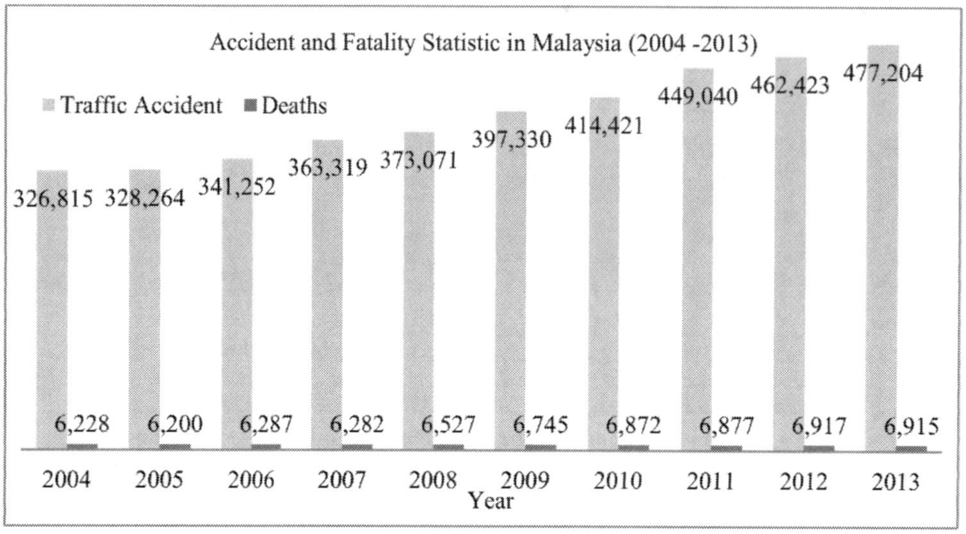

Fig. 1. Trends in reported road traffic accident in Malaysia [11].

Personality is a characteristic of individuals and concerns traits such as desire for control and sensation seeking. In the context of driving, the desire for control is one element of the control motive concept, which can lead a driver to adopt a careful driving behaviour, or a dangerous driving behaviour if the driver is in frustrated [12]. According to Dahlen et al. [13], a driver with a high sensation seeking trait likes to drive at or above the designated speed limit. This driving behaviour was reported to be the main culprit in traffic accidents [14], [15].

Taubman-Ben-Ari et al. [12] reviewed the existing scales of driving styles and believed that all these scales and definitions of driving styles can be integrated into one multidimensional concept of driving styles. They hypothesized that the existing scales of driving styles can be clustered into four general groups: reckless and careless driving style (e.g.[16]), anxious driving style (e.g. [17]), angry and hostile driving style (e.g.[14]), and patient and careful driving style (e.g.[16]). Based on their findings, they developed the Multidimensional Driving Style Inventory (MDSI) technique to assess the driving style of drivers. Originally, 80 items were used in a pilot sample in Israel, and 44 items remained as the final version of MSDI after the exploratory factor analyses were applied. This 44 items-questionnaire revealed eight clear and meaningful driving styles in Israel, which are: dissociative, anxious, risky, angry, high velocity, distress-reduction, careful and patient. Hooft van Huysduynen et al. [18] validated the developed MDSI technique by assessing the driving styles and driver profiles focussing on the drivers in the Netherlands and Belgium. Five out of the original eight factors were identified, and 36 items appeared to be stable compared to the 44 items from the original study by Taubman-Ben-Ari et al.

Therefore, in order to understand the behaviours of the drivers on the road, there is a need to study driving styles among Malaysians. Driving style denotes typical driving behaviour that includes the operational control of the vehicle's speed, tactical skills such as overtaking another vehicle, and strategic behaviour such as speeding, and that may result in traffic violations [19]. Driving itself is a complex activity that includes the interaction of personality, motivational, emotional and social factors in a dynamic environment [20]. The aim of the study presented in this paper was to define Malaysian driving styles and their relation to two personality traits which are sensation seeking and desire for control. The MDSI technique was implemented to determine the Malaysian respondents' driving style(s) and statistical analyses were applied to determine the relation with gender, age, and experience. The findings of this study will help the automotive engineers to produce the next generation of vehicles that meet the Malaysian users' level of driving pleasure and comfort. In addition, the same understanding can also be used in tackling the road accidents issue as the identification of driving styles could provide a better platform for the road safety campaigns and accident prevention measures.

2 Methodology

2.1 Procedures

A set of questionnaires was distributed through email to individuals, academic institutes, and several government and private sectors in Malaysia. In addition, the questionnaires were also circulated to several Malaysian groups in Facebook. Respondents were asked to complete a set of questionnaires which comprised of sociodemographic background questions (such as age and gender), driving experience (such as period of having the driver's license, distance of driving in a year according to the scale developed by Lajunen, Corry, Summala and Hartley [21]), the revised MDSI driving styles assessment from Hooft van Huysduynen et al. [18], the Sensation Seeking questionnaire from Zuckerman, Michael Kuhlman, Joireman and Kraft [22], and the Desire for Control questionnaire from Burger and Cooper [23].

The Driving Style questionnaire was adapted from the MSDI (44 items) by Taubman-Ben-Ari et al. [12] for Israeli drivers by Hooft van Huysduynen et al. [18], validating the MDSI for drivers in the Netherlands and proposing a questionnaire of 36 items. Participants were asked to rate each item on a seven-point scale (from "not at all" to "very much") expressing their feelings, opinions, and behaviour when driving on an average road trip.

The Sensation Seeking questionnaire developed by Zuckerman et al. [22] consists of 19 items. Response options for the items are "true" or "false. For a "yes" response, 1 point is scored. The points are summed to produce the final score (between 0 and 19). The "Desire for Control" questionnaire was developed by Burger and Cooper [23] and consists of 20 items on a seven-point scale (from "does not apply at all" to "always applies to me"). The scoring method is to average all the result with the exception that the scores for items 7, 10, 16, 19 and 20 should be inverted before summing all the scores. The methodological approach used in this study is shown in Fig. 2.

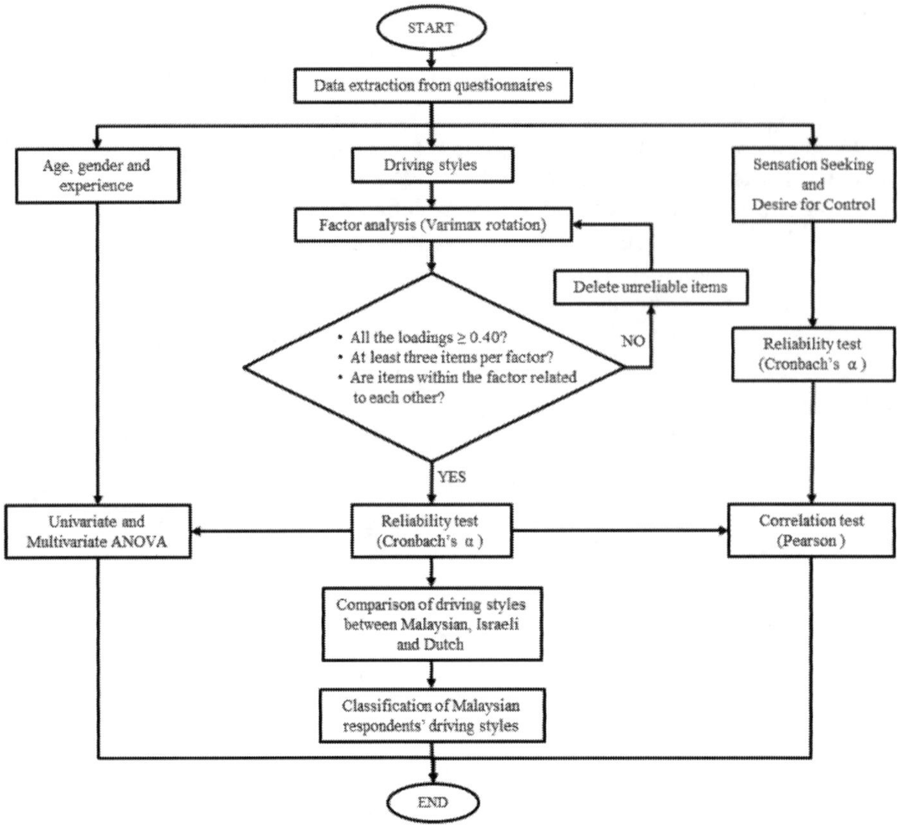

Fig. 2. Methodological approach used in the current study.

2.2 Participants

Three hundred and thirty eight (338) participants, who owned driving licenses, from various backgrounds (from university students, public servants to private sector workers) volunteered to participate is this study and completed the questionnaires through an online survey. The sample consists of 273 (80.80%) men and 65 (19.20%) women, ranging from the age of 20 to 66 years old (mean (M) = 28.00, standard deviation (SD) = 7.07). In terms of driving experience, the number varies from 1 to 47 years of owning a driving license (M = 9.11, SD = 6.77). In terms of type of driver, drivers with moderate experience (owning a driving license between 3 to 7 years) and experienced drivers (owning a driving license for more than 7 years) dominated the sample with 158 (46.75%) and 157 (46.45%) respondents, respectively, while the novice (owning a driving license for less than 3 years) drivers contributed to only 6.80% of the total respondents with 23 drivers.

3 Results and discussion

In this section, first the reliability of the Sensation Seeking questionnaire and the Desire for Control questionnaire will be considered, next, the factor analysis for the MDSI items for the Malaysian sample will be presented and the resulting MDSI questionnaire will be compared with those emerging from to the studies with Israeli and Dutch drivers. Then, the distribution of driving styles for Malaysian drivers will be presented. Finally, the relation with the sociodemographic factors (gender, age and experience) will be considered.

3.1 Internal consistency

Cronbach's α was used to assess the internal consistency of the Sensation Seeking questionnaire and the Desire for Control questionnaire. According to Field in 2009 [24], a value of 1 indicates excellent reliability, while a value of 0.7 shows good reliability and a value that is lower than 0.3 is considered unacceptable. Cronbach's α for the Sensation Seeking questionnaire was good (0.78). Therefore, the responses of all items were added up into a single score and anyone who received a score between 11 and 19 was labelled as sensation seeker. Cronbach's α for the Desire for Control was also good (0.77). A higher average score for a respondent indicates a higher desire for control.

3.2 The MDSI factors structure for Malaysian data

A factor analysis using Varimax rotation was conducted, and six factors were set as the number of fixed factors based on the six factors found in the work of Hooft van Huysduynen et al. [18]. The Kaiser-Meyer-Olkin (KMO) measure of sampling adequacy revealed a value of 0.85 which indicates a significant adequacy as a minimum value of 0.50 is usually required [25].

For the first Varimax factor analysis, there are two requirements to look at in order to filter the items in each factor. First, all the factor loadings for each of the item have to be above 0.40 [12] and secondly, at least three items are required for each of the factors for there to be any meaningful contribution [26]. Based on the first requirement, items 7, 20, 21 and 28 were excluded from the second analysis because of the low factor loadings with a value of 0.39, 0.35, 0.31 and 0.37, respectively. Additionally, by the second requirement, factor 6 was dropped since it has only two items (item 11 and 5) and, therefore, would not contribute to any significant interpretation.

From the second analysis, all the factor loadings for all the items revealed values of above 0.40. Again, in the second analysis, the requirement for each factor was, at least three items per factor and, therefore, factor 5 was omitted for the next analysis since it had only two items (items 11 and 5). Next, in each of the factors, each of the items was scrutinized with respect to its relationship with the other items to see if there are meaningful contributions towards the formulation of driving styles. The items in factor 1 were related to cautious and vigilant driving and therefore, factor 1 was labelled as "Careful" driving style for Malaysian drivers. However, item 1 and 30 did not relate coherently with the other items in factor 1 since they related to relaxing activities in driving and consequently were heterogeneous in a sense. Hence, for the third and final analysis, these two items were omitted. Factor 2 consisted of items containing words like "dangerous", "risk", "thrill" and "tailgate". Therefore, factor 2 was labelled as "Risky" driving style. In this factor, all the items were related with each other, and there was no omission of an item required.

The items in the next factor, factor 3, linked to the nervousness, distress and frustration of the driver while driving and, in addition, there were also elements like misjudged, forgetful and clumsiness. Thus, this factor was labelled as "Anxious-Dissociative" driving style. This hybrid and new driving style (the combination of two driving styles from the previous studies) will be further discussed in the next subchapter. Within this factor, item 15 was not included in the final analysis since this item was not related with the other items. In factor 4, all of the items noticeably show the driver's heated tendency toward other drivers. The kind of treatment to fellow drivers such as honking, swearing and flashing lights specified this factor as "Angry" driving style. One of the items in factor 4, item 16, was related to distraction and absent-mindedness and clearly did not relate to angry driving and, therefore, was dropped for the final analysis.

Since some items needed to be dropped after the second analysis, a final and third analysis was done to recheck the stability of each factor for the remaining items. For the third analysis, the items were pushed into four factors, and the final result of stabilized loading factors is shown in Table 1. The final scale was reduced to 24 items distributed across four factors. Next, the reliability of the scale was considered by looking at Cronbach's α for each of the factors as shown in Table 1. Examining the Cronbach's α value for factor 1 and 2 indicates good reliability (0.80 and 0.75, respectively) while factor 3 and 4 has satisfactory reliability (0.60 and 0.48, respectively) from a statistical point of view.

Table 1. Results from the final Varimax factor analysis.

Factor	Item	Final Loading
Factor 1	**Careful driving style**	
Cronbach's α = 0.80	[34] Tend to drive cautiously.	0.76
	[33] Always ready to react to unexpected manoeuvres by other drivers.	0.73
	[13] Drive cautiously.	0.73
	[8] While driving, I try to relax myself.	0.72
	[18] Base my behaviour on the motto "better safe than sorry".	0.69
	[4] Feel I have control over driving.	0.52
Factor 2	**Risky driving style**	
Cronbach's α = 0.75	[36] Enjoy the excitement of dangerous driving.	0.78
	[17] Like to take risks while driving.	0.70
	[19] Like the thrill of flirting with death or disaster.	0.68
	[24] Get a thrill out of breaking the law.	0.61
	[2] Purposely tailgate other drivers.	0.60
	[6] Enjoy the sensation of driving on the limit.	0.41
Factor 3	**Anxious-Dissociative driving style**	
Cronbach's α = 0.60	[25] Feel nervous while driving.	0.71
	[9] Driving makes me feel frustrated.	0.64
	[26] Feel distressed while driving.	0.63
	[31] Nearly hit something due to misjudging my gap in a parking lot.	0.59
	[22] Forget that my lights are on full beam until flashed by another motorist.	0.58
	[27] Intend to switch on the windscreen wipers, but switch on the lights instead.	0.53
	[32] Feel comfortable while driving. [-]	0.32
Factor 4	**Angry driving style**	
Cronbach's α = 0.48	[35] Honk my horn at others.	0.73
	[14] When a traffic light turns green and the car in front of me doesn't get going immediately, I try to urge the driver to move on.	0.68
	[3] Blow my horn or "flash" the car in front as a way of expressing frustrations.	0.59
	[23] When someone does something on the road that annoys me, I flash them with high beam.	0.58
	[12] When a traffic light turns green and the car in front of me doesn't get going, I wait for a while until it moves. [-]	0.44

Note:

- *[-] indicates negative correlation*
- *The italic items have been moved from the original factor to another factor after the final analysis.*

3.3 Comparison with Netherlands and Israel findings

The findings of this study were compared with the preceding and are shown in Table 2. In the current study, for factor 1 (careful), the main difference is the addition of items 8 and 4 in this study. Item 8, which is about relaxing oneself while driving, could be understood variedly by the diverse participants from these three countries, as this item was grouped differently for each analysis. While both of the previous studies agreed that item 4, which deals with the feeling of control over driving, was factored under the anxious driving style with negative correlation, this study found that the item sorted into careful driving style. The Malaysian study indicates that "control" has a significant and positive relationship with a careful driving style, which involves the element of planning and preparation in driving as has been discussed and found in the work of Taubman-Ben-Ari et al. [12].

Table 2. Comparison between findings of current (Malaysia) and previous two studies (Israel and Netherlands/Belgium).

Items	Current Study	Taubman-Ben-Ari et al. [12]	Hooft van Huysduynen et al. [18]
[34]	Careful	Careful	Careful
[33]	Careful	Careful	Careful
[13]	Careful	Careful	Careful
[8]	Careful	Distress-Reduction	Anxious
[18]	Careful	Patient	Careful
[4]	Careful	Anxious [-]	Anxious [-]
[36]	Risky	Risky	Risky
[17]	Risky	Risky	Risky
[19]	Risky	Risky	Risky
[24]	Risky	Risky	Risky
[2]	Risky	High-velocity	Anxious
[6]	Risky	Risky	Risky
[25]	Anxious-Dissociative	Anxious	Anxious
[9]	Anxious-Dissociative	Anxious	Anxious
[26]	Anxious-Dissociative	Anxious	Anxious
[31]	Anxious-Dissociative	Dissociative	Dissociative
[22]	Anxious-Dissociative	Dissociative	Dissociative
[27]	Anxious-Dissociative	Dissociative	Dissociative
[32]	Anxious-Dissociative [-]	Anxious [-]	Anxious [-]
[35]	Angry	Angry	Angry
[14]	Angry	High-velocity	Angry
[3]	Angry	Angry	Angry
[23]	Angry	Angry	Angry
[12]	Angry [-]	Angry [-]	Angry [-]

For the "risky" driving style, this study reveals six items and all the items, except one item (item 2 - "purposely tailgate another driver"), are in agreement with the findings of the previous two studies. All the items in the risky factor consist of elements of dangerous, thrill and sensation, which are all related to risky behaviour in driving. According to Jonah et al. [27], driving too closely to others and tailgating other drivers are strongly considered as risky driving since they would easily lead to a collision.

Four of the items found in factor 3 are identical with the items found in the previous two studies under the anxious driving style. The next three items are also similar to the previous two studies and fall under the driving style of "dissociative". As mentioned by Miller and

Taubman-Ben-Ari [28], Taubman-Ben-Ari et al. [12] in their study already reviewed the existing driving style scales and hypothesized that anxious and dissociative driving style can be clustered together. They claimed that drivers who show signs of anxiety and fear, might have less confidence in driving. Therefore, the drivers would tend to get easily distracted due to majorly focusing on the operational parts of the driving task and thus display dissociative driving behaviour.

For the fourth and last factor, the "angry" driving style, all the items are about expressing anger and irritation such as urging, honking and swearing, and relate to the treatment of fellow drivers. The current study revealed five items of which four items (items 35, 3, 23 and 32) are similar to the findings from the previous studies ([12], [18]). Only one item (item 14) is different from the finding of Taubman-Ben-Ari et al. [12], but in Hooft van Huysduynen et al. [18] item 10 was also classified under the "angry" driving style.

3.4 Malaysian driving styles distribution

Based on the results from the factor analysis, respondents were classified into their own driving styles based on the MDSI questionnaire which they have answered. In this study, there are three methods being presented in order to reveal the driving styles distributions among the respondents. The first two methods are non-refined approaches, known as *Sum Scores by Factor* and *Weighted Sum Scores* [29]. For the first method, the highest sum of raw scores determines the respondent's driving style. Each raw score is corresponding to their item loading which any raw score with a negative factor loading will be subtracted in the calculation. In addition, the score is averaging in each factor if the total items inside each factor are not equal. By using the highest sum of raw scores, the first method identified 321 (95.0%) respondents as drivers with a careful driving style, as shown in Fig. 3. For the second method, before the average scores were calculated, the score for each item was multiplied by the respective factor loading. The highest scoring result calculated by using this method indicated 330 respondents as drivers with a careful driving style. Similar to the first method, this method showed very little discriminating power as 97.6% out of the total respondents were identified as drivers with a careful driving style.

The first two methods take into account only the highest score in order to determine a respondent's driving style while leaving the scores for the other factors out of consideration. In the third method, the relative positions of the scores within the sample were taken into consideration. For example, if the driver scored relatively high for angry driving style, compared to other drivers within the sample, he or she will be classified as a driver with an angry driving style, even if his or her raw score for other driving styles were higher. Therefore, the third method which included the adjusted factor loadings was applied in this study. The same method was used by Hooft van Huysduynen et al. [18] in determining the multi-dimensional driver profiles in Netherlands and also implemented by Shahab, Terken and Eggen [32] in determining the multi-dimensional driver's personal values in driving.

The third method falls under the refined approaches where generally it is more complicated than the non-refined approaches. They are usually implemented to increase the validity and to obtain the unbiased estimation of the true factor scores by standardizing the correlation between the item and the factor by including the adjusted factor loadings. The adjusted factor loadings (calculated using Bartlett's method) were multiplied with each of the item scores before the average calculation in each factor was done. After that, the mean score and the standard deviation of the four-factor scores were calculated. Then, factor scores for each of the respondents 'driving styles were checked if they were exceeded the specific threshold, which is the sum of the mean and the standard deviation. This method identified a total of 223 (66.0%) respondents with one driving style, and the distributions of the driving styles are shown in Fig. . On the other hand, 115 respondents do not have any specific driving style due to the fact that their factor scores did not exceed any specific thresholds. Interestingly, by using the third method, it was identified that nobody has more than one driving style.

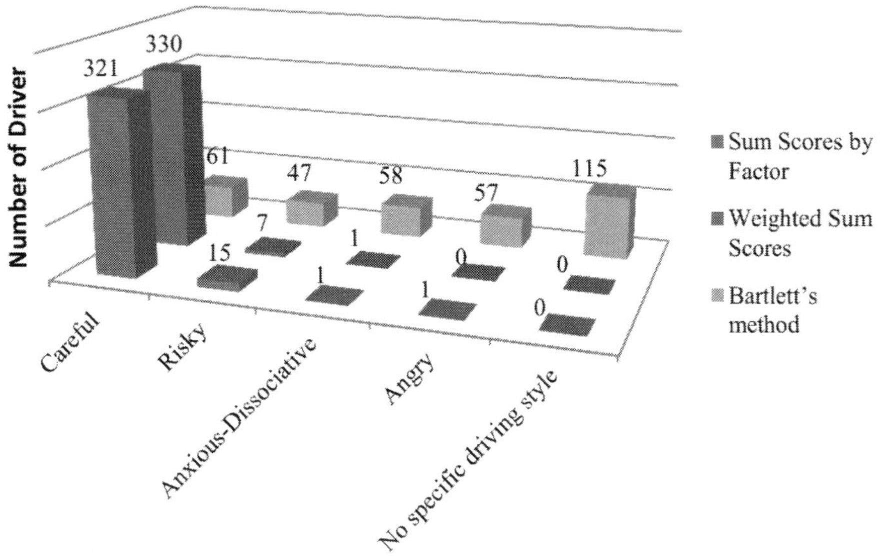

Fig. 3. Malaysian respondent's driving style determined using three different methods; sum scores by factor, weighted sum scores and Bartlett's method.

Table 3 shows the means and SDs of the raw scores for each factor for the 223 respondents with different driving styles. The columns correspond to the different driver types; the rows correspond to the Malaysian MDSI factors. As can be seen, all driver types have the highest score for careful driving, but the drivers with another driving style score relatively higher for their driving style in addition to scoring high for careful driving. This clearly demonstrates the challenge of how to assign driving styles in such a way that it gives sufficient discriminating power. The fact that they score high for careful driving as well can be understood by the fact that traffic is not a complete chaos.

Table 3. Tabulated of mean and standard deviation for Malaysian MDSI factors.

Malaysian MDSI Factors		Careful	Risky	Anxious-Dissociative	Angry
Careful	M	5.98	2.21	2.59	3.24
	SD	0.99	1.51	1.88	2.05
Risky	M	4.65	4.02	2.88	3.46
	SD	0.96	1.20	1.59	1.64
Anxious-Dissociative	M	4.82	2.56	3.68	3.32
	SD	1.03	1.13	1.26	1.48
Angry	M	5.07	2.59	2.74	4.07
	SD	1.01	1.24	1.64	1.34

3.5 Personality traits and sociodemographic factors

In addition to answering the MDSI questionnaire, the participants were also asked to fill in the Sensation Seeking questionnaire developed by Zuckerman et al. [22] and the Desire for Control questionnaire developed by Burger and Cooper [23]. The reason these two additional questionnaires were included was to investigate the correlations between driving styles and two personality traits (sensation seeking and desire for control). As stated in the literature, for example the recent work of Taubman-Ben-Ari and Yehiel [31] and Vaa [32], personality traits play an important role in determining the type of driver. Personality traits are hypothesized to be the root or the cause of why one is behaving in such way on the road. As mentioned by Vaa [32], personality traits can be viewed as dimensions of differences for any human in exhibiting their configurations of behaviour, emotions and cognition. Therefore, statistical analysis of Pearson correlations, as shown in Table below, were done to measure the relationship between the four factors (driving styles) found previously using the MDSI technique and the two personality traits of sensation seeking and desire for control. For Pearson correlations, a coefficient of +1 or -1 indicates perfect positive or negative correlations, while values of +/- 0.1 represent a small effect, +/- 0.5 represents a medium effect, and +/- 0.7 represents a large effect [24]. From Pearson correlation analyses, there are three relevant associations found, denoted by an asterisk ($P < 0.01$, highly significant) as shown in Table 4.

Table 4. Pearson correlation coefficients between MDSI factors and personality traits.

Malaysian MDSI factors	Sensation Seeking	Desire for Control
Careful	- 0.05	0.40*
Risky	0.40*	- 0.03
Anxious-Dissociative	- 0.04	- 0.24*
Angry	0.08	0.13

*$P < 0.01$

First, there is a positive and highly significant association between sensation seeking and risky driving style. As expected, risky driving is highly related to sensation seeking and this is shown by behaviours such as tailgating and driving over the limit. This correlation shows that drivers who score high for sensation seeking will have a higher tendency to be a risky driver. This result is in line with previous studies as reported by Taubman-Ben-Ari et al., [12] and Jonah [33] who also found a positive relationship between risky driving style and sensation seeking. In addition, sensation seeking is related to the opinion of the drivers about the acceleration forces experienced while driving as has been shown by the work of Lajunen and Summala [34], who found a positive correlation between sensation seeking and maximum speed and acceleration.

Second, the careful driving style has a highly positive correlation with desire for control. On the other hand, the anxious-dissociative driving style has a negative correlation with desire for control. These two correlations suggest that desire for control has a two-fold influence in driving; drivers with a high desire for control tend to be careful drivers and drivers with a low desire for control tend to be anxious-dissociative drivers.

Three MANOVA analyses were performed with the generated Malaysian MDSI factors and gender, age, and experience, to see if there are any significant effects between them. Indeed, these three elements showed significant associations as shown in Table 5.

Table 5. MANOVA for four Malaysian MDSI factors with gender, age, and experience.

	Gender	Age	Experience
Malaysian MDSI factors	$F(4,333) = 4.089$, $p < 0.05$ ($p = 0.003$)	$F(8,664) = 2.855$, $p < 0.05$ ($p = 0.004$)	$F(8,664) = 2.480$, $p < 0.05$ ($p = 0.012$)

In order to check the significant fit of each the generated Malaysian MDSI factors and the two socio-demographic data (gender, age) and experience, univariate ANOVA was conducted for this study, as shown in Table 6.

Table 6. Mean, SD and univariate ANOVA of MDSI factors with gender, age, and experience.

Malaysian MDSI Factors	Gender		Age			Experience		
	Male (n =273)	Female (n=65)	Young (n=294)	Middle-Aged (n=42)	Elderly (n=2)	< 3 years (n=158)	3 to 7 years (n=23)	> 7 years (n=157)
Careful driving style								
Mean	5.11	4.84	5.01	5.35	5.42	4.59	5.02	5.16
SD	1.39	1.41	1.38	1.50	1.06	1.47	1.39	1.37
ANOVA	$F(1,336) = 3.014$, $p \geq 0.05$, $\eta_p^2 = 0.031$		$F(2,335) = 1.754$, $p \geq 0.05$, $\eta_p^2 = 0.010$			$F(2,335) = 3.381$, $p < 0.05^*$, $\eta_p^2 = 0.020$		
Risky driving style								
Mean	2.88	2.63	2.91	2.34	2.42	2.89	2.95	2.70
SD	1.51	1.53	1.51	1.45	0.82	1.46	1.52	1.50
ANOVA	$F(1,336) = 2.281$, $p \geq 0.05$, $\eta_p^2 = 0.007$		$F(2,335) = 4.816$, $p < 0.01^{**}$, $\eta_p^2 = 0.028$			$F(2,335) = 1.451$, $p \geq 0.05$, $\eta_p^2 = 0.009$		
Anxious-Dissociative driving style								
Mean	2.96	3.01	3.01	2.71	2.36	3.11	3.06	2.86
SD	1.44	1.51	1.48	1.21	0.71	1.32	1.51	1.39
ANOVA	$F(1,336) = 2.425$, $p \geq 0.05$, $\eta_p^2 = 0.007$		$F(2,335) = 2.413$, $p \geq 0.05$, $\eta_p^2 = 0.014$			$F(2,335) = 4.769$, $p < 0.01^{**}$, $\eta_p^2 = 0.028$		
Angry driving style								
Mean	3.57	3.25	3.55	3.26	2.20	3.47	3.55	3.48
SD	1.66	1.64	1.66	1.61	1.13	1.60	1.70	1.63
ANOVA	$F(1,336) = 8.270$, $p < 0.01^{**}$, $\eta_p^2 = 0.024$		$F(2,335) = 2.255$, $p \geq 0.05$, $\eta_p^2 = 0.013$			$F(2,335) = 0.248$, $p \geq 0.05$, $\eta_p^2 = 0.001$		

* *indicates significant effect*
** *indicates very significant effect*

The univariate ANOVA revealed four important effects. Experience had a significant effect both on factor 1 (careful driving style, $F (2,335) = 3.381$, $p < 0.05$, $\eta_p^2 = 0.010$) and factor 3 (anxious-dissociative, $F (2,335) = 4.769$, $p < 0.01$, $\eta_p^2 = 0.028$). Analysing the mean values indicates that the experienced drivers (> 7 years or driving experience) are more careful than drivers with moderate experience (between 3 to 7 years) and novice drivers (< 3 years) as it showed that the higher the experience gained, the more careful the driving style would become. By analysing the mean values for anxious-dissociative driving style, it is clearly indicated that higher experience drivers scored lower compared to less experienced drivers. In terms of age, the analysis of variance exposed a very significant negative effect associated with risky driving style with $F (2,335) = 4.816$, $P < 0.01$, $\eta_p^2 = 0.028$. The younger drivers scored higher in risky driving style when compared to the older drivers. Angry driving style has a very significant dependency on gender ($F (1,336) = 8.270$, $P < 0.01$, $\eta_p^2 = 0.024$): male drivers have higher tendencies to have angry driving style compared to female drivers. This finding is in line with the work of Sullman, Stephens and Yong [35] who mentioned that male drivers in Malaysia are frequently showing their anger in driving compared to Malaysian female drivers.

4 Conclusions

In conclusion, based on this study, there are four major types of driving styles found from Malaysian drivers using the factor analysis method and MDSI questionnaire. The driving styles as revealed from the 338 respondents are careful, risky, anxious-dissociative and angry driving styles. By using Bartlett's method, 61 drivers (18.0%), 47 drivers (13.9%), 58 drivers (17.2%) and 57 drivers (16.9%) were identified as having a careful, risky, anxious-dissociative and angry driving style, respectively. The remaining 115 respondents (34.0%) had no specific driving style. The comparison with the Israel and Netherlands studies did not reveal a lot of differences other than the former identified eight driving styles and the latter identified six driving styles while in this study four driving styles were found.

Pearson correlations and ANOVA (univariate and multivariate) analyses were conducted and associations between personality traits, socio-demographic factors, experience and driving styles were presented. In general, Malaysian drivers with a risky driving style were found to score high for sensation seeking trait was also influenced by age: younger drivers tend to exhibit more risky behaviour. For drivers with a careful driving style, experience plays a major role as it was found that longer experience will mould more careful drivers. Furthermore, Malaysian drivers with a high desire for control tend to have a careful driving style. For drivers with an anxious-dissociative driving style, a negative correlation was found with desire for control and also a significant relation with experience. For the angry driving style, there was a highly significant relation with gender where men were found to have higher associations with angry driving styles when compared to women.

This study provides knowledge about the relation between sociodemographic factors (age, gender and experience), personality traits and major Malaysian driving styles. The understanding, in particular, would contribute to better campaigns in providing awareness on human behaviour on the road. In addition, effective safety measures could be designed knowing that there are four major driving styles in Malaysia. Furthermore, when a driver's driving style can be established, engineers and designers could apply personalized persuasion in order to mitigate accidents and promote safe driving among the drivers. For example, a dashboard display that provides feedback by summarizing one's driving style over one trip or a system that automatically changes the colour of the dashboard into a more calming background might be devised by drivers with particular driving styles.

In the long run, the findings of this study may contribute to the development of the driving style of autonomous vehicles. The focus of development for future automobiles should not be put solely on the technology but also the behaviour or manners of the autonomous vehicle on the road [36]. It may be assumed that users of an autonomous vehicle who score high for sensation seeking will feel discomfort unless the induced forces in riding the autonomous vehicle are about the same as if they would drive themselves. Desire for control is crucial when a fully autonomous vehicle will be released to the market. It is projected that the vehicle will be without the manoeuvring controllers like pedals and steering as this operational task will be taken over by the autonomous vehicle. As mentioned by Dudley [37], who had the chance to experience riding in Google's autonomous vehicle, his first reaction was one of surprise by the absence of the controllers in the dashboard and he immediately felt the need for some sort of controls on the car. He mentioned that riding the Google fully autonomous vehicle was like riding in one of those amusement park rides where you have no control over the vehicle and just sit and go along with the ride.

The authors would like to express their gratitude and special acknowledgements to Universiti Teknikal Malaysia Melaka (UTeM) and Ministry of Education, Malaysia (KPM) for the funding of the Ph.D. programs of Juffrizal Karjanto and Nidzamuddin Md. Yusof and also the post-doctoral program of Muhammad Zahir Hassan in the Industrial Design Department, Eindhoven University of Technology, The Netherlands.

References

1. M. M. Abdul Manan, A. Várhelyi, "Motorcycle fatalities in Malaysia," *IATSS Res.*, vol. **36**, no. 1, pp. 30–39, (2012)
2. S. Allyana, S. Mohamed, H. M. Jamil, M. Musa, N. Isah, W. S. Voon, "Impact Studies of Automated Enforcement System Implementation," (2014)
3. J. Mohd Marjan, N. Mohamed, M. R. Othman, F. Abas, F. Mohd Yusod, W. Shaw Voon, "Evaluation of the Effectiveness of OPS Bersepadu Hari Raya 2011," (2012)
4. D. J. Beirness, "Do we really drive as we live? The role of personality factors in road crashes," *Alcohol, Drugs Driv.*, vol. **9**, no. 3–4, pp. 129–143, (1993)
5. R. D. Garrity, J. Demick, "Relations among Personality Traits, Mood States, and Driving Behaviors," *J. Adult Dev.*, vol. **8**, no. 2, pp. 109–118, (2001)
6. M. J. M. Sullman, A. N. Stephens, M. Yong, "Driving anger in Malaysia," *Accid. Anal. Prev.*, vol. **71**, pp. 1–9, (2014)
7 G. J. Wilde, "The theory of risk homeostasis: Implications for safety and health," *Risk Anal.*, vol. **2**, no. 4, pp. 209–225, (1982)
8. R. Näätänen, H. Summala, "A model for the role of motivational factors in drivers' decision-making," *Accid. Anal. Prev.*, vol. **6**, no. 3–4, pp. 243–261, (1974)
9. H. Summala, "Towards understanding motivational and emotional factors in driver behaviour: Comfort through satisficing," in *Modelling Driver Behaviour in Automotive Environments: Critical Issues in Driver Interactions with Intelligent Transport Systems*, (2007), pp. 189–207
10. T. Vaa, "Modelling driver behaviour on basis of emotions and feelings: Intelligent transport systems and behavioural adaptations," in *Modelling Driver Behaviour in Automotive Environments: Critical Issues in Driver Interactions with Intelligent Transport Systems*, (2007), pp. 208–232
11. Malaysian Institute of Road Safety (MIROS), "Data Am Kemalangan Jalan Raya di Malaysia (1997 – 2014)," (2016). [Online]. Available: https://www.miros.gov.my/1/page.php?id=17. [Accessed: 16-Feb-2016]
12. O. Taubman-Ben-Ari, M. Mikulincer, O. Gillath, "The multidimensional driving style inventory - Scale construct and validation," *Accid. Anal. Prev.*, vol. **36**, no. 3, pp. 323–332, (2004)
13. E. R. Dahlen, R. C. Martin, K. Ragan, and M. M. Kuhlman, "Driving anger, sensation seeking, impulsiveness, and boredom proneness in the prediction of unsafe driving," *Accid. Anal. Prev.*, vol. **37**, no. 2, pp. 341–348, (2005)
14. J. J. Arnett, D. Offer, and M. A. Fine, "Reckless driving in adolescence: 'State' and 'trait' factors," *Accid. Anal. Prev.*, vol. **29**, no. 1, pp. 57–63, (1997)
15. M. Zuckerman and M. Neeb, "Demographic influences in sensation seeking and expressions of sensation seeking in religion, smoking and driving habits," *Pers. Individ. Dif.*, vol. **1**, no. 3, pp. 197–206, (1980)
16. D. J. French, R. J. West, J. Elander, and J. M. Wilding, "Decision-making style, driving style, and self-reported involvement in road traffic accidents.," *Ergonomics*, vol. **36**, no. 6, pp. 627–644, (1993)
17. E. Gulian, G. Matthews, A. I. Glendon, D. R. Davies, L. M. Debney, "Dimensions of driver stress," *Ergonomics*, vol. **32**, no. 6, pp. 585–602, (1989)
18. H. Hooft van Huysduynen, J. Terken, J.-B. Martens, B. Eggen, "Measuring Driving Styles : A Validation of the Multidimensional Driving Style Inventory," in *7th International Conference on Automotive User Interfaces and Interactive Vehicluar Applications*, (2015), pp. 257–264
19 M. Ishibashi, M. Okuwa, S. Doi, M. Akamatsu, "Indices for characterizing driving style and their relevance to car following behavior," *Proc. SICE Annu. Conf.*, pp.

1132–1137, (2007)
20. F. M. Poó, R. D. Ledesma, "A Study on the Relationship between Personality and Driving Styles," *Traffic Inj. Prev.*, vol. **14**, pp. 346–352, (2013)
21. T. Lajunen, A. Corry, H. Summala, L. Hartley, "Cross-Cultural Differences in Drivers' Self-assessments of Their Perceptual-motor and Safety Skills: Australians and Finns," *Pers. Individ. Dif.*, vol. **24**, no. 4, pp. 539–550, (1998)
22. M. Zuckerman, D. Michael Kuhlman, M. Joireman, H. Kraft, "Five robust questionnaire scale factors of personality without culture," *Pers. Individ. Dif.*, vol. **12**, no. 9, pp. 929–941, (1993)
23. J. M. Burger, H. M. Cooper, "The desirability of control," *Motiv. Emot.*, vol. **3**, no. 4, pp. 381–393, (1979)
24. A. Field, *Discovering Statistics using SPSS. Third Edition*. (2009)
25. B. Williams, T. Brown, A. Onsman, "Exploratory factor analysis : A five-step guide for novices," *Australas. J. Paramadicine*, vol. **8**, no. 3, pp. 1–13, (2012)
26. J. Raubenheimer, "An item selection procedure to maximise scale reliability and validity," *SA J. Ind. Psychol.*, vol. **30**, no. 4, pp. 59–64, (2004)
27. B. a. Jonah, R. Thiessen, E. Au-Yeung, "Sensation seeking, risky driving and behavioral adaptation," *Accid. Anal. Prev.*, vol. **33**, no. 5, pp. 679–684, (2001)
28. G. Miller, O. Taubman-Ben-Ari, "Driving styles among young novice drivers-The contribution of parental driving styles and personal characteristics," *Accid. Anal. Prev.*, vol. **42**, no. 2, pp. 558–570, (2010)
29. C. Distefano, M. Zhu, and D. Mîndrilă, "Understanding and Using Factor Scores: Considerations for the Applied Researcher," *Pract. Assessment, Res. Eval.*, vol. **14**, no. 20, pp. 1–11, (2009)
30. Q. Shahab, J. Terken, and B. Eggen, "Development of a Questionnaire for Identifying Driver's Personal Values in Driving," in *Proceedings of the 5th International Conference on Automotive User Interfaces and Interactive Vehicular Applications - AutomotiveUI '13*, (2013), pp. 202–208
31. O. Taubman-Ben-Ari and D. Yehiel, "Driving styles and their associations with personality and motivation," *Accid. Anal. Prev.*, vol. **45**, pp. 416–422, (2012)
32. T. Vaa, "From Gibson, Crooks to Damasio: The role of psychology in the development of driver behaviour models," *Transp. Res. Part F Traffic Psychol. Behav.*, vol. **25**, pp. 112–119, (2014)
33 B. a. Jonah, "Sensation seeking and risky driving: A review and synthesis of the literature," *Accid. Anal. Prev.*, vol. **29**, no. 5, pp. 651–665, (1997)
34. T. Lajunen, H. Summala, "Effects of driving experience, personality, driver's skill and safety orientation on speed regulation and accidents," *Traffic Transp. Psychol. Theory Appl.*, pp. 283–294, (1997)
35. M. J. M. Sullman, A. N. Stephens, M. Yong, "Anger, aggression and road rage behaviour in Malaysian drivers," *Transp. Res. Part F Traffic Psychol. Behav.*, vol. **29**, pp. 70–82, (2015)
36. M. Sikkenk and J. Terken, "Rules of Conduct for Autonomous Vehicles," in *Proceedings of the 7th International Conference on Automotive User Interfaces and Interactive Vehicular Applications*, (2015), pp. 19–22
37. D. Dudley, "Google Self-Driving Car Perfect for Elderly Drivers - AARP," *AARP*, (2015). [Online]. Available: http://www.aarp.org/home-family/personal-technology/info-2014/google-self-driving-car.html. [Accessed: 16-Feb-2016]

Simulating autonomous driving styles: Acceleration s for three road profiles

Juffrizal Karjanto[1, 2], *Nidzamuddin* Md. Yusof[1, 2*], *Jacques* Terken[1], *Frank* Delbressine[1], *Muhammad Zahir* Hassan[1, 3] and *Matthias* Rauterberg[1]

[1]Department of Industrial Design, Eindhoven University of Technology, Eindhoven, The Netherlands
[2]Faculty of Mechanical Engineering, Universiti Teknikal Malaysia Melaka, Melaka, Malaysia
[3]Faculty of Engineering Technology, Universiti Teknikal Malaysia Melaka, Melaka, Malaysia

Abstract. This paper presents a new experimental approach to simulate projected autonomous driving styles based on the accelerations at three road profiles. This study was focused on the determination of ranges of accelerations in triaxial direction to simulate the autonomous driving experience. A special device, known as the Automatic Acceleration and Data controller *(AUTOAccD),* has been developed to guide the designated driver to accomplish the selected accelerations based on the road profiles and the intended driving styles namely assertive, defensive and light rail transit (LRT). Experimental investigations have been carried out at three different road profiles (junction, speed hump, and corner) with two designated drivers with five trials on each condition. A driving style with the accelerations of LRT has also been included in this study as it is significant to the present methodology because the autonomous car is predicted to accelerate like an LRT, in such a way that it enables the users to conduct activities such as working on a laptop, using personal devices or eating and drinking while travelling. The results demonstrated that 92 out of 110 trials of the intended accelerations for autonomous driving styles could be achieved and simulated on the real road by the designated drivers. The differences between the two designated drivers were negligible, and the rates of succeeding in realizing the intended accelerations were high. The present approach in simulating autonomous driving styles focusing on accelerations can be used as a tool for experimental setup involving autonomous driving experience and acceptance.

1 Introduction

In Malaysia, the number of road accidents is steadily increasing while deaths remain static Recent studies have shown that the development of autonomous vehicle (AV) is rapidly growing, and most drivers have already accepted Advance Driving Assistance System (ADAS), which is the foundation of the autonomous driving system [1], [2]. The automakers promise that the ultimate level of automation is a fully AV where human drivers/users will no longer have to deal with the operational and tactical levels in driving

* Corresponding author: n.yusof@tue.nl

anymore [3]. Therefore, future AV users are only expected to decide where they want to go, select the destination and let the AV does the rest, from driving the car, selecting the best route, dealing with other road users and all the tasks that the human drivers used to do. In that sense, an AV is expected to behave in optimized manners that promote safety, reliability and comfort to the users. However, the acceptance of comfort may vary depending on users' perception and can be very subjective due to many factors especially driving styles [4]. Taubman-Ben-Ari et al. [5] suggested that driving styles vary across drivers and identified the most driving-specific factors into four categories; *reckless and careless* driving style, *anxious* driving style, *angry and hostile* driving style, and *patient and careful* driving style.

An AV or self-driving car contains thousands of sensors, and the way the vehicle is driven is strictly based on optimized logic (based on algorithms that analyse the inputs from sensors). In contrast, human drivers are driving based on their "targeted feeling" and heavily influenced by their own personality traits [6], [7]. Hence, this may lead to mismatched in term of driving styles preference for different types of the future AV users. The way an AV accelerates and decelerates at the intersection, handles a corner and circles a roundabout will be unfamiliar to the human drivers/users. Furthermore, since the AV drivers/users are no longer in command (no hands on the wheel, no feet on the pedals), the AV will probably accelerate and decelerate in the style that did not induce mental and physical discomfort to the human drivers/users in order to keep safety and comfort intact.

In order for an AV to be an optimized driver without sacrificing the comfort and enabling the activities other than driving for the users inside the AV, the AV is predicted and suggested to accelerate and decelerate slower than the typical human driven car and faster (or maybe the same) than the light rail transit (LRT) [8]. The LRT systems could be used as guideline such that the ways the system accelerates and decelerates in longitudinal and lateral direction are within the acceptable level of comfort for the passengers. Activities such as drinking and eating without spilling, and working or playing with electronic devices without experiencing physical discomfort like motion sickness as has been predicted would happen in the future AV riding experience [9], [10]. These activities were made possible since the induced forces felt by the users inside the LRT were minimized. In addition, the experience in riding an LRT is likely to be the benchmark used by the automotive players for their future development of AV since the LRT concept has been proven as one of the best options for the user to conduct secondary tasks while travelling in a moving vehicle [8]. On the same hand, future AV users who have had the experience travelling by train or LRT would draw mental scenarios of how traveling inside an AV should be as this new technology will also offer the kind of travelling without the need to drive.

A typical human driven car accelerates around 0.79 m/s^2 to 4.86 m/s^2 after leaving a junction or signalized intersection as stated in the work of Bogdanović and Ruškić [11]. They studied the various accelerations of cars, using digital cameras and programming, starting from the stop line at a juncture with traffic lights for ten intersections in one of the towns in Serbia. For the deceleration when approaching a junction or signalized intersection, El-Shawarby et al. [12] found that with 821 events, the range is between 1.51 m/s^2 to 7.47 m/s^2. The study was done with 60 participants in the enclosed area where it was safe to conduct the experiment. On the other hand, for the lateral accelerations, the ranges are influenced by the radius of the corner or roundabout on which the car is travelling [13]. Hugemann and Nickel (2003) [13] revealed that drivers usually produced smaller lateral accelerations with smaller radii corners while, in contrast, higher radii would usually associate with higher lateral accelerations. For the sharp turn or corner with radius in between 20 to 40 meters, the 10th, 50th and 90th percentile of the lateral acceleration found is 3.1 m/s^2, 4.1 m/s^2 and 5.3 m/s^2, respectively. Table 1 shows the list of accelerations in multi-directions for both the LRT and human driven car.

Table 1. Longitudinal and lateral accelerations and decelerations for light rail transit and human driven cars [11]–[15].

	LRT	Typical human-driven car	Upper limit in human-driven car
Longitudinal acceleration	1.34 m/s^2 or 0.14 g	2.47 m/s^2 or 0.25 g	4.86 m/s^2 or 0.5 g
Longitudinal deceleration	- 1.34 m/s^2 or 0.14 g	- 3.27 m/s^2 or -0.33 g	- 7.47 m/s^2 or -0.76 g
Lateral acceleration	1.47 m/s^2 or 0.15 g	4.10 m/s^2 or 0.42 g	5.30 m/s^2 or 0.54 g

For the vertical acceleration, Weber and Braaksma [16] stated that the measurement of discomfort, in terms of vertical force when traversing a hump, is depending on root sum of square (RSS) acceleration or peak vertical acceleration. The RSS value that indicates physical discomfort towards the motorist was at 0.17g, while peak vertical acceleration of 0.7g, is identified as heavily associated with human physical discomfort.

This paper reports the development and the execution of the experimental technique used in simulating the AV driving style in real road scenario that involves cornering, junction, and speed hump. A significant challenge to this research was to simulate the autonomous driving styles based on the selected accelerations at different road profiles. The present approach consists of experimental work that was carried out and validated using on-vehicle site testing with two designated drivers who attempted to simulate the intended accelerations using a special device which is called as the Automatic Acceleration and Data controller *(AUTOAccD)*.

The outcomes of this experiment were mapped with the results of calculated accelerations measurement from the data gained from the testing vehicle. In addition, the simulated data from the two designated drivers were tabulated and compared.

The current approach in simulating autonomous driving styles focusing on the accelerations in the triaxial direction can be used as a tool for experimental research on future autonomous driving experience and acceptance, especially in the preferred accelerations and comfort determination.

2 Methodologies

The methodology of the present study is shown in Fig. 1. The process started with the determination of driving styles to be simulated. Next, the process continued with the selection of road profiles that consist of a straight lane approaching and leaving a junction, a cornering, and a speed hump. Afterwards, the determination process of the driving styles to be simulated and followed by the determination of ranges of the accelerations for the selected driving styles based on the selected road profiles. Finally, the outcomes were examined in twofold analyses; first, was the comparison between the measured accelerations data using AUTOAccD and the calculated accelerations using the On-Board Diagnostic (OBD) data from the testing vehicle and second, was the comparison between accelerations simulated by the two designated drivers.

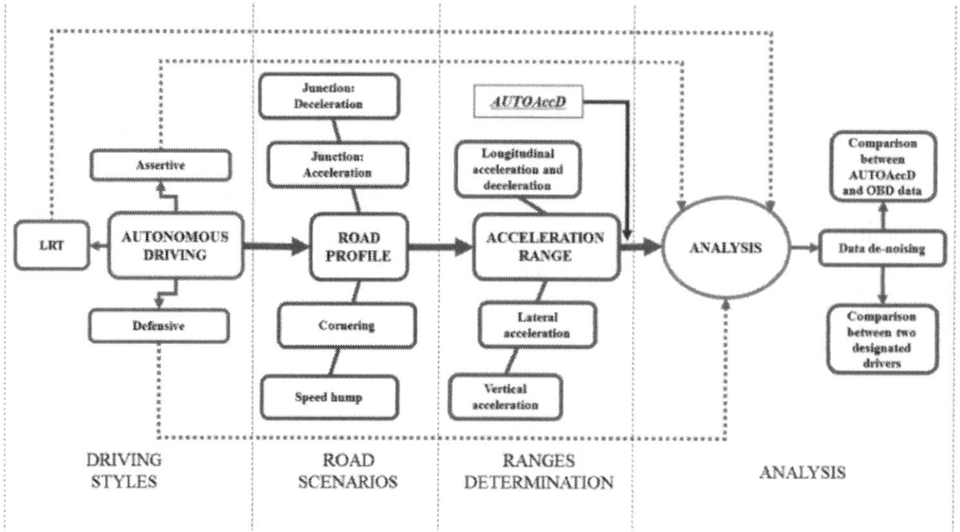

Fig. 1. The methodology of the present study.

2.1 Driving style

The first step was to determine the type of driving styles to be simulated. Elander et al. [4] defined driving style as the approach individuals select to manoeuvre their cars and something they accustom to over the years. In addition, driving styles include the selection of driving speeds in different circumstances, distance to others, the decision to overtake and tendency to violate traffic laws. Wang and Lukic [17] reviewed and summarized that driving styles are usually clustered into three categories; *mild drivers* (calm driving or economical driving style), *normal drivers* (medium driving style), and *aggressive drivers* (sporting driving style). Likewise, Taubman-Ben-Ari et al. [5] created the Multidimensional Driving Style Inventory (MSDI), which was modified from existing questionnaires (e.g., [18], [19]). Thus, driving styles vary across drivers and therefore more than one driving style for autonomous driving experience would be required [20]. In this study, three types of future autonomous driving styles of interest were selected which are defensive, assertive and LRT.

2.2 Road scenario

An Audi A3 was used as the testing vehicle for this experiment, provided by the Automotive Lab, Eindhoven University of Technology. The experimental investigation was carried out based on three different urban road profiles consisting of a junction (to access longitudinal accelerations and decelerations), speed hump (to access vertical accelerations) and corner (to access lateral accelerations). The route is located within the private compound of the Science Park campus, Eindhoven University of Technology, Eindhoven, Netherlands.

2.3 Ranges of accelerations determination

The ranges of lateral and longitudinal accelerations and deceleration to be implemented in order to simulate the LRT, defensive and assertive driving styles are presented in Fig. 2. These parameters are based on the data from LRT and typical human driven cars as reported in previous works [8], [11]–[13], [15]. For the longitudinal acceleration and deceleration, where the time plays a major role in getting the selected accelerations and decelerations, data for required time to generate the intended forces was calculated. In addition, the targeted velocity (the ending velocity in simulating acceleration or the starting velocity in simulating deceleration) would determine the required time in order to achieve the intended accelerations or decelerations and depending on where the experiment is done, the speed limit of that particular route needs to be taken into consideration.

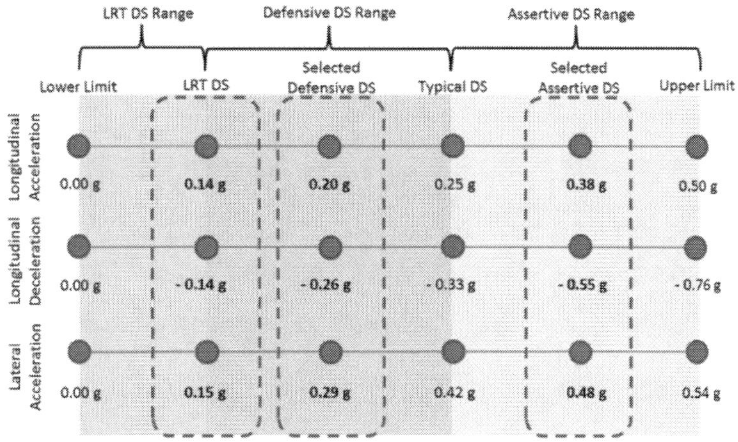

Fig. 2. The range of selection for defensive and assertive driving style for lateral and longitudinal accelerations and deceleration.

Next, the lateral acceleration, which is the centrifugal acceleration that pushes the vehicle to the outward of the curve during a cornering process, was determined. The centrifugal acceleration (a_c) formulation is shown in Eq. 1 below:

$$a_c = \frac{v^2}{r} \tag{1}$$

where v and r represent travelling velocity in the corner and radius of the corner, respectively. In order to achieve the intended lateral accelerations as has been shown in Fig. 2, the selection of velocity has to be determined based on the parameters depicted in Table 2 below.

Table 2. Required velocity to generate lateral acceleration during cornering.

Driving style	Lateral acceleration during cornering	Velocity	
		(m/s)	(km/h)
LRT	0.15 g or 1.47 m/s^2	4.79	17.24
Defensive	0.29 g or 2.84 m/s^2	6.66	23.97
Assertive	0.48 g or 4.70 m/s^2	8.57	30.85

Radius of the corner = 15.60 m

As for the vertical acceleration, the selections of velocity when approaching the speed hump for different driving styles are shown in Fig. 3 and Table 3.

DOI: 10.1051/matecconf/20179001005

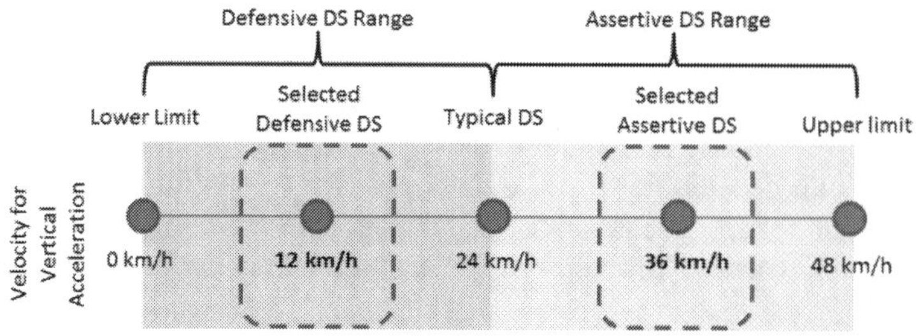

Fig. 3. The range of selection of velocity approaching the speed hump.

Table 3. Required velocity to generate vertical acceleration at the designated speed hump.

Driving style	**Velocity**		**Vertical acceleration**	
	(m/s)	**(km/h)**	**(m/s^2)**	**(g)**
Defensive	0 – 6.7	0 – 24.0	0.39	0.04
Assertive	6.7 – 13.3	24.0 – 48.0	3.72	0.38

Height of the speed hump = 0.10 m
Length of the speed hump = 4.07 m

2.4 Automatic acceleration and data controller (AUTOAccD)

A special dedicated device, the Automatic Acceleration and Data controller *(AUTOAccD)*, was specifically developed with dual capabilities: to assist the designated driver to drive according to the defined acceleration condition and also act as instrumented processor to measure and collect the vehicle dynamics data. *AUTOAccD* consists of an OBD-II adapter that is connected to Arduino board and LCD display as shown in Fig. 4. OBD is an automotive term referring to a vehicle's self-diagnostic and is able to monitor real-time engine parameters such as speed, fuel consumption and transmission of a vehicle. The *AUTOAccD* was developed based on the data logger kit from Freematics [21] which comprised of Global Positioning System (GPS) receiver, triple-axis Micro-Electro-Mechanical Systems (MEMS) accelerometers and gyro meters. In addition, the data logger was connected to a live display in the form of LCD screen. All the measured and collected data are stored in a micro Secure Digital (SD) card for analysis later.

The display was designed to assist the designated driver in maintaining a certain driving style by showing the specific ranges of lateral and longitudinal forces during the experiment. The display served as assistance and assurance device to the driver, apart from the training and his own experience in driving, in order to achieve the required acceleration at the specific location. There are three colours available to indicate the instantaneous driving style, in which LRT is indicated by purple colour while defensive and assertive are shown by green and red colours, respectively. In addition, there are two axes on which the vertical axis represents the longitudinal direction (upward movements indicates positive direction and vice versa) while the horizontal axis represents the lateral direction (rightward movements indicates positive direction and vice versa.

(a) (b)

Fig. 4. (a) The position of the *AUTOAccD* in the car; and (b) the placement of the data logger and OBD.

2.5 Data analysis

Recent developments in Micro-Electro-Mechanical Systems (MEMS) technology have enabled the development of miniaturized sensors units like accelerometers and gyro meters. However, working with data obtained from the experiments involving measurements by these devices usually comes with noise ranging from alignment errors to random noise. Hence, the noise has to be filtered for further data analysis and interpretation. Rather than using the traditional filters for de-noising such as time-frequency domain filters like Fourier transform, a time-scale domain filter known as Wavelet transform was employed. The Wavelet de-noising method was implemented because its ability to remove noises and retain the signal regardless the frequency of the signal [22], [23]. For this study, the Wavelet de-noising was done using Matlab (Wavemenu) and the selection of the appropriate filter was based on the previous literature using the same equipment. The Daubechies filter level three (Db3) was used in order to remove the high-frequency noise from the MEMS accelerometers [24].

After the de-noising process, the acceleration data from the two designated drivers were compared to check the ability of the AUTOAccD in assisting the drivers to simulate the intended accelerations. In addition, the accuracies and consistencies of the simulated acceleration will also be discussed. Next, the comparison between the acceleration data gained from the AUTOAccD and the calculated data from the testing vehicle's OBD was done to check the discrepancies between the two tools.

3 Results and discussion

The analyses started with the process of filtering or de-noising of the signal attained from the accelerometers. Wavemenu function in Matlab was used for its simplicity and its elegant visual tools that enable the users to interactively de-noise a signal and instantaneously display the result on the screen [25]. After the de-noising process, the acceleration data were analysed in terms of simulations from the two designated drivers and were compared with the calculated data that were gained from the vehicle's system.

3.1 Comparison between data measured by AUTOAccD and calculated from OBD

The comparison between acceleration data (longitudinal and lateral) gained from the triaxial accelerometers within the AUTOAccD and calculated using the speed extracted from the vehicle's OBD is shown in Table 4. For the longitudinal accelerations and decelerations, the largest variances were found in the simulation of assertive driving style where as much as 0.06 g and 0.12 g differences were recorded, respectively. Otherwise, for the defensive and LRT driving style, the differences in means between AUTOAccD and OBD were between 0.00 g to 0.03 g, which indicated higher accuracies. Meanwhile, for the lateral acceleration, between the three driving styles, AUTOAccD for assertive driving style yielded the highest difference with 0.14 g while the defensive driving style registered a difference of 0.05 g and an LRT driving style with the smallest difference with 0.01 g.

Table 5 shows the comparison between vertical acceleration data collected from the AUTOAccD and determined from the vehicle's OBD. The differences in means for each of the driving styles were 0.03 g and 0.04 g. In addition, the standard deviations recorded were between 0.00 and 0.06 which showed great consistencies among the two set of data. From Table 4 and Table 5, similar patterns can be observed in which all the measured acceleration data were higher than the calculated ones. These phenomena were parallel with the previous work of Brol and Mamala [26] who found comparable results with the measured data having the differences up to 0.10 g. The differences were contributed from the road irregularities, and some were from the engine noises. The road surface irregularities, in this study, were the cobblestones road that generated additional accelerations through the vehicle suspension system and was recorded by the accelerometers which were attached to the vehicle's body.

Table 4. Comparison between longitudinal and lateral accelerations data measured by AUTOAccD and calculated from OBD.

Type of acceleration		Assertive DS		Defensive DS		LRT DS	
		AUTO AccD	OBD	AUTO AccD	OBD	AUTO AccD	OBD
Longitudinal acceleration	Mean	0.41	0.35	0.22	0.21	0.09	0.07
	SD	0.07	0.06	0.06	0.06	0.04	0.05
Longitudinal deceleration	Mean	-0.48	-0.36	-0.32	-0.29	-0.15	-0.15
	SD	0.07	0.10	0.04	0.09	0.03	0.03
Lateral acceleration	Mean	0.54	0.39	0.31	0.26	0.12	0.11
	SD	0.12	0.05	0.03	0.02	0.04	0.03

Table 5. Comparison between vertical accelerations data measured by AUTOAccD and calculated from OBD.

Vertical acceleration	Assertive DS		Defensive DS	
	AUTO AccD	OBD	AUTO AccD	OBD
Mean	0.33	0.30	0.07	0.03
SD	0.06	0.01	0.01	0.00

3.2 Comparison between two designated drivers

For the leaving-a-junction scenario, the tabulated accelerations simulated by the two designated drivers using three driving styles (with five trials for each driving style) are presented in Fig. 5. Additionally, the means and standard deviations (SD) are presented in Table 6. For the simulated accelerations by using LRT driving style, both drivers managed to simulate the accelerations within the desired range, although the means of the drivers were quite far apart, the SD indicated that they were very consistent in simulating this driving style. For the defensive and assertive driving styles, both of the drivers had fluctuated accelerations and were less consistent compared to when simulating the LRT driving style. Nonetheless, the means for both drivers in simulating the last two driving styles were closer to the intended accelerations. For the defensive driving style, out of ten trials altogether, three of the accelerations fell outside of the desired range, whereas for the assertive driving style, all the simulated accelerations were within the intended range.

In this particular scenario, the testing car was accelerating in one smooth motion where there was no jerking or any other traffic that would impede the acceleration process. In addition, in order to successfully simulate larger acceleration forces especially for the assertive driving style, the accelerations must be done in preferably with lower transmission such as the first gear. Higher acceleration forces were usually achieved with lower gears especially when the vehicle started to accelerate from a completely static position. Higher gears usually produce acceleration not as large as the acceleration produced from the lower gears after leaving an intersection as has been shown in the previous study by Bogdanović and Ruškić [11].

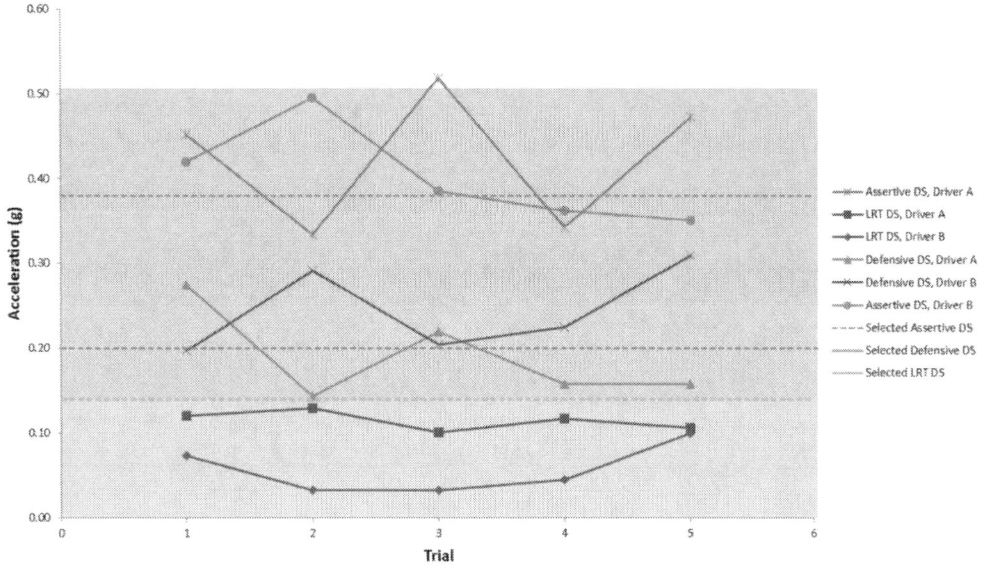

Fig. 5. Tabulated longitudinal accelerations for two designated drivers simulated with three driving styles.

Table 6. Means and SD of the three simulated driving styles for longitudinal accelerations for each driver, $n_{trial} = 5$.

	LRT DS (Targeted acc. = 0.14g)		**Defensive DS (Targeted acc. = 0.20g)**		**Assertive DS (Targeted acc. = 0.38g)**	
Driver	**Mean (g)**	**SD**	**Mean (g)**	**SD**	**Mean (g)**	**SD**
A	0.11	0.01	0.19	0.06	0.42	0.08
B	0.06	0.03	0.25	0.05	0.40	0.06

For the straight lane approaching the junction scenario, the decelerations simulated by the two designated drivers using three driving styles, with five trials each, are tabulated in Fig. 6. In addition, the means and SDs are also presented in Table 7. For this scenario, both drivers showed high consistency in simulating the accelerations although driver B actually did not manage to register any acceleration within the desired LRT range, whereas, for the defensive and assertive driving style, much higher variances were produced by both of the drivers. In addition, the attempted decelerations generally deviated quite far from the intended means. For the assertive driving style, all the simulated decelerations were within the intended range, while for the defensive driving style, both of the drivers registered half of the trials outside the intended range.

In simulating the deceleration, the drivers were instructed to decelerate, from a designated speed of 30 km/h, by adjusting the braking force according to the guides as shown on the AUTOAccD. However, in general, the simulations of deceleration are harder to generate compared to the simulations of acceleration. One of the reasons was that the drivers had to control both the clutch and brake pedal in order to achieve the designated deceleration rate and making sure the engine would not stall whereas, in the accelerations, the drivers only need to control the accelerator or gas throttle.

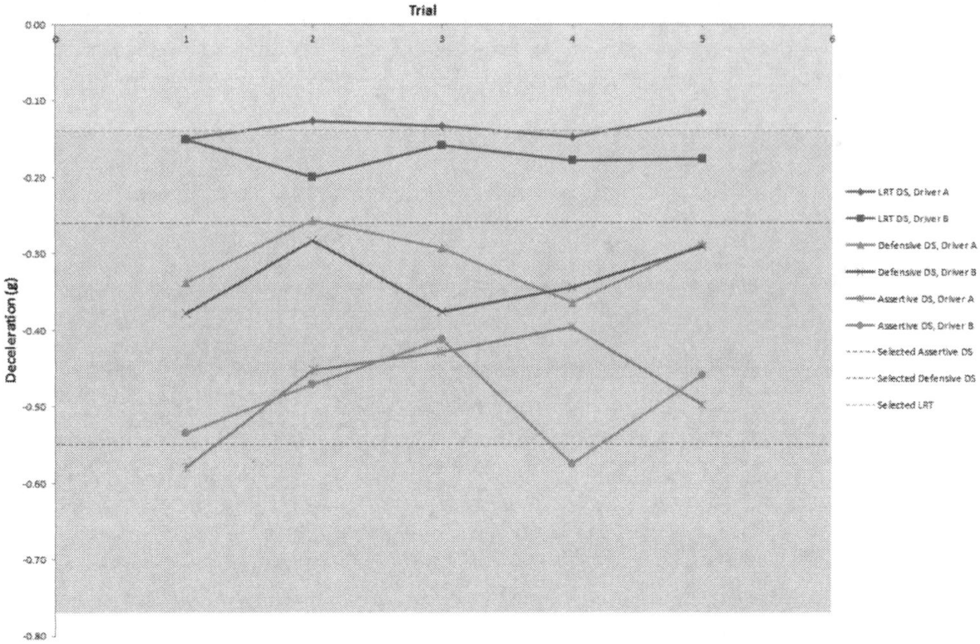

Fig. 6. Tabulated decelerations for two designated drivers simulated with three driving styles.

Table 7. Means and standard deviations of the three simulated driving styles for decelerations for each driver, $n_{trial} = 5$.

Driver	LRT DS (Targeted acc. = - 0.14g)		Defensive DS (Targeted acc. = - 0.26g)		Assertive DS (Targeted acc. = - 0.55g)	
	Mean (g)	SD	Mean (g)	SD	Mean (g)	SD
A	- 0.13	0.01	- 0.31	0.04	- 0.47	0.07
B	- 0.17	0.02	- 0.33	0.05	- 0.49	0.06

For the cornering, the lateral accelerations simulated by the two designated drivers using three driving styles are presented in Fig. 7. In addition, the means and SDs are also presented in Table 8. In the simulation of the LRT driving style, driver B had better consistencies compared to driver A as lower SD was registered. For the defensive driving style, in which both drivers achieved mean values that are higher than the intended one, both of the drivers registered good consistencies with SD of 0.03 each. Whereas for the simulation of assertive driving style, there were fluctuations in the simulated accelerations, as in this driving style the drivers had to generate accelerations that are above the average of typical human driving style. However, out of the combined ten trials from the two drivers, only once a simulated acceleration fell outside the lower limit of the intended assertive range.

 DOI: 10.1051/matecconf/20179001005

In simulating the lateral accelerations, it was crucial that the radius of the cornering was maintained to be as constant as possible. By referring back to Eq. (1), it is clear that both velocity and radius of the corner are playing a big role in determining the induced lateral acceleration [27]. In this study, the roads which have been used do not have any road markings on the surface, and therefore, both of the drivers would try to undertake the cornering by making sure the car was always approximately travelling in the middle of the road. As mentioned by Charlton [28], for more accurate results in cornering, a clearer and more accurate guideline or delineation treatment should be developed such as line marking on the road surface where the drivers can align themselves to make sure that for each trial, the radius of the travelling car in the corner stays almost the same every single time.

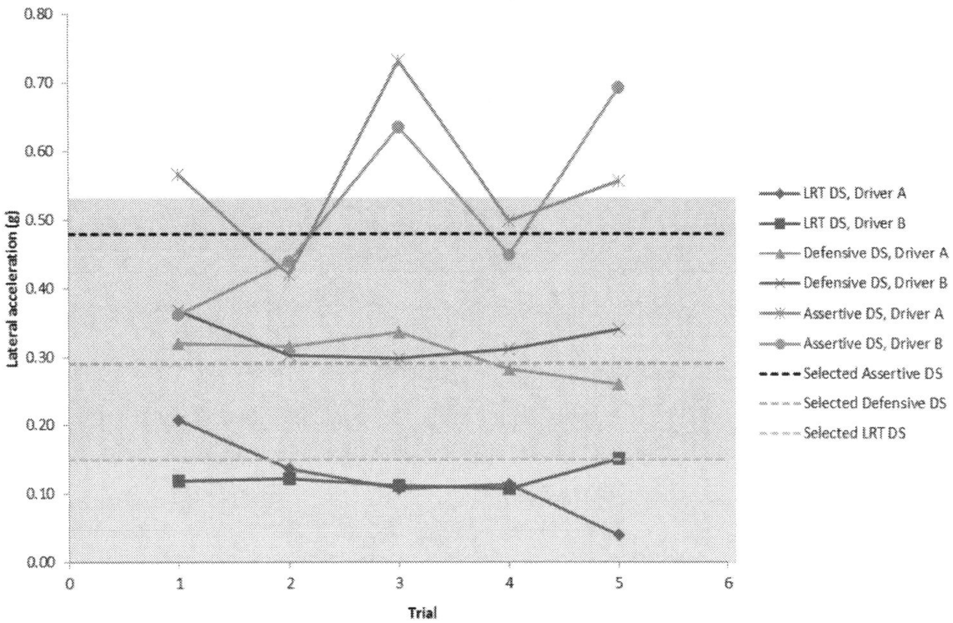

Fig. 7. Tabulated lateral accelerations for two designated drivers simulated with three driving styles.

Table 8. Means and standard deviations of the three simulated driving styles for lateral accelerations for each driver, $n_{trial} = 5$.

Driver	LRT DS (Targeted acc. = 0.15g)		Defensive DS (Targeted acc. = 0.29g)		Assertive DS (Targeted acc. = 0.48g)	
	Mean (g)	SD	Mean (g)	SD	Mean (g)	SD
A	0.12	0.06	0.30	0.03	0.47	0.11
B	0.12	0.02	0.32	0.03	0.46	0.14

For the vertical acceleration, the simulated accelerations are presented in Fig. 8. In addition, the means and SDs are presented in Table 9. For this study, both designated drivers realized higher vertical acceleration means than the intended for the simulation of defensive driving style while for the assertive driving style, they realized lower means than the intended one. In term of SDs, for the defensive driving style, both drivers managed to simulate higher consistencies as indicated by low deviation values although for the Driver A, his accelerations were sometimes lower and some other time were higher than the intended means whereas, for driver B, all the recorded values were lower than the intended means. For the assertive driving style, driver A indicated lower consistencies compared to driver B as the SD recorded were 0.08 and 0.03, respectively. From

, all of the 20 trials from both drivers were recorded within the intended ranges of interest.

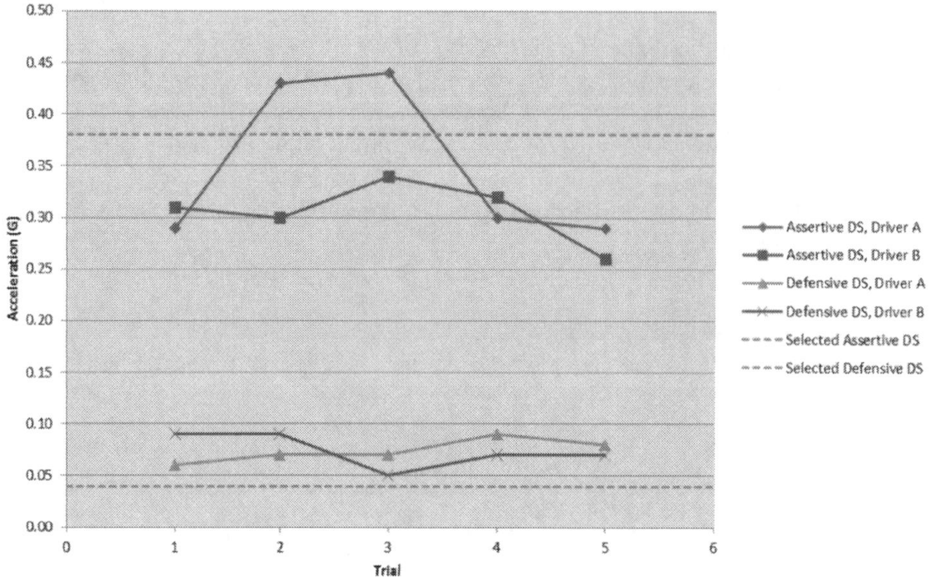

Fig. 8. Tabulated vertical accelerations for two designated drivers simulated with two driving styles.

Table 9. Means and standard deviations of the two simulated driving styles for vertical accelerations for each driver, $n_{trial} = 5$.

Driver	Defensive DS (Targeted acc. = 0.04g)		Assertive DS (Targeted acc. = 0.38g)	
	Mean (g)	SD	Mean (g)	SD
A	0.07	0.01	0.35	0.08
B	0.07	0.02	0.31	0.03

In simulating the vertical accelerations, the only parameter to be controlled was the approaching velocity. The main difference between a speed hump and a speed bump is that the former is designed for a vehicle to traverse at a low speed (25-30 km/h) without the need to slow down while the latter require a vehicle to decelerate up to 8 km/h when arriving at the tip of the bump [29]. Therefore, in this study, the intended assertive driving style required the drivers to traverse at the speed hump at a velocity of 36 km/h which is relatively high for drivers with defensive or normal driving style. In addition, the available speed hump for this experiment was relatively short at 4.07 m in length whereas a good speed hump's length is usually around 6.70 m or even extending to 9.50 m in some countries like Denmark and 12.0 m in Australia [16]. A longer speed hump would induce peak acceleration at a higher speed as usually a car has a wheelbase with a similar length to the speed hump whereas a speed bump usually has a significantly shorter length compared to a car's wheelbase. Therefore, a speed bump would generate high vertical force even at a low speed and, therefore, would definitely induce discomfort if the experiment were done using the same conditions as in this study. Since only the velocity parameter was considered in the current study, the intended accelerations were lower than measured in defensive driving style simulation because the suspension effects were not included in the initial calculations. As highlighted by Shirahatti [30], due to the vehicle suspension, when the vehicle speed was increasing, the vertical acceleration was first increasing and then decreasing at a high rate before stabilized at around 25 km/h. The study also indicated that the vehicle's vertical displacement and pitch rotation were higher at low speed and came to be lower at high speed, similar to the finding in the current study in simulating assertive driving style.

3.3 General discussion

The simulation of autonomous driving styles at three road profiles namely straight lane approaching and leaving a junction, cornering, and speed hump could be done with the help of a special device known as AUTOAccD. The experiment set up started with the determination of interested driving styles and then followed with the selection of routes and determination of ranges for acceleration in different directions. In this study, the determination of driving styles was influenced by the intention to simulate autonomous driving experience with different driving styles. Therefore, three types of driving styles were determined, namely assertive, defensive and LRT driving style. The accelerations of LRT was exclusively included as riding in an LRT could already be a mental benchmark for future AV users as the experience of using an LRT was the closest one would get in imagining using an AV. This is in line with the public opinion regarding AV reported by Schoettle and Sivak in 2014 [31], who found that most of their questionnaire respondents would like to socialize with friends and family, do work-related activities, sleeping or resting, reading, or watching television or movies inside a future AV.

In this experiment set up, the implementation of AUTOAccD resulted in two main benefits. First, this device manages to help the drivers in simulating the intended accelerations by interactively showing and clustering the live acceleration data so that the drivers could adjust the throttle position and speed in order to get the required simulations. Second, the device also acts as a data logger that measure and stores various data for further analysis. Although the AUTOAccD would help the designated drivers in simulating the intended accelerations, practise is still required as it would help the drivers to get the feeling on how to generate higher accuracy simulations with fewer deviations.

In terms of comparison between the designated drivers, most of the acceleration data have small SDs and close to the intended means. In addition, both drivers managed to simulate accelerations within the intended ranges as out of 110 trials altogether only 18 values were simulated outside the intended ranges. By comparing the three driving styles, the assertive driving styles were the hardest to be simulated regardless of the road profiles. It is evidenced by the higher SD values for assertive driving styles compare to the other two driving styles. For the defensive driving styles, in general, although low deviations were recorded the average means for the simulated accelerations are usually higher than the intended ones. For the LRT driving styles, both means and SDs for the simulated accelerations indicated variations between the two drivers. Unlike the other two driving styles which distinguished patterns were shown, for the LRT driving styles both of the drivers did not show any clear trend regarding the simulated accelerations.

A few cautions have to be taken before executing the experiment on the road such as for the lateral accelerations where the cornering radius has to be maintained in order to generate the specified centrifugal forces. For the longitudinal accelerations, a lower gear preferably the first gear is recommended in order to achieve the intended accelerations. For the longitudinal deceleration, although a specific starting velocity has been identified, the simulation of deceleration required an act of balance between clutch and brake pedal and skill to produce a continuous and smooth deceleration. On the other hand, the process of analysing the accelerometers data began with the de-noising process in order to eliminate the noises without flattening the valuable data, from the three accelerometers in the tri-axial direction.

4 Conclusions

In conclusion, this study of simulating AV driving styles could be extended with the evaluation of acceptance and preference of future autonomous driving styles such as using the subjective measurements (unipolar or bipolar comfort rating) and objective measurements (continuous physiological measures for example the galvanic skin conductance (GSR) and electrocardiogram (EKG)) [32]. The significant finding of the present work could be further extended in the area of comfort study in future AV driving styles and the effects of motion sickness where the role of the driver would change from the sole controller of the vehicle to a passive user with a much-minimized role. In addition, comfort in travelling is usually associated with the way the vehicle is controlled and driven. In the near future, where accelerations are controlled by the algorithms and machines, autonomous vehicle users who used to be the drivers may feel discomfort unless the driving styles of the vehicle are somewhat similar to their own. Hence, there is a special need to study the associations between simulated accelerations and driving styles in order to enhance the autonomous driving experience for the future AV users.

If such experiments were to be extended, experts from different backgrounds, like engineering and social sciences among others, should work together in order to solve these projected issues effectively and efficiently. The rising of AV technology brings along the benefits such as lowering the fatal accidents by taking the human out of the driving's loop and maximizing the traffic flow and usage of the road [33]. At the same time, studies of human-machine interaction and acceptance should be explored extensively because as marvellous as AV technology could be, humans are still the end users who will buy, use and experience the AV in the coming future.

The authors would like to express their gratitude and special acknowledgements to Universiti Teknikal Malaysia Melaka (UTeM) and Ministry of Education, Malaysia (KPM) for the funding of the Ph.D. programs of Juffrizal Karjanto and Nidzamuddin Md. Yusof and also the post-doctoral program of Muhammad Zahir Hassan in the Industrial Design Department, Eindhoven University of Technology, The Netherlands.

References

1. KPMG, "Self-Driving Cars: Are We Ready?," (2015)
2. KPMG, "Self-Driving Cars : The Next Revolution," (2012)
3. T. M. Gasser and D. Westhoff, "BASt-study : Definitions of Automation and Legal Issues in Germany," (2012)
4. J. Elander, R. West, and D. French, "Behavioral correlates of individual differences in road-traffic crash risk: An examination method and findings," *Psychol. Bull.*, vol. **113**, no. 2, pp. 279–294, (1993)
5. O. Taubman-Ben-Ari, M. Mikulincer, and O. Gillath, "The multidimensional driving style inventory - Scale construct and validation," *Accid. Anal. Prev.*, vol. **36**, no. 3, pp. 323–332, (2004)
6. T. Vaa, "From Gibson and Crooks to Damasio: The role of psychology in the development of driver behaviour models," *Transp. Res. Part F Traffic Psychol. Behav.*, vol. **25**, pp. 112–119, (2014)
7. H. Summala, "Towards understanding motivational and emotional factors in driver behaviour: Comfort through satisficing," in *Modelling Driver Behaviour in Automotive Environments: Critical Issues in Driver Interactions with Intelligent Transport Systems*, (2007), pp. 189–207
8. S. Le Vine, A. Zolfaghari, and J. Polak, "Autonomous cars : The tension between

occupant experience and intersection capacity," *Transp. Res. Part C*, vol. **52**, pp. 1–14, (2015)
9. C. Diels and J. E. Bos, "Self-driving carsickness," *Appl. Ergon.*, vol. **53**, pp. 374–382, (2016)
10. M. Sivak and B. Schoettle, "Motion Sickness in Self-Driving Vehicles (Report No. UMTRI-2015-12)," (2015)
11. V. Bogdanović and N. Ruškić, "The research of vehicle acceleration at signalized intersections," *Traffic Transp.*, vol. **25**, no. 1, pp. 33–42, (2013)
12. I. El-Shawarby, H. Rakha, V. Inman, and G. Davis, "Evaluation of Driver Deceleration Behavior at Signalized Intersections," *Transp. Res. Rec. J. Transp. Res. Board*, pp. 29–35, (2007)
13. D. Hugemann and D. Nickel, "Longitudinal and Lateral Accelerations in Normal Day Driving," in *6th International Conference of The Institute of Traffic Accident Investigators*, (2003)
14. Parsons Brinckerhoff Team, "California High Speed Train Program Environmental Impact Report/Environmental Impact Statement: Engineering Criteria," (2004)
15. TCRP, *TCRP Report 155 – Track Design Handbook for Light Rail Transit, Second Edition*. (2012)
16. P. A. Weber and J. P. Braaksma, "Towards a North American geometric design standard for speed humps," *ITE J. (Institute Transp. Eng.*, vol. **70**, no. 1, pp. 30–34, (2000)
17. R. Wang and S. M. Lukic, "Review of driving conditions prediction and driving style recognition based control algorithms for hybrid electric vehicles," *2011 IEEE Veh. Power Propuls. Conf. VPPC 2011*, (2011)
18. E. Gulian, G. Matthews, A. I. Glendon, D. R. Davies, and L. M. Debney, "Dimensions of driver stress," *Ergonomics*, vol. **32**, no. 6, pp. 585–602, (1989)
19. J. Reason, A. Manstead, S. Stradling, J. Baxter, and K. Campbell, "Errors and violations on the roads: a real distinction?.pdf," *Ergonomics*, vol. **33**, no. 10/11, pp. 1315–1332, (1990)
20. M. Kuderer, S. Gulati, and W. Burgard, "Learning Driving Styles for Autonomous Vehicles from Demonstration," *2015 IEEE Int. Conf. Robot. Autom.*, vol. **134**, pp. 2641–2646, (2015)
21. S. Huang, "OBD-II Telematics Advanced Kit (Arduino MEGA 2560)," (2014) [Online]. Available: http://freematics.com/pages/products/arduino-telematics-kit-3/.
22. C. Taswell, "The what, how, and why of wavelet shrinkage denoising," *Comput. Sci. Eng.*, vol. **2**, no. 3, pp. 1–11, (2000)
23. C. W. Kang, C. H. Kang, and C. G. Park, "Wavelet Denoising Technique for Improving of the Low Cost MEMS-GPS Integrated Navigation Sytems," *Inetrnational Symp. GPS/GNSS*, pp. 3–8, (2010)
24. A. M. Hasan, K. Samsudin, A. R. Ramli, and R. S. Azmir, "Wavelet-based pre-filtering for low cost inertial sensors," *J. Appl. Sci.*, vol. **10**, no. 19, pp. 2217–2230, (2010)
25. M. Misiti, Y. Misiti, G. Oppenheim, and J.-M. Poggi, *Wavelet toolbox*. (1996)
26. S. Brol and J. Mamala, "Assessment of passenger car driveability with use of two axis accelerometer mounted on car body," *J. Ultragarsas*, vol. **2**, no. 2, pp. 31–36, (2006)
27. G. Reymond, a Kemeny, J. Droulez, and a Berthoz, "Role of lateral acceleration in curve driving: driver model and experiments on a real vehicle and a driving simulator.," *Hum. Factors*, vol. **43**, no. 3, pp. 483–495, (2001)
28. S. G. Charlton, "The role of attention in horizontal curves: A comparison of advance warning, delineation, and road marking treatments," *Accid. Anal. Prev.*, vol. **39**, no. 5, pp. 873–885, (2007)
29. Institute of Transportation Engineers, "Traffic Calming: State of the Practice - Chapter

6: Legal Authority and Liability," (1999)
30. A. Shirahatti, "Investigation on effect of hump on a vehicle suspension for the comfortable ride," in *National Conference on Emerging Trends in Mechanical*, (2006)
31. B. Schoettle and M. Sivak, "Public Opinion About Self-Driving Vehicles in China, India, Japan, The U.S., The U.K. and Australia," (2014)
32. J. a. Healey and R. W. Picard, "Detecting Stress During Real-World Driving Tasks Using Physiological Sensors," *IEEE Trans. Intell. Transp. Syst.*, vol. **6**, no. 2, pp. 156–166, (2005)
33. D. J. Fagnant and K. Kockelman, "Preparing a Nation for Autonomous Vehicles: Opportunities, Barriers and Policy Recommendations for Capitalizing on Self-Driven Vehicles," in *93rd Annual Meeting of the Transportation Research Board 2014*, (2014)

Clustering of frequency spectrums from different bearing fault using principle component analysis

M.F.M Yusof[1,*], *C.K.E* Nizwan[1] , *S.A* Ong[1], and *M. Q. M* Ridzuan[1]

[1]Advanced Structural Integrity and Vibration Research Group (ASIVR), Faculty of Mechanical Engineering, Universiti Malaysia Pahang, 26600 Pekan, Pahang.

Abstract. In studies associated with the defect in rolling element bearing, signal clustering are one of the popular approach taken in attempt to identify the type of defect. However, the noise interruption are one of the major issues which affect the degree of effectiveness of the applied clustering method. In this paper, the application of principle component analysis (PCA) as a pre-processing method for hierarchical clustering analysis on the frequency spectrum of the vibration signal was proposed. To achieve the aim, the vibration signal was acquired from the operating bearings with different condition and speed. In the next stage, the principle component analysis was applied to the frequency spectrums of the acquired signals for pattern recognition purpose. Meanwhile the mahalanobis distance model was used to cluster the result from PCA. According to the results, it was found that the change in amplitude at the respective fundamental frequencies can be detected as a result from the application of PCA. Meanwhile, the application of mahalanobis distance was found to be suitable for clustering the results from principle component analysis. Uniquely, it was discovered that the spectrums from healthy and inner race defect bearing can be clearly distinguished from each other even though the change in amplitude pattern for inner race defect frequency spectrum was too small compared to the healthy one. In this work, it was demonstrated that the use of principle component analysis could sensitively detect the change in the pattern of the frequency spectrums. Likewise, the implementation of mahalanobis distance model for clustering purpose was found to be significant for bearing defect identification.

1 Introduction

In any rotating machineries, rolling element bearing is important in which it is functioning as both thrust and radial load bearer. Without bearing, the rotating shaft will be exposed to an excessive vibration which later on led to the fatigue damage. Basically, an abrupt bearing failure will precipitate massive impact to the maintenance and operational cost. Therefore, it is greatly essential to make sure that the bearing is consistently in pristine

[*] Corresponding author: fadhlan@ump.edu.my

condition while it is operating. Besides, early bearing fault detection is vital in order to prevent the failure as well as reduced the loss.

In industries, various technique can be applied for the purpose of bearing condition monitoring, and one of the common technique is vibration analysis [1, 2]. In simple cases, the vibration behavior of rolling element bearing can be analytically predicted. However, in more compounded system, vibration produced by rolling element bearing can be complex as a result from geometrical imperfection during manufacturing process, component instability as well as defect itself [3]. Apart from that, the other excitation frequency from other component or other unidentified sources might also affecting the vibration behavior. Consequently, the vibration signal produced are random and it is difficult to detect the damage-related component. Since the past several decades, the feature extraction analysis and classification technique was widely used as an approach in attempt to address on this issue. In general, the major aims of feature extraction analysis is to extract the hidden signal features among the complex signal that could lead to the detection of damage occurrence in the system that was monitored. Basically, on one hand, the feature was extracted directly from the acquired signal by determining its statistical parameter [4-7] and fundamental frequencies [3, 8] parameters. Meanwhile, on the other hand, the decomposition method such as wavelet analysis [9-12], and empirical mode decomposition [13-15] was applied before the feature is determined. The idea of decomposing the signal is to filter out all the non-related signal component which was initiated from both unidentified and unrelated sources during bearing operation.

In industrial application, it is important to identify the damage in order to assess it's fitness for service. This is vital for the purpose of maintenance planning and that is the main reason why online monitoring system is needed at the first place. Basically, for damage identification purpose, several classification technique have been applied to the extracted features. This includes, Discriminative Subspace Learning [16], Hierarchical Diagnosis Network [17], Support Vector Machine [18, 19], Extreme Learning Machine [20], and artifial neural network [21]. Despite the wide exploration on the feature extraction and classification techniques, the unavoidable problems such as low signal-to-noise ratio due to the nonlinearity of the signal still becomes a major challenge even though wide variation of feature extraction approach was taken to overcome these problems.

In this paper, the application of principle component analysis (PCA) as a pre-processing method for hierarchical clustering analysis on the frequency spectrum of the vibration signal was proposed. In common approach, PCA was implies to reduce the dimension of the complex signal before the features associated with the presence of defect was detected [22]. However, in this work, PCA was applied to identify the change in frequency spectrum pattern due to the basis that the amplitude of fundamental frequency will changed with the existence of defect. In this study, the vibration signal was acquired from the operating bearings with different condition and speed. In the beginning part, the response of vibration amplitude at the respective fundamental frequencies with the occurrence of damage will be discussed. On the next stage, the application of principle component analysis as feature extraction method and hierarchical clustering as damage identification analysis will be demonstrated.

2 Methodology

2.1 Experimental setup

The test rig for this experiment was designed to investigate failure and vibration characteristic of ball bearings. As illustrated in Figure 1, in this experiment, the shaft was driven by a variable-speed 0.37kW, 50Hz electric motor equipped with a controller in order to control the speed of the motor. A flywheel is installed at the middle of the spindle in order to apply load to the shaft and at the same time minimizing the speed oscillations of the shaft. A spring coupling was used to connect the motor and shaft to minimize shaft alignment error. The front side of the shaft (near to the motor) is fitted with tested bearing and the vibration response will be measured here while on the other side, a good bearing was fitted. In this study, a set of good bearings and another three bearings with different type of defect such as corroded, point defect and outer race defect were tested. The angular speed is set to 10%, 50% and 90% of the maximum motor speed, and shortly after the test commenced, the time response of vibration were acquired by using the Bruel & Kjær (B&K) 4506B accelerometer. The time series of vibration (acceleration) response was acquired with the sampling frequency of 20 kHz ($\Delta t = 0.039$ ms).

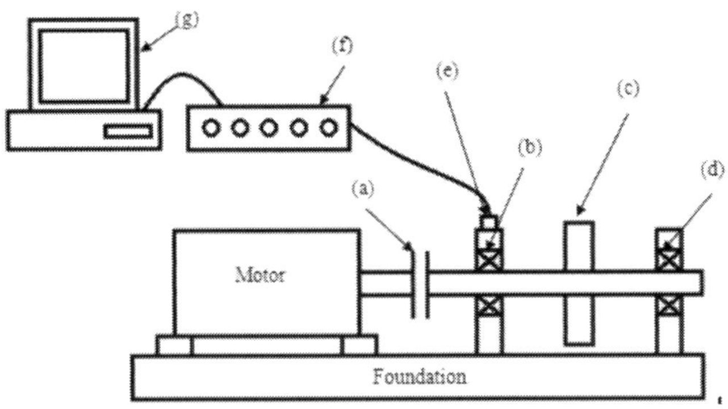

No	Part/Component
a	Spring coupling
b	Tested bearing
c	Flywheel
d	Supporting bearing
e	Accelerometer
f	DAQ
g	Analyzer

Fig. 1. Experiment setup.

2.2 Frequency spectrum clustering

In this work, 16 signals from healthy, corroded and outer race defect was selected for clustering analysis together with 12 signals from point defect bearing. Before starts the clustering process, the acquired time domain response from different types of bearings was converted into frequency spectrum through the Fast Fourier Transform (FFT) analysis. Direct observation on peak pattern at the respective fundamental frequency were made to identify the damage. Basically the fundamental frequencies were calculated based on the equation from [3]. Even though the fundamental frequency could be observed, it is important to reveal the different in structure of all the collected frequency spectrums from different bearing conditions because in some cases, the change in pattern is too small to be observed. To achieve the aims, the principle component analysis was applied. Basically, the principle component analysis implies the eigenvalue decomposition on the covariance matrix of the multiple dataset. In this work, one frequency spectrum was considered as one dataset contained n number of samples. Moreover, all the dataset was normalized by its mean value before the covariance matrix was attained. In general, the PCA process could be represent in matrix operation as shown by [23] in equation 1 to 3 whereas in equation 3, γ and $\vec{v}$ are the eigenvalue and eigenvector (principle component) respectively

$$[X] = \begin{bmatrix} x_{11} & \cdots & x_{1n} \\ \vdots & \ddots & \vdots \\ x_{m1} & \cdots & x_{mn} \end{bmatrix} \quad (1)$$

$$[C_X] = [X].[X]^T \quad (2)$$

$$[C_X].\vec{v} = \gamma.\vec{v} \quad (3)$$

Basically, the result from PCA will be represented in scatter plot to show how the numerous dataset scattered based on its pattern. To classify this numerous dataset, the hierarchical clustering approach was taken in which the distance between dataset will be measured prior to clustering process. In this work, the mahalanobis model as shown in equation 4 [24] was selected for a distance measurement due to the nature of principle components which will scattered in oval shapes when the datasets is strongly related [25].

$$d_{mahalanobis}(x, y) = \sqrt{(x-y)C^{-1}(x-y)^T} \quad (4)$$

3 Results and discussion

3.1 Frequency spectrums

In this paper, due to the same result's pattern for all rotational speed, only the result from the test with rational speed of 287 rpm was presented. As explained earlier, the time series of the vibration signal acquired from the rolling element bearings with different condition is converted into frequency domain signal and these signal were illustrated in Figure 2. Meanwhile its theoretical fundamental frequency was shown in Table 1. Based from the Figure 2, it was clear that the ball spin frequency, BSF, and the ball passing frequency outer race, BPFO are obviously appear in frequency spectrum of healthy as well as defect bearings. In contrast, the ball pass frequency inner race, BPFI are barely unseen in the frequency spectrums of the corroded bearings in which the opposite trend had been shown in the frequency spectrums of other types of bearing.

From the deeper observation, it was found that high amplitude of acceleration occur at BPFO for outer race defect bearing. In conjunction with that, among all the frequency spectrums from healthy and defects bearings, the amplitude of acceleration was higher at BPFI for point defect bearing. Based from this results, it was confirmed that the presence of specific defect will increase the amplitude of vibration at the specific fundamental frequencies. Previous findings [8] also have proven these phenomena accordingly. In contrast, the spectrums of corroded bearings shows high amplitude values for all fundamental frequencies. This is probably due to the uniform behavior of the defect itself.

Table 1. Fundamental Frequency of Rolling Element Bearing which rotates at 287 rpm.

Fundamental Frequency	Value (Hz)
Ball Passing Frequency Outer Race, BPFO	13.65
Ball Passing Frequency Inner Race, BPFI	24.58
Fundamental Train Frequency, FTF	1.707
Ball Spin Frequency, BSF	1.67

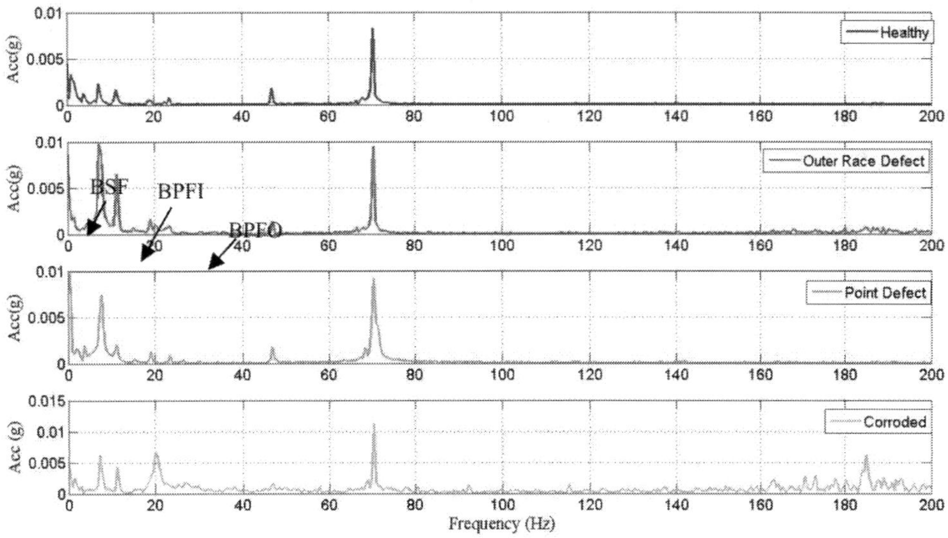

Fig. 2. Frequency spectrum for bearings that rotates 287 rpm.

3.2 Frequency signal classification

As discussed in the previous section, it is clear that the frequency spectrum from each of the bearing condition showed a significant different in its structure. To represent cluster all of these spectrums, the principle component analysis was applied to a set of frequency spectrums which consists of healthy, point defect, outer race defect and corroded bearings and the results was shown in Fig. 3. Fig. 3(a) illustrates the overall scatter plot of principle component 1 and principle component 2 while Fig. 3(b) shows the zoomed part. According to the result in both sub-figures, it was found that the principle components (PC) of frequency spectrums was scattered into four different groups. However, the scattered data form a group or population in ellipsoidal shape. This scatter trend occurs due to the similarities in the patterns of the tested dataset [26]. In other words, it is strongly believed that each ellipsoidal-shape scattered dataset is belonging to the same group of bearing types.

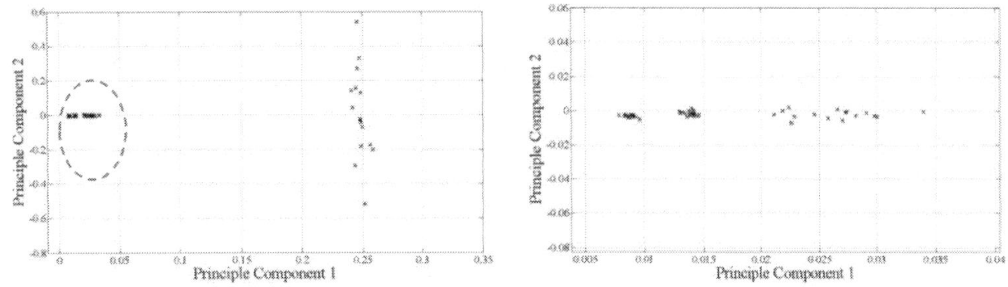

(a) (b)

Fig. 3. Classification of frequency spectrum from different bearing condition using principle component analysis (a) Overall (b) Zoomed part

To confirm the claims, clustering process in needed in which for this process, the mahalanobis distance had been calculated in order to cluster the PCs from different type of bearings. The clustering result was shown in dendrogram plot in Fig 4. According to the figures, based on mahalanobis distance, it was clear that the data have been regrouped in four major cluster. Yet, the dataset 49 and 54 was identified to be outliers. Table 2 was shown to simplify the representation of the dendrogram. As referred to the table, those dataset which belongs to corroded, outer race defect, healthy and point defect was registered to be in cluster 1, 2, 3, and 4 respectively. Meanwhile, two dataset which belongs to corroded bearing was found to be as outliers.

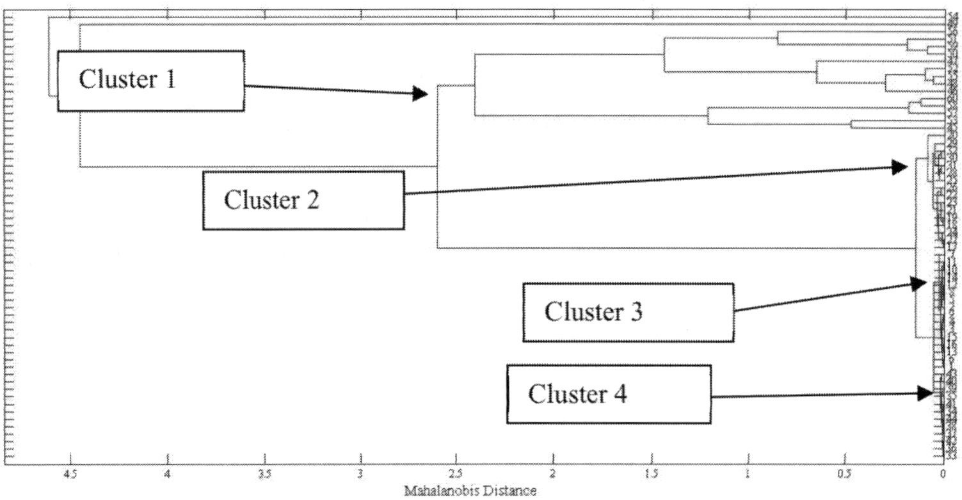

Fig. 4. Dendrogram plot of principle component 1 and 2.

Table 2. Clustering result.

Cluster 1		**Cluster 2**		**Cluster 3**		**Cluster 4**		**Outliers**	
Signal No	**Bearing Type**	**Signal No**	**Bearing Type**	**Signal No**	**Bearing Type**	**Signal No**	**Bearing Type**	**Signal No**	**Bearing Type**
56	Corroded	20	Outer Race Defect	7	Healthy	43	Point Defect	54	Corroded
51	Corroded	29	Outer Race Defect	11	Healthy	40	Point Defect	49	Corroded
59	Corroded	32	Outer Race Defect	10	Healthy	39	Point Defect		
50	Corroded	30	Outer Race Defect	14	Healthy	35	Point Defect		
47	Corroded	31	Outer Race Defect	12	Healthy	41	Point Defect		
52	Corroded	28	Outer Race Defect	8	Healthy	34	Point Defect		
55	Corroded	25	Outer Race Defect	5	Healthy	44	Point Defect		
48	Corroded	26	Outer Race Defect	2	Healthy	38	Point Defect		
46	Corroded	22	Outer Race Defect	9	Healthy	37	Point Defect		
60	Corroded	23	Outer Race Defect	4	Healthy	42	Point Defect		
58	Corroded	21	Outer Race Defect	3	Healthy	36	Point Defect		
57	Corroded	19	Outer Race Defect	15	Healthy	33	Point Defect		
53	Corroded	18	Outer Race Defect	16	Healthy				
45	Corroded	24	Outer Race Defect	13	Healthy				
		27	Outer Race Defect	6	Healthy				
		17	Outer Race Defect	1	Healthy				

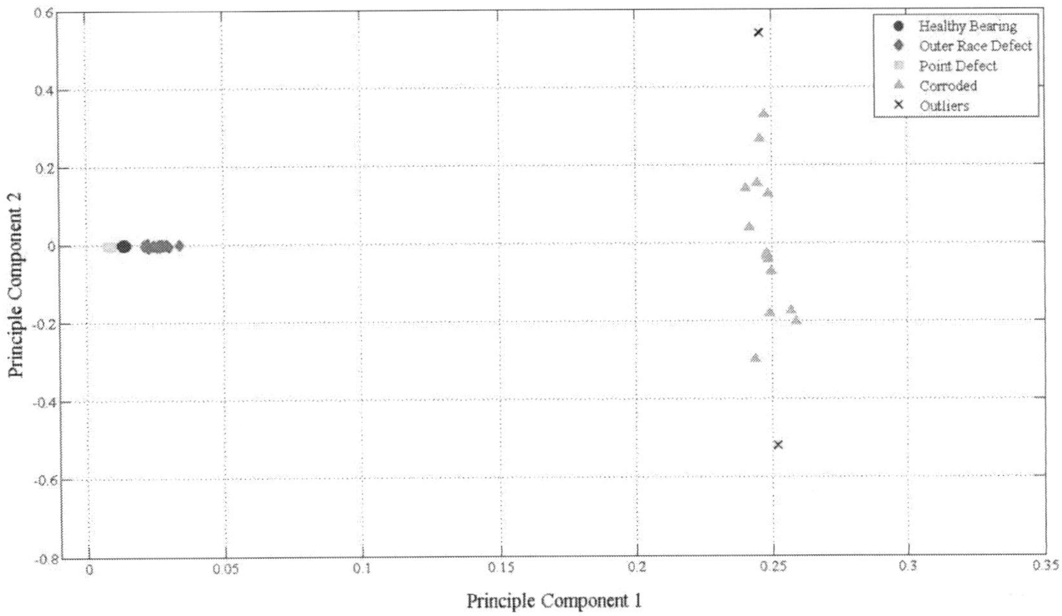

Fig. 5. Principle Components scatter plot with cluster group.

4 Conclusions

According to the results, it was found that the amplitude of vibration at Ball Passing Frequency Outer Race and Ball Passing Frequency Inner Race will increase in align with the presence of outer race defect and inner race defect respectively. Moreover, the overall amplitude of vibration spectrum was found to be uniformly increased for the case of corroded bearing due to the widespread uniform corrosion on the entire bearing. By applying principle component analysis, the change in amplitude at any of these fundamental frequencies can be detected. Meanwhile, the application of mahalanobis distance was found to be suitable for clustering the results from principle component analysis. Uniquely, it was discovered that the spectrums from healthy and inner race defect bearing can be clearly distinguished from each other even though the change in amplitude pattern for inner race defect frequency spectrum was too small compared to the healthy one. To draw the conclusion, it was demonstrated that the use of principle component analysis could sensitively detect the change in the pattern of the frequency spectrums. This was believe to give more option to detect the damage from the change in signal pattern apart from decomposing it. Likewise, the implementation of mahalanobis distance model for clustering purpose was found to be significant for bearing defect identification.

References

1. Tandon, N., A. Choudhury, A review of vibration and acoustic measurement methods for the detection of defects in rolling element bearings. Tribology International. **32**(8): p. 469-480.(1999)
2. Rai, A., S.H. Upadhyay, A review on signal processing techniques utilized in the fault diagnosis of rolling element bearings. Tribology International. **96**: p. 289-306.(2016)
3. Lacey, S.J., An Overview of Bearing Vibration Analysis. Maintenance & asset management. **23**(6): p. 11.(2008)

4. Dron, J.P., F. Bolaers, I. Rasolofondraibe, Improvement of the sensitivity of the scalar indicators (crest factor, kurtosis) using a de-noising method by spectral subtraction: application to the detection of defects in ball bearings. Journal of Sound and Vibration. **270**(1–2): p. 61-73.(2004)
5. Harvey, T.J., R.J.K. Wood, H.E.G. Powrie, Electrostatic wear monitoring of rolling element bearings. Wear. **263**(7–12): p. 1492-1501.(2007)
6. Pachaud, C., R. Salvetat, C. Fray, crest factor and kurtosis contributions to identify defects inducing periodical impulsive forces. Mechanical Systems and Signal Processing. **11**(6): p. 903-916.(1997)
7. Zhi-qiang, Z., et al., Investigation of rolling contact fatigue damage process of the coating by acoustics emission and vibration signals. Tribology International. **47**: p. 25-31.(2012)
8. Karacay, T., N. Akturk, Experimental diagnostics of ball bearings using statistical and spectral methods. Tribology International **42**((2009)): p. 836–843.(2009)
9. Choudhury, A., D. Paliwal, Application of Frequency B-Spline Wavelets for Detection of Defects in Rolling Bearings. Procedia Engineering. **144**: p. 289-296.(2016)
10. He, W., et al., Health monitoring of cooling fan bearings based on wavelet filter. Mechanical Systems and Signal Processing. **64–65**: p. 149-161.(2015)
11. Mishra, C., A.K. Samantaray, G. Chakraborty, Rolling element bearing defect diagnosis under variable speed operation through angle synchronous averaging of wavelet de-noised estimate. Mechanical Systems and Signal Processing. **72–73**: p. 206-222.(2016)
12. Wang, Y., et al., Detection of weak transient signals based on wavelet packet transform and manifold learning for rolling element bearing fault diagnosis. Mechanical Systems and Signal Processing. **54–55**: p. 259-276.(2015)
13. Dybała, J., R. Zimroz, Rolling bearing diagnosing method based on Empirical Mode Decomposition of machine vibration signal. Applied Acoustics. **77**: p. 195-203.(2014)
14. Xue, X., et al., An adaptively fast ensemble empirical mode decomposition method and its applications to rolling element bearing fault diagnosis. Mechanical Systems and Signal Processing. **62–63**: p. 444-459.(2015)
15. Zhang, X., J. Zhou, Multi-fault diagnosis for rolling element bearings based on ensemble empirical mode decomposition and optimized support vector machines. Mechanical Systems and Signal Processing. **41**(1–2): p. 127-140.(2013)
16. Zhao, M., et al., Fault diagnosis of rolling element bearings via discriminative subspace learning: Visualization and classification. Expert Systems with Applications. **41**(7): p. 3391-3401.(2014)
17. Gan, M., C. Wang, C.a. Zhu, Construction of hierarchical diagnosis network based on deep learning and its application in the fault pattern recognition of rolling element bearings. Mechanical Systems and Signal Processing. **72–73**: p. 92-104.(2016)
18. Li, X., et al., Rolling element bearing fault detection using support vector machine with improved ant colony optimization. Measurement. **46**(8): p. 2726-2734.(2013)
19. Liu, Z., et al., Multi-fault classification based on wavelet SVM with PSO algorithm to analyze vibration signals from rolling element bearings. Neurocomputing. **99**: p. 399-410.(2013)
20. Tian, Y., et al., Rolling bearing fault diagnosis under variable conditions using LMD-SVD and extreme learning machine. Mechanism and Machine Theory. **90**: p. 175-186.(2015)
21. M, S., et al., International Conference on Design and Manufacturing (IConDM2013)ANN based Evaluation of Performance of Wavelet Transform for Condition Monitoring of Rolling Element Bearing. Procedia Engineering. **64**: p. 805-814.(2013)

22. Saidi, L., J. Ben Ali, F. Fnaiech, Application of higher order spectral features and support vector machines for bearing faults classification. ISA Transactions. **54**: p. 193-206.(2015)
23. Abdi, *Encyclopedia of Measurements and Statistics*. 2007, Dallas, USA.
24. Mao, J., A. Jain, A self organizing network for hyperellipsoidal clustering (HEC). IEEE Transaction on Neural Network. **7**(1): p. 16 - 29.(1996)
25. De Maesschalck, R., D. Jouan-Rimbaud, D.L. Massart, The Mahalanobis distance. Chemometrics and Intelligent Laboratory Systems. **50**(1): p. 1-18.(2000)
26. Smith, L., *A Tutorial On Principle Components Analyis*, in *A Tutorial On Principle Components Analyis*. 2002.

Reverse engineering of B-pillar with 3D optical scanning for manufacturing of non-uniform thickness part

Md. Tasbirul Islam[1], *A.B.* Abdullah[1,*], and *Mohamad Zihad* Mahmud[2]

[1]*School of Mechanical Engineering, Universiti Sains Malaysia (USM), Engineering campus, 14300 Nibong Tebal Penang, Malaysia*
[2]*School of Aerospace Engineering, Universiti Sains Malaysia (USM), Engineering campus, 14300 Nibong Tebal, Penang, Malaysia*

Abstract. This paper presents reverse engineering (RE) of a complex automobile structural part, B-pillar. As a major part of the automobile body-in white (BiW), B-pillar has substantial opportunity for weight reduction by introducing variable thickness across its sections. To leverage such potential, an existing B-pillar was reverse engineered with a 3D optical scanner and computer aided design (CAD) application. First, digital data (i.e. in meshes) of exiting B-pillar was obtained by the scanner, and subsequently, this information was utilized in developing a complete 3D CAD model. CATIA V5 was used in the modeling where some of the essential work benches were ''Digitized Shape Editor'', ''Quick Surface Reconstruction'', ''Wireframe and Surface Design'', ''Freestyle'', ''Generation Shape Design'' and "Part design". In the final CAD design, five different thicknesses were incorporated successfully in order to get a B-pillar with non-uniform sections. This research opened opportunities for thickness optimization and mold tooling design in real time manufacturing.

1 Introduction

Major automotive manufacturers have been working to produce lightweight vehicles in order to adhere with continuous strict regulations that demand for less greenhouse gas emission and increased fuel efficiency [1]. Among various systems of a vehicle, body-in-white (BiW) is found as the most significant portion representing about 30% of the weight of a vehicle, and has tremendous potential in reducing weight of the whole vehicle [2]. Figure 1 shows the breakdown of vehicle weight by systems and components. As BiW possesses such opportunity, and has high sensitivity to structural integrity, it is the only system that is researched, designed and analyzed extensively in the studies of weight reduction technology for automobile [3]. Important components in the BiW system are passenger compartment frame, cross and side beams, roof structure, front-end structure, underbody floor structure and panels [4].

* Corresponding author: mebaha@usm.my

DOI: 10.1051/matecconf/20179001007

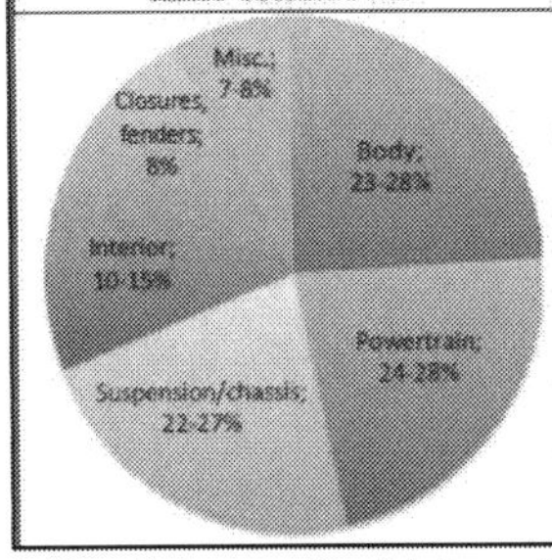

Approximate vehicle mass breakdown	System	Major components in system
	Body-in-white	Passenger compartment frame, cross and side beams, roof structure, front-end structure, underbody floor structure, panels
	Powertrain	Engine, transmission, exhaust system, fuel tank
	Chassis	Chassis, suspension, tires, wheels, steering, brakes
	Interior	Seats, instrument panel, insulation, trim, airbags
	Closures	Front and rear doors, hood, lift gate
	Miscellaneous	Electrical, lighting, thermal, windows, glazing

Fig. 1. Breakdown of vehicle weight by system and components (adopted from [3]).

BiW is mostly consists of thin-walled structural parts, which are constructed either by stamping or forming process of traditional metal sheets with uniform thickness [5]. On the other hand, non-uniform thickness section is found as one of the most effective approaches to reduce automotive part weight [6]. The most complex of all structures in a vehicle, B-pillar is located between the front and rear doors of compartment in the BiW system. Generally, it covers the electrical wiring and connections and spots for the passenger seatbelts. It also provides structural support against lateral or roof impact [2]. Figure 2 shows the B-pillar in a passenger car. For side impact safety, B-pillar is a major structural component in the BiW [7]. Having these two folded characteristics (i.e. firstly as being part of the BiW that has weight reduction opportunity, and secondly, critical to vehicle's structural integrity), the optimum design of the B-pillar must be maintain; which made the B-pillar as one of the important components in BiW system with increasing interest for further development.

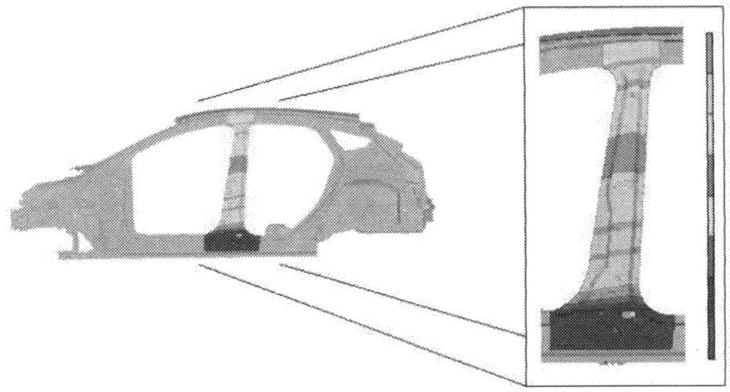

Fig. 2. B-pillar part in a typical passenger car (adopted from [8]).

Besides, traditional material alternatives and optimization techniques, manufacturing process of B-pillar is now being involved with some promising approaches such as tailor rolled blanks (TRBs) and tailor welded blanks (TWBs), where different sections of a B-pillar have non-uniform thickness, aiming to have reduced weight [8]. It is to be mentioned that TWB is an advanced manufacturing technology that consists of sheet metals that are laser welded and are characterized by different thickness and material within a single work piece [9]. On the other hand, in TRB, with a flexible rolling operation, continuous thickness transition can be made between thick and thin of a sheet metal. The thickness variation is obtained by adjusting roll gap. Research conducted by Hyrcza-Michalska et al. (2010) [10] found that using TWB for B-pillar, weight reduction was possible by 16% compared to conventional uniform thickness of the part. On the other hand, when TRB technology is employed, it was found that weight reduction of 2 kg/side was possible that consists of five different thicknesses in the B-pillar of a new BMW X5 car [11]. Another research by [8] found that weight of a B-pillar made by TRB technology with eight different thicknesses reached just over 7 kg ; a total weight saving of 1.3 kg when it is compared to conventional B-pillar design with uniform thickness. It is clear that thickness variation in different BiW system, particularly in B-pillar can significantly contribute to overall weight reduction of a vehicle.

Despite having such competitive advantages, these technologies have to follow two step processes where substantial effort and time is required to manufacture the part [12]. For instance, TRB infused B-pillar manufacturing; firstly, the TRBs are made up with the rolling process by controlling roll gap thus creating the varying thickness. After that direct hot stamping and rapid cooling is done on the TRBs, followed by laser and tool trimming for the final part. Therefore, to reduce the manufacturing time and effort, it is proposed in this research that development of a B-pillar can be made through a single process by cold forming operation where non-uniform thicknesses of the B-pillar sections will be incorporated in the tooling design of the forming operation. It is assumed to minimize the efforts of the subsequent manufacturing processes. However, to implement such idea, at first, a B-pillar with uniform thickness should be considered as a experimental object and then a reverse engineering (RE) process can be done. In this way, the actual physical object is transferred to a computer aided design (CAD) system, and further design modification can be made in order to achieve weight reduction compared with the existing one. The final modified B-pillar part will have variable thicknesses across its sections.

RE is the branch of engineering which takes advantage of an object that has already been created. The final purpose of RE is to create similar kind of identical object to the existing object considering rapid prototyping concept [13]. However, one of the essential prerequisite of this process is to get information about the physical nature of the object. In all areas in rapid prototyping, such as software, electronic components, automobile components, RE is widely being used for shorter product development time with optimized cost of the product [14]. The method is widely used and it consists a number of steps starting from capturing virtual model with a 3D optical scanner to 3D model development in CAD software [14]. The scanner coverts the physical object into point could or meshes. This kind of reverse engineering is generally being employed where access of such recorded 3D objects are confidential and identified as potential patent infringement [15]. Considering this as an important fact and to deploy rapid prototyping, RE process of the B-pillar outer part is presented in this research. This paper is organized in the following sections,

• Section 2 describes the experimental setup where the information of the physical object was transferred to meshes by 3D optical scanning,
• Section 3 highlighted the result and discussion of the research and finally,
• Section 4 concludes with achievement, limitations and future work.

2 Experimental setup

2.1 Scanning of B-pillar part

The B-pillar used in this research had complicated geometry and some free flowing surfaces. Due to relatively large size of B-pillar part, the scanning of the part was conducted in three different projections and then aligned for the complete mesh. The part had to provide with alignment points which were situated strategically for ensuring correct mesh. The original part was painted with a black colour developer spray to avoid any reflection during scanning. In this manner, the right meshes were found. The scanning of the part was conducted with the NEXTENGINE 3D optical high definition scanner that is installed in the fabrication lab at Universiti Sains Malaysia, Engineering campus. Figure 3 shows the flowchart of the process from original part to the mesh acquisition. The scan was performed in the dark space which was a prerequisite environment to avoid damages on the mesh. Finally, the meshes were found and oriented for three different faces using scanning software.

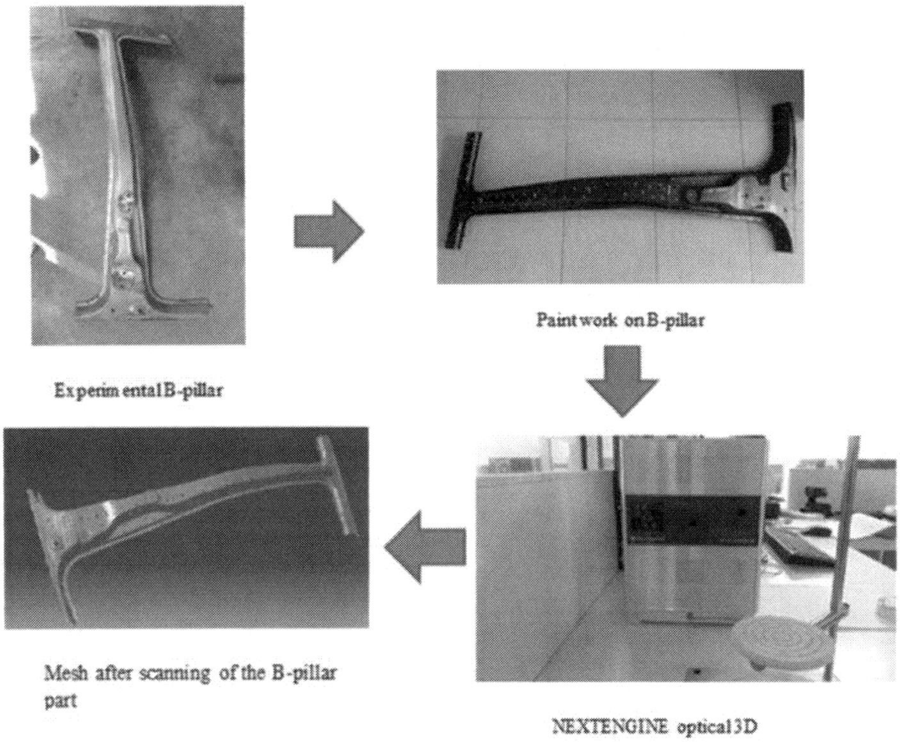

Fig. 3. Flowchart of mesh acquisition using 3D optical scanner.

3 Results and discussion

3.1 Mesh repair

After 3D scanning, the scanner software generated a (.STL) file extension that consists of meshes of the B-pillar. Initially, the obtained mesh had discontinuity, and large holes. The holes in the mesh were repaired with the fill holes, mesh smoothing, and mesh cleaning command under the ''Digitized Shape Editor'' workbench in the CATIA V5. Figure 4 shows the repaired mesh. Finally, the repaired meshes were used for developing a solid CAD model. Generally, the mesh provided a guide for placing lines and surfaces, and at the end of the process, final structure of the part will be similar to the mesh.

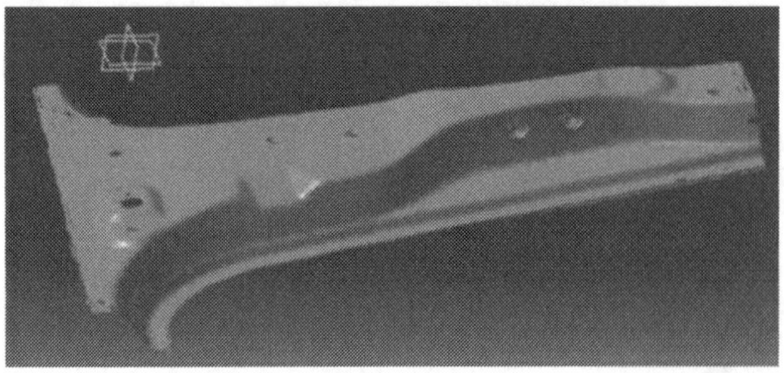

Fig. 4. Repaired mesh in Catia V5.

3.2 Reverse engineering of the surfaces

RE was started in the part by creating surfaces that exactly matched with the mesh. To create the surfaces, two important workbenches were utilized namely as ''Quick Surface Reconstruction'' and ''Wireframe and Surface Design''. The surfaces were generally identified by placing 3D curves on the mesh, and by selecting the surface activate command. As the part contained a number of different surfaces, this process was performed repeatedly until the acceptable surfaces were created. This was valid for the surfaces which have relatively free flowing shape. On the other hand, the meshes which are comparatively straight in nature and did not have complex shape were constructed by tools located at the ''Generative shape design'' workbench. Figure 5 shows the surface creation using the mesh and workbenches.

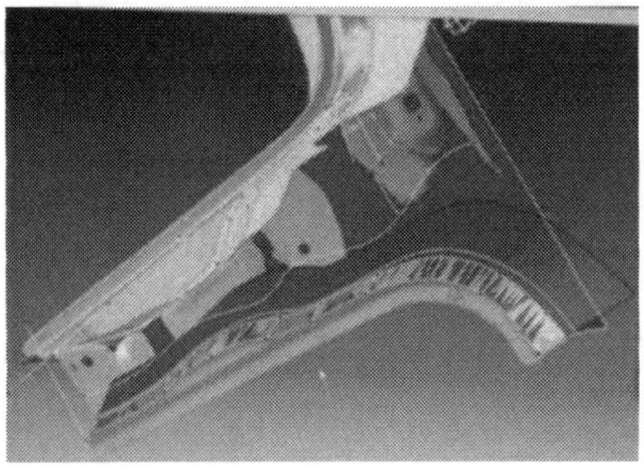

Fig. 5. Surfacing of the B-pillar part.

The surfaces were created with the points, lines, and finally with the command of multi-section surface. The created surfaces were then extended with the extend command under the ''Freestyle'' workbench. After creating and extending the surfaces, with the help of trim and sprit operation, the final surfaces according to the mesh were constructed as showed in figure 6.

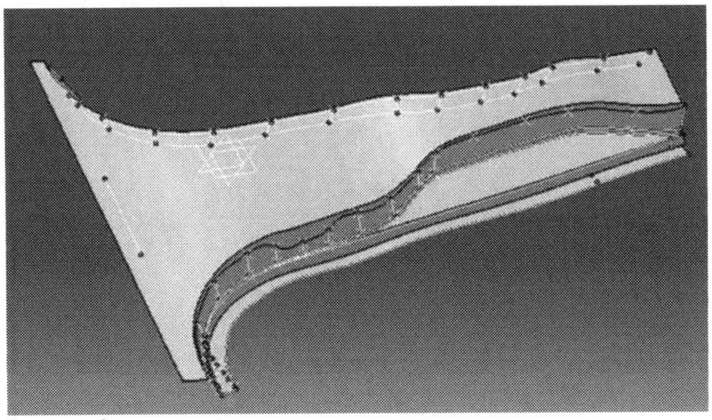

Fig. 6. Created surfaces and excess cut away.

After creating all the surfaces, the join commend was employed located under the ''Generation Shape Design'' work bench that combined all the surfaces into a single one which then provided the opportunity to create the solid part of the B-pillar. Thick surface command allowed in creating the solid CAD body from the combined surfaces. For the case of the existing B-pillar, the thickness was assigned to 1.2 mm according to the physical geometry. Fig 7 shows the solid form of the reverse engineered B-pillar part. This concluded the reverse engineering of the existing B-pillar part. In the next section, procedures for developing B-pillar part with non-uniform sections were introduced.

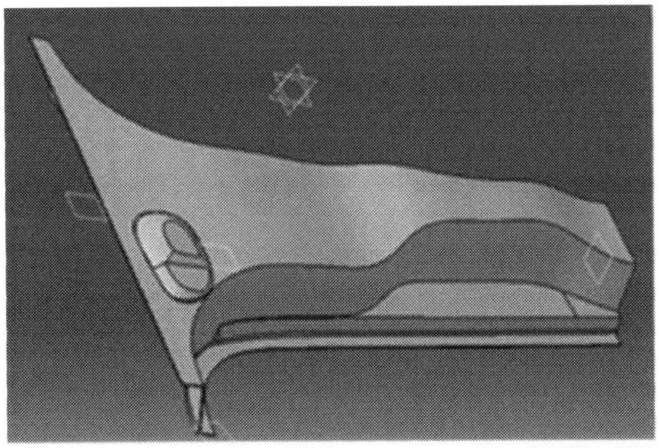

Fig. 7. B-pillar part in solid form after surface thicken command in CATIA V5.

3.3 CAD model development of non-uniform thickness B-pillar part

In this section, reverse engineered B-pillar part of the existing design went for further CAD modelling by CATIA V5 to have non-uniform thickness sections of the B-pillar. The previous design shown in figure 5 was divided into five different sections to allocate the variable thicknesses. The adjacent surfaces were given fillet as it had sharp edges, and a reference plane has been set at the top of the part. Fillet and plane operation was done under the ''Generation Shape Design'' workbench. The new planes were set to around 90 mm apart. After creating the planes, a new geometrical set for each of the sections were developed by split command using ''Generative Shape Design'' workbench. Then using ''Part Design'' workbench, the surfaces were thicken to 1.2mm, 1.6 mm, 2.2 mm, 1.4 mm and 1 mm respectively as shown in the figure 8. The primary intention was to introduce the thicknesses according to the research conducted by [11], however, at the very last section at the bottom it was limited to 1 mm rather than 1.2 mm. This occurred due to having a curved shaped socket present in the existing design of the B-pillar. Later on this will be overcome by thickness optimization using finite element analysis. With current facility at the forging laboratory, Universiti Sains Malaysia; work piece with a dimensional length of maximum 900 mm is possible to operate. So the CAD modelling here in this research was taken the total length of the part as approximately 500 mm to avoid unavoidable error.

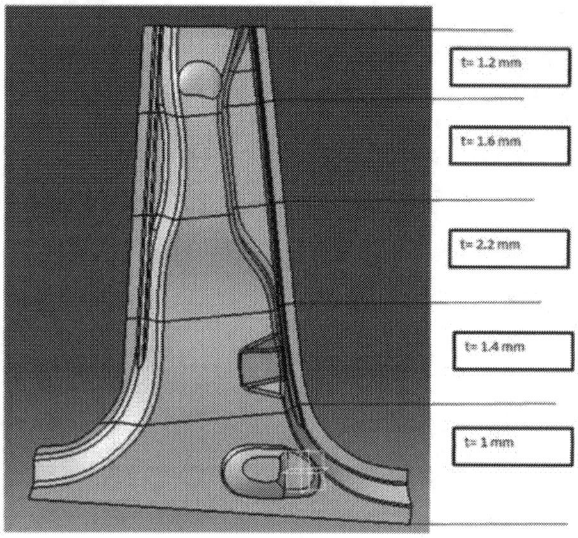

Fig. 8. Initial partition and thickness of the B-pillar part.

After providing all the necessary input for the surface thicken command, the B-pillar part was constructed as one solid body with different thicknesses across its sections. In the surface thicken command under "Part design" workbench, it was employed the second offset thickness that created the non-uniform thickness from the bottom side. These were the processes that how the surfaces of the B-pillar part became solid form from surfacing. Figure 9 shows both the front and back side of the experimental B-pillar with non-uniform sections in its solid form. In addition, figure 10 shows the section transitions with a zoom in view of the B-pillar part.

Fig. 9. Experimental B-pillar part with non-uniform sections in solid form.

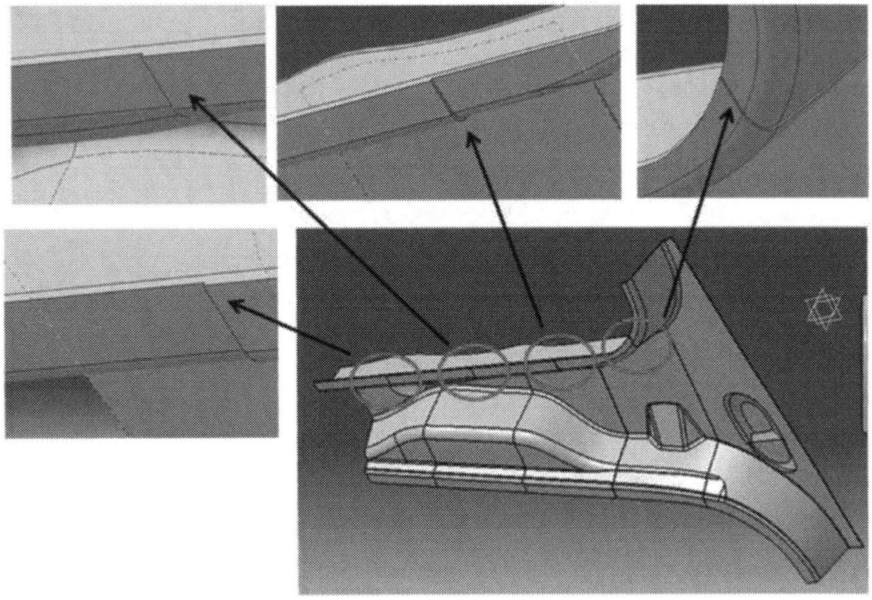

Fig. 10. Thickness transition sections of non-uniform sections of B-pillar part.

4 Conclusions

This paper described the procedure of RE operation of a complex automotive BiW component, B-pillar where an existing B-pillar part was reverse engineered with help of a 3D optical scanner and CATIA V5 as CAD application software. Subsequently, five different thicknesses across the B-pillar part were introduced in the final design to have non-uniform thickness B-pillar part. Due to existing facility at the forging lab of Universiti Sains Malaysia; where a 100 ton forging machine can only allow the part to be maximum of 900 mm in length, inhibits as major limitation of conducting RE of the B-pillar which was 1500 mm in total length. To have necessary allowance and tolerance for tooling design and for smooth machine operation, the RE of the existing B-pillar was limited to 500 mm, so that future cold forging operation can be performed. In future research mold tooling design will be developed so that the part designed in this current research can be utilized for cold forming process where the final B-pillar will have non-uniform thickness. Furthermore, research will be carried out to optimize the thickness for different sections of the B-pillar part using finite element method (FEM) that can provide optimum thickness distribution providing weight reduction opportunity.

The authors want to acknowledge Ministry of Higher Education for their sponsorship through Fundamental Research Grant Scheme (Acc. No. 203/PMEKANIK/6071308)

References

1. H.-H. Bok, M.-G. Lee, E. J. Pavlina, F. Barlat, and H.-D. Kim. International Journal of Mechanical Sciences, **53**(9): p. 744-752.(2011)
2. F. Pan, P. Zhu, and Y. Zhang. Computers & Structures, **88**(1–2): p. 36-44.(2010)
3. N. P. Lutsey. Institute of Transportation Studies.(2010)
4. R. Pathare and M. Mansour SAE Technical Paper SAE (2009)

DOI: 10.1051/matecconf/20179001007

5. G. Sun, F. Xu, G. Li, and Q. Li. International Journal of Impact Engineering, **64**: p. 62-74.(2014)
6. M. F. Adnan, A. B. Abdullah, and Z. Samad. The International Journal of Advanced Manufacturing Technology: p. 1-12.(2016)
7. P. O. Marklund and L. Nilsson. Structural and Multidisciplinary Optimization, **21**(5): p. 383-392.(2001)
8. P. Weissler. *Focus B-pillar 'tailor rolled' to 8 different thicknesses*. Available from: http://articles.sae.org/7695/. (2010)
9. K. Bandyopadhyay, S. Basak, S. Panda, and P. Saha. Materials & Design, **67**: p. 558-570.(2015)
10. M. Hyrcza-Michalska, J. Rojek, and O. Fruitos. Archives of Civil and Mechanical Engineering, **10**(4): p. 31-44.(2010)
11. M. Pfestorf. *The mixed Material Concept of the new BMW X5*. Available from: http://www.autosteel.org/~/media/Files/Autosteel/Great%20Designs%20in%20Steel/GDIS%202007/04%20-%20The%20Mixed%20Material%20Concept%20of%20the%20New%20BMW%20X5.pdf. (2007)
12. M. Merklein, M. Johannes, M. Lechner, and A. Kuppert. Journal of Materials Processing Technology, **214**(2): p. 151-164.(2014)
13. Y. Chen and C. Ng. Computers & Industrial Engineering, **33**(3): p. 481-484.(1997)
14. E. Bagci. Advances in Engineering Software, **40**(6): p. 407-418.(2009)
15. M. Paulic, T. Irgolic, J. Balic, F. Cus, A. Cupar, T. Brajlih, and I. Drstvensek. Procedia Engineering, **69**: p. 795-803.(2014)

Development of comprehensive unattended child warning and feedback system in vehicle

Norizam Sulaiman[1,*], *Kamarul Hawari* Ghazali[1], *Mohd Shawal* Jadin[1], *Amran* Abdul Hadi[1], *Muhammad Sharfi* Najib[1], *Mohd Salmizan* Mohd Zain[1], *Fatimah* Abdul Halim[1], *Suhaimi* Mohd Daud[1], *Nurdiyana* Zahed[1] and *Abdul Adam* Abdullah[2]

[1]Faculty of Electrical & Electronics Engineering, Universiti Malaysia Pahang, 26600 Pekan, Pahang, Malaysia
[2]Faculty of Mechanical Engineering, Universiti Malaysia Pahang, 26600 Pekan, Pahang, Malaysia

Abstract. The cases of children being trapped and suffocated in unattended vehicle keep increasing even though the awareness campaign on the safety of children in non-moving vehicle were carried out by government. Various methods were introduced by researchers to overcome this issue but yet to be effective. Among them were the usage of capacitive sensor, microwave sensor, pressure sensor and image sensor where most of the techniques or systems were applied on the child's seat to detect the presence of baby or infant. Thus, this research is carried out to provide a comprehensive and effective detection system to detect the presence of children including infant in unattended vehicle by using the combination of human physiological signals (voice and body odor) detectors with the temperature and motion sensors. Here, once the proposed system recognizes any signals that generated from voice, odor, motion and temperature detectors in vehicle's cabin, the system then will provide effective feedback system by sending short message to the parents first. If no response received in the specified allocation time, the system then will activate the vehicle's horn system. Finally, the system will lower down the vehicle's window to release the toxic gas and reduce the cabin temperature. The system is in prototyping stage where every design component was evaluated individually. Besides, the overall system was successfully tested where the detection and feedback system follow the instruction given by the microcontroller.

1 Introduction

In Malaysia, vehicle-related deaths for children fewer than 12 years of age keep occurring and the cases are in increasing trend [1-2]. Despite of various inventions to enhance the vehicle safety system and awareness campaign to prevent those incidents from occurring, the deaths caused by the non-moving vehicle keep increasing. When the car engine is turned-off with closed window, the temperature in the car will increase rapidly and children

* Corresponding author: norizam@ump.edu.my

might suffer heat stroke [1, 3]. A child's body overheat more easily than an adult's [1]. Most of the incidents occurred due to the negligence of the parents [2-4]. Parents or drivers, in certain situations, may leave the car without taking the children. It might also have occurred when the father or mother change his or her duty to send their children to nursery. This scenario can happen since their children taking sleep or nap, and they might not realize the presence of their children in the vehicle especially for the bigger vehicle such as multipurpose vehicle which has more than two rows of seats. Unlike vehicle in United States or European countries which requiring parents to put their children in the specified child's seat before driving as a safety matter, those rules are not abide by most of the Malaysian vehicle's drivers. It seems that the requirement to use child's seat is applied merely on the children with age below 2 years old. However, children with age ranging from 2 to 6 years old are free to move around inside the vehicle's cabin. Thus, they will be sitting, laying or napping somewhere in the vehicle's cabin. For example, it will be very hard for the multipurpose vehicles' owner to locate his or her child if the child is lie down at the back seat or at the floor between the second and last row of the seat. It is a human nature to forget something and to think driving alone. As a result, he or she will park and leave the vehicle while his or her child is left behind or still inside the vehicle [5].

Various techniques and invention had been introduced for couple of decade but yet to provide the solid solution for to cater this issue. Most of the techniques and invention were patented especially in United States and Europe. In United States and Europe, the child detection systems were developed especially for baby or infant seat and integrated with car alarm system. Here, the car security system will generate an audio alarm once the car is turned-off and the children are detected still in the car baby seat [4-5]. Among the various child detectors were introduced by researchers are Passive Infrared Red (PIR) motion sensor, RFID, RF safety pad, capacitive sensor, microwave sensor, pressure transducer, temperature detector and image detector. However, all the invention is still yet to provide the effective and comprehensive warning and feedback systems to trigger parent for the unattended child in the vehicle. Some of the inventions have drawbacks. For example, for the child seat, there were synchronization issues between child-seat-sensing device and notification device. Meanwhile, PIR required motion from the children and will not effective for the children in sleep condition [6-8]. Another invention was the use of pressure transducer on the child seat to detect the changing of air pressure due to the opening and closure of the vehicle door [9]. Here, the device was integrated with the vehicle alarm system to produce alarm once the pressure transducer activated. In addition, one of the invented techniques was the combination of the child seat sensor and temperature detector were applied to detect the present of child and to detect unsafe temperature in the vehicle cabin and trigger the vehicle alarm system [10]. The drawback for both inventions is they are designed purposely for infant which required to be placed onto child seat to trigger the car alarm system. Meanwhile, for image detection technique, it requires wide angle camera and of course, a lot of camera are required for bigger vehicle such as multi-purpose vehicle or sport utility vehicle. Another method is the use of visual cue at the car dashboard to enable driver to have a view behind the driver seat. However, this method might not practicable for bigger vehicle such as multipurpose vehicle.

In term of communication systems, it cannot be denied that the wireless communication is a practical and an effective medium to enable human to communicate with electronics device. Among them are IEEE 802.4.15, the network Global System for Mobile Communications (GSM), Global Positioning System (GPS) and Radio Frequency Identification system (RFID) were used extensively especially in vehicle safety technology [11-16]. In current vehicle technology, most of the modern vehicle already equipped with wireless anti-theft security system whereas the child detection system is installed on the vehicle baby seat and strictly applied for vehicle in US and European countries. In those countries, there is a law to put the young child or infant in the child seat when driving the vehicle [15-16]. In Asian countries especially Malaysia, the law is not fully followed and merely the expensive vehicle has the accessories for anti-theft and child detection system. Thus, it a must for every vehicle in Malaysia regardless the type of the vehicle, to be equipped with the wireless child detection system since the system is proven by researchers to enable human to communicate with the vehicle regardless the location of the vehicle unless the location cannot be reached by the GSM.

Therefore, this research will introduce a comprehensive child detection system which enhancement of the previous researches by combining human physiological signals such as voice and odor with image, temperature and motion detection systems. Besides, the system will also provide the efficient feedback system if any failure on alarm system or no response from the parent on the notification system through GSM module or no response from people who might think that the alarm sound come from the faulty alarm. The system then will be integrated with the existing vehicle security and ignition systems. The proposed system is constructed to achieve the followings objective; to design a warning system to detect the presence of children in unattended vehicle using a combination of voice, odour, motion, and temperature detection techniques. Besides, the system is created to provide the intelligent feedback system by sending Short Messaging System (SMS) to the parents, activating car's horn system and finally, moving down the vehicle's window in order to release the toxic gas and reduce the temperature inside the vehicle's cabin.

2 Experimental setup

This section will describe the overall process in designing, developing and testing the proposed system

2.1 Design of experiment

The design process begins by selecting the suitable input signals, signals processor and types of effective feedback signals. Hence, the research is divided into three studies; detection system, feedback system and selection of microcontroller to process the input signals and provide the good triggering system as illustrated by Fig. 1. In this study, 4 input signals are chosen; temperature, motion, voice and odor. In order to detect those signals, the separate detectors or recognizers are constructed. For example, temperature detector is developed to measure interior temperature of the vehicle and detect any changes in interior temperature in vehicle's cabin after the vehicle being parked. Arduino-based temperature sensor is used which can measure temperature up to 125°C. Meanwhile, motion detector is constructed using a passive infrared (PIR) sensor which become a top choice of detectors by researchers in this study due to the high accuracy, easy handling and low cost. Voice detector is developed by using the combination of Adafruit FONA mini GSM (SIM900L), EasyVR Shield and Arduino Uno R3. Here, Adafruit is used to send text message to the mobile phone through GSM module. SIM900L is a complete Quad-band GSM/GPRS in a SMT type. Meanwhile, EasyVR Shield is used to capture the voice and Arduino acts as a microcontroller to process the voice that received from the shield. In order to increase the accuracy of detecting voice, about 9 voices from difference child are captured and processed by the microcontroller.

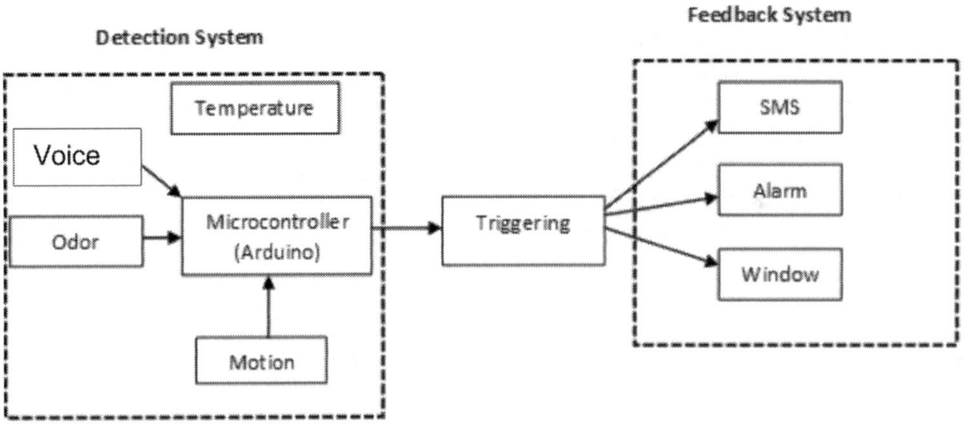

Fig. 1. Block diagram of the detection and feedback system.

The most challenging task in this research is to design the odor recognizer. Here, the odor recognizer consists of E-Nose [17]. E-nose is made up by a chamber equipped with sensor array and can be linked to microcontroller. The sample will be fed at the inlet part of the chamber and the output will be released at the outlet part of the chamber. It is a device invented by one of the member in research team to detect multiple type of odor. The E-Nose was employed to classify the Kelulut (one of bee species) honey. The intelligent detection and classification system is developed to study the odor profile and determine the type of odor as illustrated in Fig. 2. In this research, the odor of child's urine and sweat is captured and collected from 9 children. Then, the profile of those odors is created and studied. Also involved in the study is the odor from vehicle's perfume and cabin. The study requires much time in order to obtain the correct odor profile that come from human (child). The thorough study on child's odor profile is implemented to ensure the correct instruction is given to the microcontroller for the feedback system.

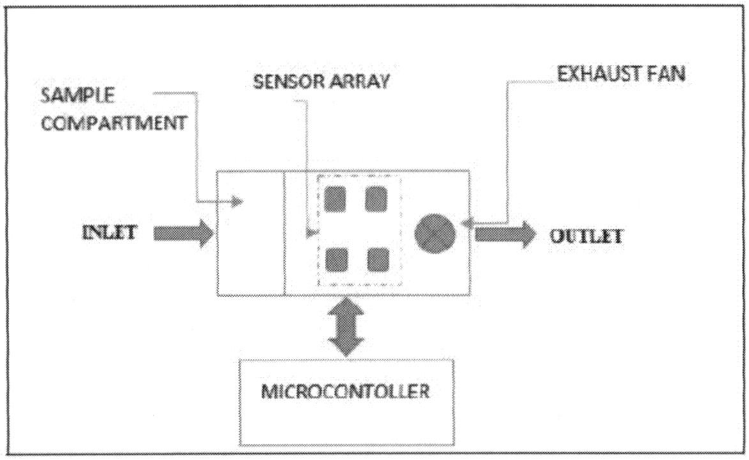

Fig. 2. Block diagram of odour detection system.

2.2 Operation of the proposed system

The proposed system will improvise the previous system by adding human physiological signals and creating the smart feedback system to detect the presence of children in the unattended vehicle. As shown by block diagram in Fig. 1, the project covers the design of the detection and feedback system which include the testing of various sensors. Hence, the experimental produce will include the chronological sequences of the system in detecting the input signals from the sensors and triggering the feedback system.

The system starts in five minutes after the vehicle is in its parking situation. The vehicle is considered to be in parking situation based on several conditions such as ignition engine is central lock is on or handbrake is locked or ignition engine is off. The alert voice message will be sent to the vehicle driver about the activation of the detection and feedback system. The system then will activate the temperature sensor first, follow by motion, voice and odor detectors respectively. However, to make the system more robust, in real implementation of the project, any detector activated will trigger the feedback system either by sending SMS to vehicle driver, activating vehicle's horn and rolling down the vehicle's window. Here, the temperature sensor will activate when the cabin temperature greater than 40°C. Meanwhile, the voice and motion detectors will take place to activate the system if detecting any movement or voice from the child. The child will start to cry or move his or her body when feeling uncomfortable due to the increasing cabin temperature. In order to make the detection system more comprehensive and since the system will take time to react based on the triggering condition of the temperature, motion and voice sensors, odor recognizer is introduced. Here, once the vehicle in its parking position with engine off, the odor detector is activated. It can immediately detect the present of the child in the vehicle's cabin based on the odor profile of the child without waiting for the other sensors to be activated. Once the odor of the child is detected, the system will trigger the feedback system immediately. The feedback system will begin by sending SMS to the vehicle driver's mobile phone through the GSM module. If no response from the vehicle driver, the vehicle horn system is activated. Finally, if no response from the vehicle driver, the feedback system will move down the window at little bit just to reduce the cabin's temperature and release the toxic gas. The system need to be reset manually by the vehicle's driver.

The full schematic of the circuit design of project is depicted by Fig. 3. As described by the diagram, 2 units of Arduino microcontroller are used to process the multiple input signals. Based on the schematic shown in Fig. 3., LCD is used to display the alert message when detecting motion and vehicle interior's temperature. The digital pins of LCD (D4, D5, D6 and D7) are connected to the digital pin of Arduino 1 (D9, D10, D11 and D12). The E and RS pins of LCD are connected to Arduino D7 and D8. Here, the temperature sensor is connected to the analog pin of Arduino (A5). Meanwhile, the transmitter and receiver pins of Adafruit GSM module is connected to the receiver and transmitter pins of Arduino Uno R3 (Digital pin 1, D1 and Digital pin 0, D0). The window motor is connected to the relay pins (GS1, GS2, GS3 and GS4). Then, the relay pins are connected to the analog pins of Arduino (A0, A1, A2 and A3). In addition, motion sensor is connected to the Arduino digital pin D6. Meanwhile, odor and voice recognizer's input signals are connected to the analog pins of another Arduino (Arduino 2). The second Arduino is attached to the EasyVR shield. The LED lights (red, yellow and green) are also installed into the detection system to indicate the status of detection process.

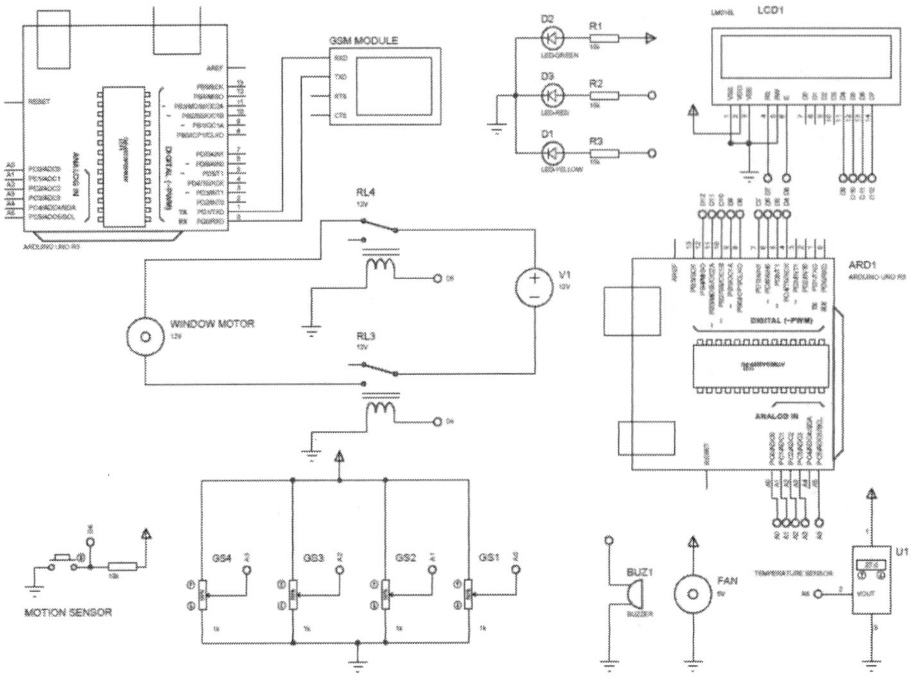

Fig. 3. The full schematic of the detection and feedback system.

3 Results and discussion

The project consists of designing the detection and feedback system using multiple detectors; temperature, motion, voice and odor. The temperature, motion and voice detection system is developed using the temperature sensor, PIR sensor and voice recognizer. Meanwhile, odor recognizer is built using E-Nose. Then, all the sensors are assembled and tested for functionality and accuracy. The prototype of temperature and motion detection system is shown in Fig. 4. When PIR detect any motion or temperature sensor detects the increasing surrounding temperature, the signals will be sent to Arduino microcontroller to process it, and then the results will be displayed on the LCD. The microcontroller also will send the processed data to the feedback system to trigger the vehicle driver by activating the alarm system.

Fig. 4. Temperature and motion recognizer.

Meanwhile, the odor detection system is built using chamber and sensor array. The odor recognizer is tested for accuracy with various odor profiles including profile of child's odor before assembling it with the other sensors. The training on the odor recognizer needs to be implemented in order to enhance the capability of that recognizer to differentiate and distinguish human's odor from other source of odor in the vehicle's cabin. The odor recognizer should able to recognize the child's odor accurately based on the child's urine and sweat. Once the odor is confirmed come from child, the detection system will trigger the feedback system.

The voice detection system is constructed using Adafruit FONA mini GSM, voice recognizer shield (EasyVR) and Arduino. Also, attached to the voice recognizer shield is the microphone to activate the system. The system is tested with 9 different child's voice and able to recognize all the voice accurately. Once the system able to recognize the child's voice, it will trigger the feedback system by sending SMS to mobile phone, activate car horn and rolling down the vehicle's window. The prototype of the voice detection system is shown in Fig 5. The example of the SMS sent to the mobile phone is elucidated by Fig. 6.

Fig. 5. Voice recognizer.

Fig. 6. Feedback system through Short Message (SMS).

The overall prototype of the detection and feedback system is elucidated by Fig. 7 (a). The prototype consists of 12V battery to provide voltage to window motor, bigger black box consists of temperature, motion, odor sensors, LCD, LEDs and Arduino microcontroller. Meanwhile, the smaller black box consists of EasyVR shield, Adafruit FONA mini GSM, Arduino microcontroller, microphone and antenna. During testing the prototype, the Arduino microcontroller is power-up through USB cable that connected to the computer. The direction of the window motor either rotating clock wise or counter

DOI: 10.1051/matecconf/20179001008

clock wise and duration of rotation will depend on the instruction given to the microcontroller. The location of window motor in the vehicle's door is shown in Fig. 7 (b).

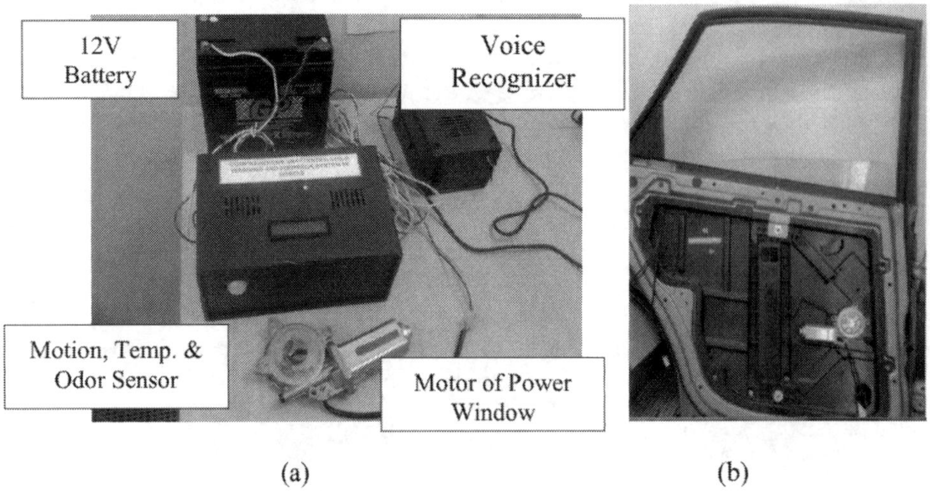

Fig. 7. The overall prototype of detection and feedback system.

4 Conclusions

In this paper, the prototype of the detection and feedback system is constructed successfully. All the sensors were tested individually and functioning correctly according to the instruction given by microcontroller. The main advantage of the system compared to the existing system is the usage of odor recognizer based on E-nose technology. The outcome of the study shows that the use of human physiological signals (voice and odor) is very effective to detect the present of the child in the unattended vehicle. The future work may include enhancement of the system by using smaller prototype such as using printed circuit board (PCB). Therefore, it can be fit nicely inside the vehicle's cabin. The next project will also include the integration of the proposed system with the vehicle's electronic system. The robustness testing of the system might be implemented as well.

The authors would like to express high appreciation to the postgraduate students and research team members for their great assistance and support during the project implementation and testing. The authors are also grateful to Universiti Malaysia Pahang for providing financial assistance through research grant, RDU140376.

References

1. Consumer Report News (2012)
2. F.R.M. Rashidi, I.H. Muhamad, *Recent Advances in Automation, Control, Modelling and Simulation* (2013)
3. A. Ranjan, B. George, *Instrumentation and Measurement Technology Conference* (2013)
4. J. Booth, G. Davis, J. Waterbor, G. McGwin, Forensic Science, Medicine, and Pathology, **6**, (2010)
5. M. Rossi, U.S. Patent 5949340 (1999)
6. N. M. Z. Hashim, H. H. Basri, A. Jaafar, M. Z. A. A. Aziz, A. Salled, A. S. Ja'afar, ARPN Journal of Engineering and Applied Sciences **9**, 9 (2014)

7. J. Morningstar, U.S. Patent 7714737 B1 (2007)
8. I., H. Muhamad, F.R.M Rashidi, *Proceeding of the Recent Advances in Automatic Control, Modelling and Simulation* (2013)
9. M. G. Burgess, U.S. Patent 6812844 B1 (2004)
10. D. C. White, U.S. Patent 7701358 B1 (2010)
11. K. Vidyasagar, G. Balaji, K. Narendra Reddy, *International Journal of Communications on Applied Electronics* **2**, 2 (2015)
12. J. K Lee, D. R. Spach, P. Srinivasan, D. P. Umstead. U.S. Patent 7321306 B2 (2008)
13. G. C. Kautz, I. La, J. Beach, U.S. Patent 34957-6605 (2010)
14. F. Mesina, IEEE Journal (2005)
15. J. Abdullah, International Journal of Engineering and Technology **3** (2011)
16. K. N. Kamil, S. I. A. Rahman, M. Gambilok, ARPN Journal of Engineering and Applied Sciences **10**, 22 (2015)
17. M. S. Najib, S. H. Azih, N. Zahed, M. F. Zahari, W. M. A. Mamat, H. Manap, ARPN Journal of Engineering and Applied Sciences **11**, 10 (2016)

Moderating effect of environmental turbulence on firm's technological innovation capabilities (TIC) and business performance in the automotive industry in Malaysia: A conceptual framework

Faridah binti Taju Rahim[1] and *Yuserrie* bin Zainuddin[1,*]

[1]Faculty of Industrial Management, Universiti Malaysia Pahang, Lebuhraya Tun Razak 26600 Gambang, Kuantan, Pahang, Malaysia

Abstract. The dimensions that constitute the technological innovation capabilities (TIC) can be a source to improve business performance. Based on the resource-based view and contingency theory, this paper tries to propose a conceptual model to further understand the moderating role of environmental turbulence in effecting of the firm's TIC on firm's business performance in the automotive industry in Malaysia. This paper then developed into concrete research hypotheses for future studies. The conceptual model will serve to guide further research in the area, which can be expected to contribute in the development of automotive industry in Malaysia. Empirical testing of this proposed model is expected in future study using survey instrument.

1 Introduction

There are increasing studies regarding how a firm can maintain their sustainable competitive advantage and improve firm performance in fast changing and unpredictable environments due to the globalization of markets, technological change, shortening product lifecycles and innovative new product development (NPD) [2–4]. One popular approach used to understand competitive dynamics is the resource-based view (RBV) of the firm. According to this view, only those firms who have the unique resources and capabilities provide the ability for competitive advantage in rapidly changing and unpredictable environments then leads to higher performance. If the firms' resources and capabilities are valuable, rare, hard to imitate, cannot be substituted and the firm also could organize and fully utilized those resources and capabilities, then they could lead to better performance [5–8].

Technological innovation capabilities (TIC) has been considered as a significant strategic resource allowing firms to attain sustainable competitive advantage when firms meet a dynamic environment [9–11]. Firms with more TIC can perform better in more

* Corresponding author: yusserie@ump.edu.my

turbulent environments as compared with firms with lesser levels of TIC. Since superior TIC can help firms receive greater efficiency gains by pioneering process innovations, and to achieve higher differentiation by innovating products in response to the changing market environment [3,12–14]. Moreover, superior TIC can accelerate the development of new product developments and adopt new processes in shorter lead time in order to reap the pioneer [15–17]. As a result, TIC can help a firm to create more value than its competitors and can receive a greater economic return above the industry average [18]. Facing competitive business environment, firms are required recurring technological innovation to continuously retain their competitiveness and to face new challenges [11,19,20]. Therefore, the firms must assimilate firm resources and TIC to maintain and improve their performance. According to Yam, Guan, Pun, & Tang (2004) in a dynamic environment, an inability to innovate eventually causes businesses to deteriorate and firms to go out of business.

In the context of internal organization, firms tend to suffer from lack of resources, especially tangible asset. However, in external circumstance, a firm challenges various environmental turbulence, such as technological turbulence and market turbulence [21]. Both technological turbulence and market turbulence are crucial dimensions of environmental turbulence [22,23]. Along with limited resources, firms need to be more adaptable to respond environmental turbulence [24,25]. Higher environmental turbulence causes reduce the innovation activity of firms [26,27] since it may raise the uncertainty and the threat of innovation investment. To constantly maintain their competitiveness, firms rely on technological innovation to deal with rapidly changing environments [11].

Many studies also have shown that technological innovation could create positive impacts and improving their competitiveness and [13,16,19,28–33]. Although studies on technological innovation are in abundance, there is nevertheless inadequately empirical evidence relating to how automotive firms improve their TIC. Even though the growing studies have been carry out to investigate the effect TIC in influencing competitive advantage and firm performance, up to now little studies has been examined the moderate effect of environmental turbulence on TIC and firm performance. Therefore, there is a need to develop a comprehensive framework for analyzing the moderating roles of two sorts of environmental turbulence: technological and market, effect on firm's TIC and firm's business performance.

Drawing upon contingency theory and the resource-based view perspectives, we suggest that environmental turbulence may moderate the relationship between TIC and business performance. This issue is important but remains mostly neglected in the existing TIC literature especially study in the automotive industry. In order to manage the high competitiveness and the various challenging business factors in the automotive industry [34] for example uncertainty about climate change policies whereby firms need to play a role in reducing global CO^2 emissions through technological innovation, therefore, it is important to create better strategic planning which strongly reflects an increase in firm performance and also need to reduce operation cost. The industry is also having experienced rapid technology changes [35] and is currently in the demand for technological innovation in order to improve their business performance. Automotive industry in Malaysia was chosen due high relevance of the automotive industry in the country's GDP. In fact, according to the Malaysian Automotive Association (MAA), production of motor vehicles for 2015 totaled 614,664 units comprising 563,883 units of passenger vehicles and 50,781 units of commercial vehicles. Sales of motor vehicles amounted to 666,674 units in 2015 consisting of 591,298 units of passenger vehicles and 75,376 units of commercial vehicles.

The article begins with a review of the literature on these areas and introduces the conceptual framework and research hypotheses in this study. Then, finally makes some concluding remarks.

2 Conceptual framework and hypothesis development

This study supports both resource-based view and contingency theory. The resources based view explains how firms use TIC as their valuable resources to achieve the goal, while contingency theory enlightens how exogenous variable influences the impact of such resources and capabilities on firm performance.

2.1 Resource-based view and TIC

Based on the assumption those firms' resources are heterogeneous and not perfectly mobile, the resource-based view (RBV) emphasizes that firms' sustainable competitive advantage is mainly attributable to their valuable, unique, inimitable, and non-substitutable resources [6]. This theory becomes a fundamental competitive advantage in which firms depend on their own tangible and intangible resources as well as their capabilities to change their temporary competitive advantage into sustainable competitive advantage. Resources are stocks of tangibles and intangibles assets partly attached to the firms [36]. However, capabilities are the complex coordinated patterns of skills and knowledge that become entrenched as organizational procedures and practices and assist firms to make the best use of its resources [3].

The concept of capabilities, derived from further research in RBV, concerns the firm skills to organize resources. Although the resources include the assets, tangible and intangible, possessed by the firm [37], the capabilities are referred to the firm ability to create a set of activities through resource deployment in order to reach a desired end ([37]; [38]). Day (1994) refers to capability as "complex of bundles of skills and gathered knowledge, implemented through organizational processes that facilitate firms to manage activities and exploit of their assets". In other words, the capabilities denote the skills allowing firms to arrange resources to achieve a desired goal.

According to Ferreira, Azevedo, & Ortiz (2011), resources are the source of the firm's capabilities, capabilities are the vital source of its competitive advantage. So as to accomplish a sustained competitive advantage, the capabilities should be valuable, unique, difficult to identify and recognize, imperfectly mobile, not easily duplicated and substitutable [6]. In this study, TIC can be considered as a subset of firm capabilities.

Applying this approach to firms, we selected the TIC which firms can exploit in order to achieve successful performance. TIC play a pivotal role to boost firm performance [13,19,20,33,41,42]. From the RBV perspective, TIC are: (1) valuable because it facilitates innovative culture [43] that may lead to produce several of new products or services and improve the operational processes that efficiently generate and deliver these goods and services to the customer which lead to provide superior value to the customers [41,44], which in turn customers may willing to pay a higher price for their products or services, therefore the firm can gain sustainable competitive advantage ; (2) rare because in any particular industry for example in automotive industry, few firms will be willing or able to undertake innovative activities [45] because product introduction into the market usually needs longer time, money and managerial activities costs [46,47]; and (3) imperfectly imitable because firm innovation cultures are socially complex [7]. Thus, TIC are a strategic resource and whose exploitation may provide a firm with a competitive advantage and superior performance [20,41,45,48,49].

Previous studies have conceptualized TIC with different approaches that results into various sets of capabilities to assessing a firm's TICs. Yam et al., (2011) has summarized three approaches which are asset approach, process approach and functional approach. According to [32] functional approach easier to understand, and it facilitates the multi-informants approach for the survey. However, the asset and process approaches are rather more complex to understand [13]. Table 1 presents a set of approaches and their elements TICs.

Table 1. Different approach and elements in assessing TICs (adapted from Yam et al. 2011).

Conceptual Approach for TICs	Elements of TICs
Asset approach - Christensen (1995) [50]	• Science research asset • Product innovation asset • Esthetics design asset
Process approach Burgelman et al. (2004,2009) [43]	Capabilities of a firm in • Resources availability and allocation • Understanding competitor innovative strategy and market • Understanding technological developments relevant to firm • Structural and cultural affecting internal innovative activities • Strategic management capability to deal with internal innovative activities
Chiesa et al. (1996) [51]	• Concept generation capability • Process innovation capability • Product development capability • Technology acquisition capability • Leadership capability • Resources deployment capability • Capability in effective use of system and tools
Functional approach (Yam et al. (2004) [32]	• Learning capability • R&D capability • Resources allocation capability • Manufacturing capability • Marketing capability • Organization capability • Strategic planning capability
Liu & Jiang (2016) [14]	• Organizational capabilities • Strategies capabilities • Human, finance, and material (HFM) resources • Knowledge resources • Fundamental research • Application R&D • Manufacturing capabilities

In this study, we carry out the functional approach of conceptualizing TICs presented by Yam et al., (2004). According to this approach, TIC results as a complex interface of activities conducted within and between different functional departments of a firm. This approach has been used in many studies recently [13,19,28,52–54]

The framework measured TIC in seven dimensions and these dimensions are described as follows. :

- Learning capability is the capacity to find, integrate, and develop new knowledge important for a firm's competitive success.
- R&D capability indicates a firm's expertise to assimilate R&D strategy, project implementation, R&D expenditure and product portfolio management.
- Resource allocation capability is the firm's ability to organize and develop its technological, financial and human resources in the innovation process.
- Manufacturing capability refers to the ability to alter R&D results into products, which meet market needs, in line with design requirement and can also be produced in batches.
- Marketing capability refer to the capacity to advertise and sell the products on the core of aware of consumer's present and future needs, customer's access methods, and competitors' knowledge.
- Organizing capability is the capacity to establish a well-established organizational structure, cultivate an organizational culture, organize the effort of all activities concerning shared objectives, and affect the speed of innovation processes through the infrastructure it generates for developmental projects.
- Strategic planning capability is the capacity to ascertain internal strengths and weaknesses and external potentials and threats, engage in different types of strategies that can adapt to environmental changes to be superior in the highly competitive environment.

2.2 Contingency theory

Contingency theory was developed in the 1960s, which claimed that there was only one way that was the best way to be coordinated. Contingency theory indicated that the most effective firm strategy and structure would vary according to the environmental demand and external condition of the firm [55,56]. Those environmental demands and external conditions which impact the firm specific under consideration are known as contingency factors. The major contingencies identified include environment, technology, age and size [57].

Hence, contingency theory considers that firm's capability to respond business environment affects firm performance [22,58]. Business environment has been recognized as one of the contingency elements in strategic management research [59]. Contingency effect can be environmental turbulence, which refers to the degrees of change in the market and/or technology within an industry [23,60]. This concept constitutes into three factors, i.e. technology turbulence, market turbulence, and competition turbulence [23]. However, this study only focuses two categories of environmental turbulence, which is technological turbulence and market turbulence because both are the most important types of uncertainty [21,22,61].

Responsiveness as crucial for firms to deal with customer needs within the technological turbulence [24]. Typically, some firms have lack of capacity to anticipate the turbulence on account of poor scanning of exact and reliable information about the industry condition [61]. Market turbulence is about the rate of dynamic composition of customers and their preferences. This causes increasing uncertainty due to changing customer's preference. Higher levels of market turbulence bring about high uncertainty due to less accurate forecast, and then firms may back off to invest in innovation activity and, as a consequence, reduce the innovation activity of firm [27]. Firm with capacity to develop their technology to respond the negative impact of marketing turbulence will be able to be better than other firms [24].

2.3 Hypothesis development

Drawing upon Resource-Based View of the firm and contingency theory, the proposed conceptual framework integrates one hierarchical component model or multidimensional constructs (TIC) that are conceptualized at second-order level and three constructs (market turbulence, technology turbulence and business performance) specified as lower-order constructs. The conceptualization of multidimensional constructs and their association with the dimensions underpinning them are derived from theory.

Our conceptual model (Fig. 1) proposes that the TIC as resources enable the firm's to enhance its business performance. Technological turbulence and market turbulence moderates the TIC– business performance relationship.

2.3.1 Relationship between TIC and business performances

TIC were viewed as a comprehensive set of elements of a firm that facilities and supports its technological innovative strategies [43] in the business environment and successful exploit of these capabilities to sustain competitiveness performance for the firm. TIC are a kind of integration of special assets or resources of the firm which comprises various assets such as technology, product, process, knowledge, experience [28,31,62,63]. TIC has become an important strategic resource that enables firms to construct performance in a dynamic environment, especially in high-tech industries such as the automotive industry, which is facing a rapidly changing environment [26,64,65].

In the theory of resource-based view, when firms have successfully created differentiating resource configurations, they could better satisfy their customers' needs, they produced more efficiently, and in the long run, they achieved competitive advantage leading to higher performance [6,66]. Superior TIC and competence can improve product and/or process innovation and accelerate the pace of new product developments in the turbulent market competition, which in turn raises a firm's competitive advantage and achieving better performance over their competitors [20,31,33]. Firms are able to increase performance when they develop new ideas, create new products or services, seek new ways of doing things and be creative in their operations [32].

This study follows an audit framework proposed by Lau, Yam, & Tang (2010) which firstly developed by Yam et al. (2004) to investigate the TIC and their impact on business performance in automotive industry. Through accumulation of TIC, firms can renovate their abilities to fit, integrate, and reorganize the internal and external resources to be better than their competitors, thereby have better performance in the changing environment. Consequently, the hypothesis is as follows:

Hypothesis 1 (H_1): Firms with superior TIC have a better performance.

2.3.2 The technological turbulence as moderating factor of TIC and business performance relationship

Environmental turbulence is considered to come with the moderating effect that changes the direction of a relationship between TIC and firm performance. This study specifies environmental turbulence into technological turbulence and market turbulence. In this study, technological turbulence refers to the rate of technological change in an industry [23]. That technological turbulence provides a moderating effect on the relationship between TIC with business performances.

Technological turbulence may moderate the strength of the relationship between TIC and business performance. When the level of technological turbulence is high, a firm can adapt by utilizing its TIC [22]. Rapid changes in technology quickly make current technology obsolete. In such a situation, firms must continually develop new technology [21] and achieve competence in the latest technologies and, thereby successfully generating product and process innovations [24]. Thus, in a highly technologically turbulent environment, firms need to depend on their TIC to exploit new technology opportunities, hence improving business performance.

Based on the aforementioned arguments, we expect that the effect of TIC on business performance will increase as technological turbulence increases than when technological turbulence is low. Therefore, we hypothesize the following:

Hypothesis 2 (H_2): Technological turbulence moderates the relationship between TIC and business performance.

2.3.3 The market turbulence as moderating factor of TIC and business performance relationship

Market turbulence indicates to the degree of change in customer preferences in an industry [21,23]. A turbulent market environment is described by constant changes in product preferences and customer demand [25], by a wide range of customer demands and needs, and by customer seeking for new products frequently [67].

Market turbulence may also influence the strength of the relationship between TIC and business performance. In a highly turbulent market, firms with greater TIC can achieve better performance. When market turbulence is increasing, firms likely require restructuring of their TIC to satisfy evolving needs of customer [24]. However, those operating in low market turbulence are less likely to require such adjustments. High market turbulence with fast changing customer needs prompts firms to learn about the changes; they also need to regularly identify reconfigurations of their resources and capabilities such as TIC [25] to address market changes in new ways and can exploit rapidly changing market demand[60]. Thus, in a highly turbulent market, firms with superior TIC can achieve better performance.

Based on the aforementioned arguments, we propose market turbulence will moderate the TICs and business performance relationship. Therefore, we hypothesize the following:

Hypothesis 3 (H_3): Market turbulence moderates the relationship between TIC and business performance.

The conceptual model is described in the Figure 1 as follows:

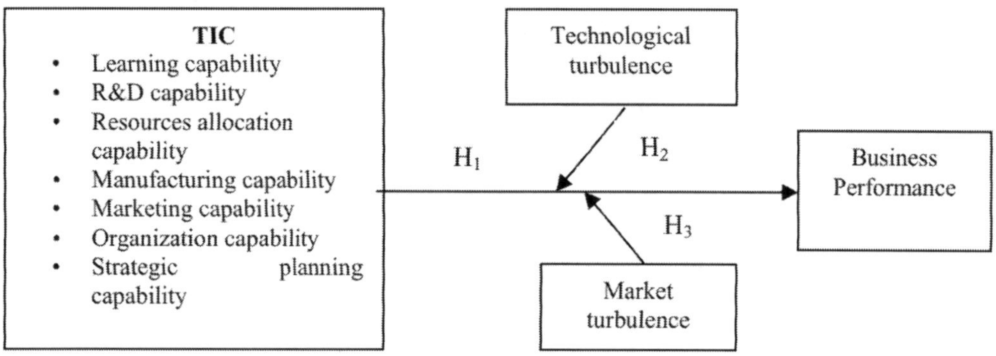

Fig. 1. Research model.

3 Conclusions

Many studies have been performed to identify the relationship of TIC and business performance. Based on previous study, this study focus on TIC which is consists of learning capability, R&D capability, resources allocation capability, manufacturing capability, marketing capability, organization capability and strategic planning capability. However, there is lack of earlier study to investigate the relationships between TIC, business performance and the moderating effects of environment turbulence on these relationships, especially in the automotive industry. A conceptual model has been recommended to study these relationships.

Our model has implications for both researchers and practitioners. For researchers, the model suggests the types of the variables that need to be included in future empirical tests of the relationship between TIC and business performance especially in automotive industry. Consequently, the model extends understanding of the moderating influence of external environment such as technological turbulence and market turbulence on the relationship between TIC and business performance.

Practitioners, on the other hand, could use the model to refine their thinking about TIC and their firm's strategic resources. The main contribution of this paper was to encourage managers to take a consideration on the relationship regarding TIC, business performance and environmental turbulence.

Based on recommended model and an earlier studied, research hypotheses are established. The next step of this study is to design a questionnaire, which will be applied for pilot study data collection in Malaysia automotive industry. We will try to conduct tests using the approaches of RBV. Therefore, the researchers as well as practitioners will have more empirical evidences associated to the fundamental theory behind the RBV, accordingly improving understanding of the relationships among TIC, and business performance and the moderating effects of environment turbulence.

References

1. N. Amira, M. Ali, M. Hanif, A. Gafar, J. Akbar, Enhancing Promotional Strategies within Automotive Companies in Malaysia. *Procedia Econ. Financ.* **7,** 158–163 (2013).
2. K. Choi, R. Narasimhan, S. W. Kim, Opening the technological innovation black box: The case of the electronics industry in Korea. *Eur. J. Oper. Res.* **250,** 192–203 (2016).
3. D. J. Teece, G. Pisano, A. Shuen, Dynamic Capabilities and Strategic Management. *Strateg. Manag. J.* **18,** 509–533 (1997).
4. C. A. Bayraktar, G. Hancerliogullari, B. Cetinguc, F. Calisir, Competitive strategies, innovation, and firm performance: an empirical study in a developing economy environment. *Technol. Anal. Strateg. Manag.* **7325,** 1–15 (2016).
5. R. M. Grant, The Resource-Based Theory of Competitive Advantage : Implications for Strategy Formulation. *Calif. Manage. Rev.* 114–135 (1991).
6. J. Barney, Firm Resources and Sustained Competitive Advantage. *J. Manage.* **17,** 99–120 (1991).
7. J. B. Barney & W. S. Hesterly *Strategic Management and Competitive Advantage: Concepts*. (Pearson Education, Inc., publishing as Prentice Hall, (2012)).
8. C. S. Lioukas, J. J. Reuer, M. Zollo, Effects of Information Technology Capabilities on Strategic Alliances: Implications for the Resource-Based View. *J. Manag. Stud.* **53,** 161–183 (2016).
9. R. Henderson, I. Cockburn, Measuring Competence? Exploring Firm Effects in Pharmaceutical Research. *Strateg. Manag. J.* 1–36 (1994).
10. F. Kong, Z. Zhang, Y. Liu, Study on the Evaluation of Technological Innovation

Capability Under Uncertainty. *2008 4th Int. Conf. Wirel. Commun. Netw. Mob. Comput.* 1–4 (2008). doi:10.1109/WiCom.2008.2999

11. Y.-L. Cheng, Y.-H. Lin, Performance Evaluation of Technological Innovation Capabilities In Uncertainty. *Procedia - Soc. Behav. Sci.* **40,** 287–314 (2012).
12. D. J. Teece, Profiting from technological innovation : Implications for integration , collaboration , licensing and public policy. *Res. Policy* **15,** 285–305 (1986).
13. R. C. M. Yam, W. Lo, E. P. Y. Tang, A. K. W. Lau, Analysis of sources of innovation , technological innovation capabilities , and performance : An empirical study of Hong Kong manufacturing industries. *Res. Policy* **40,** 391–402 (2011).
14. L. Liu, Z. Jiang, Influence of technological innovation capabilities on product competitiveness. *Ind. Manag. Data Syst.* **116,** 883–902 (2016).
15. P. Carbonell, A. I. Rodríguez-Escudero, Relationships among team's organizational context, innovation speed, and technological uncertainty: An empirical analysis. *J. Eng. Technol. Manag.* **26,** 28–45 (2009).
16. J. C. Guan, R. C. M. Yam, C. K. Mok, N. A Ma, Study of the relationship between competitiveness and technological innovation capability based on DEA models. *Eur. J. Oper. Res.* **170,** 971–986 (2006).
17. J. Prašnikar, M. Lisjak, A. R. Buhovac, Štembergar, M. Identifying and Exploiting the Inter relationships between Technological and Marketing Capabilities. *Long Range Plann.* **41,** 530–554 (2008).
18. A. J. Verdu, I. Tamayo, A. Ruiz-Moreno, The moderating effect of environmental uncertainty on the relationship between real options and technological innovation in high-tech firms. *Technovation* **32,** 579–590 (2012).
19. M. Lang, S. Hsiang, T. Nguyen, T. Vy, Mediate effect of technology innovation capabilities investment capability and firm performance in Vietnam. *Procedia - Soc. Behav. Sci.* **40,** 817–829 (2012).
20. J. Shan, D. R. Jolly, Technological innovation capabilities, product strategy, and firm performance: The electronics industry in China. *Can. J. Adm. Sci. / Rev. Can. des Sci. l'Administration* **30,** 159–172 (2013).
21. K.-P. Hung, Chou, C. The impact of open innovation on firm performance: The moderating effects of internal R&D and environmental turbulence. *Technovation* **33,** 368–380 (2013).
22. Z. Su, J. Peng, H. Shen, T. Xiao, Technological capability, marketing capability, and firm performance in turbulent conditions. *Manag. Organ. Rev.* **9,** 115–137 (2013).
23. B. J. Jaworski, A. K. Kohli, Market Orientation: Antecedents and Consequences. *J. Mark.* **57,** 53–70 (1993).
24. K. H. Tsai, S. Y. Yang, The contingent value of firm innovativeness for business performance under environmental turbulence. *Int. Entrep. Manag. J.* **10,** 343–366 (2014).
25. R. Wilden, S. P. Gudergan, The impact of dynamic capabilities on operational marketing and technological capabilities: investigating the role of environmental turbulence. *J. Acad. Mark. Sci.* 1–19 (2014). doi:10.1007/s11747-014-0380-y
26. H. Ting, H. Wang, D. Wang, The moderating role of environmental dynamism on the influence of innovation strategy and firm performance. *Int. J. Innov. Manag. Technol.* **3,** 13–16 (2012).
27. D. Jiménez-Jiménez, R. Sanz-Valle, Innovation, organizational learning, and performance. *J. Bus. Res.* **64,** 408–417 (2011).
28. G. Karagouni, I. Papadopoulos, The Impact of Technological Innovation Capabilities on the Competitiveness of a Mature Industry. *Manag. Int. Bus. Econ. Syst.* **1,** 17–34 (2007).
29. M. Lahovnik, L. Breznik, Innovation Management and Technological Capabilities as A Source of Competitive Advantage. in *Knowledge Management & Innovation*

International Conference 2013 19-21 June 2013 771–779 ((2013)).
30. Y. Liang, D. Liu, L. Zhang, Y. Zhang, Impact of Technological Innovation Capability on Business Growth: An Empirical Study for Small and Medium-Sized Enterprises. *2010 Int. Conf. E-bus. E-Government* 1177–1180 (2010). doi:10.1109/ICEE.2010.302
31. R. C. M. Yam, W. Lo, E. P. Y. Tang, K. W. Lau, Technological Innovation Capabilities and Firm Performance. *World Acad. Sci. Eng. Technol.* **42,** 1009–1017 (2010).
32. R. C. M. Yam, J. C. Guan, K. F. Pun, E. P. Y. Tang, An audit of technological innovation capabilities in chinese firms: some empirical findings in Beijing, China. *Res. Policy* **33,** 1123–1140 (2004).
33. C. Camisón, A. Villar-López, Organizational innovation as an enabler of technological innovation capabilities and firm performance. *J. Bus. Res.* **67,** 2891–2902 (2014).
34. M. C. De Stefano, M. J. Montes-sancho, T. Busch, A natural resource-based view of climate change: Innovation challenges in the automobile industry. *J. Clean. Prod.* (2016). doi:10.1016/j.jclepro.2016.08.023
35. J. Oh, S.-K. Rhee, The influence of supplier capabilities and technology uncertainty on manufacturer-supplier collaboration: A study of the Korean automotive industry. *Int. J. Oper. Prod. Manag.* **28,** 490–517 (2008).
36. B. A Wernerfelt, Resource-based View of the Firm. *Strateg. Manag. J.* **5,** 171–180 (1984).
37. R. Amit, P. J. H. Schoemaker, Strategic Assets and Organizational Rent. *Strateg. Manag. J.* **14,** 33–46 (1993).
38. C. E. Helfat, M. A. Peteraf, The dynamic resource-based view : capability lifecycles. *Stategic Manag. J.* **24,** 997–1010 (2003).
39. G. S. Day, The Capabilities of Market-Drive Organizations. *J. Mark.* **58,** 37–52 (1994).
40. J. J. Ferreira, S. G. Azevedo, rubén F. Ortiz, Contribution of resource-based view and entrepreneurial orientation on small firm growth. *Cuad. Gest.* **11,** 95–116 (2011).
41. V. M. U. Azubuike, Technological Innovation Capability and Firm ' s Performance in New Product Development. *Commun. IIMA* **13,** 43–56 (2013).
42. J. Shan, D. R. Jolly, Accumulation of technological innovation capability and competitive performance: a quantitative study in chinese electronic information industry. *Int. J. Innov. Technol. Manag.* **9,** 1–18 (2012).
43. R. A. Burgelman, C. M. Christensen, S. C. Wheelwright, *Strategic Management of Technology and Innovation*. (McGraw-Hill/Irwin, (2009)).
44. R. I. Sabir, R. M. Sabir, Managing technological innovation: China's strategy and challenges. *J. Technol. Manag. China* **5,** 213–226 (2010).
45. S. L. Newbert, Value. Rareness, Competitive Advantage and Performance: A Conceptual - Level Empirical Investigation of the Resource-Based View of the Firm. *Strateg. Manag. J.* **768,** 745–768 (2008).
46. L. Radu, Investments in Technological Innovations : a Literature Review of Organization Determinants. *Eur. Sci. J.* **1,** 52–59 (2015).
47. V. Zemlickienè, Analysis of High-Technology Product Development. *Intellect. Econ.* **5,** 283–297 (2011).
48. J. B. Barney, Strategic Factor Markets: Expectations, Luck and Business Strategy. *Manage. Sci.* **32,** 1231–1241 (1986).
49. S. L. Newbert, Empirical Research on Resource-Based View of the Firm: An Assessment and Suggestions for Future Research. *Strateg. Manag. J.* **146,** 121–146 (2007).
50. J. F. Christensen, Asset profiles for technological innovation. *Res. Policy* **24,** 727–745 (1995).
51. V. Chiesa, P. Coughlan, C. A. Voss, Development of a Technical Innovation Audit. *J. Prod. Innov. Manag.* **13,** 105–136 (1996).

52. A. K. W. Lau, R. C. M. Yam, E. P. Y. Tang, The impact of technological innovation capabilities on innovation performance An empirical study in Hong Kong. *J. Sci. Technol. Policy China* **1,** 163–186 (2010).
53. A. K. W. Lau, E. Baark, W. L. W. Lo, The effects of innovation sources and capabilities on product competitiveness in Hong Kong and the Pearl River Delta. *Asian J. Technol. Innov.* **21,** 220–236 (2013).
54. M. M. Yusr, S. S. M. Mokhtar, A. R. Othman, The effect of TQM practices on technological innovation capabilities: Applying on Malaysian manufacturing sector. *Int. J. Qual. Res.* **8,** 197–216 (2014).
55. D. Howell, C. Windahl, R. Seidel, A project contingency framework based on uncertainty and its consequences. *Int. J. Proj. Manag.* **28,** 256–264 (2010).
56. E. Fredericks, Infusing flexibility into business-to-business firms : A contingency theory and resource-based view perspective and practical implications B. *Ind. Mark. Manag.* **34,** 555–565 (2005).
57. S. C. Betts, Contingency Theory : Science Or Technology ? *J. Bus. Econ. Res.* **1,** 123–130 (2003).
58. G. Wang, W. Dou, W. Zhu, N. Zhou, The effects of firm capabilities on external collaboration and performance : The moderating role of market turbulence. *J. Bus. Res.* 1–9 (2015). doi:10.1016/j.jbusres.2015.01.002
59. D. I. Prajogo, The strategic fit between innovation strategies and business environment in delivering business performance. *Intern. J. Prod. Econ.* **171,** 241–249 (2016).
60. K. Tsai, S. Yang, Firm innovativeness and business performance : The joint moderating effects of market turbulence and competition. *Ind. Mark. Manag.* **42,** 1279–1294 (2013).
61. C. Terawatanavong, G. J. Whitwell, R. E. Widing, A. O. Cass, Technological turbulence , supplier market orientation , and buyer satisfaction. *J. Bus. Res.* **64,** 911–918 (2011).
62. J. Guan, N. Ma, Innovative capability and export performance of Chinese firms. *Technovation* **23,** 737–747 (2003).
63. M. V. Türker, A model proposal oriented to measure technological innovation capabilities of business firms – a research on automotive industry. *Procedia - Soc. Behav. Sci.* **41,** 147–159 (2012).
64. W. Wu, D. Liang, B. Yu, Y. Yang, Strategic planning for management of technology of China ' s high technology enterprises. *J. Technol. Manag. China* **5,** 6–25 (2010).
65. K. Liao, X. Deng, E. Marsillac, Factors that influence Chinese automotive suppliers ' mass customization capabilities. *Intern. J. Prod. Econ.* **146,** 25–36 (2013).
66. A. I. Ismail, R. C. Rose, J. Uli, H. Abdullah, The Relationship Between Organisational Resources, Capabilities, Systems and Competitive Advantage. *Asian Acad. Manag. J.* **17,** 151–173 (2012).
67. L. Santos-vijande, A, L. I. Santos-vijande, Innovativeness and organizational innovation in total quality oriented firms : The moderating role of market turbulence. **27,** 514–532 (2007).

Shape memory actuated sliding mechanism for rapid switching between button-patterns for adapted human-machine interaction at different automatization levels in automobiles

Manivannan Sivaperuman Kalairaj[1], and *Wei Min* Huang[1,*]

[1]School of Mechanical and Aerospace Engineering, Nanyang Technological University, Singapore 639798

Abstract. Almost all electronics and mechanical equipment have controllers to control the entire or certain tasks. These controllers respond to various types of inputs and performs the function accordingly. One type of input given to the controller is the touch and these controllers which use human touch has touch screen to act as the medium. The touch screen does not give haptic feedback to the users whereas the buttons normally take up more space and cause disturbance and confusion in recognizing the state and the function of the buttons. In this paper, we propose a simple sliding mechanism for rapid switching between different button patterns, in which the buttons remain flat in the surface when inactive and protrude out of the surface when active. The design and simulation are carried out on SolidWorks and the prototypes are fabricated and tested. This mechanism incorporated in the buttons and installed in an automobile, increases the level of automation and enhances the human-machine interaction.

1 Introduction

Nowadays automobiles are used vastly in various applications from personal to commercial and the automobiles have evolved from manual to semi or fully automated and one of the most important components that aided the evolution is the controller. The controllers which are used in automobiles are used to perform several functions ranging between turning-on to self-driving and also controllers could be either handy and portable or also fixed to the automobile. The controllers are used in automobiles to control almost all of its components and it could be that one controller to control all the functions in an automobile or a controller to control a specific function or a specific component in an automobile. Most of the controllers which are used in automobiles receives signal by human touch, and they could be in the form of touch screen or having buttons. The controllers with the buttons are bulky and also have several disadvantages like ambiguity in recognizing the functions of the buttons and also in differentiating between the active and inactive buttons. The advanced and latest controllers are touch screen where the humans can touch and interact

* Corresponding author: mwmhuang@ntu.edu.sg

with the screen to perform the functions, but this kind of interaction does not provide haptic feedback to the user which is necessary for effective communication between the human and the machine and will also adapt the design according to the necessities of the user and improve the safety of the automobile. [1]

A reconfigurable, shape-changing buttons on a vehicle control panel could be made to use shape memory alloy and/or shape memory polymer to control the movement of the buttons during active and inactive states. [2] During the inactive state, the button remains on the surface and when it is changed to the active state, there is a protrusion and this protrusion is in a hemispherical or dome shaped.[2] There is a compliant mechanism which includes single-piece flexible structure which can deform in three dimensions and transmit force to an input actuation. In another method, a mechanism could be developed to provide concealable physical buttons, while the display associated with the buttons and surroundings remains stationary. [3] In this the buttons or hard keys are concealed to the surface when not interacting and raised above the surface when needed. This motion occurs due to the presence of a motor, a threaded shaft coupled to the motor, a movable platform coupled to the threaded shaft and a hard key movably supported on the movable platform and the due to this arrangement and the movement of the motor, the hard key rises from the surface when required and this also provides haptic feedback to the end-user.[3] In this method, the space required by the setup is quite high and the mechanism is complex and the number of parts used is also large. Due to these, the power required for the actuation would also be high. In another method, the buttons rise above the surface on demand by the help of Tactus Technology. [4] In this, the button is formed with pressure in small channels inside of the substrate and small holes let the fluid in the area in-between the elastomer and the substrate. The rise occurs due to the expansion of the elastomer surface when the compressed fluid is allowed to pass through the small channels. [4] The compressed air and the micro-valves are switched using energy. This method is most suitable for touch screens. By, this method, the transparent buttons are possible. In this technology, the response time might be slow and liquid leakage could be a potential problem.

The buttons on demand are necessary as it increases the human-machine interaction and also not all the buttons are required during operation of vehicles, for example, there is no need of automatic parking buttons when the automobile is in highway at a great speed and if the automatic parking button is activated accidentally, it might induce certain problems and dissatisfaction to the users. SAE levels of a vehicle are classified based on the engagement and the performance of the features of the vehicle by the driving system. At SAE level 0 – 2, human drivers are responsible to monitor the driver environment, whereas for vehicles at SAE level 3 – 5, the automated driving system monitors the driving environment. [5, 6] For changing the different levels of automation at different points of time, an adaptation of the human machine interaction concept is necessary.

This paper presents a novel approach for the buttons on demand, in which the buttons remain concealed in the surface when the buttons are inactive and the buttons come out of the surface when they are in active. In this way, the space occupied and the ambiguity in recognizing is reduced as well as the haptic feedback is given to the user and also the mechanism used for the rapid switching between the active and inactive states of the controller is explained in details. This is achieved by sliding mechanism which could be actuated by means of such as electric switches and shape memory alloy (SMA), such as Ni-Ti alloy spring. The mechanism was designed and simulated using SolidWorks and then the prototypes were fabricated by CNC machining and laser cutting. The performance of the device was tested.

2 Design of controller with sliding mechanism for rapid switching

The controller that is designed has a linear series of seven buttons in which there are four buttons which possess the function of rapid switching of button pattern, while the other three buttons are active all the time due to its function and its necessity in the automobile. The controller was designed for the four buttons to rise or fall and possess active or inactive states, while the other three buttons are always concealed to the surface but they are active all the time. The other four buttons are designed such that they possess two positions, in which one has all the four buttons to be inactive and concealed to the surface, while in the other position, the four buttons to be active and risen above the surface. This creates two different surface patterns with respect to the state of the buttons, one pattern has all seven buttons to be concealed to the surface while the other pattern has four buttons that are alternate to the three buttons in a linear manner risen above the surface. There are seven components in the controller that play a major role in the activation process, which are top cover, bottom cover, slider, buttons, mechanochromic material, SMA (wire or spring) and a power source to activate the SMA, such as a 9 V battery. Since, there are two surface patterns obtained by the two positions of the buttons, both two-way SMA or one-way SMA could be used. In this design, we used a one-way SMA and two push buttons to activate the two different positions of the buttons and the respective surface patterns could be obtained. The design was carried out in such a way that the entire controller fits in the hand of humans and the total thickness was limited to 35 mm in which 28 mm goes inside the automobile and only 7 mm is visible and the rise of the buttons from the surface is 1.5 mm. The model of the controller using buttons on demand is shown in Figure 1.

Fig. 1. Model of the controller.

2.1 Top cover

This component acts as a cover for the buttons and the slider, so the working mechanism is not visible. It has four holes which allow four buttons to fit and these four buttons are placed, so that it is alternate with the other three buttons which are active all the time and are on the surface and will never rise or fall. The top cover should be flat, such that when the buttons are inactive and concealed to the surface, the controller looks as if it has no buttons. The top cover should have a provision for bolt or screws such that the top cover could be fixed to the bottom cover, such that all the rest of the components could be in-between the top and the bottom cover. The inner surface of the top cover should possess certain slots, such that the buttons do not rotate about its axis during the sliding mechanism and this slot further facilitates rise and fall of the buttons. The slot, which assists the rise and fall of the buttons, could be made either separately for each buttons or could be made common, such that interchangeability of buttons is facilitated. All the four edges are extended to the bottom surface, so that it provides a slot for the slider to slide with the minimal friction. A protrusion is provided on the back of the slider for pushing back and forth manually or by any other means. The top cover has a thickness of 3 mm and has an extension in each sides which increases the thickness to 5 mm. Due to the difficulty in fabricating the design using CNC machining process, a simple design was used for fabrication. The fabrication was carried out with aluminium alloy and the surface was oxidized. The top and rear view of the design of top cover in both the simple and actual design of the controller is shown in Figure 2.

Fig. 2. Actual and simplified design of top cover of the controller.

2.2 Bottom cover

The function of this component is to cover the controller and to act as a base and also this is the component which fits to the carrier in the controller, so that all the actuating components are inside the carrier. The most important function of this component is to aid the slider to slide and help in the rise and fall of the buttons. This acts as a base for sliding the slider and the material of the bottom cover was selected to ensure that the friction between the slider and the bottom cover is minimum. The bottom cover has the provision for attaching to the top cover at the edges or corners by having holes to enable bolts, so that it could be tightened with nuts or by any other screws. The bottom cover or the rear cover was fabricated with polymethyl methacrylate (PMMA) by laser cutting. The bottom cover also has a slot for allowing the protrusion on the back of the slider to slide through it and this could be used for attaching a SMA spring or wire. The design and the image of the PMMA bottom cover are shown in Figure 3.

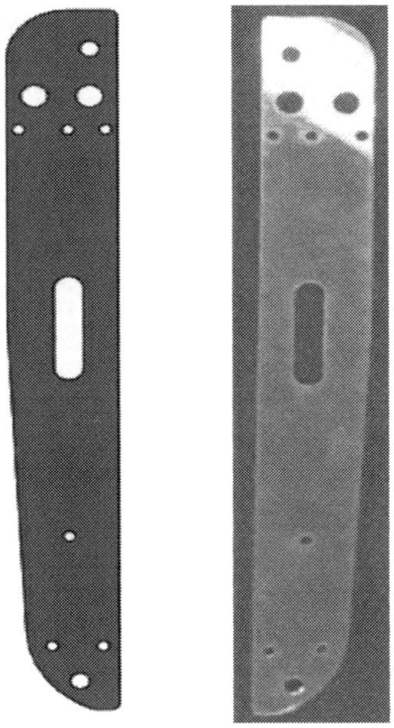

(a) (b)

Fig. 3. Bottom cover. (a) Profile design; (b) prototype made of PMMA.

2.3 Buttons

The controller has seven buttons in which three are always active but on the surface while the other four are the buttons that possess the function of buttons on demand and they are placed alternatively to those three buttons. The design of the button was made such that the top surface is flat while the bottom surface has some protrusion to enable the sliding motion of the slider and undergoes rise and fall according to the design. The design was carried out for two patterns with four alternate buttons above the surface in the first pattern and all the seven buttons are on the surface in the second pattern. To improve the visual and touch feedback protrusions based on the functions of the buttons could be made on the respective buttons. The front, rear and isometric view of the profile design of the buttons are shown in the Figure 4.

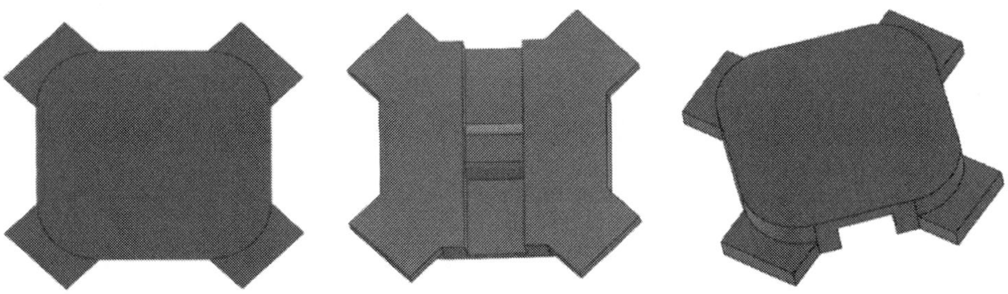

(a) (b) (c)

Fig. 4. Profile design of the Buttons. (a) Front view; (b) rear view; (c) isometric view.

2.4 Slider

The slider and the buttons are the two parts that play the most important role in this mechanism. The slider is a movable part which could in-turn help in the activation and inactivation of the buttons based on the patterns. The mechanism is designed in such a way that there are two different positions to which the slider could be moved and each position produces a different pattern of buttons on the surface. The slider was designed as a flat plate that has a protrusion along the vertical axis and the design of this protrusion determines the rise and fall of a button which in-turn gives rise to a pattern on the surface of the controller. The slider is attached with a small protrusion in the middle of its bottom surface which helps to maintain its direction and the orientation while sliding. This protrusion slides through a slot, which is cut in bottom cover and this also helps in holding the SMA spring and it is this component which indicates the position of a button (either OFF or ON). The slider was fabricated by CNC machining of aluminium alloy (Al 6061) and oxidizing its surface and is allowed to slide over the bottom cover made of PMMA. The design of the prototype of the slider is shown in Figure 5.

(a) (b)

Fig. 5. Design of slider. (a) front view; (b) rear view.

2.5 Mechanochromic material

The top surface of the controller may be enclosed with a thin surface of material called mechanochromic material. Mechanochromic material is a material that changes its colour of emission when the required amount of external mechanical force stimulus like stress or pressure is applied to it. [7, 8] When the buttons rise from the surface, the thin mechanochromic layer is stretched and this creates a localized colour change in the material which shows the difference between the active and inactive buttons in the controller. Thus, the buttons that are above the surface can be seen clearly and separately from the buttons that are concealed to the surface and are inactive. This gives additional visual feedback to the user and improves the human-machine interaction.

2.6 Shape memory alloy (SMA)

Shape memory materials (SMM) are the materials that have the ability to recover their original shapes from a deformed shape by application of certain stimulus, such as heat, light, chemical, electricity, etc. [9] The underlying effect is called the shape memory effect (SME). One of the most important and most widely used SMMs is shape memory alloy (SMA) due to its high performance and ease in development. One of the major forms of SMA is spring. This consumes less power for activation and also requires less space. SMA wire is another configuration which is more advantageous than SMA spring but still in development stages right now. The SMA spring is used in this controller to actuate the mechanism by its linear contraction and expansion. The SMA spring (NiTi) used here has a one-way SME and it was activated by an electrical power source (9 V battery) for Joule heating in order to bring the SME. The connection is made such that one terminal of the battery is connected to the middle of the SMA spring, while the other terminal is connected to the two ends of the SMA spring and by this connection the spring is made to act like two SMA springs by activating the two ends alternatively. The picture of the SMA spring connection is shown in Figure 6. The wires from the SMA spring are connected to the normally open push buttons which are connected to the battery and activating one push button, the corresponding movement of the SMA spring is obtained. The thermal behaviour of the SMA actuator is determined by the electrical heating power and the cooling mechanisms of the heat conduction and heat convection in the component. Since, the temperature range is below 200°C, heat radiation does not play a significant and need not be considered during design. [10]

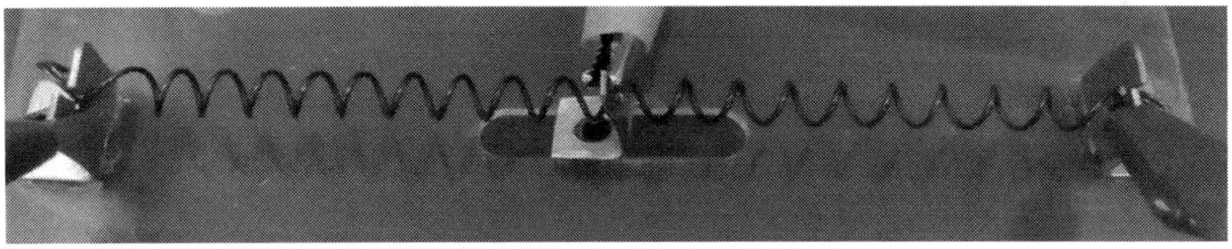

Fig. 6. Picture of SMA spring with wiring.

3 Working mechanism

The controller has two push buttons to activate the SMA spring and thus to activate the mechanism for rapid switching. When the push button responsible for making the buttons active is pushed, the current from the battery heats up the SMA spring on one side and since the SMA spring is attached to the slider, it moves the slider to one end, which aids in rising the buttons above the surface and thus resulting in activating the mechanochromic layer and the touch sensitive layer, which could be placed on the top surface. When the other push buttons are pushed, similar activation occurs but the slider is moved to the other end thus making the buttons to fall on the surface. The assembled model of the controller with rapid switching of button pattern mechanism inbuilt is shown in Figure 7.

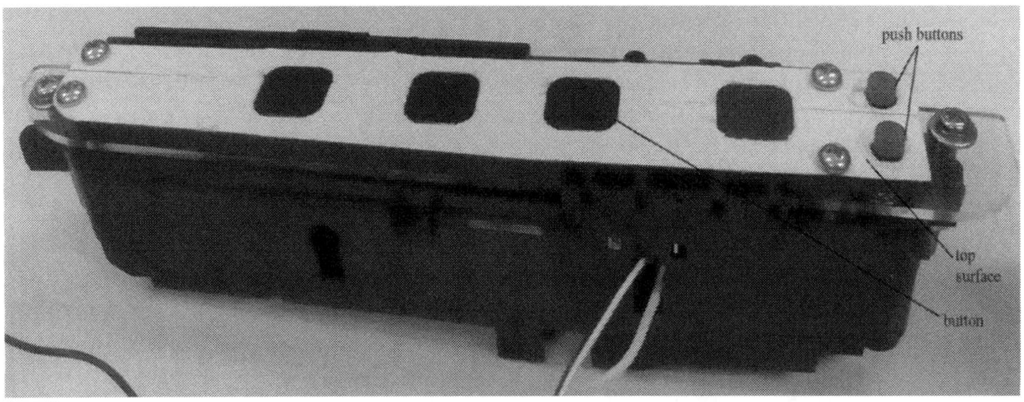

Fig. 7. Picture of the assembled controller with rapid-switching mechanism inbuilt.

 DOI: 10.1051/matecconf/20179001010

4 Results and discussion

After testing the working of the mechanism in the controller, a thin elastic cloth was used to cover the top surface of the controller to determine the activation and deactivation of the buttons. The working of the controller with the activation and deactivation of the buttons covered by a thin elastic cloth is shown in the Figure 8.

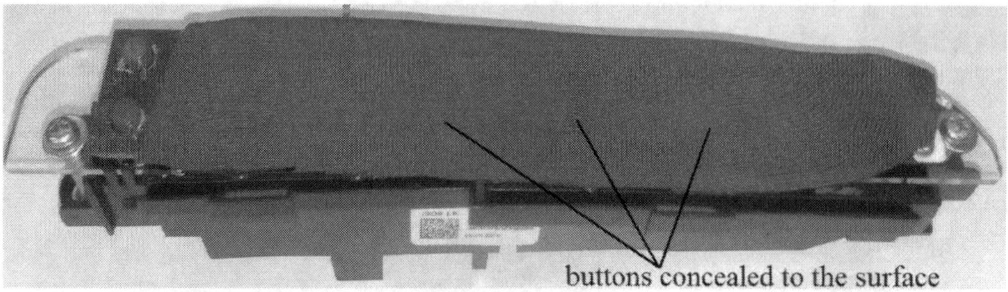

(a)

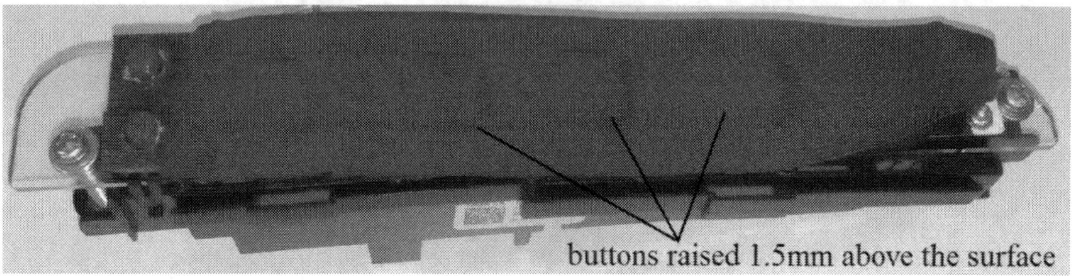

(b)

Fig. 8. Working of the controller (top surface covered with elastic cloth) : (a) Inactive state; (b) active state.

The results of the working of controller with an elastic cloth covered on the top shows that there is a considerable amount of stress developed in the cloth and so by replacing the elastic cloth with a mechanochromic material of similar thickness will produce a change in the colour on the top surface around the buttons and thus enhancing the visual interaction between the user and the machine.

In the controllers explained in the Figure 7 and Figure 8, the activation of the battery which actuates the mechanism is done with the help of push buttons or switches. This could be replaced by sensors and the processors so that the sensors (proximity) will sense the human interaction based on the movement of the hands and the processors will help in determining the right action required and the corresponding buttons responsible for that action will be actuated and will become active by providing both visual and touch feedback. Machine learning could be carried out for a particular user so that more enhanced activation of the buttons could be obtained for the specific user.

After testing the working of the mechanism in the assembled model of the controller, the efficiency of the mechanism was determined in-terms of the current required to heat up the SMA spring and the time taken for the SMA spring to activate the mechanism and the number of cycles the mechanism works with a 9 V battery. Then, the surface profile of the buttons on the top surface when they are active and inactive was determined in order to check whether the user has an enhanced visual and touch feedback on the surface, which plays the major role in this controller. The 3D profile data of the top surface of the controller, when the buttons are inactive and when the buttons are active were obtained by using a 3D scanning equipment called Talyscan 150. The surface characterization was carried out and the data when a typical button was inactive is showed in the form of meshed axonometric diagram in Figure 9. Typical 3D surface topology data of the top surface when the buttons are inactive is shown in the form of Abbott firestone curve and peak count distribution of the surface in the form of particles count histogram in Figure 10. The 2D profile was extracted for the top surface near the centre along the horizontal surface when the buttons are inactive and is shown in Figure 11. To determine the exact rise that was obtained during the working of the controller, 2D profile was extracted near the centre along the horizontal axis when the buttons are active and it is shown in Figure 12.

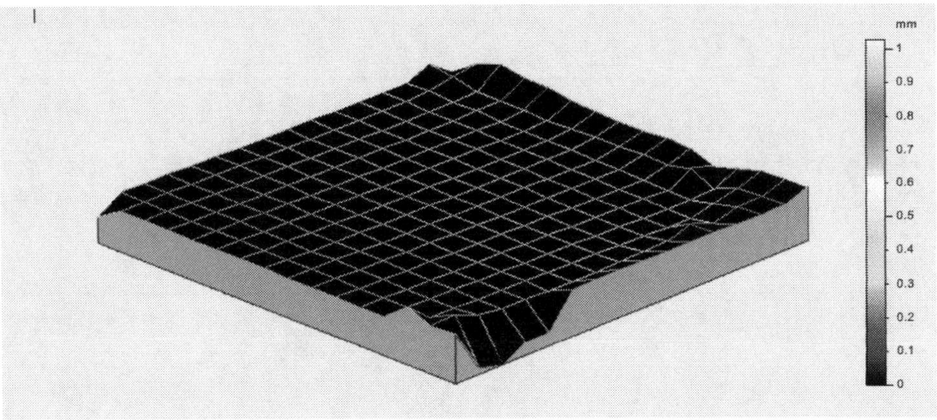

Fig. 9. Meshed axonometric diagram of the top surface when buttons are inactive.

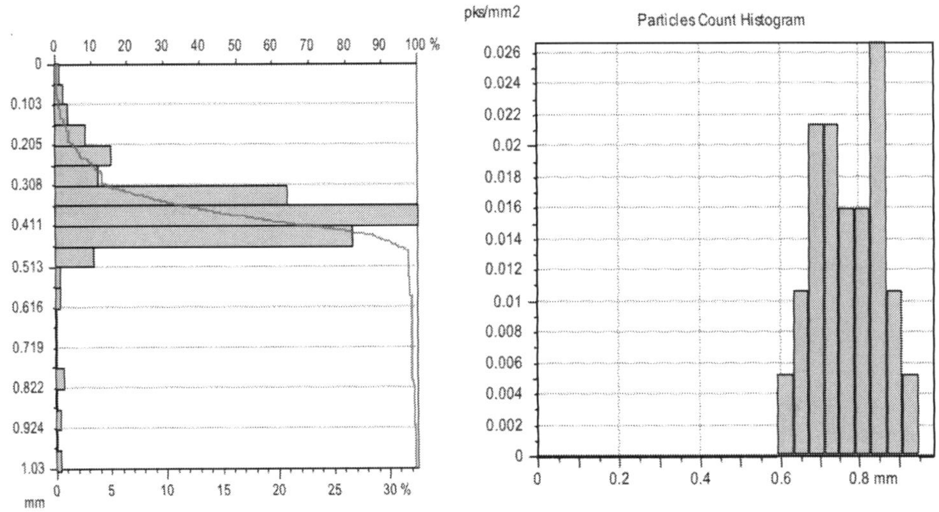

(a) (b)

Fig. 10. Surface topology when inactive. (a) Abbott-firestone curve; (b) histogram.

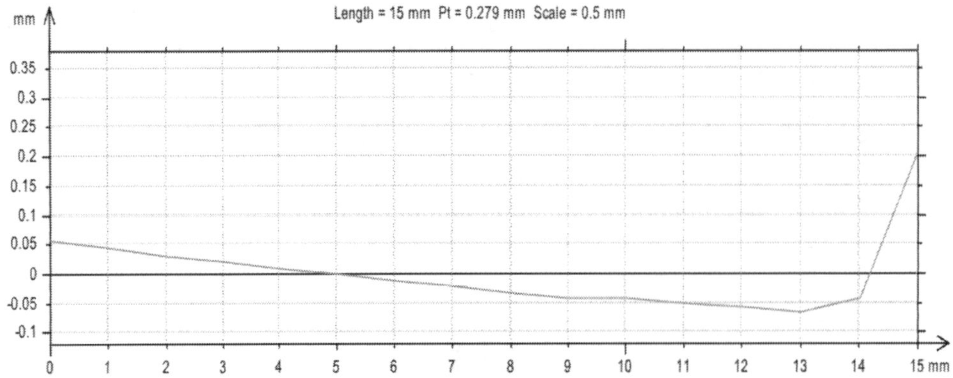

Fig. 11. 2D profile extracted near the horizontal axis when buttons are inactive.

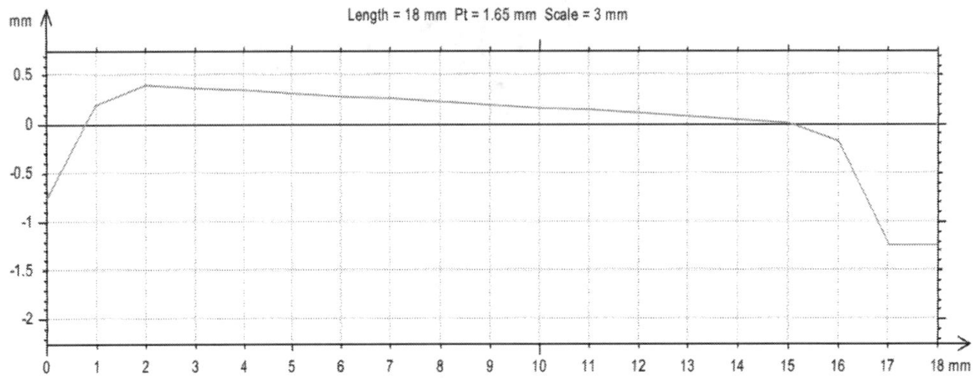

Fig. 12. 2D profile extracted near the horizontal axis when buttons are active.

4.1 Advantages of using sliding mechanism for rapid switching of button pattern

- It can be actuated manually or by motor(s) whenever required.
- Nine or more button-patterns may be achieved.
- The number of the degree of freedom of the slider is only one or two.
- The actuation and the motion of the slider and the buttons are highly reliable.
- The buttons provide haptic feedback to the users.
- Touch sensing layer may be integrated atop to provide signal feedback.
- A layer of mechanochromic material placed atop for colour change can aid the embedded word or image to appear and increase the human interaction.
- The SMA requires less power for activation (9 V) and the time required for activation is also less.

5 Conclusions

The controller with rapid switching of button patterns in an automobile along with its ability to adapt according to the necessities of the user improves the human-machine interface and also increases the safety of the user in the automobile and the environment in which the automobile is used. This controller is designed such that the activation methods could be interchanged whenever required but the most suitable activation method is using a SMA spring (NiTi) which requires only 9 V of power to activate. The addition of the protrusion in buttons according to the functions and the buttons and the use of mechanochromic layer on top of the controller along with the haptic feedback already present in the buttons will ensure both the touch and visual feedback and thus improve the human-machine interaction. In the future, the activation of the battery which activates the mechanism will be replaced by sensors and the processors so that the sensors will sense the human interaction and the processors will help in determining the right action required and the corresponding buttons responsible for that actions will become active by providing both visual and touch feedback and also suitable technology could be installed for the provision of auditory feedback, thus enhancing the human-machine interaction.

This project is supported by BMW-NTU Joint R&D Program.

References

1. S. Rodriguez Garzon, M. Poguntke, The Personal Adaptive In-Car HMI: Integration of External Applications for Personalized Use, in *Advances in User Modeling: UMAP 2011 Workshops, Girona, Spain, Revised Selected Papers*, L. Ardissono and T. Kuflik, Eds., ed Berlin, Heidelberg: Springer Berlin Heidelberg, pp. 35-46 (2012).
2. S. Strohband, A. Stoschek, M. Derse, V. Srinivasan, Reconfigurable, shape-changing button for automotive use and vehicle control panel having reconfigurable, shape-changing buttons, US Patent 20070247420 A1 (2007).
3. M. Jolda, M. McClain, A. Rudolph, S. Henze, J. Toggweiler, Sliding hard keys, US Patent, 20140257567 A1 (2014).
4. T. Technology, The Tactus Tactile layer, *http://tactustechnology.com/, USA* (2014).
5. S. o. A. Engineers, J3016: Taxonomy and Definitions for Terms Related to On-Road Motor Vehicle Automated Driving Systems, in Levels of Driving Automation for On-road Vehicles, *Society of Automotive Engineers* (2014).
6. A. Marinik, R. Bishop, V. Fitchett, J. F. Morgan, T. E. Trimble, M. Blanco, Human factors evaluation of level 2 and level 3 automated driving concepts: Concepts of operation (Report No. DOT HS 812 044), *Washington, DC: National Highway Traffic Safety Administration* (2014).
7. C. Weder, Mechanochromic Polymers, in *Encyclopedia of Polymeric Nanomaterials*, S. Kobayashi and K. Müllen, Eds., ed Berlin, Heidelberg: Springer Berlin Heidelberg, pp. 1-11 (2015).
8. G. I. Peterson, M. B. Larsen, M. A. Ganter, D. W. Storti, A. J. Boydston, 3D-Printed Mechanochromic Materials, *ACS Applied Materials & Interfaces* **7**, pp. 577-583 (2015).
9. W. M. Huang, Z. Ding, C. C. Wang, J. Wei, Y. Zhao, H. Purnawali, Shape memory materials, *Materials Today* **13**, pp. 54-61 (2010).
10. M. Kohl, Shape Memory Microactuators, *Microtechnology and MEMS, Springer-Verlag Berlin Heidelberg* (2004).

The investigation of dynamic behaviour of a structure using wave-based substructuring method

Ahmad Burhani Ahmad Basri[1] , *Muhamad Norhisham* Abdul Rani[1,*], *Mohd Azmi* Yunus[1] , *Liyana* Roslan[2] , and *Wan Imaan Izhan* Wan Iskandar Mirza[1]

[1]Center of Excellence of Dynamic and Control, Faculty of Mechanical Engineering, Universiti Teknologi MARA Kampus Shah Alam, 40450 Shah Alam, Selangor, Malaysia
[2] Faculty of Mechanical Engineering, Universiti Teknologi MARA Kampus Pasir Gudang, 81750 Masai, Johor, Malaysia

Abstract. There is an increasing need for accurate, efficient and economical methods for the investigation of the dynamic behaviour of large complex structures within the engineering community. The component mode synthesis (CMS) has been perceived by the community to be an attractive efficient method for the investigation. However, the method has substantial shortcomings, particularly in analysing a structure having a large number of interface degrees of freedom (DOFs) between substructures. This paper puts forward a method, based upon the wave-based substructuring (WBS) for the investigation of the dynamic behaviour of a structure with a large number of interface DOFs. The finite element method is used to construct the full finite element model of the structure and NASTRAN 103 is used for the normal modes analysis. A new finite element model of the structure with reduced interface DOFs is constructed based on the WBS. The measurement of the dynamic behaviour of the structure is carried out using free-free boundary conditions and an impact hammer test. The predicted results of the proposed method are then compared with those from the full finite element model and experimental counterparts. The accuracy and efficiency of the proposed method are discussed and illustrated with two different case studies.

1 Introduction

Over time, engineering structure has been developed to a certain level of large numerical model which possesses thousand numbers of DOFs. Deriving the mathematical model, especially for a complex structure that is having numbers of substructures, boundary conditions and enormous number of DOFs is a very challenging and difficult task even for the modern computer. Despite the presence of the modern computer that has been used by most of the industrial companies in increasing the work efficiency, it seems to be unable to

* Corresponding author: mnarani@salam.uitm.edu.my

perform the expected level of efficiency in terms of the iteration time. The application of conventional finite element analysis (FEA) which uses a full finite element model has been widely used and the effort of computing the mathematical model seems to be unsuitable when the computational is too expensive due to the repeated iteration solution performed.

To overcome this problem, a technique called component mode synthesis (CMS) has been introduced to enhance the computational time process. It is believed to be a very practical and efficient method for investigating the structural normal modes. The method has been widely used in the engineering field for solving complex high profile engineering structural field. In the automotive world, CMS method is used in increasing the speed of vehicle development and system-level performance assessment[1]. Not only that, in civil engineering, the method is used in the study of the bridge frame and the capability of the method towards model updating scheme[2]. A rapid analysis of the space shuttle which is used in aerospace industries has shown the success of the method in predicting the dynamic behaviour of the structure[3]. Based on the case stated, it has clearly shown the capability of the rapid CMS method in analysing the complex engineering structure.

Unfortunately, since the needs of the FEM is highly demanded by the industry and application of the method is not limited to only a few kind of structures, computing the complex engineering structure through CMS method seems to be impractical. This problem arises when the substructures are coupled with a very large interface nodes[4]. Since the efficiency of this method highly depends on the boundary nodes in which for certain structures that are still having a very large interface nodes between the substructures, the method is unable to speed up the computational time[5, 6]. Even worse, for a structure that has a limited number of the substructure, there is no other choice but only to mandatory select the large interface as the interface representation [7]. This situation has brought difficulty in using the CMS method as the reduced order scheme. Although much work has been done to date, more studies need to be conducted to find out the most reliable method that can be effectively and accurately used without highly depending on the full interface nodes.

In this paper, the work is concerned with the use of wave-based substructuring method as the reduced scheme in investigating the dynamic behaviour of a structure which is perceived able to analyse a large number of interface DOFs. It is a method, in which instead of all the interface nodes are used to be analysed as practised in CMS method, that presents a substructuring approach where the interface is written as a combination of a set of basis deformation called waves [5]. A smaller number of DOFs is used to represent the interface representation that gives a better result in terms of the computational time. As the number of waves (interface nodes) is much lower than the interface DOFs, a rapid prediction can be obtained compared to the conventional CMS method [8, 9]. Therefore, as the interface representation is smaller, easier reduction procedure and shorter time frame of the iteration process can be obtained while the accuracy of the prediction is preserved.

2 Theory of wave-based substructuring

Consider an undamped structure with no external forces applied to it, the FE matrix form will be:

$$\mathbf{M}\ddot{\mathrm{x}} + \mathbf{K}\ddot{\mathrm{x}} = 0 \tag{1}$$

The FE structure is divided into several non-overlapping substructures to obtain an assembled system. For each substructure, the DOFs, x are separated into interior DOFs, $\mathrm{x_i}$ and boundary DOFs or junction DOFs, $\mathrm{x_j}$ at the boundary [10]. The system matrices for each substructure will be:

$$\begin{bmatrix} \mathbf{M}_{ii} & \mathbf{M}_{ij} \\ \mathbf{M}_{ji} & \mathbf{M}_{jj} \end{bmatrix} \begin{bmatrix} \ddot{x}_i \\ \ddot{x}_j \end{bmatrix} + \begin{bmatrix} \mathbf{K}_{ii} & \mathbf{K}_{ij} \\ \mathbf{K}_{ji} & \mathbf{K}_{jj} \end{bmatrix} \begin{bmatrix} x_i \\ x_j \end{bmatrix} = \begin{bmatrix} f_i \\ f_j \end{bmatrix} \quad (2)$$

In the WBS approach, the method will limit the junction DOFs, x_j by reducing the interface description by introducing them to a set of basis function W, weighted with involvement factor p:

$$x_j = W \times p \quad (3)$$

The Equation (3) is substituted into Equation (2):

$$\begin{bmatrix} \mathbf{M}_{ii} & \mathbf{M}_{ij}W \\ W^T\mathbf{M}_{ji} & W^T\mathbf{M}_{jj}W \end{bmatrix} \begin{bmatrix} \ddot{x}_i \\ \ddot{p} \end{bmatrix} + \begin{bmatrix} \mathbf{K}_{ii} & \mathbf{K}_{ij}W \\ W^T\mathbf{K}_{ji} & W^T\mathbf{K}_{jj}W \end{bmatrix} \begin{bmatrix} x_i \\ p \end{bmatrix} = \begin{bmatrix} f_i \\ V^T f_j \end{bmatrix} \quad (4)$$

At this point, the set of basis functions is known as "waves". As the number of waves is lower than the junction DOFs,x_j, this substitution will reduce the size of the interface FE matrix equation.

Continuity and equilibrium condition must be translated into interface basis functions to perform the assembly for the substructures. For a rigid connection[4] between *a* and *b* substructures, the continuity and equilibrium conditions are as follow:

$$x_j^{(a)} = x_j^{(b)} \text{ and } f_j^{(a)} = f_j^{(b)} \quad (5)$$

In WBS, the assembly of two substructures, *a* and *b* will have the junction DOFs expressed as $x_j^{(a)} = W^{(a)} \times p^{(a)}, x_j^{(b)} = W^{(b)} \times p^{(b)}$, respectively. The same condition is applied to the involvement factor p:

$$p^{(a)} = p^{(b)} \text{ and } f_j^{(a)} = f_j^{(b)} \quad (6)$$

3 Academic case

In this section, there are two cases that have been investigated in order to assess the efficiency and the accuracy of the proposed method namely a normal size two-plate assembly case and 15 times bigger two-plate assembly case. The efficiency of the WBS method in computing the iteration was validated by comparing it with the full FE analysis and the conventional Craig-Bampton method and in terms of the accuracy, it was validated with the experimental counterparts. For the FE solver, MSC. Nastran [11] was used and LMS Virtual.Lab [12] as the pre- and post-processing. MATLAB [13] was used to validate the correlation of the result with the experimental result.

3.1 Case study 1: Normal size two-plate assembly

The model under the investigation is a large flat span structure that mimics a simplified aeroplane shape. The assembly structure consists of two substructures, Part A and Part B as shown in Figure 1. The model meshed into 10mm meshing size for both parts.

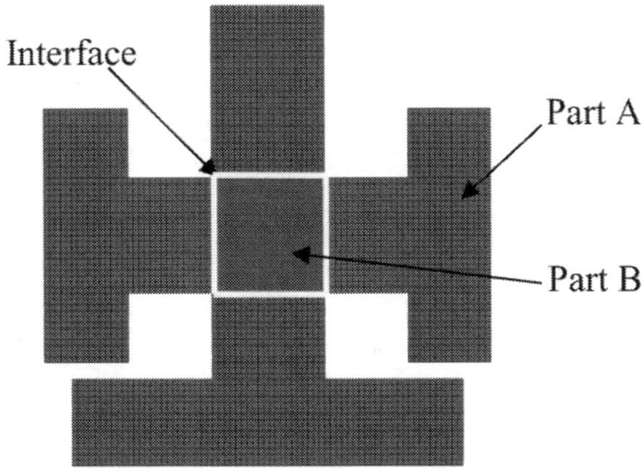

Fig. 1. Normal size structure consists of two-plate assembly (Part A:Red, Part B:Green).

The structure is a 4mm-thick aluminium made and 0.8m by 0.8m size which has the following properties:

- Young's Modulus, E = 73.58GPa
- Poisson Coefficient, n = 0.3
- Mass Density, r = $2877\text{kg}/\text{m}^3$
- Part A consists of 4168 nodes and 3870 quad elements.
- Part B consists of 441 nodes and 400 quad elements.

In between the two substructures, the interface nodes which consist of 80 coincident nodes were connected by rigid connections as shown in Figure 1(yellow line). Normal modes analysis was performed to obtain the first 16 normal modes. All the calculation was performed on an Intel(R) Core(TM) i7-3610QM CPU (2.30GHz, 8GB RAM) running on Windows 7 operating system, using MSC. Nastran v2012.

3.2 Case study 2: 15 times bigger two-plate assembly

The model under investigation is a large flat span structure that mimics a simplified aeroplane shape. The assembly structure consists of two substructures, Part A and Part B in which was increased in size to 15 times bigger than the normal structure size used in Case Study 1. The purpose of increasing the size of the structure was to increase the number of nodes and elements so that a bigger difference in terms of the computational time between full FE, WBS method and Craig-Bampton method can be obtained. The model meshed into 10mm meshing size for both parts. The same material properties were used in this case study. The geometry specifications of the structure are as follows:

- Part A consists of 875220 nodes and 870750 quad elements.
- Part B consists of 90601 nodes and 90000 quad elements.

In between the two substructures, the interface nodes which are consist of 1200 coincident nodes was connected by the rigid connection. Normal modes analysis was performed to obtain the first 16 normal modes. All the calculation was performed on an Intel(R) Core(TM) i7-3610QM CPU (2.30GHz, 8GB RAM) running a Windows 7 operating system, using MSC. Nastran v2012.

4 Results and discussion

In this study, the natural frequencies and mode shapes are the interested modal parameters to be observed which were obtained from the finite element method and experimental modal analysis. The results of modal parameters and CPU time was recorded and tabulated in Table 1. Meanwhile, two reduced order method (Craig-Bampton and WBS) were used to study the capabilities of the methods in term of efficiency computing the iteration process. The results were tabulated in Table 3.

For the first case study, all the normal mode analysis is performed in the frequency of interest range 0-120Hz. 25 number of waves are used as the interface node which has been selected from the sensitivity analysis done (see Figure 2). In order to obtain an accurate and efficient results, one should aims at using the least number of waves that is enough to represent the substructure interface deformation analogous to the full FE model normal modes behaviour of the complete structure[14].

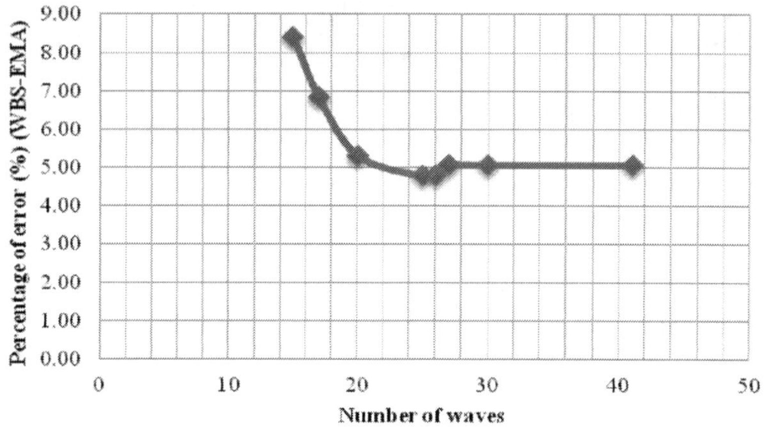

Fig. 2. Sensitivity analysis of waves. Comparison between the WBS-connected models and the EMA in terms of eigenfrequency for a different number of waves.

There are 10 elastic modes for the frequency of interest up till 120Hz. Since the number of DOFs is small in this model, the difference between the full FE and WBS method in terms of the computational time is not that significant. However, in the case of the accuracy of the method, a clear significant result is obtained. Two types of comparisons of results are shown in Table 1.

The first comparison is between the results of full FE and the test data. The comparisons of the results are tabulated in the column (IV). An error of 5.09% is obtained for this comparison. The second comparison is between the WBS model and the test data. The results of the comparisons are shown in Column (V) in Table 1. A lower percentage of error with a value of 4.79% is obtained. In terms of the MAC analysis, the average MAC value is 0.92 which indicates that the results have a very good agreement in terms of the mode shapes between the numerical and experimental work (see also Table 2).

It was found that the method not recommended to be used for a small structure since setting the reduction process for the substructure might take some time. It can be said that the method used was able to accurately maintain and significantly improved the level of accuracy. This indicates that WBS offers a benefit for reduction where the accuracy of the prediction is maintained even a smaller number of DOFs is used to analyse the structure. In order to get a better view on the efficiency of the WBS method, the second case study was performed.

 DOI: 10.1051/matecconf/20179001011

Table 1. Comparisons of natural frequencies obtained from the full FEM, WBS and EMA for the normal size structure.

Mode	(I) Full FE (Hz)	(II) WBS (Hz)	MAC	(III) EMA (Hz)	(IV) Error between I & III (%)	(V) Error between II & III (%)
1	16.38	16.38	0.9	16.27	0.67	0.67
2	17.05	17.09	1.0	17.15	0.57	0.34
3	31.25	31.25	1.0	31.29	0.14	0.12
4	31.77	31.78	0.8	31.52	0.79	0.82
5	37.33	37.35	1.0	37.87	1.43	1.39
6	46.71	46.71	0.8	46.77	0.12	0.12
7	73.82	73.83	1.0	74.23	0.56	0.54
8	90.21	90.23	1.0	90.56	0.39	0.36
9	110.41	110.43	1.0	110.17	0.22	0.23
10	118.88	118.88	0.7	118.65	0.20	0.20
CPU Time (s)	2	1		**Total Error**	5.09	4.79

Table 2. Comparison between the measured and predicted mode shapes of the structure.

Measured mode shape	Numerical mode shape	Measured mode shape	Numerical mode shape
 Mode 1	 Mode 1	 Mode 2	 Mode 2
 Mode 3	 Mode 3	 Mode 4	 Mode 4
 Mode 5	 Mode 5	 Mode 6	 Mode 6
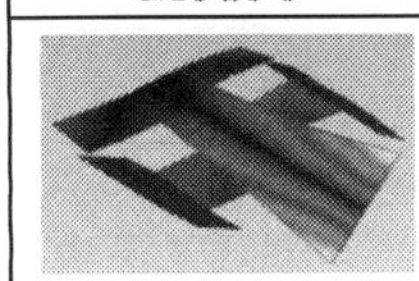 Mode 7	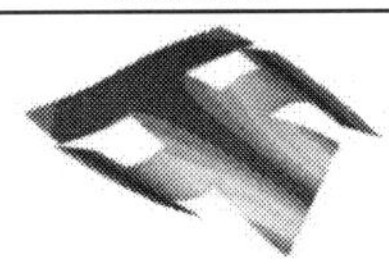 Mode 7	 Mode 8	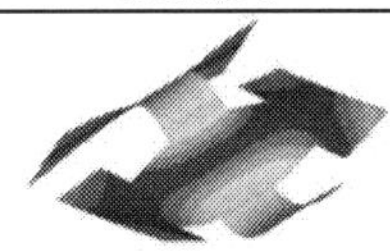 Mode 8
 Mode 9	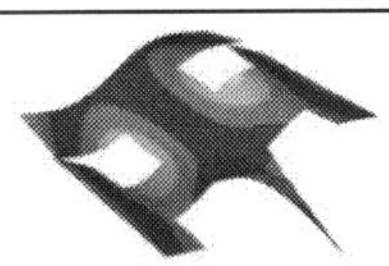 Mode 9	 Mode 10	 Mode 10

For the second case study, the structure has increased the size up to 15 times than the normal size as in Case Study 1. A total of 16 number of normal modes are computed. The results shown in Table 3 consist of three comparisons between the full FE, WBS and Craig-Bampton method.

Table 3. Comparisons between Full FE results and WBS results when the structure size was 15 times. bigger.

	Full FE	Craig-Bampton	WBS
Number of component normal modes	16	16	16
Number of Nodes	964621	1200	36
Reduction iteration time (s)		8907	1688
Assembly iteration time (s)		207	123
Total CPU Time (s)	11349	9114	1811
Time reduction (%)		19.7	84

In full FE analysis, there are 964621 numbers of nodes which took 8585 seconds computational time for the analysis to complete. This situation shows the need to use an efficient method to calculate the iterative solution since it is taking too much time.

As the Craig-Bampton method was applied, instead of using all the nodes to perform the analysis, 1200 interface nodes are used to analyse the structure. It is found that the computational time has been reduced to 9114 seconds which indicated 19.7% time reduction from the full FE analysis. This result shows a huge lacking in efficiency still exists in the CPU time especially during the reduction iteration time. In addition, to able the reduction of the part A become achievable, the substructure (Part A) has to be decomposed into several new substructures. Part A was decomposed into four new substructures and analyse them separately. This situation has brought a new burden in reduction process since it will take more time to prepare a new substructure for an analysis.

As the WBS method is applied, the desired efficiency is achieved, as shown in Table 3. The notable achievement is gauged via the reduction in computational time between the full FE, Craig-Bampton and WBS analysis. Only 36 number of nodes is used to analyse the structure which gave the necessary time to compute the structure just by only 1811 seconds. This result clearly shows a significant reduction in the expenditure time by 84% time reduction from the full FE and 5 times quicker than the Craig-Bampton method. Plus, during the reduction process, it takes lesser time compared to Craig-Bampton method and a single substructure (Part A) was reduced without having to decompose it into several new substructures. Through the results obtained, a strong evident indicates that the WBS model is successfully, efficiently and accurately used to analyse a structural normal mode in comparison with the full FE model and Craig-Bampton model.

5 Conclusions

A wave-based substructuring approach for investigating the dynamic behaviour of a structure is presented. The efficiency and accuracy of WBS have been demonstrated in two different case studies. It has been shown that the WBS has accurately and efficiently predicted the modal parameters of the structure over the full FE and Craig-Bampton method. The significant advantages of the WBS method are that it is able to perform the investigation of the dynamic behaviour of the structure in the most efficient way. It is suggested that the proposed method is used when predicting a complex structure with a large number of iteration required.

A similar degree of accuracy can be obtained through the proposed method as obtained from the full FE model. Even the structure is analysed through the reduction model in

WBS, the accuracy of the predicted is maintained as the same, while a rapid iteration process is used. This situation is proven that the WBS has the same capability as the full FE model in accurately predict the dynamic behaviour of a structure.

Authors gratefully acknowledge the Malaysia Ministry of Higher Education (MOHE) and Research Management Institute (RMI) of University Teknologi MARA (UiTM) for providing financial support for this study through the research acculturation grant scheme (RAGS)- 600-RMI/RAGS 5/3 (21/2015). Authors would also like to express their appreciation for helpful comments, suggestions and technical support given by Mr David Starbuck and Mr Mohamad Azwan Abdul Khair.

References

1. M. Bennur, "Superelement, Component Mode Synthesis, and Automated Multilevel Substructuring for Rapid Vehicle Development," *SAE Int. J. Passeng. Cars - Mech. Syst.*, vol. **1**, no. 1, pp. 2008–01–0287, (2008).
2. S. Weng, Y. Xia, Y.-L. Xu, H.-P. Zhu, "Substructure based approach to finite element model updating," *Comput. Struct.*, vol. **89**, no. 9–10, pp. 772–782, (2011).
3. R. R. J. Craig and C. J. Chang, "Substructure Coupling for Dynamic Analysis and Testing," (1977).
4. S. Donders, B. Pluymers, P. Ragnarsson, R. Hadjit, W. Desmet, "The wave-based substructuring approach for the efficient description of interface dynamics in substructuring," *J. Sound Vib.*, vol. **329**, no. 8, pp. 1062–1080, (2010).
5. A. Maressa, D. Mundo, S. Donders, W. Desmet, "A wave-based substructuring approach for concept modeling of vehicle joints," *Comput. Struct.*, vol. **89**, no. 23–24, pp. 2369–2376, (2011).
6. S. Donders, R. Hadjit, M. Brughmans, L. Hermans, W. Desmet, "A wave-based sub-structuring approach for fast vehicle body optimisation," *Int. J. Veh. Des.*, vol. **43**, no. 1/2/3/4, p. 100, (2007).
7. E. Balmès, "Use of Generalized Interface Degrees of Freedom in Component Mode Synthesis," *Proc. 14th Int. Modal Anal. Conf.*, pp. 1–7, (1996).
8. J. M. Mencik, D. Duhamel, "A wave-based model reduction technique for the description of the dynamic behavior of periodic structures involving arbitrary-shaped substructures and large-sized finite element models," *Finite Elem. Anal. Des.*, vol. **101**, pp. 1–14, (2015).
9. M. A. Tournour, N. Atalla, O. Chiello, F. Sgard, "Validation, performance, convergence and application of free interface component mode synthesis," *Comput. Struct.*, vol. **79**, no. 20–21, pp. 1861–1876, (2001).
10. S. Donders, R. Hadjit, M. Brughmans, L. Hermans, W. Desmet, "Speeding up vibro-acoustic modification predictions using a Wave-Based Substructuring approach and ATV technology," *Forum Acusticum Budapest 2005 4th Eur. Congr. Acustics*, pp. 161–166, (2005).
11. "MSC. MSC/Nastran 2012." (2012).
12. "LMS International. LMS Virtual.Lab 13.5." (2015).
13. "The MathWorks Inc. MATLAB." (2013).
14. S. Donders, "Computer Aided Engineering Methodologies For Robust Automotive NVH Design," Katholieke Universiteit Leuven, (2008).

DOI: 10.1051/matecconf/20179001012

The investigation of the dynamic behaviour of a complex assembled structure using the frequency response function based substructuring method

Wan Imaan Izhan Wan Iskandar Mirza[1], *Muhamad Norhisham* Abdul Rani[1,*], *Mohd Azmi* Yunus[1], *Liyana* Roslan[2] and *Ahmad Burhani* Ahmad Basri[1]

[1]Dynamics and Control Laboratory, Faculty of Mechanical Engineering, Universiti Teknologi MARA (UiTM), 40450 Shah Alam, Malaysia
[2]Faculty of Mechanical Engineering, Universiti Teknologi MARA Kampus Pasir Gudang, 81750 Masai, Johor, Malaysia

Abstract. The frequency response function based substructuring method (FRF-BSM) for modelling and investigating dynamic behaviour of engineering structures has received much attention in recent years among modal analysts. However, the accuracy and efficiency of the predicted dynamic behaviour of the structures via the method is often found to be different from the test data. The discrepancy is believed to be the result of the coupling types used in the modelling. This paper aims to investigate the potential candidates of coupling types for FRF based substructuring in predicting the dynamic behaviour of a complex assembled structure which consists of a large flat span and two simplified aircraft pylons. Modal tests are performed to measure the dynamic behaviour of the assembled structure and its components. The finite element method is used for constructing analytical models of the assembled structure. The FRF-BSM is then used for the assembly of the span and pylons, and also to predict the dynamic behaviour of the assembled structure using rigid and elastic coupling. The comparison of results revealed that elastic coupling has demonstrated better capabilities to represent the bolted joints in the test structure, which may due to the coupling calculated is reasonably representing the stiffness of the bolted joints.

1 Introduction

In the field of structural dynamics, dynamic substructuring method has played an important role and has continued to be a great value since five decades ago. Dynamic substructuring is a process which divides an assembled structure into subsystems which can be analyzed or tested individually and then combining them by an assembly procedure to predict the response of the build-up structure [1-3]. This method has been used to evaluate the dynamic behaviour of a large complex structure and to give possibility to combine the analytical

* Corresponding author: mnarani@salam.uitm.edu.my

with the experimental parts. It also allows combining and sharing subsystems from different project groups. In recent times, de Klerk [4] has summarised the history, review, classification of techniques and has provided a general framework for dynamic substructuring. In general, the dynamic substructuring is divided into two different classes which are the time-domain and frequency-based method [5]. FRF based substructuring (FBS) is the process that predicts the FRF of a combined structure (coupled system) on the basis of free interface FRF of the substructures (uncoupled system). The equations for the FRF"s of the combined system are derived by balancing the forces and enforcing continuity at the interface [6,7]. The FBS represents the dynamic behaviour of the coupled system with FRFs, rather than set of discretized mass and stiffness equations.

By using the FBS method, the FRFs of the subsystem model can be either experimentally measured or analytically derived. This enable to combine the FRF's from the test data with the analytical model in order to predict the FRF of the whole combine system. This is the main advantage of this method. Allen [8-10] and Mayes [11, 12] presented several approaches about coupling the analytical model and some experimental issues related to FBS especially during the measurements of rotational degree of freedoms at the coupling. Neglecting the rotational degree of freedom will affect the FRF of the assembled system, as presented by Manzato recently in [13]. Hence, for this study, the rotational degree of freedom for the subsystems were included to archive a high degree of accuracy.

Nevertheless, when it comes to analytical modelling the dynamic behaviour of a structure with multiple joints, the results calculated from this method may differ from the test data. This is believed to be a result of the inappropriate coupling types used to represent the joints. Many researchers have developed FBS schemes for the particular case studies. However, little attention has been given on the effects of the coupling type used in modelling and predicting the dynamic behaviour of a structure. Nevertheless, modelling bolted joints is becoming very difficult because of the properties of the joint itself which is complex to be understood [14, 15]. In addition, there is no universal method to model dynamics of joints so much so that each application has been investigated on case by case basis [16]. For dynamic substructuring method, the properties of coupling need to be considered and understood in order to obtain a satisfactory level of accuracy in dynamic behaviour prediction. Although each subsystem is modelled flawlessly, the invalid assumption in coupling type to assemble the subsystems may lead to discrepancies in dynamic behaviour prediction of coupled system.

This work is concerned with the study of appropriate coupling types used for multiple bolted joints in a structure. The appropriate coupling type is presented for the investigation of the dynamic behaviour of the structure. Bolted joints have been chosen for this study as these joints are commonly used in most engineering application and are also used by modal analysts in their research related to FBS in [8, 10, 12]. For the dynamic substructuring model, the FRFs of the subsystems were derived based on the analytical model of the subsystems and are assembled together using rigid and elastic type of coupling.

1.1 Theoretical explanation of FRF coupling

Consider a system(S) consists of subsystem (A) and (B) which the DOFs are classified either as internal or coupling DOFs as shown in Figure 1. Both subsystems are independent of one another before the coupling. The c-DOFs must be the same for both subsystems.

 DOI: 10.1051/matecconf/20179001012

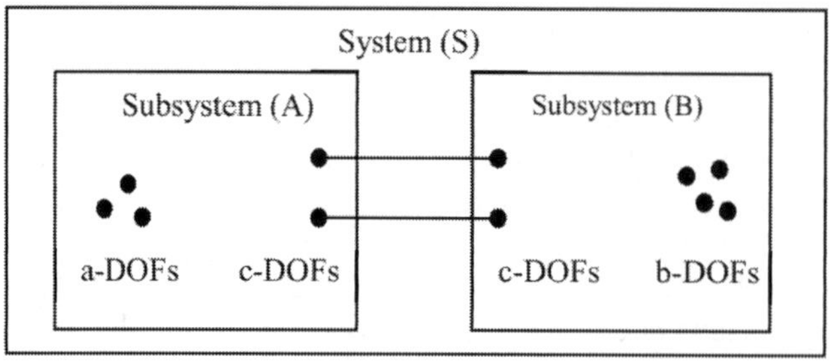

Fig. 1. Schematic diagram of systems (s), subsystem (A) and subsystem (B).

The equation of motion for each subsystem in the frequency domain is:

$$\{X\}_n = [H]_{nn}\{F\}_n \tag{1}$$

where $\{X\}_n$ is the complex displacement, $[H]_{nn}$ is the complex admittance matrix in form of displacement, force $\{F\}_n$ is the applied force vector and the subscript n is the total number of DOF for each subsystem.
The basic governing relationship for subsystem A:

$$\begin{Bmatrix} \{X^A\}_a \\ \{X^A\} \end{Bmatrix}_n = \begin{bmatrix} [H^A]_{aa} & [H^A]_{ac} \\ [H^A]_{ca} & [H^A]_{cc} \end{bmatrix} \begin{Bmatrix} \{F^A\}_a \\ \{F^A\}_c \end{Bmatrix}_n \tag{2}$$

and for subsystem B:

$$\begin{Bmatrix} \{X^B\}_b \\ \{X^B\}_c \end{Bmatrix}_n = \begin{bmatrix} [H^B]_{bb} & [H^B]_{bc} \\ [H^B]_{cb} & [H^B]_{cc} \end{bmatrix} \begin{Bmatrix} \{F^B\}_b \\ \{F^B\}_c \end{Bmatrix}_n \tag{3}$$

For the rigid connections between subsystems A and B, compatibility implies that

$$\{X^A\}_c = \{X^B\}_c = \{X^C\}_c \tag{4}$$

and the force equilibrium requires that

$$\{F^A\}_c = \{F^B\}_c = \{F^C\}_c \tag{5}$$

The FRFs of the system can be defined as

$$\begin{Bmatrix} \{X^S\}_a \\ \{X^S\}_c \\ \{X^S\}_b \end{Bmatrix}_n = \begin{bmatrix} [H^S]_{aa} & [H^S]_{ac} & [H^S]_{ab} \\ [H^S]_{ca} & [H^S]_{cc} & [H^S]_{cb} \\ [H^S]_{ba} & [H^S]_{bc} & [H^S]_{bb} \end{bmatrix} \begin{Bmatrix} \{F^S\}_a \\ \{F^S\}_c \\ \{F^S\}_b \end{Bmatrix}_n \tag{6}$$

The c-DOFs from the partitioned equation of equation (2) and (3) are derived and can be equated and solved for the connection force as

$$\{\tilde{F}^A\}_c = [[H^A]_{cc} + [H^B]_{cc}]^{-1} [[H^B]_{cb}\{F^B\}_c - [H^A]_{ca}\{F^A\}_a + [H^B]_{cc}\{F^S\}_c] \tag{7}$$

with $\{\tilde{F}^A\}_c$ is the internal transmitted force at the interfaces of subsystem A at the coupling. The equation derived above can be used to determine the coupled system FRFs on the basis of the uncoupled FRFs of the individual components. While for the elastic coupling, a joint can be described analytically as follows;

$$\{F\} = [Z]\{X\} \tag{8}$$

where F is the force vector, Z is the impedance matrix at the coupling and X is the displacement vector. The detail derivation of the coupling with joints can be found in [17].

2 Description of the test structure

The structure under the investigation was a mimicked version of an aircraft. It consists of three major substructures which are a unit of a large span plate and two units of a mimicked version of aircraft pylons. The large span substructures was made of an aluminium sheet with 4mm thickness while each of the pylon consists of seven other components which were assembled together using bolted joints. The three major substructures were assembled together using eight bolted joints as shown the Figure 2. To ensure the characteristic of bolted joints are similar to each other, the same amount of torque force has been applied. In this study, the assembled structure was numerically and experimentally investigated.

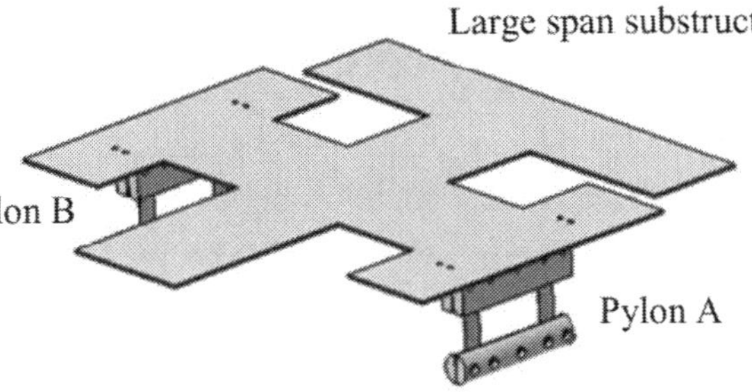

Fig. 2. Assembled structure.

3 FRF based substructuring method

In this investigation, the FRFs of the subsystems were derived from the numerical model. Prior to performing the FRF based substructuring assembly analysis of the assembled structure, each of the subsystem was numerically solved under normal mode solutions. The updated material properties were used to the subsystems to minimize components error. The desired excitation and response points are derived in the form of FRF for each of the coupling points. In general, there are two methods to derive the FRFs for the numerical model as discussed in [18]. The first method is by derivation of the FRFs from the numerical model of the subsystems from the eigenvalues. The second method is by performing an FRF synthesis based on a finite number of the modal parameters of the subsystem. Since the natural frequencies and mode shapes of the components were already obtained from the normal mode solution of the numerical model, the second method was used to obtain the FRFs. For FRF synthesis method, the synthesized FRF matrix $H_{syn}(\omega_k)$ and mode shapes are expressed by:

$$\mathrm{H_{syn}}(\omega_k) = \sum_{i=1}^{N} \frac{\{\emptyset\}_i\{\emptyset\}_i^T}{\left(\omega_{n_i}^2 - \omega_k^2\right) + j2\xi_i\omega_k\omega_{n_i}} \tag{9}$$

where N is the number of calculated modes, $\{\emptyset\}_i$ is the ith mass normalised mode shapes, ω_{n_i} is ith natural frequency and ξ_i is the ih modal damping ratio.

The subsystems with synthesized FRFs were assembled in Virtual.Lab software as shown in Figure 3. The frequency of interest was set between 1-30 Hz with 0.5 Hz frequency step. For this case study, two types of coupling have been used which are the rigid and elastic types of coupling in order to identify which type of coupling is suitable to be used for multiple bolted joints. The FRF at the reference point obtained from the FBS analysis was compared with the measured FRF at the reference point for validation process.

Fig. 3. FBS setup for the assembled structure.

4 Elastic coupling modelling for bolted joints

This way of modelling presented in this section is one of most used in Nastran software, as it is simple but sufficiently accurate. Bolted joints consist of axial, shear and rotational stiffness for the six degrees of freedom (DOF) on each nodes. For the translational DOFs, assume K_1 is the value for the axial stiffness, K_2 and K_3 are the shear stiffness of the bolted joint. The axial stiffness K_1 is obtained from the equation derived based on the Hooke's Law:

$$K_1 = \frac{AE}{L} \tag{10}$$

where A, E and L is the cross-sectional area, Young's modulus and the length of the bolt respectively. The shear stiffness of K_2 and K_3 are calculated using the shear flexibility by Huth formulation:

$$K_2 = K_3 = \frac{1}{c} \tag{11}$$

Which C is the shear flexibility which acquire from equation:

$$C = \left(\frac{t_1 + t_2}{2d}\right)^a \left(\frac{b_1}{t_1 E_1} + \frac{b_2}{t_2 E_2} + \frac{b_1}{2t_1 E} + \frac{b_2}{2t_2 E}\right) \tag{12}$$

Where subscripts t_1 , E_1 and t_2 , E_2 are the thickness and Young's modulus of two plates. b_1 and b_2 are the coefficients depending on the joint plate material, and a is the fasterner type coefficient.

For the rotational stiffness of K_5 and K_6, the values can be calculated using;

$$K_{5-6} = \frac{1}{\varepsilon}\left(Max(K_2, K_3)\frac{L^2}{4}\right) \tag{13}$$

Usually, rotational stiffness around the fastener K_4 can be fixed to 100.

5 Test setup of the assembled structure

The structure was tested in free-free boundary conditions by using an impact testing method. To simulate the free-free boundary conditions, the test structure was hanged to the test rig using four nylon strings and springs. From the initial finite element result of the assembled structure, the frequency bandwidth, excitation point and also the mode shapes of the structure has been determined. The frequency under the investigation was between 1-30 Hz where the rigid body modes was not included. An impact hammer was used to excite the structure and 3 rowing accelerometers were used to acquire the dynamic data. The response signals and load were interpreted via LMS SCADAS analysis. The modal parameters from the FRF were extracted from the resonance amplitude of FRF data. The initial results of the finite element model were used for mode pairing purposes.

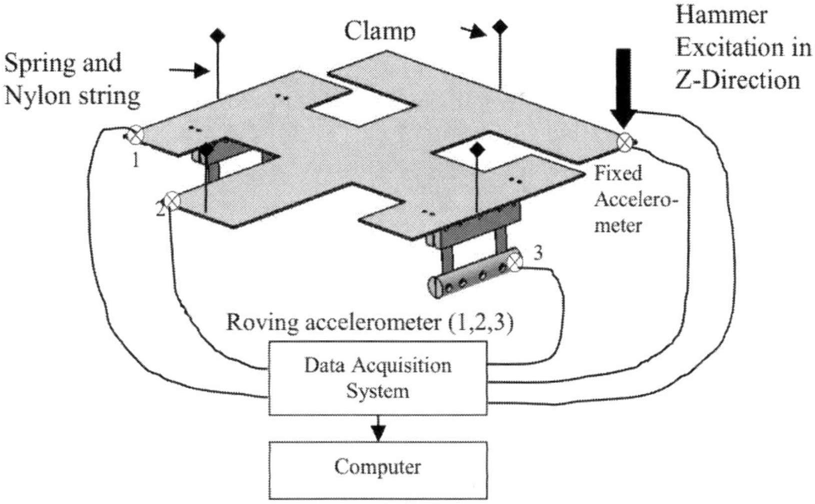

Fig. 4. Test setup for the assembled structure.

6 Results and discussion

Prior to performing the assembly analysis via FBS method for the structure with bolted joints, the dynamic behaviour of each of the subsystem were numerically and experimentally investigated. The measured results were used to reconcile the finite element model of the subsystems in order to minimize the discrepancies within the test data. The percentage error of the predicted result from the finite element model of the subsystems are summarize in Table 1.

Table 1. Natural frequencies and percentage errors of the updated substructures of large flat span, Pylon A and Pylon B.

Mode	Large Span (Hz)	Error with test data	Pylon A (Hz)	Error with test data	Pylon B (Hz)	Error with test data
1	16.33	0.18%	31.14	0.54%	32.05	5.18%
2	17.00	0.59%	37.29	5.72%	38.58	4.21%
3	31.35	0.19%	115.12	0.73%	120.31	1.43%
4	31.63	0.35%	695.01	4.69%	710.39	1.17%
5	37.79	0.19%	715.10	0.50%	733.54	1.81%
		1.50%		12.19%		13.80%

It can be seen in particular that the total errors for the large span in Table 1 are 1.5 percent, which is the significance achievements in this investigation. This is due to the large flat span was a simple structure without any joints. In contrast, the total percentage errors for both pylon are in high values which are 12.19% and 13.80% for pylon A and pylon B respectively. The high errors were contributed by the second and fourth modes for pylon A while by the first and second modes for the pylon B. It is believed that the source of the discrepancies might as a result of the invalid assumption in finite element modelling of the bolted joints at the subsystems, as discussed in [19, 20]. However, since the numerical results of the two Pylons are above 30 Hz, the finite element models seems to be reasonable as the modes of interest for the assembled system in this study are no more than 30Hz.

On the other hand, the dynamic behaviour of the assembled structure was calculated using the FBS method. For the first investigation of the coupling in FBS, the reconciled model of the subsystems were combine together using rigid type of coupling. The FRF calculated from the use of rigid coupling was compared with the test FRF as presented in Figure 5. The drive point FRF was obtained from both test and finite element model reference point as shown in Figures 3 and 4. Figure 5 shows that amplitude of the rigid coupling FRF is ahead of the resonance peak of the test FRF. This indicated that the numerical model was posed with higher stiffness in comparison with the test structure. This is as the result of the rigid type of coupling at which can be considered as infinitely stiff. Therefore, it may be reasonable to assume that rigid type of coupling seem to lack the capability to bolted joints with an acceptable level of accuracy.

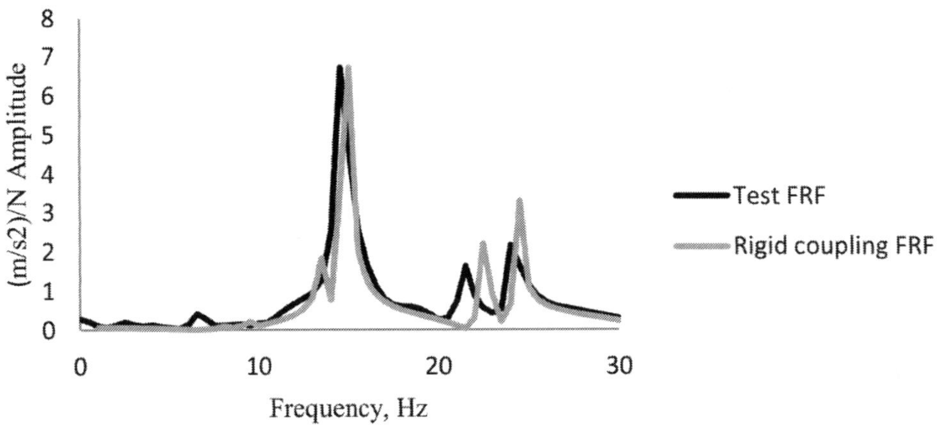

Fig. 5. Comparison of test and rigid coupling FRFs.

In order to reduce the stiffness at the joints, elastic type of coupling has been used to combine the structure. Benefitting from the capability of the elastic type of coupling in predicting the dynamic behaviour of the structure, the stiffness value for all the translations and rotations can be included at the coupling. The stiffness of the coupling are calculated by using the Huth formulation as stated before. The FRF calculated from the using of elastic coupling was compared with the test FRF as presented in Figure 6. The comparison of both FRF seems to be more acceptable compared with the rigid type of coupling. . This may due to the fact that elastic coupling calculated is reasonably representing the stiffness of the bolted joints. Obviously, the elastic type of coupling by using the Huth formulation has better capabilities to represent the bolted joints in the test structure. The comparisons of the FRFs of the measurement, rigid and elastic type of coupling are presented in Figure 7 for better view.

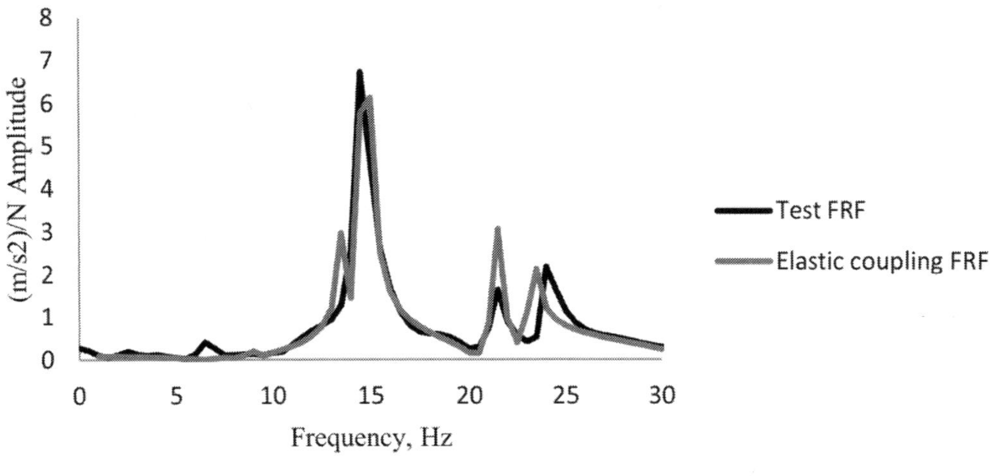

Fig. 6. Comparison of test and elastic coupling FRFs.

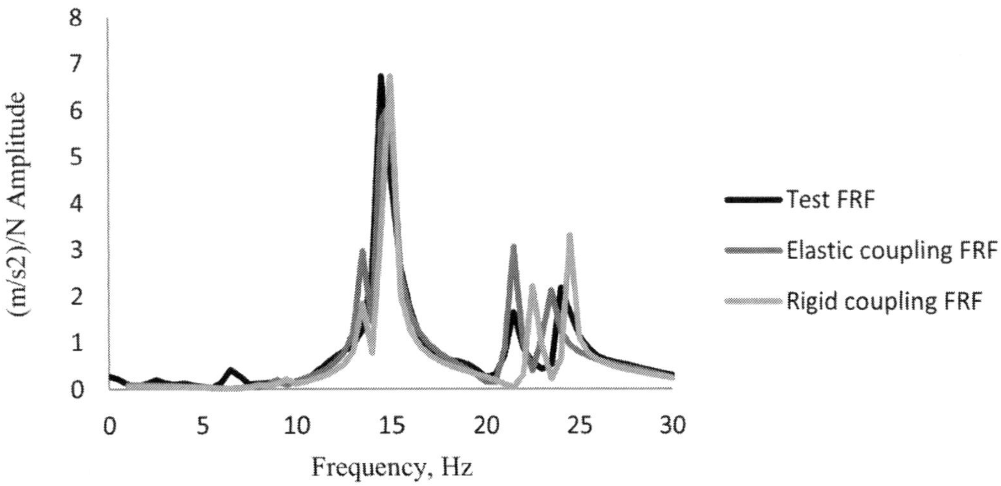

Fig. 7. Comparison of test data, rigid and elastic coupling FRFs.

7 Conclusions

The dynamic behaviour of a simplified model of an aircraft structure with multiple bolted joints was investigated numerically and experimentally. The evaluation of dynamic behaviour of bolted joints via FBS analysis by using two types of coupling which are the rigid type and elastic coupling has been performed by the authors. The test result has been carefully measured and the comparison of results in term of FRFs from the measurement and FBS method is presented. The comparison of results revealed that elastic coupling has demonstrated better capabilities to represent the bolted joints in the test structure. This may due to the fact that elastic coupling calculated is reasonably representing the stiffness of the bolted joints. The study reveals that there are high possibilities to obtain a high degree of accuracy of dynamic behaviour prediction from the use of elastic coupling in combining a numerically derived and experimental model of a structure via FBS method.

Authors gratefully acknowledge the Malaysian Ministry of Higher Education (MOHE) and Research Management Institute (RMI) of Universiti Teknologi MARA (UiTM) for providing financial support for this study through the research acculturation grant scheme (RAGS) (600-RMI/RAGS 5/3 (159/2014). They would like to express their heart-felt appreciation for the constant support and help given by Prof. Dr. Hadariah Bahron, Assoc. Prof. Dr. Siti Akmar Abu Samah, Mr Fauzi, Mr Mohamad Azwan Abd Khair and Mr Mohd Hakimi of SIRIM.

References

1. Y. Xia, S. Weng, Y.-L. Xu, "A Substructuring Method for Model Updating and Damage Identification," *Procedia Eng.*, vol. **14**, pp. 3095–3103, (2011).
2. P. Soleimanian, M. H. Sadeghi, and A. Tizfahm, "Substructure-Joint Based Approach to Damped FE Model Updating of Complex Assembly Structure," vol. **464**, pp. 1169–1174, (2012).
3. L. Baeza, a. Roda, J. C. O. Nielsen, "Railway vehicle/track interaction analysis using a modal substructuring approach," *J. Sound Vib.*, vol. **293**, no. 1–2, pp. 112–124, (2006).
4. D. De Klerk, D. J. Rixen, S. N. Voormeeren, "General Framework for Dynamic Substructuring: History, Review and Classification of Techniques," *AIAA J.*, vol. **46**, no. 5, pp. 1169–1181, (2008).
5. A. Williams, C. Chipman, P. Avitabile, "Modal and frequency based substructuring using rotational DOF conidetations," *26th Int. Modal Anal. Conf. (IMAC XXVI)*, no. 1, (2008).
6. V. Paulo Antdnio Urgueira, "Dynamic Analysis of Coupled Structures Using Experimental Data," *Techniques*, no. October, (1989).
7. D. J. E. M. Imregun, D.A. Robb, "Structural Modification and Coupling Dynamic Analysis using FRF Data," in *Proceedings of the Fifth International Modal Analysis Conference (IMAC V)*, (1987).
8. M. S. Allen, R. L. Mayes, "Comparison of FRF and Modal Methods for Combining Experimental and Analytical Substructures," *Modal Anal. Conf. 2007 A Conf. Expo. Struct. Dyn.*, p. 19, (2007).
9. M. S. Allen, H. M. Gindlin, R. L. Mayes, "Experimental modal substructuring to estimate fixed-base modes from tests on a flexible fixture," *J. Sound Vib.*, vol. **330**, no. 18–19, pp. 4413–4428, (2011).
10. M. S. Allen, R. L. Mayes, E. J. Bergman, "Experimental modal substructuring to couple and uncouple substructures with flexible fixtures and multi-point connections," *J. Sound Vib.*, vol. **329**, no. 23, pp. 4891–4906, (2010).
11. R. Mayes, P. S. Hunter, T. W. Simmermacher, "Combining Lightly Damped Experimental Substructures with Analytical Substructures," *Sem Imac Xxvi*, (2008).
12. R. L. Mayes, P. S. Hunter, T. W. Simmermacher, M. S. Allen, "Combining Experimental and Analytical Substructures with Multiple Connections," in *26th IMAC: Conference and Exposition on Structural Dynamics 2008: IMAC XXVI*, (2008).
[13. B. Manzato, S. Risaliti, E. Napoli, C. Tamarozzi, T. Peeters, "A Review Of Frequency-Based Substructuring Methods And Their Applicability To Engineering Structures," in *Proceeding of International Conference of Structural Engineering Dynamics (ICEDyn)*, (2015).
14. K. Karpanan Nakalswamy, "Experimental and numerical analysis of structures with bolted joints subjected to impact load," (2010).
15. H. Ouyang, M. J. Oldfield, J. E. Mottershead, "Experimental and theoretical studies of a bolted joint excited by a torsional dynamic load," *Int. J. Mech. Sci.*, vol. **48**, no. 12, pp. 1447–1455, (2006).
16. S. Bograd, P. Reuss, A. Schmidt, L. Gaul, M. Mayer, "Modeling the dynamics of

mechanical joints," *Mech. Syst. Signal Process.*, vol. **25**, no. 8, pp. 2801–2826, (2011).

17. W. Liu, D. J. Ewins, "Substructure Synthesis Via Elastic Media Part II : Coupling Analysis."
18. K. Cuppens, P. Sas, L. Hermans, "Evaluation of the FRF based substructuring and modal synthesis technique applied to vehicle FE data," *Proc. ISMA 25*, pp. 1143–1150, (2001).
19. M. A Yunus, H. Ouyang, M. N. A. Rani, "Finite Element Modelling and Updating of Structure of Sheet Metal with Bolted and Welded Joints," *Appl. Mech. Mater.*, vol. **70**, pp. 7–11, (2011).
20. M. N. Abdul Rani, D. Stancioiu, M. a. Yunus, H. J. Ouyang, H. Deng, S. James, "Model Updating for a Welded Structure Made from Thin Steel Sheets," *Appl. Mech. Mater.*, vol. **70**, pp. 117–122, (2011).

Response surface reconciliation method of bolted joints structure

Mohd Azmi Yunus[1,2,*], *Sharifah* Nazri[1,2] , *Mohamad Norhisham* Abdul Rani[1,2]
Azela Tormodi[1,2] and *Salmiah* Kasolang[1,2]

[1]Faculty of Mechanical Engineering,Universiti Teknologi Mara,40450 Shah Alam, Selangor, Malaysia
[2]Structural Dynamics Analysis Validation, Universiti Teknologi Mara,40450 Shah Alam, Selangor, Malaysia

Abstract. Structural joining methods such as bolted joints are commonly used for the assembly of structural components due to their simplicity and easy maintenance. Understandably, the dynamic characteristic of bolted joined structure is mainly influenced by the properties of their joints such as preload on the bolts and joints stiffness which alter the measured dynamics response of the structure. Therefore, the need to include the local effect of the bolted joints into the numerical model of the bolted joined structure is vitally important in order to represent the model accurately. In this paper, a few types of connector elements that can be used to represent the bolted joints such as CBAR, CBEAM and CELAS have been investigated numerically and experimentally. The initial numerical results of these element connectors are compared with the experimental results in term of natural frequencies and mode shapes. The comparative evaluation of numerical and the experimental data are performed in order to provide some insights of inaccuracies in the numerical model due to invalid assumption in the numerical modelling such as geometry, material properties, and boundary conditions. The discrepancies between both results (numerical and experimental data) are then corrected using the response surface reconciliation method (RSRM) through which the finite element model is altered in order to provide closer agreement with the measured data so that it can be used for subsequence analysis.

1 Introduction

In recent years, finite element (FE) models have become an important analysis tools for structural dynamics to produce numerically the dynamic behaviour of structures in mechanical systems. However, the finite element (FE) model of a structure is commonly constructed on the basis of design engineering and may not truly represent of an actual structure [1]. As a result, the numerical predictions from a finite element (FE) model

* Corresponding author: mayunus@salam.uitm.edu.my

largely differ from the results of the real structure. These discrepancies originate from the uncertainties in simplifying assumptions of structural geometry, materials properties or boundary conditions [2]. In order to reduce these discrepancies, the uncertain parameters of a finite element (FE) model should be updated to leads the better predictions of the responses of an actual structure [3]. Many model updating techniques have been tested. Most techniques are based on the minimization of structural parameters to minimize an error function between the measured and numerical responses. There are two methods of model updating which are direct and iterative methods [4].

Direct methods of updating are directly update the finite element (FE) model properties commonly one step procedure while the iterative methods of updating uses the sensitivities of the parameters to find the changes and involving solution of an optimization problem [5]. Jaishi and Ren use the sensitivity updating methods that require complicated constructions of sensitivity matrices because the finite element (FE) models should be tuned and recomputed during optimization process [6]. Moreover, for a large finite element (FE) model, the evaluation of performance was repeated and local gradients are not only computational intensive but may also have result convergence difficulty when nonlinear complex structures are involved. Furthermore, each of the iteration in the iterative method needs to go back to run finite element (FE) with any parameter updated. Thus, in the field of model updating, computational efficiency is a major issue in the development of more applicable and practical approaches. Hence, an alternative technique for efficient computation of response of complex structures by overcoming the drawbacks of iterative method is Response Surface Methodology. Response Surface Method (RSM) is based on meta-modelling has emerged as an alternative solution to such problems to achieve balance. The primary advantages of the response surface method (RSM) in the updating method are its ease of implementation and low computational cost [7].

Response surface method has shown attractive ability in modifying finite element (FE) model parameters. Guo and Zhang present that the response surface method gave similarly accurate predictions while requiring much fewer number of finite element analyses compared with the sensitivity-based model updating [8]. Besides that, Zhang et al proposed a model updating technique based on generic algorithm and response surface methodology [9]. The present paper emphasizes the development of the connector element to represent as a bolted joined structure. As well, it is aimed to determine the better correlation between the experimental and finite element (FE) for bolted joined structure when updating the finite element (FE) model using response surface methodology (RSM).

2 Experimental work

The simplified model structure of bolted joined was developed. The completed bolted joined structure consists of two different types of components namely as plate and u-shape. The u-shape is made from aluminium T60-6134 with the thickness of the top is 15mm and the thickness of below 10mm. The plate is made of mild steel with the thickness of 1mm. There are in total seven bolted joints on the assembled structure. The experimental modal testing was performed to the bolted structure in free-free boundary conditions and the interest frequencies from 1~1000 Hz was used in the experiment. An impact hammer (PCB Model 086C02) and roving accelerometers (Dytran 3032A) were employed used in the measurement of the natural frequencies and mode shapes of the test structure. Then, the modal parameters from experimental work are obtained via the LMS data acquisition system. The finite element (FE) analysis results were employed to provide guidance on determining the frequency bandwidth of the testing, the locations of the excitation points, excitation directions and also response measurement points. As a result the model structure was set up as showed in Figure 1 in which four springs and nylon strings were used to simulate free-free boundary conditions of the structure.

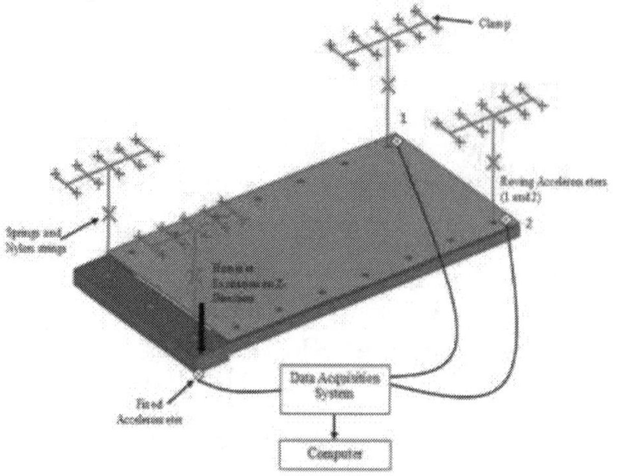

Fig. 1. Schematic diagram of bolted joined structure for experimental work.

3 Finite element model and analysis

The finite element models used in this study were developed using Altair HyperWork software as shown in Figure 2. Then, MSC PATRAN/NASTRAN Solution 103 and 200 were used for normal modes and sensitivity analysis respectively. In order to represent the bolted joints in the finite element model of the bolted joined structure, three types of connectors elements were used namely CBAR, CBEAM and CELAS. The material properties of mild steel and aluminium T60-6134 are used in the finite element model with nominal values as tabulated in Table 1.

Fig. 2. Simplified CAE model of the bolted joined structure.

Table 1. Details of structure FE model.

Properties (Mild Steel)	**Value**	**Properties (Aluminum)**	**Value**
Young's Modulus	210 GPa	Young's Modulus	69 GPa
Poisson's Ratio	0.3	Poisson's Ratio	0.33
Density	7900kg/m^3	Density	2700kg/m^3

In this research, the MSC NASTRAN SOL103 was used for normal modes analysis. The governing equation of motion for the homogeneous system with absence damping is given as;

$$\mathbf{M}\ddot{\mathbf{x}}(t)+\mathbf{K}\mathbf{x}(t)=0 \tag{1}$$

Where **M, C** and **K** are represent as mass, damping and stiffness matrices of the system respectively. Then, $\ddot{\mathbf{x}}$, $\dot{\mathbf{x}}$ and **x**, known as vectors of accelerations, velocities and displacement respectively. The damping value in physical system can be disregarded for most engineering system due to its amount is so small [4]. The eigen-equation is applied in order to determine dynamic characteristics which are modes shapes and natural frequencies. It can be written as follows;

$$(\mathbf{K}-\omega^2\,\mathbf{M})\emptyset= 0 \tag{2}$$

where eigenvalues ω^2 are natural frequencies squared and $\emptyset$ are corresponding modes.

The initial natural frequencies and mode shapes of the bolted structures are computed using Altair HyperWork with free-free boundary condition to determine the frequencies and mode shapes of the structures.

4 Response surface based on finite element model updating

Response surface methodology is a combination of mathematical and statistical techniques useful for the modelling and analysis engineering problems in which a response of interest is influenced by several parameters and the objective is to optimize this response [10]. Furthermore, the choices of response surface function type are important. First order, the mathematical is able to describe the true relationship between input parameters and output responses. Usually, a low-order polynomial in some region of the independent variables is employed. If the response is well modelled by a linear function of the independent variables, the approximating function is the first order model as

$$y=\beta_o+\sum_{i=1}^{\eta}\beta_i x_i \tag{3}$$

Second order, this expression should contain the minimum number of items the increase of which will considerably increase the number of design points. For most engineering problems, a second-order polynomial function with interaction effects is adequate for representing physical systems:

$$y=\beta_o+\sum_{i=1}^{\eta}\beta_i x_i+\sum_{i=1}^{\eta}\beta_{ii}x_i^2+\sum_{i=1}^{\eta}\sum_{j=1+1}^{\eta}\beta_{ij}x_i^2 x_j+\varepsilon \tag{4}$$

In, Eq. 1 and Eq. 2 y is the response variable, β_0 is the constant term, β_i is the coefficient of the linear term, β_{ii} is the coefficient of quadratic single term, β_{ij} is the coefficient of the quadratic cross product term. The x_i, x_j, x_k terms represent the main effect of parameter i and the interaction effect between i and j, respectively. [11].

5 Results and discussion

The first ten frequencies of the bolted joined structure were calculated numerically using finite element method (FEM). The accuracy of the finite element models of the bolted joined structure with three types of element connectors was investigated. The result as shown in Table 2 consists of three types of comparisons with different types of connector element. The comparisons were made between test data and initial finite element data that show in summation of relative error for first ten modes. These relative errors might happen due to the deficiencies in the FE modelling. Non-linearity was one of the reason that the relative error which caused by the location and nature of joining especially in structure that made from metal [14]. On top of that, the assumptions input parameters (Young's modulus, Poisson's ratio, density and thickness) also influence the frequency of the structure.

As can be seen from the Table 1, the CBEAM connector element has shown the lowest error (10.55 %) connector compared to CBAR and CELAS element. The result also shows that the higher error is CELAS connector element which is 80.89% and followed by CBAR is 10.64% respectively. Therefore, CBEAM connector element was used to represent bolted joints of the assembled structure in the finite element model.

Table 2. Comparison of connector element.

Mode	Test Data (Hz)	CELAS		CBAR		CBEAM	
		FE (Hz)	Error (%)	FE (Hz)	Error (%)	FE (Hz)	Error (%)
1	123.15	163.25	32.57	124.35	0.98	124.24	0.89
2	197.31	174.12	11.75	197.62	0.16	197.55	0.12
3	205.62	209.95	2.11	208.71	1.50	208.66	1.48
4	227.52	279.59	22.89	223.36	1.83	223.29	1.86
5	303.44	288.59	4.89	295.02	2.77	294.96	2.79
6	418.93	390.95	6.68	404.70	3.40	404.64	3.41
Total Error (%)			80.89		10.64		10.55

On the other hand, the response surface model updating was performed on the basis of the first six measured frequencies by minimising the objective function (Eq. 4). The updated values of the natural frequencies of the finite element model of bolted joints are shown in Table 2. From the table, it is clearly shown that, the total error of the updated natural frequencies is slightly reduced from 10.55 percent to 9.62 percent and MAC values are above 0.60.

Table 3. Three comparisons of the results between the measured, the initial finite element model and updated model of the bolted structure (CBEAM).

Mode	I	II	III	IV	V	VI	VII
	Test Data (Hz)	Initial FE (Hz)	Error (%) [I-II/I]	Initial FE MAC	Updated FE (Hz)	Error (%) [I-V/I]	Updated FE MAC
1	123.15	124.24	0.89	0.98	122.14	0.82	0.97
2	197.31	197.55	0.12	0.82	196.39	0.47	0.84
3	205.62	208.66	1.48	0.93	205.35	0.13	0.88
4	227.52	223.29	1.86	0.83	238.02	4.61	0.94
5	303.44	294.96	2.79	0.83	310.27	2.25	0.97
6	418.93	404.64	3.41	0.63	424.55	1.34	0.91
Total Error(%)			10.55			9.62	

6 Conclusions

The bolted assembly structure was analysed by experimentally and numerically. The relative error was determined by comparing the natural frequencies and model shapes of the experimental result and numerical data. A replacement model of the bolted structure was updated using response surface method in reducing the value of relative error in term of the natural frequencies. It can be concluded that the cause of the error was contributed by the invalid assumption of the finite element model. Therefore, the continuous study in modelling bolted joint was recommended. It may be included the effect of deformation happen while joining process in further study in modelling bolted joints structure.

The authors would like to thank Kementerian Pendidikan Malaysia (KPM) and Research Management Centre (RMC) UiTM for the Fundamental Research Grant Scheme (FRGS) with the file number 600-RMI/FRGS 5/3 (78/2014) for the sponsorship provided in completing this paper. The authors would like to express appreciation to En. Azwan bin Abd. Kadir, technician, Structural Dynamic and Validation (SDAV), UiTM for their technical assistance and laboratory facilities.

References

1. T. Marwala, Finite-element-model Updating Using the Response-surface Method, *Finite element model Updat. Using Comput. Intell. Tech. Appl. to Struct. Dyn.*, pp. 103–125, (2010).
2. W.-X. Ren, H.-B. Chen, Finite element model updating in structural dynamics by using the response surface method , *Eng. Struct.,* vol. **32**, no. 8, pp. 2455–2465 (2010).
3. J.E. Mottershead and M.I. Friswell, Model Updating in structural dynamics: A survey, *J Sound Vibration* ,pp.167-347, (1993).
4. B. a. Zárate, J. M. Caicedo, Finite element model updating: Multiple alternatives *Eng. Struct.,* vol. **30**, no. 12, pp. 3724–3730 (2008).
5. S. Chakraborty, A. Sen, Adaptive response surface based efficient Finite Element Model Updating, *Finite Elem. Anal. Des.,* vol. **80**, pp. 33–40, (2014).
6. B. Jaishi, W.X. Ren, Structural finite element model updating using ambient vibration test results , *J. Struct. Eng.* , ASCE 131 (4) ,pp 617-628, (2005).
7. Q. Rui, H. Ouyang, H. Y. Wang, An efficient statistically equivalent reduced method on stochastic model updating , *Appl. Math. Model.*, vol. **37**, no. 8, pp. 6079–6096, (2013).
8. S. G. Shahidi, S. N. Pakzad, Response surface model updating for nonlinear structures, *Proc. IMAC-XXXI Conf. Expo. Struct. Dyn. SEM*, Orange County, CA, USA;, vol. **18015**, (2013)
9. L. M. Zhang, Q. Fei, Q. T. Guo, Dynamic finite element model updating using meta-model and genetic algorithm, *Proc., IMAC-XXIII Conf. Expo. Struct.* Dyn. SEM, Orlando, FL, USA, (2005).
10. D.C. Montgomery, Design and Analysis of Experiments: Response surface methods and designs. *New Jersey: John Wiley and Sons*, Inc. (2005).
11. W. X. Ren, S. E. Fang, M. Y. Deng, Response surface based finite element model updating using structural static responses, *J Eng Mech-ASCE*, vol. **137**, no. Montgomery 2004, pp. 248–257, (2011).

Effect of heat-treatment on the hardness and mechanical properties of Boron Alloyed Steel

Mohammad Raffik bin Khiyon[1], *Salwani* Mohd Salleh[1,*]

[1]Faculty of Mechanical Engineering, Universiti Malaysia Pahang (UMP), 26600 Pekan, Pahang, Malaysia

Abstract. In an automotive industry, hot stamped, die quenched structural components have been widely used to provide extra protection against crash intrusion. Boron alloyed steel exhibit limited ductility, but it also promotes improvement in impact performance. This study analyzed the effect of cooling rate on the hardness and energy absorption. Self-quenched specimens were heated to 850°C and cooled in air of room temperature, water at room temperature and cold water. Vickers hardness test and tensile test was then carried out to analyze the effect of different quenching rate. Self-quenched specimens were compared to the properties of the die-quenched specimens obtained from commercial automobile body. Result shows that boron steel with the highest cooling rate has the highest value of hardness but low in strength.

1 Introduction

Steel manufacturers produce alloy boron steels that have undergone heat treatment, including quenching or quenching and tempering at low temperatures to attain high resistance to dynamic loads [1]. The adding of boron significantly increases the hardenability and strength of low and medium carbon steel was already observe in the year 1930s.

The demand in automotive industry to reduce the vehicle weight while considering the safety have rapidly increase the manufacturing of lightweight body parts from ultrahigh-strength steels (UHSS). The process of forming this ultrahigh-strength steel is limited by low formability and considerable spring back. To overcome this problem, boron has been used as an alloying element for alloy steel. It is a vital material that plays an important role as it has strong and hard properties [2]. Thus, when the strength is increased, the size of the steel can be decreased and the weight will become lighter.

The desired mechanical properties and microstructure suitable for the specific purpose can be reached, by selecting the parameters of heat treatment of these steels [3,4]. This would allow them to extend their range of applications. However, the different

* Corresponding author: salwani@ump.edu.my

DOI: 10.1051/matecconf/20179001014

microstructures can cause different mechanical properties, require careful decision making to define the application of these steels in many states of heat treatment. Thus, this study the effect of heat-treatment on the hardness and mechanical properties of boron alloyed steel.

2 Experimental setup

2.1 Specimens

Specimens are made of boron alloyed steel. This boron alloy steel grade 22MnB5 belongs to the product category of quenched and tempered steels, and features outstanding strength after heat treatment. The materials strength is propagated by adding a small fraction of boron to the carbon, manganese and chromium composition. Table 1 below shows the chemical composition of boron alloyed steel 22MnB5 that has been used in this research.

Table 1. Chemical composition of 22MnB5.

Chemical Element	Percentage (%)
Carbon, C	0.250
Silicon, Si	0.400
Manganese, Mn	1.350
Phosphorus, P	0.023
Sulphur, S	0.010
Aluminium, Al	0.080
Nitrogen, N	0.010
Chromium, Cr	0.250
Boron, B	0.004

Different types of specimens tested are notated as: Blanked specimens (specimen-B); Die-quenched specimens (specimen-HPF), Blanked specimens that are heated and cooled at Room Temperature (specimen-RT); Blanked specimens that are heated and cooled in room temperature water (specimen-WRT); and Blanked specimens that are heated and cooled in cold water (specimen-WC). A total of 9 specimens were used in this heat treatment. First, all the specimen was heated in the furnace up to 900 °C for 2 hours. Secondly, the initial temperature of the specimen was taken by using infrared thermometer (range -50 °C – 1150 °C). Then, three of the specimen was taken out and cooled at room temperature. Another three specimens was taken out and cooled in water at room temperature (25 °C) for 5 seconds and the final temperature of the specimen was taken. Then, last three specimens was taken out and cooled in cold water (5 °C) for 5 seconds and the final temperature of the specimen was taken. The temperature was taken to calculate the cooling rate for each type of quenching process in both type of water.

2.2 Experimental procedures

Vickers test was done based on the ASTM E384 to determine the hardness of each specimen. Since the material is in form of sheet, which is used for hot stamping, it is important to study the behaviour of the rolled sheet using tensile test. For the tensile testing, boron steel plate were cut into dimension as specified in ASTM 08M. All of the specimens were tested subjected to three velocities of the top platen plate, 3 mm/min, 6mm/min and 9 mm/min.

3 Results and discussion

There are two types of testings that was conducted in this research which is the Hardness Test (Vickers) and Tensile Test. Both test conducted based on the ASTM. Both tests consist of control specimen of 22MnB5 with different type of treatment that has been done.

3.1 Hardness test result

Vickers Hardness test is carried out to obtain the hardness. Pyramid diamond indenter is pressed at 7 points for an interval of 0.2 mm for each of specimen types. The graph for this test were summarized in Figure 1 below.

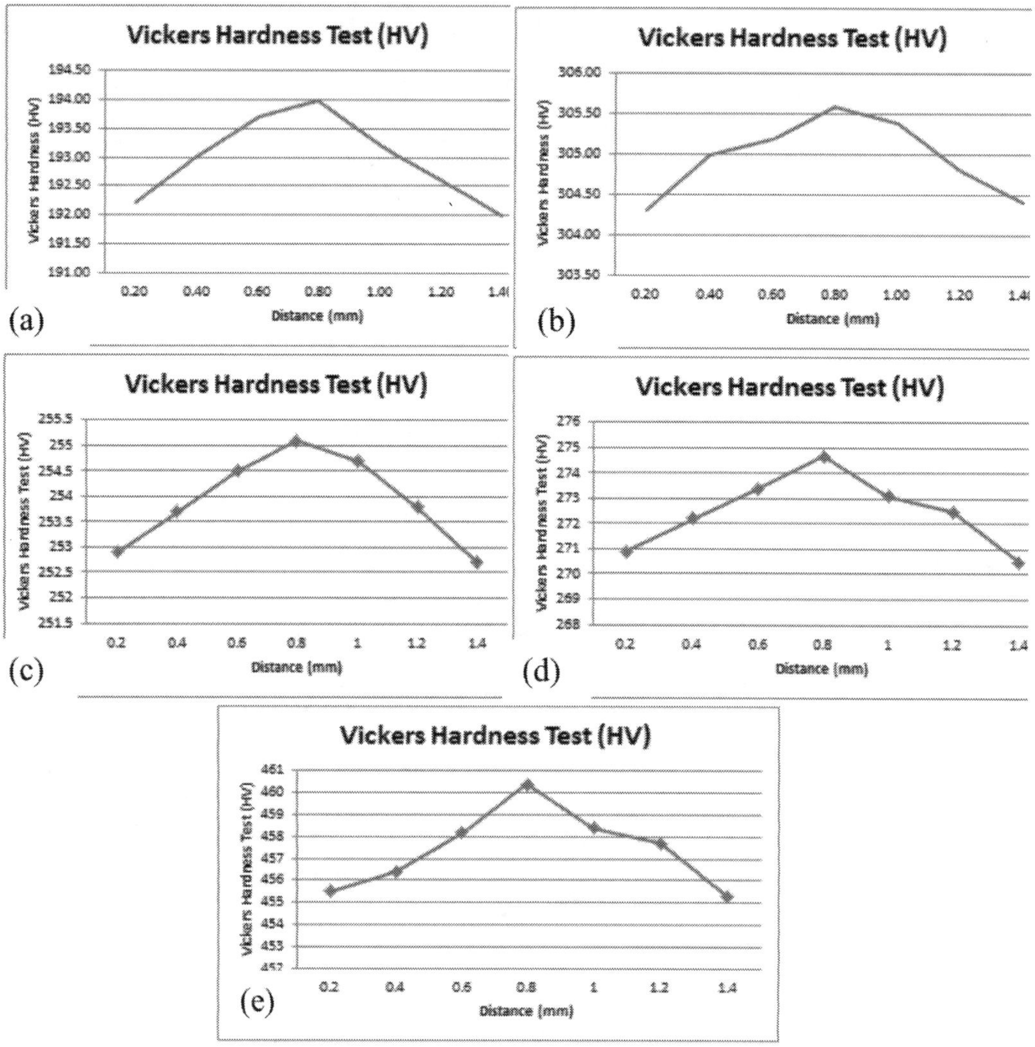

Fig. 1. Vickers Hardness test result for (a) specimen-B (b) specimen-HPF (c) specimen--RT (d) specimen-WRT and (e) specimen-WC.

Based on Fig. 1, the average Vikers Hardness for each of the specimens are: 192.96 HV (specimen-B); 304.19 HV (specimen-HPF), 253.91 (specimen-RT); 272.4HV (specimen-WRT); and 457.93HV (specimen-WC). The Vickers hardness test results for 22MnB5 shows that the highest value for hardness is for specimen-OWC that is and the lowest value is for specimen-O. Specimen-O has the lowest Vickers Hardness value because the microstructure for this steel contains a mixture of pearlite phase, ferrite and small amount of carbide. It has no martensitic in the microstructure. For other types of 22MnB5, it has martensitic in the microstructure because of the process of annealing. The hardness of the steel depends on the amount of martensitic on it and it also depends on the grain size, the smaller the grain size, the higher the value of hardness and strength. The size of the grain size depends on the cooling rate of the specimen. The faster the cooling rates, the smaller the size of the grain size. Specimen-OWC has the fastest cooling rate which is 122.6 °C/s. The fine grain of the martensitic structure of these steels creates their characteristic features of high toughness, strength.

3.2 Tensile test result

In this analysis, properties that are directly measured via a tensile test are ultimate tensile strength, and Young's modulus. This tensile test was done based on ASTM E8/E8M – 13 which is the Standard Test Methods for Tension Testing of Metallic Materials. The reading for tensile test was taken with three different velocities which is 3, 6 and 9 mm/min. The graph for all results was summarized in Table 2

Based on all the results obtained, the best steel is 22MnB5 HPF. 22MnB5 HPF has good Tensile Stress and good Young Modulus compare to other 22MnB5. For as-received 22MnB5 and 22MnB5 RT, it has good Young Modulus but poor Tensile Stress. For 22MnB5 WRT, it has good Tensile Stress but poor Young Modulus and for 22MnB5 CW it has poor Tensile Stress and Young Modulus. Poor Young Modulus will make the steel becomes less flexible and become more to brittle. Brittle materials cannot absorb the impact applied on it and it will break easier. Poor Tensile Stress means that the steels are not strong enough. Weak materials cannot hold the high Tensile Stress applied on it and it also will break easier.

Table 2. Stress-strain curves and fractured specimens.

Specimen types	Stress-strain curve	Fractured specimen
specimen-B	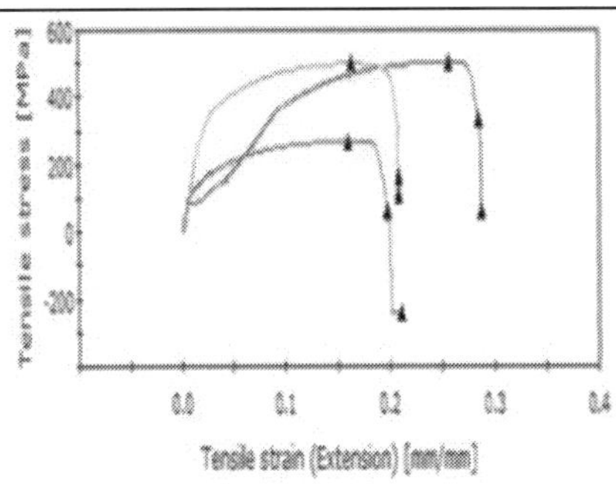	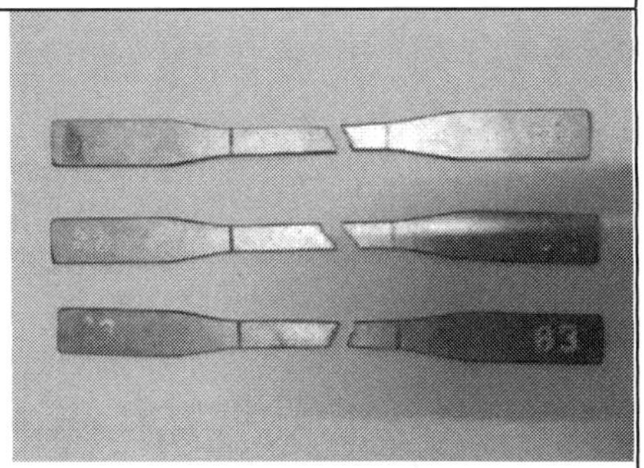

specimen-HPF	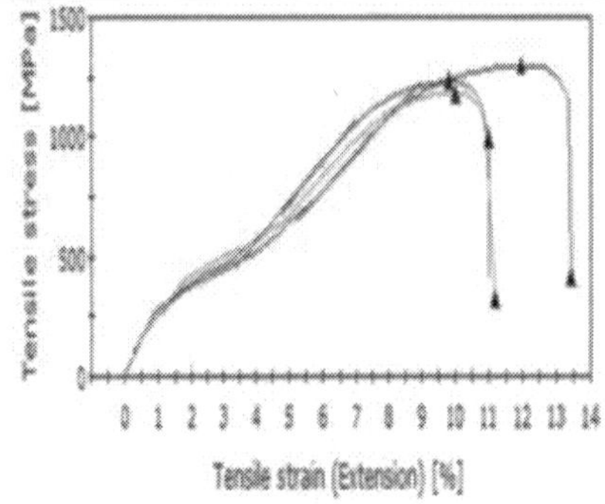	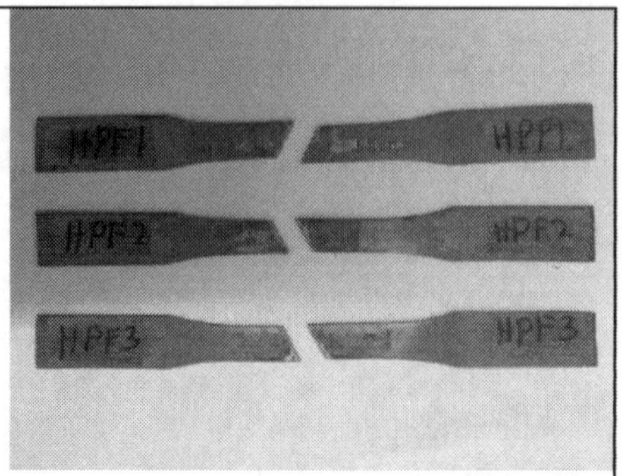
specimen-RT	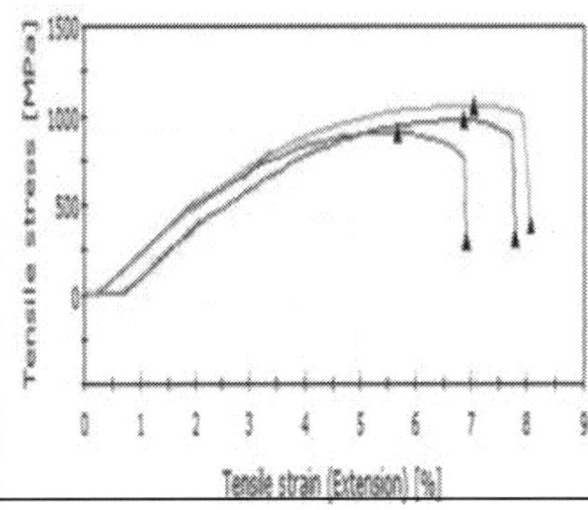	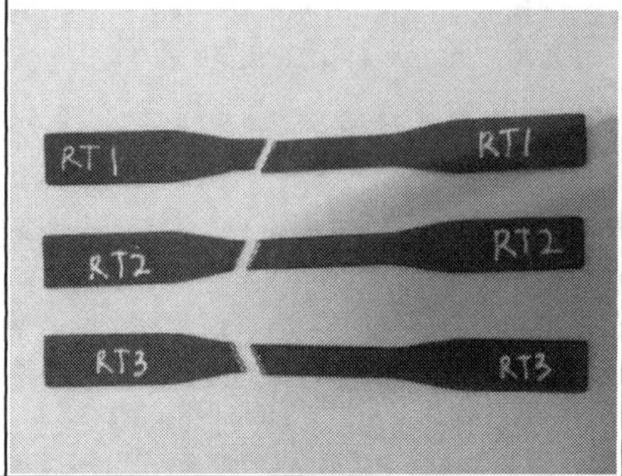
specimen-WRT	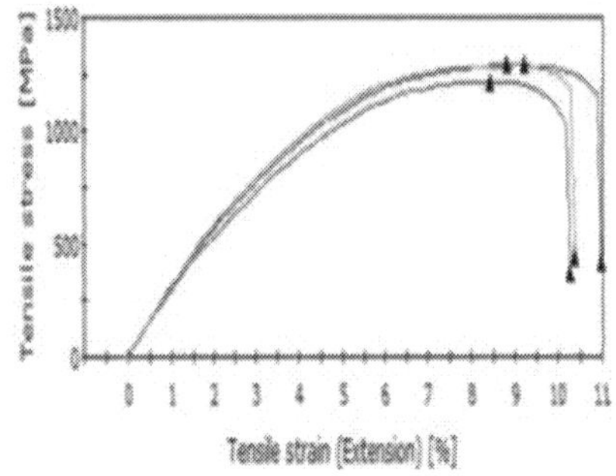	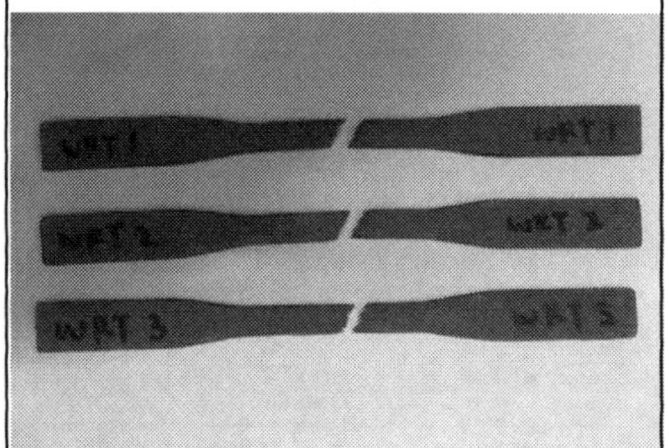
specimen-CW	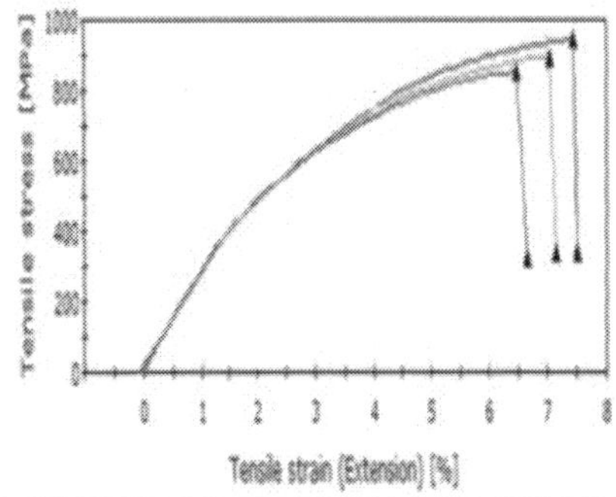	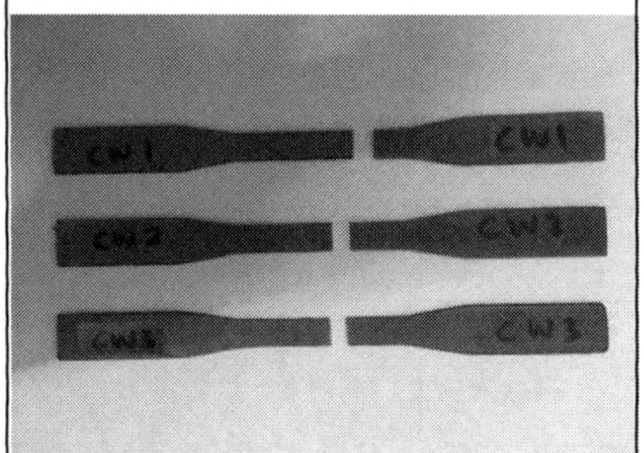

Legend: 3 mm/min 6 mm/min 3 mm/min

4 Conclusions

As conclusion, the heat treatment done to the 22MnB5 with different type of quenching process has portrayed an encouraging result, particularly in the hardness of the 22MnB5. The higher the cooling rate of the quenching, the smaller the size of the grain size. Hence, it will increase the hardness of the steel. When the cooling rate is very high, it will increase the strength of the steel but it will reduce the toughness and the ductility of the steel. Higher cooling rate tends to make the steel become less flexible and more brittle.

The authors would like to thanks Universiti Malaysia Pahang for providing laboratory facilities and financial assistance under project no. RDU1303114.

References

1. H. So, D. Fasmann, H. Hoffmann, R. Golle, M. Schaper, An investigation of the blanking process of the quenchable boron alloyed steel 22MnB5 before and after hot stamping process, Journal of Materials Processing Technology 212 (2012) 437– 449.
2. M. Naderi, V. Uthaisangsuk, U. Prahl, W. Bleck, A Numerical and Experimental Investigation into Hot Stamping of Boron Alloyed Heat Treated Steels, steel research int. 79 (2008) No. 2.
3. N. Aziz, S.N. Aqida, Optimization of quenching process in hot press forming of 22MnB5 steel for high strength properties for publication, Materials Science and Engineering 50 (2013).
4. P. Hu, D. Shi, L. Ying, G. Shen, W. Liu, The finite element analysis of ductile damage during hot stamping of 22MnB5 steel, Materials and Design 69 (2015) 141–152.

 DOI: 10.1051/matecconf/20179001015

State of art of cooling method for dry machining

Kaaliarasan Ramachandran[1,*], *Balakumaran* Yeesvaran[1], *Kumaran* Kadirgama[1,2], *Devarajan* Ramasamy[1,2], *Saiful Anwar* Che Ghani[1] and *Keeran* Anamalai[1]

[1]Faculty of Mechanical Engineering, Universiti Malaysia Pahang, 26600 Pekan, Pahang, Malaysia
[2]Automotive Engineering Centre, Universiti Malaysia Pahang, 26600 Pekan, Pahang, Malaysia

Abstract. The usage of cutting fluid in the machining operations will not only poses health risk to the workers but also creates environmental challenges associated with fluid treatment and disposal. The objective of this review paper is to identify present cooling method in dry machining and novel feature than can be included in the cooling system for a better heat removal which tend to increase machining performance. From the review study, it shows that cooling method by internal manner is most widely been researched because researcher believes that it has a promising future in adaptive machining to produce a contamination free products which supports the green manufacturing concept. The suggested novel feature is ecologically desirable and it will become as a necessity for improving the performance of dry machining in the near future.

1 Introduction

Machining can be performed under three different condition; wet machining, near dry machining and dry machining. Traditionally, in wet machining, cutting fluid is used to reduce cutting temperature, reduce cutting force, minimize formation of buildup edge and improve tool life [1, 2]. In USA alone, volume of cutting fluid that is been discharged is nearly 115 million liters per year [3] and cost to retreat the cutting fluid with additive is USD 15 per gallon [3]. Study in year 1994 at Germany, records usage of cutting fluid is 1.15 million which is responsible of 7% - 17% for cutting fluid and 2% to 4% for cutting tool from overall cost [4]. This clearly shows that the cost of cutting fluid exceeds the cost of cutting tool and overhead cost [5]. Meanwhile, cutting fluid also creates environmental issues and causes health injuries to humans [6, 7]. Exposure of cutting fluid causes skin cancer and evaporation of cutting fluid at cutting zone is dangerous to human respiratory system [1, 6]. Cutting fluid cannot be used for long term as it is tend to wear out after mixing with the chips [4]. Wet machining decreases the cutting fluid efficiency because of two factors; limitation of cutting fluid to reach cutting zone and chip movement during machining [4, 8]. Near dry machining has only small effectiveness in cooling cutting tool and ineffective to machine difficult materials [4].

In these cases, dry machining is much more preferable as it saves cost, environmentally friendly and does not harm the human health [5]. The main limitation in dry machining is the heat generation at cutting tool specifically; cutting zone. During machining, friction and

[*] Corresponding author: kaaliarasan@gmail.com

chip deformation at cutting interface causes thermal load which causes high temperature at cutting tool [1, 2]. High temperature is because of the high friction and adhesion formation at cutting tool [1]. Heat generated during machining causes two changes in the cutting tool; reduces mechanical and wear resistance [4]. It is proven that the deformation and friction converts 99% input energy to heat [9]. This leads to the tool wear subsequently which reduces the tool's life span [2]. Formation of tool wear causes the workpiece subjected to poor surface roughness and less dimensional accuracy [1, 2]. There are two methods to control the heat generated during machining; limitation of cutting parameter and usage of cutting fluid [4]. Every combination of cutting tool and workpiece has its own optimum cutting parameter to produce good surface finish product and prolongs the tool life. But in the first method the work productivity is affected because of the machining limitation [4]. In the second method, cutting fluid is used to reduce the cutting zone temperature. Meanwhile, friction coefficient in dry machining can be reduced by two methods; coating layer at cutting tool and material selection of cutting tool [10]. By reducing temperature at cutting tool during dry machining operating, performance of the machining can be increased. Dry machining will only be acceptable if the machining performance results exceeds or equals the performance of the wet machining [4].

Researchers are finding alternate solution to eliminate the usage of cutting fluid and encourage dry machining method which supports the green manufacturing concept [2, 4]. Some of the cooling method for dry machining are internally cooled, indirect cooling, solid lubricant and heat pipe technology [2, 4, 6, 10-12]. For time being, internal cooling system is the most promising cooling system to improve dry machining's performance [1, 13]. Indirect cooling which uses cryogen is effective but very costly [6]. Heat removal by using heat pipe is more attractive as no external flow source and external power supply is used in the cooling system [12]. This paper is about the cooling method and introduces novel features to improve existing dry machining performance.

2 Cooling method in dry machining to improve machining performance

Cooling a cutting tool by internal manner is the most effective solution for dry machining [2]. L. E. A. Sanchez [4] come out with an idea to make a chamber in the tool holder in which cutting fluid will be channeled through the inlet and exits through the outlet. Cutting tool will be placed on top of the cavity whereby the heat generated during machining will be carried away by the cutting fluid which flows inside the chamber of tool holder [4]. The heated cutting fluid is then carried away to condenser. Copper board which has a high value of thermal conductivity is placed in between to increase the heat transfer rate. Copper board addition will promote for a better heat transfer rate since it has large surface area. The main challenge in this design is to make the internal chamber inside the tool holder. The schematic diagram of the internally cooled tool for turning as shown in Figure 1.

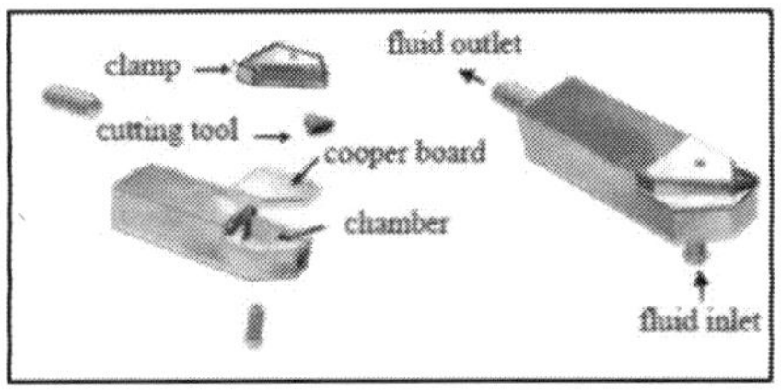

Fig. 1. Assembly of internally cooled tool for turning [4].

Sun [1] approach by inverting own cooling method by combination of "bottle-cap insert" and chambered adaptor. Cutting tool is designed and machined to have a "bottle-cap insert" look as shown in Figure 2(a). Adaptor is designed with chamber and two micro holes at the corner as shown in Figure 2(b). As the cutting fluid is channeled through the micro hole, the chambered section in the adaptor forms a cooling tube with a triangle section. By channeling the cutting fluid at the tip of the cutting tool, the heat transfer rate increases and improves the tool life. This type of cooling system is most suitable for customized square shape cutting tool. Sun [1] proves that the internal cooling actually helps to decrease the cutting temperature which improves the tool life. The diameter of the internal structure is the most important factor when designing an internal cooling system [1, 13].

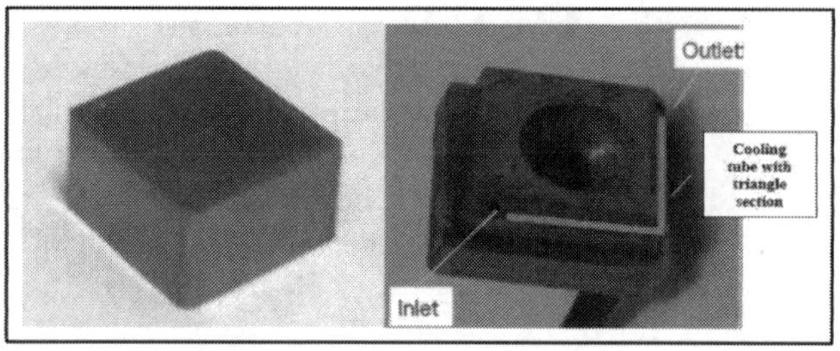

(a) (b)

Fig. 2(a). Final look of "Bottle-cap insert" [1].

Fig. 2(b). Chambered adaptor with micro holes which represent inlet and outlet of cutting fluid [1].

Y Isik [14] finds that heat transfer can be increased by placing the internal cooling system closer to the cutting tool [14]. Internally cooled tool (ICTool) is a reliable achievement in the internally cooling system. Pure cooling water is used as the cutting fluid because it has high heat capacity value and does not pollute the environment [13, 14]. In this design, as shown in Figure 3, 2mm holes of cooling channel is made at the pocket of tool holder. Cutting tool will be placed at the top of the tool pocket to directly absorb the heat transfer during machining [14]. Cutting fluid will be supplied to the cooling channel through the inlet and outlet of the pipe situated at the sides of the cutting tool. Additives are included in the cooling pure water to prevent corrosion on the tool holder [13, 14].

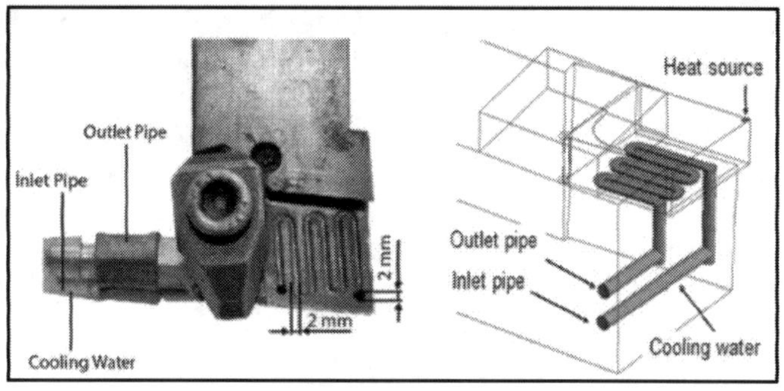

Fig. 3. Parts and exploded view of Internally cooled tool (ICTool) [14].

Another approach of the design is by using cryogen as an indirect cooling method of cutting tool. Thermal conductor plate is placed between the cutting tool and tool holder groove to increase the heat removal. The indirect cryogen is supplied to the cutting tool by having a modified micro cooling channel at the sides of the cutting tool [7]. Fins can be added inside of the grooves, to increase the heat transfer efficiency. The required flow rate of cryogen to cool down the cutting tool during the machining is less than 1% compared to the amount of the standard cutting fluid required to cool the same insert during the same machining operation [2, 7]. Figure 4 shows the schematic diagram of indirect cooling system in the cutting tool.

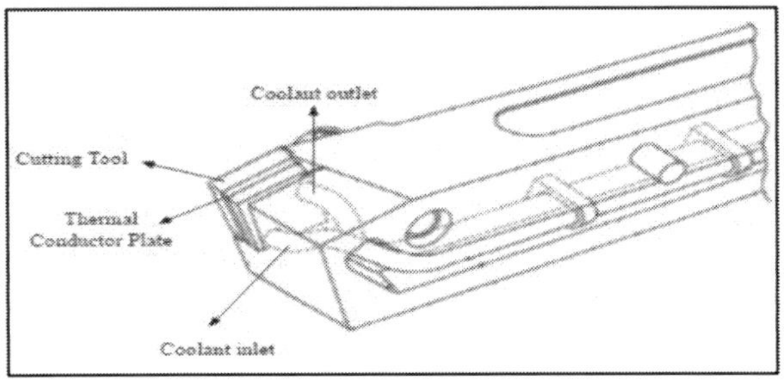

Fig. 4. Parts and exploded view of indirect cooling system in cutting tool [7].

Liang [12] describes that the usage of heat pipe as a cooling system to reduce the thermal elongation of tool holder which reduces the tool wear. Heat pipe is attached to the tip of the cutting tool. A modified cutting tool with a hollow structure is made to hold the heat pipe at the sides. Placement of the heat pipe absorbs the heat generated during machining. This cooling does not require any external flow medium or power source. [12] Heat transfer efficiency of the cooling system can be increased by considering the lengths, diameters, vacuums and materials of the heat pipe [12]. Figure 5 shows the placement of the heat pipe at the cutting tool.

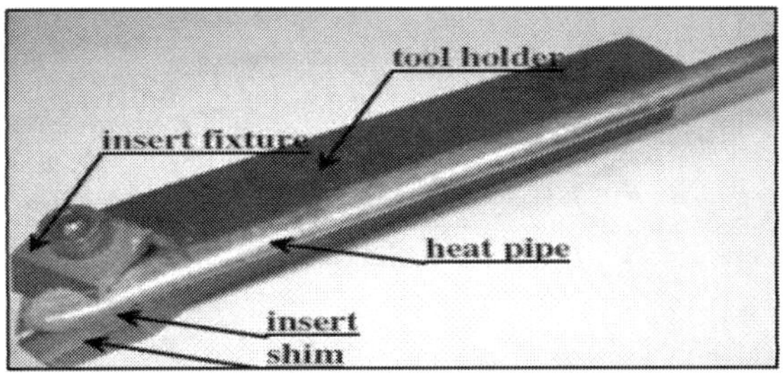

Fig. 5. Placement of heat pipe on cutting tool [12].

Jianxin [15] suggests the use of a solid lubricant to encounter the friction coefficient problem during machining. Micro hole is made at the cutting tool to place the solid lubricant onto it. Micro hole can be made at two locations; rake face or flank face. Solid lubricants at the rake face results in a lower flank wear than solid lubricant at flank face [15]. Studies show that solid lubricant placed at the rake face records the lowest friction coefficient and least tool wear value [10, 15]. Figure 6 shows the self-lubricated tool system in the dry machining.

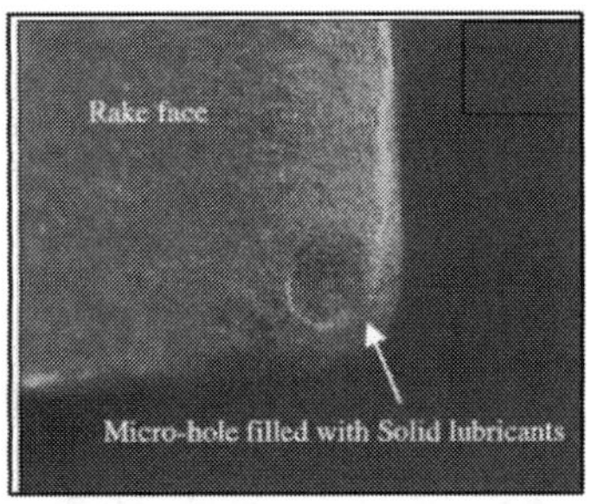

Fig. 6. Location of micro hole in self-lubricated tool system [15].

 DOI: 10.1051/matecconf/20179001015

Solid lubricant, Calcium Difluoride (CaF_2) will be included in fabricating the ceramic cutting tool by hot pressing process [10]. About 10% of solid lubricant is included into the ceramic matrix [10]. Friction coefficient at tool-chip interface can be reduced as the solid lubricant produces self-lubricating film on it [10]. The formation of the self-lubricating film as shown in Figure 7 smears on the cutting tool reduces friction coefficient, tool wear and increases life span of tool [10]. This can be achieved as easy shearing along basal plane of the hexagonal crystalline structure at the ceramic matrix [10].

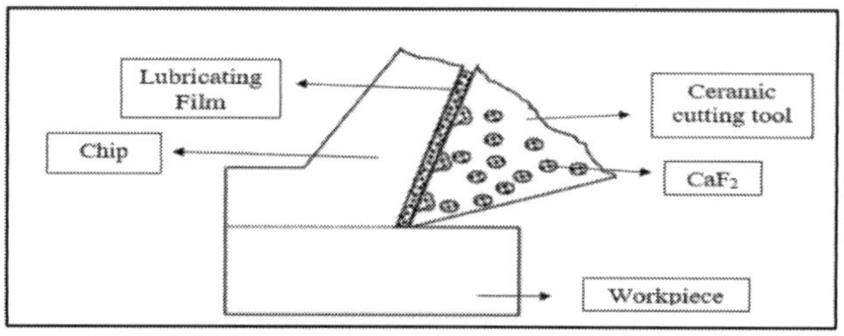

Fig. 7. Location formation of lubricating film on tool-chip interface [10].

3 Comparison of cooling methods for dry machining

Tables 1 and 2 show a complete comparison of cooling system for dry machining. Generally, the main drawback of internally cooled cooling method is the complex internal cooling design which is difficult to fabricate but it is very efficient in heat absorption during machining. Meanwhile, indirect cooling is the most efficient in heat removal but it requires high setup cost. Cooling by solid lubricant is effective but the strength of the cutting tool need to sacrificed. Finally, heat pipe technology is an innovative cooling method without the presence of power supply but it has low heat transfer effectiveness.

Table 1. Complete comparison of cooling system for dry machining (internally cooled).

Cooling Method	Design	Cooling Medium	Pro	Con
Internally cooled	Internally cooled tool for turning	R-123 cutting fluid [4]	• Simple and low cost [4] • Efficient heat exchanger [4] • Improvement in the heat removal [4] • Environmental and economic advantages [4]	• Complex design • Difficult to machine • HCFC leakage pollutes environment [4]
	Internally cooled "bottle-cap" smart cutting tool	Pure water [1]	• Low cost and easy to use [1] • Substantial cost reduction [1] • Friendly environmental machining [1]	• Custom made cutting tool • Time consuming to produce cutting tool
	Internally Cooled Tool (ICTool)	Cooled pure water [14]	• Green cooling method [14] • Efficient in heat removal [14] • Better surface quality and extended tool life [14]	• Complex cooling system • Difficult to machine cooling channel

Table 2. Comparison of dry machining cooling system for other methods.

Cooling Method	Design	Cooling Medium	Pro	Con
In direct cooling	Indirect cooling of cutting tool	Cryogen [7]	• High heat removal capacity [4] • High heat transfer effectiveness [7] • Decreases the usage of the cutting fluid [2]	• Complex micro cooling channel design • High setup cost [4] • Cryogen is expensive and not reusable [16]
Solid lubricant	Self-lubricated tool in dry machining	Molybdenm disulfide (MoS_2) [15]	• Reduces friction coefficient [10, 15, 17] • Controls heat generation [17] • Improves flank wear [15]	• Reduces physical property and strength of cutting tool
	Self-made cutting tool with solid lubricants	Calcium Difluoride (CaF_2) [10]	• Improves tribology [10, 18-20] • Usable in high temperature applications [10]	• Study on the solid lubricant before fabricating the cutting tool.
Heat pipe technology	Dry turning assisted by heat pipe cooling	Heat pipe: Height: 4mm Length: 150mm Thickness: 0.65mm [12]	• No external power supply [12] • Reduces the thermal elongation of tool holder [12] • Zero usage of cooling medium [12]	• Low heat transfer effectiveness [7] • Groove need to be made at the cutting tool, which will reduce the mechanical strength.

4 Future of cooling system to improve performance of dry machining

In the near future, the cooling system for the dry machining can be designed with more complex structure with the developed technologies like the 3D printers and Selective Laser Melting (SLM). Conversely, new type of cutting fluid with higher heat capacity can be used to accelerate the heat transfer rate. It will be a positive impact to the environment and green technology if biodegradable cutting fluid can be used for the cooling purposes. Besides, it is very possible with the development of Nano cellulose. Absence of cutting fluid in dry machining causes the chips not to be washed away. An integrated air system can be installed together with the cooling system at the tool holder to clear away the chips. This is feasible as most of the laboratories are equipped with air compression system in the recent era. Temperature and force can be monitored during machining with the innovation of the smart cutting tool. Smart cutting tool which is integrated with sensors can be used to predict the changes in the cutting zone and determine the life span of the cutting tool.

5 Conclusions

Dry machining with a proper cooling method can be used to achieve better machining performances than the wet machining by maximizing the heat removal rate, focusing heat removal at the tip of the cutting tool. From the review, internal cooling method is most widely researched compared to other cooling methods. This is because of the easiness and record promising results of the internal cooling method result compared to the other methods. Researches and developments of cooling method need to be widen up to improve the dry machining in term of the surface roughness, tool wear and life span of the cutting tool. Development of the cooling method is aided by the modern technologies and innovations all over the world.

The authors would like to be obliged to Universiti Malaysia Pahang for providing laboratory facilities and financial assistance under project no. RDU1403134.

References

1. X. Sun, R. Bateman, K. Cheng, S. Ghani, Design and analysis of an internally cooled smart cutting tool for dry cutting. *Proceedings of the Institution of Mechanical Engineers, Part B: Journal of Engineering Manufacture* **226**, 585-591 (2012).
2. A. Kromanis, G. Pikurs, , G. Muiznieks, K. Kravalis, V. Gutakovskis, Design of internally cooled tools for dry cutting. in *9th International DAAAM Baltic Conference: INDUSTRIAL ENGINEERING* 24-26 (2014).
3. S.Y. Hong,, M. Broomer, Economical and ecological cryogenic machining of AISI 304 austenitic stainless steel. *Clean Products and Processes* **2**, 157-166 (2000).
4. L.E. Sanchez, V.L. Scalon, G.G. Abreu, Cleaner machining through a toolholder with internal cooling. in *3rd International Workshop Advances in cleaner production. Brazil* (2011).
5. P. Sreejith, B. Ngoi, Dry machining: machining of the future. *Journal of materials processing technology* **101**, 287-291 (2000).
6. N. Dhar, S. Paul, A. Chattopadhyay, The influence of cryogenic cooling on tool wear, dimensional accuracy and surface finish in turning AISI 1040 and E4340C steels. *Wear* **249**, 932-942 (2001).
7. J.C. Rozzi, W. Chen, E.E. Archibald, Indirect cooling of a cutting tool. (Google Patents, 2011).
8. K. Seah, X. Li, K. Lee, The effect of applying coolant on tool wear in metal machining. *Journal of materials processing technology* **48**, 495-501 (1995).
9. A. Balaji, G. Sreeram, I. Jawahir, E. Lenz, The effects of cutting tool thermal conductivity on tool-chip contact length and cyclic chip formation in machining with grooved tools. *CIRP Annals-Manufacturing Technology* **48**, 33-38 (1999).
10. D. Jianxin, C. Tongkun, Y. Xuefeng, L. Jianhua, Self-lubrication of sintered ceramic tools with CaF 2 additions in dry cutting. *International Journal of Machine Tools and Manufacture* **46**, 957-963 (2006).
11. A. Kromanis et al. Design of Internally Cooled Tools for Turning. *Rigas Tehniskas Universitates Zinatniskie Raksti* **35**, 93 (2013).
12. L. Liang, Y. Quan, Z. Ke, Investigation of tool-chip interface temperature in dry turning assisted by heat pipe cooling. *The International Journal of Advanced Manufacturing Technology* **54**, 35-43 (2011).
13. B.C. Ghani, S. Anwar, Brunel University School of Engineering and Design PhD Theses (2013).
14. Y. Isik, Using internally cooled cutting tools in the machining of difficult-to-cut materials based on Waspaloy. *Advances in Mechanical Engineering* **8**, 1687814016647888 (2016).

15. D. Jianxin, S. Wenlong, Z. Hui, Design, fabrication and properties of a self-lubricated tool in dry cutting. *International Journal of Machine Tools and Manufacture* **49**, 66-72 (2009).
16. T. Stefánsson, Application of Cryogenic Coolants in Machining Processes: State-of-the-art Literature Study and Experimental Work on Metal Matrix Composite. (2014).
17. U.M.R. Paturi, Y.R. Maddu, R.R. Maruri, S.K.R. Narala, Measurement and Analysis of Surface Roughness in WS 2 Solid Lubricant Assisted Minimum Quantity Lubrication (MQL) Turning of Inconel 718. *Procedia CIRP* **40**, 138-143 (2016).
18. R. Westergård, A. Åhlin, N. Axen, S. Hogmark, Sliding wear and friction of Si3N4-SiC-based ceramic composites containing hexagonal boron nitride. *Proceedings of the Institution of Mechanical Engineers, Part J: Journal of Engineering Tribology* **212**, 381-387 (1998).
19. Y. Wang, F. Worzala, A. Lefkow, Friction and wear properties of partially stabilized zirconia with solid lubricant. *Wear* **167**, 23-31 (1993).
20. A. Gangopadhyay, S. Jahanmir, Friction and Wear Characteristics of Silicon Nitride-Graphite and Alumina-Graphite Composites©. *Tribology transactions* **34**, 257-265 (1991).

Techno-economic and sensitivity analysis for grid-connected renewable energy electric boat charging station in Terengganu

N. A. S. Salleh[1,*] *and W. M. W.* Muda[1, 2]

[1]School of Ocean Engineering, University Malaysia Terengganu
[2]Special Interest Group of Eastern Corridor Renewable Energy

Abstract. In order to encourage the eco-friendly technologies in transportation sector, the reliance on fuel need to be reduced and the use of renewable energy (RE) technology as energy source are widely explored by researchers. Thus, this study focus on the feasibility of developing grid-connected renewable energy electric boat charging station for the fishermen in Terengganu using simulation-based method by HOMER software. Five year solar radiation and wind speed data were collected at Universiti Sultan Zainal Abidin (UNISZA) weather station. For load profile, the information about fishing activities and the amount of subsidy spent by the government were obtained from the interview session with the fishermen and validated with Lembaga Kemajuan Ikan Malaysia (LKIM). The results acquired are compared between grid-only and grid-connected RE systems in term of net present cost (NPC), operational cost and payback period. A sensitivity analysis is done to find the minimal Feed-in Tariff (FiT) rate that can be implemented in order to encourage the use of RE system in this sector. Then, the relationship between FiT and NPC, payback period and emission of pollutants are analyzed. At current FiT rates RM 0.813/kWh, hybrid grid-PV system manages to achieve its optimal in generating high income from selling the power to the grid with convincing amount of electricity production and short payback period. It is concluded at minimum RM 0.56/kWh of FiT, the grid-connected RE system is possible to be developed because its performance shows better outcome compared to the grid-only system.

1 Introduction

The heavy dependency on fuels by conventional boat (CB) to generate the motor normally affecting the water ecosystem [1]. The engine cooling water and waste water contains oils and fuel residue which is pumped out into the sea causing negative impacts to underwater organism. Besides the harmful waste, CB also emit exhaust gases and noise pollution from the engine [2, 3]. Similar to other electric vehicle, electric boat (EB) can be

* Corresponding author: gsk2357@pps.umt.edu.my

used to replace the use of CB in reducing the abovementioned problems. Inspired by the rapid growth of electric vehicles nowadays, EB is widely used in a few countries such as Australia and Canada [4]. According to Minami [5], the use of electric motor gives a lots of advantages in fishing activities such as quietness, less vibration and lack of toxic during fishing. The reliance on fuel is replaced with battery which is needed to be charged by directly connected to the grid. However, if the charging demand is not well-controlled, it can damage the grid [6]. Therefore, the encouragement to use renewable energy (RE) in operating grid-connected charging station need to be proposed to control the dependency on grid [7].

As Malaysia is a tropical climate country, the potential to exploit RE technologies is high. Malaysia shows high solar power generation with approximately 4000-5000 Wh/m^2 of daily average solar radiation [8, 9]. Terengganu has been identified with the capability to generate wind power electricity [10]. With possible amount of solar irradiance and wind speed, Terengganu is one of high potential states in developing RE technology. Unfortunately, statistics from Sustainable Energy Development Authority (SEDA) shows that Terengganu is still far behind compared to western region in implementing this technology [11].

Kuala Terengganu faces the South China Sea which makes fisheries sector is one of important source of economic development. Results from interview session with LKIM Chendering officer, a total amount RM 1,228,325.70 of fuel subsidy was spent by the government in 2014 for 242 fishermen in Kuala Terengganu Utara area [12]. Other than that, fishermen also accepting numbers of incentive such as "e-Dana", "e-kasih", and monthly allowance to accommodate the needs on fuel. But with high cost of living and uncertain oil rates nowadays, the distribution of subsidies for the fisherman is still not enough. Mentioned by Hayati in [13], most of the fishermen are poor. Thus, solutions are needed to help them out with these struggles.

Over the years, few energy policies and programs have been introduced by the government to encourage the use of RE power application in industries, community and individuals sector. The government began to drawing up initiatives in RE field since 1996 in the 7th Malaysia Plan (7MP). Since then, the next Malaysia Plan covering the areas of RE up to now in the 10th Malaysia Plan (10MP) [14]. New targets and goals constantly introduced to improve the quality control and energy management in Malaysia. Regarding to this effort, government introduced Feed-in Tariff (FiT) scheme which the implementation is managed by SEDA in 10MP where the RE system installation is accessed to the TNB's grid and setting a favorable price per unit of RE sold [11, 15, 16].

Researchers, for example in [9, 17, 18] are already striving to minimize the gap in terms of power system technology development in Terengganu from the rural to the urban areas. The progress from literatures can be seen that Terengganu has its own potential to be successful in generating power from renewable energy. However, statistics of the development for renewable energy technology in Peninsular Malaysia are quite slow [11]. Therefore, due to big investment cost involved to develop RE technology [19], economic analyses on the proposed system need to be highlighted.

This paper also is an extension of the proceeding "Feasibility Study of Optimization and Economic Analysis for Grid-connected Renewable Energy Electric Boat Charging Station in Kuala Terengganu" [20]. In [20], FiT rate RM 1.49/kWh obtained from [21] was used to conduct the optimization and simulation in HOMER to analyze the cost-benefit of the system. However in this paper, the current FiT rate (for January 2015) which is RM0.813/kWh (include the bonus) is used. Since the FiT rates keep decreasing every year, the sensitivity analysis is also done to find the minimal FiT rates for the proposed system in gaining profitable revenue in RE technologies.

2 Hybrid system configurations

Renewable energy components i.e. PV modules and wind turbine are combined with non-renewable components which is the grid in HOMER to produce a hybrid renewable energy system. Fig 1 shows the configuration of grid-connected RE charging station in HOMER. The load represented in this system is battery for EB which is needed to be charged by the hybrid renewable energy charging station. Since the proposed system is a grid-connected system, when electrical production is insufficient, energy are purchased from the grid. But, if there is excess electricity, it is sold to the grid by following FiT scheme which is fixed by SEDA. The HOMER software is used in this study to determine the optimal configuration based on the optimization, simulation and sensitivity analysis.

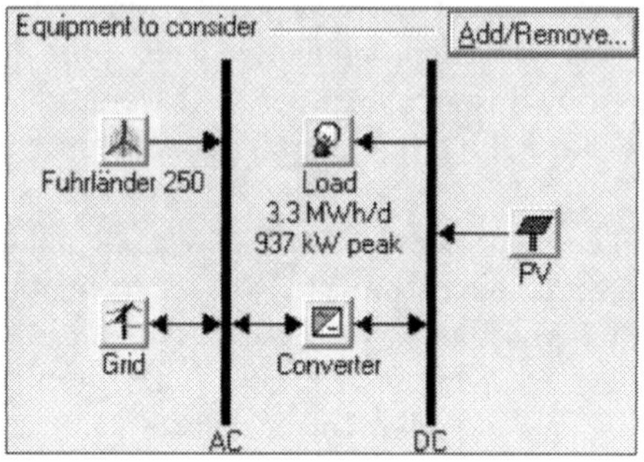

Fig. 1. Configuration of grid-connected hybrid RE system in HOMER.

Kuala Terengganu Utara (KTU) is selected as a case study area with latitude and longitude are 5.31 and 103.12 respectively [9]. The cost and specification for each component used in the configuration such as the PV module, wind turbine, converter and grid are needed to calculate the feasible cost of system configuration. Some input parameters are also required such as solar radiation inputs (kWh/m^2/day), wind speed (m/s) and load profile (kW) as daily usage of power by the load.

2.1 Profile load/ load demand

Similar to [20], the information on fisheries activities of 16 out of 242 fishermen were obtained by doing interview session. Then, the data was validated with LKIM Chendering [12] by conducting a short dialogue session. In addition, LKIM is providing information on the annual amount of fuel subsidy and government incentive for 2014. Summarization of data can be seen in Table 1 of [20].

Approximately, 45kW of electric motor is needed for EB to replace 60hp engine power from conventional boat. Inspired by [5], a 20 kWh Li-ion battery is chosen to function as electrical supplier for the electric motor. The operating hours to charge the EB is from 2:00 p.m. to 12:00 a.m. To make a convenience charging station for all the fishermen in KTU, 25 units of charging station are needed. Assuming 25 units of charging station are operating simultaneously in charging 20kWh loads, thus, the charging demand of the load in a day for the system is 500kW for 10 hours after the working hours. Fig 2 shows the load profile of the proposed system.

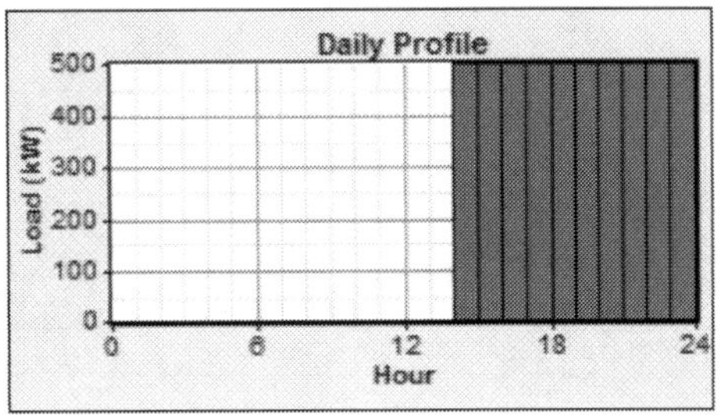

Fig. 2. Load profile of March – October.

The load profile starts to be at peak after 2:00 p.m. as the fisherman are home from their fisheries activities to recharge their battery boat. But, during the Northeast Monsoon from November to February, the load profile becomes null because the fishermen not be able to go out for fishing. There is no charging activities occur for four months. This explains the average electricity used by the load is 3.3 MWh/day, as calculated by HOMER.

2.2 Renewable energy resources

Solar radiation data was collected from UNISZA weather station at Kuala Terengganu [22]. Fig 3 illustrates a plotted annual solar radiation and the clearness index of case study area calculated by HOMER with the following daily radiation data entered. The range of solar radiation is from 2.493 kWh/m^2/day to 6.447 kWh/m^2/day and the average of solar radiation is 4.472 kWh/m^2/day. According to Fig 3, the solar irradiance from March to October is high; meanwhile it is low from November to February due to Northeast Monsoon season.

Fig 4 shows the average monthly wind speed data measured at 25m above the sea level. The average wind speed ranges from 2.174 m/s to 4.175 m/s with an annual average of 2.605 m/s. the highest wind shown to occur in December – February, because of the Monsoon season. The wind speeds over the year; a Weibull parameter (k) is assumed 2.00 and represented in Fig 5 with autocorrelation factor of 0.85. It is measured based on hour-to-hour randomness of wind speed [23].

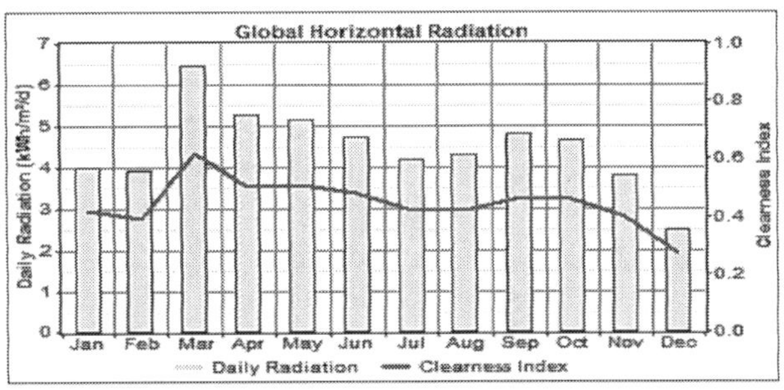

Fig. 3. The annual solar radiation and clearness index of Kuala Terengganu.

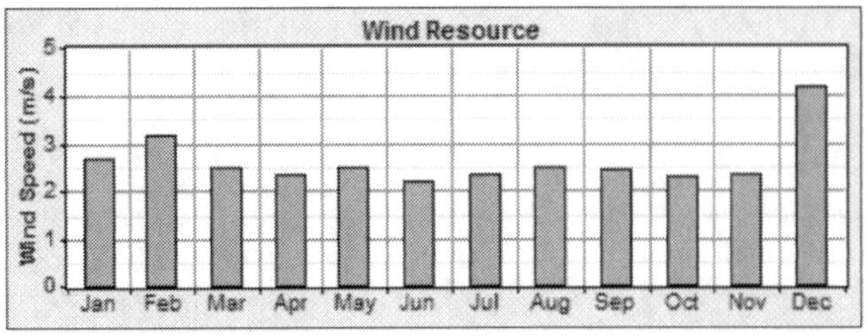

Fig. 4. Monthly average wind speed data.

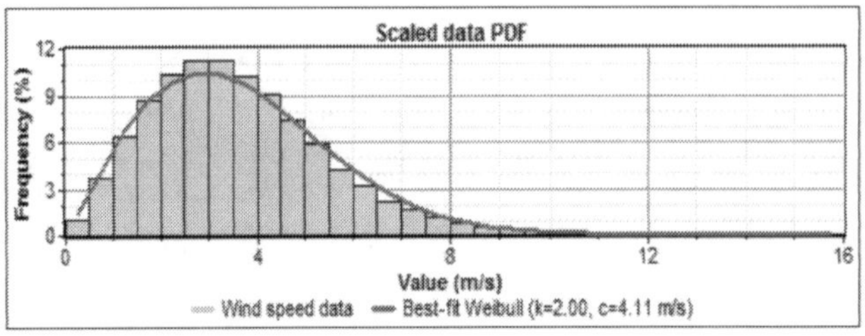

Fig. 5. Weibull probability function of wind speed.

2.3 Hybrid system component assessments

In a power system, a component generates, delivers and converts energy to feed the needs of the system. Thus, in this system, PV module, wind turbine and converter are selected to fill the requirement of charging station. Compared to [20], battery was not included in the configuration because it contributes to high NPC. Moreover, the system done well without battery storage because any excess electricity generated by RE sources can be sold directly to the grid which is more cost-effective than store it in the battery. Specifications of the components can be referred to [20].

2.3.1 Grid utility and feed-in tariff (FiT)

This project uses the Tariff C1; Medium Voltage General Commercial Tariff [24] with the commercial rate and maximum demand are RM 0.365/kWh and RM 30.30/kW/month, respectively. By referring [11] for January 2015, the FiT rate for solar PV with a capacity 1MW to 10MW is RM 0.547. For the bonus FiT rates, the proposed system eligible to receive RM 0.166 as it is used as building materials. The PV modules are installed at the rooftop of the boat dock. In addition, PV module and converter was assembled locally for grid compatibility in Malaysia. Thus, additional RM 0.050 included for each bonus criteria. Therefore, the summation of basic rates and bonus rate in 2015 made the FiT rate for grid-connected solar PV is set at RM 0.813/kWh. In addition, the purchase capacity from the grid is set up to 5000 kW and sale capacity from the grid is set to 1000 kW only. This is due to the fact that the amount of electricity sold to grid should not exceed the maximum capacity used by load to prevent the effect of reverse power flow [25].

3 System analyses

A few analyses on grid-connected RE hybrid system in term of economic feasibility are carried out based on net present cost (NPC), operating cost and payback period from optimization results obtained in HOMER. Comparison between investment cost of developing the charging station and fuel subsidy given by the government are also highlighted in this study.

3.1 Net present cost (NPC)

The NPC assesses all the cost throughout 25 years system lifetime including the initial costs, replacement costs, operation and maintenance (O&M) and grid rate costs. Total NPC is calculated using the Eq. (1) to Eq. (3) from [21]. Department of Statistics Malaysia in [26, 27] reported the inflation rate of 2015 in Malaysia is 0.90% and the nominal interest rate is 3.25%. Therefore, the annual real interest rate is 2.33%.

$$NPC = \frac{C_{ann,tot}}{CRF(i,N)} \tag{1}$$

$$CRF(i,N) = \frac{i(1+i)^N}{(1+i)^N - 1} \tag{2}$$

$$i = \frac{i' - f}{1+f} \tag{3}$$

3.2 Operating cost

The annual O&M, annualized replacement cost minus the annualized salvage value and net grid purchases are included in total operating cost of the system. The salvage value is the cost of remaining in a component at the end of project lifetime. Meanwhile, the net grid purchases are obtained from grid purchases minus grid sales.

3.3 Payback period

The Payback Period indicates the length of time for a project to generate enough cash flow to pay back the initial investment. Payback period is the ratio of the extra first cost, ΔP, to the annual savings, S, as in equation (4) [28].

$$Payback\ period = \frac{Extra\ first\ cost, \Delta P\ (\$)}{Annual\ saving, S\ (\$/yr)} \tag{4}$$

4 Results and discussion

All possible configurations of components are obtained from optimization result in HOMER. The orders of configurations are ranked based on the NPC from the most to the least cost-effective. Fig 6 is a list of optimization results for hybrid renewable energy charging station for 25 years project lifetime with FiT rates RM 0.813/kWh.

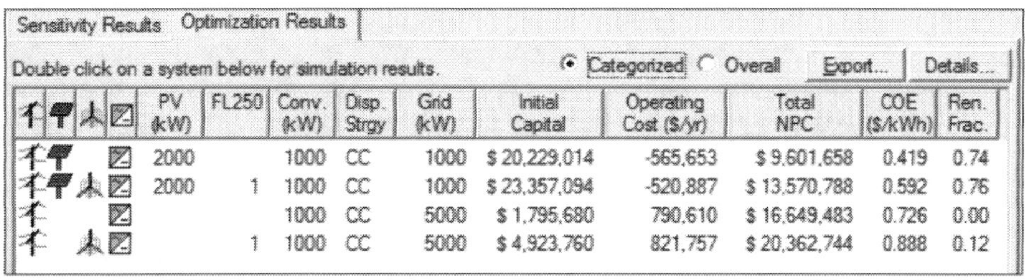

	PV (kW)	FL250	Conv. (kW)	Disp. Strgy	Grid (kW)	Initial Capital	Operating Cost ($/yr)	Total NPC	COE ($/kWh)	Ren. Frac.
	2000		1000	CC	1000	$ 20,229,014	-565,653	$ 9,601,658	0.419	0.74
	2000	1	1000	CC	1000	$ 23,357,094	-520,887	$ 13,570,788	0.592	0.76
			1000	CC	5000	$ 1,795,680	790,610	$ 16,649,483	0.726	0.00
		1	1000	CC	5000	$ 4,923,760	821,757	$ 20,362,744	0.888	0.12

Fig. 6. Hybrid system optimization result.

The result shows that the grid-PV system is the most feasible grid hybrid RE system to be developed with lowest NPC at RM 9,601,658 with cost of electricity (COE) is RM 0.419/kWh. The 0.74 RE fraction is produced from 2000 kW of PV with combination of 1000kW converter and 1000kW grid purchased. Meanwhile, grid-wind-PV system also feasible to be developed with addition of one unit wind turbine from grid-PV configuration. The NPC and COE for this system is RM 13,570,788 and RM 0.592/kWh, respectively. It is noticeable that both grid-PV and grid-wind-PV system producing negative operating cost. The negative sign indicates both systems earn more income by selling excess electricity to the grid than purchasing electricity from the grid.

The existing system which is grid-only system becomes the third based on the NPC value at RM 16,649,483 with COE is RM 0.726/kWh. This system might not invest much in capital cost since the system only needs to be connected to grid utility system with 1000kW converters. The least optimal configuration is grid-wind system with a very high NPC which is RM 20,362,744 and the COE RM 0.888/kWh. Since purchasing electricity from the grid is more cost-effective than using the wind turbine, the system chooses only one unit wind turbine and 5000 kW from the grid although the sizing range of wind turbine is set up to 10 units. The summary of all system performance at FiT rates RM 0.813/kWh are shown in Table 1.

Table 1. The performance of grid-only and grid-connected RE systems at FiT rates RM 0.813/kWh.

Performance	Systems			
	Grid-only system	**Grid-PV system**	**Grid-wind-PV system**	**Grid-wind system**
Grid purchase (kWh/yr)	1,436,381	918,103	890,243	1,393,448
Electrical production (kWh/yr)	-	2,609,669	2,796,430	186,761
Grid sales (kWh/yr)	-	1,880,463	2,016,457	143,829
Total income (RM/yr)	-	1, 528, 816.42	1, 639, 379.54	116, 932.98
Total emission (kg/yr)	913, 654	-612, 138	-716, 362	794, 857
Payback (yr)	-	13.2	14.2	42.1

From the table, it is clearly that the dependency on the grid can be reduced when the grid is hybrid with RE sources as the grid purchase capacity is reduced for all three configurations. The electrical production by RE sources is highest for hybrid grid-wind-PV as two RE sources are used to generate electricity. As more electricity generated, more excess electricity can be sold to the grid as shown in grid sales. The total income is simply obtained by multiplying the grid sales with the FiT rate. Although the grid-wind-PV system is able to generate more income compared to grid-PV system, the cost of wind turbine is so expensive that contribute to higher NPC.

Since all systems are grid-connected, the emission still occurs. The negative total emissions occur for systems with negative operating cost which are grid-PV and grid-wind-PV system. This is because HOMER simply multiplies the net grid purchases by the emission factor for each pollutant. So, for configurations with grid purchase lower than grid sale, the net grid purchase is negative. As expected, the hybrid renewable energy sources reduced the pollutant emission compared to grid-only systems.

The payback period is the length of time it takes for a project to generate enough cash flow to pay back the initial investment. From equation (4) [28], the payback period of the system can be determined. This system project lifetime is 25 years. In Table 1, the grid-PV system has shortest period to a return of investment with 13.2 years at rates RM 0.813/kWh

 DOI: 10.1051/matecconf/20179001016

followed by grid-wind-PV systems and grid-wind with 14.2 years and 42.1 years, respectively. However, the grid-wind system is still not being able to precede grid-only system even though the wind energy technology is promising energy sources in Terengganu [29]. Thus, to find the least FiT that can be used to get the most optimal hybrid grid-RE configuration, sensitivity analysis was done.

4.1 Sensitivity analysis of hybrid renewable energy system

In this analysis, the FiT rates of solar PV that have been implemented by SEDA in the previous years were used as sensitivity values which are {0.813, 0.86, 0.991, 1.03, 1.14 and 1.24} in unit RM/kWh.

4.1.1 NPC verses FiT rates

The increments of the FiT rates generally reduce the NPC values as more income can be generated from selling electricity to the grid. This is clearly shown in Figure 8 for grid-PV and grid-wind-PV systems. The significant reductions occur in NPC value as the percentage of grid sales for both systems are above 60%. However, only slight changes in NPC value for grid-wind system as the percentage of grid sale for this configuration is 11 %. The FiT doesn't affect the value of NPC for grid-only system as it only supplies energy to the load.

From Fig. 7, the minimal value of FiT rates in gaining profitable revenue is unknown for all three configurations as the points do not intersect with grid-only system. So, Figure 8 and Figure 9 are used to predict the minimal value of FiT for all configurations.

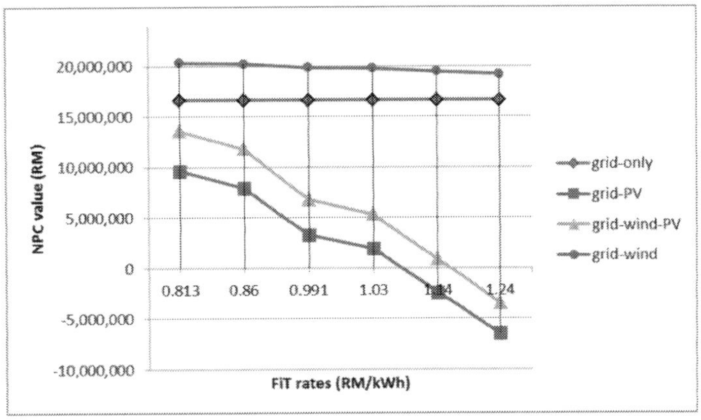

Fig. 7. FiT rates versus NPC value.

From Fig 8, the minimal FiT for grid-wind system is RM 2.20/kWh. From Fig 9, the minimal FiT for grid-wind-PV and grid-PV systems are RM 0.72/kWh and RM0.56/kWh, respectively. Above the values, the grid-RE systems are more optimal compared to the grid only system.

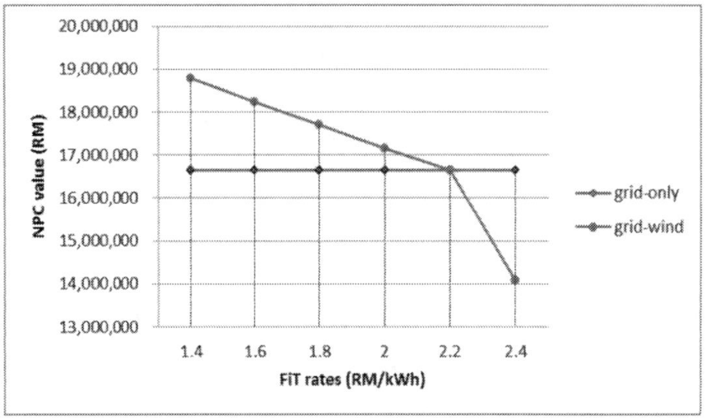

Fig. 8. Possible FiT rates for grid-wind system.

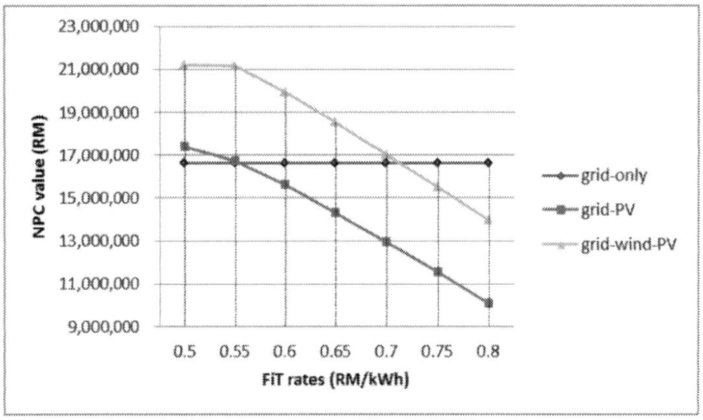

Fig. 9. Minimal FiT rates for grid-PV system and grid-wind-PV system.

4.1.2 Emission of pollutants verses FiT rates

Emission of pollutants do not effect directly with the FiT. If the size of PV remains the same for different FiT, the net grid purchase is also remaining the same. As the emissions of pollutants are obtained by multiplying the net grid purchase with the emission factor, the emissions also remain. However, with the changes in FiT, HOMER calculates that the optimal size of PV is 2500 kW for FiT RM 1.14/kWh and RM 1.24/kWh. As a result, grid sales for these FiT become higher and net grid purchases become more negative. So, the emission of pollutants also decrease for both grid-PV and grid-wind-PV systems as shown in Fig 10. However, no changes happened in the optimal size of wind turbine for grid-wind system, thus, the value of emission of pollutants for this configuration doesn't change with the FiT and remain at the same value as in Table 1.

4.1.3 Payback period verses FiT rates

Fig 11 shows the payback decrease with the increase of FiT. It is true as more income can be generated from selling electricity to the grid, the payback period becomes shorter. Since the grid-wind system produces the payback period longer than the project lifetime, it is not included in the results.

 DOI: 10.1051/matecconf/20179001016

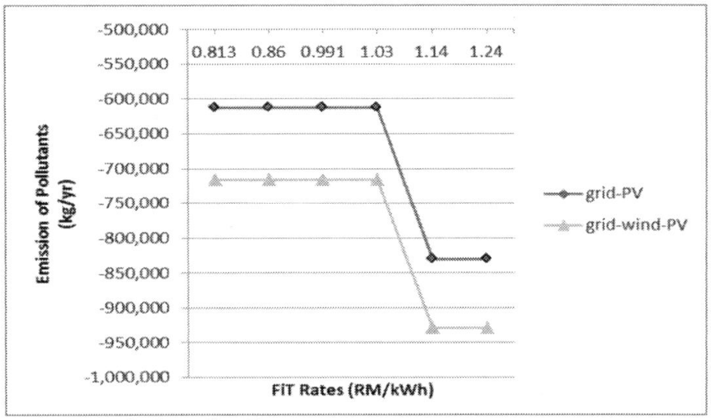

Fig. 10. FiT rates versus emission of pollutants.

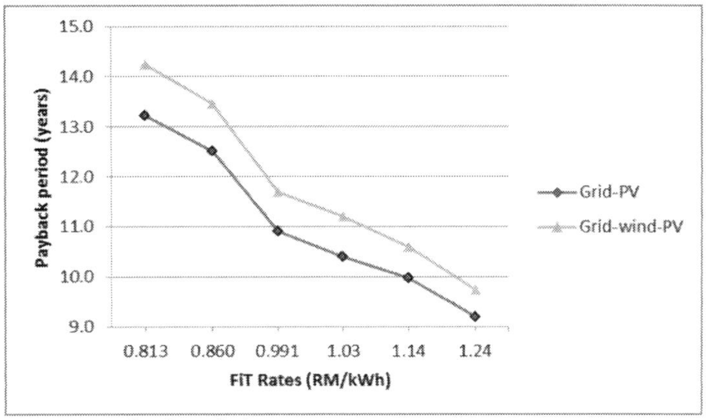

Fig. 11. FiT rates versus payback period.

4.1.4 Malaysia subsidy scheme

Government has allocated an amount of subsidy for the fishermen. Based on [12], in 2014, the amount of fuel subsidy received by the fishermen at KTU area was RM 1,228,325.70. However, from the results in Table 1, the most optimal system which is grid-PV system is able to generate RM 1,528,815/year, higher than the subsidy spent by the government. So, for long term investment, the development of the proposed system is more beneficial than fuel subsidy. Moreover, the fund in other sector such as education and medical sector can be raised when the subsidy is reduced.

5 Conclusions

From HOMER simulation, Kuala Terengganu has potential in developing RE system and the result of economic analysis with four different configurations which are the grid-only, grid-PV, grid-wind-PV and grid-wind systems have been presented. The grid-PV system is the most feasible configuration using the current FiT rates at RM 0.813/kWh. To encourage the use of PV, the FiT rate for the system should not be reduced below RM0.56/kWh. Since the wind turbine is expensive and the wind speed in the selected area is low, the FiT rate should be increased up to RM 2.20/kWh in order to get the benefit of the grid-wind and grid-wind-PV system.

References

1. P. Suuronen, F. Chopin, C. Glass, S. Løkkeborg, Y. Matsushita, D. Queirolo, D. Rihan, Low impact and fuel efficient fishing—Looking beyond the horizon. *Fisheries Research* **119–120**, 135-146 (2012)
2. M. Ferry, W. W. Nik, MFAhmad, F. Gaspersz, M. Manuputty, Comparative study of hybrid catamaran versus diesel monohull boat as ferry for short distance routes. *Indonesian Journal of Naval Architecture* **1**, 29-36 (2013)
3. J. Nóbrega, T. C. Dan, Rubanenco, Electric propulsion applied for research vessel. *Renewable Energy and Power Quality Journal*, (2013)
4. J. V. a. C. Rast, "2015 Top markets report recreational transportation," *A Market Assessment Tool for U.S. Exporters* (International Trade Administration, U.S. Department of Commerce, 2015)
5. S. Minami, K. Koizumi, T. Hanada, N. Matsuda, K. Ishizu, J. Nishi, T. Fujiwara, Performance of a newly developed plug-in hybrid boat. *Electric Vehicle Symposium and Exhibition (EVS27), 2013 World*, 1-4 (2013)
6. Vítor Monteiro, J. C. Ferreira, J. L. Afonso, Smart battery charger for electric mobility in smart grids. *1st International Conference on Smart Grids and Battery IT Systems*, 1-6 (2012)
7. M. A. A. F. Yousif, E. Alfegi, N. Asim And K. Sopian, Simulation of hybrid PV-"3 phase grid" electric vehicle charging system. *7th International Conference on Renewable Energy Sources (RES '13)*, 310-315 (2013)
8. S. Mekhilef, M. Barimani, A. Safari, Z. Salam, Malaysia's renewable energy policies and programs with green aspects. *Renewable and Sustainable Energy Reviews* **40**, 497-504 (2014)
9. W. M. W. Muda, C. A. Fong, Simulation-based method to evaluate pv-wind hybrid renewable energy system in Terengganu. *Energy and Sustainability V* **186**, 12 (2015)
10. R. Ali, I. Daut, S. Taib, A review on existing and future energy sources for electrical power generation in Malaysia. *Renewable and Sustainable Energy Reviews* **16**, 4047-4055 (2012)
11. SEDA, Sustainable Energy Development Authority Malaysia Accessed on: 14th August 2015
12. LKIM, Lembaga Kemajuan Ikan Malaysia. Accessed on: 16th May 2015
13. N. H. Sa'at, Social mobility among the coastal community: A case study in kuala terengganu. *Kajian Malaysia* **29**, 95–123 (2011)
14. S. C. Chua, T. H. Oh, Review on Malaysia's national energy developments: Key policies, agencies, programmes and international involvements. *Renewable and Sustainable Energy Reviews* **14**, 2916-2925 (2010)
15. *Tenth Malaysia plan 2011-2015* (2010).
16. A. S. Abdullah, M. P. Abdullah, M. Y. Hassan, F. Hussin, Renewable energy cost-benefit analysis under Malaysian feed-in-tariff. *Research and Development (SCOReD), 2012 IEEE Student Conference*, 160-165 (2012)
17. T. Khatib, A. Mohamed, K. Sopian, Optimization of a PV/wind micro-grid for rural housing electrification using a hybrid iterative/genetic algorithm: Case study of Kuala Terengganu, Malaysia. *Journal of Energy and Buildings* **47**, 321-331 (2012)
18. N. A. b. A. Razak, M. M. bin Othman, I. Musirin, Optimal sizing and operational strategy of hybrid renewable energy system using homer. *Power Engineering and Optimization Conference (PEOCO), 2010 4th International*, 495-501 (2010)
19. T. Khatib, K. Sopian, H. A. Kazem, Actual performance and characteristic of a grid connected photovoltaic power system in the tropics: A short term evaluation. *Energy Conversion and Management* **71**, 115-119 (2013)

20. N. A. S. Salleh, W. M. W. Muda, S. S. Abdullah, Feasibility study of optimization and economic analysis for grid-connected renewable energy electric boat charging station in Kuala Terengganu. *2015 IEEE Conference on Energy Conversion (CENCON)*, 510-515 (2015)
21. R. N. S. R. Mukhtaruddin, H. A. Rahman, M. Y. Hassan, Economic analysis of grid-connected hybrid photovoltaic-wind system in Malaysia. *Clean Electrical Power (ICCEP), 2013 International Conference*, 577-583 (2013)
22. UniSZA, Weather Station. Accessed on: 10th February 2015
23. M. S. Ngan, C. W. Tan, Assessment of economic viability for PV/wind/diesel hybrid energy system in southern Peninsular Malaysia. *Renewable and Sustainable Energy Reviews* **16**, 634-647 (2012)
24. TNB, Tenaga Nasional Berhad. Accessed on: 30th March 2015
25. R. Hara, H. Kita, T. Tanabe, H. Sugihara, A. Kuwayama, S. Miwa, Testing the technologies. *IEEE Power and Energy Magazine* **7**, 77-85 (2009)
26. Global economics research. Accessed on: 22nd December 2015
27. MEIH, Department of Statistic Malaysia: Electricity Generation. Accessed on: 6th May 2015
28. M. M. Gilbert, in *Renewable and efficient electric power systems*. (John Wiley and Sons, New York, 2004), chap. Chapter 5: Economics of Distributed Resources, pp. 240-254.
29. A. Albani, M. Z. Ibrahim, Y. K.H., The feasibility study of offshore wind energy potential in Kijal, Malaysia: the new alternative energy source exploration in Malaysia. *Journal of Energy Exploration & Exploitation* **32**, 329–344 (2014)

Design and development of coil casing MRF brake system

R. Siti Lydia[1,*] , *R.* Mokhtar[1], *A.B.* Muhamad Husaini[1] and *B.J.* Norhaniza[1]

[1]System Engineering and Energy Laboratory, Section of Manufacturing, Universiti Kuala Lumpur Malaysian Spanish Institute, Kulim Hi-Tech Park, 09000 Kulim, Kedah.

Abstract. Nowadays magneto-rheological brakes has been introduce to overcome the drawback of conventional braking such as bulky, leakage and late build up pressure. Magnetic field is a main parameter need to be maximized in order to increase the efficiency of the brake. In MRF brake system the magnetic field directly affected by coil casing shapes. This paper focused on the study of the effect of coil casing shapes to the magnetic field distribution in MR brake. Finite Element Method will be used to analyze the magnetic field distribution produced by the coil. Five coils shape that different angle from 50° to 90° used in this study. As the result magnetic field was change quadratically by increasing the angle size. The best angle is between 60° to 70°. The peak value is at 70°. If more than that the magnetic field will drop. The significant of this study is to get the most efficient angle for the electromagnet casing for the MR brakes.

1 Introduction

Nowadays there are two types of brake that are widely used. It is drum brake and disc brake. There are several problems these two types of brake. Drum brake has lower breaking efficiency and heat generation in drum is brake more [1]. The excessive heat can cause the brake fade [2]. Then disc brake being introduced to the industry, with several advantages compared to the drum brake. Although the disc brake is much better compared to the drum brake, but it also has its disadvantages. One of it, is, the efficiency of disc brake is too high which may cause skidding or wheel lock. Furthermore, during wet condition disc brake tended to decrease its efficiency. Considering the disadvantages of conventional brake system, thus there is a requirement to design a new braking system which is minimizing the effect mention.

Magneto rheological fluid is types of smart fluid that can change its rheological properties. A rapid change from liquid to semi solid state can occurs when a magnetic field induced to the fluids as shown in Fig. 1.

Magneto rheological fluid is a combination of magnetic particles suspended in a fluid medium. There are three component that made up MR fluid, iron particle, base fluid and additives. The sizes of the iron particle in the magneto rheological fluid are between 1-10 μ-meters. The metal particle usually made up from carbonyl iron or powder iron or iron/

* Corresponding author: sitilydia@unikl.edu.my

DOI: 10.1051/matecconf/20179001017

cobalt alloy, to achieve high magnetic saturation. When the magnetic field is applied the micro size particle will line up and a chain like structure will be form [3]. The viscosity of the fluid will be increase due to the alignment of the particles [4]. The base fluid commonly use is, hydrocarbon oil, mineral oil and silicon oil. The base fluid will act as the carrier fluid in which the metal is suspended. The base fluid should low in viscosity and not vary to the temperature. For the additive it should be high viscosity material such as grease or others thixotropic additives. The additives will act as suspending agent, thixotropers, friction modifier and anti-corrosion component [5].

(a) Magneto rheological fluid without magnet field (b) Magneto rheological fluid with magnetic field

Fig. 1. Magneto Rheological Fluid behavior with or without magnet.

In recent year researcher has start developed MR brake [6], has comes out with Design and Consideration for an Automotive Magneto rheological Brake as shown in Fig. 2. A prototype of MR brakes is then built and tested and the experimental result show a good correlation with the finite element simulation predictions. However, the braking torque generated is still far less than conventional hydraulic brakes has design the brake using a Small Steel Roller and MR Fluid [7]. As the result shows the designed MR Brake have about 200% larger torque than the conventional one.

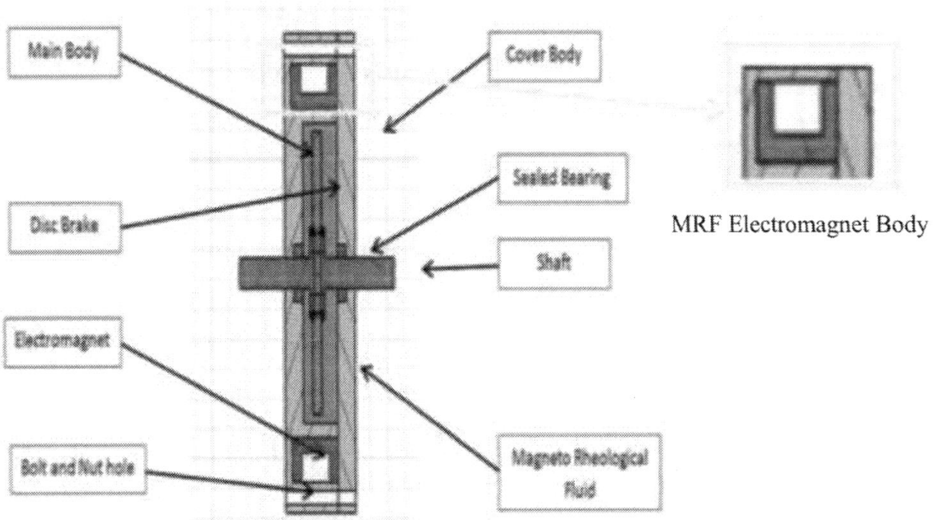

Fig. 2. Magneto Rheological Brake cross sectional area.

Figure 2 shows the cross sectional area for the MR brakes. The MR brake consists of a disc that being immersed into MR fluid in a closed contained and all the parts are made up of steel. The steel has been choosing because it has good electromagnetic conductivity. The highlighted part is the electromagnetic body which will be surrounding by copper wire to generate magnetic field to activate the MR fluid. The objective of this study is to analyze the effect of electromagnet body shapes on magnetic field induced. The material for each of the component is constant. The parameter that will be study is the angle of the electromagnet body and varies from 50° to 90°. In order to design MRF brake system the magnetic circuit need to be analyzed. The variation of magnetic field inside the system is compulsory to study. Due to complicity in geometry the numerical method need to be used. In this paper will focusing on the shapes of the electromagnet. In which angle the electromagnet will give much greater magnetic field result.

2 Methodology

In this section will explain about the Magneto Rheological Brake electromagnet body design. Five different angles has been test, 50°-90°, for the electromagnet body as shown in Fig. 3. FEMM software was used to analyze the magnetic field produced by the electromagnet.

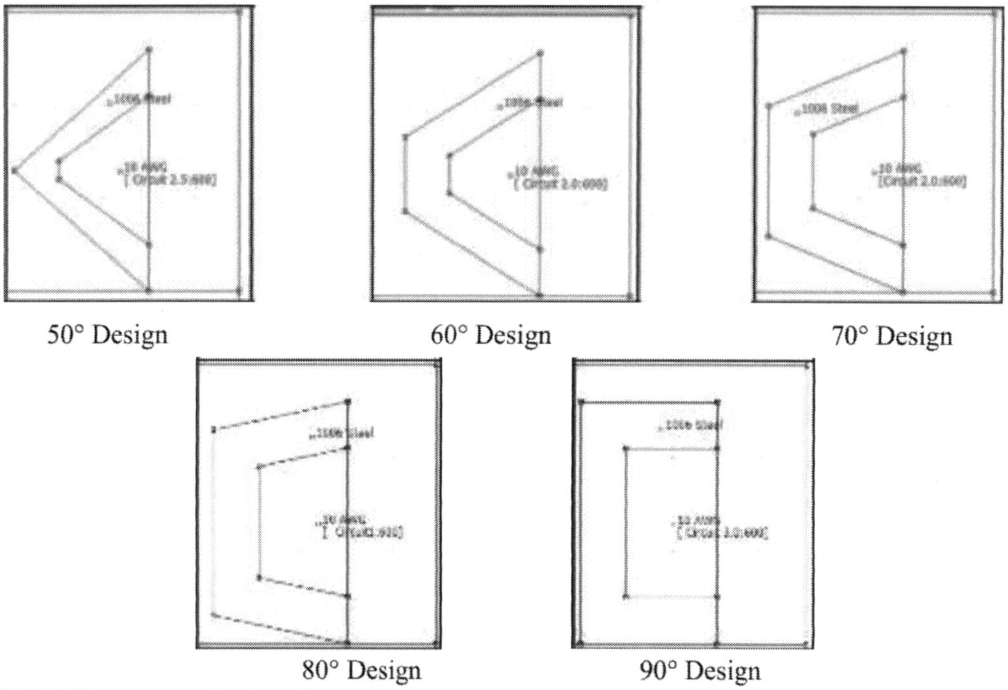

Fig. 3. Electromagnet body design.

FEMM software used to discretize the geometry and solve the Maxwell equation numerically.The examined MRF Brake is characterized by the axial symmetry hence the field problem can be solved in the two-dimensional formulation in the Cartesian coordinate system. The computational domain for the magnetic field analysis is at the electromagnet body design with several different angles. Magnetostatic problems are problems in which the fields are time-invariant. In this case, the field intensity (H) and flux density (B) must obey:

$$\nabla X H = J \tag{1}$$

$$\nabla . B = 0 \tag{2}$$

subject to a constitutive relationship between B and H for each material:

$$B = \mu H \tag{3}$$

 DOI: 10.1051/matecconf/20179001017

If a material is nonlinear (e.g. saturating iron or alnico magnets), the permeability, μ is actually a function of B:

$$\mu = \frac{B}{H(B)} \tag{4}$$

Flux density is written in terms of the vector potential, A, as:

$$B = \nabla X A \tag{5}$$

Now, this definition of B always satisfies (3.0). Then, (4.0) can be rewritten as:

$$\nabla X\left(\frac{1}{\mu(B)}\nabla X A\right) = J \tag{6}$$

For a linear isotropic material (and assuming the Coulomb gauge, $\nabla .A = 0$) reduces to :

$$\frac{1}{\mu}\nabla^2 A = J \tag{7}$$

In any numerical method geometry model is the main input, same goes to the Finite Element. Figure 4(a) shows CAD model for MRF valve used in this research work. In order to reduce the computational time, the geometry was converted into 2 dimensional. Figure 4(b) shows the two dimensional valve after meshing process. Figure 5 shows the Magneto rheological brake assemble.

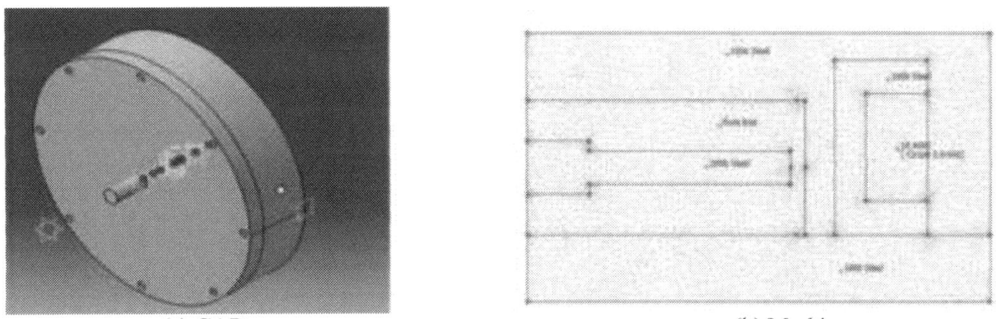

(a) CAD (b) Meshing

Fig. 4. CAD model for MR Brake and MRF Brake after mesh.

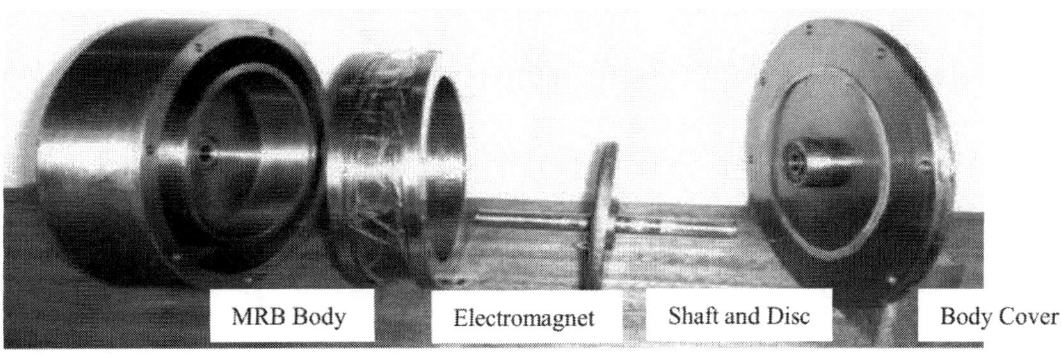

Fig. 5. Magneto rheological brake assemble.

3 Results and discussion

The result from the FEMM can be shown in color contour for magnetic flux. The color will present the magnetic flux density and magnetic field strength. The plot value will be define along a define contour. From the result, it can define the B (magnitude of Flux Density). Figure 6 shows the result of the density plot for 50°, 60°,70° and 90° designs. From the result, the purple region will show the highest Tesla value. Design 50°, 60° and 70° has larger region with highest tesla value compared to the 90°. However when the tesla value at point 1 will show the highest value for tesla comes from 70° degree design with value of 1.30498 Tesla.

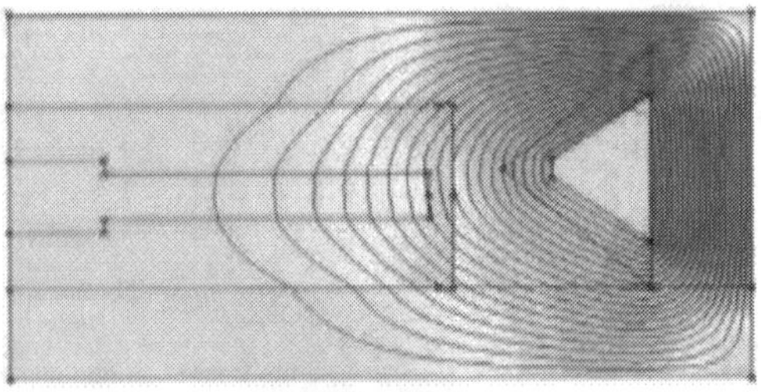

50° Design

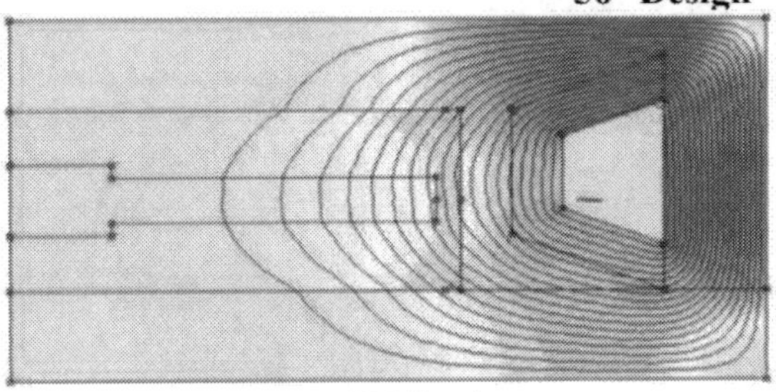

70° Design

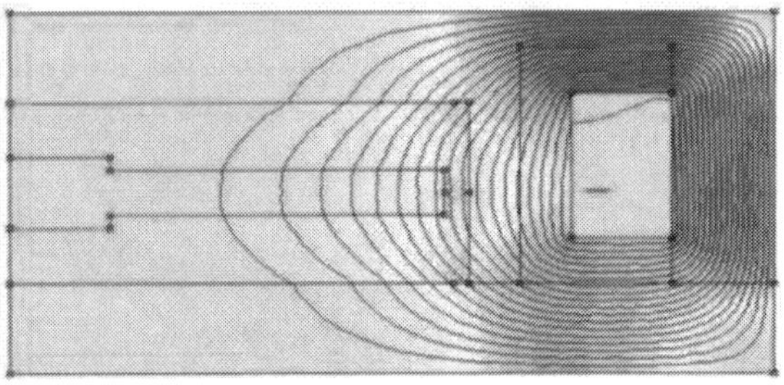

90° Design

Fig. 6. Magnetic Flux Density contour.

Magnetic flux density is the amount of magnetic flux per unit area of a section, perpendicular to the direction of flux. From Fgiure 7, there are slightly increasing value of normal flux density (B.n) from 1.29024 tesla to 1.30303 tesla when the angle being increase from 50° to 60°. From the density plot, the 60° design has larger region that produce higher value of tesla. When the design changed to 70° the value has reach to 1.30498 tesla, the highest value compared to the others design. The magnetic flux density plot shows that the 70° angle will allow the magnetic flux to leading to the MR fluids more efficiently. The flux density slightly decreases when the angle changes to 80° and 90°. As shown by the density plot, there are too small magnetic flux region being produces in the design.

The graph above shows the result of effect of current to the Tesla value. The test has been don at the point 1.The current has been manipulated from 0.5 to 2.5 ampere. As the result from the three design chosen, 50°,70° and 90° it shows that the value of Tesla is increase due to the increasing current value. Although all the value is increasing, but the higher value comes from 70° design. From the lower current supply, 0.5A, to the highest current, 2.5A, 70° design shows the higher value. This may cause by the geometry of the design that has good magnetic flux flow to the point 1 compared to the others.

 DOI: 10.1051/matecconf/20179001017

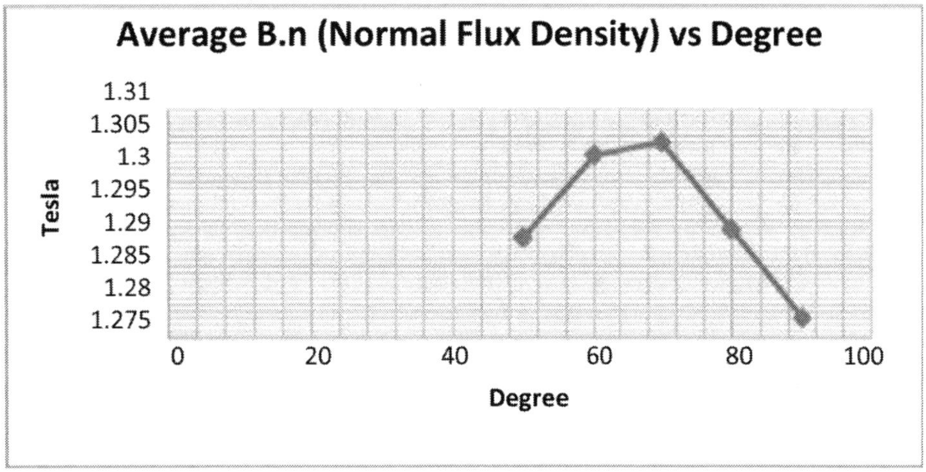

Fig. 7. B.n vs. Degree.

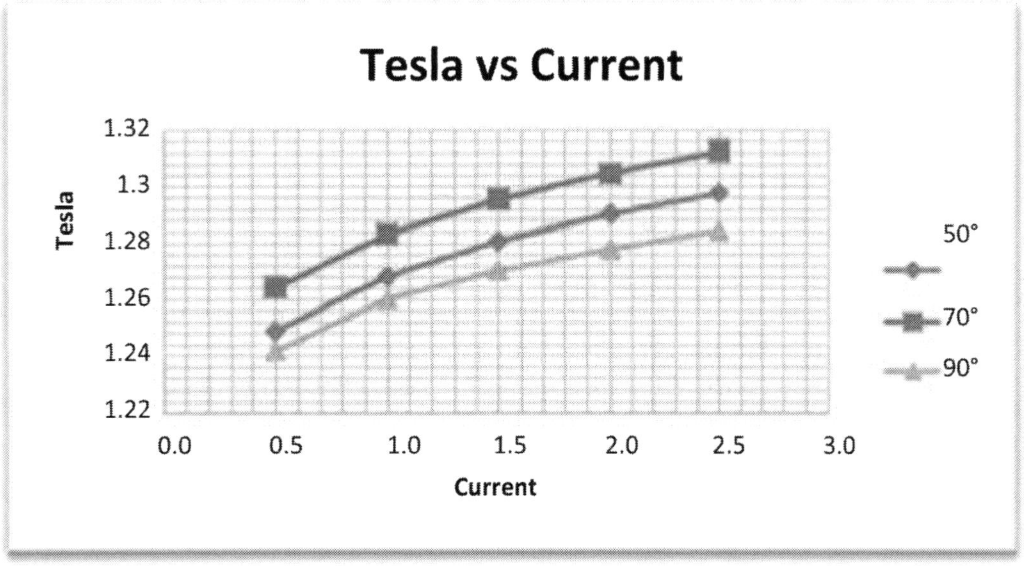

Fig. 8. Tesla vs. Current.

4 Conclusions

FEMM is the software is a solver to solve the governing equation for magnetic field distribution inside the domain. From the experiment above five different angle of design has been analysis using the FEMM. As the result, the good range for the electromagnet body design is between 50° to 70°. If the design goes far than that the value of the magnetic field will be drop. As shown on graph 1. Although the supply current been manipulated, the 70° design still give the high value up to 1.31275 Tesla at 2.5 ampere.

References

1. Sriram T. Mutalik & Tapan S. Kulkarni "*Disc Oriented in Drum Brakes*". International Journal of Emerging Technology and Advance Engineering, **2**, Issues 10, (2012)

2. Kannan M. Munisamy, Ramel Shafik "*Disk Brake Design for Cooling Improvement Using Computational Fluid Dynamics (CFD)*".IOP Conf. Series: earth And Environmental Science 16 (2013)
3. Luo Yiping, Xu Biao, Ren Hongjuan "*Summarization of magneto-rheological fluid and its application in engineering*". American Journal of Mechanics and Applications. **2**, No. 1, 2014 pp. 1-5. doi: 10.11648/j.ajma.20140201.11
4. M. Kciuk, R. Turczyn " *Propeties and Application of Magneto Rheological Fluid*". Journal of Achievement in Material and Manufacturing Engineering, **18**, (2006).
5. Deepak Baranwal, Dr. T.S. Deshmukh "*MR-Fluid Technology and Its Application-A Review*". International Journal of Emerging Technology and Advance Engineering, **2**, (2012).
6. Kerem Karakoc, Edward J.Park, Afzal Suleman "*Design consideration for an automotive magnetorheological brake*".(2008).
7. Tran Hai NAM, Kyoung Kwan AHN "*New approach to designing an MR brake using a small steel roller and MR fluid*". Journal of Mechanical Science and Technology,**23** 1911~1923 (2009).

Effect of varying geometrical parameters of trapezoidal corrugated-core sandwich structure

*N.Z.M.*Zaid[1,*], *M.R.M.*Rejab[1,2], *A.F.*Jusoh[1], *D.*Bachtiar[1,2], *J.P.*Siregar[1,2]

[1]Faculty of Mechanical Engineering, Universiti Malaysia Pahang, 26600 Pekan, Pahang, Malaysia
[2]Structural Material & Degradation Focus Group, Faculty of Mechanical Engineering, Universiti Malaysia Pahang, 26600 Pekan, Pahang, Malaysia

Abstract. Sandwich structure is an attractive alternative that increasingly used in the transportation and aerospace industry. Corrugated-core with trapezoidal shape allows enhancing the damage resistance to the sandwich structure, but on the other hand, it changes the structural response of the sandwich structure. The aim of this paper is to study the effect of varying geometrical parameters of trapezoidal corrugated-core sandwich structure under compression loading. The corrugated-core specimen was fabricated using press technique, following the shape of trapezoidal shape. Two different materials were used in the study, glass fibre reinforced plastic (GFRP) and carbon fibre reinforced plastic (CFRP). The result shows that the mechanical properties of the core in compression loading are sensitive to the variation of a number of unit cells and the core thickness.

1 Introduction

Composite laminate with single skin, made from carbon, glass and other fibres may be strong but lack of stiffness due to their relatively low thickness[1]. Sandwich structures comprising of face-sheets and cores are broadly used in transportation vehicles and civil infrastructure due to their high stiffness/strength-to-weight ratio [2–10]. Sandwich structures have numerous topologies in different material- e.g composite corrugated-core, metallic core, or metallic textile core [10–14]. Sandwich structures have different core material including the shape. The foam cores are preferably used when the waterproof, sound and heat insulation qualities of cores are required. Moreover, the foam cores are the cheaper among core materials and can offer some advantages in sandwich construction. Furthermore, the honeycomb core sandwich structure comprise of a thick layer core separate with thin stiff layers (skins) and they are describe as cellular cell because of the shape like bee honeycomb. Honeycomb is lightweight, high flexural stiffness and can support classical loadings like tension and bending. Applications of corrugated-core sandwich structure have been used in aerospace and automotive industries, marine ship and civil due to their high strength to weight ratio[6][9][15]. Previously, Mohammadi et al. [16] proposed analytical formulation for trapezoidal corrugated panels. The formulation was based on energy approach and castigliano's theorem. The analytical was compared with FEM results. Investigation of large displacement behaviour of aramid/epoxy trapezoidal

* Corresponding author: nzakiahmz@gmail.com

 DOI: 10.1051/matecconf/20179001018

corrugated laminates under tensile loading transverse to the corrugation direction was conducted by Thill et al. [7]. Rejab and Cantwell [9] studied compressive properties of triangular corrugated-core sandwich structures based on three dissimilar material both experimental and numerically. From the study, indicate that thick CFRP corrugations offer properties that are similar to those exhibited by aluminium honeycombs. In this paper, sandwich structure with trapezoidal corrugated-core is loaded in compression direction and the effects of geometrical parameters are discussed. The trapezoidal corrugated specimens are made of carbon fibre reinforced plastic (CFRP) and glass fibre reinforced plastic (GFRP).

2 Experimental setup

2.1 Core fabrication

The corrugated-core specimens were fabricated using a trapezoidal machined mild steel mould. By using epoxy resin carbon fibre plain weave fabrics were hand-layup and cured for 24 hours in room temperature. Ply of the carbon fibre are count to obtained desired thickness. Figure 1 shows the trapezoidal corrugated-cores mild steel mould. The cores were cut to test specimen size using electrical saw.

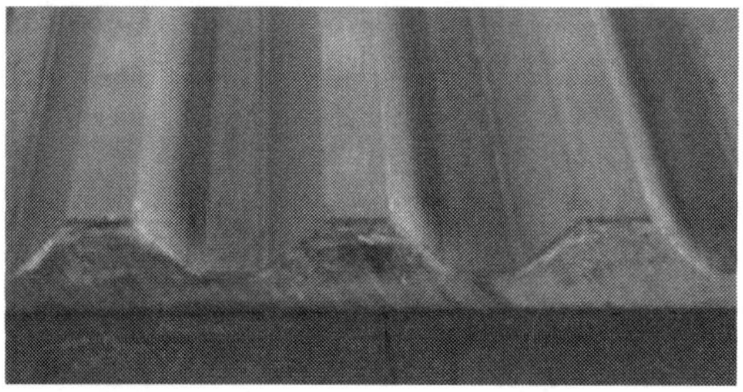

Fig. 1. Trapezoidal corrugated-core mould.

2.2 Core geometry

The length of the trapezoidal corrugated-core investigated in this paper is 40 mm, 120 mm and 200 mm. Which is 40 mm is for 1 unit cell, 120 mm is for 3 unit cell and 200 mm is for 5 unit cell. The width of compression test and tensile test specimen are 25 mm, respectively. Furthermore, thickness of tensile test specimen and thickness of compression test specimen is tabulated in Table 1.

2.3 Core material characterization

The corrugated-core is made of different plies of plain weave carbon fibres with epoxy resin. To evaluate the mechanical properties of the carbon fibre reinforce polymer, a total of five specimens were manufactured and were end tabbed using aluminium and Araldite 420AB adhesive for tensile test. The length, width and thickness of the tabbed sections are 250 mm, 25 mm and 2 mm individually.

3 Experimental works

3.1 Static compression test

A compression test tested on each specimen by using a testing machine, Instron series 3369, load capacity 50 kN as shown in Figure 2. The specimen was placed and a uniform compression was then applied at rate of 2 mm/min. Each specimen have core ID, for example CF1U3P (CF is stand for carbon fibre, 1U stand for one unit cell and 3P is stand for three plies). Table 1 below shows material used, core ID, number of unit cell, number of plies, average core width, average core length and average core height. Tests were repeated three specimens for one core ID. The data of load-displacement was recorded until the specimen fully crushed by compression.

Table 1. Compression test parameter for corrugated composite core.

Material	Core ID	No of unit cell	No of plies	Average core width (*w*)	Average of core length (*l*)	Average core height (mm)
CFRP	CF1U3P	1	3	25	40	13
	CF2U3P	2	3	25	80	13
	CF3U3P	3	3	25	120	13
	CF1U4P	1	4	25	40	14
	CF2U4P	2	4	25	80	14
	CF3U4P	3	4	25	120	14
	CF1U5P	1	5	25	40	16
	CF2U5P	2	5	25	80	16
	CF3U5P	3	5	25	120	16
GFRP	GF1U2P	1	2	25	40	14
	GF2U2P	2	2	25	80	14
	GF3U2P	3	2	25	120	14
	GF1U3P	1	3	25	40	17
	GF2U3P	2	3	25	80	17
	GF3U3P	3	3	25	120	17
	GF1U4P	1	4	25	40	18.5
	GF2U4P	2	4	25	80	18.5
	GF3U4P	3	4	25	120	18.5

Fig. 2. Static compression test setup on CFRP specimen.

 DOI: 10.1051/matecconf/20179001018

4 Results and discussion

4.1 Compression test

Compression testing on the trapezoidal corrugated-cores displayed that the behaviour of different core were significantly difference.

4.2 Mechanical bBehaviour of corrugated-core

The failure processes in CFRP trapezoidal corrugated-core are shown in Figure 3, the early failure start with debonding at the end of right and left specimen. Following failure include delamination, fibre breaking and debonding (III-VI).

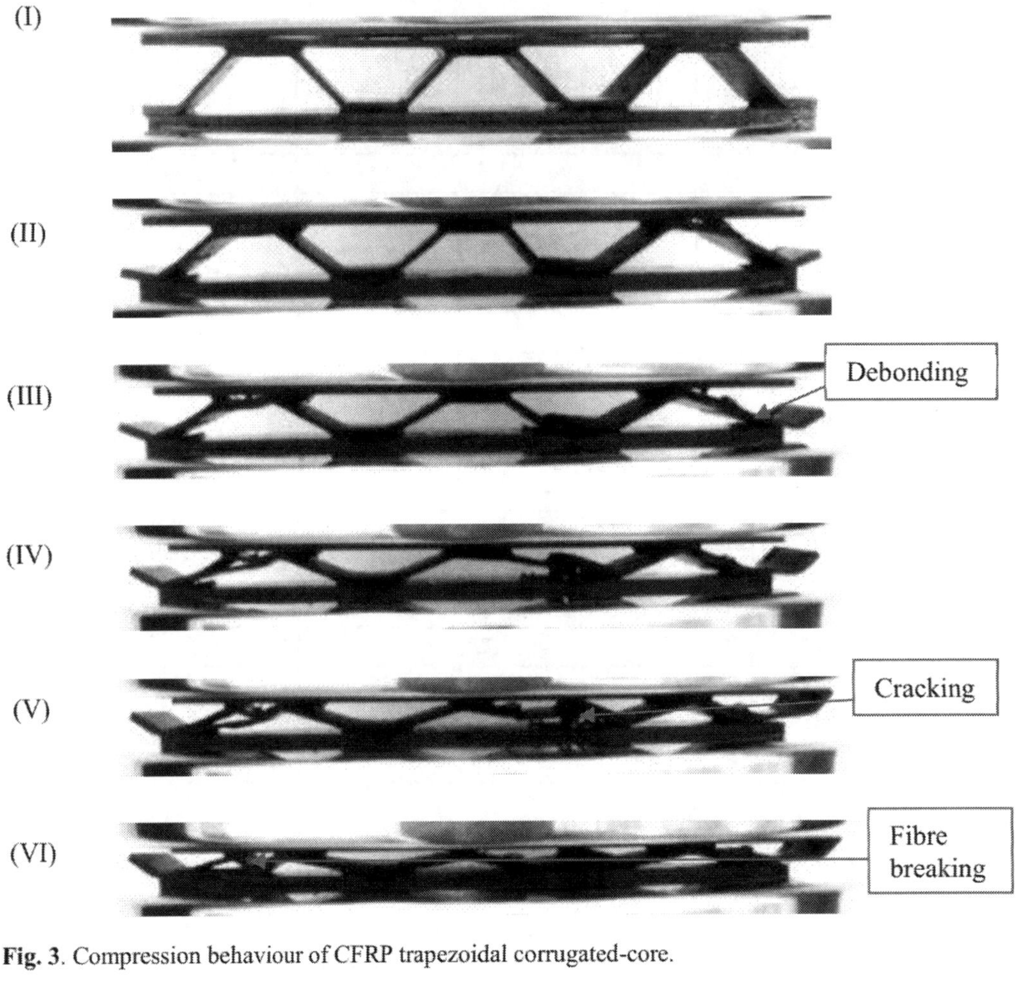

Fig. 3. Compression behaviour of CFRP trapezoidal corrugated-core.

The failure processes in GFRP trapezoidal corrugated-core are shown in Figure 4, the initial failure start with debonding at the end of right and left specimen. Following failure include delamination, fibre breaking and debonding (III-VI). The stress strain curve in Figure 5 for GFRP trapezoidal corrugated-core shows that the compressive stress increases to 3.36 MPa at 0.0589 mm of displacement and drop to 1.38 MPa at 0.6 mm. When the cell

DOI: 10.1051/matecconf/20179001018

wall buckled or debonding at left and right, the force dropped progressively in image (I-II). Following failure include delamination, fibre cracking and debonding (III-V).

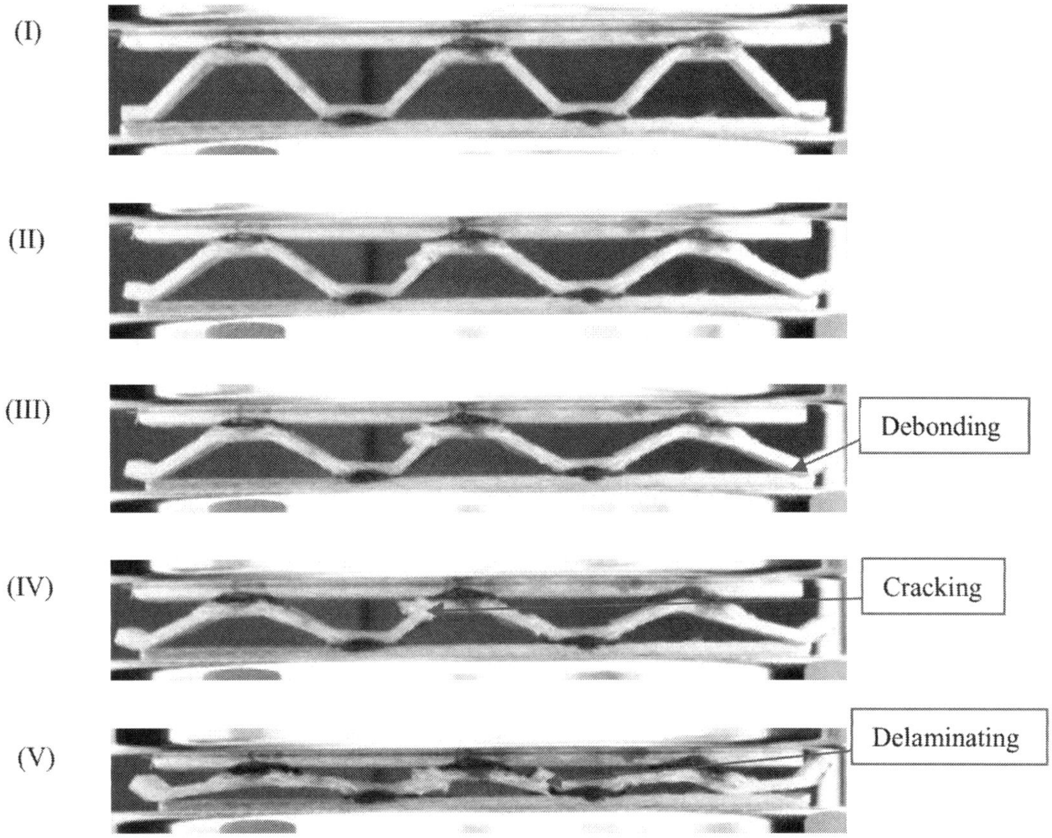

Fig. 4. Compression behaviour of GFRP trapezoidal corrugated-core.

4.3 Effect of varying no of plies and varying no. of unit cell

In an attempt to validate more precisely, test were taken on three specimens for GFRP and CFRP. The effect of varying number of unit cell and varying wall thickness on compression strength of GFRP and CFRP trapezoidal corrugated-core sandwich structure is shown in Figure 6 and Figure 7. Obviously, the compression strength of (CFRP) is much higher than GFRP structure due to the fact that carbon fibres have higher mechanical properties than GFRP. Figure 6 and 7 shows effect of varying the number of unit cell for GFRP and CFRP from one to three unit cells. It shows that the higher number of unit cell, it influences the composite strength. For the effect of cell wall thickness, Figure 7 shows the higher the wall thickness, the higher the compression strength.

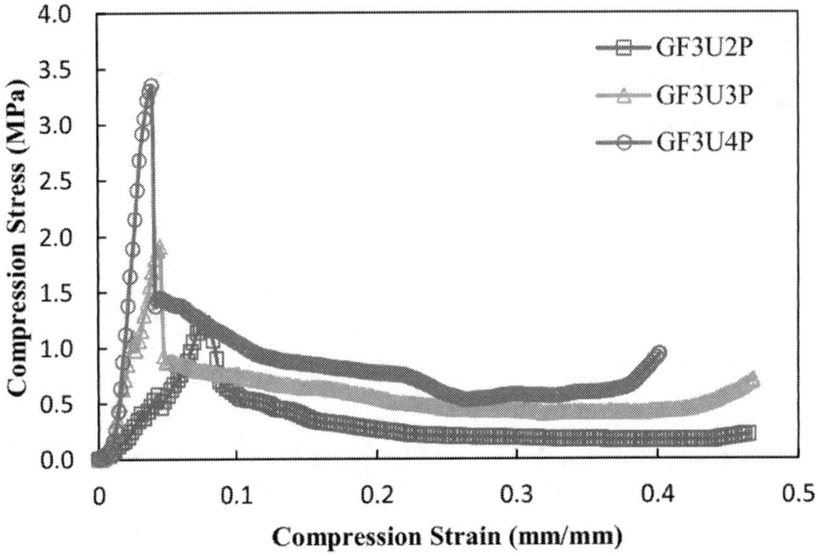

Fig. 5. Compression Stress Strain Curve for a GFRP trapezoidal corrugated-core.

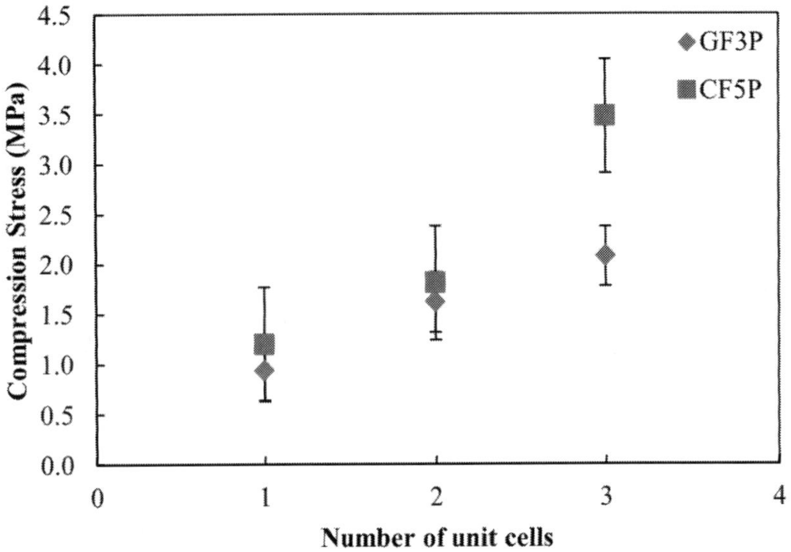

Fig. 6. Comparison between glass fibre four plies (GF3P) with carbon fibre 5 plies (CF5P).

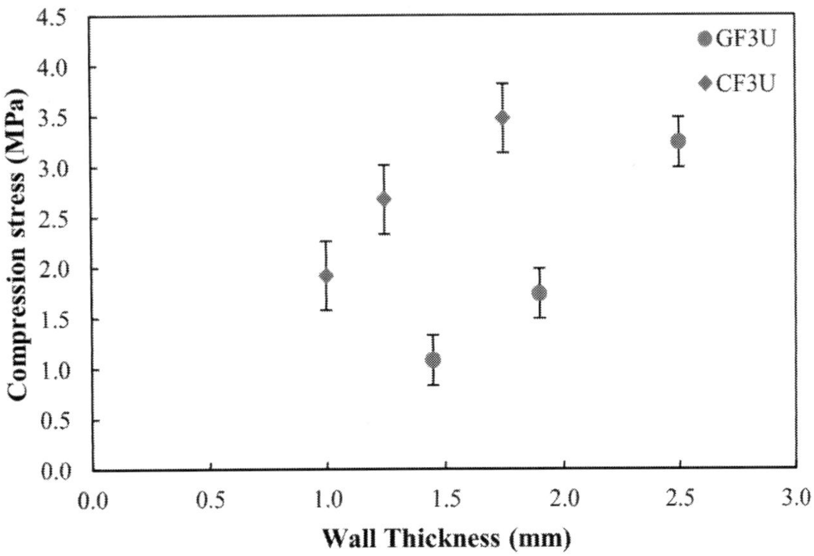

Fig. 7. Comparison between glass fiber three unit cells (GF3U) with carbon fiber three unit cell (CF3U)

5 Conclusions

Trapezoidal corrugated-core was fabricated using a 45° profiled mould, was used to manufacture range of lightweight sandwich structures. A 45° corrugation angle was designated since it represents an optimal configuration for all combinations of bending, shear and stretching stiffness. The compressive behaviour and resulting failure mechanism in structures based on two different materials have been investigated experimentally. An investigation of corrugation during testing indicated that initial failure was dominated by instabilities as the cell walls begin to buckle. In contrast, the composite fibre exhibited fibre fracture, delamination and debonding. From the finding the effects of varying the number of unit cell dominate by CFRP is 3.48 MPa higher than GFRP that 2.08 Mpa at three unit cell. It shows that the higher number of unit cell, it influences the composite strength. For the effect of cell wall thickness, shows the higher the wall thickness, the higher the compression strength. Compression strength CFRP and GFRP are 3.48 MPa and 1.74 MPa at thickness 1.75 mm and 1.90 mm respectively. For future work, the testing will be extended to correlation under bi-axial loading.

References

1. D. Bannister, "An introduction to core materials," *Reinf. Plast.*, vol. **58**, no. 2, pp. 32–37, (2014).
2. J. Zhang, P. Supernak, S. Mueller-Alander, and C. H. Wang, "Improving the bending strength and energy absorption of corrugated sandwich composite structure," *Mater. Des.*, vol. **52**, pp. 767–773, (2013).
3. B. Castanié, Y. Aminanda, C. Bouvet, and J.-J. Barrau, "Core crush criterion to determine the strength of sandwich composite structures subjected to compression after impact," *Compos. Struct.*, vol. **86**, no. 1–3, pp. 243–250, (2008).
4. H. Herranen, O. Pabut, M. Eerme, J. Majak, M. Pohlak, J. Kers, M. Saarna, G. Allikas,

and A. Aruniit, "Design and Testing of Sandwich Structures with Different Core Materials," vol. **18**, no. 1, (2012).
5. H. S. Ji, W. Song, and Z. J. Ma, "Design, test and field application of a GFRP corrugated-core sandwich bridge," *Eng. Struct.*, vol. **32**, pp. 2814–2824, (2010).
6. G. Bartolozzi, M. Pierini, U. Orrenius, and N. Baldanzini, "An equivalent material formulation for sinusoidal corrugated cores of structural sandwich panels," *Compos. Struct.*, vol. **100**, pp. 173–185, (2013).
7. C. Thill, J. A. Etches, I. P. Bond, K. D. Potter, P. M. Weaver, and M. R. Wisnom, "Investigation of trapezoidal corrugated aramid/epoxy laminates under large tensile displacements transverse to the corrugation direction," *Compos. Part A Appl. Sci. Manuf.*, vol. **41**, no. 1, pp. 168–176, (2010).
8. M. Yazici, J. Wright, D. Bertin, and A. Shukla, "Experimental and numerical study of foam filled corrugated core steel sandwich structures subjected to blast loading," *Compos. Struct.*, vol. **110**, pp. 98–109, (2014).
9. M. R. M. Rejab and W. J. Cantwell, "The mechanical behaviour of corrugated-core sandwich panels," *Compos. Part B Eng.*, vol. 47, pp. 267–277, (2013).
10. G. Bartolozzi, N. Baldanzini, and M. Pierini, "Equivalent properties for corrugated cores of sandwich structures: A general analytical method," *Compos. Struct.*, vol. **108**, no. 1, pp. 736–746, (2014).
11. S. Kazemahvazi and D. Zenkert, "Corrugated all-composite sandwich structures. Part 1: Modeling," *Compos. Sci. Technol.*, vol. **69**, pp. 913–919, (2009).
12. K. Mohan, T. H. Yip, S. Idapalapati, and Z. Chen, "Impact response of aluminum foam core sandwich structures," *Mater. Sci. Eng. A*, vol. 529, pp. 94–101,(2011).
13. C. Kiliçaslan, M. Güden, I. K. Odaci, and A. Taşdemirci, "Experimental and numerical studies on the quasi-static and dynamic crushing responses of multi-layer trapezoidal aluminum corrugated sandwiches," *Thin-Walled Struct.*, vol. **78**, pp. 70–78, (2014).
14. C. Kiliçaslan, M. Güden, I. K. Odaci, and A. Taşdemirci, "The impact responses and the finite element modeling of layered trapezoidal corrugated aluminum core and aluminum sheet interlayer sandwich structures," *Mater. Des.*, vol. **46**, pp. 121–133, (2013).
15. I. Dayyani, S. Ziaei-Rad, and H. Salehi, "Numerical and Experimental Investigations on Mechanical Behavior of Composite Corrugated Core," *Appl. Compos. Mater.*, vol. **19**, no. 3–4, pp. 705–721, (2011).
16. H. Mohammadi, S. Ziaei-Rad, and I. Dayyani, "An equivalent model for trapezoidal corrugated cores based on homogenization method," *Compos. Struct.*, vol. **131**, pp. 160–170, (2015).

Comparison performance of split plug-in hybrid electric vehicle and hybrid electric vehicle using ADVISOR

Muhammad Ikram Mohd Rashid[1,*], *Hamdan* Daniyal[1] and *Danial* Mohamed[2]

[1]Sustainable Energy & Power Electronics Research Group(SuPER), Faculty of Electrical and Electronics Engineering, Universiti Malaysia Pahang, 26600 Pekan, Pahang, Malaysia
[2]Automotive Engineering Centre(AEC), Universiti Malaysia Pahang (UMP), 26600 Pekan, Pahang, Malaysia

Abstract. Electric vehicle suffers from relatively short range and long charging times and consequently has not become an acceptable solution to the automotive consumer. The addition of an internal combustion engine to extend the range of the electric vehicle is one method of exploiting the high efficiency and lack of emissions of the electric vehicle while retaining the range and convenient refuelling times of a conventional gasoline powered vehicle. The term that describes this type of vehicle is a hybrid electric vehicle. Many configurations of hybrid electric vehicles have been designed and implemented, namely the series, parallel and power-split configurations. This paper discusses the comparison between Split Plug-in Hybrid Electric Vehicle(SPHEV) and Hybrid Electric Vehicle(HEV). Modelling methods such as physics-based Resistive Companion Form technique and Bond Graph method are presented with powertrain component and system modelling examples. The modelling and simulation capability of existing tools such as ADvanced VehIcle SimulatOR (ADVISOR) is demonstrated through application examples. Since power electronics is indispensable in hybrid vehicles, the issue of numerical oscillations in dynamic simulations involving power electronics is briefly addressed.

1 Introduction

A hybrid electric vehicle draws its propulsion power from an (Internal Combustion Engine)ICE and an electric motor, which can also operate as a generator to charge the onboard electrical energy storage device. Depending on the connectivity of the power sources to the load (wheels), the HEV may have one of the following drivetrain configurations: series hybrid, parallel hybrid or combined series-parallel hybrid [1].

A series-hybrid Fig. 1 only has the electric motor turning the drive-shaft while the ICE serves as an on-board electric generator to charge the batteries and/or to power the motor directly ('series' connection between the ICE the electric motor, resulting in an electric transmission of power to the wheels) [1].

* Corresponding author: mikramump@gmail.com

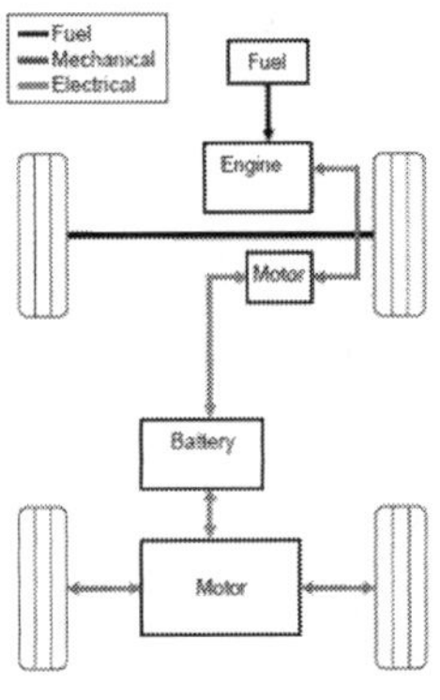

Fig. 1. Series-hybrid electric vehicle [2].

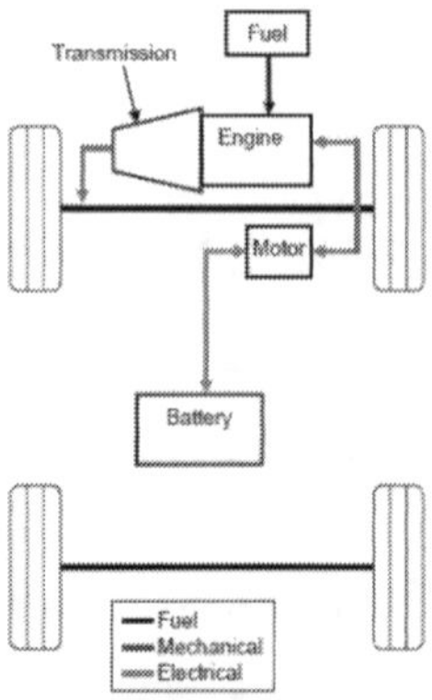

Fig. 2. Parallel-hybrid electric vehicle [2].

Instead, a parallel-hybrid vehicle Fig. 2 has the ICE turning the drive-shaft, just like a conventional vehicle, but the drive-shaft can also be turned by the electric motor ('parallel' connection of the ICE and motor to the drive shaft via some form of mechanical coupling and transmission) [1].

A hybrid vehicle typically has the motor-generator onboard the chassis, as in most production hybrids of today, the parallel "through the road" hybrid uses one motor for propulsion assist or to recharge the battery by loading the rear axle with the motor. In contrast, the series hybrid provides power to the wheels by a series path of energy conversion through the engine to the generator, battery, then the rear motor. Note that the series hybrid and through-the-road parallel hybrid can still propel the vehicle without the engine on, sourcing power for the rear motor from the battery [2].

In this paper addressed about comparison between modelling split hybrid electric vehicle and hybrid electric vehicle and simulate it using ADVISOR. All simulation was done by using ADVISOR and the results were discussed through this paper.

2 HEV modeling using ADVISOR

ADvanced VehIcle SimulatOR (ADVISOR) is a modelling and simulation tool developed by U.S. National Renewable Energy Laboratory (NREL) [4,5]. It can be used for the analysis of performance, fuel economy, and emissions of conventional, electric, hybrid electric, and fuel cell vehicles. The backbone of the ADVISOR model is the Simulink block diagram shown in Fig. 3, for a parallel HEV as an example. Each subsystem (block) of the block diagram has a Matlab file (m-file) associated with it, which defines the parameters of that particular subsystem. The user can alter both the model inside the block as well as the m-files associated with the block to suit the modelling needs. For example, the user may need a more precise model for the electric motor subsystem. A different model can replace the existing model as long as the inputs and the outputs are the same. On the other hand, the user may leave the model intact and only change the m-file associated with the block diagram. This is equivalent to choosing a different make of the same component (for example choosing a 12-Ah battery manufactured by Hawker-Genesis instead of a 6-Ah battery manufactured by Caterpillar). ADVISOR provides modelling flexibility for a user.

ADVISOR models fit empirical data obtained from the component testing to simulate a particular subsystem. In general, the efficiency and limiting performances define the operation of each component. For example, the ICE is modelled using an efficiency map that is obtained via experiments. The efficiency map of a Geo 1.0 L (43 kW) engine is shown in Fig. 4. The maximum torque curve is also shown in this map. The engine cannot perform beyond this maximum torque constraint. Maximum torque change is another constraint to the engine subsystem. In other words, the model considers the inertia of the component in the simulation.

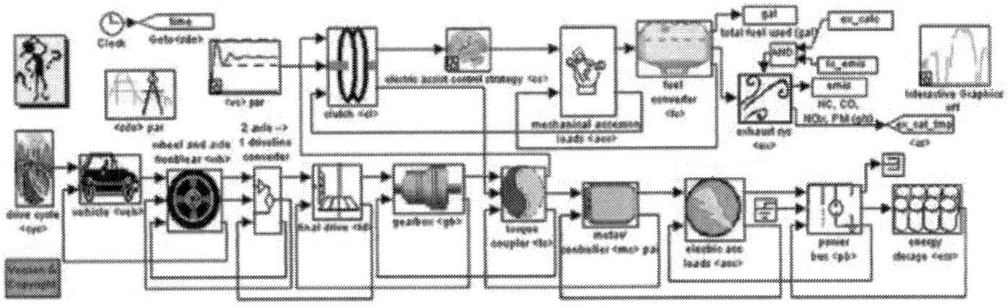

Fig. 3. Block diagram of parallel HEV in ADVISOR [3].

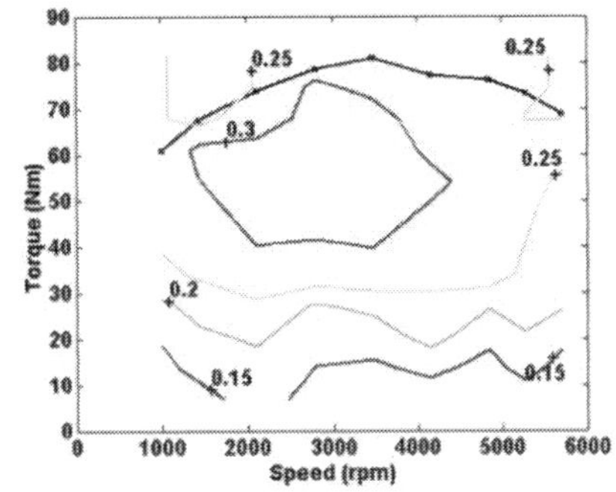

Fig. 4. Geo 1.0 L (43 kW) SI engine efficiency map model [3].

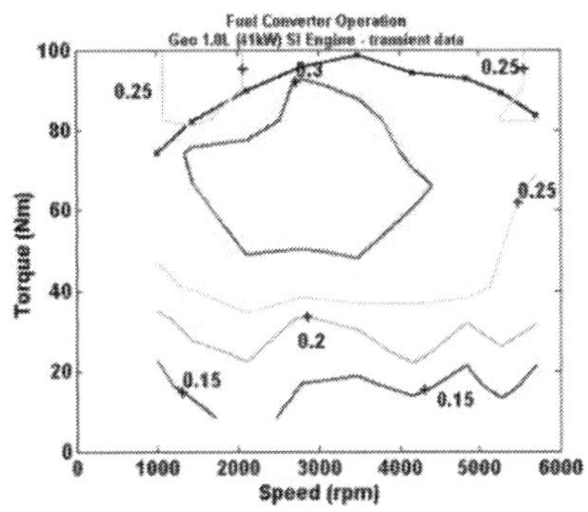

Fig. 5. Geo 1.0 L engine scaled to give a maximum power of 50 kW by linear alteration of torque characteristics.[3].

The program also allows for the linear scaling of components. For an ICE, this means linear scaling of the torque to provide the required maximum power. This type of scaling is valid only in the neighborhood near the actual parameter where the efficiency map for a slightly larger or smaller component would not change drastically. Scaling of the Geo ICE is shown in Fig. 5 so that the ICE gives a maximum power of 50 kW instead of the nominal 43 kW.

In the latest version of ADVISOR, the functionality of the software was improved by allowing links to other software packages such as Ansoft Simplorer [6] and Synopsys Saber [7]. These powerful software packages allow for a more detailed look at the electric systems of the vehicle.

As an application example, ADVISOR is used to simulate a hybrid battery/ultracapacitor energy storage system. More extensive applications can be found in [8], where ADVISOR is used to model hybrid fuel cell/battery powertrain and hybrid fuel cell/ultracapacitor powertrain and simulate their fuel economy and performance. The concept of using a hybrid energy storage system consisting of a battery and an ultra-capacitor (UC) is well known and well documented in literature [9,10]. The ultracapacitor provides and absorbs the current peaks, while the battery provides the average power required for the electric motor. This arrangement of hybrid energy storage in an HEV extends the life of the battery and allows the motor to operate more aggressively. Simulating such a system in ADVISOR allows the user to visualize the fuel economy benefit. At the same time, the program allows the user to design the best control strategy for the battery/ ultra-capacitor hybrid to improve the battery life and the overall system performance. Finally, the size of the components can be optimized and, thus, the cost and weight of the system can be reduced.

The default battery model in ADVISOR operates by requesting a specific amount of power from the battery as decided by the vehicle control strategy. Depending on the amount of power that the battery is able to supply, the battery module will send out the power available from the battery to the other subsystems. Due to the hybrid backward/forward simulation method of ADVISOR, the amount of power that the batteries are able and required to supply in a given time step is calculated in a single iteration. From this value, the battery model calculates the battery variables like current, voltage, and the battery temperature.

However, a hybrid battery/ultracapacitor energy storage system cannot be modeled within ADVISOR using the above default battery model. So, have to replace the energy storage model with a more complex model. Fortunately, the subsystem model in ADVISOR can be altered as long as the types of inputs and outputs to the rest of the vehicle are not altered. In the simulation, need to replace the battery model by a model of a combination of a battery and an ultra-capacitor connected to a local control strategy unit that splits the power demand between the battery and the ultra-capacitor. Detailed information about the control strategy is available in [9]. The block diagram representation of the system is shown in Fig. 6.

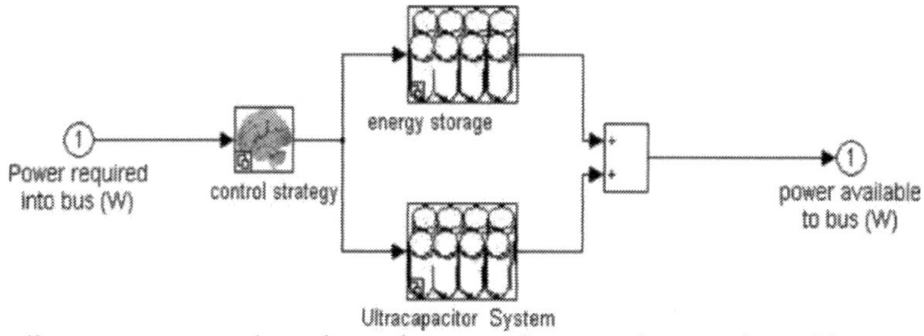

Fig. 6. Block diagram representation of new battery subsystem that consists of battery and ultra-capacitor. Input/output relation with rest of the system is left unchanged [3].

The use of the model described gives the user a way to quickly and easily simulate the battery/ultra-capacitor subsystem in a vehicle environment. It allows the user to observe the benefit of using the ultra-capacitor on the fuel economy of the vehicle as well as the benefit to the battery by making the battery state of charge more even and by reducing the peaks of the battery current that the battery has to accept. It also allows the user to validate the system whether it operates as efficiently if the battery size were reduced. Finally, the user can optimize the battery/ultra-capacitor control strategy (in other words, how the power demand will be split) without having to think about the complexities of designing the power electronics to make this control system feasible. In addition, the system can be optimized before any system is built and the system cost and possible savings can be easily calculated at the early design stage. Once the control strategy is optimized, the actual dc/dc converter with the required control strategies can be integrated into the simulation using Saber or Ansoft Simplorer software [9].

3 Propose System

The overall objective of this paper is to define a performance for a Split PHEV shown in Fig. 7 by simulates powertrain of the system. Fig. 8 shows block diagram of the system. The results of these simulations should define how the vehicle can decrease fuel consumption, while maintaining low vehicle emissions. Because of the hybrid system, just operating an engine in its regions of high efficiency does not guarantee efficient vehicle operation. These results will not give the specific powertrain commands necessary to enable complete vehicle operation, and are meant only to define a literal strategy; that is, an understanding as to why the vehicle should operate in a certain way under the given conditions.

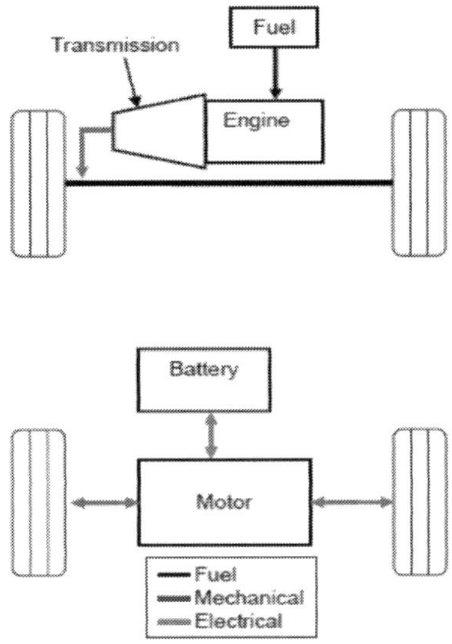

Fig.7. Split Plug-in Hybrid Electric Vehicle [2].

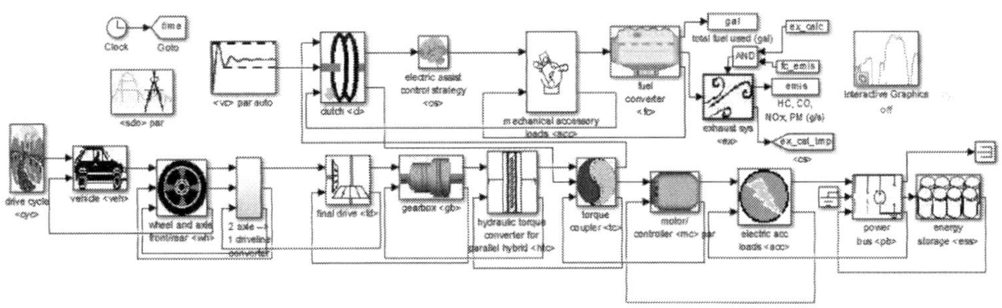

Fig. 8. Block diagram of Split PHEV in ADVISOR.

4 Simulation Setup

4.1 Simulation configuration

The system configuration was shown in Tables 1 and 2. Small car Produa Kancil was used in this simulation with default component and rear wheel system was installing the 10kW electric vehicle conversion kit (Figs. 9 and 10).

 DOI: 10.1051/matecconf/20179001019

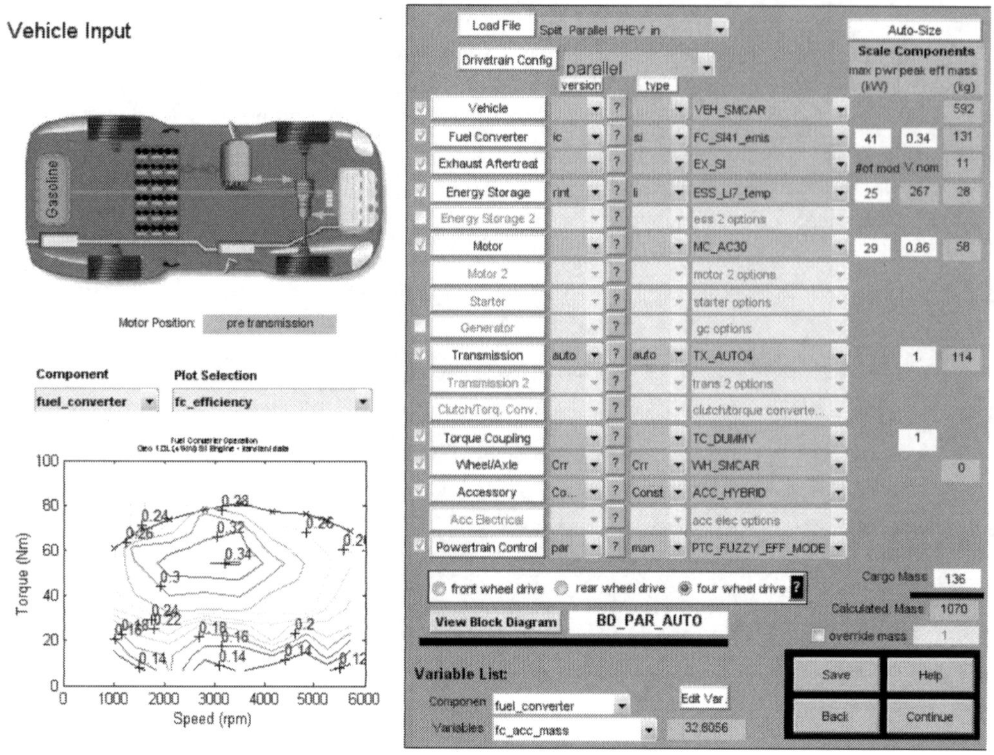

Fig. 9. Propose System: Split Plug-in hybrid.

Table 1. Split PHEV configurations.

Component	Description
Engine	850CC Gasoline Engine
Motor	10kW AC Motor
Battery	LifePO4 96V 60Ah
Transmission	4 Speed on the front and Rear Axle (Wheel distance 1385mm)Ratio: 10.6:1
Control Strategy	Propelling, Shifting and Breaking
Weight	550kg

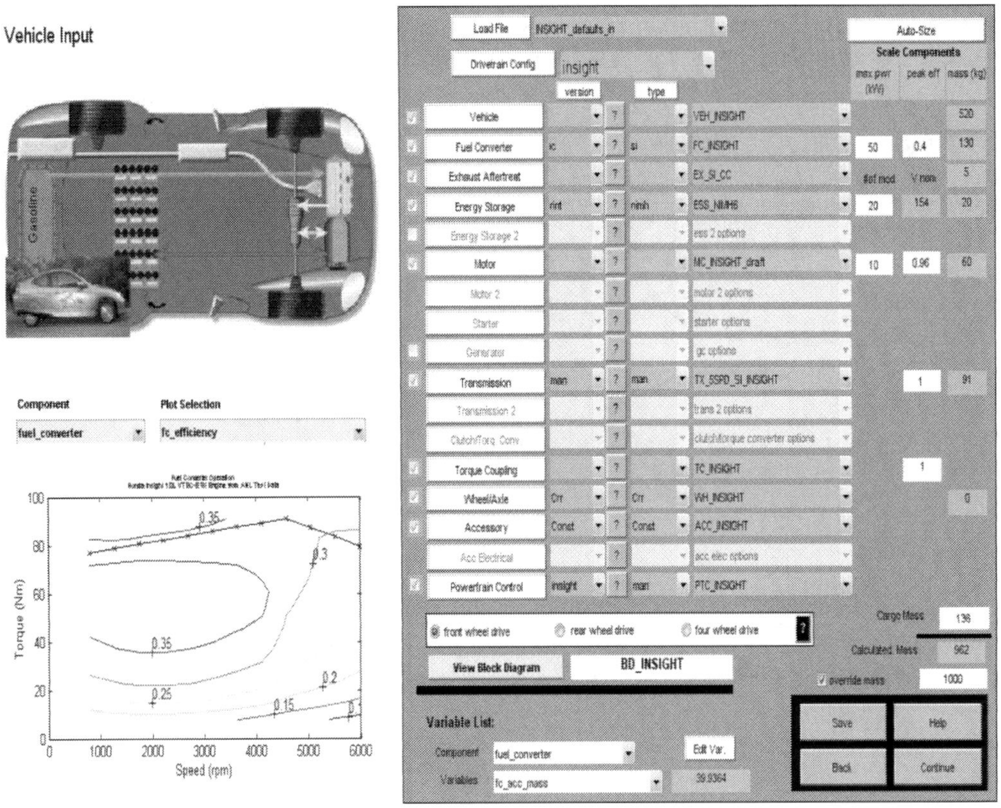

Fig. 10. Honda Insight Hybrid.

Table 2. Honda Insight Hybrid configuration.

Component	Description
Engine	995CC Gasoline Engine
Motor	10kW Permanent Magnet
Battery	Nickel Metal Hydride (Ni-MH) 144V 6.5Ah
Transmission	Front-Wheel Drive Ratio: 3.208
Control Strategy	Propelling, Shifting and Breaking
Weight	891kg

5 Results and discussion

The Graphs in Fig 11 shown the performance of propose split PHEV in ADVISOR. The patent of driving showed in first graph how drive cycle graph perform in speed versus distance. In the Fig 11 shown summary results for the propose system. Fuel consumption for this proposes vehicle is approximately 35km/L. The distance for this test is about 12km on city highway road.

Fig 12 shown comparison performance between propose PHEV and Honda Insight HEV. The results shown that the propose PHEV have better performance compare to Honda Insight. States of Charge (SOC) for propose PHEV is red line and Honda Insight HEV is blue line. Honda Insight HEV used more fuel per km compare to propose PHEV.

Honda Insight fuel consumption approximately 29km/L. The propose PHEV lead about 6km/L compare to Honda Insight base from the ADVISOR simulation.

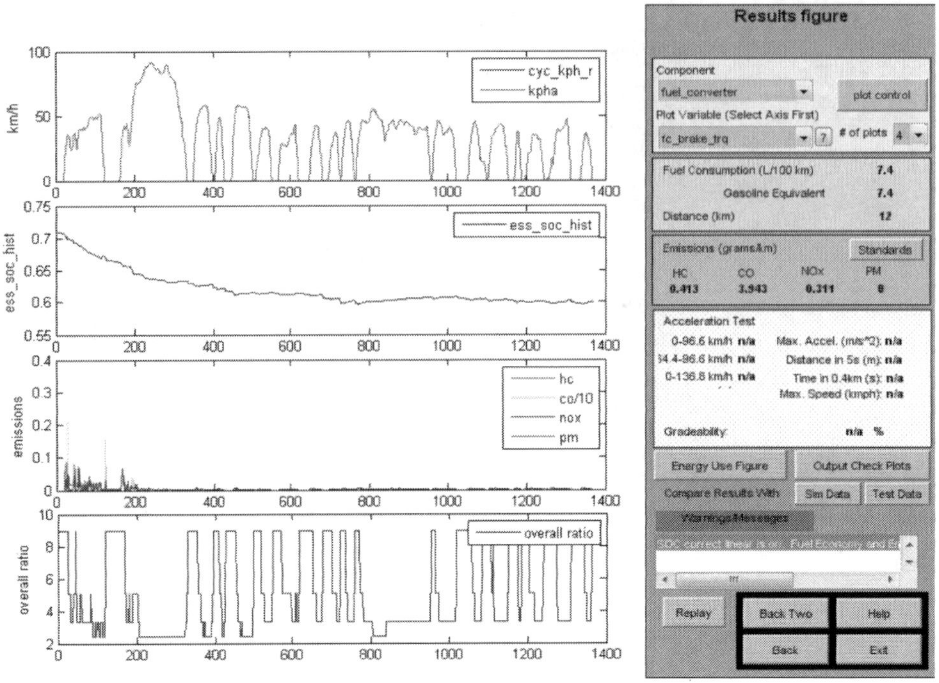

Fig. 11. Simulation result for Split Plug-in Hybrid.

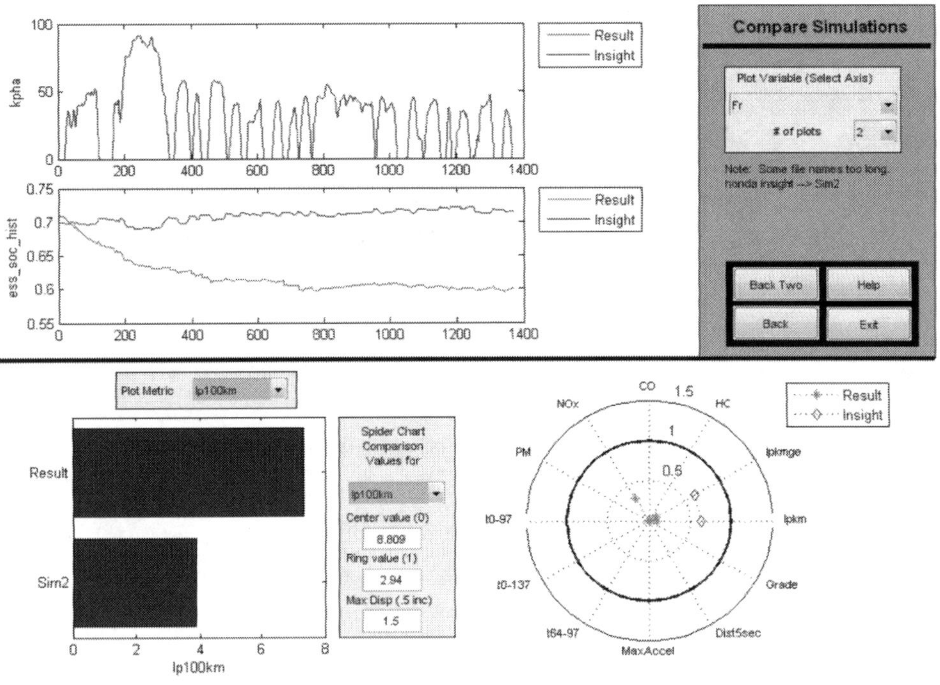

Fig. 12. Simulation results comparison for Split Plug-in Hybrid and Honda Insight Hybrid

6 Conclusions

In this paper, a dynamic model for the split PHEV was developed and compared with Honda Insight HEV. The results show that propose PHEV have better performance compare to Honda Insight HEV. These models include detailed representations for the split hybrid system, the vehicle ICE and electric motor. The developed models were simulated in ADVISOR. Using the proposed vehicle controller, the vehicle can be operated in different modes of operation depending on the driving conditions. The transient response of each mode is also studied in this paper. By applied the modelling and working in different modes of operation, the engine operates in more efficient areas of the efficiency map which implies a higher efficiency in comparison to the conventional drive-trains.

The developed platform can be used for real-time simulation of the drive-train and also for hardware-in-the-loop simulation of the vehicle if implemented on a real-time simulator such as ADVISOR.

The authors would like to be obliged to Universiti Malaysia Pahang for providing laboratory facilities and financial assistance under project no. RDU150340.

References

1. Z.A. Saiful, N. Saad, S.Mohd, and A. Rashid A. Aziz. "Split-Parallel In-Wheel-Motor Retrofit Hybrid Electric Vehicle". IEEE International Power Engineering and Optimization Conference (PEOCO), Melaka, Malaysia: 6-7 (2012).
2. J. B. Steven "Hybrid Electric Vehicle Control Strategy Based on Power Loss Calculations". Virginia Polytechnic Institute and State University Master of Science Mechanical Engineering (2006).
3. D. W. Gao, Senior Member IEEE, Chris Mi, Senior Member IEEE, and Ali Emadi, Senior Member IEEE. Modeling and Simulation of Electric and Hybrid Vehicles". Gao et al.: Modelling and Simulation of Electric and Hybrid Vehicles, **95**, No. 4, Proceedings of the IEEE (2007).
4. K. B. Wipke, M. R. Cuddy, and S. D. Burch. Nov. "ADVISOR 2.1: A user-friendly advanced powertrain simulation using a combined backward/forward approach". IEEE Trans. Vehicular Technol., **48**, pp. 1751–1761 (1999).
5. T. Markel, A. Brooker, T. Hendricks, V. Johnson, K. Kelly, B. Kramer, M. O'Keefe, S. Sprik, and K.Wipke.." ADVISOR: A systems analysis tool for advanced vehicle modeling". J. Power Sources, vol. **110**, no. 2, pp. 255–266 (2002)
6. Ansoft Simplorer Website. [Online]. Available: http://www.ansoft.com/.
7. Saber Website. [Online]. Available: http://www.synopsys.com/saber/.
8. W. Gao. "Performance comparison of a hybrid fuel cell Battery powertrain and a hybrid fuel cell Ultracapacitor powertrain". IEEE Trans. Vehicular Technol., vol. **54**, no. 3, pp. 846–855 (2005).
9. A. C. Baisden and A. Emadi. "An ADVISOR based model of a battery and an ultra-capacitor energy source for hybrid electric vehicles". IEEE Trans. Vehicular Technol., vol. **53**, no. 1 (2004).
10. M. I. M. Rashid, H. Daniyal. "Modelling And Simulation Of Split Plug-In Hybrid Electric Vehicle Using Advisor". ARPN Journal of Engineering and Applied Sciences, Vol. **10**, No. 21, pp. 9860–9865 (2015).

MATEC Web of Conferences **90**, 01020 (2017) DOI: 10.1051/matecconf/20179001020
AiGEV 2016

Correlation of numerical and experimental analysis for dynamic behaviour of a body-in-white (BIW) structure

N.A.Z. Abdullah[1], *M.S.M.* Sani[2,*] , *M.M.* Rahman[1,2] and *I.* Zaman[3]

[1]Advance Structural Integrity and Vibration Research (ASIVR), Faculty of Mechanical Engineering, Universiti Malaysia Pahang, 26600 Pekan, Pahang, Malaysia
[2]Automotive Engineering Centre, Universiti Malaysia Pahang, 26600 Pekan, Pahang, Malaysia
[3]Department of Engineering Mechanics, Faculty of Mechanical and Manufacturing Engineering, Universiti Tun Hussein Onn Malaysia, 86400 Parit Raja, Batu Pahat, Johor, Malaysia

Abstract. In order to determine the reliability of data gathered using computational version of finite element analysis, experimental data is often used for validation. In case of finite element analysis, it can sometimes be considered as inaccurate especially when subjected to complex and large structure such as body-in-shite. This is due to difficulties that might occur in modelling of joints, boundary conditions and damping of the structure. In this study, a process of comparison and validation of model based test design with modal testing results was conducted. Modal properties (natural frequencies, mode shapes, and damping ratio) of a body-in-white (BIW) structure were determined using both experimental modal analysis (EMA) and finite element analysis (FEA). Correlation of both sets of data was performed for validation. It appeared that there was significant value of error between those two sets of data. The discrepancies that appear after correlation was then reduced by performing model updating procedure. The results presented here may demonstrate the effectiveness of model updating technique on improving the complex structure such as BIW structure.

1 Introduction

The trustworthiness of structural mechanical numerical analysis such as finite element analysis (FEA) can always be validated by using experimental analysis data. The difference of values gathered through experiments indicates the level of accuracy for the existing finite element model used in numerical prediction analysis. It is very important to ensure that the existing finite element model that was created during design process is a reliable model and able to provide accurate prediction of structural behavior and performance before the actual structure undergo mass production in manufacturing area. As stated in many previous studies, finite element modeling is the most elementary instruments used to mock up the complex system or structure [1-5]. On the other hand, the body-in-white (BIW) structure,

* Corresponding author: mshahrir@ump.edu.my

which consists of several sheet metal components welded together, and other numerous types of joints, dents and curvy profile on the structure, can be classified as one complex structure as well. Modeling an exact structure of BIW in finite element can be tricky and very hard to achieve. Thus, the goal of modeling in this case is to come up with a representative model that is simple enough for mathematical manipulation yet capable for describing, inducing, and reasoning complicated phenomena. By using principle from finite element method, a complicated structure, BIW for instance, can be divided into many small elements. The mass and stiffness matrices of each element can be assigned easily. Then the global mass and stiffness matrices of the structure can be assembled using these element matrices by considering connectivity and all the boundary condition [6].

In order to validate the existed finite element model whether it is accurate enough to replicate the response that will be produced by actual structure, many researchers had suggested the usage of modal data gathered both through FEA and experimental modal analysis (EMA) [3, 7–9].The modal properties of the structure in subject which include the natural frequencies, damping ratios, and mode shapes will be obtained through both FEA and EMA. These two sets of data will be correlated in order to calculate the level of discrepancies between them.

While discrepancies between prediction analysis and experimental on the real structure is unavailable due to existence of many local effects that are not taken into account by FE modeling when predicting frequencies and modes, model updating is a method widely suggested and frequently applied when it comes to improve the correlation between FE model and the experimental data [10]. Increase of interest in the field of model updating has been developed throughout the years and quite a number of researchers has demonstrate the effectiveness of model updating procedure in reducing the error between prediction and actual structure [5, 11-15]. In addition, not a few numbers of studies has contributed to the development and improvement of model updating technique [16-20].

Yet, very few studies have done the correlation of dynamics properties and perform globalize model updating procedure on a very complex structure such as BIW. It can be seems from literature survey that most study only demonstrate the procedure of localize model updating on various types of joints, which was shown using simple beam and plate. Thus, the present study deals with correlation of experimental modal data with the data gathered through numerical production of a BIW structure which can be considered as a complex structure and demonstrate the procedure of model updating on the correlated data.

2 Finite element modelling and analysis

BIW structure, which is the foundation of vehicle's shape, can be considered as a complex structure. Due to this fact, modeling the structure accordingly to the actual structure might be difficult. Therefore, major simplification in modeling the structure was carried out when constructing the finite element model of the BIW. Most of the indentations and curvatures on the surfaces of the BIW structure, as well as the available joints on the structure, were all neglected during modeling process.

Computer aided design (CAD) model was produced first before being imported into a computer aided engineering (CAE) software, MSC.Nastran/Patran, in order to convert the model into a finite element model. the BIW is modeled according and based on five basic modules consisted in BIW structure, which are the floor and underbody, dash panel assembly, front structure, body sides, and roof assembly. The whole BIW model was modeled as surface structure where the thickness for each surface created will be assigned in properties manager in MSC.Nastran/Patran software.

The finite element model of the BIW was created by using 30911 elements of tria3-shaped elements while applying hybrid mesher. Equivalence of nodes was then done with equivalence tolerance to be 0.005 m, in order to eliminate the coinciding nodes between each surface. The BIW model was assigned to three different shell properties with the same material properties of steel, which details are shown in Table 1. All those different thickness was assigned according to condition of each surface on actual structure. Thicker properties were assigned due to existing of multilayer and stiffening substructure on each surface. Since the complexness of the BIW structure was ignored on modeling process, the assigned thick properties aims to represent the stiffness of the surface.

In general, floor and underbody, body side, and dash panel surface were assigned with thicker properties while roof and front structure surface were assigned with thinner properties.

Table 1. Nominal value of material properties assigned to BIW model.

Properties		Nominal Value
Material properties (Steel)	Young's modulus	200 GPa
	Poisson ratio	0.3
	Density	7860 kg/m^3
Shell properties	Thickness 1	0.012 m
	Thickness 2	0.008 m
	Thickness 3	0.006 m

Neither boundary condition nor external forces were applied to the model as the model was let to be in free-free boundary condition for calculation of modal properties using SOL 103 in MSC.Nastran/Patran, which is the solution for the normal modes analysis. The solution of the equation of motion for natural frequencies and normal modes requires a special reduced form of the equation of motion. The reduced equation of motion in matrix form is as shown in Eq. (1)

$$[M]\{\ddot{u}\}+[K]\{u\}=0 \quad (1)$$

where $[M]$ is the mass matrix, $[K]$ is the stiffness matrix and u is the vector of the variable describing the motion. Equation 1 was solved by assuming a harmonic solution of the form as Eq. (2)

$$\{u\}=\{\phi\}\sin\omega t \quad (2)$$

where $\{\phi\}$ is the eigenvector or mode shape and ω is the circular natural frequency. This harmonic solution is the key to the numerical solution of the problems and indicates that all the degrees of freedom of the vibrating structure move in asynchronous manner. The structural configuration does not change its basic shape during motion and the only thing that changes is its amplitude. If the differentiation of the assumed harmonic solution is performed and substituted into the equation of motion, the simplification of the equation will be represented as Eq. (3).

$$([K]-\omega^2[M])\{\phi\}=0 \quad (3)$$

As Eq. (3) was reduces to an eigenvalue problem, it will be represented as Eq. (4)

$$[K]-\omega_i^2 M\{\phi\_i\}=0 \quad (4)$$

where ϕ is the eigenvector (mode shape) corresponding to its eigenvalue and the eigenvalue λ_i is related to the natural frequency as Eq. (5)

$$f_i=\omega_i/2\pi \quad (5)$$

where f_i is the i-th natural frequency; and $\omega_i=\sqrt{\lambda_i}$. The constructed finite element model as viewed in graphic interface of MSC.Nastran/Patran software is as shown in Figure 1. Meanwhile, the computed first 5 mode shapes obtained in FEA is shown in Figure 2.

Fig. 1. Finite element model of the BIW structure.

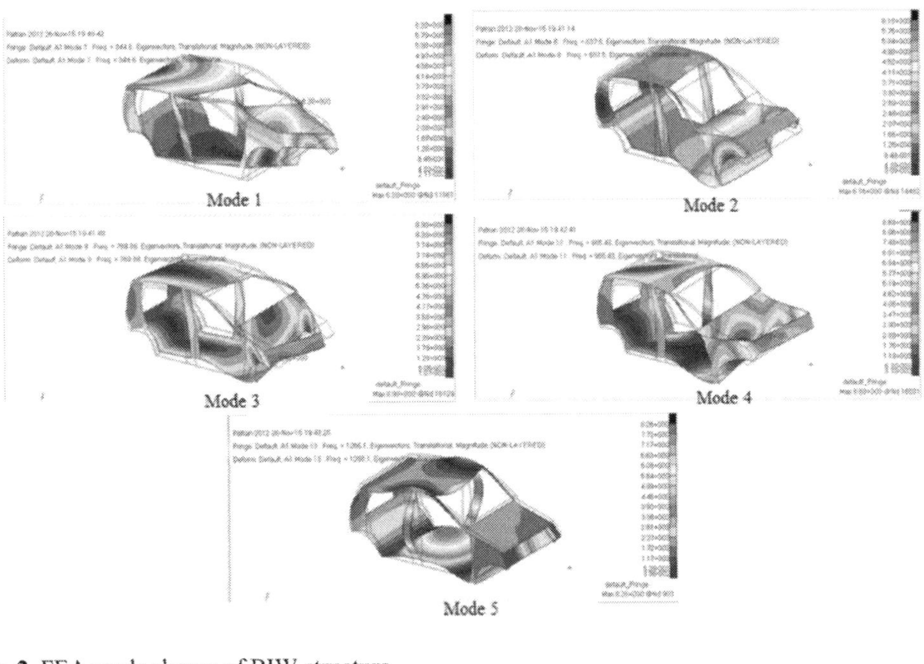

Fig. 2. FEA mode shapes of BIW structure.

3 Experimental modal analysis

Experimental modal analysis (EMA) or sometimes called as modal testing, is the process of extracting dynamic characteristics of a system, machinery or structure experimentally. Carrying out EMA on a structure or system has the advantage of having modal characteristic defined from actual measurements. An important property of modes or the measured frequency response functions from a modal testing can be used to describe the structure's dynamic or modal properties, which are the natural frequencies, damping ratios and mode shapes.

There are several types of response domain (whether frequency or time domain) that can be gathered via EMA in order to extract those modal properties. However, the most typical data used for parameter extraction, and was used in this study, are the gathered frequency response functions (FRFs), which use excitation input and the corresponding output of the test structure. The method of excitation is the impact hammer test. Impact hammer testing is one of the most frequently used methods in modal. Apart from the ability to compute FRF measurements in an FFR analyzer, impact hammer testing is also a fast, convenient, and low cost way of finding the modes of machines and structure [21].

Before conducting the impact hammer testing, an experimental model of BIW was created by using post-processing software. The model consists of lines and points which virtually represent the geometric shape of the BIW structure (refer Figure 3). Data gathered through numerical analysis was used as guide to perform the experiments in way of indicating the best location for accelerometers on the BIW structure. In order to perform the impact hammer test, the BIW structure was hanged from a test rig by using towing cable in order to put the structure under free-free boundary condition (see Figure 4). The measurements were made using a modal analysis software and several other equipment such as PCB 086D20 impact hammer with medium soft tip attached, 4-channel NI DAQ device, and a tri-axial PCB accelerometer. Roving accelerometer method was adopted for the testing procedure, where one knocking point and 59 measurement points was assigned on the structure. Roving accelerometer test was done by creating initial disturbance on the BIW structure at one fixed position while the single tri-axial accelerometer was roved around other measurement points. Nodal points were ensured to be avoided when choosing the impact and measurement point. The vibrational response was measured by using the 4-channel NI DAQ device. Curve fitting procedure available in the modal analysis software was used to extract the modal properties of the BIW structure from the computed FRFs.

The computed FRF based on all measurement points is shown in Figure 5. The estimated modal parameters were extracted by applying curve fitting method to the FRF graph in order to obtain a set of experimental derived data. The outcome of curve fitting, which is a set of modal parameters, which consist of natural frequencies, damping and residues, for each of the identified modes within the range of frequency of interest, is shown in Figure 6. After curve fitting process was completed, the modal parameters are stored in a Shape Table as residue mode shapes. The mode shapes was then displayed through the created experimental model. The computed modal parameters are shown in Figure 6. According to the computed FRF graph, the natural frequency for the first mode is 7.84 Hz.

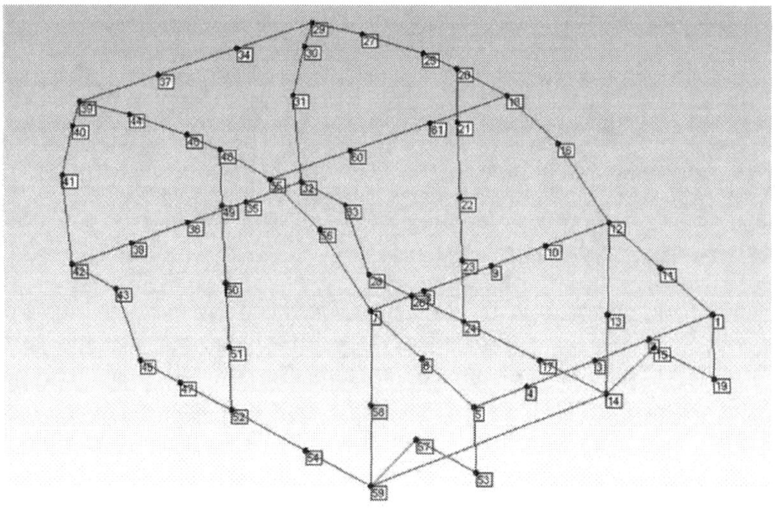

Fig. 3. Experimental model of BIW structure.

Fig. 4. Hanging BIW structure under free-free boundary condition.

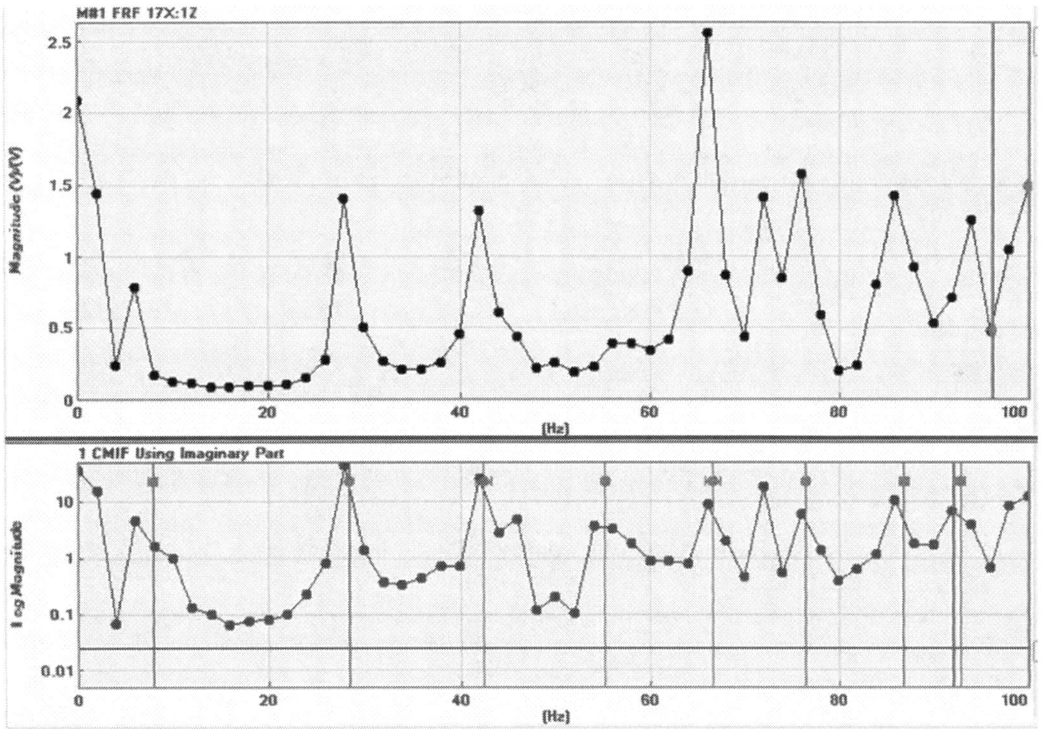

Fig. 5. Computed FRF for all measurement points.

 DOI: 10.1051/matecconf/20179001020

	Select Mode	Frequency (Hz)	Damping (Hz)	Damping (%)	Residue Mag	Residue Phs (deg)
1	Yes	7.84	0.453	5.77	2	-30.7
2	Yes	28.4	0.162	0.571	8.36	-36.8
3	Yes	42.4	0.816	1.93	14.6	-172
4	Yes	55.1	0.344	0.624	3.78	-170
5	Yes	66.4	0.815	1.23	21.1	-44.4
6	Yes	76.3	0.143	0.188	5.46	172
7	Yes	86.8	0.466	0.537	15.4	174
8	Yes	92.8	0.47	0.506	8.06	176

Fig. 6. Computed modal data from FRF graph.

4 Correlation and finite element model updating

Correlation of data that was obtained through finite element analysis and modal testing was conducted in order to analyze the discrepancies existed between those two sets of data. In addition, correlation of data is essential in order to have accurateness estimation on the existing BIW model. Table 2 exhibits the correlation of natural frequencies of BIW structure that was gathered via experimental and numerical (finite element) analysis. Value of discrepancies between those two sets of data was calculated by accepting the value obtained through experiment as the actual value. As illustrated, the percentage of error that exists is significantly high especially in second, third and fourth mode.

Table 2. Correlation of natural frequencies between EMA and FEA.

Mode	Experimental Natural Frequency (Hz)	Numerical Natural Frequency (Hz)	Percentage of Error (%)
1	7.84	7.41	5.48
2	28.40	22.78	19.79
3	42.40	38.65	8.84
4	55.10	50.07	9.13
5	66.40	66.01	0.59
6	76.30	72.89	4.47
		Total Error	48.30

In order to reduce the existing discrepancies, model updating procedure was applied on the finite element model of BIW, thus improving the model to have better correlation with the actual structure. Several updating parameters (Young's modulus and all the thicknesses used in finite element analysis) were considered to be included in model updating procedure.

Approximation subject to the simple first-order Taylor series expansion was used in SOL 200 in MSC.Natran in order to change the vector λ of eigenvalues based on the vector θ of structure updating parameters as stated in Eq. (6).

$$\lambda_{(i+1)}=\lambda_i+[S_i](\delta\theta) \quad \textbf{(6)}$$

where $[S_i]$ is a sensitivity matrix of i-th iteration, which signify the rate of change of the structural eigenvalues λ_i with respect to changes in $\delta\theta$. The expression for the eigenvalue sensitivity S_i can be stated as Eq. (7).

$$S_i=(\delta\lambda_i)/\delta\theta=\phi_i^T[\delta K/\delta\theta-\lambda_i(\delta M/\delta\theta)]\ \phi_i \quad \textbf{(7)}$$

The objective function for the prediction error is defined as Eq. (8).

$$g(x)=\sum W((\omega_i^e/\omega_i^a)-1)^2 \quad \textbf{(8)}$$

where ω_i^e and ω_i^a are the experimental and finite element natural frequencies respectively and W is the weighting factor for each mode.

The number of updating parameters was kept to be less than number of modes to be updated in order to avoid ill-conditioning problem in updating procedure. The sensitivity analysis was carried out in the first place in the interest of ensuring that the selected updating parameters were meaningful. The prediction of objective function g(x) in updating procedure was set to obtain the minimized value. Therefore, the procedure was kept going until convergence was accomplished when the values of g(x) is sufficiently small.

Comparison of error in initial finite element results to the experimental results and the error after model updating was performed on the finite element data was illustrated in Table 3. Data from this table can be compared with the data in Table 2 which clearly shows the notable changes in the natural frequency values. The most striking result to emerge from the data is that total percentage of error obtained after applying model updating procedure reduced significantly when compared to the initial results. Although the error for mode 5 in the updated data is greater than the error from the initial model, other modes show noteworthy reduction of error. Thus, the updating procedure can be considered as successful.

On the other hand, the deviations of the updating parameters assigned initially on the BIW structure to the updated value were visible in Table 4. The process of updating achieved convergence after 13 iterations. The finding highlights that Young's modulus, thickness 2 and thickness 3 shows greater increment. This is owing to the reason that these parameters shows higher sensitivity value during sensitivity analysis compared to thickness 1.

Table 3. Comparison of discrepancies between initial and updated results.

Mode	EMA Frequency (Hz)	Initial FEA Natural Frequency (Hz)	Initial Percentage of Error (%)	Updated Numerical Natural Frequency (Hz)	Percentage of Error after Model Updating (%)
1	7.84	7.41	5.48	8.25	5.23
2	28.40	22.78	19.79	25.25	11.09
3	42.40	38.65	8.84	41.94	1.08
4	55.10	50.07	9.13	53.79	2.38
5	66.40	66.01	0.59	70.16	5.66
6	76.30	72.89	4.47	77.76	1.91
		Total Error	48.30	Total Error	27.36

Table 4. Changes of updating parameters from the initial values.

Parameter	I	II	Deviation (%)
	Initial Value	Updated Value	=|(II-I)/I|×100
Young's Modulus (GPa)	200	226.34	13.17
Thickness 1 (m)	0.012	0.0121	0.83
Thickness 2 (m)	0.008	0.0084	5.00
Thickness 3 (m)	0.006	0.0049	18.33

5 Conclusions

This study was undertaken to correlate the experimental modal data to the data gathered from finite element analysis of a BIW structure. A finite element model of the BIW structure was produced and the percentage of errors between those two sets of data was obtained and the study was furthered into performing model updating procedure in order to reduce the existing percentage of errors. The updating procedure is considered as parameter identification and strikes to improve numerical prediction to be as closely as possible to the actual structure.

Four parameters were chose as the updating parameters and correlation was done based on the first six measured natural frequencies. The results of this investigation show that discrepancy is unavoidable when constructing model for a complex structure such as BIW because of inaccuracies in parameters assumption and simplification in process of modeling. However, this inconsistency between the prediction model and the actual structure can be improved by carrying out model updating procedure.

It is recommended that further research be undertaken by including the joint during modeling process. Therefore, procedure of model updating can be carried out at the joint element and joint properties can be chosen as one of the updating parameters. At the end, the updated finite element model structure can possibly exhibit more accurate and reliable numerical prediction model for future research.

This work was supported and funded by FRGS fund provided by the Ministry of Higher Education (MOHE) under RDU 130149.

References

1. M. Imregun and W. J. Visser, "A review of model updating techniques," *Shock Vib. Dig.*, 1991.
2. M. I. Friswell and J. E. Mottershead, *Finite element model updating in structural dynamics*, vol. 38. 1995.
3. D. J. Ewins, "Model validation: Correlation for updating," *Sadhana*, vol. 25. pp. 221–234, 2000.
4. A. Bagchi, "Updating the Mathematical Model of a Structure Using Vibration Data," *J. Vib. Control*, vol. 11, no. 12, pp. 1469–1486, Dec. 2005.
5. N. A. Husain, H. H. Khodaparast, A. Snaylam, S. James, G. Dearden, and H. Ouyang, "Finite-element modelling and updating of laser spot weld joints in a top-hat structure for dynamic analysis," *Proc. Inst. Mech. Eng. Part C J. Mech. Eng. Sci.*, vol. 224, no. 4, pp. 851–861, Apr. 2010.
6. O. C. Zienkiewicz and R. L. Taylor., *The Finite Element Method Vol. 1*. 1994.
7. M. I. Friswell, J. E. Mottershead, and H. Ahmadian, "Finite-element model updating using experimental test data: parametrization and regularization," *Philos. Trans. R. Soc. London, Ser. A (Mathematical, Phys. Eng. Sci.*, vol. 359, pp. 169–186, 2001.
8. G. Banwell, S. Mohr, S. Rothberg, and J. Roberts, "Using experimental modal analysis to validate a finite element model of a tennis racket," *Procedia Eng.*, vol. 34, pp. 688–693, 2012.
9. E. Reynders, "System Identification Methods for (Operational) Modal Analysis: Review and Comparison," *Arch. Comput. Methods Eng.*, vol. 19, no. 1, pp. 51–124, 2012.
10. N. A. Z. Abdullah, M. S. M. Sani, M. M. Rahman, and I. Zaman, "A review on model updating in structural dynamics," *IOP Conf. Ser. Mater. Sci. Eng.*, vol. 100, no. 1, p. 012015, Dec. 2015.
11. N. A. Husain, H. H. Khodaparast, A Snaylam, S. James, M. Sharp, G. Dearden, and H. Ouyang, "Modal testing and finite element model updating of laser spot welds," *J. Phys. Conf. Ser.*, vol. 181, p. 012024, Aug. 2009.
12. J. E. Mottershead, M. Link, and M. I. Friswell, "The sensitivity method in finite

element model updating: A tutorial," *Mech. Syst. Signal Process.*, vol. 25, no. 7, pp. 2275–2296, Oct. 2011.
13. Q. Yuan, "Dual approaches to finite element model updating," *J. Comput. Appl. Math.*, vol. 236, no. 7, pp. 1851–1861, 2012.
14. C. Zang, S. C. Ma, and M. I. Friswell, "Structural model updating with an improved updating parameter selection method," *Proceedings of the International Conference on Noise and Vibration Engineering ISMA 2012*, 2012.
15. S. N. Zahari, M. S. M. Sani, N. A. Husain, M. Ishak, and I. Zaman, "Dynamic Analysis Of Friction Stir Welding Joints In Dissimilar Material Plate Structure," *J. Teknol.*, vol. 78, no. 6–9, 2016.
16. Q. Yuan, "A constrained matrix least-squares problem in structural dynamics model updating," *J. Comput. Appl. Math.*, vol. 280, pp. 367–376, May 2015.
17. H. Sun and R. Betti, "A Hybrid Optimization Algorithm with Bayesian Inference for Probabilistic Model Updating," *Comput. Civ. Infrastruct. Eng.*, vol. 30, no. 8, pp. 602–619, Aug. 2015.
18. J. Zapico-Valle, R. Alonso-Camblor, M. P. González-Martínez, and M. García-Diéguez, "A new method for finite element model updating in structural dynamics," *Mech. Syst. Signal Process.*, vol. 24, pp. 2137–2159, 2010.
19. S. Adhikari and M. I. Friswell, "Distributed parameter model updating using the Karhunen–Loève expansion," *Mech. Syst. Signal Process.*, vol. 24, no. 2, pp. 326–339, Feb. 2010.
20. W. Wang, J. E. Mottershead, A. Ihle, T. Siebert, and H. Reinhard Schubach, "Finite element model updating from full-field vibration measurement using digital image correlation," *J. Sound Vib.*, vol. 330, no. 8, pp. 1599–1620, Apr. 2011.
21. B. J. Schwarz and M. H. Richardson, "Experimental modal analysis," in *CSI Reliability Week, Orlando, FL*, 1999, pp. 1–12.

 DOI: 10.1051/matecconf/20179001021

Finite element modelling and updating of friction stir welding (FSW) joint for vibration analysis

Siti Norazila Zahari[1], *Mohd Shahrir* Mohd Sani[1,2,*] and *Mahadzir* Ishak[2]

[1]Advanced Structural Integrity and Vibration Research (ASIVR), Faculty of Mechanical Engineering, Universiti Malaysia Pahang, 26600 Pekan, Pahang, Malaysia
[2]Automotive Engineering Centre, Universiti Malaysia Pahang, Malaysia

Abstract. Friction stir welding of aluminium alloys widely used in automotive and aerospace application due to its advanced and lightweight properties. The behaviour of FSW joints plays a significant role in the dynamic characteristic of the structure due to its complexities and uncertainties therefore the representation of an accurate finite element model of these joints become a research issue. In this paper, various finite elements (FE) modelling technique for prediction of dynamic properties of sheet metal jointed by friction stir welding will be presented. Firstly, nine set of flat plate with different series of aluminium alloy; AA7075 and AA6061 joined by FSW are used. Nine set of specimen was fabricated using various types of welding parameters. In order to find the most optimum set of FSW plate, the finite element model using equivalence technique was developed and the model validated using experimental modal analysis (EMA) on nine set of specimen and finite element analysis (FEA). Three types of modelling were engaged in this study; rigid body element Type 2 (RBE2), bar element (CBAR) and spot weld element connector (CWELD). CBAR element was chosen to represent weld model for FSW joints due to its accurate prediction of mode shapes and contains an updating parameter for weld modelling compare to other weld modelling. Model updating was performed to improve correlation between EMA and FEA and before proceeds to updating, sensitivity analysis was done to select the most sensitive updating parameter. After perform model updating, total error of the natural frequencies for CBAR model is improved significantly. Therefore, CBAR element was selected as the most reliable element in FE to represent FSW weld joint.

1 Introduction

Friction stir welding (FSW) is rapidly used to join thin sheet metal with similar and dissimilar materials in the application of shipbuilding, aerospace, railway and automotive [1-3]. FSW seems to be prominent technique as its enable to join dissimilar materials in solid state while avoiding drawbacks of fusion welding due to the large difference in terms

* Corresponding author: mshahrir@ump.edu.my

of chemical composition and physical properties between the components to be joined. FSW joints not only provide connection between sheet metal but also significantly influence the global structural behaviour of complete structures. Therefore, it is crucial to understand the dynamics characteristics of the welded joints which can be achieved via computational and experimental work. Even though prediction task through computational methods are widely used in predicting the behaviour of welds on a complete structure, developing a numerical model of the weld itself is a complex issue. This is mainly because of the existence of many local effects that are not taken into account by FE modelling when predicting frequencies and modes.

Recent years, many works regarding on FSW in dissimilar materials such as dissimilar alloy [4-6] focusing more on microstructure and mechanical properties of the weld itself. There have been several studies in the literature reporting on modelling of FSW [7-11] concentrated in simulation of the process and limit capacity analysis. Modelling for these types of analysis required detailed mesh in order to work out a smooth stress field within and around FSW weld joint. On the contrary for vibration analysis, only simplified model which represent the stiffness of FSW weld and enable predict their influence for entire structure [12]. To the author's best knowledge, there is no reported work on modelling the dynamic behaviour of a structure with FSW welded structure. For the dynamic analysis of FSW welded structures, a major requirement of FSW finite element models is to accurately predict the dynamic characteristics of welded structures with a small number of degrees of freedom. Previous research has reported not taking into consideration the effect of welded joints in their mode [13]. For instance, Sani [13] investigated on identification of dynamics modal parameter for car chassis. A rigid connection was assumed at the welded joint in the car chassis.

The primary goal of this study is to present an appropriate way on how to model FSW welded joints between dissimilar materials using finite element method. In this paper, AA6061 and AA7075 Al alloy used as the base material due to its significant application in automotive, rail transportation and aerospace industries.

First, the welded joint will be modelled using equivalent technique as an initial model and result from numerical analysis will be compared using nine set of experimental results in order to choose the most ideal set of FSW welded plate. Then, three types of modelling technique using rigid body connection element, single bar element and spot weld connection; CWELD will be discussed. All types of FE modelled will undergo normal mode analysis and the predicted result will be validated using experimental modal analysis (EMA) to select the most reliable model to represent FSW joint. Model updating will be carried out for selected to improve correlation between experiment and numerical counterparts.

2 Description of structure

2.1 Material specification and welding process

Nine specimens consists of two different series of aluminium alloy plates; AA6061 and AA7075 with the same dimensions (200 mm × 100 mm × 2 mm) as illustrated in Figure 1 jointed together by FSW. The joint configuration for this paper is square butt joint, where its being joined together using tool pin that transverse along longitudinal length at the centre point between two plates. Tool steel with material AISI H13 with cylindrical pin profile was used as welding tool [14]. The tool consists of shoulder and pin with diameter 17.7 mm and 5.80 mm for each. Backing plate and parallel bar were made of mild steel used to support and hold the specimen during welding process.

DOI: 10.1051/matecconf/20179001021

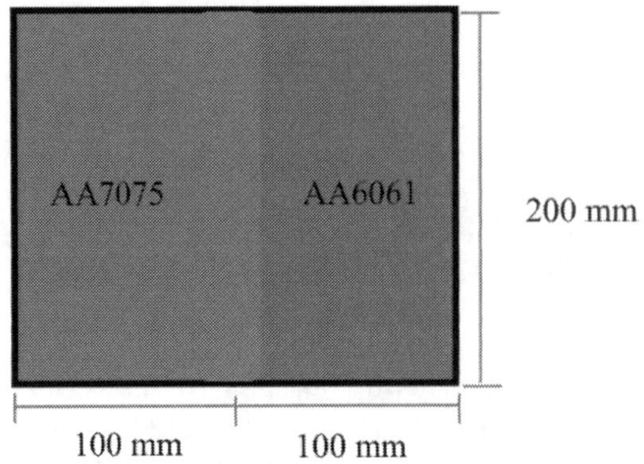

Fig. 1. Two plates with dissimilar material welded by FSW.

FSW process was carried out using vertical milling machine. For this study, plate AA6061 was placed on advancing side [4, 15] due to its higher mechanical strength and tool pin was positioned at the center of joint line. Process parameters used in this study are tabulated in Table 1. Figure 2 below show the specimen after being welded.

Table 1. Welding parameters.

	Rotational Speed (rpm)	Welding Speed (mm/min)	Tilt Angle (°)
A	900	30	0
B	1000	30	1
C	1100	30	2
D	900	40	0
E	1000	40	1
F	1100	40	2
G	900	50	0
H	1000	40	1
I	1100	50	2

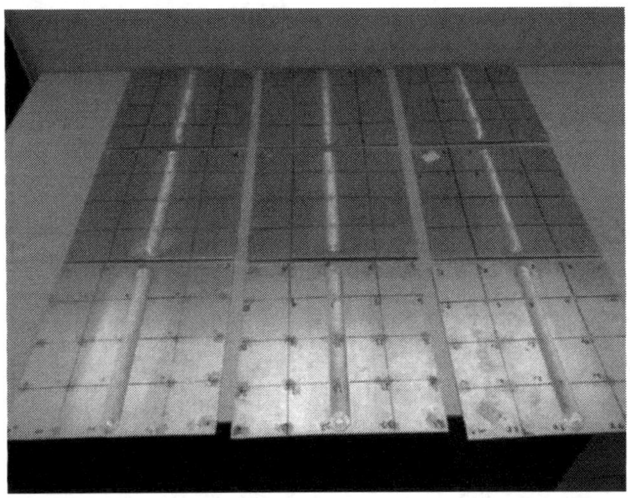

Fig. 2. Nine set of specimens.

 DOI: 10.1051/matecconf/20179001021

3 Experimental modal analysis

Modal testing was executed to extract modal parameters such as natural frequency and mode shapes experimentally. Impact hammer testing used as excitation method in this EMA. The plate was supported by a sponge in order to achieve free-free boundary conditions. The experiment setup is illustrated in Figure 3 below.

Fig. 3. Experimental modal analysis setup.

The welded plates were divided into 25 small grids points to obtain frequency response function (FRF). A PCB Piezotronic uniaxial accelerometer with sensitivity of 97.17mV/g was used in this experiment and it roved to each of 25 points shown in this test. The structure is excited by an impact hammer fixated at point 14 on the specimen. The excitation force and the response accelerations are acquired and analyzed using the FFT analyzer. Then the FRFs are calculated. Modal parameters (natural frequency and mode shapes) extracted using curve fitting method in MEscope software and results tabulated in the Table 2 below as validation of numerical result.

4 Finite element model and analysis (FEA) for initial model

A commercial finite element software package namely, MSC Patran was used to model the specimen with FSW joint and MSC Nastran was utilized to generate its natural frequency and mode shapes using normal mode analysis (SOL103). A FE model of the structure shown in Figure 1 is built using four noded shell elements (CQUAD4). Neither constraints nor load were assigned to create free-free boundary state. Minimum frequency of 1Hz being set to avoid the solver from calculating the six rigid body motions that having frequency less than 1 Hz. The materials for two flat plates are assigned according to their properties. Material AA6061 have density, ρ of 2700 kg/m3 with Young modulus, E of 69 GPa. The density for material AA7075 is 2820 kg/m3 and Young modulus of 72 GPa. Initial model using equivalence nodes had been developed in order the select the most optimum specimen that can be used later for another FSW modelling technique.

4.1 Material specification and welding process

When the structure consists of more than one surface, use of equivalence becomes mandatory in order to connect the mesh. The equivalence command in MSC Patran plays a role as remover of redundant node between two surfaces. Figure 4 present the details on equivalence nodes in connecting the two surfaces and represent the rigid connection between them and neglecting the joint connection that exist.

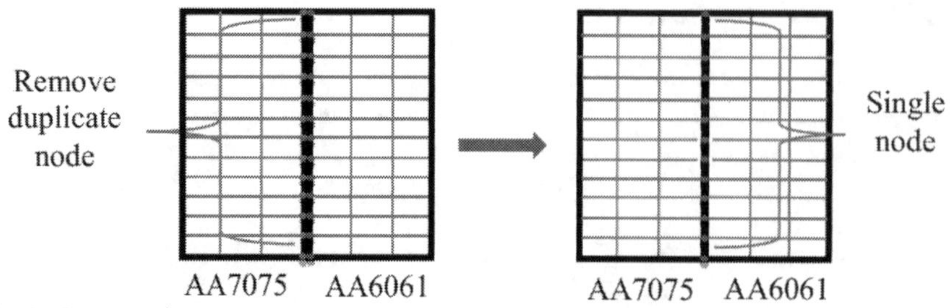

Fig. 4. Equivalence node.

4.2 Finite element analysis (FEA) of initial model

Normal mode analysis (SOL 103) in MSC NASTRAN was performed to compute the modal data of the FSW welded structure.

4.3 Comparison between EMA and FEA for initial model

Natural frequency from the experimental are compared with the natural frequency of FEA and being listed in Table 2 and Table 3 show the percentage error of nine set of specimen and Figure 5 presents the mode shape extracted from numerical and test analysis.

Table 2. FEA vs EMA for nine set of specimen (natural frequency).

Natural Frequency (Hz)									
FEA	EMA (SET)								
	A	B	C	D	E	F	G	H	I
160.12	144	178	154	158	143	162	154	143	168
232.19	190	235	240	203	232	208	200	217	260
296.01	298	260	316	308	308	286	329	301	326
406.47	398	402	364	426	432	398	396	345	363
413.80	458	422	444	450	437	472	538	424	436

Table 3. Percentage of error between FEA and EMA.

Natural Frequency (Hz)									
FEA	EMA (SET)								
	A	B	C	D	E	F	G	H	I
160.12	11.19	10.04	3.97	1.34	11.97	1.16	3.97	11.97	4.69
232.19	22.21	1.20	3.25	14.38	0.08	11.63	16.10	7.00	10.70
296.01	0.67	13.85	6.33	3.89	3.89	3.50	10.03	1.66	9.20
406.47	2.13	1.11	11.67	4.58	5.91	2.13	2.64	17.82	11.98
413.80	9.65	1.94	6.80	8.04	5.31	12.33	23.09	2.41	5.09
Total Error	**45.85**	**28.14**	**32.02**	**32.23**	**27.16**	**30.75**	**55.83**	**40.86**	**41.66**

Table 3 indicates that set E have the minimum total error between FEA and EMA. These findings provide evidence that Set E represent the ideal joint condition compare to other set of specimen. Therefore, Set E will undergo other modelling technique to find the most suitable method to represent this specimen.

 DOI: 10.1051/matecconf/20179001021

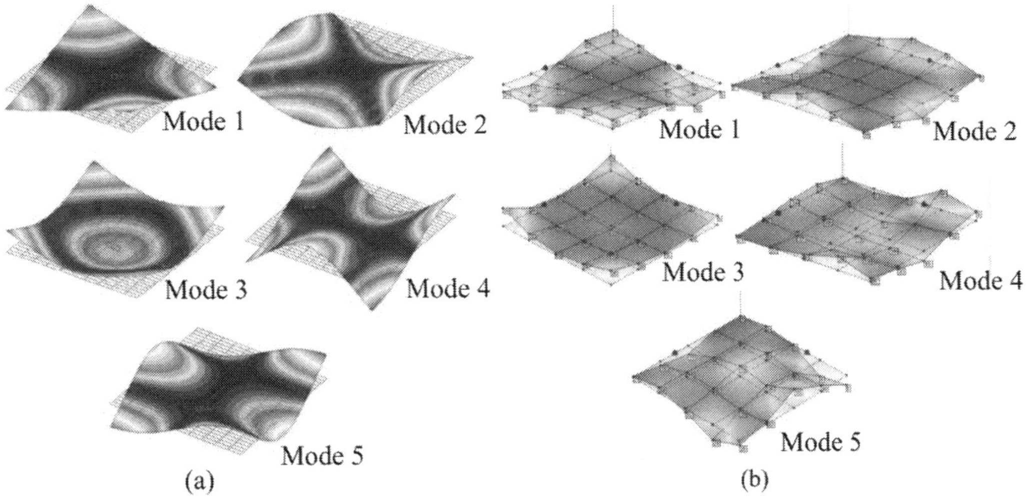

Fig. 5. Mode shapes from (a) FEA and (b) EMA.

5 Different approach of finite element model technique of FSW

This section discusses the details on the other technique for modelling the FSW weld joints for Set E using rigid body element (RBE2), one dimensional bar element and CWELD element.

5.1 Rigid Body Element (RBE2)

Rigid body with independent degree of freedom (DOF) at one grid and dependent DOF at an arbitrary number of grids called RBE2. These elements rigidly weld multiple grids to other grid.

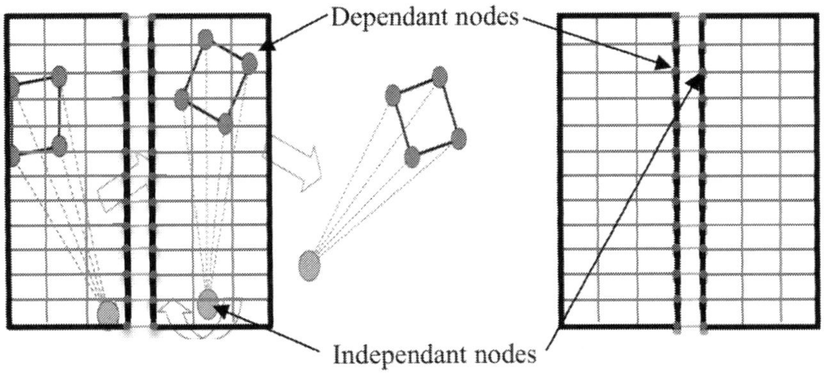

Fig. 6. RBE2.

In modelling the welded flat plate, 31 RBE2 elements being assigned with the middle nodes of each plate attached to one another in a straight line [16]. The first five frequencies of the FE model with RBE2 weld joint representative are tabulated in Table 4 (column III) and the FE results are validated by comparing them with their experimental counterparts.

 DOI: 10.1051/matecconf/20179001021

5.2 One dimensional (1D) bar element (CBAR)

31 curves had been assigned to connect the two plates as shown in Figure 7. Bar element defined as a type of properties for curve. The diameter for bar is 0.018m and its geometrical properties; Young modulus, density and Poisson ratio had been set same as base material.

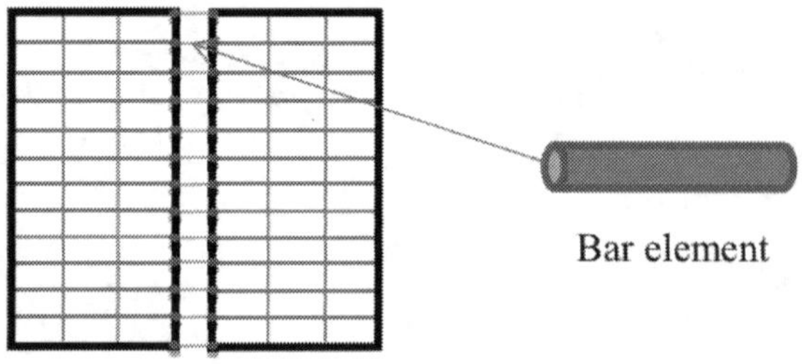

Fig. 7. CBAR element.

5.3 Spot weld connector element - CWELD

The CWELD element (Figure 8) is developed using a two noded special shear flexible beam type element with 12 DOFs (six for each node) and each node is connected to its corresponding patch with constraints from the Kirchhoff shell theory [17].

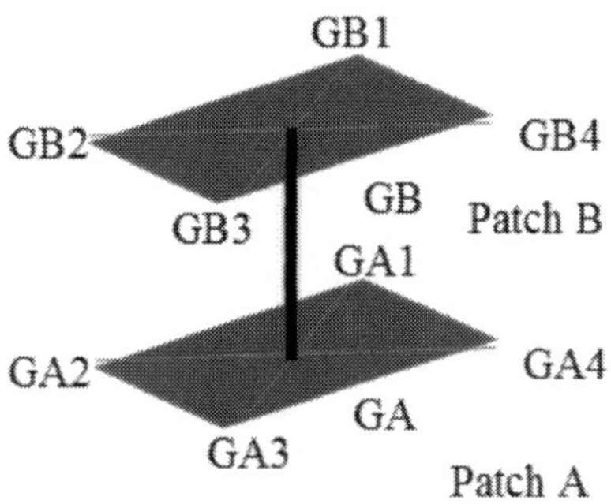

Fig. 8. CWELD element.

In modelling the FSW weld joints, 16 CWELD elements are employed to represent continuous weld at the center of adjacent edge of two plates. The property of CWELD element defined on a PWELD entry. The diameter of spot weld being set approximately 18 mm equals to diameter of tool shoulder that been used in FSW.

5.4 Comparison of EMA and FEA and selection of elements representing FSW joints

The three types of model are compared in order to choose the most reliable to represent FSW joints. Table 4 summarized the percentage of error of RBE2, CBAR and CWELD while Table 5 illustrated the prediction of mode shape.

From the result, it shows that CBAR suffering of underestimation of natural frequency meanwhile CWELD element and RBE2 capable to reduce the percentage error. However based on Table 5 for prediction of mode shapes, mode swapping problem occurred for mode 4 and 5 when utilizing the CWELD element. In contrast, the mode shapes are successfully predicted by RBE2 and CBAR element.

Table 4. Percentage of error between EMA and FEA for 3 types of modelling.

Modes	Natural Frequency (Hz)						
	EMA	RBE2	Percentage error (%)	CBAR	Percentage error (%)	CWELD	Percentage error (%)
1	143	160.15	11.99	159.78	11.73	160.15	1.29
2	232	232.15	0.06	218.79	5.69	249.31	7.46
3	308	295.96	3.91	281.15	8.72	333.86	8.40
4	432	406.39	5.93	380.31	11.97	413.51	4.28
5	437	413.68	5.34	413.18	5.45	450.60	3.11
Total Error			**27.23**		**43.56**		**24.54**

Table 5. Mode shapes for EMA and FEA.

Mode shapes (MS)	Types of modelling			
	EMA	RBE2	CBAR	CWELD
MS 1	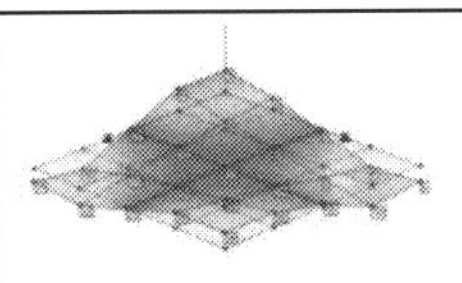	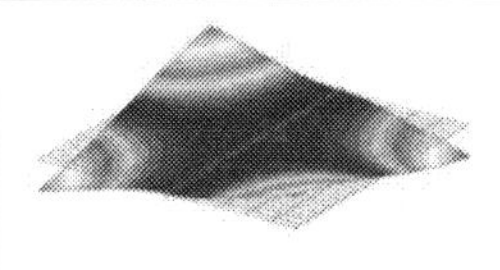		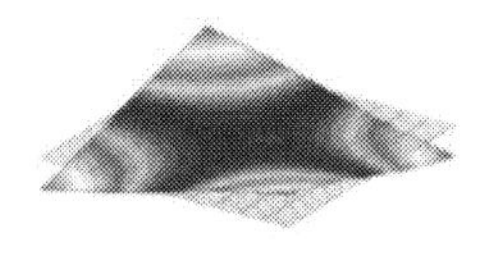
MS 2	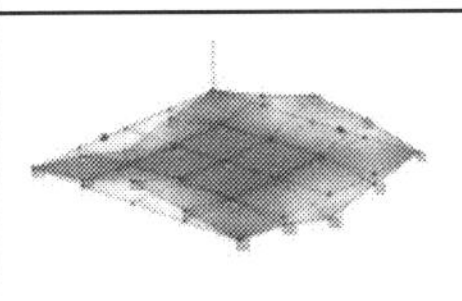	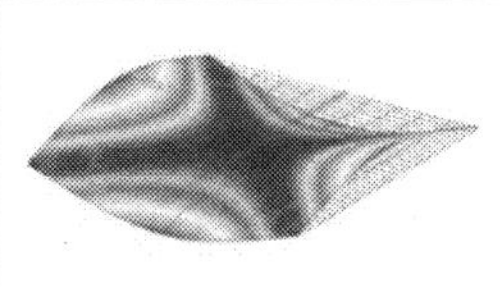	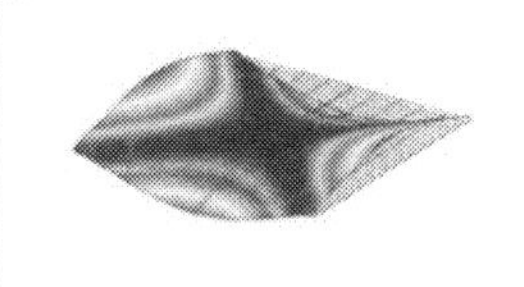	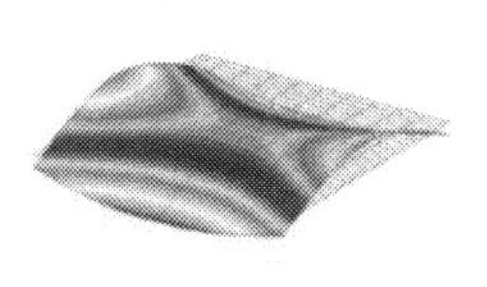
MS 3	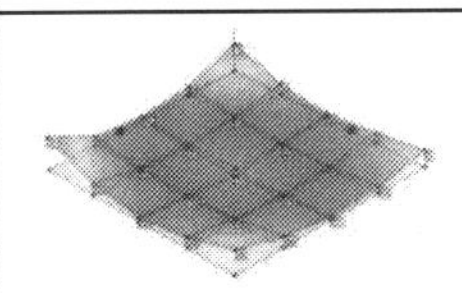		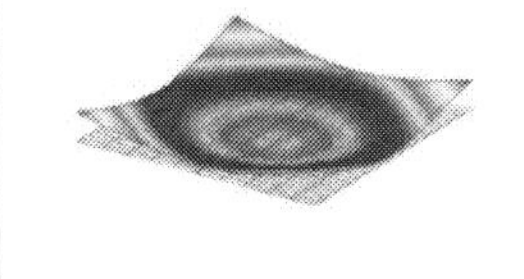	
MS 4	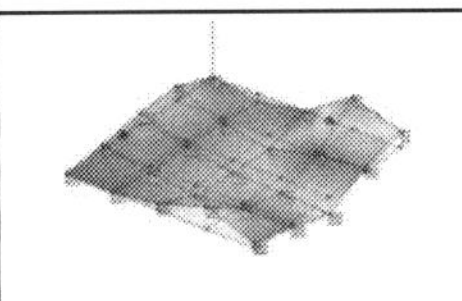	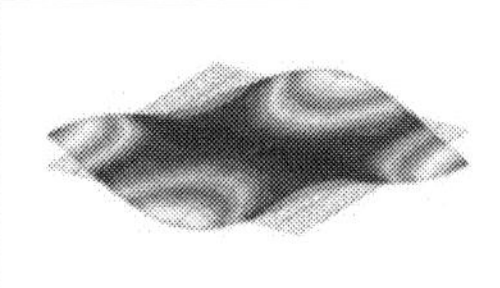	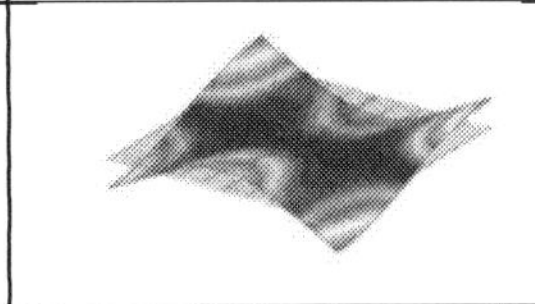	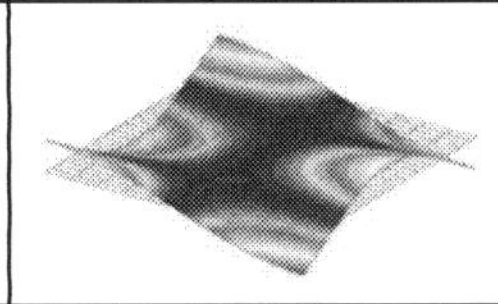
MS 5	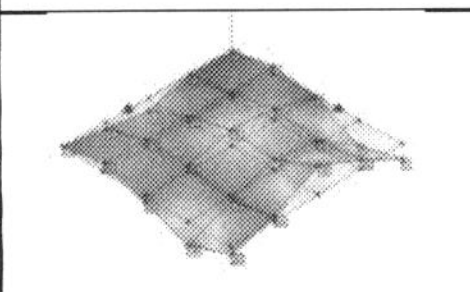	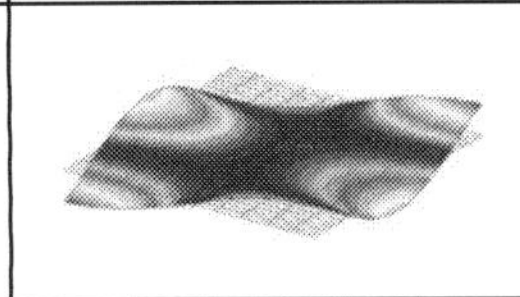	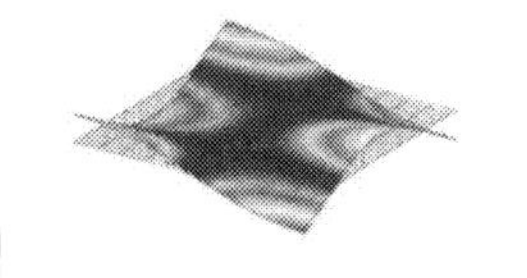	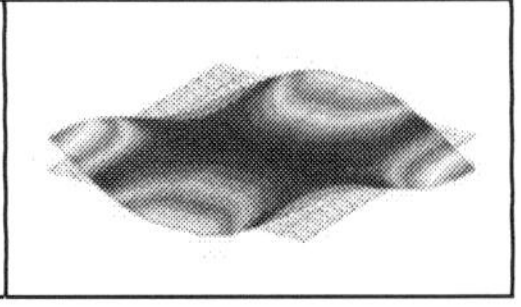

With regards to modelling joints, all types of model can be modelled when congruent mesh was applied to the model. There is no problem occurred in this study because the FE model was meshed using CQUAD4 for both plates therefore all meshing are congruent and simply coincides with each other. For model updating purpose, RBE2 that represent weld joints do not possess both geometrical and material properties that can be used to update compare to CBAR and CWELD that contains parameters in modelling of weld that can be used to update. From these comparisons, it is found that the characteristic of CBAR is much more appealing compare to other two models, so the CBAR element is selected to represent FSW butt joint modelling and will be proceed to model updating method to improve the correlation between numerical and experimental counterparts.

5.5 FE Model updating of welded structure

Before proceed to model updating, it is essential to make sure the geometrical properties of FE model was well defined because FE model updating only correct the errors that exist from the uncertainties and assumptions of modelling parameters. There are several methods which the model can be parameterized for updating. Seven parameters had been chosen and undergo the sensitivity analysis using SOL200 in MSC Nastran software to select the suitable parameter to update. The sensitivities of the first five natural frequencies were tabulated in Table 6 and comparisons to be made in order to choose the most sensitive parameters.

Table 6. Sensitivity analysis for six parameters.

Output Type	Young Modulus (E_{7075})	Young Modulus (E_{6061})	Poisson Ratio (ν_{7075})	Poisson Ratio (ν_{6061})	Diameter CBAR (D)	Density (E_{7075})	Density (E_{6061})
NF 1	40.19	39.72	-8.95	8.5	-0.0008	-41.06	-38.67
NF 2	54.90	54.51	-8.48	8.58	-23.46	-52.19	-47.75
NF 3	70.42	70.20	27.87	28.94	-26.85	-69.53	-62.07
NF 4	94.92	95.28	-9.08	9.06	-43.51	-87.94	-82.46
NF 5	102.52	104.12	-11.16	10.63	-0.0005	-102.35	-100.32

Based on sensitivity data, Young's Modulus and density for both plates are the most sensitive among all six parameters. But only one of these parameters is selected for updating because of their direct relation in calculation of natural frequency. Modulus of elasticity was chosen because it provides more uncertainties than the density. Diameter of CBAR despite not being sensitive enough is also selected for its contribution for modes 2, 3 and 4. Same goes to Poisson ratio for both materials that sensitive to mode 3 and 5. But to avoid ill conditioning problem, only four parameters was chosen. Therefore only one Poisson ratio was selected due to the same level of sensitivities. In conclusion, four parameters; Young Modulus of AA7075 and AA6061, Poisson ratio for AA6061 involved in model updating. Updating is performed by minimizing the error function and executed on the basis of the first five modes of measured frequencies. Table 7 displays the error between FEA and EMA after updating.

Table 7. Percentage of error between FEA and EMA after updating.

Modes	Experiment (Hz)	Initial FEA		Updated FEA	
		FEA (Hz)	Error (%)	FEA (Hz)	Error
1	143	159.78	11.73	159.77	11.85
2	232	218.79	5.69	231.79	0.09
3	308	281.15	8.72	300.08	2.57
4	432	380.31	11.97	406.22	5.97
5	437	413.18	5.45	415.25	4.98
Total Error			43.56		25.34

From the table, it can be concluded that the natural frequency improved significantly and total error become decreased. It is proved that the modulus of elasticity, Poisson ratio for both plates and diameter of CBAR which are considered as the parameter that significantly affect the behavior of the model. It is also found that not only material properties of the structure influenced the natural frequency of the structure but geometrical properties of CBAR element that used to model the weld joint also plays an important part that improving the frequency. The changes in value for updated parameters are shown in Table 8.

Table 8. Updated value of parameters.

Parameter	I	II	Deviation (%)
	Initial value	Updated Value	= |(II-I)/I|
Young's Modulus, E_{AA7075}[GPa]	72	68.4	0.95
Young's Modulus, E_{AA6061}[GPa]	69	75.3	1.09
Poisson Ratio, υ_{AA6061}	0.33	0.35	1.06
Diameter CBAR (m)	0.018	0.004	0.25

6 Conclusions

In summary, this study has deals with some of the technique on modelling the FSW weld joints for prediction of dynamic behaviour. Three types of modelling using RBE2, CBAR and CWELD were developed and the suitability of these three modelling was compared using correlation between natural frequency and mode shapes extracted from EMA and FEA for set E. CBAR element was selected as a reliable model for FSW joints due to its accurate prediction of mode shapes and contains an updating parameter for weld modelling compare to other weld modelling. The model updating was executed in order to improve the correlation between measured counterparts. Before proceed to model updating, sensitivity analysis had been done to make sure the most sensitive parameters will be chosen for model updating. Results show that Young modulus, Poisson ratio for both material and diameter of CBAR are the most sensitive parameter for updating. After perform model updating, total error of the natural frequencies for CBAR model is improved significantly.

The author would like to greatly acknowledge the support by focus group of Advanced Structural Integrity of Vibration Research (ASIVR) and Universiti Malaysia for providing all the equipment used for this project and special thank for Knowledge Transfer Programme (KTP) and Ministry Of Education for financial assistance support.

References

1. B. Gibson, et al. Friction stir welding: process, automation, and control. *Journal of Manufacturing Processes*. **16**(1): p. 56-73 (2014).
2. R. Nandan, T. DebRoy, and H. Bhadeshia. Recent advances in friction-stir welding–process, weldment structure and properties. *Progress in Materials Science*. **53**(6): p. 980-1023 (2008) .
3. D.Lohwasser, and Z. Chen. *Friction stir welding: From basics to applications*. 1st ed., United Kingdom: Woodhead Publishing Limited (2009).
4. J. Guo, J, et al. Friction stir welding of dissimilar materials between AA6061 and AA7075 Al alloys effects of process parameters. *Materials & Design*. **56**: p. 185-192 (2014).
5. P. Sadeesh, et al. Studies on friction stir welding of AA 2024 and AA 6061 dissimilar metals. *Procedia Engineering*. **75**: p. 145-149 (2014).
6. R. Kesharwani, S. Panda, and S. Pal. Multi Objective Optimization of Friction Stir Welding Parameters for Joining of Two Dissimilar Thin Aluminum Sheets. *Procedia Materials Science*. **6**: p. 178-187 (2014).
7. D.M. Neto and P. Neto. Numerical modeling of friction stir welding process: A literature review. *International Journal of Advanced Manufacturing Technology*. **65**(1-4): p. 115-126 (2013).

 DOI: 10.1051/matecconf/20179001021

8. C.C Tutum, and J.H. Hattel. Numerical optimisation of friction stir welding: review of future challenges. *Science and Technology of Welding and Joining*. **16**(4): p. 318-324 (2011).
9. J.E. Gould, and Z. Feng. Heat flow model for friction stir welding of aluminum alloys. *Journal of Materials Processing and Manufacturing Science*. **7**: p. 185-194 (1998).
10. H. Schmidt, J. Hattel, and J. Wert. An analytical model for the heat generation in friction stir welding. *Modelling and Simulation in Materials Science and Engineering*. **12**(1): p. 143 (2004).
11. X. He, F. Gu, and A. Ball. A review of numerical analysis of friction stir welding. *Progress in Materials Science*. **65**: p. 1-66 (2014).
12. N.A. Husain, et al. Finite-element modelling and updating of laser spot weld joints in a top-hat structure for dynamic analysis. *Proceedings of the Institution of Mechanical Engineers, Part C: Journal of Mechanical Engineering Science*. **224**(4): p. 851-861 (2010).
13. M.S.M.Sani, et al. Identification of Dynamics Modal Parameter for Car Chassis. *IOP Conference Series: Materials Science and Engineering*. **17**(1): p. 012038 (2011).
14. V. Infante, et al. Study of the fatigue behaviour of dissimilar aluminium joints produced by friction stir welding. *International Journal of Fatigue*.
15. B. Fu, et al. Friction stir welding process of dissimilar metals of 6061-T6 aluminum alloy to AZ31B magnesium alloy. *Journal of Materials Processing Technology*. **218**: p. 38-47 (2015).
16. S.N. Zahari, et al. Dynamic analysis of friction stir welding joints in dissimilar material plate structure. *Jurnal Teknologi (Sciences & Engineering)*. **78**(6-9): p. 57-65 (2016).
17. J.Fang, et al. *Weld modelling with MSC. Nastran*. in *Second MSC Worldwide Automotive User Conference, Dearborn, MI, USA*. (2000).

MATEC Web of Conferences **90**, 01022 (2017) DOI: 10.1051/matecconf/20179001022
AiGEV 2016

The study of stability, combustion characteristics and performance of water in diesel emulsion fuel

Zulkifli Syafiq[1], *Othman* Fahmi[1] , *Omar*. I Awad[1] and *Abdullah* Adam[1,2,*]

[1]Faculty of Mechanical Engineering, Universiti Malaysia Pahang, 26600 Pekan, Pahang, Malaysia
[2]Automotive Engineering Center, Universiti Malaysia Pahang, 26600 Pekan, Pahang, Malaysia

Abstract. A single cylinder diesel engine study of water in diesel emulsions was conducted to investigate the stability effect of emulsion fuel on three different fuel blends and the water emulsification effect on the engine performance. Emulsified fuels contained 2% of surfactant including Span 80 Tween 80 and tested 10 HLB number. The blends also varied of 5%, 10% and 15% of water in diesel ratios namely as BSW5, BSW10 and BSW15. The fuel blends performance was tested using a single cylinder, direct injection diesel engine, operating at 1860 rpm. The results on stability reveal that high shear homogenizer yields more stability on emulsion fuel than mechanical stirrer and ultrasonic water bath. The engine performance results show that the ignition delay and peak pressure increase with the increment of water percentage up to 15%. However, the results indicate the increment of water percentage is also shows a significant decrease in engine power.

1 Introduction

Limitations of fossil fuel resources and restrictions on environmental pollution have directed research programs worldwide to introduce the new method that not only ensures rationalization of fuel consumption but also keep emissions at low levels from different combustion devices. Particular attention is indeed given to both petrol and diesel engines as they represent a significant contributor to both fuel consumption and environmental pollution. As the demand for fossil fuel increase, there is a concern for the effect of burnt fossil fuel that will pollute the environment. Alternative fuels including biodiesels as well as the use of fuel additives are among recent methodologies that aim at reducing the utilization of fuels and/or improving engine performance while meeting environmental regulations. The mixing of water-fuels emulsions is considered as promising ways to realize improved fuel economics and usage of fuels. The technology of water fuel emulsion has been stated as the most universal and effective method that enable simultaneous reduction of the engine smoke level and nitrogen oxides (NO_X) without the necessity for engine modifications [1].

* Corresponding author: adam@ump.edu.my

The interest in water-in-diesel emulsions derives from the fact that water in the form of micrometer-sized droplets exerts some positive effects on the combustion of the fuel. The main reason for a stronger interest in incorporating water in diesel than in gasoline is that the high combustion temperature and the high pressure that exist in diesel engines is particularly suitable for the concept. An experimental on water addition emulsified droplet combustion reveal the combination of high-pressure environment and a high boiling temperature fuel is emphasis on improving smoke reduction . Therefore, water–oil emulsions are particularly suitable for applications in diesel engines, where high-pressure combustion exists, a relatively non-volatile fuel is used, and smoke emission problem exists [2]. In addition, using emulsion diesel can lead to other advantages such as additional momentum in jet behavior which may assist better mixing of fuel and air. Moreover, the presence of tiny water particles may lead to micro explosion, which also enhance further fuel atomization [3]. Use of diesel emulsions has been shown to give several interesting effects, such as (i) reduced NOx emission, lower soot and particulate contents in the exhaust, and (ii) improved combustion efficiency [4-6].

Emulsion stability study is to measure the separation time for water and fuel. Research about emulsion stability reveals that emulsion instability leads to changes of emulsion properties [7-9]. This is the major challenge to prepare emulsion itself. The first challenge depends on the emulsion preparation method or the technology used to deform the dispersed phase into fine droplets, while the second is solved or retarded by the proper selection of the type and concentration of emulsifier to reduce the interfacial tension between two immiscible liquids.

The proper preparation method is needed to verify the efficiency the water fuel emulsion including the mixer/blender types and the surfactant's HLB number. Gouvêa et al. found that the chemical reaction rates can be enhanced by using ultrasonic technique [10]. The ultrasonic waves are of mechanically vibrating types and propagate through a transmission medium causing cavitation. Side-by-side, the motion of ultrasonic wave creates an alternate positive and negative pressure waves. Both of these methods forms a strong stirring affect and mixes the two immiscible fluids. Therefore, the ultrasonic technique would be used beneficially to prepare water–diesel emulsions [11-14]. The mechanical stirrer machine, the mostly used emulsifying machine, stirs up the mixture of continuous and dispersed phase with the presence of surfactant [15]. With the increase in the stirrer speed, the mean droplet diameter is observed to be reduced almost linearly[16]. In addition, the high shear homogenizer is used to blend the emulsion to ensure the mixture in a homogenous state [17]. It is used to allow optimal mixing and homogenization under higher viscosity and delivers more reduction in droplets size.

Surfactants are compounds that lower the surface tension or interfacial tension between two liquids. Surfactants may act as foaming agents, wetting agents and emulsifier. To increase the kinetic stability, surfactant act as emulsifier needed to stabilize the emulsion. The emulsifier has the ability to dissolve with oil (lipophilic or hydrophobic) and water (hydrophilic) with different degree of solubility expressed by the hydrophilic–lipophilic balance (HLB). HLB value an important to develop emulsification [18, 19]. The most popular emulsifiers used for water-diesel emulsification are those based on aliphatic hydrocarbons as sorbitan esters known as span. While for oil-in-water emulsifications, emulsifiers based on polysorbitan esters known as tween. Researchers reported that the applications of a mixture of span and tween lead to better results [20, 21]. This study is to investigate emulsion with Span 80 Tween 80 and HLB 10. The purpose of the current work is to create emulsified diesel-water fuel with three different type of mixer and study the effect of 5%, 10% and 15% percentage of water content in water diesel emulsion on diesel engine performance.

2 Experimental setup

Table 1 shows the commercial non-ionic surfactants Span 80 (hydrophobic) and Tween 80 (hydrophilic) adopted in this work as they were expected to stabilize emulsions effectively. The formula for blend Span 80 and Tween 80 to any given HLB value shown in equation 1 and 2.

$$\%(A) = (HLB\text{-}HLB_B)\times 100\,/\,HLB_A\text{-}HLB_B \quad (1)$$

$$\%(B) = 100 - \%(A) \quad (2)$$

Table 1. Physicochemical properties of surfactants.

Properties		Span80	Tween80
Density	kg/m	994	1060
Molar mass	g/mol	428.6	1310
Hydroxyl value	mgKOH/g	193-210	65-80
HLB value	-	4.3	15.0
Content in 10 HLB (%)	-	46.7	53.3

To stabilize the emulsions, 2% by volume surfactant mixture consisting of Span 80 and Tween 80 (HLB 10) were blended by a mechanical stirrer, ultrasonic water bath machine (FB15051 Fisherbrand) and high shear homogenizer. The mixer were all set to operate for 10 minutes to make sure there are no impurities to achieve the maximum homogenization. Then the emulsified fuel was kept in 100ml lab container and observation been done to measure the emulsion stability of the fuel in the room temperature.

The engine used in this experiment was the YANMAR TF120M single cylinder, four strokes with water cooled diesel engine. The engine specifications are listed in Table 4. A schematic diagram of the experimental setup is shown in Figure 1. The engine is equipped with two fuel tanks, air intake system and data acquisition system. The engine works at a constant speed of 1860 rpm. Three blends emulsion fuel with water content of 5%, 10% and 15% (namely, BSW5, BSW10, and BSW15) were used in this experiment. The engine was set at 0%, 25%, 50%, 75% and 100% engine load at injection angle 18^0 BTDC. The engine was tested repeatedly prior to actual testing in order to ensure all instruments are in the stable condition and the engine testing was done according to the SAE J1349 standard.

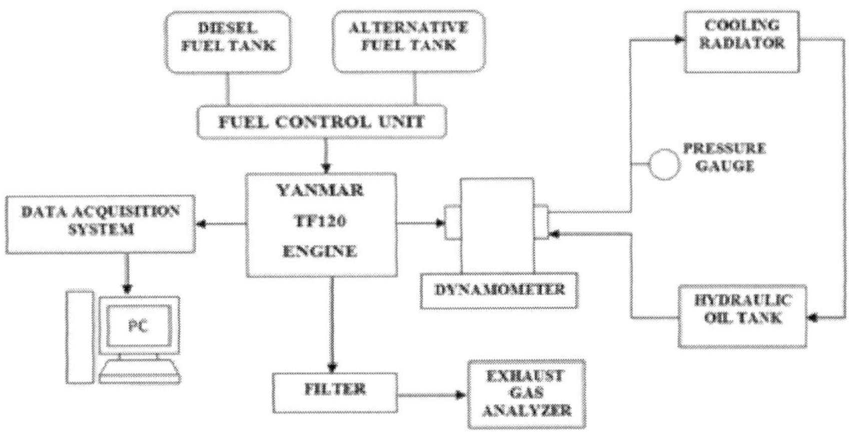

Fig. 1. Schematic diagram of engine testing.

Table 2. Engine specification.

Description	Specification
Engine model	YANMAR TF120M
Engine type	Single cylinder, horizontal, diesel 4 stroke cycle
Combustion system	direct injection
Number of cylinders	1
Bore x stroke (mm)	92 x 96
Displacement (L)	0.638
Injection timing	18° BTDC
Compression ratio	17.7
Continuous output (HP)	10.5 HP at 2400 rpm
Rated output (HP)	12 HP at 2400 rpm

3 Results and discussion

The effect of emulsion stability on the three different mixers including mechanical stirrer, ultrasonic water bath and high shear homogenizer were shown in Figure 2, 3 and 4 respectively. Emulsion stability is the capability of an emulsion to conserve the emulsifying layer after being kept motionless or heated constant temperature at a preset time. In overall, the effect of water addition in the form of emulsion fuel reveals that the increasing separation length of the emulsion fuel. The ultrasonic water bath shows the emulsion produced is the fastest to conserve the emulsifying layer than mechanical stirrer and high shear homogenizer. The faster time taken of separation of water from emulsion show the ultra-sonication is significantly not in a homogenized state. The separation length of the emulsion produced by the mechanical stirrer and high shear homogenizer show similar value, but emulsion of high shear homogenizer take longer time to be in a stable state. It is due to the principle of the mixer of the high shear homogenizer that able to subdivision of particles into subtle sizes to create an emulsion [22].

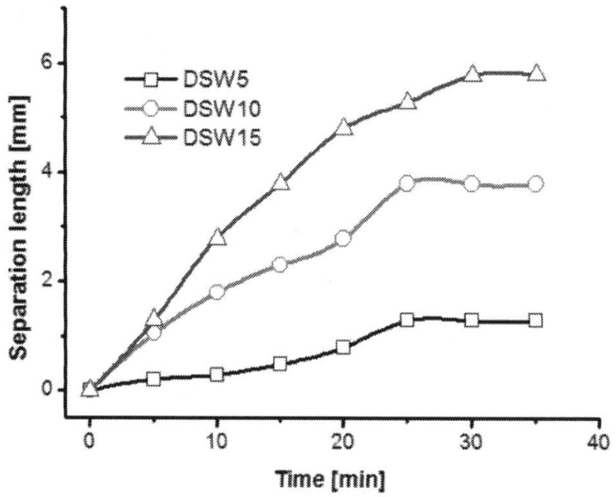

Fig. 2. Separation rate of emulsified fuel using mechanical stirrer.

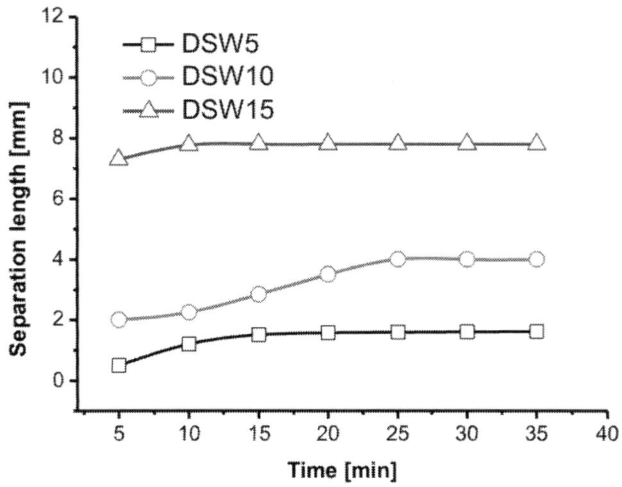

Fig. 3. Separation rate of emulsified fuel using ultrasonic water bath.

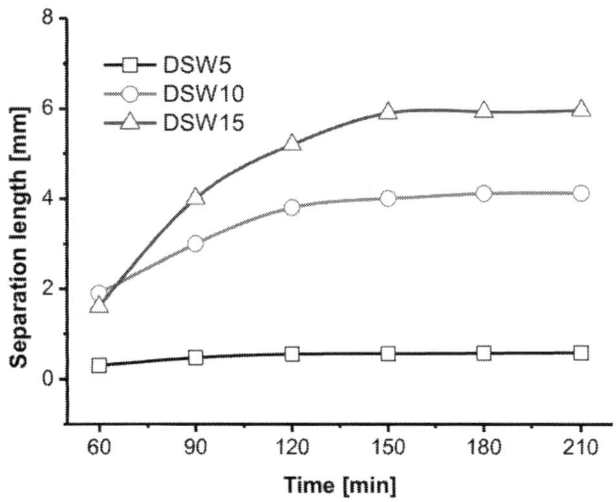

Fig. 4. Separation rate of emulsified fuel using high shear homogenizer.

The effect of the emulsion on the power curve has been illustrated in Figure 5. It is observed that with the increase in load condition, the power slightly increases for the three blended emulsions. The average power curve decreases when the increase the addition water in emulsion fuel. This is because of water in emulsion influence combustion of the fuel. However, at 100% engine load, the power increases as the water in emulsion increase. This is due to the engine reached the limit so that the speed will change (slow down) or a lower gear will be used to multiply the engine torque by a greater factor (and consequently cause the engine speed to increase).

Figure 6 shows the effect addition of water in emulsion to the location of the peak pressure in unit CAD. As, expected, the crank angle degree increase as increase the water content in emulsion. This is due to the effect of ignition delay to start of the combustion. By increasing the water content in emulsion, the vaporization of water in emulsion increase and increase the compression ratio of the engine then affect the peak pressure of in the engine.

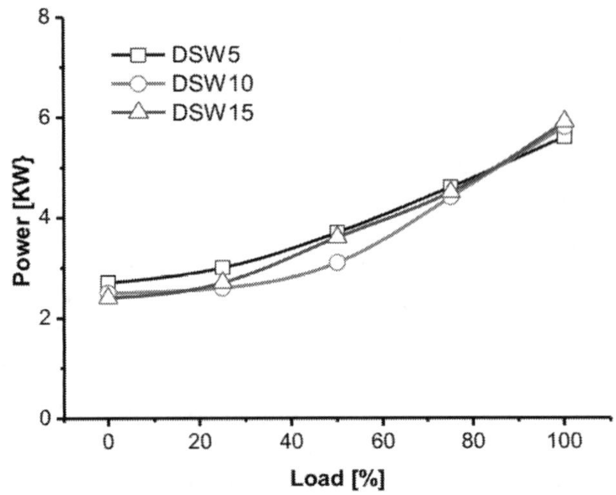

Fig. 5. Engine power curve of different blend concentration.

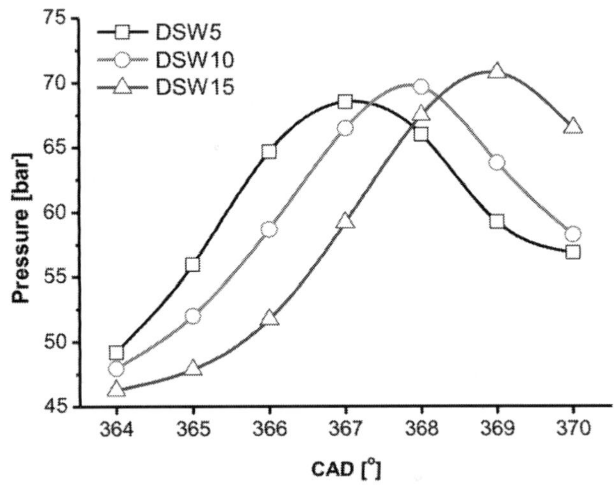

Fig. 6. Combustion peak pressure of various emulsion blends.

Ignition pressure is the start of ignition point at which the pressure increases suddenly from the graph pressure-crank angle degree. Ignition delay is defined as the period from when the start of injection (SOI) to the start of combustion (SOC). So, the ignition delay angle is the difference between the two angles in degree. The variation of the pressure with crank angle degree (CAD) for different emulsion is shown in Figure 7. It is clear that as the percentage of water in emulsion increase, the ignition delay increase. It is due to the heat absorption by water vaporization in the fuel jet. The ignition delay period is when the fuel that has been injected into the cylinder is undergoing chemical and physical preparation for combustion. Thus, by adding water into emulsion fuel, it requires less compression (negative) work due to the longer ignition delay during the compression stroke [23].

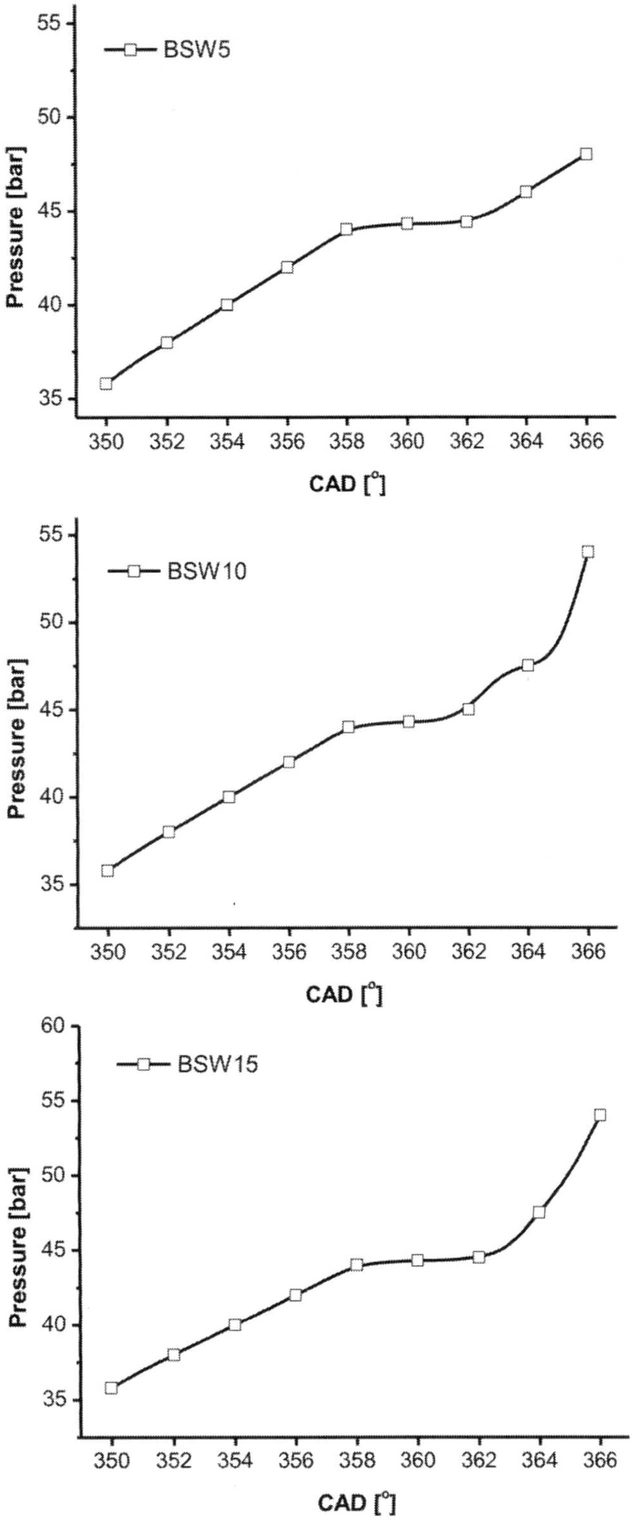

Fig. 7. Comparison of ignition delay of various emulsion blends.

4 Conclusions

The research project summaries the specific issues related to emulsion preparation, the types of emulsifying agents or surfactants and the HLB value. In this paper, an attempt has been made to choose the optimum mixer to blend the water-diesel by adding surfactant. By measuring the emulsion stability, the high shear homogenizer produce the optimum homogenous state for the emulsion compared to the mechanical stirrer and ultrasonic water bath. The applications of emulsified fuel, in the form of water-diesel have found to have influenced the engine combustion. Water-diesel emulsions added by surfactant Span 80 Tween 80 (HLB 10) were tested on diesel engine. The comparisons were made in different water content (5%, 10% and 15%) of emulsion. It may conclude that the addition of water up to 15% in the form of emulsion improves the combustion efficiency of the diesel engine. Addition the water, the ignition delay and peak pressure increase. However, the power curve decreases when increasing the water content.

Malaysian Ministry of Higher Education and Universiti Malaysia Pahang is greatly acknowledged for the technical and financial supports under RAGS grant (RDU121418).

References

1. A. M. Attia, A. Kulchitskiy, Fuel, **116**, 703 (2014)
2. D. Cook, C. Law, Comb. Sci. & Tech., **18**, 217 (1978)
3. O. Armas, R. Ballesteros, F. J. Martos, J. R. Agudelo, Fuel, **84**, 1011 (2005)
4. M. E. A. Fahd, Y. Wenming, P. Lee, S. Chou, C. R. Yap, Appl. Enrg., **102**, 1042 (2013)
5. P. Janulis, Rnwble. Enrg., **29**, 861 (2004)
6. M. Abu-Zaid, Enrg. Conv. & Mgmt., **45**, 697 (2004)
7. H. Schubert, R. Engel, Che. Eng. R&D, **82**, 1137 (2004)
8. Y. Xie, J. Chen, S. Zhang, K. Fan, G. Chen, Z. Zhuang, M. Zeng, D. Chen, L. Lu, L. Yang, Int. J. of Pharm., **500**, 110 (2016)
9. X. Zhang, J.-y. Wu, J. Niu, Solar Enrg. Mat. & Solar Cells, **147**, 211 (2016)
10. D. P. Gouvêa, V. D. Bareño, J. Bosenbecker, B. B. Drawanz, P. D. Neuenfeldt, G. M. Siqueira, W. Cunico, Ultrasncs. Sonoche., **19**, 1127 (2012)
11. X. Liu, J. Wu, J. of Acoust. Soc. America, **125**, 1319 (2009)
12. W. Lauterborn, C.-D. Ohl, Ultrasncs. Sonoche., **4**, 65 (1997)
13. A. Tezel, S. Mitragotri, Biophy.J., **85**, 3502 (2003)
14. L. Wolloch, J. Kost, J. of Controlled R., **148**, 204 (2010)
15. C.-Y. Lin, K.-H. Wang, Fuel, **82**, 1367 (2003)
16. B. K. Debnath, U. K. Saha, N. Sahoo, Rnwble. & Stnble. Enrg. Rvws., **42**, 196 (2015)
17. S. M. Kumar, A. Kerihuel, J. Bellettre, M. Tazerout, P. of Int. of Mech. Eng., **219**, 583 (2005)
18. C. P. Fernandes, M. P. Mascarenhas, F. M. Zibetti, B. G. Lima, R. P. Oliveira, L. Rocha, D. Q. Falcão, R. B. F., **23**, 108 (2013)
19. X. Jin, D. A. Streett, C. A. Dunlap, M. E. Lyn, Bio. Cntrl., **46**, 226 (2008)
20. A. Lif, K. Holmberg, Adv. in Clld. & Int. Sc., **123**, 231 (2006)
21. Z. Fu, M. Liu, J. Xu, Q. Wang, Z. Fan, Fuel, **89**, 2838 (2010)
22. J. Taghinia, M. Rahman, K. Tim, T. Siikonen, Che. Eng. R&D, **106**, 327 (2016)
23. N. Samec, B. Kegl, and R.W. Dibble, Fuel, **81**, 2035 (2002)

 DOI: 10.1051/matecconf/20179001023

Engine performance testing using variable RON95 fuel brands available in Malaysia

Aizuddin Fahmi Mohd Riduan[1,*] , *Noreffendy* Tamaldin[1,2] , and *Ahmad Kamal* Mat Yamin[1,2]

[1]Faculty of Mechanical Engineering, Universiti Teknikal Malaysia Melaka, Hang Tuah Jaya, 76100 Durian Tunggal, Melaka, Malaysia.
[2]Centre for Advanced Research on Energy, Universiti Teknikal Malaysia Melaka, Hang Tuah Jaya, 76100 Durian Tunggal, Melaka, Malaysia.

Abstract. There are various gasoline fuel producers available in Malaysia. The effects of fuel variations from different manufacturers on vehicle performance have always been a debate among users and currently the facts still remains inconclusive. Hence, this study focuses on analyzing various RON95 fuel brands available in the Malaysian market and finding the differences towards engine performance. In terms of engine output, the important data of power (hp) and torque (Nm) will be gathered by using an engine dynamometer. Another data that would also be taken into account is the knocking where the relative knock index can be measured in percentage using the knock sensor accelerometer. Results have shown that the performance of different fuel brands tested are indeed different albeit by only a small margin even though all fuels are categorized with the same octane rating. The power and torque results also imply that both are influenced by the amount of vibration generated due to engine knocking. Based from the overall outcome, consumers would not need to only focus on a certain type of gasoline brand as all differentiates the engine performance marginally.

1 Introduction

In Malaysia, most all of the fuel brands that sell gasoline would offer octane ratings of RON95 and RON97. Various gasoline fuel brands with different octane ratings are sold at a fixed price which is regulated by a managed float system that was implemented on 1st of December 2014 to cut out all the fuel subsidies. As of 1st April of 2016, the fuel prices for both RON95 and RON97 were set at RM1.70/L and RM2.05/L respectively. The most common query among users about the difference in fuel brands is whether or not all of them are the same or unlike when used in their vehicles even though the gasoline used has the same octane rating and price point. Hence, this research examines certain RON95 fuel brands available in Malaysia and discuss the differences between all tested gasolines.

Gasoline is a type of chemical mixture that consists of numerous hydrocarbons. It varies from one refinery to another and there no ideal reference point that can be used to portray the quality and purity of gasoline as a whole [1]. In this study, octane rating defines the

* Corresponding author: lucky_din_v1@yahoo.com

characteristics of the gasoline. Octane rating can be determined by three methods which are Research Octane Number (RON), Motor Octane Number (MON), and Anti-Knock Index (AKI). The difference between all octane ratings does not necessarily mean the fuel would have difference fuel performance when tested on. It is just simply a method of measurement that can be categorized through a standard experiment procedure.

RON is measured by testing fuel in an engine that has a variable compression ratio under controlled conditions. MON in the other hand is a step up test from RON where the fuel will be tested with a higher engine speed, preheated fuel mixture, and also with a variable ignition timing to increase knock resistance load of the fuel. Both of those octane ratings are still measured in a standardized single cylinder, variable compression ratio (from 4:1 to 18:1), internal combustion engine (Cooperative Research Fuel-CFR engine), according to the standard methods ASTMD2699 and D2700, respectively [1 – 2]. The engine is kept at a constant operating speed, where 600 rpm for RON and 900 rpm for MON, and the compression ratio is increased until the onset of knocking. [1]. The RON accounts for fuel performance under low severity engine operation while the MON for more severe operation that might be incurred at high speed or high load [1]. Anti-Knock Index meanwhile is just a formula of (RON + MON)/2 where is sometimes can also be known as Pump Octane Number (PON) and Road Octane Number (RdON). The difference between all of the types of measurement is the fuel sensitivity towards the experiment however the similarity between all is that all the results were compared to a standard mixture of iso-octane and n-heptane. As an example a fuel with an octane rating of 95 means it is similar to a standard fuel of 95% iso-octane and 5% heptane. Octane quality for a fuel is essential to avoid knocking phenomena. The absence of knocking will able combustion phasing of a spark ignition (SI) engine to be tuned optimally. It allows engine to work on high-load operating points more efficiently. Furthermore, engine compression ratio (CR) can also be increased which in return improves the whole range of engine speed and load thermodynamic efficiency [3].

Typically, it is conceived that a higher octane number increases engine performance. This thought has attracted the general public to use gasoline of a higher octane for their vehicles [4]. Adding oxygenates such as tetra-ethyl lead (TEL), methyl tertiary butyl ether (MTBE), or tert-amyl methyl ether (TAME) to fuels can increase efficiency of combustion and ignition, stabilizing fuel mixture, preserve the motor from erosion, as well as reducing harmful emissions. Adding nanomaterials into the liquid fuel as catalyst will accelerate the burning rate and promote clean burning. As an example, cerium oxide (CeO2) nanoparticles was tested as a gasoline additive to improve combustion [5]. One other research suggests that addition of alcohol to gasoline such as butanol can help reduce the knocking effect of the engine [6]. Another study that also tempered with the gasoline blend was done by David A. Rothamer and Jamie H. Jennings where they mixed the fuel with 2, 5–Dimethylfuran and ethanol to test the knocking propensity [7]. More examples can be seen where a research conducted an experiment of testing different fuel octane numbers, where the differences of the fuel can be characterized by different content of olefins and oxygen [8]. Based on the studies reviewed, it is shown that adding additives can play a major part in contributing towards engine efficiency which would explain different fuel brands of the same octane rating can variably affect engine performance.

To determine the difference between tested fuels, another quantity that will be look upon is the knocking of an engine. Knocking has been a pivotal problem for fuel efficient and high compression ratio SI engines development for more than 60 years [10]. Knocking phenomena takes place when end gas auto-ignition is ahead of the propagating flame. When auto-ignition happens in the combustion chamber, the chemical energy contained in the end gas is released rapidly. At a certain magnitude, shock waves propagation occurs across the cylinder causing it to resonate at its natural frequencies. This will result to high frequency shock waves striking the cylinder walls and inducing permanent damage [9]. These pressure waves are carried through the engine structure to the surrounding air leading to the combustion 'ping' sound be heard in close range. [11]. Pressure waves resulted from knocking happens at frequencies that are acoustic vibration modes of the chamber geometry. Amplitude of the oscillations linked immediately with each resonant mode after the occurrence of knock depending on the initial boundary conditions. The oscillations frequencies depend on the shape and size of the chamber, auto-ignition location, resonant mode, and the sound average speed. Vibration modes can be radial and circumferential, and the range of frequencies are between 6 – 25 kHz for automotive sized engines [11]. Whatever reasons that might lead to knocking, normally it is tripped by rapid combustion and release of unequally distributed energy in the combustion chamber [12].

Modern SI engines requirement nowadays are directed towards low fuel consumption and reduced pollutant emissions. A simple approach in obtaining better thermal efficiencies for SI engines is increasing the CR. At part load, higher thermal efficiency can be achieved by raising the CR or by means of supercharging (concept of downsizing) [13, 16]. In the last decade, small sized SI engines working under high loads have become more favourable for car manufacturers. The big advantage of such engines is the increase in efficiency and reduction of CO_2 emissions [14]. Nevertheless, for every great improvement of engines thermal efficiency, one major challenge faced from SI engine is the underlying problem of combustion knock [15]. A higher compression ratio contributes to greater unburned end gas pressure and temperatures, which develops a spontaneous ignition from the portion of unburned end gas mixture. Still, this end gas auto-ignition also produces high-frequency oscillations and shockwaves and of the in-cylinder pressure, which consequently cause engine knocking [16]. Hence, it has been a key technical obstruction for modern SI engines to increase power per litre extensively while limited to knock [15].

According to one research [18], these can be the consequences of prolong effect by knocking:

- Piston rings breakage
- Cylinder head, piston crown, and top land erosion
- Melting of piston
- Limiting engine CR or acceleration performance of vehicle
- Greater air pollution
- Lower engine efficiency
- Increase in fuel consumption
- Structural damage to engine in the long run
- Source of intolerable noise

Prediction of knocking is very difficult and requires specific information of the occurring process within the combustion chamber. In SI engines, the combustion phenomenon can be proceed as normal or unnatural depending on engine operating conditions [17]. Detection of knock are commonly obtained from data gathered by either cylinder pressure sensors or accelerometers. Because of its simplicity, vibration measurement using accelerometer is widely used in industry. Even so, parasitic noise related to engine running can affect the quality of knock measurement when using this method [19].

There are variety of ways in measuring knocking without experimenting on an engine to preserve the engine life. One of which is by large eddy simulation (LES) [20 – 21] which can be very helpful to cut down cost of full scale testing. Another simulation method is done by a parametric modelling approach where vibration signal of the cylinder block is modelled by auto regressive moving average (ARMA) parametric model [23]. Other simulation that use equations to study the effects of knocking can be seen in this research [24] in which a pressure wave equation of ICE is derived and modelled and then coupled with KIVA-3V codes with reduced kinetics mechanism. The oscillation pressure induced by auto-ignition was accurately simulated using the model. One other technique of determining knocking is by using the prediction method where using a model to mimic the combustion chamber by means of two zones, which are one in front and one behind the flame front. It is performed by conjuring up a theoretical model using equations to imitate the combustion of SI engines [22]. Although there are many methods in measuring knock either through simulation or prediction, theoretical studies no matter how accurate, cannot hold up with the actual test as a lot of deciding factors can affect the results. Hence, the scope of this research will only focus on the vibration at one point which could be consider the engine relative knock index detected by a knock sensor accelerometer.

Concerning this research, there are some examples of other studies that has been conducted similar type of testing approach. One relatable research [25] that can be found done in Malaysia was a study on part-load performance and emission of a SI engine fuelled with RON95 and RON97 gasoline. It was conducted due to the rapid rise and fall of the country fuel prices and determining whether using gasoline of a higher octane is beneficial in terms of economy. Another study [26] which was also similar, researched on the effects of RON and fuel system on SI engine performance and emissions by using Saudi Arabian RON91 and RON95 fuel with port injection and direct injection system. They also worked on the effects of varying spark timing on the performance and emission characteristics using the same mentioned fuel RONs [27]. Other studies that tried to examine between two fuel grades (RON90 & RON95) was performed Ali Alahmer and Wail Aladayleh [28], where they found using octane grades higher than requirement will decreased the engine performance but will have lower NOx and CO emission concentration for the higher RON. They also found that using a higher octane can incur higher noise. There was also an investigation [29] carried out on finding the effects of varying spark timing at different octane ratings on engine performance and emission characteristics where the research tested five different types of RON (91, 93, 95, 97 and 98). Hence, this paper will study upon various fuel from different gasolines manufacturers with the same octane number to determine whether or not all are different or the same without altering any other variables such as engine tuning or using other fuel RON.

2 Methodology

In this research, three different brands of fuel, Brand A (A95), Brand B (B95), and Brand C (C95) were tested. All of the mentioned brands were categorized in the same octane rating of RON95. The process of testing the fuels performance was done using Land & Sea's DYNOmite Engine Dynamometer **(Fig. 1)**. Engine testing parameters conducted were only to focus on performance of the engine where data collected from each run were power (hp) and torque (Nm) which determines engine performance of each tested gasoline brands. To determine the contributing factor towards performance figures, engine knocking data was also gathered by the knock sensor which was placed at the side of the engine block as can be seen in **Fig. 2**. This positions the knock sensor axis perpendicular to the thrust side of the piston skirt where the knocking signals are typically most useable near to the cylinder walls. The aforementioned location provides a strong "rattle" as detonation slams the piston's skirts against its bore. Ideal placement of the sensor can be vary by engine design, hence the mounting point of the knock sensor was determined by its original factory standard location and it is considered as the relative knock index (%) at that attached position.

 DOI: 10.1051/matecconf/20179001023

Fig. 1. Land & Sea's DYNOmite engine dynamometer.

Fig. 2. Knock sensor position at the side of the engine block.

To conduct the experiment, certain procedures were needed to be in checked. The fuel tank was first drained using a fuel pump to completely remove the previous fuel that was used for testing and the test fuels were also filled by using the same pump to avoid any spillage. Before the test run for actual data collection, the fuel must run at least 15 minutes to ensure the engine is circulating only the test fuel to avoid any unnecessary data errors. The engine temperature before initiating the run must at least be at a minimum of 70°C and must not exceed 100°C while the test is carried out to ensure the optimum working temperature of the engine. The absorber temperature in the other hand must not exceed 60°C to prevent it from overheating. Five runs were recorded using automated sweep test where the engine speed starts from 2000 RPM and ends at 6000 RPM with increment of 100 RPM every second at wide open throttle. All data collected were stored using the engine dynamometer data acquisition electronics.

3 Results and discussion

From the graphs shown, it can generally be seen that all three fuel brands indeed have differences between one another in terms of engine performance. From Fig. 3, engine

 DOI: 10.1051/matecconf/20179001023

power when tested with gasolines of different brands, fuel A95 shown to be the best among the three with the highest power output of 79.58 hp at 5400 rpm. At that particular engine speed also, both fuel B95 and C95 have nearly the same high power yield with 78.10 hp and 78.17 hp respectively. However from Fig. 4, at engine speed of range between 4500 rpm to 5000 rpm, the differences between the three fuels can be seen clearly. A95 have a better overall power output at that range while B95 is the lowest. There is also need to mention that all tested brands have the same trend of descending power yield when referred to the graph where the rate of power decrease for fuels A95, B95, and C95 are -1.076 hp/s, -1.074 hp/s and -1.118 hp/s respectively.

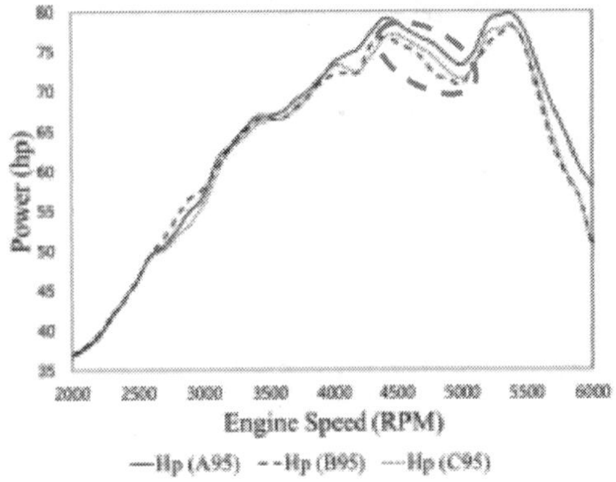

Fig. 3. Power vs. Engine Speed of RON95 fuel brands.

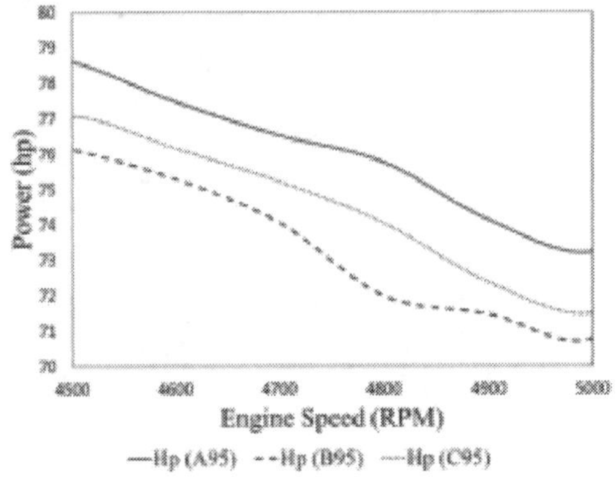

Fig. 4. Enlarged view of engine power at 4500 – 5000 rpm range.

In Fig. 5, results show that A95 and B95 almost have the same highest magnitude value of torque with 140.4 Nm and 139.8 Nm both at 3300 rpm. C95 is slightly lower than the former mentioned with 138.7 Nm at 3200 rpm. Closer examination with referred to Fig. 6, at speed range of engine from 4500 rpm to 5000 rpm, the differences between gasoline RONs can be clearly analyzed. In that specific range, fuel A95 again have a better overall torque figures compare to both B95 and C95. The trend of the torque output between the three brands is also the same with decreasing torque yield across the mentioned range. This is further supported by the rate of decrease of torque between A95, B95, and C95 where the results are -4.04 Nm/s, -3.94 Nm/s, and -4.08 Nm/s respectively.

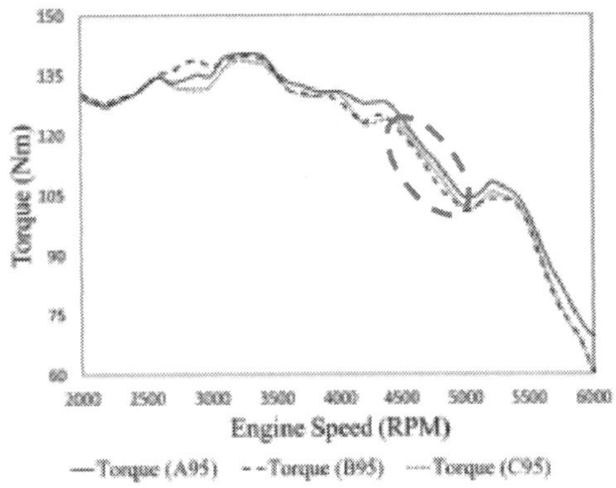

Fig. 5. Torque vs. Engine Speed of RON95 fuel brands.

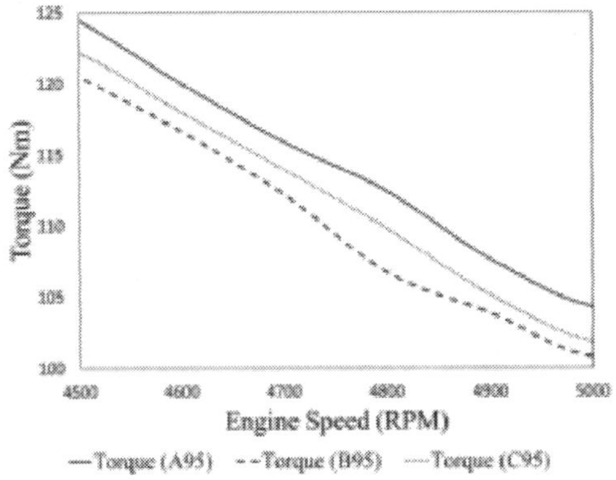

Fig. 6. Enlarged view of engine torque at 4500 – 5000 rpm range.

Based on the performance figures mentioned, it needs to be stated that when the power and torque output is high, the relative knock index is reversely proportionate to it. This is evident from Fig. 8, where when A95 has the highest overall power and torque output at

 DOI: 10.1051/matecconf/20179001023

engine speed range from 4500 rpm to 5000 rpm, the relative knock index of A95 is lower overall across that engine speed scope. Furthermore, at the same speed range of engine, the lowest performance output among the three which is B95 has the highest overall knocking effect while fuel C95 remains constant in the middle from performance figures to relative knock index. Therefore, what can be said from this analysis is that the knocking effect or relative knock index in this case, does indeed contribute to the overall engine performance.

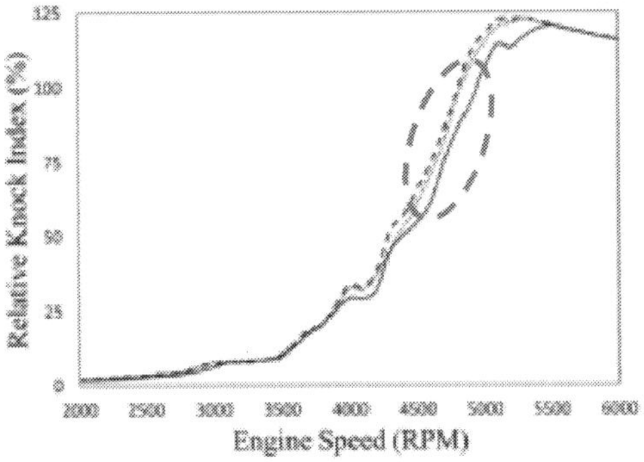

Fig. 7. Relative Knock Index vs. Engine Speed of RON95 fuel brands.

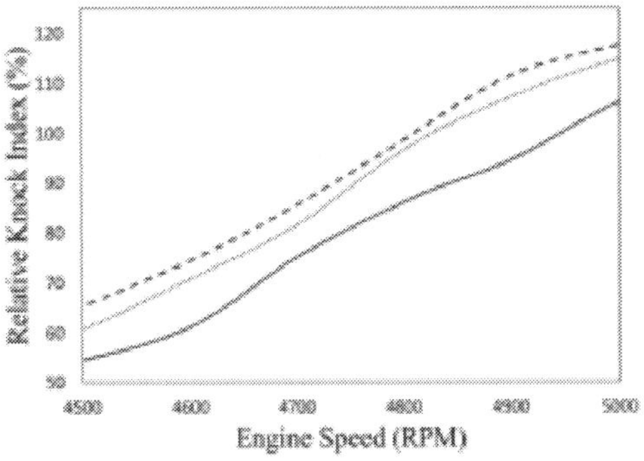

Fig. 8. Enlarged view of relative knock index at 4500 – 5000 rpm range.

Nevertheless, one important issue that needs to be address in this matter is whether or not the differences between the gasoline producers of the same RON is significant or marginal. To do this, fuel C95 is set as the baseline to compare it between the other two fuel brands. From Fig. 4, fuel A95 only have at most the highest power increase of only 2.3%, while fuel B95 at most decrease by 2.7%. For torque difference, fuel A95 only have a maximum increase of only 2.4%, while B95 decreases to only at a maximum of 2.7% which can be seen in Fig. 6. However, for relative knock index with referred to Fig. 8, fuel A95 performed relatively better when compared to the baseline with a maximum decrease to 11.7%, while B95 only increase at most by 4%.

4 Conclusions

From all the results gathered, it can be concluded that all tested gasoline brands of the same octane rating does indeed effect the overall performance of the engine but only by a small gap. Moreover, with the power and torque output having a difference of less than 3%, it shows that the fuel manufacturers really follows the guidelines in maintaining the quality of the fuel to a certain octane rating. In addition to that, it can be put to rest of the debate of which brands is better when comparing to one and another as proved from the results that all fuel brands when tested for engine performance gives an output of relatively similar. Hence, the general public would not need to only focus a certain fuel brand to use as all are in the same octane rating and performs roughly the same. From this paper on, further studies would be conducted to determine if the emissions would have a similar effect as well when tested with the same experimental procedures.

The authors would like to give formal recognition to Universiti Teknikal Malaysia Melaka (UTeM) for providing laboratory facilities to conduct the engine testing experiment for this research. Further credit is given towards UTeM for supporting this study financially under the RAGS grant with project code RAGS/1/2014/TK06/FKM/B00071.

References

1. N. Nikolaou, C.E. Papadopoulos, I.A. Gaglias, and K.G. Pitarakis. "A new non-linear calculation method of isomerisation gasoline research octane number based on gas chromatographic data." *Fuel* 83.4-5 (2004).
2. L. Guan, X.L. Feng, Z.C. Li, and G.M. Lin. "Determination of octane numbers for clean gasoline using dielectric spectroscopy." *Fuel* 88.8 (2009).
3. N. Rankovic, G. Bourhis, M. Loos, and R. Dauphin. "Understanding octane number evolution for enabling alternative low RON refinery streams and octane boosters as transportation fuels." *Fuel* 150 (2015).
4. C. Sayin, I. Kilicaslan, M. Canakci, and N. Ozsezen. "An experimental study of the effect of octane number higher than engine requirement on the engine performance and emissions." *Applied Thermal Engineering* 25.8-9 (2005).
5. S. S.Kish, A. Rashidi, H. R. Aghabozorg, and L. Moradi. "Increasing the octane number of gasoline using functionalized carbon nanotubes." *Applied Surface Science* 256.11 (2010).
6. H. Wei, D. Feng, M. Pan, J. Pan, X. Rao, and D. Gao. "Experimental investigation on the knocking combustion characteristics of n-butanol gasoline blends in a DISI engine." *Applied Energy* 175 (2016).
7. D. A. Rothamer, and J. H. Jennings. "Study of the knocking propensity of 2,5-dimethylfuran–gasoline and ethanol–gasoline blends." *Fuel* 98 (2012).
8. T. Cerri, G. D'Errico, and A. Onorati. "Experimental investigations on high octane number gasoline formulations for internal combustion engines." *Fuel* 111 (2013).
9. C. Rahmouni, G. Brecq, M. Tazerout, and O. Le Corre. "Knock rating of gaseous fuels in a single cylinder spark ignition engine." *Fuel* 83.3 (2004).
10. C. Hudson, X. Gao, and R. Stone. "Knock measurement for fuel evaluation in spark ignition engines." *Fuel* 80.3 (2001).
11. S. Szwaja, K. Bhandary, and J. Naber. "Comparisons of hydrogen and gasoline combustion knock in a spark ignition engine." *International Journal of Hydrogen Energy* 32.18 (2007).
12. A. D. Gaeta, V. Giglio, G. Police, and N. Rispoli. "Modeling of in-cylinder pressure oscillations under knocking conditions: A general approach based on the damped wave equation." *Fuel* 104 (2013).
13. E. Galloni. "Dynamic knock detection and quantification in a spark ignition engine by means of a pressure based method." *Energy Conversion and Management* 64 (2012).
14. J. Rudloff, J.-M. Zaccardi, S. Richard, and J.m. Anderlohr. "Analysis of pre-ignition in highly charged SI engines: Emphasis on the auto-ignition mode." *Proceedings of the Combustion Institute* 34.2 (2013).
15. G. Shu, J. Pan, and H. Wei. "Analysis of onset and severity of knock in SI engine based on in-cylinder pressure oscillations." *Applied Thermal Engineering* 51.1-2 (2013).
16. N. Kawahara, E. Tomita, and Y. Sakata. "Auto-ignited kernels during knocking combustion in a spark-ignition engine." *Proceedings of the Combustion Institute* 31.2 (2007).
17. S. S. Merola, and B. M. Vaglieco. "Knock investigation by flame and radical species detection in spark ignition engine for different fuels." *Energy Conversion and Management* 48.11 (2007).
18. X. Zhen, Y. Wang, S. Xu, Y. Zhu, C. Tao, T. Xu, and M. Song. "The engine knock analysis – An overview." *Applied Energy* 92 (2012).
19. G. Brecq, J. Bellettre, and M. Tazerout. "A new indicator for knock detection in gas SI engines." *International Journal of Thermal Sciences* 42.5 (2003).

20. A. Robert, S. Richard, O. Colin, L. Martinez, and L. De Francqueville. "LES prediction and analysis of knocking combustion in a spark ignition engine." *Proceedings of the Combustion Institute* 35.3 (2015).
21. G. Lecocq, S. Richard, J. B. Michel, and L. Vervisch. "A new LES model coupling flame surface density and tabulated kinetics approaches to investigate knock and pre-ignition in piston engines." *Proceedings of the Combustion Institute* 33.2 (2011).
22. E. Moses, A. L. Yarin, and P. B. Yoseph. "On Knocking Prediction in Spark Ignition Engines." *Combustion and Flame* 101.3 (1995).
23. M. M. Ettefagh, M. H. Sadeghi, V. Pirouzpanah, and H. Arjmandi Tash. "Knock detection in spark ignition engines by vibration analysis of cylinder block: A parametric modeling approach." *Mechanical Systems and Signal Processing* 22.6 (2008).
24. J. Pan, G. Shu, and H.Wei. "Research on in-cylinder pressure oscillation characteristic during knocking combustion in spark-ignition engine." *Fuel* 120 (2014).
25. T. I. Mohamad, and H. G. How. "Part-load performance and emissions of a spark ignition engine fueled with RON95 and RON97 gasoline: Technical viewpoint on Malaysia's fuel price debate." *Energy Conversion and Management* 88 (2014).
26. S. Binjuwair, T. I. Mohamad, A. Almaleki, A. Alkudsi, and I. Alshunaifi. "The effects of research octane number and fuel systems on the performance and emissions of a spark ignition engine: A study on Saudi Arabian RON91 and RON95 with port injection and direct injection systems." *Fuel* 158 (2015).
27. S. Binjuwair, and A. Alkudsi. "The effects of varying spark timing on the performance and emission characteristics of a gasoline engine: A study on Saudi Arabian RON91 and RON95." *Fuel* 180 (2016).
28. A. Alahmer, and W. Aladayleh. "Effect two grades of octane numbers on the performance, exhaust and acoustic emissions of spark ignition engine." *Fuel* 180 (2016).
29. C. Sayin. "The impact of varying spark timing at different octane numbers on the performance and emission characteristics in a gasoline engine." *Fuel* 97 (2012).

 DOI: 10.1051/matecconf/20179001024

The effect of fiber laser parameters on microhardness and microstructure of duplex stainless steel

Ghusoon R. Mohammed[1,2,*], *M.* Ishak[1], *S.* N. Aqida[1], and *Hassan* A. Abdulhadi[1,2]

[1]Faculty of Mechanical Engineering, University Malaysia Pahang, Pekan 26600, Pahang, Malaysia
[2]Middle Technical University, Baghdad, Iraq

Abstract. An investigation was implement to study the influence of laser power, and speed of the welding on hardness, microstructure, and penetration of laser welding bead on plate duplex stainless steel, which is not exhibited so far. A fiber laser was selected for welding duplex stainless steel sheet with 2 mm thickness. Then, optical microscope (OM) was used in the morphologic observation of cross section, penetration depth, and bead width. Microhardness of the welded sheet was measured using Vickers hardness. Profiles of hardness and microstructure were utilized to discriminate welding line and to propose superior welding parameters. The experimental results displayed that, a good quality of duplex steel welds can be acquired when a suitable fiber laser welding parameters were selected. It was found that microhardness profiles showed a rise in the hardness of the weld and heat-affected zones as the solidification process proceeds rapidly. Additionally, the crystal solidification process induced by the fiber laser welding was schematically clarified and systematically exposed.

1 Introduction

Typically, duplex stainless steels is a grade of virtually an equal proportion of two phases austenite-ferrite. Duplex stainless steels offer a few favorable benefits. They consolidate the mechanical property and corrosion resistance points of interest of both the austenitic and the ferritic grades. The austenite conferring extensive corrosion resistance and strength while the ferrite giving impedance to chloride stress corrosion and toughness [1].

Various welding and joining techniques are adopted in products as well as structures. Among such, laser welding accelerates power saving, automation as well as systematization. Noted as a technology, it achieves high precision and quality, large flexibility and welding rates, used in various applications. Laser welding's heat source may be a LD (diode laser), YAG laser, LD pumped solid laser, disc and fiber laser, and CO2 laser (carbon dioxide gas laser) [2]. In recent years, a rapid growth has been existed in fiber laser application with the flexibility of fiber, resulting in high beam quality and efficiency by miniaturization, where it is used as a heat source with high power and power density. For instance, using a 10 k fiber laser with high power, attaining a large beam quality of a few mm mrad is possible, with about 25% high efficiency, high compactness and about $1 MW/mm^2$ of power densification [3].

The duplex stainless steel has been explored to offer a composition of pitting and stress corrosion cracking resistance and strength, comparing with the 300 series austenitic

* Corresponding author: ghusoon_ridha@yahoo.com

DOI: 10.1051/matecconf/20179001024

stainless steel [4]. Such particular properties have led to the various industrial application uses, such as oil and gas transmission lines, chemical process plant piping and marine environment structure[5,6]. Because welding is adopted extensively in such applications, it is critical for the good implementation of such materials of engineering to understand the factors affecting the ability of welding of duplex stainless steels.

In general, the balance ration of ferrite/austenite has a remarkable impact on the corrosion and mechanical properties [7,8]. The best property of such steels is realized while in the microstructure almost equal ferrite and austenite proportions exist (Fig.1). Within base metals, such phase balance is achieved through a proper combination of solution heat treatment and composition. Nevertheless, the balance control of austenite/ferrite in welds is not easy as among base metals [9-12].

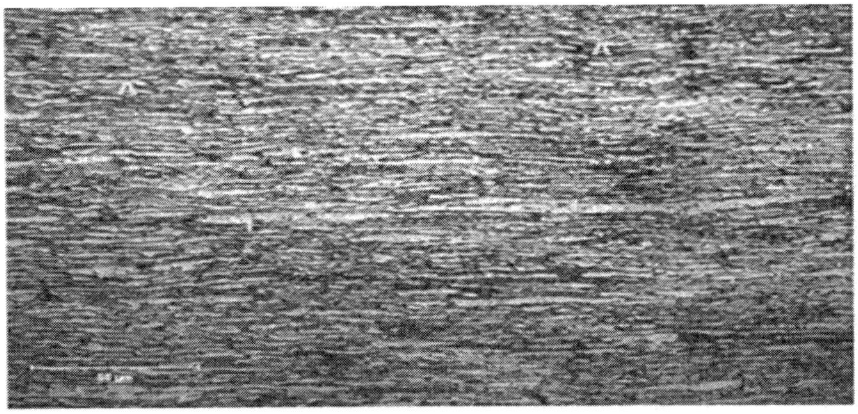

Fig. 1. The microstructure of duplex stainless steel base metal showing Ferrite (F) and Austenite (A) phases.

Previously, researches have proved that the microstructure of weld and heat affected zones is a function of the cooling rate of high temperature [13-15]. As a result, the austenite/ferrite balance within fusion zone is expected to be influenced by welding speed and laser power [16,17]. But the research region has got relatively few explorations, more efforts are needed for in-depth understanding of the influence of various laser welding parameters on duplex stainless steel properties. Presently, this research is about the effect laser beam parameters and its influence on fusion zone and microstructure of duplex steel welded.

Laser welding process provides a lot of good qualities on welded metal. However, the rapidity and perfection of the weld is a future problem. Currently, a lot of researchers focus on welding of cans, domestic equipment, aircrafts and cars [18-20]. Welding depth and shape are affirmed with the way that the energy of welding is used on the joint. The input energy of laser welding is under the control of the following parameters: Laser power, welding speed, keyhole shielding gas and focus spot size. In addition, the repetition of welding performance depends on the material preparation, joint fit-up and laser beam to joint alignment. All the above-mentioned parameters are the inputs to welding process which are controlling the output quality of the joint [21].

Stainless steels may be classified into the austenitic, duplex, martensitic and ferritic family. The duplex family provides a suitable blend of toughness, weldability and economy are frequently chosen where both corrosion and strength properties are essential [22]. DSSs are more, in fact, requesting than the grades of standard austenitic stainless steel, like type 316 and 304.

During welding, avoiding secondary phases and maintaining the good austenite and ferrite balance is important. Typically, laser welding is a process of a high cooling rate that is so far, causing a problem in the welding of duplex stainless steels. The fast cooling results in undesired alterations in the microstructure of the weld metal. That's considered in the case of welding duplex steel as laser welding of duplex stainless steels can be challenge exceptionally whether filler metal did not utilize [23]. Besides, further effort is needed for preferable comprehension the effectiveness of welding parameters (welding speed, laser power) on the forming of weld bead; the evolution of weld microstructure; the allocation of the essential alloying component; and from, the variance in microhardness crossed the weld. Thus, in the present paper, the impact of input power on weld joint shape, the progression in the microstructure, and alteration in the microhardness throughout weld regions will investigate.

2 Experimental procedure

In this research, the material adopted is SAF 2205 duplex stainless steel; 2mm plate thickness, which chemical composition is presented in Table 1. Bead on plate weld was achieved with fiber laser welding process in pulse wave mode with maximum output power of 2 kW. Figure 2 shows the configuration of fiber laser welding joint.

The weld specimens were prepared with dimension (75 × 50 × 2 mm), and a fixture was used to prevent distortion. Table 2 show the parameters of fiber laser welding that used (P: Peak power, S: welding speed, Heat input: Q, Pd: Pulse duration, Df: defocusing). In addition, Argon was used as a shielding gas with a flow rate 20 l/min.

After welding process, the samples were inspected and sectioned transverse to the direction of welding. The samples were prepared for the metallographic inspection using standard technology. The size and microstructure of fusion zone were inspected using optical microscope. Hardness was carried out for fiber laser welded having complete penetration.

After welding process, both macro-and microscopic examinations were carried out for cross sections taken from welded specimens. The welds were cut and prepared as metallographic samples to evaluate, the integrity of the weld beads. Microscopic examinations were carried to clarify the microstructure of WM, HAZ and BM including the extent of ferrite and austenite phases.

The specimens were prepared for such metallographic examination using standard grinding, polishing and etching techniques. Specimens were ground with emery paper ranging in fineness from 320 up to 1200 equivalent mesh, and then polished with alumina suspension. Finally, polished specimens were etched in a Modify Kalling's solution (5 mg $CuCl_2$, 100 ml ethanol, 100 ml HCl) then, cleaned using distilled water with and dried by air. Microhardness Vickers tester was implemented to measure the hardness by applying a standard load of 500gf for a time of 10 s as a dwell period.

Table 1. Chemical composition (wt. %) of duplex stainless steel used.

Cr	Ni	Mn	C	Si	P	S	Mo	Cu	N
22.07	5.54	1.146	0.03	0.587	0.026	0.003	3.01	0.20	0.16

Table 2. Process parameters of fiber laser welding.

P (kW)	S (m/min)	Pd (ms)	Df (mm)	Q (J/mm)
1,1.2,1.4,1.6	0.16	4	+2.5	240,288,336,384
1.4	0.08,0.09, 0.11,0.16	4	+2.5	336

DOI: 10.1051/matecconf/20179001024

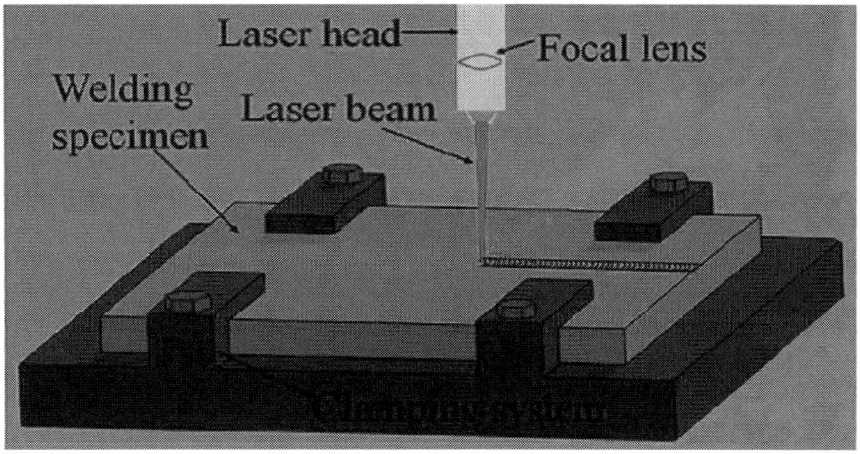

Fig. 2. Schematic of the laser welding configuration.

3 Results and discussion

3.1 Influence of laser power

The influence of laser power was explained with the welding speed of 0.11m/min and defocusing distances of +2.5 mm. In Fig. 3, the cross section of laser bead on plate welds was performed with a welding speed of 0.11 m/min. The characteristic of fiber laser welding is affected by numerous significant ingredient. These features are the penetration depth, width of the welded joint and the final appearance of the weld surface [24].

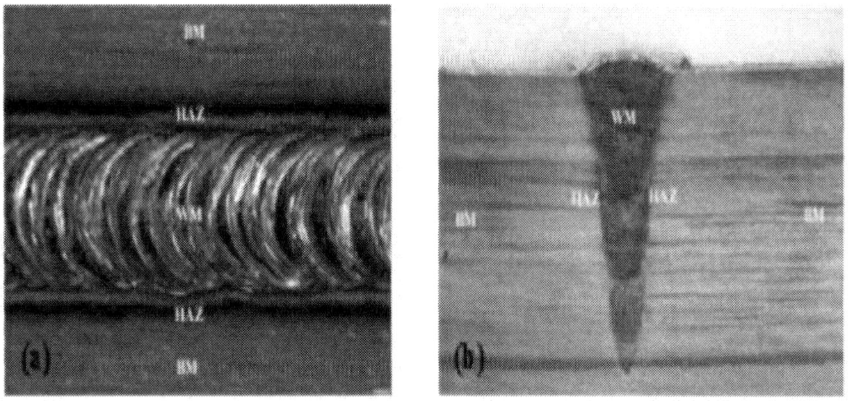

Fig. 3. Macrographs of the fiber laser welded present the various welding regions (a) the top surface appearance and (b) sectional appearance.

Fig.4(a-d) shows the surface of the fiber laser welds under different laser welding parameter compounds and the influence of fiber laser welds on the surface appearances under laser power of 1.4 KW and speeds of 0.08, 0.09, 0.11, 0.16 m/min sequentially. The average results of the hardness are present in Table 3 for duplex steel 2205 at three value (1.2, 1.4, 1.6 KW) of heat inputs. An improvement was noticed in the weld zone and HAZ Hardness of SAF 2205 with the increase of the energy from 288 to 384 J/mm. The profile of the Microhardness of different laser welding power across section of the bead on plate duplex stainless steel was illustrates in Fig.5. Additionally, it was found that, the relation between the peak power and the penetration was almost liner as can be notice in Fig.6.

 DOI: 10.1051/matecconf/20179001024

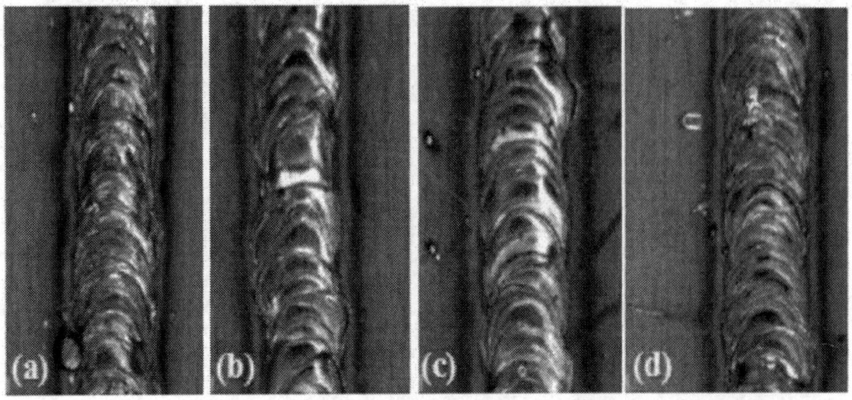

Fig. 4. The surface appearances of the fiber laser weld with laser power of 1.4 KW and speeds of (a) 0.08 m/min, (b) 0.09 m/min, (c) 0.11 m/min, (d) 0.16 m/min.

Table 3. Hardness of different zones of bead on plate duplex stainless steel.

Heat input (KW)	BM	HAZ	WM
1.2	266.1	274.1	313.9
1.4	266.2	284.1	322.6
1.6	267.4	290.3	329.6

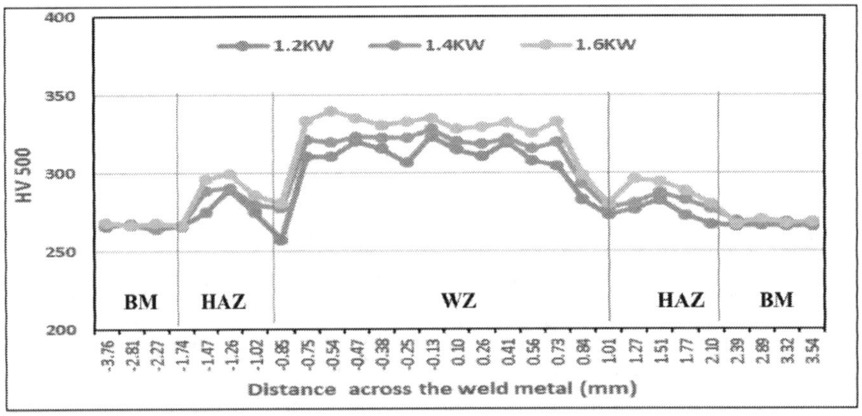

Fig. 5. Microhardness profile of different laser welding power across section of the bead on plate duplex stainless steel.

Results of different welding parameters are exhibited in (Fig.8). As to distinct sets of welding parameters, cross sections of the weld were examined. It was showed that appropriate weld bead profiles were accomplished with 0.11, 0.16, 0.08 and 0.09m/min welding speed. On top of the weld metal area, at the speed of 0.16 m/min the undercut was observed

3.2 Influence of laser welding speed

The speed of welding and laser power is the parameter determining the energy of weld (J/mm of weld length) and as a consequence, the property, and shape of the weld. Considering the influence of welding speed on the quality of the weld, 0.09 m/min was the chosen speed. No evidence of defects exists, like undercut or porosity. The weld width is about 0.75 mm, and the HAZ width is about 0.1mm. The joint may be divided into numerous discrete metallurgical areas, the HAZ (heat affected zone), the BM (base metal) and the FZ (fusion zone) in the center. The zone of fusion includes grains that are coarse, columnar and dendritic, aligned in a perpendicular direction to the boundary of fusion zone.

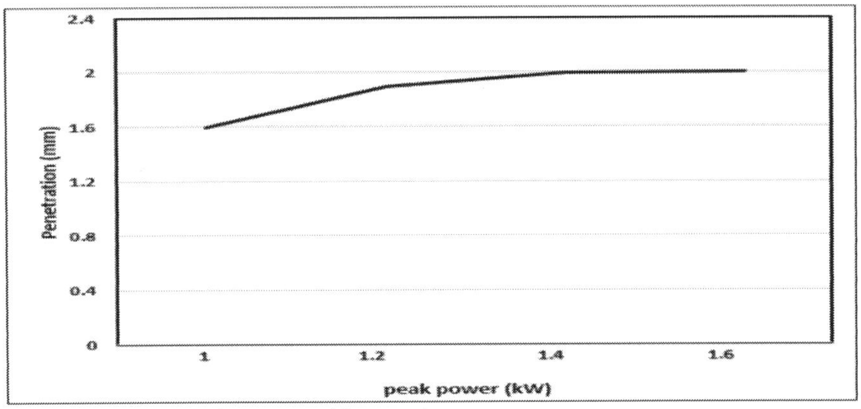

Fig. 6. Relation between peak power and Penetration.

The maximum heat flow direction and the boundary of fusion zone are perpendicular, therefore the grains are likely to grow most rapidly there, leading to columnar grains in the zone of fusion (Fig.7). The zone of weld metal is featured as a regular and narrow shape devoid of defects like shrinkage and solidification. Taken at higher magnification, the best micrographs indicated that the structure that was corresponding consisted of elongated primary ferrite grains.

During welding, the orientation of grain approximately succeeds the direction of the heat flux. Also, a little proportion of austenite was visible, which mainly had precipitated from the boundary of ferrite grain and interior of grain, in finer form. Observed as a faded area on macro-view of welded joint (Fig.7), the narrow heat-affected zone was not corresponding to any important modification of structure on a micro-scale. The noticeable alteration was barely discerned in the high-temperature heat affected zone adjacent to the fusion line, in which the indication of developing decomposition of the austenite phase occurred in an area from the bead of weld [14,25]

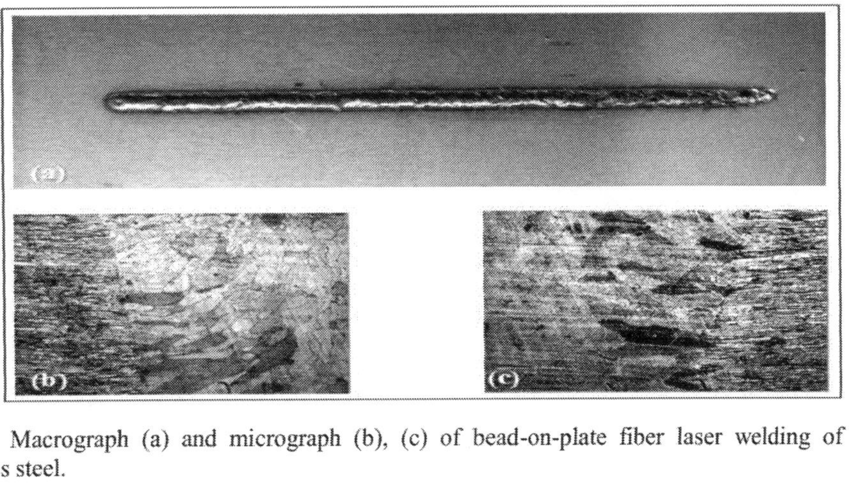

Fig. 7. Macrograph (a) and micrograph (b), (c) of bead-on-plate fiber laser welding of duplex stainless steel.

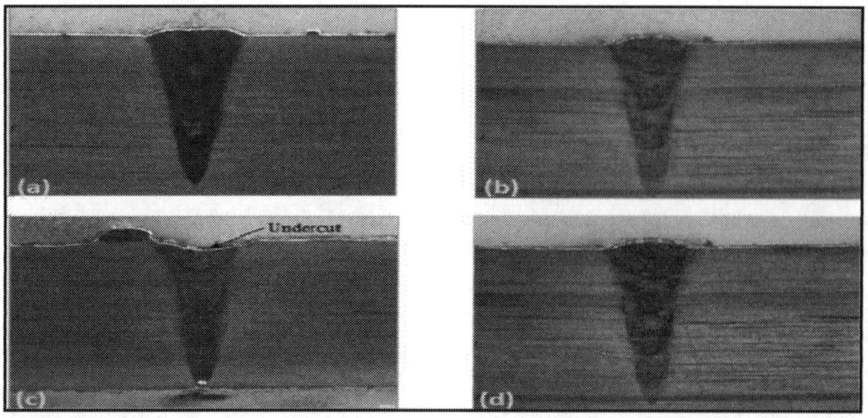

Fig. 8. Effect of different welding speed (a) 0.08, (b) 0.09, (c) 0.16, and (d) 0.11 m/min on the depth and width of the weld bead.

4 Conclusions

In this research and based on the outcomes accomplished in, the following conclusions can be represented. First of all, the welds produce by the fiber laser parameters highlighted a uniform and thin weld metal zone without any defects like shrinkage and solidification. Also, the structure of weld metal comprised an extended ferrite grains besides a limited measure of austenite, fundamentally place at the border of the grain.

Fiber laser welding resulted in a pronounced reduction in the size of FZ and a weld profile that is acceptable. However, the minimum FZ size produced relied on the input of heat which is a function of LBW parameters.in addition, fiber laser welding is expected to result in an enormous difference in the balance of ferrite-austenite of FZ comparable to the BM due to high cooling rate.

Lastly, hardness profile of duplex stainless steel is affected by the heat input, an increase in the heat input results in the decrease of hardness.

References

1. M. Yousefieh, M. Shamanian, A. Saatchi, JISR, International, **18**(9), 65-69 (2011)
2. W. Steen, K.G. Watkins, J. Mazumder, Springer SBM, (2010)
3. S. Katayama, H. Nagayama, M. Mizutani, Y. Kawahito, WI, **23**(10), 744-752 (2009)
4. A.-M. El-Batahgy, A. Tsuboi, , SFC, **8**(3), 189-203 (2014)
5. K.A. Chandler, , Elsevier, (2014)
6. J.K.L. Lai, C.H. Shek, K.H. Lo, B. Sci. Publishers, (2012)
7. S. Tavares, V. Terra, J. Pardal, M.C. Fonseca, JMS, **40**(1), 145-154 (2005)
8. V.S. Moura, L.D. Lima, J.M. Pardal, A.Y. Kina, R.R.A. Corte, S.S.M. Tavares, MC, **59**(8), 1127-1132 (2008)
9. B. Deng, Z. Wang, Y. Jiang, T. Sun, J. Xu, J. Li, Corr Sci, **51**(12), 2969-2975 (2009)
10. J. Dobránszky, J. Ginsztler, MSF, Trans Tech Publ, pp 2119-2122 (2007).
11. V. Muthupandi, P.B. Srinivasan, S. Seshadri, S. Sundaresan, Mat. Sci.Eng: A, **358**(1), 9-16 (2003)
12. I.F. KoleniČ, D.-I.L. Kovac, D.-I.D. Drimal, WW, **55**(5-6), 19-25 (2011)
13. A.H.I. Mourad, A. Khourshid, T. Sharef, Mat. Sci.Eng: A, **549**, 105-113 (2012)
14. B.R. Gnanasundaram, M. Natarajan, MT, **48**(5), 761-763 (2014)
15. J. Pekkarinen, V. Kujanpää, PP, **5**, 517-523 (2010)

16. A.-M. El-Batahgy, A.-F. Khourshid, T. Sharef, , M Sci App, **2**(10), 1443-1451 (2011)
17. M.M.A. Khan, L. Romoli, M. Fiaschi, G. Dini, F. Sarri, J Mat. Proc. Tech., **212**(4), 856-867 (2012)
18. W. Duley, Laser processing and analysis of materials, Springer Sci. & BM (2012)
19. J. Dutta Majumdar, I. Manna, , IMR, **56**(5/6), 341-388 (2011)
20. G. Moskvitin, A. Polyakov, E. Birger, WI, **27**(7), 572-580 (2013)
21. Z. Zhang, H. Jing, L. Xu, Y. Han, L. Zhao, Mat.& Des, **109**, 670-685 (2016)
22. I.M. Association, T. Stainless, *Practical guidelines for the fabrication of duplex stainless steels,* 3rd Edition., IMOA, London, UK (2014).
23. F. Mirakhorli, F.M. Ghaini, M. Torkamany, JMEP, **21**(10), 2173-2176 (2012)
24. Y. Zhao, Y. Zhang, W. Hu, X. Lai, , *Opt.& Las.En.*, **50**(9), 1267-1273 (2012)
25. V. Amigó, V. Bonache, L. Teruel, A. Vicente, WI, **20**(5), 361-366 (2006)

Identification of damage based on frequency response function (FRF) data

M. S. A. Sulaiman[1], *M. A.* Yunus[1,*], *A. R.* Bahari[2], *M. N.* Abdul Rani[1]

[1]Center of Excellence (CoE) of Structural Dynamics Analysis and Validation (SDAV), Faculty of Mechanical Engineering, Universiti Teknologi MARA (UiTM), 40450 Shah Alam, Selangor, Malaysia.
[2]Faculty of Mechanical Engineering, Universiti Teknologi MARA (UiTM), 23200 Bukit Besi, Dungun, Terengganu, Malaysia

Abstract. Mechanical joints, particularly fasteners such as bolted joints have a complex non-linear behaviour. The non-linearity might emerge from the material, geometry or by the contacts in the joints. However, damage to a structure can be happened either their connections or the material of components. The effect of damage can change the dynamic properties of the structure such as natural frequencies and mode shapes and structural performance and can cause premature failure to structure. This paper presents a damage detection method using a vibration based damage detection method based on the frequency response function (FRF) data. A combination of numerical model and physical bolted jointed structure of damaged and undamaged structure will be investigated. The validation is employed to detect the presence of damage in the structure based on the frequency response function (FRF) data from the parameter values used in the benchmark model and damaged model. The comparisons of the undamaged and damaged structure of the FRF have revealed the damaged structure was shifted from the undamaged structure. The effect of the FRF between undamaged and damaged structure is clearly affected by the reduction of stiffness for the damaged structure.

1 Introduction

A complex jointed structure, such as the automotive structure is produced using a mix of thin metal sheets and are joined by a thousand number of joints. The reliability arrangement of the structure extremely depends on these joints [1, 2]. The bolted joint is a typical sort of fastener for jointed the structure that is commonly used in the automotive structure. Complex jointed structures with involved with joints have its own dynamic properties, and the dynamic properties of the structure will alter due to the addition of stiffness in the structure.

For instance, mechanical joints, particularly fasteners such as bolted joints have a complex non-linear behaviour. The non-linearity might emerge from the material, geometry

* Corresponding author: mayunus@salam.uitm.edu.my

or by the contacts in the joints. However, damage to a structure can be happened either their connections or the material of components. Understandably, it also will reduce vehicle performance and can cause premature failure of the vehicle structure. Therefore, to overcome the presence of damage is vital to avoid the additional problem of the structure, such as significantly affects the performance, safety and dynamics properties of the structure due to the reduction of stiffness in the structure. For detecting the damage on the structure, it has several types of damage identification methods; non-destructive methods and destructive method [1]. The visual examination has usually been utilised in damage detection method because it is easy to perform for clear structural damage. However, it is problematic when the structure becomes complicated given the limited capability of conventional visual assessment.

Normally, the non-destructive methods have wide variety methods for examples using strain gauges, penetrating liquids, ultrasound, visual inspection, etc. [3]. However, some of these methods are localised methods, it has a disadvantage in requiring the area of the damage to be known from the earlier and the area of the structure being investigated is readily accessible by a machine or labour, which makes the examination procedure practically difficult to perform, expensive and time-consuming [4].

Typically, there are other option methods like vibration-based damage detection method which can locate the damage region more efficiently [5]. Besides, the result of these methods can be utilised for further damage appraisal [6]. These methods are widely used in structural applications to identify the dynamic properties of the structure and frequency response function (FRF). The purpose of the damage identification is to offer data in the presence of the damage, area, and the level of damage in the structure [7]. Thus, it is important to monitor and detect the damage on the structure to reduce the cost and increase its reliability [8].

Frequency response function (FRF) data are one of option methods for vibration-based damage detection method. Essentially an FRF is a numerical representation of the relationship between the input and output system. They may also be identified regarding magnitude and phase. Besides, it also covers the data of both natural frequencies and mode shapes [3], [9], [10]. This paper presents a damage detection method using a vibration based damage detection method based on the frequency response function (FRF) data. A combination of numerical model and physical bolted jointed structure of damaged and undamaged structure will be investigated, and the result will be validated. The reconciliation procedure is employed to detect the presence of damage in the structure and disparity in certain parameter values from the parameter values used in the benchmark model is used to localise and quantifies the damaged at the particular joint/structure.

2 Frequency response function (FRF)

Figure 1 is shown the relationship between excitation input and the output signal of a measuring point on mechanical system of frequency response function (FRF). The concept of FRF is at the basis of modern experimental modal analysis. The experimental FRF data are generally acquired from vibrational testing.

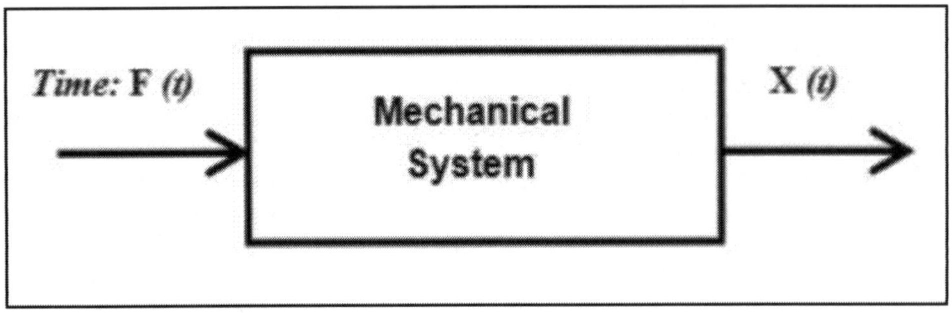

Fig. 1. Block diagram of FRFs [11]

$$\{\mathbf{X}(\omega)\} = [\mathbf{H}(\omega)]\{F(\omega)\} \tag{1}$$

where,
$\{\mathbf{X}(\omega)\}$ = output force
$\{F(\omega)\}$ = input force
$[\mathbf{H}(\omega)]$ = response model (the ratio of output by input)

Therefore, the relationships between the response model $\mathbf{H}(\omega)$ to modal model and spatial can be arranged and expressed by;

$$\mathbf{H}(\omega) = \left(-\omega^2\mathbf{M} + i\omega\mathbf{C} + \mathbf{K}\right)^{-1} \tag{2}$$

where,
M = mass
C = damping
K = stiffness

Since the input force to the structure and dynamic response of the structure are acquired from physical measurement, it is hypothetically conceivable to get a mathematical description of the structure through experiment.

3 Experimental setup

3.1 Experiment modal analysis

The experimental modal analysis (EMA) is a method to define the dynamic properties of the structure from on vibration testing, such as natural frequencies, mode shapes and damping ratios [4]. Essentially, it is carried out according to both input and output experimental data by the impulse response functions (IRF) in the time domain and frequency response functions (FRF) in the frequency domain. These method has developed since the beginning era for the digital Fast Fourier Transform algorithm (FFT) analyser in the early 1970's [11]. Vibration testing is to check the result from finite element model proposed for the experimental data.

3.2 Experimental procedure

The structure under investigation is a simplified model structure of u-shape components made from Aluminium 6061-T4 and for the thin metal plate is made of mild steel. Before the experimental arrangement, every one of this procedure must consider getting the reliable result. It is the hanging orientation of the structure, the number of measuring points or nodes of the structure, the excitation method and the number of the accelerometers. The initial idea of the procedure was determined by finite element analysis (FEA) results. The experimental model was set-up in free-free boundary conditions as shown in Figure 2.

In this experiment, four nylon strings with springs were used to simulate the conditions. Based on finite element analysis, the frequency range of interest in this study was 0 Hz to 600 Hz. Besides, roving accelerometer method and impact hammer were used to excite the experimental model. LMS SCADAS analysis interpreted the load and response signals. The data used in this research consists of two stages; the benchmark structure and damaged structure. It is because to simulate the present of damage to the structure. The damage was presented to the structure by cutting on the thin metal plate as shown in Figure 3.

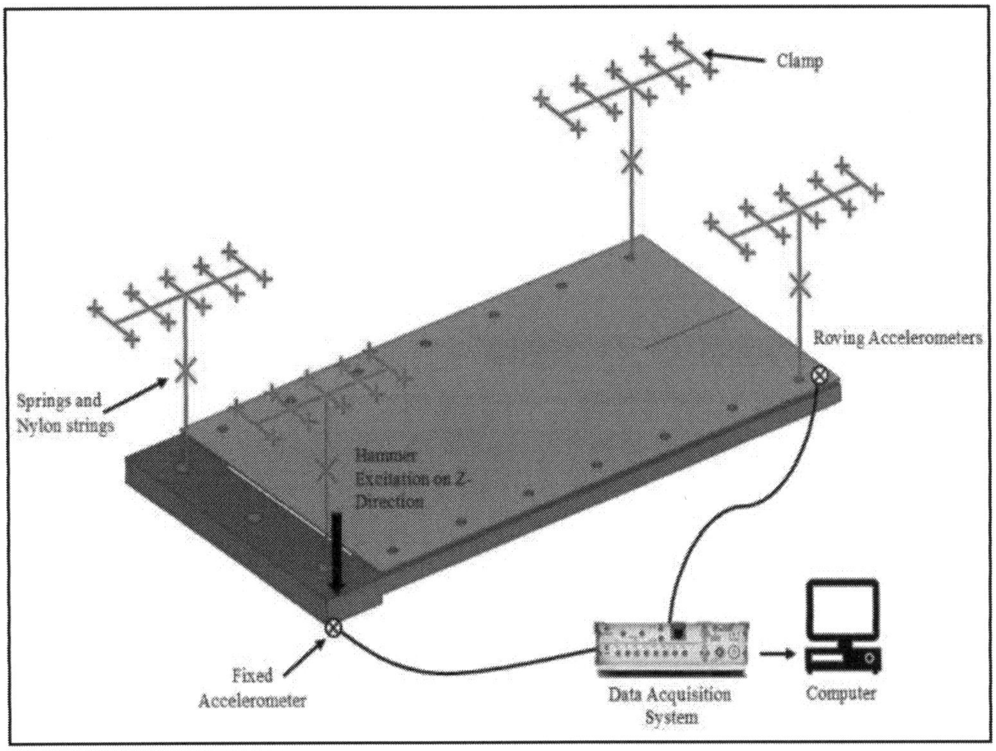

Fig. 2. Schematic diagram of the simplified model tests setup.

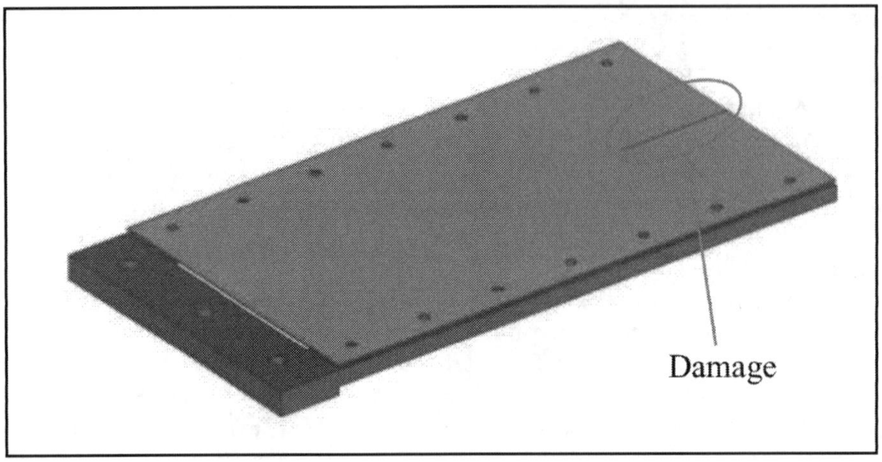

Fig. 3. Damaged structures.

4 Finite element analysis

The simplified model of the car hood was constructed using finite element analysis. The simplified model consists two components, u-shape component and thin metal plate as shown in Figure 4. It will be assembled by a number of bolted joints. The FE model for the u-shape component is made from Aluminium 6061-T4 and for the thin metal plate is made of mild steel. Table 1 and Table 2 showing the detail of FE model for u-shape component and thin metal plate. For Table 3, it is only showing the detail of the FE model for the joints.

Table 1. Detail of the FE model for the u-shape component.

	No. of Element	Properties	Value
Type of element CTETRA	52309	Young's Modulus	68.9 GPa
		Poisson's Ratio	0.33
		Density	2700 kg/m^3

Table 2. Detail of the FE model for the thin metal plate.

Type of element	No. of Element	Properties	Value
CQUAD4	5603	Young's Modulus	210 GPa
CTRIA3	252	Poisson's Ratio	0.30
		Density	7900 kg/m^3

Table 3. Details of the FE model for the joints.

Type of Elements	No of Elements
RBE2	98

The simplified structure was discretized into two elements, solid element for u-shape component and shell element for the thin metal plate. Meanwhile, the model of the jointed structure was constructed using rigid connector elements (RBE2) to represent the bolted joints. Therefore, since the only rigid body element (RBE2) was used, there is no mass apply to the joint and mechanical properties. Though it is a 1D element, the idea of using this model is to form a look like bolted joints as shown in Figure 5. It is the combination of the rigid body element to form a spider-web like connecting the node at the edge of the hole. It was modelled using Altair HyperMesh software and Nastran SOL103 was used to compute the dynamic properties of the structure. The value of frequencies of interest is set between 1Hz to 600Hz and only first ten modes were recognised.

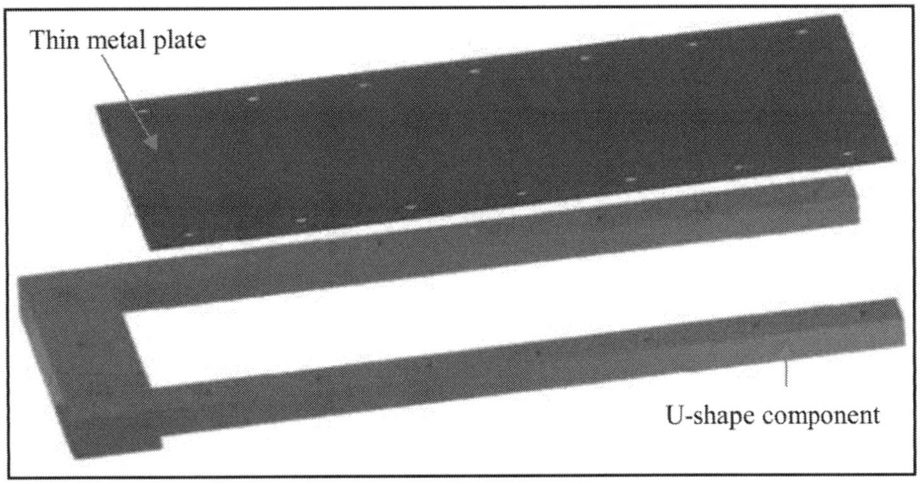

Fig. 4. The simplified model of the car hood.

Fig. 5. The FE model of the connector element to represent a bolted joint.

5 Results and discussion

The identification of natural frequencies of the structure has always been a challenging task and difficult. In this study, the measured data for natural frequencies of undamaged and damaged will be compared with the predicted result to confirm the accuracy of the model [4]. Table 4 and Table 5 shows the result of natural frequencies obtained from the experiment and predicted by comparison the results between undamaged and damaged structure.

The results in Table 4 shows the comparisons of the measured data and predicted result based on the natural frequencies that were obtained from the undamaged structure. Meanwhile, in column IV of Table 1 shows the total results of the percentage error is 27.54 percent. These are clearly presented that the least error was in mode five with MAC value 0.83 and the mode seven shown with the highest error of 6.26 percent. Meanwhile, for Table 5 the highest error was in mode nine and the mode three shown the least of 0.01 percent with a total of MAC value 0.87. Based on Table 4 and Table 5, it shows the changes in the natural frequencies of the undamaged and damaged structure. The changes in the natural frequencies achieved from the measured data are shown the presence of damaged in the structure.

Table 4. Comparisons of measured and predicted natural frequencies for undamaged structure.

I	II	III	IV	MAC
Mode No.	**Test Data (Hz)**	**Undamaged (Hz)**	**Error (%)**	**Value**
1	120.17	126.82	5.53	0.88
2	193.16	198.66	2.85	0.83
3	203.95	210.53	3.23	0.85
4	220.02	224.03	1.82	0.79
5	295.64	295.76	0.04	0.83
6	409.56	405.83	0.91	0.78
7	454.33	482.79	6.26	0.80
8	513.50	500.28	2.57	0.73
9	524.44	528.30	0.74	0.70
10	562.17	542.01	3.59	0.84
Total Error (%)			27.54	

Table 5. Comparisons of measured and predicted natural frequencies for damaged structure.

I	II	III	IV	MAC
Mode No.	**Test Data (Hz)**	**Damaged (Hz)**	**Error (%)**	**Value**
1	98.49	102.85	4.43	0.90
2	147.22	150.22	2.04	0.83
3	189.16	189.18	0.01	0.87
4	197.76	197.60	0.08	0.82
5	216.58	218.94	1.09	0.87
6	292.64	285.91	2.30	0.79
7	399.28	400.44	0.29	0.83
8	411.83	396.15	3.81	0.74
9	426.34	473.46	11.05	0.72
10	487.30	493.78	1.33	0.84
Total Error (%)			26.43	

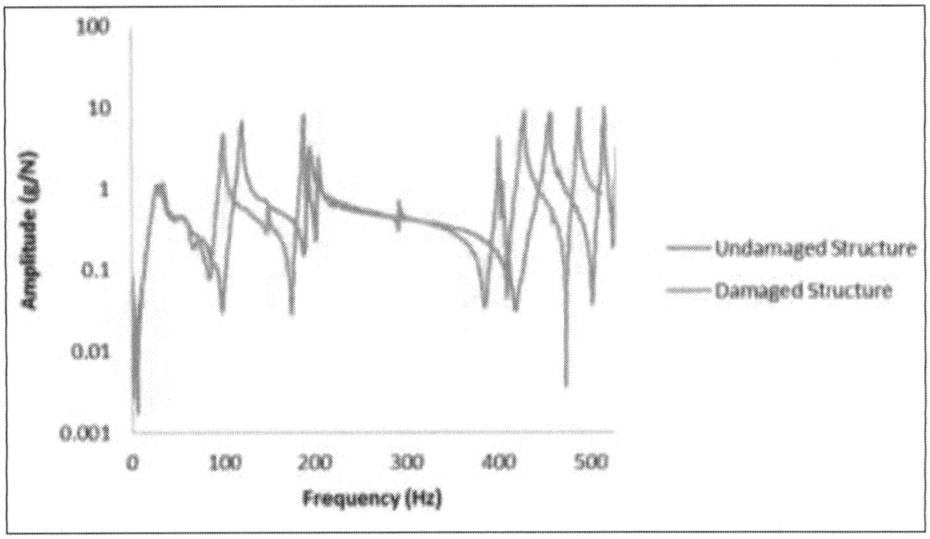

Fig. 6. Comparisons of measured data of the frequency response function (FRF) for undamaged structure and damaged structure.

For instance, the natural frequencies for mode one of the undamaged structure (Table 4) and damaged structure (Table 5) were diminished from 120 Hz to 98 Hz because of the loss of stiffness caused by the damaged in the structure. Meanwhile, Figure 6 shows the super-impose on the FRF's of damaged and undamaged structure that was obtained from the vibration testing. Therefore, the identification of the presence of damage of the structure was carried out by mapping the FRF for undamaged and damaged structure. It clearly can be seen that the FRF data for the damaged structure was shifted from the undamaged structure due to the loss of stiffness in the structure due to the existence of damage to the structure. The effect of the FRF between undamaged and damaged structure is clearly affected by the reduction of stiffness for the damaged structure [12]. Therefore, if the crack becomes bigger, a more stiffness reduction in the structure will lead to a larger shift of the peak in the FRF.

6 Conclusions

The dynamic properties of the both bolted joined structure, undamaged and damaged structure were investigated experimentally and numerically. Moreover, combinations of the experimental and predicted data are shown in relatively in good agreement. This paper has shown the presence of damage structure in a bolted joined structure based on the frequency response function (FRF) data from the parameter values used in the benchmark model and damaged model. It has been shown that identification of damage based on frequency response function (FRF) data has been successfully and accurately using the proposed technique with guidance from the predicted data. The comparisons of the undamaged and damaged structure of the FRF have revealed the damaged structure was shifted from the undamaged structure. The effect of the FRF between undamaged and damaged structure is clearly affected by the reduction of stiffness for the damaged structure.

The research has been funded by the Kementerian Pendidikan Malaysia (KPM) and Research Management Centre (RMC) UiTM for the Research Acculturation Grant Scheme (RAGS) with the file number 600-RMI/RAGS 5/3 (165/2014). Besides, the authors would like to express thankfulness to the technician of the Center of Excellence (CoE) Structural Dynamics Analysis and Validation (SDAV) UiTM by Mohamad Azwan Abd Khair for their technical assistance and laboratory facilities.

References

1. M. A. Yunus, M. N. A. Rani, H. Ouyang, H. Deng, S. James, "Identification of Damaged Spot Welds in a Complicated Joined Structure," J. Phys. Conf. Ser., **305**, p. 012057, (2011)
2. R. A. Ibrahim, C. L. Pettit, "Uncertainties and Dynamic Problems of Bolted Joints and Other Fasteners," J. Sound Vib., **279**, no. 3–5, pp. 857–936, (2005)
3. R. P. C. Sampaio, N. M. . Maia, J. M. M. Silva, "Damage detection using the frequency response function curvature method," J. Sound Vib., **226**, no. 5, pp. 1029–1042, (1999)
4. K. A. Ramsey, S. M. Systems, S. Jose, "Experimental Modal Analysis , Structural Modifications and FEM Analysis on a Desktop Computer mathematical dynamic models of structures so that they could," Sound Vib., no. February, pp. 1–10, (1983)
5. L. Wang, S. T. Lie, Y. Zhang, "Damage detection using frequency shift path," Mech. Syst. Signal Process., **66–67**, pp. 298–313, (2016)
6. S. W. Doebling, C. R. Farrar, M. B. Prime, "A Summary Review of Vibration-Based Damage Identification Methods," Shock Vib. Dig., **30**, no. 2, pp. 91–105, (1998)

7. Y. Xia and H. Hao, "Statistical Damage Identification of Structures with Frequency Changes," J. Sound Vib., **263**, no. 4, pp. 853–870, (2003)
8. M. N. Abdul Rani, D. Stancioiu, M. a. Yunus, H. J. Ouyang, H. Deng, S. James, "Model Updating for a Welded Structure Made from Thin Steel Sheets," Appl. Mech. Mater., **70**, pp. 117–122, (2011)
9. M. K. Yoon, D. Heider, G. J.W., R. Jr., C. C.P., R.M., "Local damage detection using a global fitting method on mode shape data," Proc. Int. Modal Anal. Conf. - IMAC, **1**, pp. 231–237, (2001)
10. R. P. Bandara, T. H. T. Chan, D. P. Thambiratnam, "Frequency response function based damage identification using principal component analysis and pattern recognition technique," Eng. Struct., **66**, pp. 116–128, (2014)
11. B. J. Schwarz , M. H. Richardson, "Experimental Modal Analysis," CSI Reliab. Week, **35**, no. 1, pp. 1–12, (1999)
12. K. Dems, J. Turant, "Structural damage identification using frequency and modal changes," Bull. Polish Acad. Sci. Tech. Sci., **59**, no. 1, pp. 5–10, (2011)
13. A. Shahdin, J. Morlier, H. Nieman, Y. Gourinat, "Correlating low energy impact damage with changes in modal parameters: diagnosis tools and FE validation", Structural Health Monitoring 10(**2**) 1, (2010)
14. Y. Zhao, C. Yang, L. Cai, W. Shi, Z. Liu, "Surface Contact Stress-based Nonlinear Virtual Material Method for Dynamic Analysis of Bolted Joint of Machine Tool," Precis. Eng., **43**, pp. 230–240, (2016)
15. A. A. S. M. Darwish, Sh. M. Elseufy.Advanced, "Finite Element Analysis of an Automobile Engine Finite Element Analysis of an Automobile Engine Hood," no. MARCH 2013, pp. 1–8, (2016)
16. L. Majkut, "Acoustical diagnostics of cracks in beam like structures," Arch. Acoust., **31**, no. 1, pp. 17–28, (2006)
17. N. Rieger, "Relationship between finite element analysis and modal analysis.," Sound Vib., pp. 16–31, (1986)
18. Stephen A. Dyer, *Survey of Instrumentation on Measurement*. Wiley, (2004)
19. W. Kuntjoro, *An Introduction to the Finite Element Method*. McGraw-Hill Education (Asia, 2005)
20. N. S. Gokhale, *Practical Finite Element Analysis*. (2008)
21. D. Hutton, *Fundamental of Finite Element Analysis*. Mc Graw Hill, (2004)
22. He, J. and Z.F. Fu, *Modal analysis*. (Butterworth-Heinemann, (2001)

DOI: 10.1051/matecconf/20179001026

Finite element simulation of twist forming process to study twist springback pattern

M. N. Nashrudin[1] , and *A. B.* Abdullah[1,*]

[1]Metal Forming Research Laboratory, School of Mechanical Engineering, Universiti Sains Malaysia, 14300 Nibong Tebal, Pulau Pinang, Malaysia. Phone: 604-5996332, Fax: 604-5941025

Abstract. Springback is one of the most common defects found in the metal forming of automotive parts. There are three conditions which can be considered as springback i.e. flange angle change, sidewall curl and twist springback and among them, twist springback is the most complicated problem. This study will focuses on the development of finite element simulation model of the twist forming process. The main aim of this project is to investigate the parameters that may affect the twist springback. Few parameters including twist angle, hardening constant and thickness are explored using finite element (FE) software ANSYS Workbench (16.0). The rectangular mild strips are used to form the twist forming. The standard material properties and stress-strain curve of mild steel had been used to get the springback prediction. The results of springback were measured by the difference of the bending angles before and after unloading process. The results were then be validated with the research made of Dwivedi et al., (2002). The results show that the springback angle reduces as the thickness of strips are increased and also as the angle of twist increases.

1 Introduction

In automotive industries, metal forming is the main process used in manufacturing process. The raw material is deformed plastically to obtain required shape and size. Examples of the process are forging, rolling, extrusion, sheet metal forming and hydro forming. Usually the primary shape of a car is created by plastically deforming the material. The dimension must be precise to assemble another parts into it. In every industry, quality and productivity are vital issues as it is related to the cost of manufacturing. Every forming process, there are defects can be occurred and one of it is springback. The defects need to be controlled.

Springback occurs after forming that change the part to the undesirable shape [1]. The cause of it is due to the disappearance of the elastic deformation after removing the external load and the final shape is different from the die shape [2]. It happens toward the end of plastic deformation[3]. The effects of springback are it can bring the undesirable shape change, leads to some difficulties in assembly, decrease the material mechanical properties

* Corresponding author: mebaha@usm.my

and causes geometrical and unreliability of sheet metal after forming [2, 3]. The springback also affected by the applied tension, the bending radius and the mechanical properties of material [5]. There are 3 general types of springback which are angular change, sidewall curl and twist springback [4]. Among of them, twist springback is the most complicated and it will be focused in this study.

Twist springback involved twisting and bending that considered as one of the major geometrical defect in strip metal forming. It is tendency of twisted part return to its original shape due to the elastic recovery when unloading [6]. The unbalanced elastic-plastic deformation and residual stress after the forming process are the causes of the twist springback [7]. When the twist keeps increasing, the non-recoverable plastic deformation increases and the recoverable elastic deformation keeps decreasing [8]. Springback in angle of twist depends on the point on the curve where unloading is initiated and the shape of the elastic unloading line [9]. There are many automotive parts that can exhibit twist springback such as B pillar and twist rail [10].

In this study, the development of finite element simulation model of twist forming process will be focused. The parameters that may affect twist springback are the main investigation of this project. The result will be validated with the research made of Dwivedi et al., (2002) [8]. From the previous research by [8], an analysis for twist forming process has been made using experimental method. The parameters studied for this method will be compared by using finite element simulation method on this research. Then validation can be made that finite element simulation method is relevant to be used to predict springback percentage. All parameters will be used same as the previous research by Dwivedi et al., (2002).

2 Methodology

2.1 Finite element simulation

In this study, twist forming process was simulated using finite element (FE) software ANSYS Workbench (16.0). Figure 1 shows the model and boundary condition of mild steel strip that undergo twist forming.

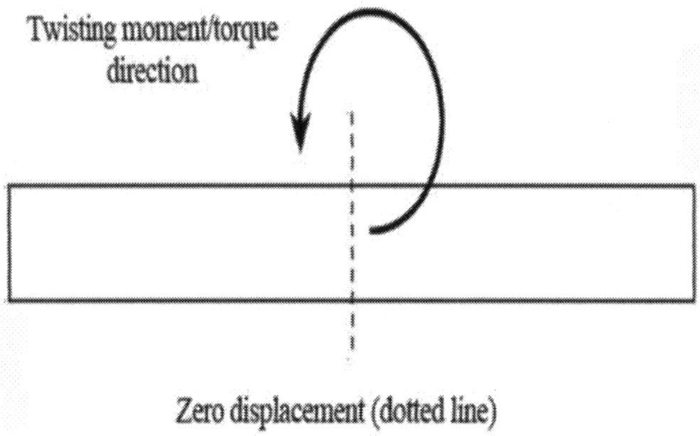

Fig. 1. Model and boundary conditions for simulation process.

The rectangular mild steel strips were modeled in SolidWorks with thickness of 2.2 and 3.2 mm, width of 25.4 mm and length of 178 mm. Then the strips were modeled in ANSYS to undergo simulation for springback prediction. The center of the strip are fixed and

DOI: 10.1051/matecconf/20179001026

twisting moment or torque are applied at both ends. Table 1 list the material properties of mild steel used for simulation process.

Table 1. Material properties of mild steel used.

Material properties	Value
Elastic modulus	210 GPa
Poisson's ratio	0.29
Modulus of rigidity	82 GPa
Density	7800 kg/m^3
Yield strength	200 MPa

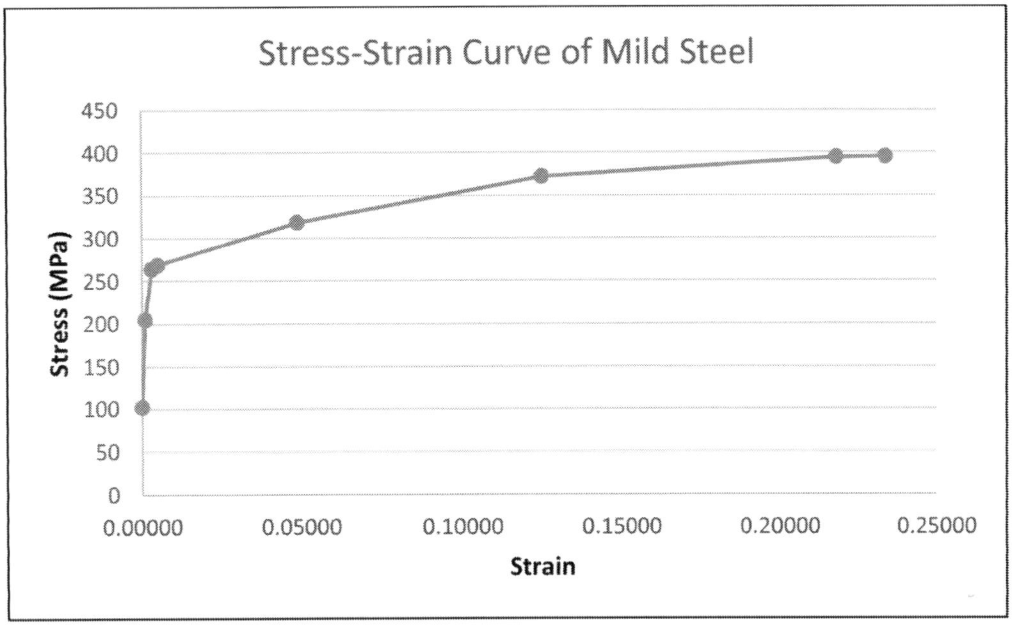

Fig. 2. Stress-strain curve of mild steel.

2.2 Analysis of springback measurement

From the front view, the angle for both before and after of springback are shown in Figure 3. The value of springback angle can be expressed as the change of bending angle [11].

$$\Delta\Theta = \Theta_o - \Theta_f \quad (1)$$

Where Θ_o and Θ_f are the bending angles before and after unloading process, respectively.

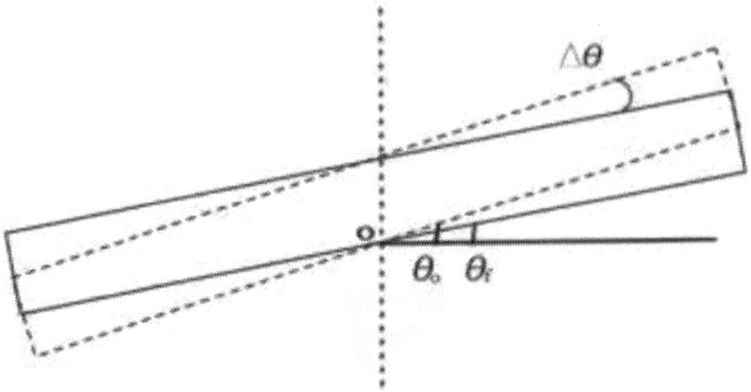

Fig. 3. Angle before and after springback.

3 Results and discussion

In this study, simulations were made to compare the result with the result published by Dwivedi et al., (2002). The parameters that have been used were different twist angle, the different thickness of strips and the strain hardening index, n of 1/3 for mild steel with material properties stated on Table 1. The results show the percentage difference between the both results are quite large but the curve patterns are similar. Pattern of springback in percentage against thickness are shown in Figure 4, while relationship of springback in percentage against twist angle in Figure 5 and 6.

Based on the result, the springback angle reduces as the thickness of strips are increased. Higher plastic zone is created at higher sheet thickness and result to lower springback angle. The value of angle of twist also gives significant effect to springback angle. The springback angle will decrease as the angle of twist increases. Twist angle is more sensitive to process parameters and material direction as well as the materials. Torque slows down when the twist increases and result to lower springback angle.

State that the result is below 10% and 36% of deviation for the thickness of 2.2 mm and 3.2 mm respectively. This can be concluded that the validation is reliable since the percentage difference are quite low for 2.2 mm thickness. From research by [12], the deviation below 10% is considered low in springback cases. For 3.2 mm thickness, the deviation is quite large because of the twisting moment goes beyond the plastic region of the section. It may be reduced if more accurate material properties for stress-strain curve is obtained. Additional study about FE simulation also may improve the results.

DOI: 10.1051/matecconf/20179001026

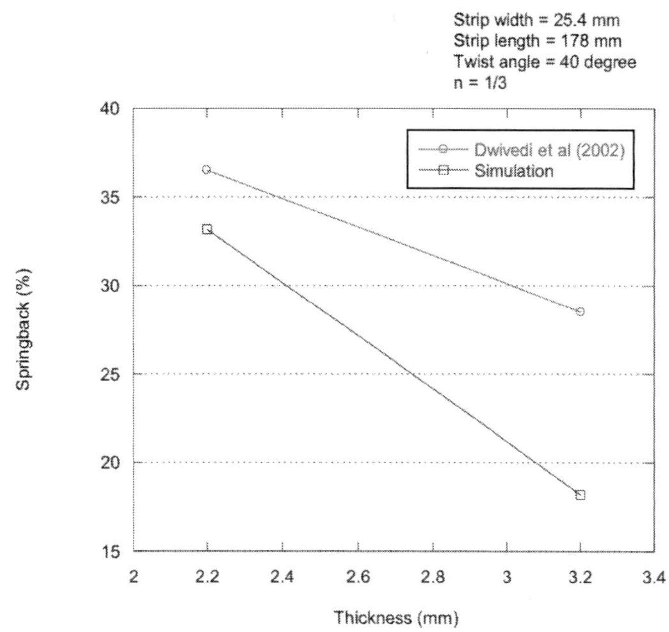

Fig. 4. Springback percentage against thickness for angle of twist of 40°.

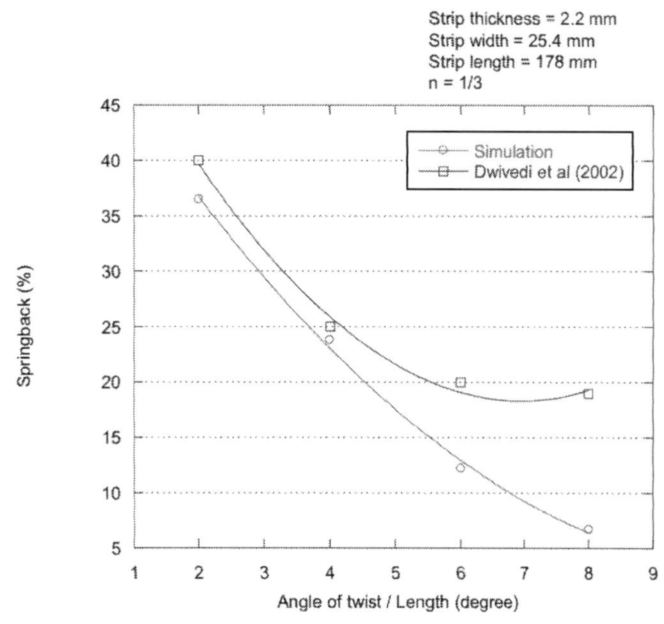

Fig. 5. Springback percentage against angle of twist for strip thickness of 2.2 mm.

 DOI: 10.1051/matecconf/20179001026

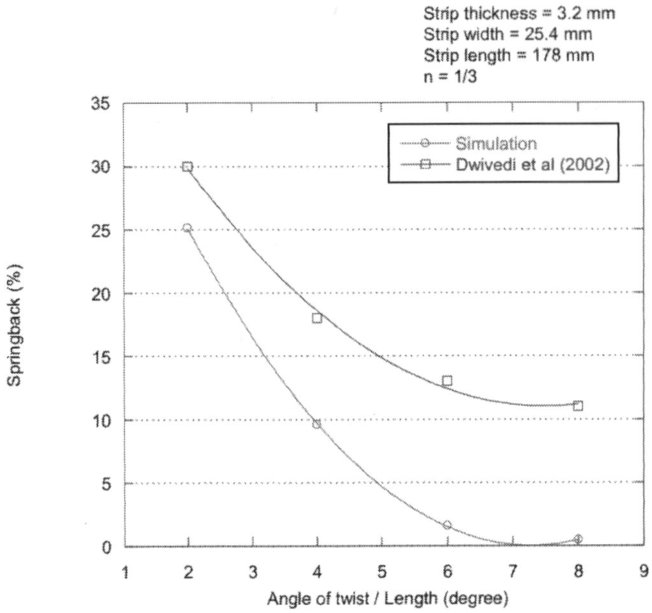

Fig. 6. Springback percentage against angle of twist for strip thickness of 3.2 mm.

4 Conclusions

As a conclusion, the parameter of thickness of strips and angle of twist gave significant effect to the twist springback angle based on the results. The validation is accepted since the curve patterns are same and the deviation is low i.e. between the results for 2.2 mm thickness. For future work, the effect of another parameters to the twist springback angle will be explored using simulations and experiments.

The authors want to acknowledge Ministry of Higher Education for their sponsorship through Fundamental Research Grant Scheme (Acc. No. 203/PMEKANIK/6071308).

References

1. X. Xue, J. Liao, G. Vincze, J. Sousa, F. Barlat, and J. Gracio, "Modelling and sensitivity analysis of twist springback in deep drawing of dual-phase steel," *Mater. Des.*, **90**, pp. 204–217, (2016).
2. W. Jiang, B. Yang, X. Guan, and Y. Luo, "Bending and twisting springback prediction in the punching of the core for a lattice truss sandwich structure," *Acta Metall. Sin. (English Lett.,*. **26**, no. 3, pp. 241–246, (2013).
3. S. Xu, K. Zhao, T. Lanker, J. Zhang, and C. T. Wang, "Springback prediction, compensation and correlation for automotive stamping," *AIP Conf. Proc.*, **778** A, pp. 345–352, (2005).
4. W.Y.D. Yuen, "Springback in the Stretch-Bending of Sheet Metal with Non-Uniform Deformation," *J. Mater. Process. Technol.*, **22**, pp. 1–20, (1990).
5. M. Ueda, K. Ueno, and M. Kobayashi, "A study of springback in the stretch bending

DOI: 10.1051/matecconf/20179001026

of channels," *J. Mech. Work. Technol.*, **5**, no. 3–4, pp. 163–179, (1981).

6. A. B. Abdullah and Z. Samad, "Measurement of Twist Springback on AA6061-T6 Aluminum Alloy Strip - A Preliminary Result," *Appl. Mech. Mater.*,. **699**, pp. 44–48, (2014).
7. X. Xue, J. Liao, G. Vincze, and J. Gracio, "Modelling of mandrel rotary draw bending for accurate twist springback prediction of an asymmetric thin-walled tube," *Procedia Eng.*, **81**, pp. 2177–2183, (2014).
8. J. P. Dwivedi, S. K. Shah, P. C. Upadhyay, and N. K. Das Talukder, "Springback analysis of thin rectangular bars of non-linear work-hardening materials under torsional loading," *Int. J. Mech. Sci.*, **44**, no. 7, pp. 1505–1519, (2002).
9. N. K. T. J. P. Dwivedi, A.N. Singh, Sohan Ram, "Springback analysis in torsion of rectangular strips," *Pergamon Journals Ltd*, **28**, no. 8, pp. 505–515, (1986).
10. H. Li, G. Sun, G. Li, Z. Gong, D. Liu, and Q. Li, "On twist springback in advanced high-strength steels," *Mater. Des.*, **32**, no. **6**, pp. 3272–3279, (2011).
11. J. Liao, X. Xue, M. G. Lee, F. Barlat, and J. Gracio, "On twist springback prediction of asymmetric tube in rotary draw bending with different constitutive models," *Int. J. Mech. Sci.*, **89**, pp. 311–322, (2014).
12. A. B. Abdullah, M. S. Salit, Z. Samad, K. H. MTandoor, and N. A. Aziz, "Twist springback measurement of autonomous underwater vehicle propeller blade based on profile deviation," *Am. J. Appl. Sci.*, **10**, no. 5, pp. 515–524, (2013).

Influence of rear-roof spoiler on the aerodynamic performance of hatchback vehicle

See-Yuan Cheng[1,2,*], and *Shuhaimi* Mansor[3]

[1]Centre for Advanced Research on Energy, Universiti Teknikal Malaysia Melaka, Hang Tuah Jaya, 76100 Durian Tunggal, Melaka, Malaysia
[2]Faculty of Mechanical Engineering, Universiti Teknikal Malaysia Melaka, Hang Tuah Jaya, 76100 Durian Tunggal, Melaka, Malaysia
[3]Faculty of Mechanical Engineering, Universiti Teknologi Malaysia, 81310 UTM Skudai, Johor, Malaysia

Abstract. Rear-roof spoiler is commonly used for improving the aerodynamic performance of road vehicles. This study aims to investigate the effect of strip-type rear-roof spoiler on the aerodynamic performance of hatchback vehicles. The main parameter of study was the inclination angle of the spoiler. A computational fluid dynamics (CFD) method was used. The numerically obtained results were compared to the experimental data for validation of the CFD method. The spoiler effectively reduced the aerodynamic lift at positive inclination angle by causing the surface pressure near the roof-spoiler junction to increase. However, its effect is unfavourable when configured at negative angle due to the downward accelerating flow that causes the surface pressure around the roof-spoiler junction to drop. Although the aerodynamic lift was found to decrease with the spoiler angle, this was accompanied by drag increment.

1 Introduction

A properly designed rear spoiler can reduce the rear-axle lift of a vehicle. This effect is important because a vehicle with its rear-axle lift lower than the front exhibits better stability, i.e. requires minimum driver intervention to maintain a straight path [1]. In addition, when a vehicle makes a turn, it needs sufficient traction to provide the centripetal force for it to pass the curve without slip. Hence, the use of spoiler will enable a vehicle to achieve greater downforce (negative lift), and thus better traction without significant increase in its weight.

There are two types of rear spoiler design: strips and free-standing wing. The effectiveness of wing-type spoiler has been reported in numerous studies. For instance, Tsai et al. [2] investigated numerically five different spoiler configurations. The spoilers are of inverted airfoil type and were mounted on the trunk of a simplified car model based on the HONDA S2000. Although all the cases had shown reduction in lift, however, only one out of five configurations succeeded in producing a negative lift coefficient value.

* Corresponding author: cheng@utem.edu.my

Meanwhile, Daryakenari et al. [3] has shown a similar trends. Its numerical results indicate that up to 75% reduction in lift coefficient for a passenger car model can be achieved by properly manipulating the inclination angle of its flat-plate-type spoiler, i.e. at 50° relative to the ground. Similarly, Kodali and Bezavada [4] has observed a decrease of 80% lift coefficient in the case of a simplified-passenger-car model mounted with an inverted-airfoil type spoiler.

As for strip-type rear spoilers, Menon et al. [5] has investigated the influence of spoiler on the aerodynamic drag and crosswind stability of a simplified hatchback model, namely, an Ahmed model at 35° slant angle. Although these two performances are important, the study did not cover aerodynamic lift performance which is important for driving stability. Moreover, extensive literature search found that the coverage of this particular topic – strip-type rear spoiler's influence on lift performance – is scarce. To fill the gap, hence, the main objective of the present paper is to investigate the effect of strip-type rear-roof spoiler on the aerodynamic lift performance of simplified hatchback model.

2 Hatchback model and spoiler configurations

The Ahmed body which represents simplified road vehicle geometry in the form of a bluff body was adopted. The slant angle was at 35°, which is typical for most hatchback cars. Figure 1 shows the Ahmed model fitted with a rear-roof spoiler configured at 0° pitch angle. Also shown in the figure are the profiles of the spoiler at various pitch angles investigated in the present study, namely, -15°, 0°, 5°, 10°, and 15°. For all configurations, the length of the spoiler was fixed at 66.6 mm, which corresponds to 30% of the length of the slant section. The trailing edge of the spoiler was filleted (5 mm radius) for avoiding highly skewed cells during meshing. For the same reason, the angle between the slant section and the rear face of the spoiler was maintained at right angle for all configurations. For details of the Ahmed model dimensions, the reader is referred to Ahmed [6].

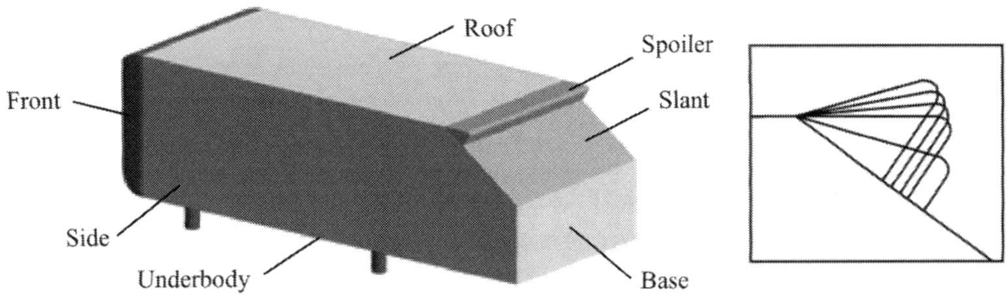

Fig. 1. Side view of Ahmed model fitted with a rear-roof spoiler (left), and the spoiler configurations (right).

3 Computational method

3.1 CFD settings

The present study employed a numerical simulation method to investigate the influence of pitch angle of rear spoiler on the aerodynamics forces of a simplified hatchback model. All results were obtained using the commercial finite-volume solver ANSYS Fluent 16. The Reynolds-averaged Navier-Stokes (RANS) approach was used. The governing equations solved in the simulations are the time-averaged continuity and momentum equations for an incompressible Newtonian fluid:

DOI: 10.1051/matecconf/20179001027

$$\frac{\partial u_i}{\partial x_i} = 0 \tag{1}$$

$$\rho u_j \frac{\partial u_i}{\partial x_j} = \frac{\partial}{\partial x_j}\left[-p\delta_{ij} + 2\mu S_{ij} - \rho u_i' u_j'\right] \tag{2}$$

where, u is the time-averaged velocity, ρ is the fluid density, p is the time-averaged static pressure, μ is the fluid viscosity, S_{ij} is the mean rate of strain tensor, and the quantity $-\rho u_i' u_j'$ is the Reynolds stresses which represent the effects of turbulence. To close equation (2), the Reynolds stresses term must be modelled. The chosen turbulence model was the widely used two-equation model, namely, k-epsilon realizable model. The enhanced wall treatment (EWT) option of the simulation package was used. The EWT automatically imposes the wall function to model the boundary layer profile for estimation of the surface friction on the wall surfaces when the first grid point is in the log-layer range. The steady, pressure-based solver was utilized to achieve steady-state simulations. All reported results were obtained using a second-order node-based upwinding discretization scheme.

The inlet boundary condition was set as uniform flow with inlet velocity U = 40 m/s and the turbulence intensities of 0.2%. The corresponding Reynolds number (Re) was 768,000 based on the model height. As for the outlet boundary, the pressure outlet at zero gauge pressure was imposed. The side and top walls of the domain was defined as symmetry boundary condition. Meanwhile, the ground and model surfaces were set as no-slip wall.

The computational domain resembles a rectangular box. Since the model is symmetric and the flow is steady, all simulation cases were run for half of the flow domain with a symmetry plane placed at the centreline location. The cross sectional area of the half flow domain was 1738 mm x 1129.5 mm (height x width). The corresponding blockage ratio was less than 1.5%, which is well within the typically accepted range of 5% in automotive aerodynamic testing [7]. The upstream and downstream extends of the domain were $1.4l$ and $11.4l$, respectively.

3.2 Meshing

The computational domain was decomposed into unstructured and prismatic cells (see Figure 2). The latter was employed around the model and the ground for improving the boundary layer resolution. The result of grid convergence study indicates that the mesh is sufficiently refined at around 315,000 nodes and 890,000 cells. The first prismatic cell layer thickness around the model surface was at 0.5 mm. The corresponding y+ ranges from around 1 to 57.

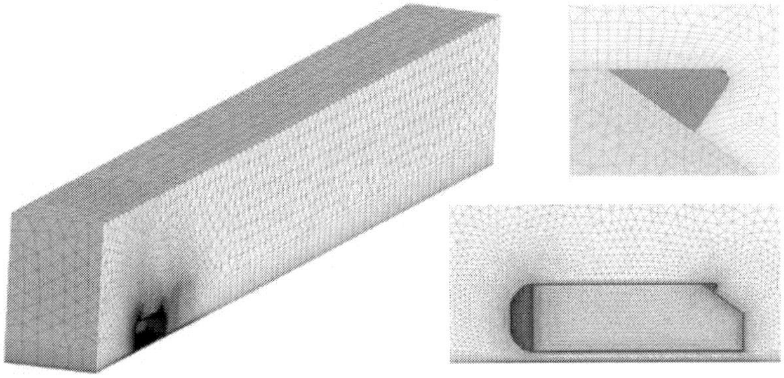

Fig. 2. Numerical cells of the simulation domain (left), close-up of prismatic cells around the rear-roof spoiler (top right), and mesh density distribution around the Ahmed body (bottom right).

3.3 Validation

Validation of the numerical method was carried out by comparing the numerically obtained results of Ahmed model without the spoiler to the experimental results found in the literature. The Re was at 768,000 (based on the model height) which was consistent with the experimental Re of Lienhart et al. [8]. Figure 3 shows the flow features at the rear section of Ahmed model along the centreline reproduced by the present simulation. As depicted, the flow separates from the roof-backlight junction. It does not reattach at the slanted edge, but continues downstream and joins the separation bubble trailing the vertical base of the model. Qualitatively, this flow pattern is consistent with the experimental results reported in the literature (e.g. [6, 8, 9]).

Table 1 compares the drag coefficient C_d values obtained from the present study to the experimental data by Lienhart et al. [8]. As shown, the total C_d value was in excellent agreement with the experiment, i.e. percentage difference of 2.68%. The drag breakdown shows that the C_d of the base and slant sections were well predicted. Good prediction of C_d in these two sections is important because the influence of the spoiler was expected to be most prominent at the rear section.

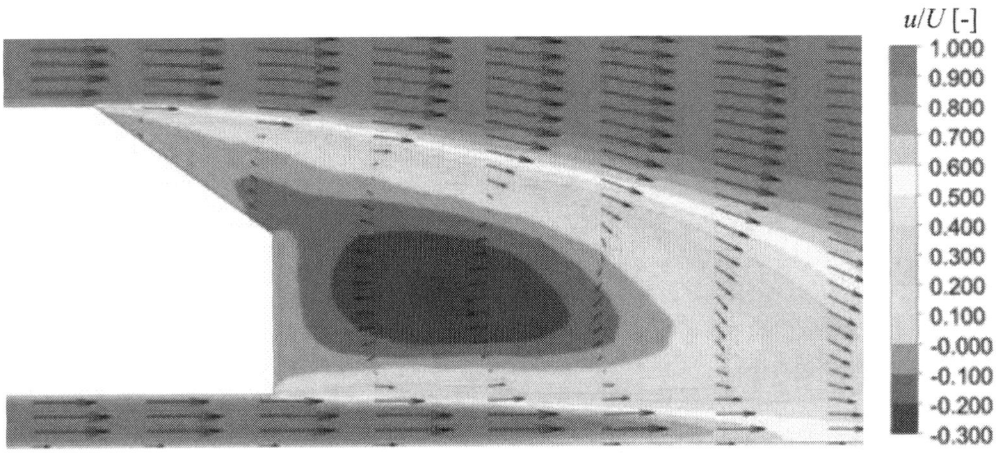

Fig. 3. Distribution of streamwise velocity component along the symmetry plane at the rear section of Ahmed model.

Table 1. Comparison of the experimental C_d.

Source	CFD	Experiment [8]	Percentage diff. (%)
Slant	0.095	0.097	2.08
Base	0.094	0.096	2.11
Front	0.039	0.015	88.89
Viscous	0.049	0.055	11.54
Total	0.265	0.258	2.68

4 Results and discussion

4.1 Effects of rear spoiler angle

Figure 4 shows that the lift coefficient C_l decreased almost linearly with pitch angle for the range of pitch angle tested. In addition, it is evident that the benefit of fitting a spoiler only occurred at positive pitch angle. At negative pitch angle, relatively higher lift was found, which is deemed unfavourable to driving stability.

Figure 5 compares the distribution of static pressure coefficient C_p (colour contour) and streamwise velocity component (vectors) around the rear section of the model at different spoiler configurations. In general, the flow at the upper body separated near the trailing edge of the roof when without the spoiler, whereas, it separated near the end of the spoiler for all spoiler angles when a spoiler is used. Besides, when a spoiler was configured at zero or positive pitch angle, it prevented the flow from accelerated near the roof-backlight junction. Thus, it resulted in a higher static pressure around the junction. As for the spoiler with negative pitch angle, augmentation of flow occurred near the junction, and was accompanied by significant drop in static pressure.

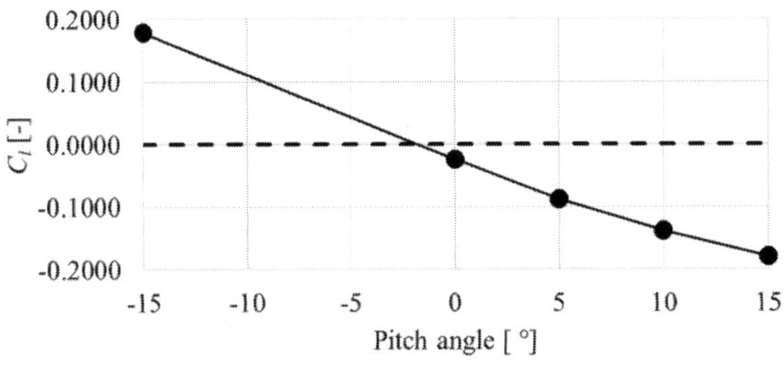

Fig. 4. Graph of C_l against rear spoiler angle; Dashed-horizontal line is without spoiler.

Meanwhile, as depicted in Figure 5, the spoiler angle also has an effect on the size of the separation bubble behind the model. As shown, the height and length of the separation bubble are found to increase with larger spoiler angle.

Figure 6 compares the effect of spoiler angles on the C_p distribution along the centreline of the model. As expected, the C_p at the front section is almost identical in all cases. However, from the midsection onward, deviations in C_p are gradually becoming more apparent along the upper body. The 0° and 15° cases exhibited higher C_p, with the latter being more pronounced, as compared to the case without a spoiler. Whilst, the -15° case shows a significant drop in C_p particularly around the rear end of the roof, and the value peaked at the roof-spoiler junction.

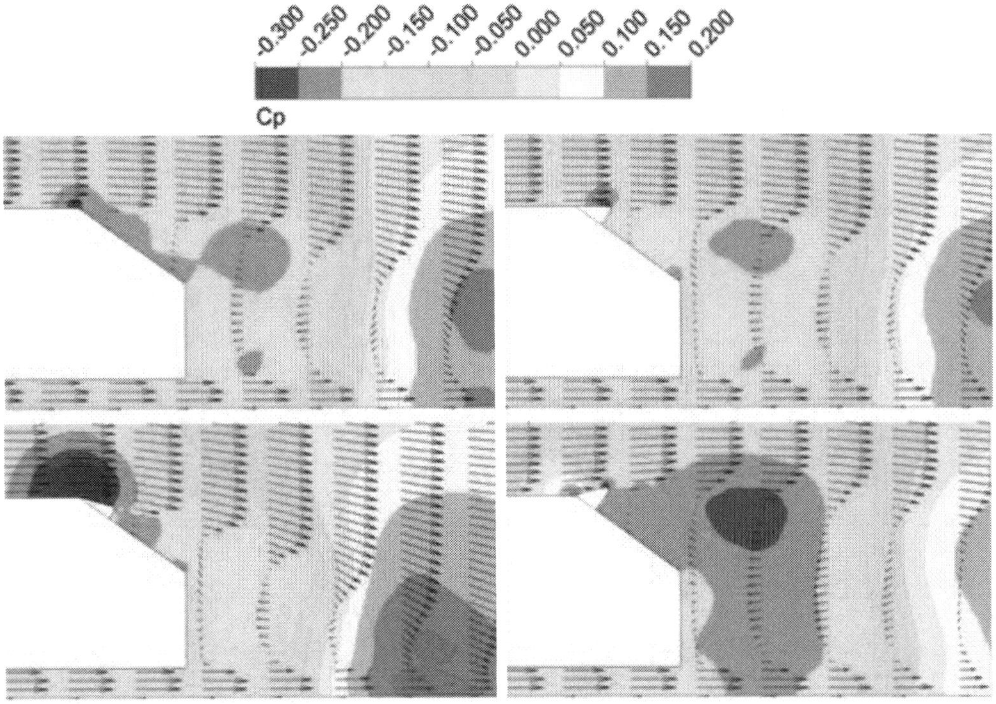

Fig. 5. Vectors of streamwise velocity component and C_p distribution along the symmetry plane at the rear section of Ahmed model; Without spoiler (top left), with 0° spoiler (top right), 15° (bottom right), and -15° (bottom left).

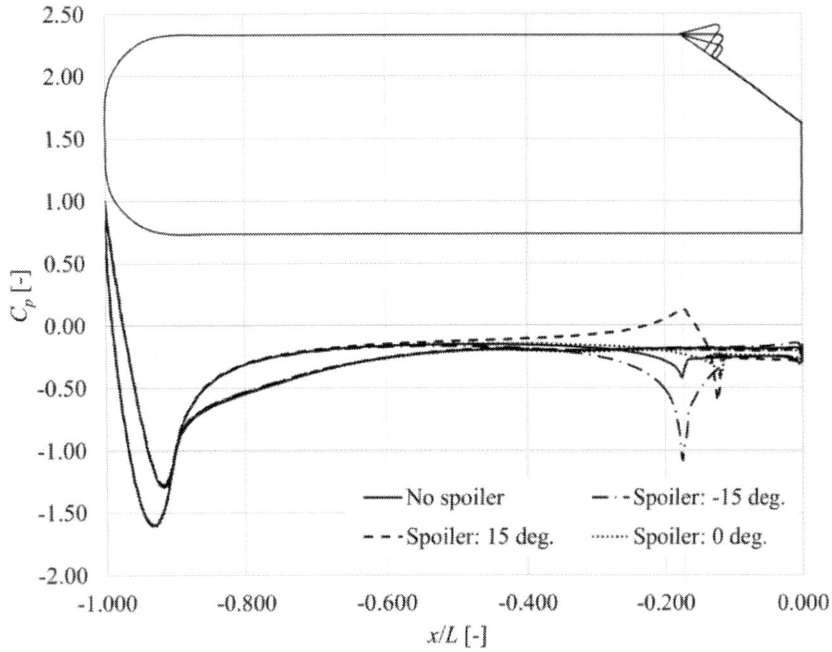

Fig. 6. C_p distribution along the centerline of four cases: Without spoiler (solid curve), with 0° spoiler (dotted curve), 15° (dashed curve), and -15° (dashed-dotted curve).

In general, the spoiler affects the aerodynamic lift of the model mainly by altering the flow near the rear end of the roof where the spoiler was mounted (see Figure 6 and 7). At 0° inclination angle, the relatively higher surface pressure near the rear end of the roof has caused significant reduction in C_l (by about 488%) as compared to the case without the spoiler. In addition, the C_d has reduced by about 3% (see Figure 8). Note that this positive combination – reduction in both the drag and lift – is highly appreciable as it is often the case that the element that increases the downforce will also increase drag (e.g. [2-4]).

At 15° pitch angle, significant increase in the surface pressure near the roof-spoiler junction may be evident (see Figure 7). This high pressure region has resulted in the dramatic drop in C_l, by about 2937%. However, it was accompanied by about 11% C_d increment. Hence, this configuration might not be suitable in situations where fuel economy is important.

At negative inclination angle (i.e. the -15° case), the airflow around the roof-spoiler junction accelerated downward with the corresponding drop in the model's surface pressure around the region. Therefore, the C_l has increased (by about 2737%).

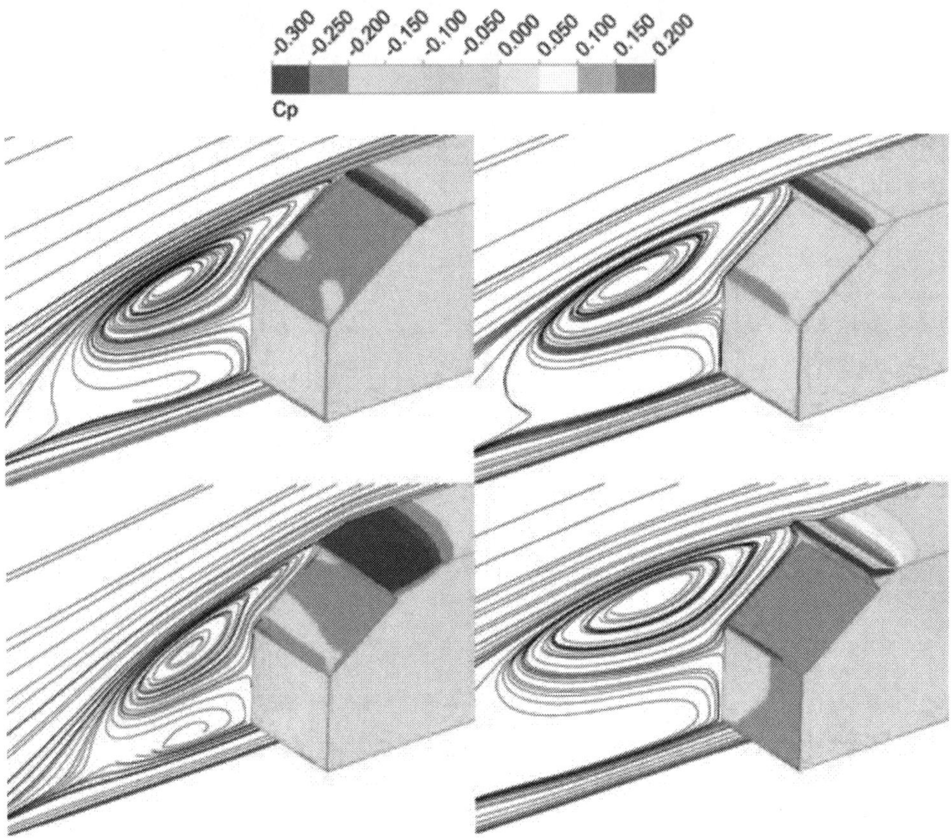

Fig. 7. Surface pressure distribution around the rear section of the model and the streamlines along the symmetry plane; Without spoiler (top left), with 0° spoiler (top right), 15° (bottom right), and -15° (bottom left).

Figure 8 shows that in general, the reduction in lift is accompanied by drag penalty. Therefore, lift reduction practice has to be carried out with care so as not to introduce significant drag increment in cases where drag performance is important such as in hybrids and EVs. This could be the reason why the use of rear-roof spoiler at high pitch angle (above 5°) is scarce in commercial vehicles as their need for downforce is not as crucial as in race cars.

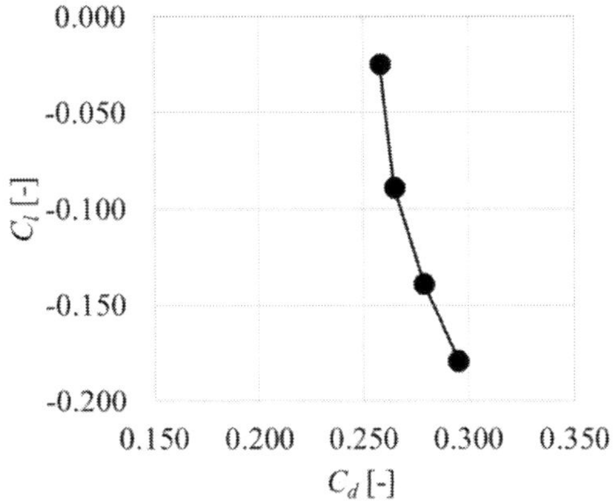

Fig. 8. Graph of C_l against C_d of the model fitted with the strip-type rear-roof spoiler.

4.2 Body part contribution to C_l reduction

Figure 9 compares the percentage change of C_l per body part as the pitch angle of the spoiler increases from 0° to 5°, and from 5° to 15°. As depicted, the decrease in C_l was mainly come from the spoiler and roof. This result supports the qualitative discussions pertaining to Figure 7.

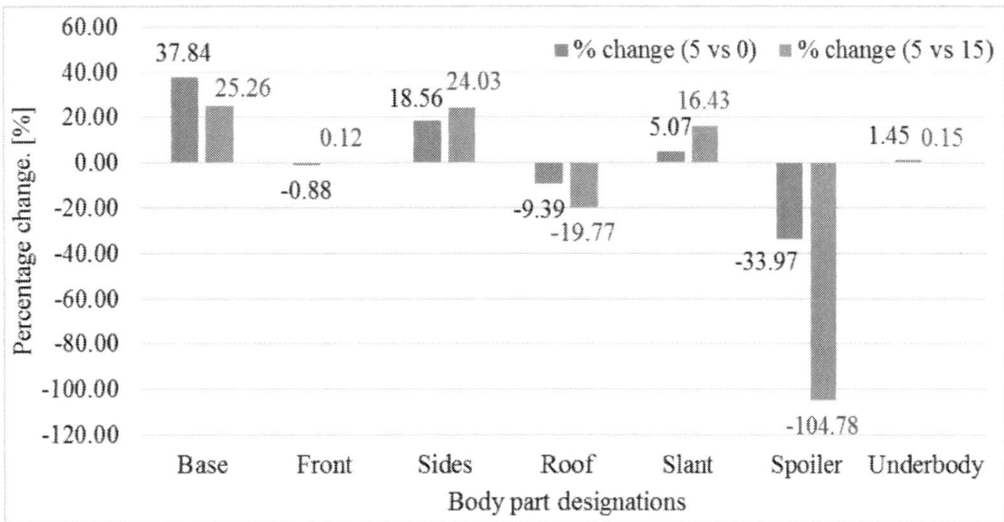

Fig. 9. Percentage change in C_l per body part as the pitch angle of the spoiler increases from 0° to 5° (left bar), and from 5° to 15° (right bar).

5 Conclusions

This paper investigated the effect of strip-type rear-roof spoiler on the aerodynamic performance of hatchback vehicles by a RANS-based CFD method. The results show that the aerodynamic lift was to decrease nearly linearly with the inclination angle of the spoiler, and was accompanied by the increment in aerodynamic drag. However, it is possible to achieve reduction in both the lift and drag when the spoiler is configured at 0° - parallel to the roof. In contrast, at negative spoiler angle, the use of rear spoiler was counterproductive to both the lift and drag reductions. In addition, when a spoiler is used, apart from the spoiler itself, the main body part that has contributed to the lift reduction was the model's roof, particularly at the rear end where the spoiler was mounted.

The present results were obtained from stationary simulations in which the motion of vehicle body was not considered. In practice, motion of vehicle body is very common and could change the inclination angle of the spoiler when the motion mode is of pitching. Hence, it helps to provide insight into how each spoiler configurations would perform under a more realistic driving condition if certain motion mode could be incorporated in the flow simulation. For this purpose, the method introduced by Cheng et al. [10, 11] could be employed in the future studies.

The authors would like to thank Universiti Teknikal Malaysia Melaka (UTeM) and Ministry of Higher Education for supporting this research under FRGS FRGS/1/2015/TK03/FKM/02/F00273.

References

1. J. Howell, and G. Le Good, The influence of aerodynamic lift on high speed stability, SAE Paper No 1999-01-0651 (1999)
2. C. H. Tsai, L. M. Fu, C. H. Tai, Y. L. Huang, and J. C. Leong, Computational aero-acoustic analysis of a passenger car with a rear spoiler, J. Appl. Math. Model. **33**, No. 9, pp. 3661-3673 (2009)
3. B. Daryakenari, S. Abdullah, R. Zulkifli, E. Sundararajan, and A. Mohd Sood, Numerical study of flow over ahmed body and a road vehicle and the change in aerodynamic characteristics caused by rear spoiler, Int. J. Fluid Mech. Res. **40**(4), pp. 354-372 (2013)
4. S. P. Kodali, and S. Bezavada, Numerical simulation of air flow over a passenger car and the influence of rear spoiler using CFD, IJATP, **01**(1), pp. 6-13 (2012)
5. D. P. Menon, G. S. Kamat, Y. S. Mukkamala, and P. S. Kulkarni, To improve the aerodynamic performance of a model hatchback car with the addition of a rear roof spoiler, *16th Annual CFD Symposium*, August 11-12, Bangolore (2014)
6. S. R. Ahmed, An experimental study of the wake structures of typical automobile shapes, J. Wind Eng. Ind. Aerodyn, **9**(1-2), pp. 49-62 (1981)
7. W-H. Hucho, and G. Sovran, Aerodynamics of road vehicles, Annu. Rev. Fluid Mech, **25**, pp. 485-537 (1993)
8. H. Lienhart, C. Stoots, and S. Becker, Flow and turbulence structures on the wake of a similified car model (Ahmed model), *DGLR Fach. Symp. der AG ATAB*, Stuttgart University (2000)
9. G. Vino, S. Watkins, P. Mousley, J. Watmuff, and S. Prasad, Flow structures in the near-wake of the Ahmed model, J Fluid Struct, **20**(5), pp. 673-695 (2005)
10. S.Y. Cheng, M. Tsubokura, T. Nakashima, T. Nouzawa, Y. Okada, A numerical analysis of transient flow past road vehicles subjected to pitching oscillation, J Wind Eng Ind Aerod, **99**, pp. 511-522 (2011)
11. S.Y. Cheng, M. Tsubokura, T. Nakashima, Y. Okada, T. Nouzawa, Numerical quantification of aerodynamic damping on pitching of vehicle-inspired bluff body, J Fluid Struct, **30**, pp. 188-204 (2012)

Prediction of available power being generate in small hydropower system at Sungai Perting Bentong Pahang

*M.*Razi[1], *M.A.*Yusuff[1,*], *B.T.*Tee[1], *K.A.*Zakaria[1]

[1]Faculty of Mechanical Engineering, Universiti Teknikal Malaysia Melaka, 76100 Durian Tunggal, Melaka, Malaysia

Abstract. Tremendous studied had been conducted on small hydropower system based on run-of-river schemes as an alternative renewable energy. Small hydropower system can be classified based on electricity generated between 1MW to 10MW. This system is normally being applied in rural area for providing the consumer electricity demand. Basically the researches to date are more focusing on the large scale of hydropower rather than the small scale hydropower technology. Therefore, this study is aimed to focus on predicting the available power generated by the small hydropower system specifically for the river stream in peninsular Malaysia. The water flow rate is measured by using ultrasonic level sensor located at the intake of the small hydropower system. The water flow rate is important data to be used in predicting the power output of the power house. The result shows that, the power outputs are depending on the fluctuation of water flow rate and the electricity being generated is more than 1MW. This finding can be used as the benchmark for daily and monthly monitoring process of the system efficiency or target output.

1 Introduction

Peninsular Malaysia is blessed with an abundance numbers of water resources and due to the extreme topography of the mountain; most streams have very high slopes. This combination of stream flow and high stream gradient or slope is ideal for the run-of-river hydropower development. This kind of small hydropower development has the least environmental impact and is generally less capital intensive than typical hydropower plants built in conjunction with high dams and large amounts of water storage [1]. Besides, small hydropower system is one of the significant renewable energy alternatives that have been greatly developed in the past few decades due to its small scale, lower investment, and the promotion of local economic development especially for rural application in generating electricity [2–6].

* Corresponding author: mohd.asri@utem.edu.my

The small hydropower system can be developed in a reservoir or run-of-river type. In For run-of-river type, there is no water storage for power generation, but the water is diverted from the main river by weir. For small capacities of hydropower system, run-of-river types are highly recommended. A typical run-of-river small hydropower system for electricity generation is consisted of the following basic components: water intake structure (e.g., weir and settling tank), penstock, turbine, mechanical power transmission system to generator, generator, and electricity transmission system to load centers, and control system [7]

The power production of small hydropower plant is a dynamic process which indicates the maximum generation capacity under certain meteorological and hydrological conditions. It is also easily being influenced by hydrology, climate, and installation capacity. Therefore predicting the power production of a small hydropower system is a non-linear and multi-factor complicated problem. Due to the less value of the installed capacity in a single plant, it is hard to predict its power production along with strong uncertainty and fluctuation [4]. Besides, the important of predicting the production output in a small hydropower plant is to maintain its sustainability development. To obtain the power production estimation, the value of the net head and water flow of the system are required. Net head is difference between the gross head and the losses due to friction and turbulence in piping, while the water flow is the quantity of water in cubic meter per second [8].

The power production of small hydropower plant shows great spatiotemporal diversity, and thus, it is difficult to establish a commonly used model. These data can be served as a source of information for the model development and to build a procedure to simulate the operation of hydrological and hydraulic systems that can be used for optimizing the marketing of the renewable energy technology. The power production depends a lot on the environmental conditions for example, precipitation, weather changes and etc. In Malaysia generally, small hydropower plant based on run-of-river types is less implemented, therefore it has some challenges in predicting the power production and becoming complex task. So, prediction of performance output is necessary in providing adequate information for further research. This paper aim to study the power generated by small hydropower system based on run-of-river located in Peninsular Malaysia, specifically at Sungai Perting, Bentong, Pahang.

2 Experimental setup

The present experiment aims to study the power output generated by the 5MW grid-connected small hydropower plant in Sungai Perting, Bentong, Pahang with net head of 159.6 m. In order to obtain the result, ultrasonic level sensor was installed to measure the water flow rates (m^3/s) of the river which is crucial variable needed in calculating the power generation by the small hydropower system. The Signature Flow Meter which is connected through TIENET 310 Ultrasonic Level Sensor was specially designed for open channel flow monitoring application. The flow meter sensor was installed at water intake of the river for each pairs of the water channel. This device is a non – contact liquid level measurement work by measuring the time interval between the transmissions of the river surface sound pulse to the sensor. The function of signature flow meter is to process the data obtained from the level sensor in terms of water flow rates numerical values. In this research, the data of water flow rate were used as crucial parameters to calculate the power prediction that were obtained hourly for whole day in January 2016 which is equivalent to 744 hours. Figure 1 (a) shows Ultrasonic Level Sensor and Figure 1 (b) shows Signature Flow Meter.

(a) (b)

Fig. 1. Equipment used in this experiment.

3 Results and discussion

Statistical analysis will be performed in evaluating the result. The water flow rates obtained were in form of continuous random variables. Power generated by the small hydropower system can be calculated by using Equation (1) where, ρ (kg/m^{3}) is a density 1000 kg/m^3, g (m/s^2) is the acceleration due to gravity 9.81 m/s^2, Q (m^3/s) is the flow rate, and H (m) is the net head available at the inlet to the turbine and η is the overall energy conversion efficiency [9]. The overall efficiency for small hydropower system electricity generation is lower (< 80%) than for mechanical power supply system because of generator losses in the former [10]. Therefore, in this analysis, the net head and overall efficiency used are 159.6 m and 80% or 0.8 respectively.

$$P = \rho g H Q \eta \tag{1}$$

Based on the calculation by using Equation (1), the average power output generated in January 2016 was 4.25 MW. Figure 2 shows the scattered plot for exploited power in the small hydropower system. Based on the graph, the data are randomly distributed. The maximum and minimum value was spotted away from the overall linear pattern distribution and basically the exploited power fluctuation is directly proportional to the flow rates of the stream as shown in Figure 3. The fluctuations in the graph are caused by wet and dry weather which also influenced the increment of precipitation [11]. The level of precipitation will be lesser during dry weather compared to wet weather.

Figure 4 shows the probability density function for the generated power output. The probability density function is a curve that represents the probability distribution of a continuous random variable of the generated power output. The curve indicates the regions of higher and lower probabilities for the random variable values. From the graph, it shows that the probability of the generated power is denser between 2.0 MW – 5.0 MW. It means this area of distribution it has the highest probability of the power generated in MW by the system. The graph was skewed to the left and the mean value was located to the left of the peak value. The peak value of the graph shows the mean of the power generation from the small hydropower system which is 4.25 MW.

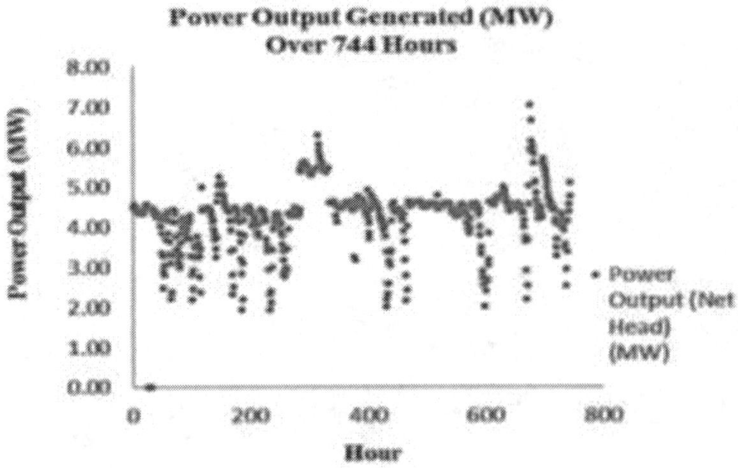

Fig. 2. Scatter Plot of Generated Power Output (MW) in January 2016.

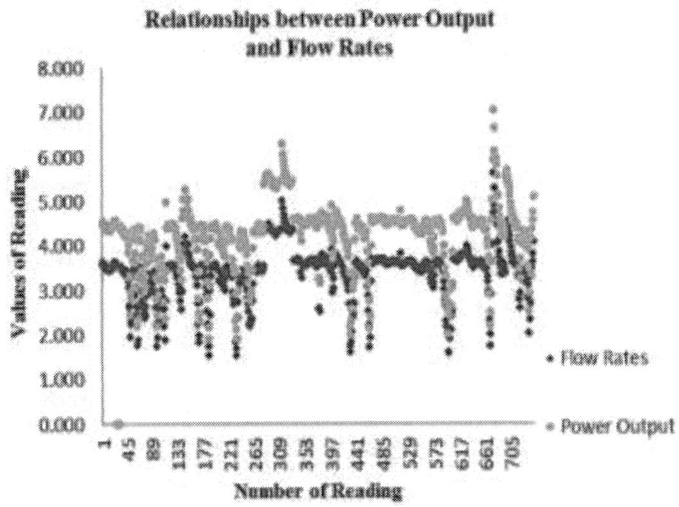

Fig. 3. Relationship between Power Generation (MW) and Water Flow Rates (m^3/s) in January 2016.

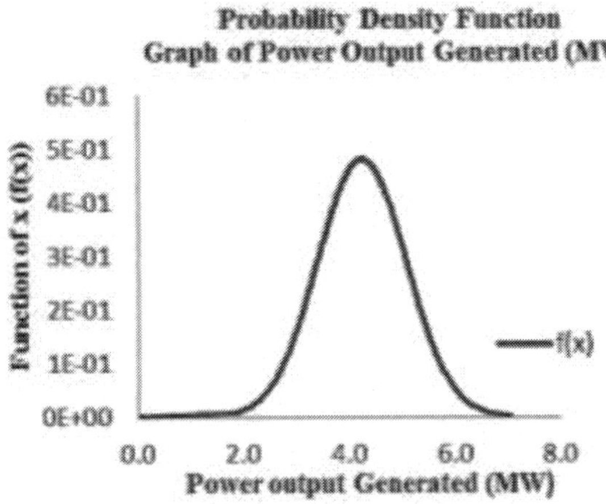

Fig 4. Probability Density function for Generated Power Output (MW).

4 Conclusions

This paper provides adequate information on predicting power production by small hydropower system based on run-off-river in Peninsular Malaysia Rivers. Besides, the prediction of power production for such small power plants is an essential tool to ensure ongoing power supply for user demand, to plan for reserve power supply and transaction between power plants especially in rural area application. The power production of small hydropower can be ranged from few kW until thousands of MW and much reliable as a clean energy sources compared to other types of power plant. In this study, the mean value of generated power obtained in January 2016 was 4.25 MW. The highest production value was around 7 MW and the least value was around 2 MW. The current findings from this work is hoped to be one of information sources in designing or choosing potential sites for small hydropower plant installation. Those present and further findings will provide mass benefit to the rural electrification generation in fulfilling the increasing energy demand.

The authors would like to thank Ministry of Higher Education of Malaysia and Universiti Teknikal Malaysia Melaka (UTeM) for providing support and financial assistance through the Fundamental Research Grant Scheme award (FRGS/2/2014/TK06/FKM/03/F00235).

References

1. Heitz, Leroy F., Shahram Khosrowpanah. "Prediction of Flow Duration Curves for Use in Hydropower Analysis at Ungaged Sites in Pohnpei," FSM. Water and Environmental Research Institute of the Western Pacific, University of Guam, (2010)
2. Yuksel, I. "As a renewable energy hydropower for sustainable development in Turkey," Renew. Sustain. Energy Rev. 14, 3213–3219, (2010)
3. Dudhani, S., Sinha, A.K., Inamdar, S.S. "Assessment of small hydropower potential using remote sensing data for sustainable development in India," Energy Policy, **34**, 3195–3205, Water 2015, 7 4819, (2006)
4. Bakken, T.H. Sundt, H. Ruud, "A Development of small versus large hydropower in Norway comparison of environmental impacts," Energy Procedia, **20**,185–199, (2012)
5. Bøckman, T.; Fleten, S.E.; Juliussen, H.J. "Investment timing and optimal capacity choice for small hydropower projects," Eur. J. Oper. Res, 190, 255–267, (2008)
6. Kosnik, L. "The potential for small scale hydropower development in the US," Energy Policy, **38**, 5512–5519, (2010)
7. Kaunda, Chiyembekezo S., Cuthbert Z. Kimambo, and Torbjorn K. Nielsen. "Potential of small-scale hydropower for electricity generation in Sub-Saharan Africa." *ISRN Renewable Energy* (2012)
8. Kosa Preeyaphorn, Thanatchai Kulworawanichpong, Rerkchai Srivoramas, Avirut Chinkulkijniwat, Suksun Horpibulsuk, and Neung Teaumroong. "The potential micro-hydropower projects in Nakhon Ratchasima province, Thailand." Renewable Energy **36**, no. 3, 1133-1137, (2011)
9. Kaunda, C. S., Kimambo, C. Z., Nielsen, T. K. "Potential of small-scale hydropower for electricity generation in Sub-Saharan Africa". ISRN Renewable Energy, (2012).
10. Paish, O. Micro-hydropower: status and prospects. Proceedings of the Institution of Mechanical Engineers, Part A: Journal of Power and Energy, 216(**1**), 31-40 (2002).
11. Sarkkula, J., et. al "Origin, fate and role of Mekong sediments. Detailed Modelling Support Project, Mekong River Commission, Vientiane, Laos PDR", (2010).

 DOI: 10.1051/matecconf/20179001028

12. Li G., Liu, B. Li, S.Z. "An overview of large scale small hydropower in Yunnan Power Grid: situations, challenges and measure," In Proceedings of World Environmental and Water Resources Congress, Cincinnati, OH, USA,(2013)
13. Amcorp Properties Berhad, Amcorp Perting 5MW Mini Hydro Power Plant.

A review on failure characteristics of polymer gear

Wafiuddin Md Ghazali[1], *Daing Mohamad Nafiz* Daing Idris[1,*], *Azizul Helmi* Sofian[2], *Januar Parlaungan* Siregar[1] and *Ismail Ali* Abdul Aziz[1]

[1]Faculty of Mechanical Engineering, Universiti Malaysia Pahang, 26600 Pekan, Pahang, Malaysia
[2]Faculty of Chemical & Natural Resources Engineering, Universiti Malaysia Pahang, 26300 Gambang, Kuantan, Pahang, Malaysia

Abstract. Polymer are now widely used as substitute material for steel gear in low load devices. Its failure differs from gears made of steel, thus it is important to categorize the failures shown by polymer gears. Several previous studies noted that wear detection, microstructure surface condition monitoring, weight loss and temperature detection can be used in detecting failure of polymer gear. This article reviews the failure detection method mentioned above. Other researcher works were studied and their findings were extracted in order to identify the methods they used. The most common method used was wear detection and it was supplemented by other methods such as microstructure surface condition monitoring. Failures shown by polymer can be concluded to be tooth breakage, tooth deformation, material removal and surface fatigue.

1 Introduction

Polymer are chosen to replace steel gear in low load devices due to the technical advantage and economic gains. The advantages of polymer gear is ability to operate with small amount of lubrication, light in weight material and low operating noise [1] compared to metal gear. However, considerations such as; working environment, strength, weight, material properties and cost [2] must be carefully considered when deciding to replace steel with polymer. Standards for polymer gear like AGMA [3] and British Standard [4] can be referred when designing the gears.

Polymer gear failures are different from steel because of the material properties of polymer are totally different. An example of polymer gear failure is melting of material which does not occur for steel gears. These type of failure can be categorized under thermal damage. This article reviews four types of failure characteristic; wear, microstructure surface condition monitoring, weight loss and thermal damage which is shown in Fig. 1. These characteristic are often studied together as they complement one another. The other two characteristics is not discussed in this article.

Wear detection is the most commonly used method [5] to determine failure as it covers a variety of conditions such as cracks, breakage and debris formation [6]. Microstructure

* Corresponding author: daingnafiz@ump.edu.my

surface condition monitoring provides a ready and quick method to inspect the surface failure [7] which are not visible during wear detection. Weight loss failure characteristic are only suitable for gears made from pure polymer as reinforcements may affect the weight of gear. In thermal damage characteristic, temperature detection has a vital role to determine at what temperature damage starts to occur.

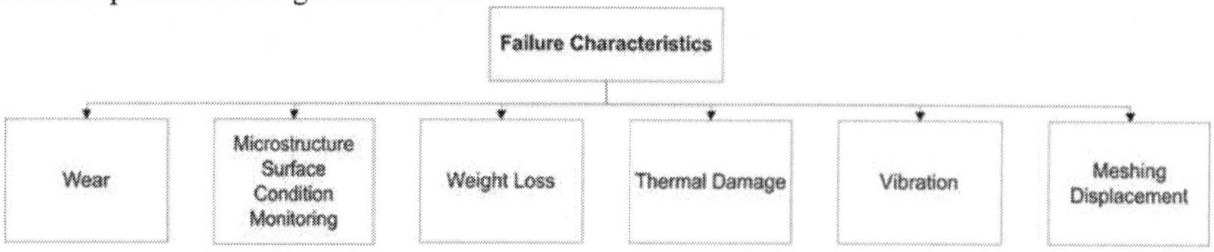

Fig. 1. Types of failure characteristics for polymer gear.

2 Failure characteristics

2.1 Wear

There are many types of failure that can be categorized under wear, such as crack or breaking, tooth thickness reduction and debris formation. Each review will include all known wear formation of polymer gear.

2.1.1 Wear debris formation

Acetal gear were found to have different failure compared to Nylon gear as reported by K. Mao et al [8]. Polymer gear wear can be divided into three stages; running in, linear and final rapid wear as shown in Table 1. The wear debris size increases as the gear approaches final wear period. When Acetal gears were tested in the high range load, 10 – 16.1 N.m, the wear debris formed immediately after the test started. When Nylon gears were tested at high load, 10 N.m, it fractures after going through running in and linear wear period. The gear made from Acetal failed by melting and in Nylon by fracture as shown in Fig. 2.

The same result was also obtained by W. Li et al [9] in their research where the test gears were paired with different materials. Acetal gear started to melt at load torques higher than 9 N.m and fracture occurs when load is 10 N.m and above. However, the wear performance improved when it is paired with dissimilar material where Acetal as the driver gear and Nylon as the driven gear, this pair showed the highest performance from other pairs which can be seen in Fig. 3.

Table 1. Stages of polymer gear wear.

Phase	Explanation
Running in wear	Occurs for a short time but the amount of wear is high
Linear wear	Low amount of wear can be seen but is progressive
Final rapid wear	High wear rate but small amount of debris, indicating debris is due to deformation undergone by the polymer gear caused by thermal effects

 DOI: 10.1051/matecconf/20179001029

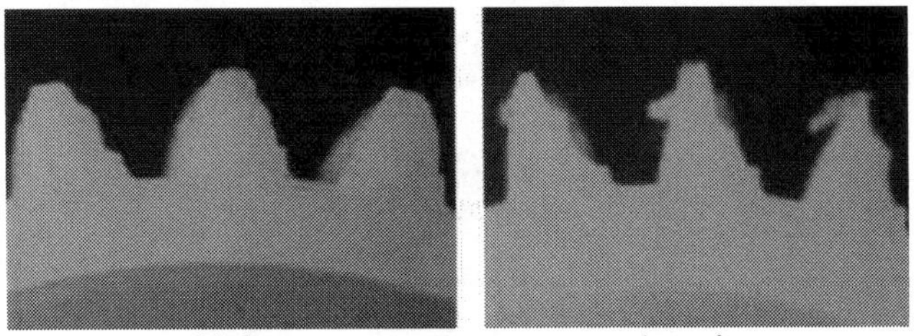

Nylon Acetal

Fig. 2. Wear on polymer gears with module 2 mm and 30 gear tooth [8].

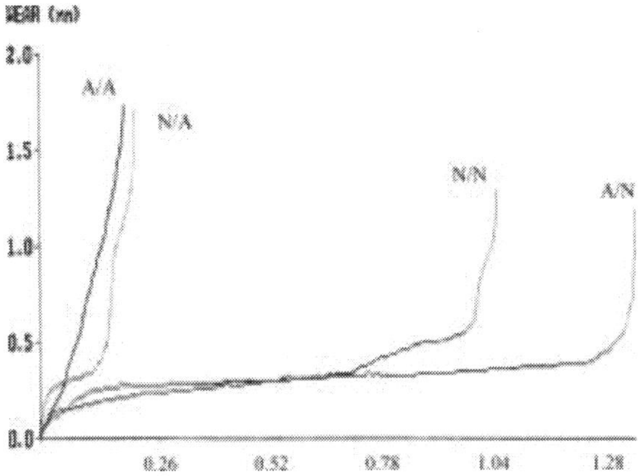

Fig. 3. Results on Nylon/Acetal (N/A), Acetal/Nylon (A/N), Acetal/Acetal (A/A) and Nylon/Nyon (N/N) gear pairs [9].

2.1.2 Tooth thickness reduction

Wear rates for Acetal gear produced by machine cut and injection moulding are independent to the manufacturing process as reported by K. Mao et al [10]. The gears undergo testing at load 6 – 9 Nm at 100 rpm and undergo wear in three phases. The running-in and linear phase produced little wear debris, but in the rapid wear period, wear debris increased and so does the wear rate. After about 33% of tooth thickness removed, the gear started to fail. An incremental step loading (load is incrementally added without changing the test gears) was used and it was compared to the conventional procedure (test gears is changed for every load value). The result obtained showed that the incremental step loading test produces adequate result and can be completed within hours compared to conventional testing which takes up weeks. It was also noted that bending occurred when the material is momentarily melted causing it to jump out of mesh. The repeated motion of sliding at addendum and dedendum region produced heat caused by the friction of the tooth surface leading to adhesive wear.

The various types of failure in unreinforced and reinforced Nylon 66 gear was studied by S. Senthilvelan and R. Gnanamoorthy [11] using tooth thickness and weight loss measurement technique. Unreinforced and reinforced Nylon 66 gears were meshed with a stainless steel (SS316) gear. Fig. 4 shows the deformation of teeth region undergone by the glass reinforced Nylon 66 gear. Reinforced gears showed a uniform material loss compared to unreinforced gears because glass fiber have better adhesion to the matrix compared to carbon fibre. Wear of tooth flank region in glass reinforced fibre is caused by softening of material and scraping by opposing stainless steel gear tooth. The wear occurred is due to the low thermal resistance of the material. In the case of carbon fibre reinforced gear, no appreciable tooth deformation was present due to high stiffness and good thermal resistance

DOI: 10.1051/matecconf/20179001029

of the material. This result was obtained at the test parameter of 1000 rpm rotational speed and loads ranging from 1.5 N.m to 3 N.m.

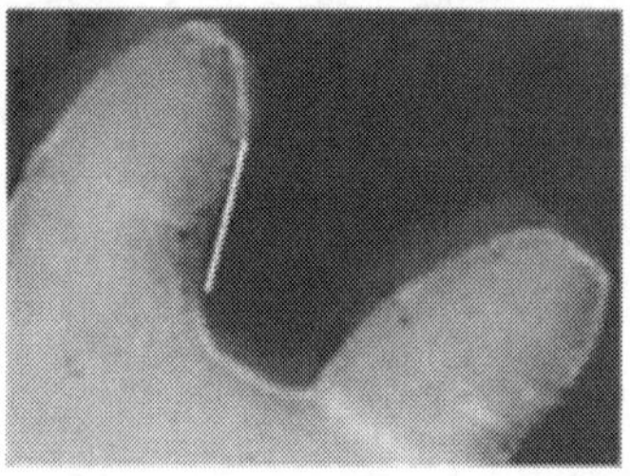

Fig. 4. Tooth thickness reduction due to scraping of steel gear tooth [11].

2.1.3 Cracks

Cracks often occur at the root of the tooth and will propagate causing tooth breakage. The effect of rotational speed on the performance of unreinforced and glass reinforced Nylon 6 was studied by S. Senthilevan and R. Gnanamoorthy [12]. The glass fibre reinforced gear showed improvement in mechanical strength and thermal deformation. They noted that the performance of gears was influenced by the load applied. The performance was only influenced by speed at the higher load condition. Gear root tooth crack and tooth wear were observed occurring at lower load, 8 MPa for both materials. When the load is higher, plastic deformation occurs on the unreinforced gear and at 15 MPa deformation starts to occur in glass fibre reinforced gear. At low stress levels, gear tooth root cracking and tooth wear was the main factor of failure, and in the higher stress level, deformation of material at high temperature causes failure. Modification on gear tooth made from Nylon 6 was reported by H. Imrek [13] and the failure for each design was studied. The tooth was modified as seen in Fig. 5 so that the single mesh area was increased thus reducing the load and temperature of the area. This reduces the wear rate and improves the overall teeth temperature. The unmodified gear showed cracking at the pitch area in Fig. 5 while in modified gear, cracking occurred at tooth roots.

Modification of gear tooth was also studied by H. Duzcukoglu [14] where holes are introduced to the tooth body of the gear. This serves as a cooling mechanism and to improve the heat distribution. Gears with modification shown smooth wear transition compared to unmodified gears. As the tooth load increases, the tooth profile wear becomes more noticeable at the tooth root and tooth tip region. It was concluded that wear occurs due to softening and detachment of material from the contact area and with the modification, the damage is delayed with the help of increased heat transfer from the gear. The effect of different hub type on the spur gear performance was studied by S.

Senthilvelan and R. Gnanamoorthy [15] using gears made from Nylon 66 and reinforced with carbon fibre. The hub was made from Nylon 66 and in cylindrical or spline shape. At 15 MPa of bending stress, both unreinforced and reinforced gears showed wear at tooth surface and flank. When the bending stress is at 20 MPa, the gear fitted with cylindrical hub failed at the gear and hub joint at 2 x 10^5 cycle and the gear with spline hub showed wear characteristics such as cracks at tooth root region. The failure between circular hub and gear is caused by the joint failure meanwhile in spline hub and gear the failure is caused by the gear tooth. The failure for unreinforced and reinforced gear were the same in the spline hub.

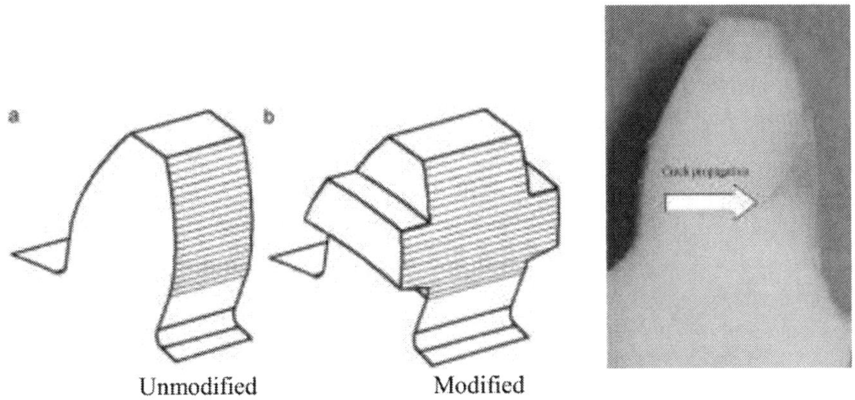

Fig. 5. Gear profile models, arrow indicates crack propagation [13].

2.2 Microstructure surface condition monitoring

This method is used to detect micro crack or deformation on the gear surface which are not visible with naked eye. With the introduction of composite polymer, this method becomes more important as it is capable to inspect the fibre structure and alignment.

Scanning electron micrograph (SEM) was used by A. R. Breeds et al [16] in order to study the surface condition. Three major interest region were dedendum, addendum, and pitch line. With the help of SEM, they were able to detect large pits or scoops of material were removed at the dedendum, smooth surface due to wear at addendum caused by sliding and rolling motion of gears and the formation of a ridge at pitch line caused by rolling. An SEM examination around the gear tooth pitch and root areas were conducted by K. Mao et al [10] to determine whether wear occurs at that region. From the gear mesh theory, there is nearly zero friction around the pitch point, however the images from SEM showed otherwise. This shows that SEM can also be a reliable method to detect failure in polymer gears. Fig. 6 shows the difference of wear occurring at the tip and pitch point of the gear.

The effect of fibre orientation was analyzed using SEM by S. Sentilvelan and R. Gnanamoorthy [11]. A perpendicular aligned fibre orientation showed better performance where it helps slowing the crack growth, thus improving the gear life. In the glass reinforced gear, matrix nylon material was found adhered to the protruded glass fibre on the fracture surface. The cracked surface showed few cavities and nearly flat. This is due to the better adhesion of glass fibre and nylon matrix. Molten smeared layers were also seen on the surface. A high number of cavities were observed in the carbon reinforced gear cracked surface due to the poor adhesion between carbon fibres and nylon matrix. Molten material was absent due to better thermal properties of the carbon fibre reinforced Nylon 66. The fibre orientation is influenced by the gate location and tooth geometry. Both glass and carbon reinforced show poor wear resistance compared to unreinforced gear at the investigated condition. Molten material can be seen on the unreinforced Nylon 66 in Fig. 7 and on glass fibre reinforced Nylon 66 in Fig. 8. There was no molten material present on carbon fibre reinforced Nylon 66. However, cavities are present on both glass fibre and carbon fibre reinforced Nylon 66 in Fig. 8.

The surface of loading tooth was inspected by H. Duzcukoglu et al [17] to determine the presence of transverse crack. It was found that the transverse cracked occurred due to thermal softening caused by accumulated heat. These cracks shall merge and grow resulting removal of material in the shape of flakes. The possibility of controlling wear by applying coatings on tooth flanks were studied by K. Dearn et al [18]. Five types of coating were used; PTFE, boron nitride, molybdenum disulphide and graphite to protect the gear. SEM was used to study the surface of each gear with different coatings. PTFE and graphite provide most optimum protection as it lowers the friction between gear tooth, reduces running temperature and subsequently the wear of gears. However, it is possible that the coating will lose its effectiveness as the protection film wears over time.

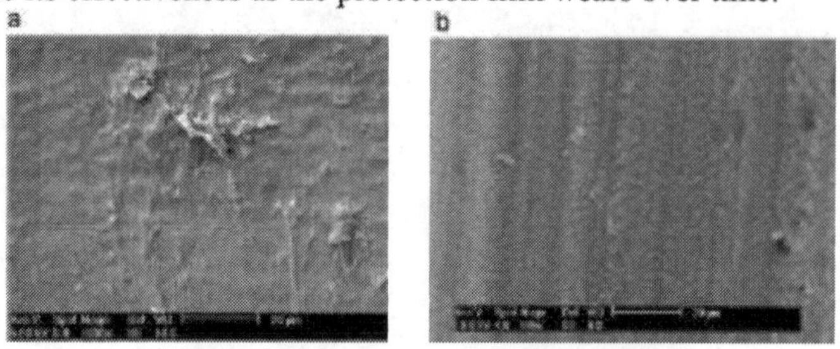

Wear at tip Wear at pitch

Fig. 6. SEM image of wear on polymer gear occurring at tip and pitch [10].

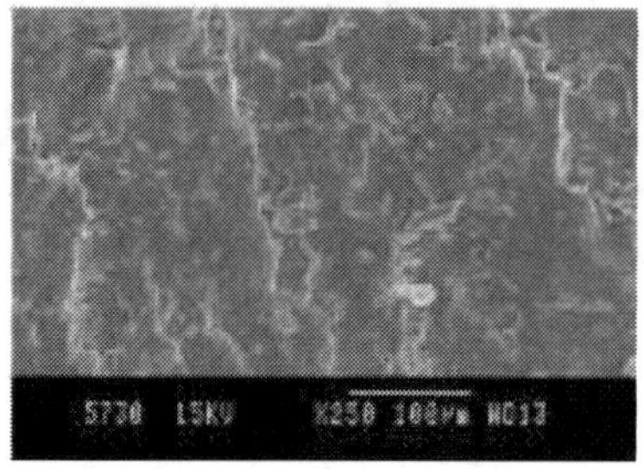

Fig. 7. Surface condition of unreinforced Nylon 66 [11].

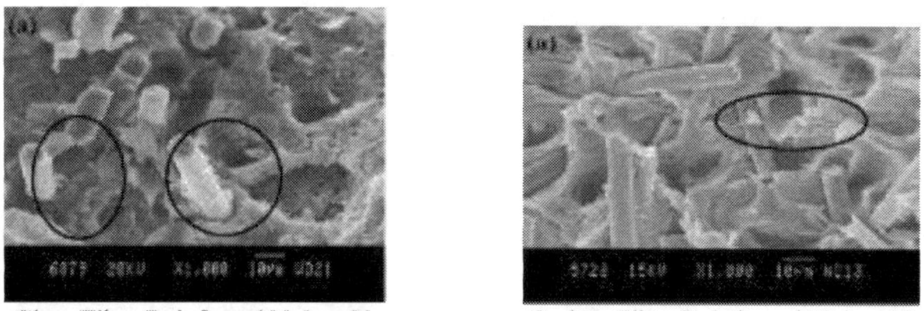

Glass Fibre Reinforced Nylon 66 Carbon Fibre Reinforced Nylon 66

Fig. 8. Surface condition of polymer gear [11].

2.3 Weight loss

This characteristic was found to be acceptable if gears were made using pure polymer. However, if it is made from composite, it become less reliable as the weight is affected by the composition of fibre and moisture or water presence. This was shown by N. A. Wright and S. N. Kurenka where they introduced a pair of control gear in their research [19]. They stated that the weight loss from running test can be considered as one of the method to determine failure, if only the material used does not have a high affinity for water. They noted that Polyamide 66 absorbs a significant amount of water, therefore the introduction of control gears and it was placed on the drive gearbox. They calculated the weight loss by subtracting the weight loss by control gear from the total measured weight loss by the test gear.

The weight loss method was used by C. H. Kim [20] to determine the wear volume of both Nylon and Acetal pinion. The pinion had three design, a solid gear tooth body, a drilled hole on the gear tooth body and a hole inserted with steel pin. The hole type and insert type showed less wear rate than the solid one. In the Nylon gear, hysteric heat loss was decreased by the hole in the tooth, while in the steel pin type, heat is absorbed and distributed by the pin. Both design led to decrease in wear rate and degradation of Nylon material. In the case of Acetal pinion, the variation of cross section increased the specific wear rate. The decrease of cross section area led to deformation and plastic flow on the Acetal pinion. This will lead to severe wear due to interference and severe contact between the Acetal pinion and steel driver gear. The wear rate in Nylon pinion decreased by over 30% and an increase in service life by 415%. While the Acetal pinion, it causes increase in wear, therefore this method can only be applied to visco-elastic material only.

The wear resistance of carbon nanotube reinforced Acetal gear was studied by S. Youseff [21] by determining the weight loss of the gear. It was then compared to results from previous research [22]. The results showed that the average wear resistance of Acetal reinforced with carbon nanotube compared to Acetal improved significantly. Spur gear improved by 28%, helical gear by 35%, bevel gear by 44% and lastly worm gear up to 47%.

2.4 Thermal damage and temperature detection using thermal camera or temperature sensor

In this method, temperature of the gears is taken during or after they were tested. Some researcher also used data acquisition system to record the operating temperature. This failure detection method is essential as different loads will influence the running temperature and affect the material properties when it reaches the glass transition temperature or the melting temperature.

The failure mode of polymer and polymer composite was found to be different as shown by S. Sentilvelan and R. Gnanamoorthy [11]. The gears were made from Nylon and reinforced with carbon or glass fibre. They also found that the surface temperature of unreinforced gear was higher compared to reinforced gear. In the reinforced Nylon, carbon reinforced had a lower temperature than glass reinforced. The reinforced gears lower temperature was contributed by a better tooth stiffness, lower friction and good thermal properties. A high tooth stiffness prevents tooth deflection which contribute to less unwanted contact between tooth surface which causes heat. The improved heat dissipation ability increased the gears life considerably. The introduction of cooling holes was reported by H. Duzcukoglu [17] in order to decrease thermal damage. Three design of gears were studied, first is unmodified, the second gear had a hole drilled at the pitch point of the gear tooth and the third design have holes at the pitch point and on the body of the tooth as seen in Fig. 9. The temperature was detected using a non-contact infrared temperature sensor and recorded on a PC by using data acquisition system. The first design failed at the vicinity of the pitch diameter, caused by softening. This was due to the gear inability to emit heat which was accumulated during the running process. As the load increases, the thermal damage also increased. This causes the material to soften and severe tooth deformation occurs. In the second design, partial thermal softening at the pitch region and tooth root region was observed. The amount of thermal damage was reduced by using this design, however, there is still damage on the surface of the loading tooth. For the third design, only thermal damage initiation was observed at the high load, 18.1 N.m. The heat produced in each design is from the friction between the driver and driven gear. The result from heat produced affecting the gear tooth can be seen in Fig. 9.

DOI: 10.1051/matecconf/20179001029

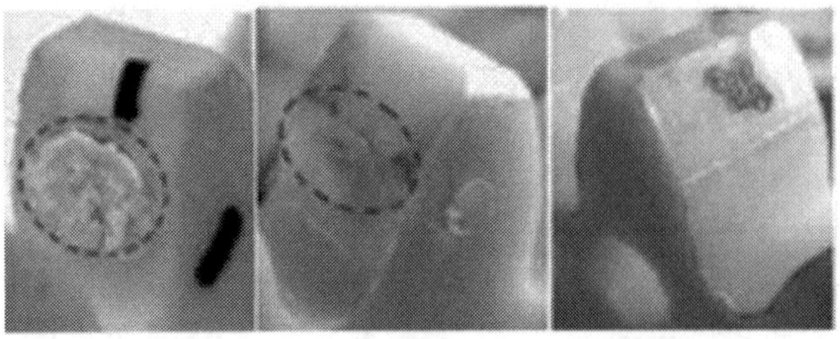

Fig. 9. Tooth condition for each design when the load is at 6.1 N/mm [17].

A design in which an internal hole or steel pin inserts are introduced to the tooth body was presented by C. H. Kim [20] to improve heat transfer process and stress concentration. Temperature of the tooth surface was measured and investigated using a non-contact type temperature sensor. Three load value was used, which are 9.8N/mm, 19.6 N/mm and 29.4 N/mm. At first load value, the hole type pinion had the lowest temperature value but have a higher fluctuation. The steel insert pinion has a slightly higher temperature, but the temperature maintained. When the load is 19.6 N/mm, the insert type pinion showed better performance than the others. At the highest test load, fracture can be seen from all types of design. the insert type took the longest time before failing followed by hole type and lastly solid type. It can be noted that the decrease in tooth temperature will result in better life and reduction of wear.

Polymer gear can fail in two typical ways, fatigue or sudden melting as reported in the research by A. Pogacnik and J. Tavcar [23]. A new multilevel accelerated testing procedure was proposed by the authors and the results which are life span and gear temperature were compared with a calculation procedure. The temperature was recorded using a thermal camera and the materials were PA 6, PA 6 with 30% glass fibres and Polyacetal. The maximum gear temperatures and load levels are different for every pair of materials. PA6/PA6 pair generated the highest temperature due to the high coefficient of friction. POM/PA6 pair gives the lower temperature due to lower coefficient of friction. The melting of gears was a consequence of overload and an increase in temperature. By avoiding problematic material combination, the failure due to thermal characteristic can be avoided.

The effect of different surface roughness was studied by J. Mertens and S. Sentilvelan [24] where three different value of coefficient of friction studied. Three stainless steel gear with coefficient of friction 3.8-4.1μm, 2.5-2.8μm and 1.9-2.2 μm was mated with polypropylene gear. The surface temperature of test gear was measured using a non-contact infra-red sensor. The frictional values of the surface are influenced by the hardness and micro geometry of the stainless steel gear. When a polymer slides on steel, adhesion and deformation occurs, contributing to the friction between those two surface. At a higher load, the surface interaction will increase, causing the friction, wear and temperature to increase which can be seen in Fig. 10. Polymer gear will generate more heat when meshed with surfaces having a high friction coefficient thus affecting the performance of the polymer gear. It can be seen that Gear A have the highest friction followed by B and C which relates to the higher temperature produced by A and followed by B and C at each load.

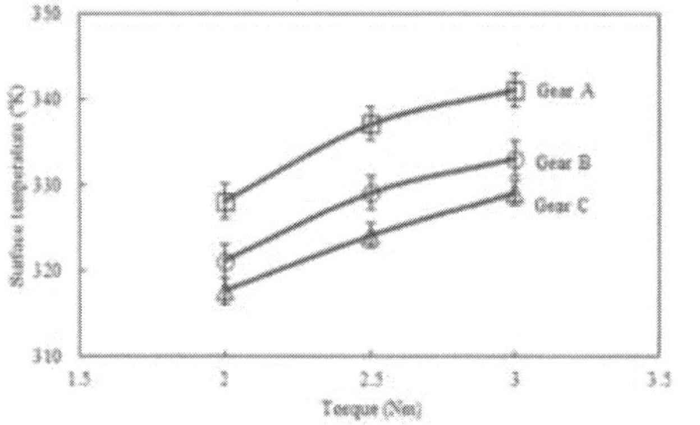

Fig. 10. Comparison of surface temperature for test gears [25].

3 Conclusions

From the reviewed articles, it can be concluded that detection of wear, microstructure surface condition monitoring, weight loss and thermal damage and temperature detection are the most prominent method alongside vibration and meshing displacement in order to detect failure in polymer gears. Most of the failures occurred due to the limitation of material, such as the load handling capability and thermal properties. In order to optimize the usage of polymer gear in applications, the operating parameters such as load and running temperature must be calculated beforehand so that the working environment of the polymer gear is the most optimum.

The authors would like to express their gratitude to Universiti Malaysia Pahang and Automotive Excellence Centre (AEC), UMP for their financial support under grant number RDU150386.

References

1. R. L. Mort, *Machine Elements in Mechanical Design*: Pearson Education South Asia Pte Ltd, (2006).
2. J. L. Elmquist. Deciding When to Go Plastic. *Gear Technology*. 46 - 47 (2014).
3. A. G. M. Association, "Standard 1106-A97. Tooth Proportions for Plastic Gears," ed. Washington DC: American Gear Manufacturers Association.
4. B. S. Institution, "BS 6168. Specification for non-metallic spur gears," ed. London, (1987).
5. K. M. Marshek and P. K. C. Chan, "Qualitative analysis of plastic worm and worm gear failures," *Wear,* vol. **66**, pp. 261-271, 1981/02/16 (1981).
6. H. J. Watson, "5–Wear of gears," *Tribology,* **2**, pp. 212-216, 1969/11/01 (1969).
7. B. A. Shaw, J. T. Evans, T. F. Page, "Scanning electron acoustic microscopy imaging of sub-surface microcracks produced in gear fatigue," *Journal of Materials Science Letters,* **13**, pp. 1551-1554, (1994).
8. K. Mao, W. Li, C. J. Hooke, and D. Walton, "Friction and wear behaviour of acetal and nylon gears," *Wear,* **267**, pp. 639-645, 6/15/ (2009).
9. W. Li, A. Wood, R. Weidig, and K. Mao, "An investigation on the wear behaviour of dissimilar polymer gear engagements," *Wear,***271**, pp. 2176-2183, Jul 29 (2011).

10. K. Mao, P. Langlois, Z. Hu, K. Alharbi, X. Xu, M. Milson, *et al.*, "The wear and thermal mechanical contact behaviour of machine cut polymer gears," *Wear,* **332–333**, pp. 822-826, 5// (2015).
11. S. Senthilvelan and R. Gnanamoorthy, "Damage Mechanisms in Injection Molded Unreinforced, Glass and Carbon Reinforced Nylon 66 Spur Gears," *Applied Composite Materials,* **11**, pp. 377-397, (2004).
12. S. Senthilvelan and R. Gnanamoorthy, "Effect of rotational speed on the performance of unreinforced and glass fiber reinforced Nylon 6 spur gears," *Materials & Design,* **28**, pp. 765-772, // (2007).
13. H. İmrek, "Performance improvement method for Nylon 6 spur gears," *Tribology International,* **42**, pp. 503-510, 3// (2009).
14. H. Düzcükoğlu, "Study on development of polyamide gears for improvement of load-carrying capacity," *Tribology International,* **42**, pp. 1146-1153, 8// (2009).
15. S. Senthilvelan and R. Gnanamoorthy, "Selective Carbon Fiber Reinforced Nylon 66 Spur Gears: Development and Performance," *Applied Composite Materials,* **13**, pp. 43-56,(2006).
16. A. R. Breeds, S. N. Kukureka, K. Mao, D. Walton, and C. J. Hooke, "Wear behaviour of acetal gear pairs," *Wear,***166**, pp. 85-91, 1993/06/15 (1993).
17. H. Düzcükoğlu, R. Yakut, and E. Uysal, "The Use of Cooling Holes to Decrease the Amount of Thermal Damage on a Plastic Gear Tooth," *Journal of Failure Analysis and Prevention,* **10**, pp. 545-555, (2010).
18. K. D. Dearn, T. J. Hoskins, D. G. Petrov, S. C. Reynolds, and R. Banks, "Applications of dry film lubricants for polymer gears," *Wear,*. **298–299**, pp. 99-108, 2/15/ 2013.
19. N. A. Wright and S. N. Kukureka, "Wear testing and measurement techniques for polymer composite gears," *Wear,* **251**, pp. 1567-1578, 10//(2001).
20. C. H. Kim, "Durability improvement method for plastic spur gears," *Tribology International,* **39**, pp. 1454-1461, 11// (2006).
21. S. Yousef, T. A. Osman, A. Abdalla, and G. Zohdy, "Wear Characterization of Carbon Nanotubes Reinforced Acetal Spur, Helical, Bevel and Worm Gears Using a TS Universal Test Rig," *JOM,* pp. 1-8, 2014/12/23(2014).
22. S. Yousef, A. Khattab, M. Zaki, and T. A. Osman, "Wear Characterization of Carbon Nanotubes Reinforced Polymer Gears," *Nanotechnology, IEEE Transactions on,* **12**, pp. 616-620, (2013).
23. A. Pogačnik and J. Tavčar, "An accelerated multilevel test and design procedure for polymer gears," *Materials & Design,* **65**, pp. 961-973, 1// 2015.
24. A. J. Mertens and S. Senthilvelan, "Effect of Mating Metal Gear Surface Texture on the Polymer Gear Surface Temperature," *Materials Today: Proceedings,*. **2**, pp. 1763-1769, // (2015).
25. A. J. Mertens and S. Senthilvelan, "Effect of mating metal gear surface texture on the polymer gear surface temperature," *Materials Today-Proceedings.* **2**, pp. 1763-1769, (2015).

Theoretical modelling of a beam with attached spring-mass-damper system

Syaiful Azmirul Mohd Rozlan[1], *Izzuddin* Zaman[1,*], *Bukhari* Manshoor[1], *Amir* Khalid[1], *Shiau Wei* Chan[2], and *Mohd Shahrir* Mohd Sani[3]

[1]Faculty of Mechanical and Manufacturing Engineering, Universiti Tun Hussein Onn Malaysia, 86400 Batu Pahat, Johor, Malaysia
[2]Faculty of Technology Management and Business, Universiti Tun Hussein Onn Malaysia, 86400 Batu Pahat, Johor, Malaysia
[3]Automotive Engineering Centre, Universiti Malaysia Pahang, 26600 Pekan, Pahang, Malaysia

Abstract. Vibrations are always undesirable, wasting energy besides producing noise. In this case, beams which are prominent component in most engineering having no exemption from the vibration effect when imposed by dynamic loading. One of the approach to attenuate vibration of a structure is by having a spring-mass-damper (SMD) system or typically known as vibration neutralizer attached to the vibrating structure. This method is more promising as it does not contribute significant additional energy to the structure. The work presented in this paper describes the frequency response (FRF) of a simply supported beam with an attached SMD system. A mathematical model of a beam was at first developed in the study which was further derived to include the attachment of SMD system. In order to transform the derived equations into a form of graph that can be analysed, Matlab® software was used. The outcome from Matlab® shows that the attachment of SMD onto beam attenuates its vibration significantly. The result also displays a good resemblance FRF when compared with numerical finite element analysis of Ansys®. It is expected that the theoretical derivation demonstrated in this paper provide a helpful reference to future researchers who endeavour to find equations of a simply supported beam with an attached SMD system as well as for a vibration control study.

1 Introduction

A beam is a typically horizontal structural member applied to support vertical loads such as floors, panels, roofs and decks. It is capable of withstanding load primarily by resisting against bending which is caused by the external loads, own weight, span and external reaction when the bending force is induced into the beam [1-4]. There are few support configuration used for beam, for example cantilever, fixed supported and a common one is simply supported [4]. Beam models are usually made to simplify its application such as in structures of automobile frame, bridge, aircraft wings, propeller blades and others

* Corresponding author: izzuddin@uthm.edu.my

engineering structure. This is because beam models are much easier to be analysed [5,6]. Figure 1 shows some of the common examples of beam application.

Fig. 1. Potential of beam's applications.

Although beam is a popular structure employed in engineering applications, even it cannot escape from the effects of vibration phenomena when a dynamic load is applied [7,8]. With the current trend of using dynamic systems, the exposure of structure to vibration could not be avoided and it would become worse if the structural vibration of beam is not controlled. In other words, vibrations is always not needed because induce uncomfortable noise, lead to excessive deflections and eventually failure of structure [9]. In order to overcome the problem, few methods have been developed for the past 20 years and one of them is by adding vibration neutralizer.

In general, vibration neutralizer consists of spring, mass and damper systems – which is the term used in this study. A vibration neutralizer system has been widely used to suppress or attenuate vibrations in linear, nonlinear vibrating mechanical systems and many structures. The applications of this device can be found particularly in bridges, building structures, aeroplanes, machine tools and others engineering systems [10-16]. A previous study of Hao et al. [14] investigated about tuned vibration neutralizer to suppress of hand-arm vibration in electric grass trimmer. From the experimental results, it shows that the vibration of the electric grass trimmer with attached neutralizer was reduced almost 95% along the frequency band. In 2013, Acar & Yilmaz [15] proposed a new design of the passive dynamic vibration neutralizer made of a string under variable tension. The author used finite element method to verify the natural frequencies of the system and validate data outcome from experiments. A similar study was performed by Rubio et al. [16] who used finite element to investigate the performance of vibration neutralizer attached to a boring bar.

Although it has been proved through the experimental study and finite element analysis that vibration neutralizer can reduce the structural vibration [17-19], none have addressed the theoretical background of vibration absorber attached to the structure in depth. Therefore, this paper strive to present details derivation of a mathematical model which allows the calculation of frequency response function of a simply-supported beam with an attached vibration neutralizer (abbreviated as SMD–spring mass damper system). The outcome of this study will provide a helpful reference to future researchers who endeavour to find an equation for a simply-supported beam with attached vibration neutralizer.

2 Mathematical modelling

This section describes the derivation of a mathematical model which allows the calculation of frequency response function of a simply supported beam with an attached vibration neutralizer. At first, the mathematical equation for dynamic response of a simply-supported beam is derived, followed by a derivation of SMD system and after that the derivation of a simply-supported beam with attached SMD.

2.1 Simply-supported beam

Figure 2 shows the schematic of a simply supported beam. The response of beam *w(x)* can be determined when beam subjected to vertical load *F(x)*.

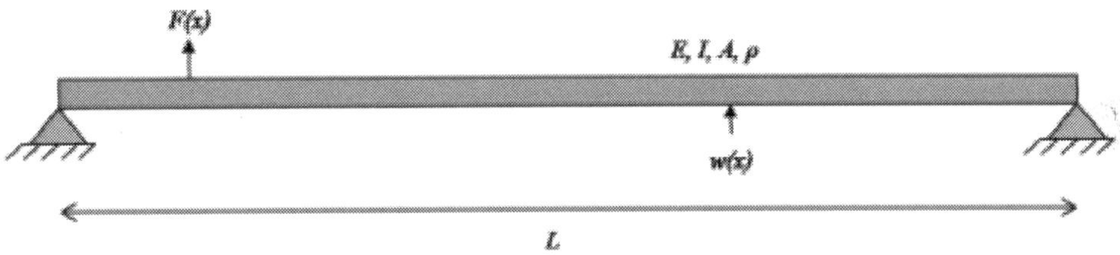

Fig. 2. Schematic of a simply supported beam.

Assuming the Young's modulus E, cross sectional area A, area moment of inertia I and density ρ are constant along the beam length L, the equation of motion for a simply-supported beam can be expressed by [20]:

$$EI\left(\frac{\partial^4 w(x,t)}{\partial x^4}\right) + \rho A \frac{\partial^2 w(x,t)}{\partial t^2} = F(x,t) \tag{1}$$

Therefore, the total normal response of beam is given by the summation of superposition of the individual Eigen solutions, or modes given by Eq. (2):

$$w(x,t) = \sum_{n=1}^{\infty} W_n \cdot \psi_n e^{j\omega_n t} \tag{2}$$

Where, W_n is the modal amplitude, ψ_n is the mode shape of a simply supported beam, ω_n is the natural frequency and n is the mode number. The mode shape ψ_n of a simply supported beam is given by:

$$\psi_n = \sqrt{2}\left(\frac{n\pi}{L}x\right) \tag{3}$$

Thus, the natural frequencies of the nth mode of a simply-supported beam can be calculated from Eq. (4):

$$\omega_n = \left(\frac{n\pi}{L}\right)^2 \sqrt{\frac{EI}{\rho A}} \tag{4}$$

Where the area moment of inertia of beam is obtained from Eq. (5):

$$I = \frac{bh^3}{12} \tag{5}$$

By neglecting the exponential time varying term, the total response of simply-supported beam incorporating the viscous damping ζ can be simplified as in Eq.(6):

$$\omega(x,t) = \frac{2F}{\rho AL}\sum_{n=1}^{\infty}\frac{(\sin(n\pi x_1)/L)\sin(n\pi x/L)}{\omega_n^2 - \omega^2 + j2\zeta\omega\omega_{mn}} \tag{6}$$

2.2 Spring mass damper (SMD) system

Figure 3 shows the schematic of spring mass damper system. The general equation of motion applied for spring mass damper system is given by Eq. (6) [12]:

$$m\ddot{x} + c\ddot{x} + kx = F \tag{7}$$

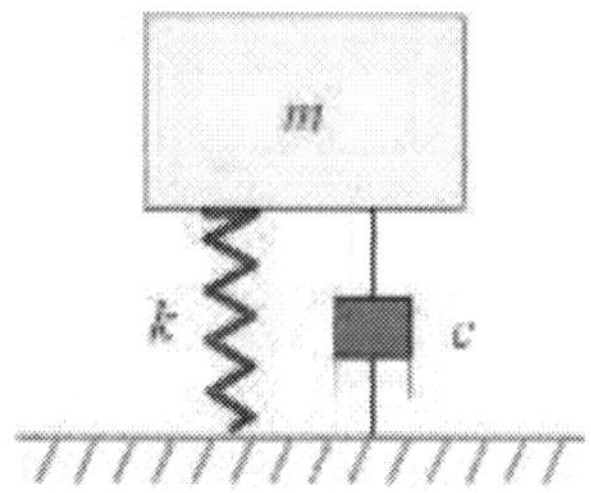

Fig. 3. Schematic of spring mass damper system.

Taking the Laplace transform of a general second order differential equation with initial conditions, the transfer function of the system is:

$$\frac{x(s)}{F(s)} = \frac{1}{ms^2 + cs + k} \tag{8}$$

By applying the following definitions $\zeta = c/2\sqrt{km}$ and $\omega_n = \sqrt{k/m}$, Eq. (8) can be further simplified and rewrite as in Eq. (9):

$$\frac{x(s)}{F(s)} = \frac{1/m}{s^2 + 2\zeta\omega_n s + \omega_n^2} \tag{9}$$

Where, ω_n is the undamped natural frequency and ζ is the damping ratio.

2.3 Simply-supported beam with attached SMD system

Figure 4 shows a simply-supported beam with attached a single degree of freedom SMD system subjected to a harmonic load *F(x)*.

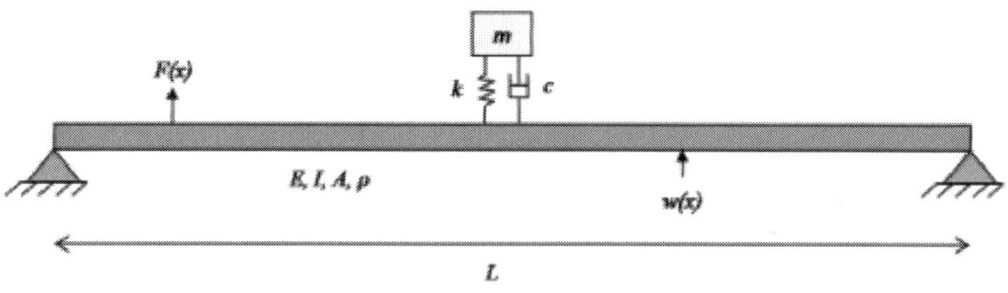

Fig. 4. Schematic of a simply-supported beam attached with a SMD system.

The equation of motion of a SMD system attached to a simply supported beam is given by Eqs. (10) and (11):

$$m\dot{y} + c(\dot{y} - \dot{w}) + k(y - w) = 0 \tag{10}$$

$$-c(\dot{y} - \dot{w}) - k(y - w) = F_e e^{st} \tag{11}$$

Where y is the response of absorber mass, w is the response of a beam at the point of attachment of the spring-damper and F_e is a force of spring damper system imparts on the beam.

Let say $y = Ye^{st}$ and $w = We^{st}$, therefore Eqs.(10) and (11) reduce to:

$$ms^2Y + csY + kY - csW - kW = 0 \tag{12}$$

$$-csY - kY + csW + kW = F_e \tag{13}$$

Putting into matrix form, Eqs. (12) and (13) are simplified as:

$$\begin{bmatrix} ms^2 + cs + k & -(cs+k) \\ -(cs+k) & cs+k \end{bmatrix} = \begin{bmatrix} Y \\ W \end{bmatrix}\begin{bmatrix} 0 \\ F_e \end{bmatrix} \tag{14}$$

And solving for W gives:

$$W = \frac{F_e(m_e s^2 + c_e s + k_e)}{m_e s^2(c_e s + k_e)} \tag{15}$$

Therefore, the receptance at a point of attachment of spring-mass damper system on the beam is given by Eq. (16):

$$\beta_{22} = \frac{m_e s^2 + c_e s + k_e}{m_e s^2(c_e s + k_e)} \tag{16}$$

By modifying Eq.(6), the response of the beam at x_1 with attached spring mass damper system at x_2 can be calculated using receptance method. The displacement equation for a beam with attached one spring-mass-damper system is simplified as follows:

$$\omega_1(x) = \left[\alpha_{11} - \frac{\alpha_{21}^2}{\alpha_{22} + \beta_{22}}\right]F(x) \tag{17}$$

Where,

$$\alpha_{11} = \frac{2}{\rho AL}\sum_{n=1}^{\infty}\frac{\sin^2(k_n x_1)}{\omega_n^2 - \omega^2 + j2\zeta\omega\omega_n} \tag{18}$$

$$\alpha_{21} = \frac{2}{\rho AL}\sum_{n=1}^{\infty}\frac{\sin(k_n x_1)\sin(k_n x_2)}{\omega_n^2 - \omega^2 + j2\zeta\omega\omega_n} \tag{19}$$

$$\alpha_{22} = \frac{2}{\rho AL}\sum_{n=1}^{\infty}\frac{\sin^2(k_n x_2)}{\omega_n^2 - \omega^2 + j2\zeta\omega\omega_n} \tag{20}$$

Where ω_n is the natural frequency of beam, ζ is the viscous damping of beam, n is the mode number and k_n is given by:

$$k_n = \frac{n\pi}{L} \tag{21}$$

The full solution method involves inverting the mass matrix and solving inhomogeneous Eq.(22) directly.

$$\left\{-\omega^2[M] + j\omega[C] + [K]\right\}\{w\} = \{F\} \tag{22}$$

Where $[M]$ is structural mass matrix, $[C]$ is structural damping matrix, $[K]$ is structural stiffness matrix, $\{w\}$ is nodal displacement vector and $\{F\}$ is applied load vector.

3 Results and discussion

The normalized displacement of a simply-supported beam was plotted against frequency by using Matlab®. The solution method that employed for the harmonic analysis was mode superposition method. This method uses natural frequencies and mode shapes from the modal analysis to characterize the dynamic response of a structure [21,22]. Tables 1 and 2 describe the properties of the beam and SMD that used in the study.

Table 1. Properties and values used in beam analysis.

	Length	Cross Section Area	Height	Young's Modulus	Density	Viscous Damping
Symbol	L	A	h	E	ρ	ζ
Value	1.0	4.0×10^{-3}	0.02	210	7850	1.0
Units	m	m^2	M	GPa	kg/m^3	%

Table 2. Properties and values used in SMD.

	Mass	Spring	Damping
Symbol	m	k	C
Value	1.0	1000	6.3
Units	kg	N/m	Ns/m

Figure 5 displays the frequency response function (FRF) and phase angle of the normalized displacement of simply supported beam. The mathematical equation was plotted using Matlab® while numerical model of Ansys® [23] was used to validate this equation. In order to obtain an accurate result, 500 modes were defined in Matlab® for the frequency range of 10 to 1000 Hz. Obviously it can be seen that both theoretical and numerical graphs of displacement and phase show no difference. Table 3 indicates the percentage difference obtained between theoretical and numerical Ansys® are too small and can be negligible. Therefore demonstrate both equation and numerical model developed in the study are accurate.

Table 3. Resonance frequency (Hz) of simply supported beam.

Mode n^{th}	Theoretical Matlab®	Numerical Ansys®	Error (%)
1	47.0	46.9	0.1
2	187.8	187.5	0.2
3	421.6	421.7	0.1
4	749.3	749.7	0.1

DOI: 10.1051/matecconf/20179001030

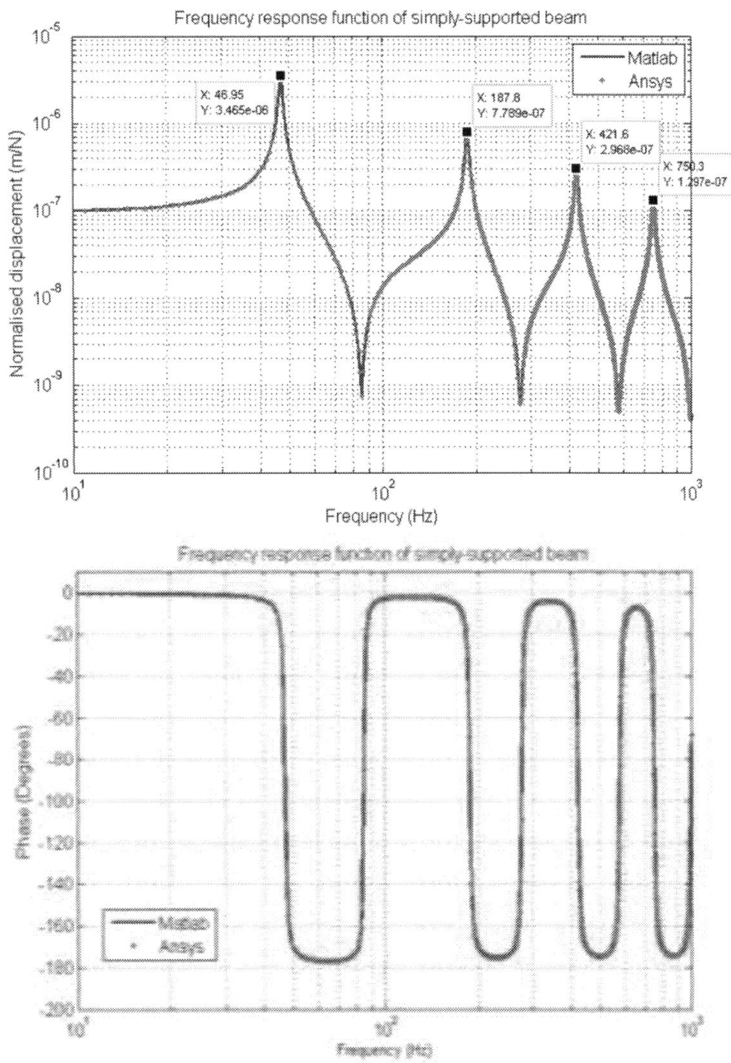

Fig. 5. A comparison of the magnitude and phase of the receptance of a beam.

Figure 6 shows the total displacement and phase of beam with an attached SMD system. Again it shows that the theoretical equation plotted by Matlab® is almost identical with numerical of Ansys®. It is found that by adding SMD to a beam has significantly reduce the displacement amplitude about 30% and this was agreed with our previous findings [16-20]. Thus indicates that the derivation of mathematical equation of a simply supported beam with attached SMD was successful.

DOI: 10.1051/matecconf/20179001030

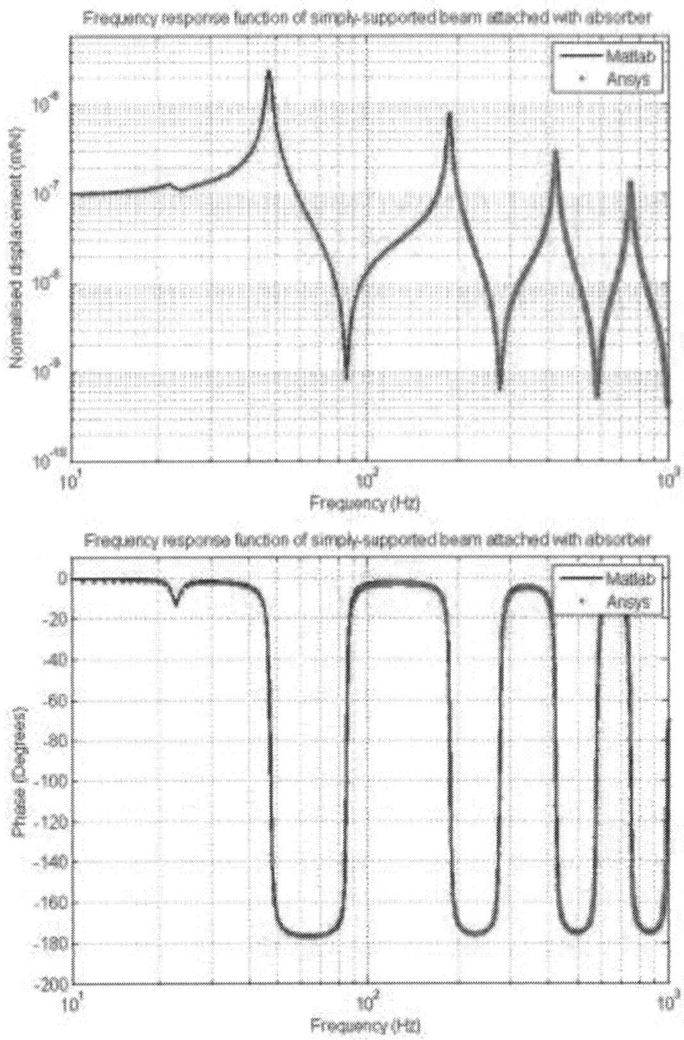

Fig. 6. A comparison of the magnitude and phase of the receptance of simply supported beam with SMD attached.

4 Conclusions

In this paper, it was attempted to predict the vibration response of a simply-supported beam with and without an attached SMD system by deriving from the mathematical model. The application software such as Matlab® was used to transform the model into the form of FRF graph. Later, in order to prove this mathematical model, numerical analysis of Ansys® was carried out. Results found that the difference are too small which can be neglected, thus prove the theoretical model prediction.

Special thanks to Research, Innovation, Commercialization and Consultancy Management, Universiti Tun Hussein Onn Malaysia and Ministry of Higher Education Malaysia for the financial support under Fundamental Research Grant Scheme (FRGS) vote 1546 and Incentive Grant Scheme for Publication (IGSP), vote U247.

References

1. R. Gonçalves, M. Ritto-Corrêa, D. Camotim, *Computational Mechanics* **46**, 759-781 (2010).
2. G. Piccardo, G. Ranzi, A. Luongo, *Mathematics and Mechanics of Solids* **19**, 900-924 (2014).
3. T. P. Vo, J. Lee, *Engineering Structures* **30**, 1958-1968 (2008).
4. M. Aydogdu, *Physica E: Low-dimensional Systems and Nanostructures* **41**, 1651-1655 (2009).
5. H. Ma, X.-L. Gao, J. Reddy, *Journal of the Mechanics and Physics of Solids* **56**, 3379-3391 (2008).
6. B. Wang, J. Zhao, S. Zhou, *European Journal of Mechanics-A/Solids* **29**, 591-599 (2010).
7. M. Şimşek, T. Kocatürk, *Composite Structures* **90**, 465-473 (2009).
8. A. E. Alshorbagy, M. Eltaher, F. Mahmoud, *Applied Mathematical Modelling* **35**, 412-425 (2011).
9. M. S. Qatu, M. K. Abdelhamid, J. Pang, G. Sheng, *International Journal of Vehicle Noise and Vibration* **5**, 135 (2009).
10. R. I. Wright, M. R. F. Kidner, *Journal of Vibration and Control* **10**, 1221-1237 (2004).
11. I. Zaman, M. M. Salleh, B. Manshoor, A. Khalid, S. Araby, *Applied Mechanics and Materials* **629**, 191-196 (2014).
12. M. J. Brennan, *Shock and Vibration* **13**, 531543 (2006).
13. J. Q. Sun, M. R. Jolly, M. A. Norris, *Transactions of the American Society of Mechanical Engineers* **117**, 234-242 (1995).
14. K. Y. Hao, L. X. Mei, Z. M. Ripin, *International journal of industrial ergonomics* **41**, 494-508 (2011).
15. M. A. Acar, C. Yilmaz, *Journal of Sound and Vibration* **332**, 231-245 (2013).
16. L. Rubio, J. Loya, M. Miguélez, J. Fernández-Sáez, *Mechanical Systems and Signal Processing* **41**, 691-704 (2013).
17. I. Zaman, A. L. M. Tobi, B. Manshoor, A. Khalid, N. A. M. Amin, *ARPN Journal of Engineering and Applied Sciences* **11**, 2308-2313 (2016).
18. H. L. Sun, P. Q. Zhang, X. L. Gong, H. B. Chen, *Journal of Sound and Vibration* **300**, 117-125 (2007).
19. I. Zaman, M. M. Salleh, M. Ismon, B. Manshoor, A. Khalid, M. S. M. Sani, S. Araby, *Applied Mechanics and Materials* **660**, 511-515 (2014).
20. I. Zaman, M. M. Salleh, M. Ismon, B. Manshoor, A. Khalid, M. S. M. Sani, S. Araby, *MATEC Web of Conferences* **13**, 03003 (2014).
21. E. Reynders, *Archives of Computational Methods in Engineering* **19**, 51-124 (2012).
22. I. Zaman, K. A. Kamarudin, M. T. M. Nordin, in *International Conference on Construction and Building Technology (ICCBT)*, Kuala Lumpur, Malaysia (2008).
23. M. M. Salleh, I. Zaman, *ARPN Journal of Engineering and Applied Sciences* **11**, 2336-2339 (2016).

Prevalence of distracted pedestrians while crossing: a study of Malaysia's situation

Mohd Syazwan S.[1,2,*], *Baba* Md. Deros[2], *Nur Zarifah* H.[1] , *Aqbal Hafeez* A.[1], and *Noor Faradila* P.[1,2]

[1]Crash Safety Engineering Unit, Vehicle safety and Biomechanics Research Centre Malaysian Institute of Road Safety Research, Kajang, Selangor, Malaysia
[2]Faculty of Engineering, Department of Mechanical and Materials Engineering, National University of Malaysia, Bangi, Selangor DE, Malaysia

Abstract. Pedestrian's crashes account for approximately 7% of road death in Malaysia. Even though the percentage is smaller compared to motorcycle-related crashes, amount of road crashes involving pedestrians in Malaysia is a serious concern. The objective of this study is to establish the common type of distraction for Malaysian. Crashes among pedestrians usually highly recorded at straight road followed by inter-section road type. For that, study focusing on the type of distraction among pedestrians while crossing the road by using video recording method. Two type of road were selected observational field survey – signalized and non-signalized intersection. A video recording method was used to analyse the behaviour of pedestrians crossing with or without distraction. Approximately 375 samples were observed while crossing with various type of distraction such as eating, texting, mobile phone talking and others. The study found the highest distraction type observed among pedestrians is usage of electronic equipment-mobile phone with 84.8% followed by other type of distraction – smoking and talking with more that 5%. Mobile phone distraction cause a significant effect in term of time to cross which average time to cross was 14.77 second. The overall results provide a prevalence information that is useful in the development of countermeasures aimed at improving pedestrians' safety.

1 Introduction

Globally, pedestrians constitute 22% of all road deaths, and in some countries this proportion is as high as two thirds [1]. In ASEAN country alone the magnitude of pedestrian fatalities was rather high especially in Myanmar, Philippines, and Singapore with more than 15 percent, whereas Malaysia, Thailand, Indonesia, Cambodia, Brunei and Indonesia had the rate in between 8 to 13 percent. It is noted that Lao DPR, Vietnam and Timor-Leste data are unavailable might due to insufficient accident data. It is noted that if compared with motorcyclist fatalities, pedestrian deaths are considered low for each country.

* Corresponding author: syazwan@miros.gov.my

In Malaysia, road injuries and fatalities are major issues, with more than 6,000 involved in fatal accident and over 10,000 in minor and major incident yearly. Pedestrian fatalities are averaging approximately more than 500 deaths each year. Records consistently have shown that fatal road traffic injuries among elder (<60 years old) are increasing from 24.4% (2006) to 44.2% (2013) of the total pedestrian's fatalities. In term of incident occurred, it is statistically shown that straight road contributed more than 60% fatalities followed by junction-type of road [2]. Since the number of pedestrian's fatalities in Malaysia create media attention, this issue need to be addressed. Table 1 describes the fatalities involving pedestrian in the Southeast Asia countries in the year of 2015, as reported by the World Health Organization (WHO) [3].

Table 1. Vehicles, population, road traffic deaths and proportion of pedestrian casualties by country in Southeast Asia, 2015.

Southeast Asia Country	Registered Vehicles (Count)	Population for 2015 (Count)	Reported Road Traffic Fatalities (Count)	Deaths by Pedestrian Category (Proportion)[1]
Laos DPR	1,439.481	6,769,727	910	10%
Malaysia	**23,819,256**	**269,716,965**	**6,915**	**7 %**
Myanmar	4,310,112	53,259,018	3612	26%
Philippines	7,690,038	98,393,574	1,513	19%
Singapore	974,170	5,411,737	159	27 %
Thailand	32,476,977	67,010,502	14,059	8 %
Vietnam	40,790,841	91,679,733	9,156	-
Cambodia	2,457,569	15,135,169	1950	13%
Brunei	304,432	390,056	54	9.2 %
Indonesia	104,211,132	249,865,610	26,460	21 %
Timor-Leste[2]	63,553	1,132,979	74	-

Note:
- Data Not Available
[1] Proportion of pedestrian deaths per reported road traffic fatalities
[2] Not ASEAN country
Source: WHO, 2015 [3]

1.1 Factors related to pedestrians-vehicle accident

There are many factors associated with pedestrians-vehicle accident as discussed in literatures available worldwide. This paper primarily reviewed Zegeer et al. [4]'s key factors that were derived from nearly 200 pedestrian safety studies from 2000 to 2010, including national and international articles published in Transportation Research Records, Accident Analysis and Prevention, and other transportation-related journals. The authors classified the pedestrian crash factors into five categories: (1) Driver factors; (2) Pedestrians factors; (3) Vehicle factors; (4) Roadway/ Environmental factors; and (5) Demographic/Social/Policy factors. For this study, the focus is on the pedestrian factors specifically on pedestrian's distraction [4]. Figure 1 provides an overview of factors associated with pedestrian crash risk and/or severity.

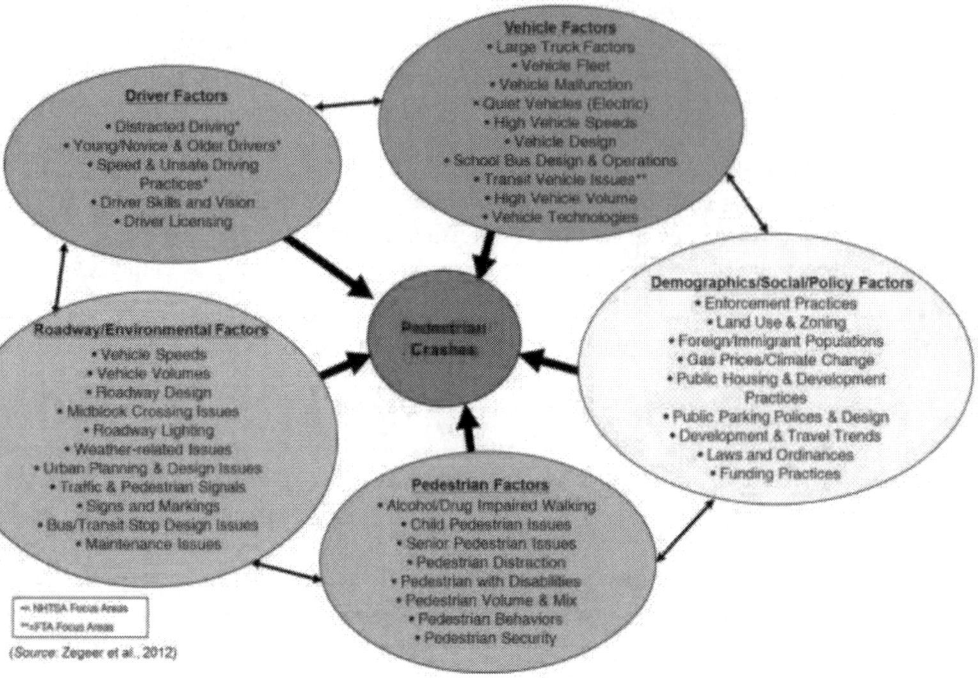

Fig. 1. Illustration of factors associated with pedestrian crash risk and/or severity [4].

1.2 Pedestrian distraction

Pedestrian distraction is not new issue in road safety. Study by Bungum et. al, [5] revealed that 15% of pedestrian's accident occurred because they lacked attention. Distraction has negative effects on traffic behavior. Most studies estimate that distraction plays a role in the occurrence of about 5-25% of the car crashes [6]. Pedestrian's distraction can be defined as those wearing headphones, talking on a cell phone, eating, drinking, smoking, or talking with another pedestrians as they crossed the street [7]. Those examples potentially increase due to lift up of lifestyle and itis aligned with advance telecommunication. Distracted pedestrians usually at high risk during crossing which the process needs cognitive (e.g., road user focuses more on phone conversation rather than walking activity), cautionary behavior (e.g., looking before crossing a road), auditory (e.g., road user listens to the music via headphone) and motor coordination judgment to minimize the risk [8].

Due to distractions, it is shown that it can significantly affect crossing speed, increase conflict, less attention to environment and many more errors which possibly end up with accident [9-12]. Pedestrian's distraction while crossing is a major issue in developed countries and those countries proactively take this issue and come out with various intervention that suit to their cultures and laws.

To date, with the best knowledge of author, there is no research being conducted in this area to present the situation of distracted pedestrians while crossing in Malaysia. This study aims to determine the type of distraction among Malaysian pedestrian's distraction while crossing for both signalised and non-signalised crosswalks. Moreover, this study wants to establish the common type of distraction for Malaysian. For that, an observational study is proposed for this study.

2 Methodology

This short, cross-sectional study was conducted for a month and a sampling was carried out in Capital of Malaysia, Kuala Lumpur. This study, based on observation survey was conducted to determine the characteristics of pedestrians crossing. The four selected areas (Central Market and Puduraya – non-signalised; Bukit Bintang and Sogo – signalised) were selected to represent the two type of crosswalks; signalized and non-signalized. Kuala Lumpur is selected due to high density of population and also high capacity of pedestrians walking activity.

Data collection in every area was carried out by an assigned team member at pre-determined fixed observation points with high pedestrians walking volume. For this observation method, a video recording by using video recorder was used and it was set-up at the strategic location on site which enabled to capture the best view of pedestrian activities. Pilot testing was done prior to the actual data collection at the Kajang Train Station to ensure this method is correct and the data is sufficient to be analyzed. In term of crossing time, the data collected using time embedded in recorded video.

This study used stratified random sampling technique for field observation data collection. Targeted sample include all pedestrians who crossing the area while doing other activity such as eating, texting, listen to music, mobile phone talking and other type of distraction which passed through the observation points. All variables from the observed samples were recorded in the video and transferred all variable needed into observational form for analysis. The observations were conducted during daytime - two interval time (7am – 9am and 12pm – 2pm) on weekdays. This observation lased for a week. The reason to select in the particular interval time is due to high volume of pedestrians. For this study, a certain pedestrians behavior not take into account such as disabled, running and large group.

Fig. 2. Example of signalized (left) and non-signalized (right) pedestrians crossing.

3 Results and discussion

In average, 1,455 pedestrians were observed crossing the junctions per period during a week data collection. From that, a total of 375 samples were collected both at signalized and non-signalized control type. Out of these, 275 samples were collected at signalized crosswalk, while the rest, 103 samples were observed at non-signalized crosswalk. A sample of 375 pedestrians in this study consisted 226 (60.3%) males and 149 (39.7%) females. In term of estimated age of the sample, it shows that distracted crossing pedestrians are made of 191 (50.9%) for 21 to 30 year olds, 155(41.3%) for 31 to 40 year olds, 24(6.4%) for 41 to 50 year olds and 5(1.3%) for more than 51 year olds. Table 2 shows distracted pedestrian characteristics. Furthermore, 68.3% distracted pedestrians were observed using designated area to cross. During the observation period, it was revealed that 84.8% pedestrians were distracted with mobile phone usage followed by others type of distraction. The lowest type of distraction recorded was drinking and eating with 4.5%.

Table 2. Pedestrians characteristics.

Characteristics	**n**	**%**
Estimated age group		
21 -30 years	191	50.9
31-40 years	155	41.3
41-50 years	24	6.4
>50 years	5	1.3
Gender		
Male	226	60.3
Female	149	39.7
Type of distraction		
Drinking/eating	17	4.5
Mobile phone usage	318	84.8
Reading	18	4.8
Other –e.g. Smoking, talking	22	5.9
Crossing at Designated crossing(zebra cross)		
Yes	256	68.3
No	119	31.7

In term of gender distribution, it shows that use of mobile phone is among males is higher as compared to females with 59.7% and 40.3%, respectively (Table 3). For female, they were more distracted by reading with 66.7%.

Table 3. Distribution of pedestrian's distraction according to gender.

Pedestrians distraction	**Male**	**Female**
Mobile phone usage	190 (59.7%)	128
Drinking/eating	8 (47%)	9
Reading	6 (33.3%)	12
Others – smoking, talking	22 (100%)	0

In this study, six type of pedestrian behaviours were recorded for analysis. The behaviours accounted for pedestrians waiting before crossing until arrived on the other side of road. For mobile phone distraction, it shows that pedestrians often crossing the road before car completely stopped – 55.1% and the distracted pedestrians while using mobile phone revealed that almost half of them walking not in straight path with 49.7%. In addition, for eating/drinking distraction among pedestrians, the most typical behaviour that was observed is crossing while vehicle still in moving with 82.3%. Table 4 gives a summary of the pedestrian behaviours according to type of distractions.

Several types of pedestrian's distraction are also associated with risk of their behaviour. Form the analysis, it shows that only three type of behaviours were associated with the type of distractions observed. The analysis revealed that pedestrians drinking/eating (OR=6.375, 95% CI 0.835 to 48.69), reading (OR=0.221, 95% CI 0.083 to 0.587) and other type of distractions (OR=8.534, 95% CI 1.133 to 64.283) aresignificant as these pedestrians were observed not looking left and right prior to crossing. Furthermore, pedestrians crossing while using mobile phone (OR=1.975, 95% CI 1.092 to 3.573) and drinking/eating (OR=4.413, 95% CI 1.247 to 15.612) were significantly not walking straight during crossing. Table 5 shows the result of odd insecure pedestrian behaviour by type of distractions.

Time to cross

Table 6 shows the distracted pedestrian's crossing time. The mean crossing time for all type distraction is 14.55 seconds across the 3-lane road. In term of average time, mobile phone usage recorded the highest time with 14.55 second followed by other type of distractions with 14.0 second. The lowest average time recorded in this study is pedestrians with reading distraction with 12.77 second.

Table 4: Pedestrians behaviour according to type of distraction

Pedestrians behaviour		**Mobile phone usage**	**Eating/drinking**	**Reading**	**Others e.g smoking, talking**
Crossing at designated area	Yes	218	11	15	12
	No	100 (31%)	6 (35.3%)	3 (16.6%)	10 (45.4%)
Observe right and left prior to crossing	Yes	228	16	7	21
	No	90 (28.3%)	1(5.8%)	11 (61.1%)	1 (9%)
Look left or right during crossing	Yes	171	9	9	16
	No	147 (46.2%)	8 (47%)	9 (50%)	6 (27.2%)
Crossing after car complete stop	Yes	149	3	14	4
	No	169 (53.1%)	14(82.3%)	4 (22.2%)	18 (81.8%)
Walk straight during crossing	Yes	160	14	9	15
	No	158(49.7%)	3 (17.6%)	9(50%)	7 (31.8%)

This study aims to establish type of pedestrian's distraction in capital of Malaysia-Kuala Lumpur. The result discovered that the most common type of distraction among pedestrians is using the electronic device – mobile phone which accounted for 84.8%. While other type of distractions only accounted less than 6 percent. This finding similar with study by Liberty Mutual Insurance [13] whereby most of pedestrians distracted by mobile electronic device activity such as taking on the phone, texting/hand operate or listening to music. From this observational study, it seems that growth of mobile technology suggested influencing human walking behavior. Statistics shows that up to end of 2015 there are 27.8 million mobile phone users in Malaysia [14]. This statistics shows that will increase potential risk to road user if there is no action taken by government or other related parties. Previous study by Nasar J.L et al. [8] found that distracted walking injuries in the United States are rising fast, with 1,506 recorded in the US emergency rooms in 2010, up from 256 in 2005. The teenagers were especially likely to be affected. Other countries also reported the same significant findings whereby increment of injuries among pedestrians due to mobile phone distraction [8].

Furthermore, in term of pedestrian's behavior while crossing, it is discovered that most of distracted mobile phone users were crossing the road even the vehicles were not completely stop with 53.1% followed by diagonal crossing with 49.7%. The result is consistent with the findings of Bungum et. al, [5] whereby mobile phone distraction may cause less pay attention to traffic both before and during crossing. As the result using talking on the phone – auditory distracted, it can cause pedestrians to miss significant objects in their environment and appear to exhibit unsafe behavior (failure to look right and left, wait on curb for light turn to green before cross [15].

Table 5: Insecure pedestrian behaviour by type of distractions

Insecure Pedestrians behaviour			
	OR	**95% CI**	**p_value**
Not Crossing at designated area:			
Phone usage	0.917	0.504 to 1.671	0.778
Drinking/eating	0.847	0.305 to 2.344	0.747
Reading	2.407	0.683 to 8.478	0.159
Others – e.g smoking, talking	0.838	0.275 to 2.551	0.755
Not Observe right and left prior to crossing:			
Phone usage	1.336	0.687 to 2.598	0.392
Drinking/eating	**6.375**	**0.835 to 48.69**	**0.041**
Reading	**0.221**	**0.083 to 0.587**	**0.001**
Others – e.g smoking, talking	**8.534**	**1.133 to 64.283**	**0.013**
Not Look left or right during crossing:			
Phone usage	1.271	0.716 to 2.254	0.412
Drinking/eating	0.93	0.351 to 2.465	0.884
Reading	0.821	0.319 to 2.118	0.684
Others – e.g smoking, talking	2.314	0.885 to6.051	0.079
Not crossing after car complete stop:			
Phone usage	0.662	0.37 to 1.183	0.162
Drinking/eating	**0.245**	**0.069 to 0.868**	**0.019**
Reading	**4.510**	**1.456 to 13.970**	**0.005**
Others – e.g smoking, talking	**0.25**	**0.083 to 0.755**	**0.008**
Not Walk straight during crossing :			
Mobile phone usage	**1.975**	**1.092 to 3.573**	**0.023**
Drinking/eating	**4.413**	**1.247 to 15.612**	**0.012**
Reading	0.889	0.345 to 2.292	0.807
Others – e.g smoking, talking	1.991	0.792 to 5.001	0.136
*Note: Bold p-values are significant at 0.05 level			

One of the main reason pedestrian's accident while crossing is due to long time to cross safely. In other studies, it was found that, distracted pedestrians was associated with the slower crossing speed [8, 10]. In our study, the mean speed for males are slower as compared to females. It is might be due to uneven samples distribution. Previous research on pedestrian injuries due to mobile phone usage shows that crossing with low speed and changing lane will increase exposure to risk [10]. Additionally, for other road users such as driver, distraction also has a big issue which it potentially poses hazard to other road users. Study by Dzubak, C. M [16] has identified that brain researchers have identified "reaction-time switching costs which is a measurable time when the brain is switching its attention and focusing from one task to another. Research studying the impact of talking on cell phones while driving has identified slowed reaction time to potential hazards are tangible, measurable and risky.

This paper provides valuable insight on the safety impacts of pedestrian's distraction, the severity and demographics of pedestrian's distraction problems, and potential causes of distracted crossing. The findings from this study are recommended to be used for developing proper countermeasures and educational campaigns to alleviate the traffic safety problems caused by distracted crossing, especially for young populations.

Table 6: Average crossing time for type of distraction

Pedestrians behaviour	**Average time to cross (second)**	**p_value**
Distraction		
Mobile phone usage	14.77	**0.000**
Drinking/eating	12.83	0.924
Reading	12.77	0.296
Others – e.g smoking, talking	14.0	**0.000**
*Note: Bold p-values are significant at 0.05 level		

4 Conclusions

Pedestrian distractions such as mobile phone usage, eating or drinking, reading and others seem to impair with some behaviors while crossing the road. This study shows that mobile phone usage is the highest percentage in term of type of distraction in accordance with advance technology. While this study only focuses on one area in Malaysia, the findings could be an indicator of a similar problem occurring in other areas throughout the country. However, the method need to be strengthen and it is suggested that further studies covering not for intersection crossing only. Since the usage of mobile phone is the highest numbers of distracted recorded, it is also suggested the further analysis need to be done - focus on electronic device distraction. To highlight this kind of problem, the relevant parties should conduct an intervention which highlight the danger of any kind of distractions while crossing.

The authors would like to appreciate the special support provided by The Malaysian Institute of Road Safety Research (MIROS) and National University of Malaysia (UKM). Special thanks also due to Zulyahikem Zakaria, Azhar Hamzah, Zulhaidi Jawi and Mohd Hafzi Isa for their assistance in this project.

References

1. World Health Organization (WHO). "Pedestrians Safety: A Road Safety Manual for Decision-Makers and Practitioners". Geneva, Switzerland: World Health Organization. (2013)
2. RMP (2010) Statistical Report of Road Accidents Malaysia 2010. *Royal Malaysia Police*, Traffic Branch, Bukit Aman, Kuala Lumpur (2010)
3. World Health Organization (WHO). *Global Status Report On Road Safety* (2015).
4. Zegeer, C.V., Bushell, M., "Pedestrians crash trends and potential countermeasures from round the world". *Accident Analysis and Prevention*, 2012:44(**1**):3-11 (2012)

5. Bungum, T.J., Day, C., Henry, L.J., "The association of distraction and caution displayed by pedestrianss at a lighted crosswalk". *J. Commun. Health* 30 (**4**), 269–279,(2005)
6. SwOV Fact sheet, *The Hague, the Netherlands*, September Retrieved from http://www.swov.nl/rapport/Factsheets/UK/FS_Distraction.pdf. (2013)
7. Amy Williamson, alexia Lennon," Pedestrians self-reported exposure to distraction by smart phones while walking and crossing the road" *Proceeding of the 2015 Australasian Road Safety Conference* (2015)
8. Nasar JL, Troyer D. "Pedestrians injuries due to mobile phone use in public places". *Accident Analysis & Prevention* (2013)
9. Nasar J, Hecht P, Wener R. "Mobile telephones, distracted attention, and pedestrians safety". *Accident analysis & prevention*. 2008;40(**1**):69-75(2008)
10. Hatfield J, Murphy S., "The effects of mobile phone use on pedestrians crossing behaviour at signalised and unsignalised intersections," *Accident analysis & prevention*. 2007;39(**1**):197-205 (2007)
11. Byington K, Schwebel D., "Effects of mobile Internet use on college student pedestrian injury risk," *Accident Analysis & Prevention*. 2013;51:78-83 (2013)
12. Ferguson, R.W., Xu, Z., Green, A., & Rosenthal, K.M,"Teens and Distraction: An In-Depth Look at Teens' Walking Behaviors". *Washington, DC: Safe Kids Worldwide* (2013)
13. Libert Mutual Insurance, "Pedestrians Safety Survey Infographic", (2013)
14. MCMC," Communications and Multimedia : Pocket Book of Statistics, Q4 2015". *Malaysian Coomunications and Multimedia (MCMC)*, (2015)
15. J. Mwakalonge, S. Siuhi and J. White, "Distracted walking: examining the extent to pedestrians safety problems[J]", *Journal of Traffic and Transportation Engineering (English Edition)* (2015)
16. Dzubak, C. M. Multitasking: The good, the bad, and the unknown. *Association for the Tutoring Profession.Orlando, FL* (2007)

Effect of fiber laser parameters on laser welded AZ31B Magnesium alloys

Naqiuddin Mat Salleh[1], *Mahadzir* Ishak[1,*], and *Fadhlur* Rahman Romlay[1]

[1]Faculty of Mechanical Engineering, Universiti Malaysia Pahang, 26600 Pekan, Malaysia.

Abstract. Recently, the usage of Magnesium (Mg) alloys has been hugely applied in the industrial application such as in automotive, marine, and electronic due to its advantages of recyclability and lightweight. This alloys required low heat input to be weld since it is easily evaporated due to the Magnesium Oxide (MgO) at the surface and it also possesses lower melting point compared to steel. Laser welding is more convenient to weld Mg alloys due to its high power and lower heat input. AZ31B was selected since it has strong mechanical properties among others Mg alloys due to the major alloying elements; Aluminium (Al) and Zinc (Zn). Low power fiber laser machine with wavelength of 900 nm was used in this experiment. The intention of this work was to investigate the effect of low power fiber laser parameters and effect of shielding gas on weld penetration and microstructure. Another aim in this work was to produce the joint for this thin sheets metal. Penetration depth and microstructure evaluation were emphasized in the analysis section. Bead-on-Plate (BOP) and laser lap welding was conducted on AZ31B with thicknesses of 1.0 mm and 0.6 mm for feasibility study using pulsed wave (PW) mode. Defocusing features was used in order to find better focal position, which has less occurrence of evaporation (underfill). The effect of different angle of irradiation was also investigated. Two types of shielding gases, Argon (Ar) and Nitrogen (N_2) were used in order to study the effect of shielding gas. Lastly, the effect of pulsed energy on penetration types and depth of BOP welded samples was investigated. Focus point was found at focal length of 156 mm with 393.75 μm. For BOP experiment, higher pulsed energy used contributes to melt through defect. Meanwhile, N_s shielding gas proved to be better shielding gas in laser welding the AZ31B. Higher angle of irradiation could reduce the underfill defect. Fillet Lap joint of similar metal was successfully done where 2.0 J of pulsed energy reveals better weld joint compared to 2.4 J.

1 Introduction

In mechanical engineering field, joining product has become a great demand where it is important to make two or more of similar or different parts to become one product with good characteristics. There have several types of joining especially in mechanical engineering fields such as bolt and fasteners, adhesive, solder, rivet, and welding method.

* Corresponding author: mahadzir@ump.edu.my

In metal joining, welding method was usually selected to produce strong permanent joints compared to others. However, laser welding promising a unique ways compared to the conventional welding that have been applied industries long before. As a means known in industrial especially in manufacturing line, fibre laser welding was proved to be best among the laser welding method in terms of saving overall operating cost, easily automated, and allows higher welding speed with excellent weld bead [1]. This process already becomes an important manufacturing process especially in joining metals due to its advantages as a bonding process. The applications of laser welding have been discussed by G.V. Moskvitin, et., al, [2] where the laser welding application can produce a powerful laser beam, concentrated heat source, and allowing to obtain deep and narrow penetration for the high speed process. Laser light is monochromatic and collimated (single wavelength and parallel), so that beam can be focused to very small diameters where the photon can produce enough laser energy to melt the metal in short time.

For past few years, there has been tremendous usage of fibre laser in welding metals such as steels, aluminium alloys, magnesium alloys, and titanium alloys [3-9]. Among of this metals, Magnesium alloys was the attractive metal in automotive, aerospace, and marine industries due to their lower density and high specific strength with high corrosion resistance [10]. Magnesium is the sixth most abundant element on Earth's surface and represents about 2.5% of its composition [11]. This non-ferrous metal have attracted huge attraction recently as a new kind of degradable biomaterial where it has advantages of density almost same to that of human bone [12]. This means that this material was having a good usage although to human being. In aerospace, magnesium alloys also was highly recommended as it has lightest density among other materials and they are good for engineering application by replacing engineering plastics since magnesium alloys are stiffer, more recyclable, and less costly to produce [13]. The trending laser welding on magnesium alloys was attracted on AZ31B magnesium alloys where it has the properties of Medium-strength alloy, weldable, and good formability [13]. In recent years, AZ31B was successfully welded using laser welding method by some researcher. AZ31B-H24 wrought magnesium alloys with thickness of 1.7 mm was used in the studies of effect of shielding gas and laser wavelength using CO_2 laser type and Nd:YAG laser type for BOP test [14]. Nd:YAG laser was similarly with fibre laser since the laser beam was conducted by the fibre core with same wavelength. However, maximum 2 kW average power in continuous wave (CW) mode was needed to weld 1.7 mm thick AZ31B. Fibre laser welding with 4 kW using two weld pass method was successfully done to 1.5 mm thick of AZ31B by applying overlapped joint [7] where also CW welding mode was used with the 1 kW average power. However, they still need to use laser average power up to 3 kW in order to perform preheating (first pass). AZ31B was then successfully joined using Nd:YAG laser welding types with fillet lap configuration [8]. They focused on pulsed wave (PW) laser mode with pulsed energy of 1.8 Joules to make similar fillet lap joint of 0.3 mm thick AZ31B. So far, there was still lack of research in laser welding of thin sheets AZ31B magnesium alloys where these products could become a huge significant in many industries especially for transportation (automotive) since it will increase the percentage of fuel economy in future.

The intention of this present work carried out was to find out the possible laser welding behavior to obtain sound weld of thin sheets (0.6 mm thick) AZ31B using low power fibre laser welding machine with 200 W average powers with 900 – 1200 nm wavelength. Pulsed wave (PW) was more focused in this experimental work and the important parameters which plays major roles were laser pulsed energy, types of shielding gas, and angle of irradiation. The effects of these parameters were investigated in order to investigate the effect toward weld penetration and its microstructure. As well, an attempt to produce similar joint for AZ31B magnesium alloys have been conducted where the parameters were selected form the BOP welded result.

DOI: 10.1051/matecconf/20179001032

2 Experimental setup

The experimental works start with finding the focus points on AZ31B magnesium alloys surface since different metals produce different heat absorptions when subjected to the laser beams since different heat transfer, *Q* occurred compared with other metals. Material used in this research works was thin sheet AZ31B with thickness range from 0.6 to 1.0 mm. One laser spot was shot on the metals surface using the range of pulsed power from 80 to 100 % (1.6 – 2.0 kW) with wavelength of 900 nm. The focal lengths used were ranged from 150 to 160 mm. Beam diameter from the laser spot was measured at each focal length. Next, BOP welding was conducted on 1.0 mm thickness of AZ31B where the pulsed energy, *Ep* was studied. In this work, 0 degree angle of irradiation (AOI) with the absent of shielding gas were used. The effect of pulsed energy was studied based on the macrostructure of the weld cross section. Calculation for laser pulsed energy was made based on *Eq.* (1):

$$Ep = Pp \times \Delta t \qquad (1)$$

where Pp is peak power in Watt (W) , and Δt is pulse width in milliseconds (ms). After that, AZ31B samples was BOP welded using two different types of shielding gas which were Argon (Ar), and Nitrogen (N_2) gas with flow rates of 25 L/min. Previous researcher found that N_2 helps preventing the formation of pores in the WZ in non-ferrous metal compared to using Ar gas despite being use angle of irradiation up to 40° inclination [15]. Another motivation using N_2 gas is it was inflammable gas which can reacts with high performance since AZ31B is a flammable metal. The parameters such as pulsed energy (1.7 J), welding speed (3 mm/s), angle of irradiation (3°), was used at defocused position of +26 mm (FL= 130 mm). The comparison between the welded samples cross section was studied and defect in the microstructure was inspected in this experiment. Next to that, an experiment to study the effect of angle of irradiation (AOI) also has been operated. Using same parameter as before, range of AOI selected was 1° to 5°. AZ31B sheets with thickness of 1.0 mm were used in this experiment with N_2 shielding gas at flow rate 25 L/min. Figure 1 shows the schematic of depth measurement at the weld cross-section. From the cross section, the underfill defect and weld depth were measured for each sample from different AOI.

After the BOP welds have been performed, an attempt on joining the thin sheet metals (AZ31B) of 0.6 mm thick have been done. Fillet lap joint configuration has been conducted using pulsed wave (PW) mode with N_2 shielding gas at flow rate 25 L/min. Figure 2 shows the experimental setup for fillet lap joint. As shown in Figure 2(a), AOI used in this fillet lap weld was 3° with the angle of shielding gas nozzle at 30°. From the schematic view in Figure 2(b), this preliminary test used defocused position at +26 mm from the focused point which at beam diameter of 524 μm. The centre of beam was focused on upper sheet, meanwhile another half of the beam diameter focused on lower sheet. Primarily, single fillet lap joint was made and the parameters selected was 2.0 and 2.4 J of pulsed energy with controlled parameters of welding speed and angle of irradiation with the value of 3 mm/s and 3°, respectively. Pulsed width used was 2 ms with the Pulsed Repetition Rate of 60 HZ also was applied. At 2.0 J, 1000 Watt (50%) of laser peak power was used. Meanwhile, 1200 W (60%) of peak power used for pulsed energy of 2.4 J.

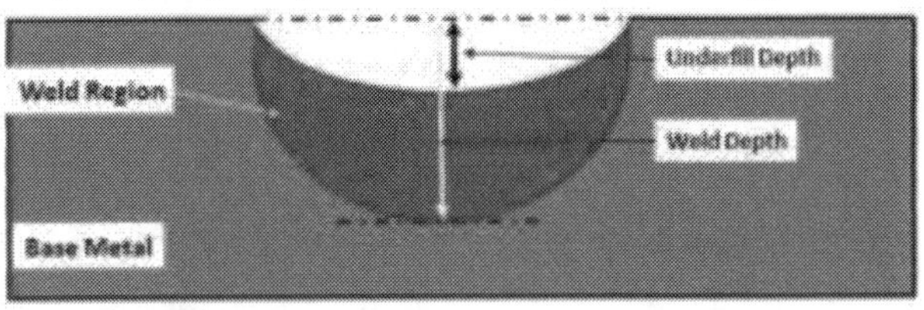

Fig. 1. Schematic view for depth measurements.

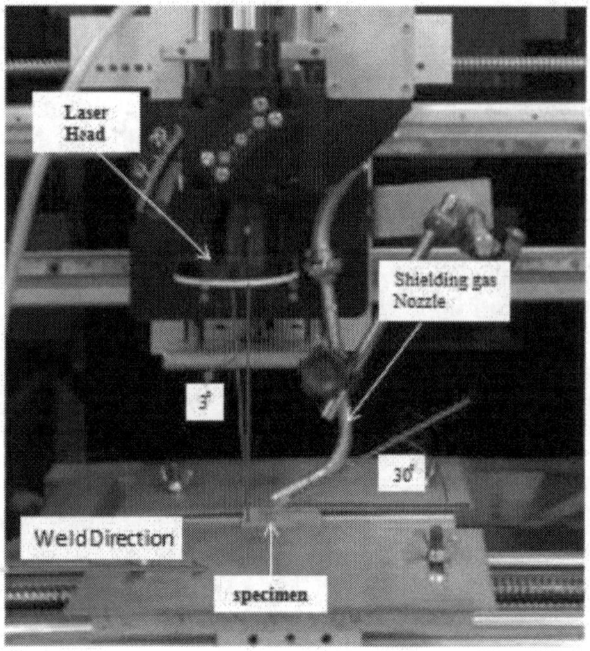

(a)

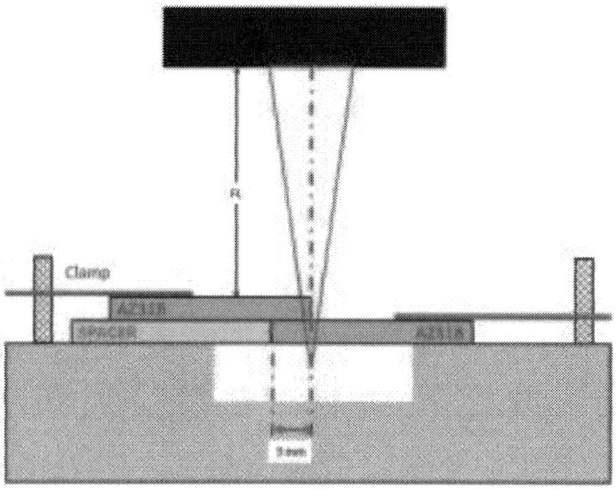

(b)

Fig. 2. Experimental setup for fillet lap welding (a) real view (b) schematic view.

3 Results and discussion

3.1 Focus point experiment

Table 1 shows the result in finding smallest beam diameter which represents the focus points of laser beam on AZ31B magnesium alloys surface. From the tabulated data, a graph of spot diameter versus focal length with the range of 1 mm between them was constructed as shown in Figure 3. From the graph below, the focus point for material AZ31B was at focal length, FL = 156 mm with beam diameter of 393.75 μm. So that, the focused point is set to 0 (FP= 0) at focal length 156 mm. It was concluded that the focus point of the laser beam on AZ31B was at FL = 156 mm with approximately 400 μm of beam width.

Table 1. Focus point results of focal lengths.

Focal length (mm)	Spot diameter (μm)
150	467.5
151	445
152	432.5
153	442.5
154	421.25
155	428.5
156	393.75
157	393.8
158	417.5
159	424.94
160	428.23

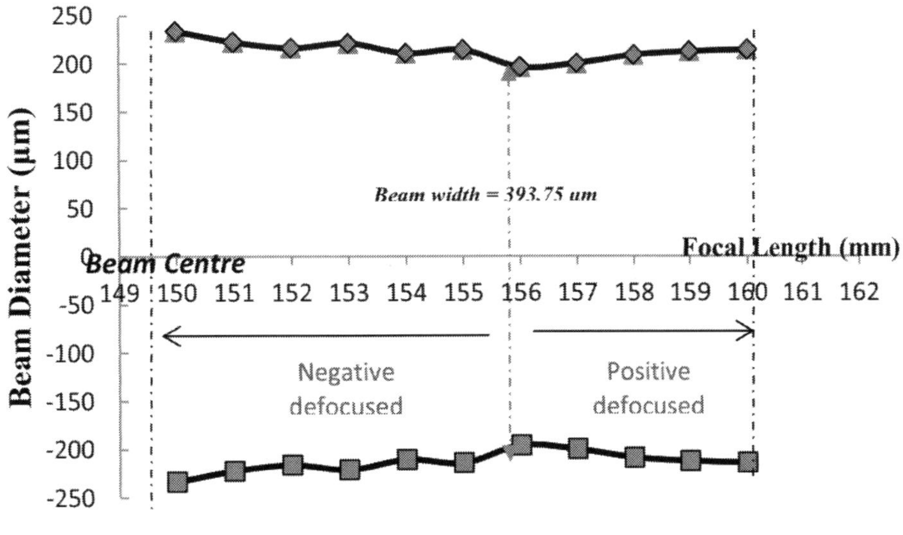

Fig. 3. Graph of beam diameter versus focal length.

3.2 Bead-on-plate (BOP) experiment

Firstly, the results from the investigation of different usage of pulsed energy were discussed. It was shown that increasing in pulsed energy contributes to the burnt through defects since metals are evaporated with high energy absorption in investigating the pulsed energy without using shielding gas and angle of irradiation.

DOI: 10.1051/matecconf/20179001032

Ep,(J)	Cross section	Results
0.5	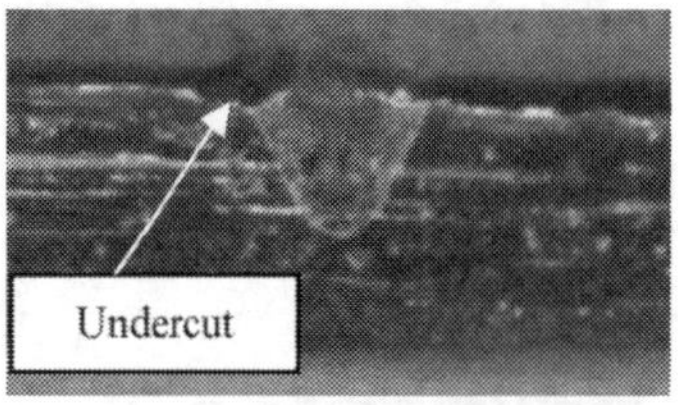	Half Penetration with small undercut.
0.7		Full penetrate, undercuts and burnt through present.
0.9	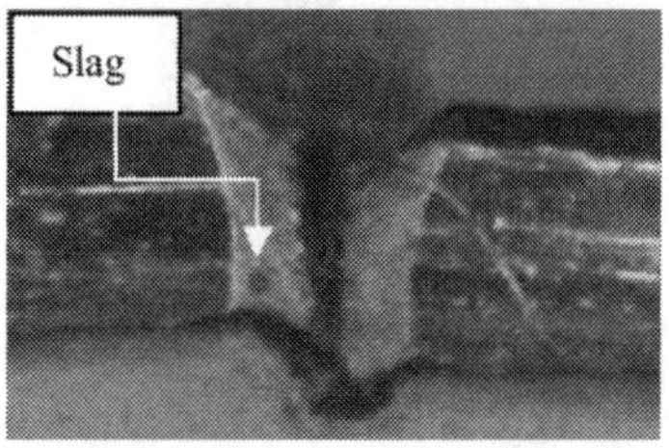	Sample distorted and slag inclusion presents.
1.2	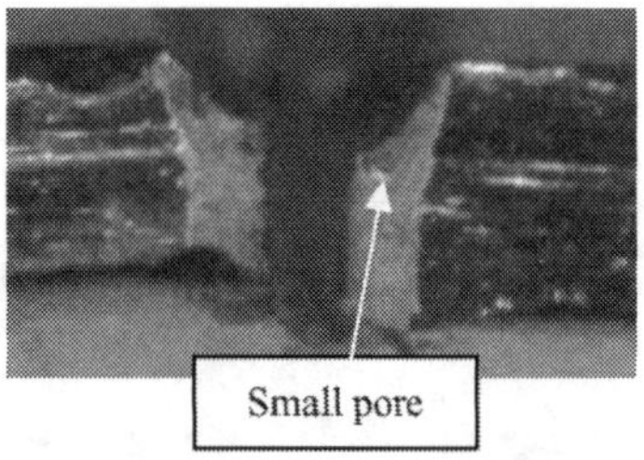	Larger burnt through area, a lot of metal evaporation, small pore observed.
1.4	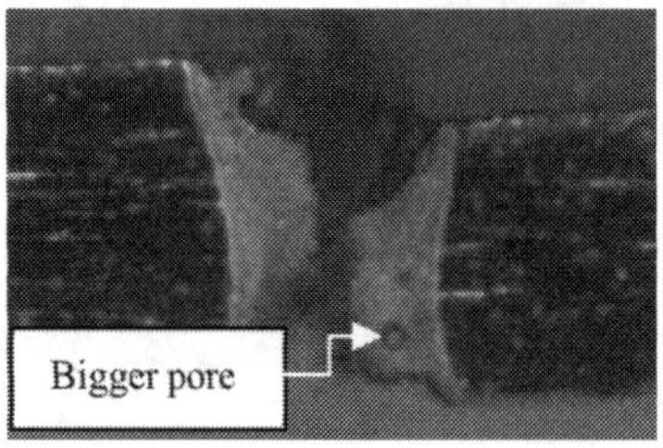	Large burnt through area, bigger pore formations.

Fig. 4. BOP welded AZ31B with different pulsed energy.

From figure above, it was found that AZ31B was evaporated with the increasing pulsed energy. This factors also was caused by zero usages of shielding gas where the metals was easily evaporated due to this metals was easily burned. From observation, only 0.5 J pulsed energy shows better weld result compared to others weld products with small undercut defects. For sample with 0.9 J pulsed energy, slag inclusion presented in the weld due to the improper samples cleaning before the experiments was conducted. Porosity was observed starting from the pulsed energy of 1.2 J and became bigger at 1.4 J. Pores presented in the weld due to the high rates of burning factors during welding process. For the different shielding gas experiment, Figure 5 shows the comparison between samples welded with Ar and N_2 shielding gas with same value of flow rates. It was observed that sample which laser welded with Ar gas produces porosity in the weld region while pores do not present in the weld region to sample that welded with N_2 gas. However, both samples experienced hot crack at the centre of weld zone (WZ) due to the rapid cooling after being laser welded. Due to the same laser energy absorption on the metal surfaces, underfill defect still showed up due to the metal evaporations occurred during welding process.

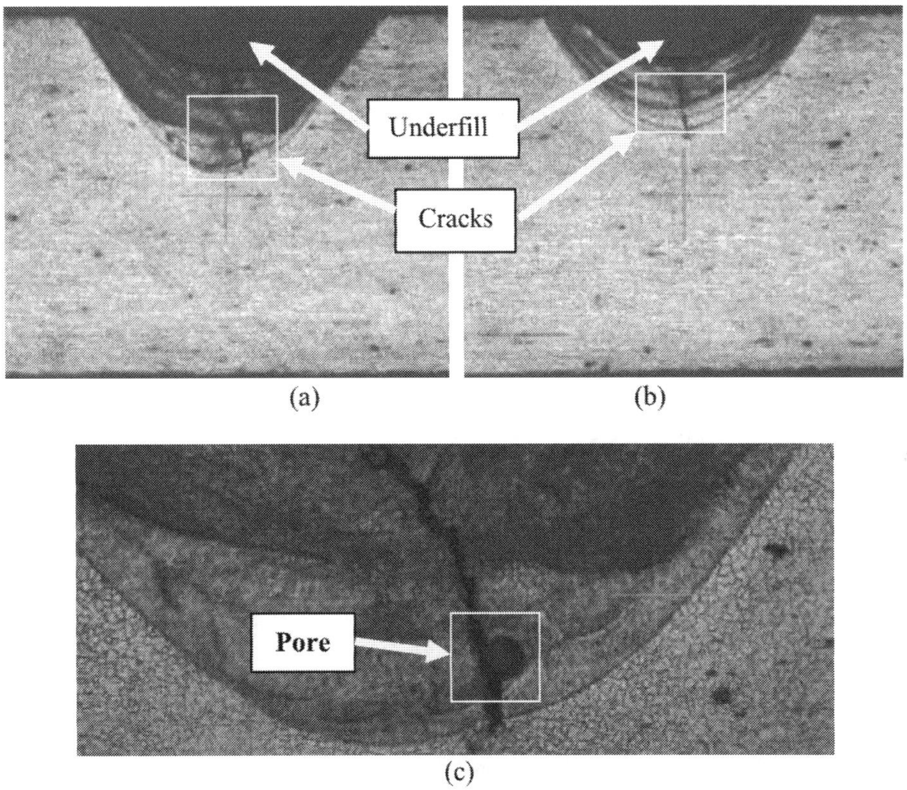

Fig. 5. Cross section of sample welded with (a) Argon (b) Nitrogen (c) welded zone of sample welded with Ar, 50 X magnifications.

As mentioned in methods section, an experiment to study the effect of inclination angle of irradiation (AOI) also has been operated. Using same parameter as before, range of AOI selected was 1° to 5°.

AOI, (°)	Cross section microstructure	Results
1	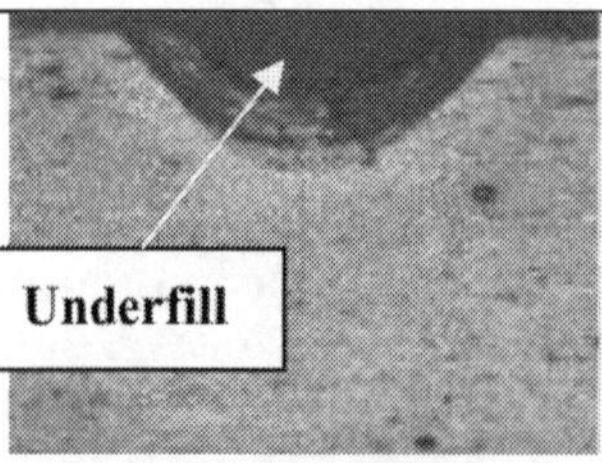	Weld Depth= 0.333 mm Underfill depth = 0.143 mm
2	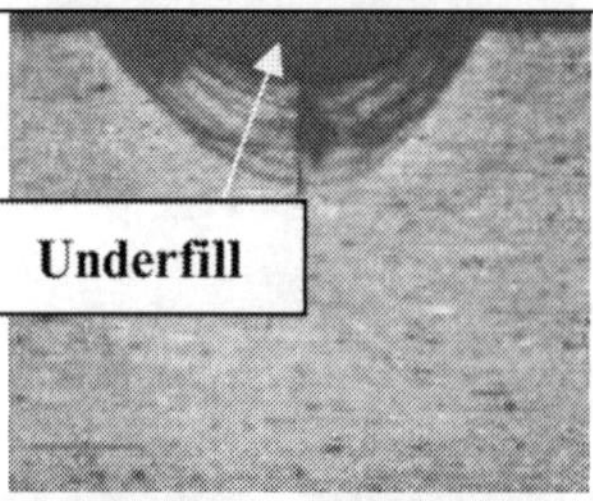	Weld Depth= 0.326 mm Underfill depth = 0.095 mm
3	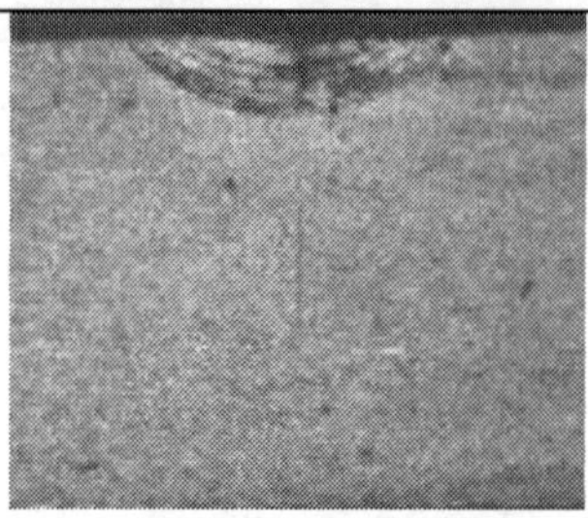	Weld Depth= 0.160 mm Underfill depth = 0.00 mm
4	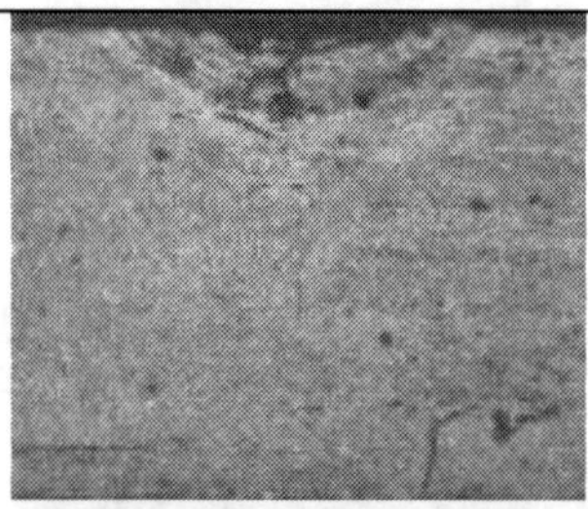	Weld Depth= 0.155 mm Underfill depth = 0.00 mm
5	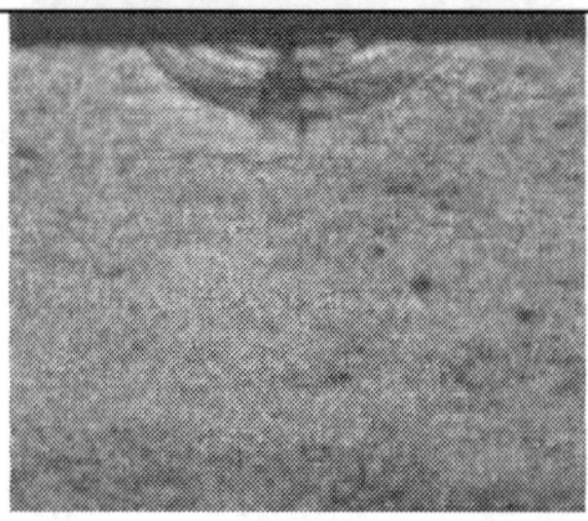	Weld Depth= 0.150 mm Underfill depth = 0.00 mm

Fig. 6. Microstructure of BOP welded AZ31B with different AOI.

Figure 6 above shows the microstructure with 5 x magnifications for each AOI used. It was observed that the smallest angle of irradiation, the underfill depth became more. It was observed that when using the AOI bigger than 3°, the underfill defect was successfully eliminated although the penetration depth was reduced by 0.05 mm. Porosity also was not observed in the microstructure as N_2 gas was used. Nitrogen was proved to be better selection as the shielding gas to weld Mg alloys compared to Argon gas.

3.3 Lap joint of thin sheet AZ31B

Similar weld of AZ31B was successfully welded with fillet lap joint method where the centre of the laser beam with defocused position was used to produce the weld. Figure 7 shows the macrostructure image of fillet lap joint with different value of pulsed energy. From this figure, it was observed that fillet lap weld when using pulsed energy at 2.0 J produced better weld joint compared to 2.4 J pulsed energy. It was proved that 2.0 J was suitable to make fillet weld joint at defocused of +26 mm. Crack was not observed at the FZ for sample welded with 2.0 J compared to that welded with 2.4 J. Figure 8 shows the microstructures of a small portion in sample from Figure 7 (a).

From Figure 8 (a), it was observed that the grain boundaries were transform into smaller equiaxed dendritic grains where it represents the α-Mg grain structure. In heat affected zone (HAZ) region, the hexagonal structures have bigger grain size with 16.5 μm compared to its structure in the unaffected BM with 8.0 μm as shown in Figure 9 (b) where there was a double increments due to more heat absorbs since HAZ is a region of base material near the interface to the fusion zone. From previous research, the basic shape of BM in AZ31B was hexagonal closely packed (hcp) with the α-Mg and β-phase of $Mg_{17}(Al,Zn)_{12}$ precipitates where α-Mg grain size was about 7.7 μm. Meanwhile, β-phase of $Mg_{17}(Al,Zn)_{12}$ ellipsoidal particle with maximum length of 200 nm [16]. Typically, β-$Mg_{17}(Al,Zn)_{12}$ precipitates were coarsening with the increasing grain size of α-Mg which affected by the heat input. From Figure 8 (a) also, it was found that fusion zone (FZ) was characterized by α-Mg grain phase but with smaller grain sizes with value of 4.2 μm. In FZ also, voids presented due to insufficient shielding gas flow rates supplied during laser welding process.

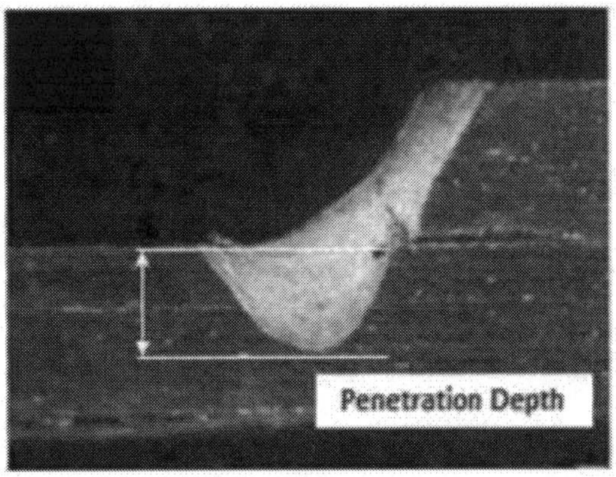

(a)

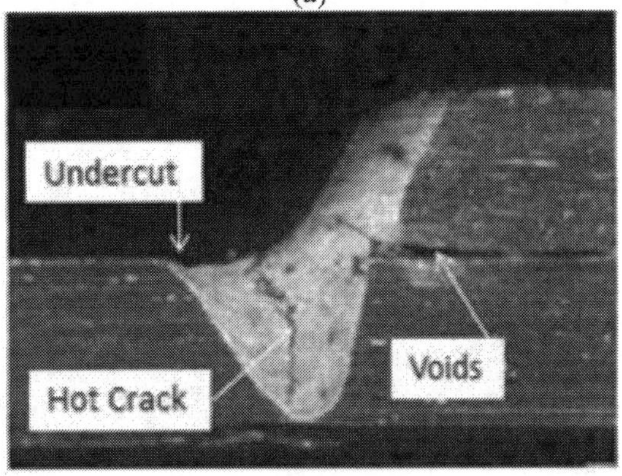

(b)

Fig. 7. Cross-section images of fillet welded with (a) Ep = 2.0 J (b) Ep = 2.4 J.

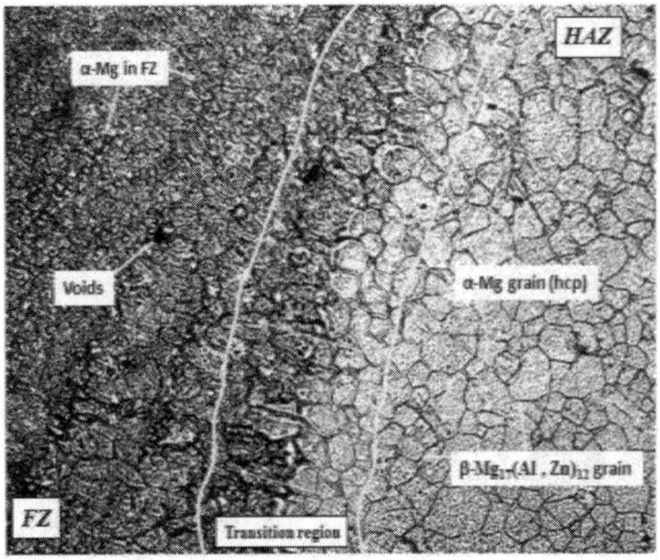

(a)

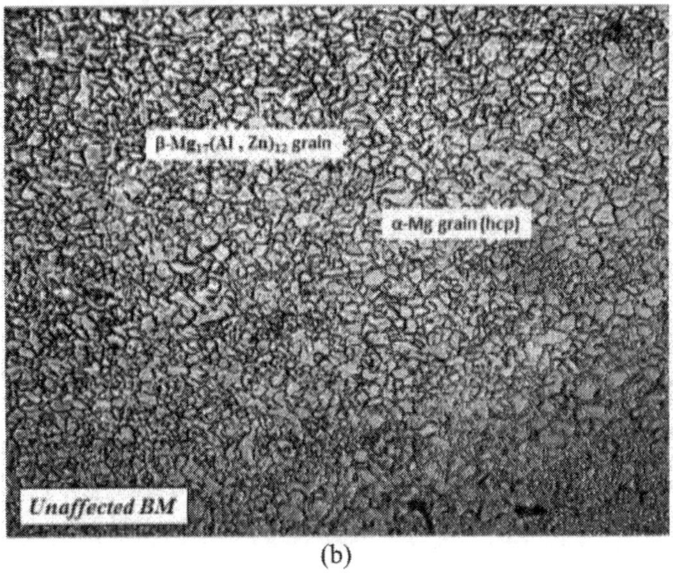

(b)

Fig. 8. Microstructure image of (a) FZ and HAZ (b) unaffected BM.

4 Conclusions

The influences of low power fibre laser welding parameters, such as pulsed energy, angle of irradiation, and types of shielding gas, on the weld penetration and microstructure of BOP and fillet lap welding of AZ31B magnesium alloys with thickness of 0.6 mm using pulsed low power fibre laser have been investigated. The following conclusions can be drawn for this present research. Firstly, lower pulsed energy could give better result in making the weld joints. So that, low power laser machine can be used to weld this thin sheets AZ31B. Secondly, increasing the angle of irradiation (AOI) can eliminates the underfill defects in weld joint since it reduce the metal evaporation during laser welding processes. Thirdly, fusion zone (FZ) was characterized with smaller α-Mg grain structure which relatively smaller compared to BM where at this region, the grain also consists of β-phase of $Mg_{17}(Al,Zn)_{12}$ precipitates which the grain size relatively higher than that of α-Mg grain. Lastly, pulsed wave (PW) was successfully created the fillet lap joint of 0.6 mm thick of AZ31B with fewer defects observed. In addition, this joint could be improved with the aid of design of experiment (DOE) such as response surface methods (RSM).

The author would like to thank the technical staff of Universiti Malaysia Pahang for all of the work by providing laboratory facilities within which the experiments were conducted. Also, financial support by the Ministry of Education through Universiti Malaysia Pahang for research grant RDU140118 is gratefully acknowledged.

References

1. E. Assunção, L. Quintino, and R. Miranda, "Comparative study of laser welding in tailor blanks for the automotive industry," *The International Journal of Advanced Manufacturing Technology,* vol. 49, pp. 123-131, 2009.
2. G. V. Moskvitin, A. N. Polyakov, and E. M. Birger, "Application of laser welding methods in industrial production," *Welding International,* vol. 27, pp. 572-580, 2013.
3. J. Ahn, L. Chen, C. M. Davies, and J. P. Dear, "Parametric optimisation and microstructural analysis on high power Yb-fibre laser welding of Ti–6Al–4V," *Optics and Lasers in Engineering,* vol. 86, pp. 156-171, 2016.
4. R. Lai, Y. Cai, Y. Wu, F. Li, and X. Hua, "Influence of absorbed nitrogen on microstructure and corrosion resistance of 2205 duplex stainless steel joint processed by fiber laser welding," *Journal of Materials Processing Technology,* vol. 231, pp. 397-405, 2016.
5. I. Bunaziv, O. M. Akselsen, A. Salminen, and A. Unt, "Fiber laser-MIG hybrid welding of 5mm 5083 aluminum alloy," *Journal of Materials Processing Technology,* vol. 233, pp. 107-114, 2016.
6. M. Harooni, B. Carlson, and R. Kovacevic, "Detection of defects in laser welding of AZ31B magnesium alloy in zero-gap lap joint configuration by a real-time spectroscopic analysis," *Optics and Lasers in Engineering,* vol. 56, pp. 54-66, 2014.
7. M. Harooni, J. Ma, B. Carlson, and R. Kovacevic, "Two-pass laser welding of AZ31B magnesium alloy," *Journal of Materials Processing Technology,* vol. 216, pp. 114-122, 2015.
8. M. Ishak, K. Yamasaki, and K. Maekawa, "Lap Fillet Welding of Thin Sheet AZ31 Magnesium Alloy with Pulsed Nd:YAG Laser," *Journal of Solid Mechanics and Materials Engineering,* vol. 3, pp. 1045-1056, 2009.
9. M. Ishak, K. Yamasaki, and K. Maekawa, "Lap Fillet Laser Welding of AZ31B Thin Sheet Magnesium Alloy using Silver Nanoparticles," *Journal of Solid Mechanics and Materials Engineering,* vol. 4, pp. 51-62, 2010.
10. Z. D.T. YAN Yong, QIU Cheng, ZHANG Wen, "Dissimilar friction stir welding between 5052 aluminum alloy and AZ31 magnesium alloy," *Transaction of Nonferrous Metal Society China,* vol. 20, pp. 619-623, 2010.
11. X. Cao, W. Wallace, J. P. Immarigeon, and C. Poon, "Research and Progress in Laser Welding of Wrought Aluminum Alloys. II. Metallurgical Microstructures, Defects, and Mechanical Properties," *Materials and Manufacturing Processes,* vol. 18, pp. 23-49, 2003.
12. X.-N. Gu and Y.-F. Zheng, "A review on magnesium alloys as biodegradable materials," *Frontiers of Materials Science in China,* vol. 4, pp. 111-115, 2010.
13. S. Fleming, "An Overview of Magnesium based Alloys for Aerospace and Automotive Application," MASTER OF ENGINEERING IN MECHANICAL ENGINEERING, Faculty of Rensselaer Polytechnic Institute, Rensselaer Polytechnic Institute, 2012.

14. H. Hiraga, T. Inoue, S. Kamado, and Y. Kojima, "Effects of the shielding gas and laser wavelength in laser welding magnesium alloy sheets," *Welding International,* vol. 16, pp. 442-450, 2002.
15. S. Katayama, H. Nagayama, M. Mizutani, and Y. Kawahito, "Fibre laser welding of aluminium alloy," *Welding International,* vol. 23, pp. 744-752, 2009.
16. R. S. Coelho, A. Kostka, H. Pinto, S. Riekehr, M. Koçak, and A. R. Pyzalla, "Microstructure and mechanical properties of magnesium alloy AZ31B laser beam welds," *Materials Science and Engineering: A,* vol. 485, pp. 20-30, 2008.

Assessing retro-reflective markers (RRMs) usage on heavy vehicles with respect to MS 828:2011

Mohd Amirudin M.R[1,*], *Ahmad Noor Syukri* Z.A[1], and *Kak* D-Wing[1]

[1]Malaysian Institute of Road Safety Research (MIROS), 43000 Kajang, Selangor, Malaysia

Abstract. Malaysia is continuously developing and enhancing the safety system and policies for road users. One of the essential elements of safety is a good marking and thus the visibility of vehicles on the road. Since 2011, MS 828:2011 has been gazetted to guide industry on the right specification to follow with regards to RRMs. However, the RRMs that being used by heavy and long vehicle including in the current market were not up to the standard. Thus, the objectives of this study is to identify the current status of RRMs in Malaysia besides to measure the photometric values of current RRMs and to determine the usage rate of substandard RRMs on heavy vehicles and current market. Market surveillance and sample testing were conducted to determine on current RRMs usage and to analyse the RRMs with respect to MS 828:2011. As a result, 62% of RRMs condition on heavy vehicles are still not in a good condition or covered by dust. Also, only 4% of the RRMs sample complied to MS 828:2011. From this paper, it can be concluded that majority or 96% of the RRMs in current market and have been used by current heavy vehicles are substandard markings.

1 Introduction

Every year more than 6000 people are killed in road accidents in Malaysia. Approximately 1000 of these fatalities are car occupants and motorcyclists, killed in collision involving rigid and articulated type's trucks and rear-end crashes are one of the frequently occurring types of crashes [1]. Malaysia is continuously developing and enhancing the safety system and policies for road users to reduce the number of crashes and fatalities on the road. One of the essential elements of safety is a good marking and thus the visibility of vehicles on the road. In order to better visibility of long, heavy goods vehicles and their trailers, etc., these vehicles have to be marked with effective RRMs. This marking has to be from highly reflective materials which can guarantee better visibility during night and poor weather conditions.

Since 2011, MS 828:2011 Malaysian Standard has been gazetted to guide industry on the right specification to follow with regards to RRMs. However, industry readiness was compromised at the time of implementation in terms of product compliance and test

* Corresponding author: mohdamirudin@miros.gov.my

facility. Fast forward to 2014, the test facility was finally ready and the industry was eager to obtain certification for their products.

The implementation of MS 828:2011 was scheduled on 1st July 2014 for all new trucks and trailers. A 6-month grace period was given for all fleet operators to implement such installation and for awareness purposes. However, the implementation for old heavy and long vehicles that was scheduled by early 2015 will be implemented by 1st July 2016 stage by stage.

Currently, the RRMs that being used by heavy and long vehicle including in the current market were not up to the standard with respect to colorimetric, photometric, dimension and etc. Before full implementation, this paper will assess the compliant RRMs usage on heavy goods vehicles towards MS 828:2011 in Malaysia. Objectives of this study are to identify the current status of RRMs in Malaysia with respect to MS 828:2011 besides to measure the photometric values of current RRMs. The usage rate of substandard RRMs on heavy goods vehicles and current market also will be determined.

1.1 Rear-end crashes and conspicuity issues in Malaysia

Based on data from fatal crashes in Malaysia involving heavy commercial goods vehicles (HCGV), the most common types of crash configurations for HCGVs are head-on collisions (33%) and rear impact collisions (28%) as shown in Figure 1 [2]. Furthermore, for rear-end collisions involving HCGVs, the amount of light was not significant as the proportion of crashes that occurred during daylight and in the different categories of dark were 51% and 49%, respectively [2].

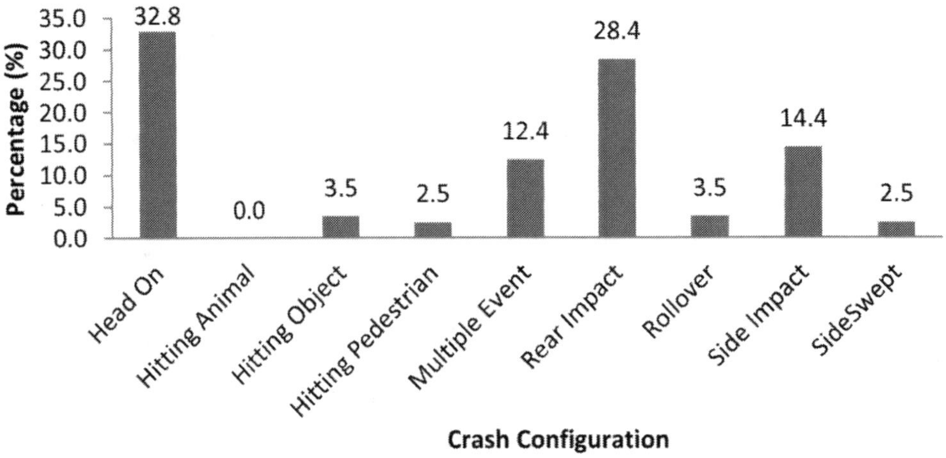

Fig. 1. Investigated HCGV fatal cases by crash configuration [2].

1.2 Conspicuity markings

In rear-end crashes, conspicuity plays a significant role, especially at night. To improve heavy goods vehicle visibility, heavy vehicles and their trailers need to be equipped with conspicuity markings, such as RRMs, and maintained in a good condition. The purpose of these markings are to increase the visibility of heavy goods vehicles to other motorists, especially in low light or and bad weather conditions. These markings brightly reflect the light emanating from the headlights of other motorists to warn them that they are closing in on a heavy vehicle. The alternating red and white on the rear of a HGV flags it as a heavy trailer and at the same time helps other road users gauge their distance and rate of approach [3].

In the United Kingdom (UK), the Road Vehicles Lighting Regulations (RVLR) (1989) require certain motor vehicles with a gross vehicle weight exceeding 7.5 tonnes to be fitted with retro-reflective rear marker plates in compliance to UNECE Regulation 70.00 (R70). The latest version of this regulation is R70.01, however in the UK R70.00 is still permitted.

Some vehicle operators also fit conspicuity markings to their vehicles. These are vertical and horizontal lines of retro-reflective material which emphasise the length and shape of the vehicle and also warn other drivers of their presence. UNECE Regulation 104 (R104) provides a technical specification for these conspicuity markings and their use on vehicles in the UK is optional. UNECE Regulation 48 (R48) is a European regulation that sets out harmonised installation requirements for vehicle lights and reflectors. Recently R48 was amended to mandate R104 specification conspicuity markings on certain new heavy goods vehicles and trailers [4].

1.3 RRMs in Malaysia

In Malaysia, all heavy goods vehicles, especially lorries and trailers, are required to use RRMs to provide other drivers with information and warnings. Two types of rear and side marking are specified by Malaysian Standard (MS) 828:2011 as shown in Figure 2 [5]. The dimensions of the plate, retro-reflective and coloured areas, and the proportions of the letters shall comply with the regulations currently in force in Malaysia.

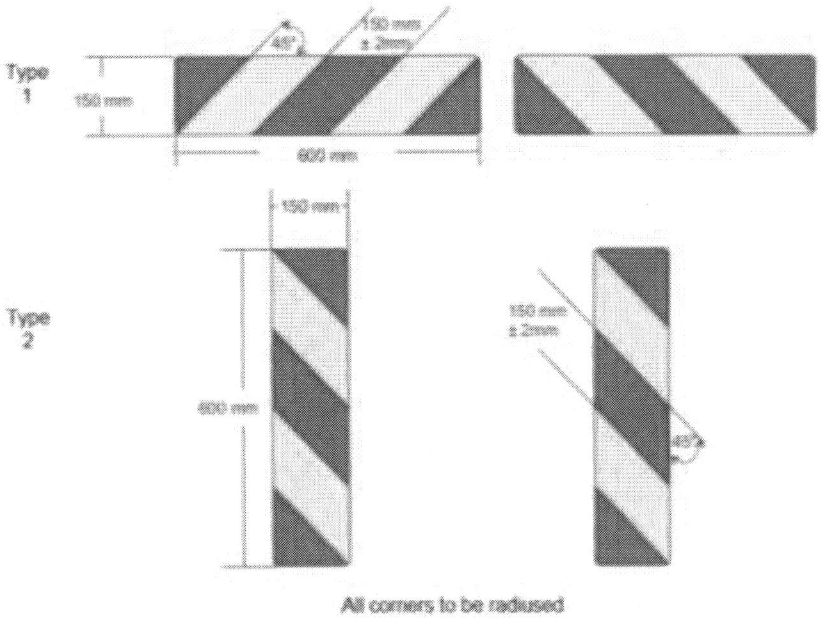

Fig. 2. Types of rear and side marking specified in Malaysia [5].

R104 provides a technical specification for high quality markings, which can be used to identify the contours of large vehicles. Meanwhile, MS 828:2011 specifies the requirements for marking with retro-reflective materials to be fitted to the rears and sides of heavy and long vehicles and trailers. MS 828:2011 was developed with reference to R70 and R104.

DOI: 10.1051/matecconf/20179001033

The specified requirements include design, colorimetric, photometric, physical and mechanical properties of retro-reflective markers

Also, Table 1 below shows the minimum values of the coefficient of retro-reflection R' or photometric value for, yellow, white and red [5]. When the marking was achieved the minimum of photometric value as in Table 1, that marking was useful and were intended to make the vehicle more conspicuous to other drivers, so that they are aware of the presence of a large, potentially slow moving vehicle especially on the road.

Table 1. Minimum photometric values for yellow and red colour [5].

Observation angle, α^o	**Entrance angle, β^o**				
α = 0.33° (20')	β_1	0	0	0	0
	β_2	5	30	40	60
Colour					
Yellow		300	130	75	10
White		450	200	90	16

Observation angle, α^o	**Entrance angle, β^o**				
α = 0.33° (20')	β_1	0	0	0	0
	β_2	5	20	30	40
Colour					
Red		120	60	30	10

1.4 Effectiveness of RRMs

The usage of retro-reflective tapes on trucks was effective and significantly reduced side and rear impacts [3]. The effectiveness of retro-reflective tape was assessed in four categories of dark conditions, namely "dark-not-lighted," "dark-lighted," "dusk" and "dawn," and it was found to reduce overall side and rear impacts into heavy trailers by 29%. In "dark-not-lighted" conditions the use of retro-reflective tape reduced side and rear impacts by 41% [3].

In a study of night-time and day time accidents between a test group comprising 1000 vehicles equipped with contour markings and a control group of 1000 vehicles without such measures. After 2 years of testing the markings, 95% of night-time collisions could have been avoided if trucks in the control group had retro-reflective visibility markings. Increased truck visibility due to reflective outline markings saw a 41% reduction of rear-end crashes and a 37% decrease in side impacts [6]. Also the risk of an accident between truck and car was 30 times greater if the truck was without conspicuity markings [6].

 DOI: 10.1051/matecconf/20179001033

2 Methodology

2.1 Market surveillance

Fifty sample of RRMs were purchased randomly at hardware and accessories shops. All the shops are located in Klang Valley Area. The purchased price of all the samples were recorded inclusive of Government Service Tax (GST). From these 50 sample, two samples were approved by SIRIM on the compliant towards MS 828:2011 and to be used as the references sample. Then, the dimension of these sample were measured.

2.2 Sample testing

Testing was carried out to profile the photometric values of RRMs sample for both yellow and red colour respectively by using photometer equipment as shown in Figure 3. All the sample from market surveillance were tested. The photometer was able to measure retro-reflectivity value in lux unit.

In addition, the sample from real world (lorries) were also collected and compared with the sample from current market. Those sample were collected randomly from 100 lorries which were stopped at Sungai Buloh Rest Area. The old RRMs on the lorries were exchanged with the new RRMs that complied to MS 828:2011. The old RRMs were cleaned and used as the sample for testing to get the photometric values reading. The height of those samples from the ground clearance for each lorries was also measured besides the dimension of the sample itself.

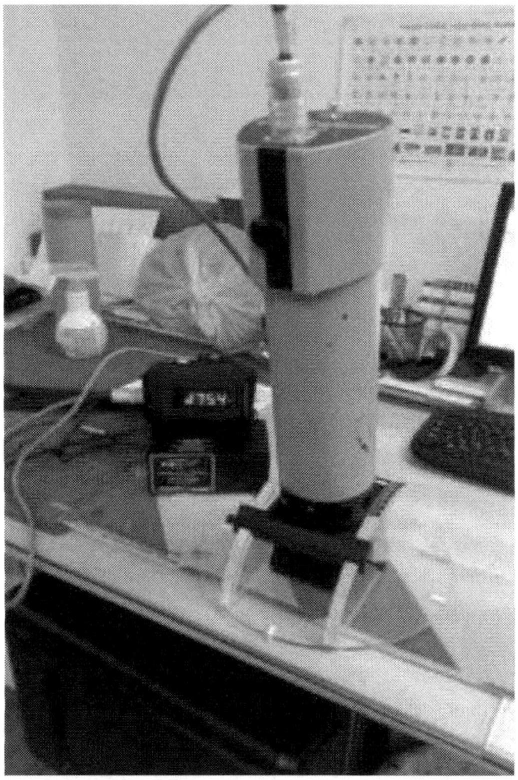

Fig. 3. Photometer equipment for testing.

The testing was conducted twice for two different types of sample; current market and real world. Before conducting the test, the photometer was calibrated to get the accuracy of the reading. After that, the angle of the photometer was adjusted to 5° as required by the MS 828:2011 [5] in order to get the actual reading of photometric values of all samples. The testing for each samples including both colour was repeated for three times in order to get the average reading.

3 Results and discussion

3.1 Market surveillance and sample testing

3.1.1 Sample price

From the market surveillance, the average price of the sample is about RM26.30. The lowest price is RM8.50 and the highest price is RM 85.00. The different between the highest and lowest price is about RM77.50. However, the two highest prices; RM65.00 and RM85.00 are the reference RRMs that complied to MS 828:2011.

3.1.2 MS 828:2011 compliant (dimension)

Dimension of the sample is another requirement for the compliance to MS 828:2011. Figure 4 shown the percentage of RRMs dimension that comply to the standard. Only 20% of installed RRMs complied to the standard. This shows that the awareness level on MS 828:2011 especially among lorry owners and fleet operators is low and need to be improved.

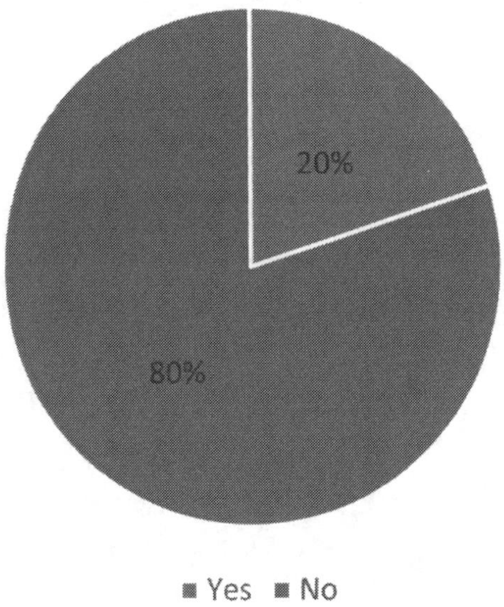

Fig. 4. MS 828:2011 compliant (dimension).

3.1.3 MS 828:2011 compliant (photometric value)

Photometric value is one of the specifications that has been considered on the compliant towards MS 828:2011. Table 1 above shown the minimum photometric values to be achieved for both yellow and red colour are 330 and 120 respectively. However, Figure 5 and 6 below show that only 4% of the RRMs sample that have been analysed complied to MS 828:2011 which means that the other sample are counterfeit markings. The usage of counterfeit markings may not fully functional to reflect the light emanating from the headlights of other motorists to warn them that they are closing in on a heavy vehicle. This is dangerous to the other road users who are travelling at night time and they are at risk to be involved on the rear end crashes especially with the lorries and trucks that are using non-compliance markers.

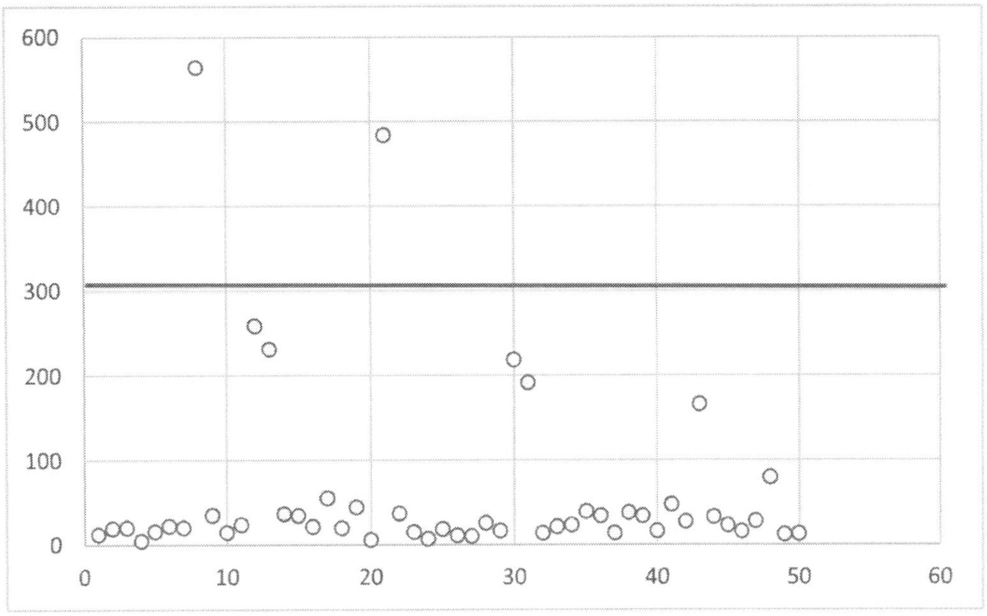

Fig. 5. MS 828:2011 compliant (Photometric value of yellow colour).

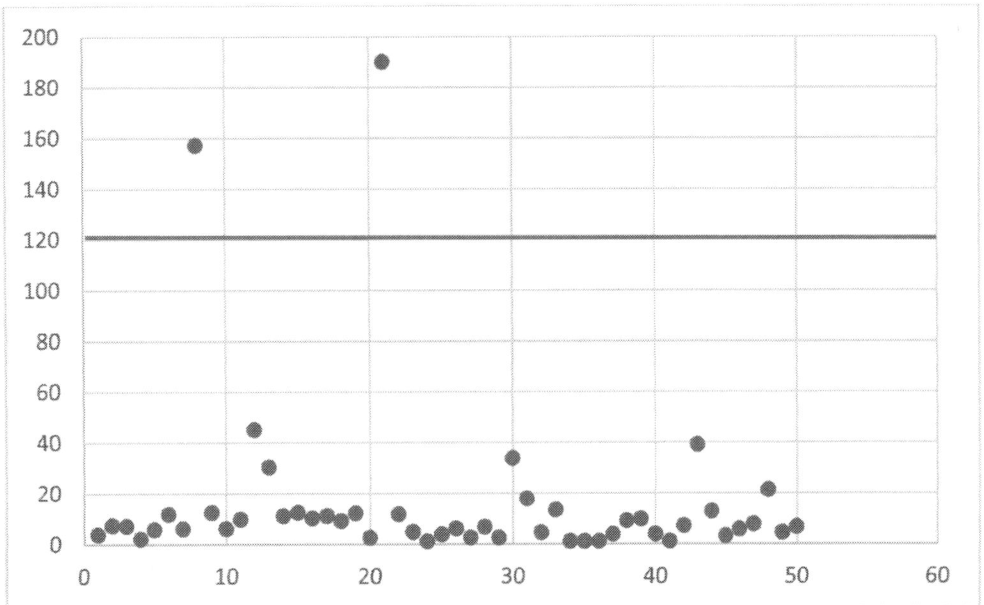

Fig. 6. MS 828:2011 compliant (Photometric value of red).

3.2 Real world sample testing

At Sungai Buloh Rest Area, 100 sample are collected from both lorries and trucks. Table 2 shows the type of lorries involved for sample collection and the condition of old RRMs sample which collected from the trucks and lorries.

Table 2. RRMs condition on trucks and lorries.

Sample, N	Lorry Type	RRMs Condition		
100	Rigid	Good	Covered by Dirt	Missing Segment
	50 (50.0%)	10 (20.0%)	33 (66.0%)	7 (14.0%)
	Articulated			
	50 (50.0%)	8 (16.0%)	29 (58.0%)	13 (26.0%)

3.2.1 MS 828:2011 compliant (dimension)

Figure 7 shows the number of RRMs installation compliant to the dimension from MS 828:2011. Only 19% of installed RRMs are found to be compliant and this percentage is almost same with the percentage from current market sample which is only 20%.

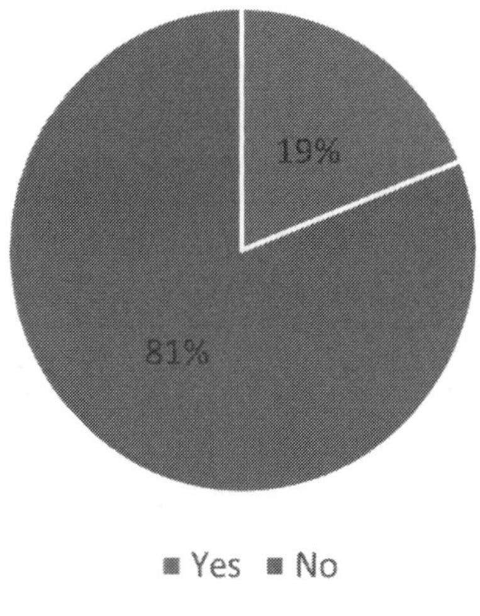

Fig. 7. MS 828:2011 compliant (dimension).

3.2.2 MS 828:2011 compliant (photometric values)

Besides the height of ground clearance and marker dimension, the photometric values are the other for the compliance towards MS 828:2011. However, both Figure 8 and 9 show that only 2% of yellow colour of the RRMs sample that have been analysed complies to

MS 828:2011 while none of red colours complies to the standard. None of the samples from real world complies to the MS 828:2011 or on the other word, all the sample are counterfeit markings. The usage of counterfeit markings may not fully functional to reflect the light emanating from the headlights of other motorists to warn them that they are closing in on a heavy vehicle. This is dangerous to the other road users who are travelling at night time and they are at risk to be involved on the rear end crashes especially with the lorries and trucks that are using non-compliance markers.

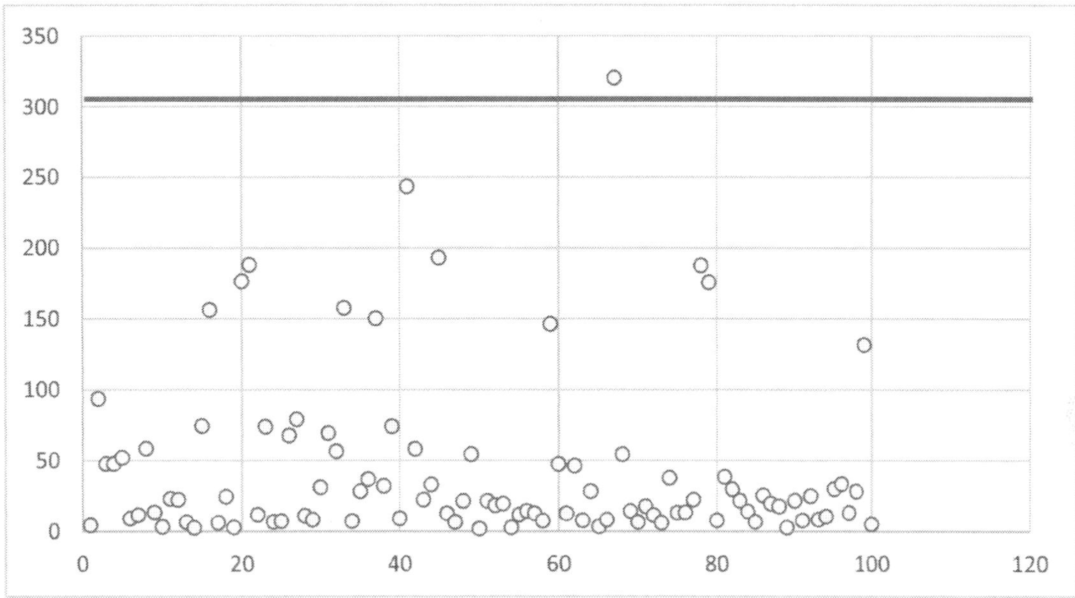

Fig. 8. MS 828:2011 compliant (Photometric values for yellow colour).

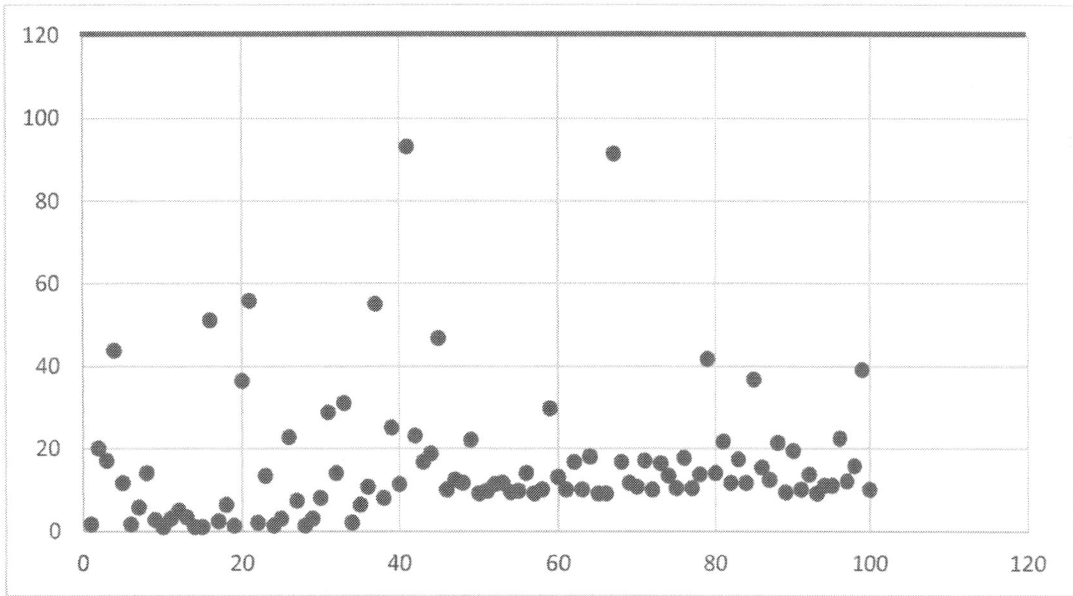

Fig. 9. MS 828:2011 compliant (Photometric values for red colour).

4 Conclusions

In this paper, several conclusions are outlined as follows:

- From the market surveillance, the average price of the sample is about RM26.30 while the lowest price is RM8.50 and the highest price is RM 85.00.

- 62% of the RRMs on both types of lorries are found to be covered by dirt. When the RRMs are not in good condition such as covered by dirt, the marking will not be effective thus will not fully function to reflect the markers.

- For the sample from current market, only 4% of the sample complies to MS 828:2011 while for the sample from lorries (real world), none of the sample complied to MS 828:2011.

- After the analysis is carried out, it is found that the usage rate for substandard RRMs on current market and real world are 96% and 100% respectively. It shows that the usage rate of substandard markers especially on the current market and real world in Malaysia is very high and need to be enforced by related agencies.

References

1. M. M. Huzaifah, *Rear End Crash Compatibility Between Cars and Trucks in Malaysia: A Preliminary Study*, presented at the The 8th Int. Forum of Automotive Traffic Safety (INFATS), Wuhu, China (2010)
2. A. N. Syukri, *Heavy Commercial Goods Vehicle Crashes in Malaysia,* Malaysian Institute of Road Safety Research (MIROS), Malaysia (2013)
3. C. Morgan, *The Effectiveness of Retro Reflective Tape on Heavy Trucks,* National Highways Traffic Safety Authority, Washington DC Report No. DOT HS **809 222** (2001)
4. D. Babic, *Impact of Heavy Vehicle Visibility on Traffic Safety,* presented at the The International Scientific Conference - ZIRP 2014, Zagreb, Crotia (2014)
5. Department of Standards Malaysia and SIRIM Berhad, *Malaysian Standard (MS828:2011), Road Vehicles - Rear and Side Marking - Specification (Second Revision),* ed. Malaysia (2011)
6. P. D.-I. Schmidt-Clausen, *Contour Marking of Vehicles,* H.J. Laboratory of Lighting Technology, Darmstadt University of Technology Final Report FO 76 / 00, **8** (2000)
7. Amirudin M.R, *Rear-end Markers (REMs) Issues involving Trucks and Lorries in Malaysia*, Malaysian Institute of Road Safety Research (MIROS) (to be published)
8. A Berces, *Improving Road Safety by Increased Truck Visibility*, 3M Traffic Safety Systems Division, ARRB Conference, 25th, 2012, Perth, Western Australia, Australia (2012)
9. M.R.J. Baldock, A.D. Long, V.L. Lindsay, A.J. McLean, *Rear End Crashes*, Report No. CASR018, Adelaide, Australia: Centre for Automotive Safety Research, the University of Adelaide (2005)
10. Clare Lawton, John Richardson, Ruth Welsh, *Assessment of The Safety Benefit of Retro Reflective Markings On HGVs and Buses*, Report for Department for Transport USA, May (2005)

MATEC Web of Conferences **90**, 01034 (2017) DOI: 10.1051/matecconf/20179001034
AiGEV 2016

Real time emergency auto parking system in driver lethargic state for accident preventing

Mohammed Hayyan Alsibai[1,*], *Syafiq Fauzi* Kamarulzaman[2] , *Hossam Adden* Alfarra[1] and *Yasir Hashim* Naif[1]

[1]Faculty of Engineering Technology, Universiti Malaysia Pahang, 26300, Kuantan, Pahang, Malaysia
[2]Faculty of Computer Systems & Software Engineering, Universiti Malaysia Pahang, 26300, Kuantan, Pahang, Malaysia

Abstract. This paper is presenting a safety driving and accident preventing system which uses a vision sensor to detect driver drowsiness and lethargic states. The system notifies the driver in dangerous situations. Moreover, in case the driver is unable to conduct safe driving, emergency parking system is to be activated. The system comprises two stages. First is a drowsiness detection stage which uses a smartphone or a tablet computer as a processing unit. The second stage is the vehicle emergency parking control system which uses a microcontroller unit (MCU). The MCU is connected to an alarm system, hazard lights and a vehicle control interface. The experiment results showed realistic real time responses. Drowsiness detection processing time average is about 480 ms / frame. Alarming system is responding perfectly within 500 ms. Simulation results illustrate the effectiveness of the developed schemes for the auto parking system in real time. The average time from drowsiness detection to fully parking, if the vehicle is moving at the speed of 100 km/s, is about 15s.

1 Introduction

Traffic accidents are serious global problem. Studies show that sleep and fatigue are involved in around 10 percent of accidents [1, 2]. According to National Highway Traffic Safety Administration (NHTSA), annually from 2009 to 2013, on average there were over 72,000 police-reported crashes involving drowsy drivers. Therefore, it is suggested by many road safety related entities to develop driver alarm and assistance systems to reduce the number of accidents. For example, Malaysian Institute of Road Safety (MIROS) introduced what is called "research pillars". Pillar 3 is "Safer Vehicles" [3].

Driver fatigue detection has been studied during the last two decades by many researchers. The causes of fatigue can be physical, physiological, or psychological [4]. Unfortunately, there is no standard fatigue measurement, because the direct measures are few. A recent comprehensive review on various commercial and academic literatures is available in [5]. A quick summary is presented in this introduction. Most of the research methodologies can be divided into four categories: Methodologies which are focusing on the driver's current state, driver performance, a combination of the driver's current state

* Corresponding author: mhdhayyan@gmail.com

and driver performance and methodologies focusing on vehicle's current state [5]. Furthermore, few commercial products are already available in markets. For example, the Lumeway eye alert Distracted Driving and Fatigue Warning Systems, Toyota driver attention monitor, DADSTM (Driver Alertness Detection System^{TM}), Fatigue Monitor, and many other systems. Few systems use methods depending on inputs from devices which are not visual sensors, like [6]. But most of the algorithms are depending on visual information.

The outcome of most of the systems is an audible alarm. Few number of researchers suggested different ways for overcoming drowsiness. For example, Singh H. et al. [7] proposed a system which gives warning output in form of sound and seat belt vibration. The alarm is deactivated manually. In 2009, Mercedes-Benz unveiled a system called "Attention Assist" [8]. The system monitors the driver's fatigue level and issues a visual and audible alarm. The significant feature in this system is the linking with the vehicle's navigation system. This allows the system to tell the driver where coffee is available. In 2016, the latest version of "Attention Assist" not only makes the driver aware of behaviour that may be influenced by drowsiness, but also helps driver to correct minor errors [8].

The practical use and efficiency of the devices in preventing accidents are still under inspection. United Kingdom royal society for the prevention of accidents (RoSPA) published a literature review on driver fatigue and road accidents [9]. The study investigated number of technical devices. The study concluded that such devices may prove beneficial, but there are concerns that drivers would rely on them instead of managing themselves for safety. The study raised the question: "Drivers are normally aware that they are sleepy, so why is a device necessary to tell them so?" Therefore, more efficient actions should take place to achieve the main goal of preventing and reducing accidents. In this paper a new emergency auto parking system is proposed. To the best of our knowledge, there are studies for autonomous cruise control system [10], but no studies are available for parking on road side in case of emergency. The proposed system is activated when the driver is unable to conduct safe driving. The purpose of this system is to automatically control the speed and the steering of a vehicle to park it safely on the emergency lane.

Smart systems for vehicle parking are not new field of research. They are attracting a great deal of attention recently. Parking systems in general deals with two categories: Garage-parking problem and parallel-parking problem [11]. Figure 1 (a) and (b) shows the difference between garage-parking and parallel-parking. The target of such systems is to help the driver to park the vehicle in a limited parking space [12]. The proposed system in this paper can be categorized to another category which is emergency parking. Figure 1 (c) shows the emergency stopping lane which is the destination for vehicle parking in emergency cases.

2 Emergency parking system

Figure 2 shows the flowchart of the proposed system. The system comprises two stages. The first stage is drowsiness detection, which uses the built-in camera of a smartphone. The video frames captured by the camera are processed by an Android application (App). The output of this stage is received by the second stage, namely, the vehicle emergency parking control system. The second stage uses a microcontroller unit (MCU). The MCU is connected to an alarm system, hazard lights and vehicle control interface.

(a) (b) (c)

Fig. 1. Parking systems are divided to (a) garage-parking, (b) parallel-parking and (c) highway emergency stopping lane.

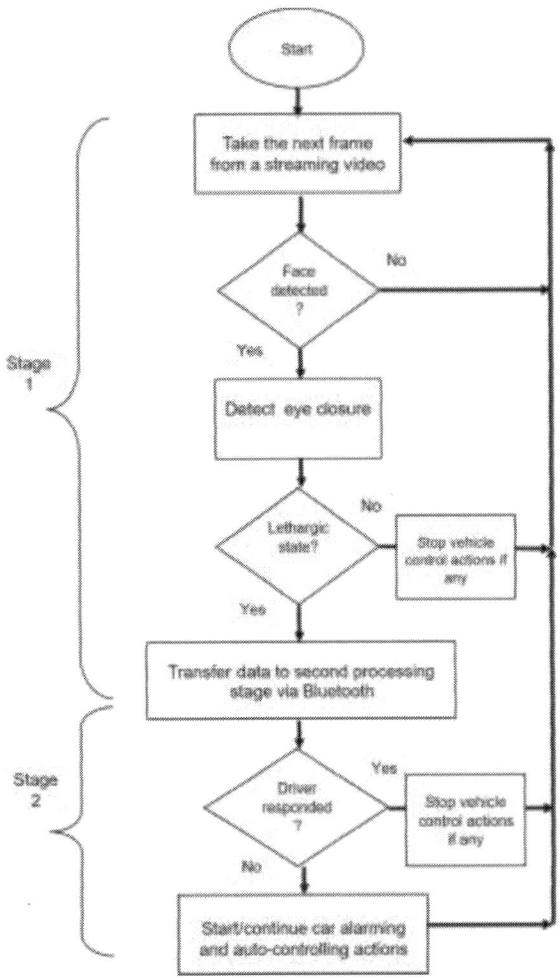

Fig. 2. System flowchart.

2.1 Drowsiness detection

There is no standard fatigue measurement system. Most measures are of the outcomes of fatigue rather than of fatigue itself [5]. Moreover, there is no study which links the effect of fatigue and driver's ability to drive. Consequently, no robust model or algorithm for risk assessment based on fatigue is currently available. In the proposed system, fatigue is evaluated by monitoring drowsiness effect on driver's eyes. A smartphone camera is used to capture driver face. An Android App is used to process the video frames captured by the camera.

Smartphones and tablet computers are increasingly becoming popular among researchers and developers. Researchers are recently using these devices for powerful controlling applications, like robot operating [13]. Image processing and pattern recognition applications performance on smartphone platforms are still slower than the performance on personal computers. However, the performance is reaching acceptable levels [14]. Therefore, smartphone can be considered as efficient processing unit for fatigue detection.

To monitor drowsiness effect on driver's eyes, the first step is face detection. Face is detected by applying Viola and Jones method [15]. The method works by applying masks which are rectangles contain a small number of Haar wavelets. Haar descriptor for one mask defines a weak classifier. A set of weak classifiers is used to decide whether there is a face in the window occupied by the masks. Slide sets of masks through the image; decide whether a given window contains a face or not. The Haar feature-based cascade classifiers using an AdaBoost classifier are applied on driver face images of the video stream. AdaBoost, which is an adaptive machine-learning meta-algorithm, cascades weak classifiers to build a strong classifier. The cascade function is trained using a set of positive images (images of faces) and negative images (images without faces).

After detecting the face, the region which includes the face, is extracted. The next step is eye detection. The area of the face that contains the eyes is analyzed using the Haar cascade algorithm again. To detect eyes, the cascade function is trained using two classes: Images of open eyes, and images of closed eyes. Eyes data are used to calculate eye blinking frequency. The blinking frequency parameter is expressed as the percentage of number of images with closed eyes to the total number of images at a fixed interval of time. If the blinking frequency is greater than 80%, the driver is considered asleep and the state is considered as a state of emergency. Driver state data are sent to the second stage via the Bluetooth module.

2.2 Parking system prototype and simulation

The diagram of the second stage processing unit is shown in Figure 3. The unit receives eye closure data from the smartphone via Bluetooth module. If closed eyes are detected and the state is evaluated as state of emergency, the MCU fires the alarm signal. Hazard lights and the automatic control system of the vehicle are activated if the driver does not respond within 3 seconds. The driver may respond by manually pushing a button that is connected to the second stage processing unit to turn off the alarm and the automatic vehicle control system. A prototype of the system is built using Arduino MCU, which is connected to four light-emitting diodes (LEDs), a buzzer, a push-button switch, and a motor as shown in Figure 4.

As long as the driver is not responding, the second stage processing unit will continuously determine braking and speed-reduction scenarios based on vehicle speed and the distance of any other vehicle in its rear. To determine braking and speed-reduction scenarios, parameters such as steering angle, throttle input, and forward velocity are

DOI: 10.1051/matecconf/20179001034

considered as shown in Figure 5. The control system of braking, speed reducing, and steering control is simulated using MATLAB–Simulink. The simulation is performed according to a proportional-derivative (PD) controller and the mathematical model shown in Figure 6 and Equations (1) and (2).

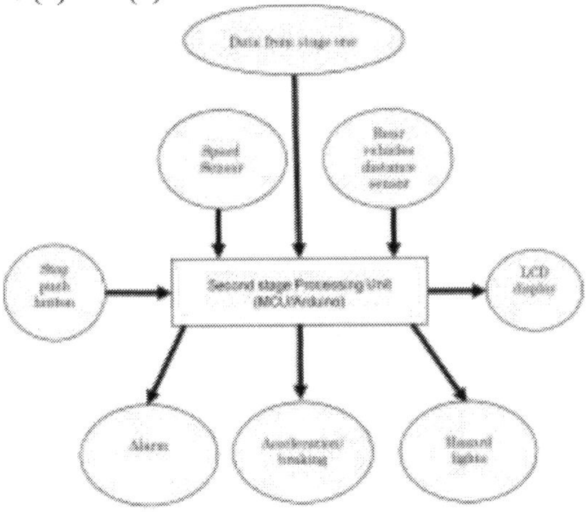

Fig. 3. Second stage system diagram.

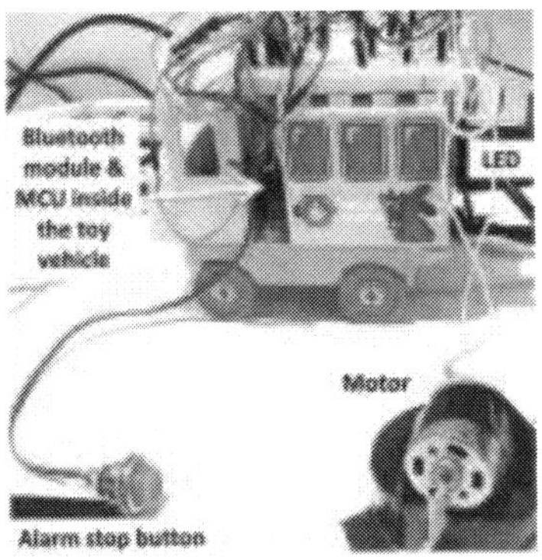

Fig. 4. System prototype.

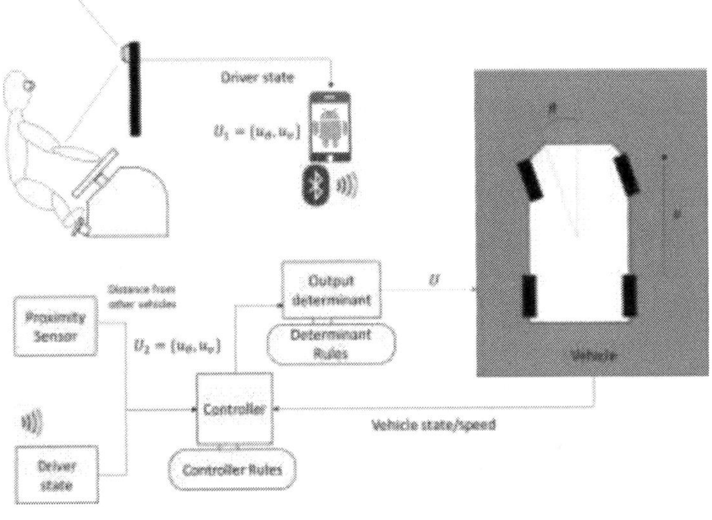

Fig. 5. Speed and steer controlling system diagram.

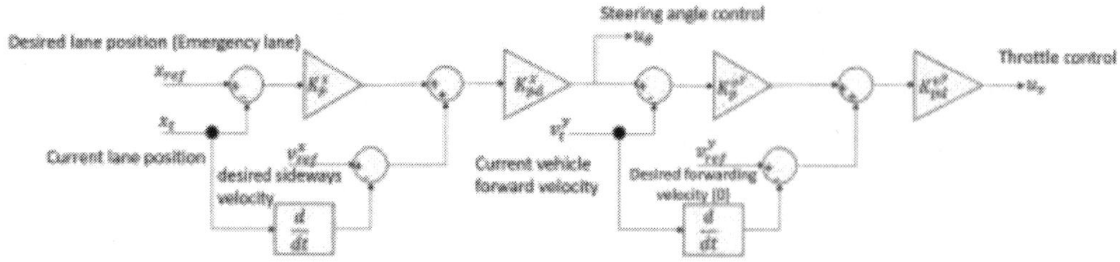

Fig. 6. Braking, speed reducing and steer control mathematical model.

$$u_\theta = k^x_{pd}\left(k^x_p\left(x_{ref} - x_t\right) + \left(v^x_{ref} - \frac{dx_t}{dt}\right)\right) \tag{1}$$

$$u_v = k^{v^y}_{pd}\left(k^x_p\left(k^x_{pd}\left(k^x_p\left(x_{ref} - x_t\right) + \left(v^x_{ref} - \frac{dx_t}{dt}\right)\right)\right) - v^y_t + \left(v^y_{ref} - \frac{dv^y_t}{dt}\right)\right) \tag{2}$$

where u_θ is the steering angle input, u_v is the throttle input, x_t is the current lane position according to road width, x_{ref} is the desired lane position according to road width, v^y_t is the current vehicle forward velocity, and v^y_{ref} is the desired vehicle forwarding velocity.

The steering angle input, u_θ relates to the desired vehicle position according to the lane width, x_{ref}; the current vehicle position according to lane width, x_t; and the desired vehicle sideways velocity, v^x_{ref}. The lane width x relates to average travelling speed of vehicle on each lane of the road. The lane width x may include slow lane, fast lane and emergency lane.

The throttle input, u_v is related to the steering angle input, u_θ, current vehicle forward velocity, v^y_t and final vehicle forward velocity, v^y_{ref}. When the steering angle input, u_θ is changed towards the emergency lane, the vehicle forward velocity, v^y_t gradually decreases until the vehicle arrives at the desired location which is the emergency lane. The current vehicle forward velocity v^y_t is to be determined by a speed sensor. Furthermore, the desired vehicle forward velocity, v^y_{ref} relates to changes of steering angle input, u_θ wherein if the vehicle is moving straight with no changes of steering angle input, the desired vehicle forward velocity is equal to "0". The value of the desired vehicle forward velocity v^y_{ref} is targeted to be "0" which means the controlled vehicle stops on the emergency lane. Steering angle input is taken into account to make sure that the vehicle forward velocity gradually decreases according to distance of the controlled vehicle from the emergency lane.

Based on the gain coefficients of the PD controller, $k^x_{pd}, k^{v^y}_p$, $k^{v^y}_{pd}$ and k^x_p which vary according to vehicle physical specifications, the steering angle input, u_θ and throttle input, u_v are computed, wherein the steering angle value of "0" indicates that the vehicle is in a straight position, negative values indicate that the vehicle is moving to the left, and positive values indicate that the vehicle is moving to the right.

For example, in case the desired location according to the lane width is the emergency lane, the steering angle input, u_θ changes from "0" to a negative value. When the steering angle input, u_θ is changed, the desired vehicle forward velocity, v^y_{ref} is reduced accordingly due to the negative value of the steering angle input until the value of the steering angle input changes back to "0". The braking scenario is based on reduction of velocity of the vehicle wherein the steering angle input value, u_θ is equal to "0".

3 Results and discussion

The proposed system was tested on several video sequences captured for different people. The temporal sampling rates are 30 frames per second. When the system detects closed eyes and state of emergency, alarm system is fired. If the driver does not respond within three seconds, the electrical motor is started in the prototype.

Face and eye correct detection rate was 95%. After detecting the face, eyes are extracted from the face to determine eye closure according to the described algorithm. An example of detecting face and eyes is shown in Figure 7. Drowsiness detection processing time average is about 480 ms / frame. After drowsiness detection the data are sent via Bluetooth module to the second stage. The second stage processing unit triggers the alarm signal within 500 ms.

Further tests were done using the MATLAB simulation. The MATLAB simulation results illustrated the effectiveness of the developed schemes for the automatic emergency parking system. It showed real-time response for the testing cases. Figure 8 shows snapshots of the vehicle parking simulation system. The system is taking in consideration the surrounding vehicles. Other vehicle positions and speed data are fed to the controller to allow it to control the vehicle safely. In reality, such data are collected using a Hall Effect speed sensor and a rear ultrasonic or laser distance sensor. Simulation results illustrate the effectiveness of the developed schemes for the auto parking system in real time. The average time from drowsiness detection to fully parking, if the vehicle is moving at the speed of 100 km/s, is about 15 s. The Processing times for the entire emergency parking system actions are summarized in Table 1.

Table 1. Processing times for the entire emergency parking system actions.

Action	Processing time
Face detection	380 ms
Eye and drowsiness detection	100 ms
Firing alarm after drowsiness detection	500 ms
Waiting time before activating the emergency parking system	3 s
Average time from drowsiness detection to fully parking if the vehicle is moving at the speed of 100 km/s	15 s
TOTAL	19 s

The complete process requires about 19 seconds to park the vehicle safely on the emergency lane. This system is activated in case the driver is unconscious or unable to conduct safe driving.

The successfully simulated system is to be applied on the all-terrain vehicle (ATV) shown in Figure 9. The main constraint for applying it on the prototype ATV is designing a steering control system to handle the ATV steering stem. Various systems and theories are investigated currently to design the mechanical steering system.

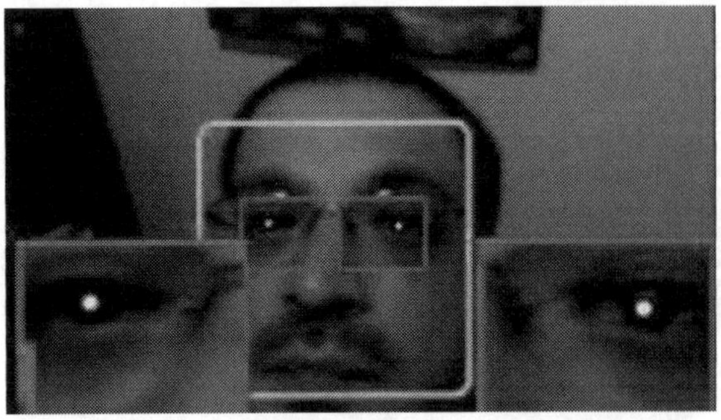

Fig. 7. Face and eye detection.

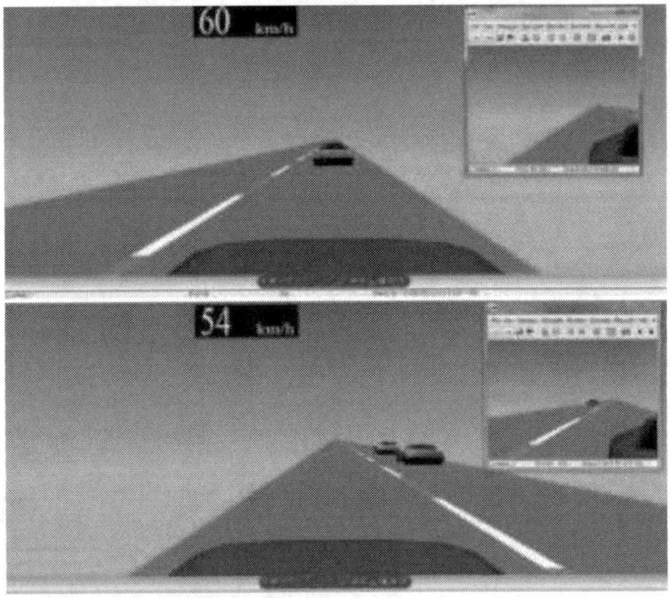

Fig. 8. Snapshots of the vehicle parking simulation system.

Fig. 9. The prototype vehicle with the speed sensor.

4 Conclusions

A safety driving assistance system is presented. The system notifies the driver in dangerous situations and activates an emergency parking system in case the driver is unable to conduct safe driving. The alarm system responds perfectly. The simulation results illustrate the effectiveness of the developed schemes for the automatic emergency parking system. The complete process requires about 19 seconds to park the vehicle safely on the emergency lane. This time is considered acceptable for preventing, or at least reducing, the sleep and fatigue related accidents.

Next step is to design a mechanical automatic steering system to apply the model on the ATV. The effectiveness of the detection method and the risk evaluation method in relation to fatigue are not discussed. The study is focusing on eye closure as an important parameter for fatigue detection. More robust evaluation methods for detecting driver fatigue and linking it with its effect on driver's ability to drive is needed. Moreover, it is recommended to improve the classification method by detecting three or more levels of eye closure: open, half open, and closed. Other parameters like gaze duration, looking direction, yawning, blinking frequency, and vehicle movement are suggested to be involved in fatigue and risk evaluation.

The authors would like to be obliged to University Malaysia Pahang for providing laboratory facilities and financial assistance under project no. RDU1403151.

References

1. US Department of Transportation, an Evaluation of Emerging Driver Fatigue Detection Measures and Technologies (2009).
2. R. O. Phillips, F. Sagberg and T. Bjørnskau, Fatigue in operators of land- and sea-based transport forms in Norway, Risk Profiles Fatigue in Transport Report IV", Project Report, Institute of Transport Economics, Oslo, TØI (report 1440/2015).
3. MIROS (Malaysian Institute of Road Safety Research), Five research pillars, www.miros.gov.my/ (Accessed October 2016)
4. A. Frank et al., The exploration of physical fatigue, sleep and depression in paramedics: a pilot study, Australasian Journal of Paramedicine, **9(1)**, pp. 1– 33 (2012)
5. M. H. Alsibai and S. Abdul Manap, A Study on Driver Fatigue Notification Systems, ARPN Journal of Engineering and Applied Sciences, **11 (18)**, pp. 10987 - 10992 (2016)
6. L. W. Li et al., Multi-sensor soft-computing system for driver drowsiness detection, Proceedings of Online conference on soft computing in industrial applications, pp. 1–10 (2012)
7. H. Singh, J.S. Bhatia and J. Kaur. Eye tracking based driver fatigue monitoring and warning system, International Conference on Power Electronics (IICPE). India. pp. 1 - 6 (2011)
8. Mercedes Benz Tech. Center: http://techcenter.mercedes-benz.com/en/attention-assist/detail.html, (Accessed October 2016)
9. Royal society for the prevention of accidents, Driver fatigue and road accidents a literature review and position paper (2001)
10. H. L. Man, G. P. Hyung, H. L. Seok, S. Y. Kang, S. L. Kil., An adaptive cruise control system for autonomous vehicles, International Journal of Precision Engineering and Manufacturing, **14 (3)**, pp 373-380 (2013)
11. S. L. Tzuu-Hseng and S. J. Chang, Autonomous Fuzzy Parking Control of a Car-Like Mobile Robot, IEEE transactions on systems, man, and cybernetics—part a: systems and humans, **3**, pp. 451-465 (2003)
12. N. E. Cavalcanti et al., Development Control Parking Access Using Techniques Digital Image Processing And Applied Computational Intelligence, IEEE Latin America Transactions, **13(1)**, pp. 272 -276 (2015)

13. S.-Y. Juang, J.-G. Juang, Remote control of a mobile robot for indoor patrol. Appl. Sci. **6 (82)**, URL http://www.mdpi.com/2076-3417/6/3/82 (2016)
14. M. H. Alsibai, H. Manap, A. A. Abdullah, Enhanced face 323 recognition method performance on android vs windows platform. ARPN Journal of Engineering and Applied Sciences, **10 (23)**, pp. 17479-17485 (2015)
15. P. Viola and M. Jones, Rapid object detection using a boosted cascade of simple features, IEEE Conf. on Computer Vision and Pattern Recognition, pp. 511-518 (2011)

Magnetic flux gradient observation during fatigue crack propagation: A case study of SAE 1045 carbon steel used for automotive transmission parts

S.R. Ahmad[1,*], *A.* Arifin[1] , *S.* Abdullah[1] and *S.A.N.* Mohamed[1]

[1]Department of Mechanical and Materials Engineering, Faculty of Engineering and Built Environment, Universiti Kebangsaan Malaysia, 43600 UKM Bangi,Selangor, Malaysia

Abstract. The objective of this study is to evaluate the application of the metal magnetic memory (MMM) technique for investigations on fatigue crack propagation in a ferromagnetic material. Fatigue failure caused by stress concentration is serious in practical engineering. However, early fatigue damages cannot be detected by using traditional nondestructive testing (NDT) methods. Therefore this paper study about NDT method called metal magnetic memory (MMM) that has potentials for evaluating the fatigue damage at the early damage and critical fracture stages. While its capacity to evaluate the distribution of self-magnetic leakage field signals on the component's surface is well-established, there remains a need to scrutinize the physical mechanism and quantitative analysis aspects of this method. To begin with, a fatigue test involving a loading of 7kN was conducted on a SAE 1045 carbon steel specimen. This material is frequently used in the manufacturing of automotive transmission components that include the axle and spline shaft. MMM signals were measured along a scanning distance of 100 mm and analysed during the propagation stage. Other than revealing that the value of the magnetic flux gradient signals *dH(y)/dx* increased in tandem with the crack length, the results also led to the detection of the crack growth location. It was anticipated that the *dH(y)/dx* value will also exhibit an upward trend with a rise in the fatigue growth rate of *da/dN*. A modified Paris equation was utilized to correlate *dH(y)/dx* with da/dn through the replacement of the stress intensity factor range ΔK. This resulted in the log-log plot of *da/dN* versus *dH(y)/dx* portraying an inclination similar to the log-log plot of *da/dN* versus ΔK. A linear relationship was established between dH(y)/dx and ΔK with the R^2 value as 0.96. Players in the automotive industry can benefit from the disclosure that *dH(y)/dx* can effectively replace ΔK for the monitoring of fatigue crack growth behaviour.

1 Introduction

* Corresponding author: shahridhwan.jpk@gmail.com

The development of fatigue-safe components is among the most crucial factors in the automotive industry domain. This is attributed to the fact that fatigue failure can lead to accidents, which in turn could result in injuries and loss of life. Automotive transmission components are among those often beset with biaxial fatigue, and SAE 1045 carbon steel is among the most scrutinized materials in this area [1]. The axle and spline shaft are two examples of automotive transmission parts made of SAE 1045 carbon steel where fatigue is the primary cause of failure. This is especially so on the stress raisers [2, 3]. A wide range of studies have been conducted on the fatigue behaviour of this material in the context of automotive components. Pals & Stephens [4] ascertained the fatigue behaviour of SAE 1045 steel subjected to extremely elevated tensile mean stress for un-notched, mildly notched, and sharply notched test specimens under dissimilar heat treatments. Firat [5] executed a simulation that involved the combined bending and torsion fatigue of a SAE 1045 notched shaft where the design was fashioned to resemble the distinctive geometric and manufacturing features of automotive transmission shafts. Ince & Glinka [6] on the other hand, investigated the fatigue life of a SAE 1045 notched shaft under multi axial loadings.

Failure deterrence is a significant aspect of automotive engineering. Generally, an inspection for flaws in components by way of non-destructive testing (NDT) is the standard procedure for making certain that a component is in good working order [7]. However, conventional NDT procedures such as ultrasonic testing, radiographic testing, magnetic particle testing, penetrant testing, and eddy current testing are only designed to uncover developed defects. As such, these procedures fall short when it comes to initial fatigue damage prevention [8]. This dilemma can be overcome with the employment of the MMM technique. This technique focuses on identifying the stress concentration zones in order to provide an early indication of damage to a ferromagnetic material [9]. It was first introduced in the 50th international welding conference by the Russian researcher Dubov. The principle of MMM is the magneto-mechanical effects. While equipment working, ferromagnetic material is having self-magnetization phenomenon. When the work load is off the magnetic field is kept down. The residual magnetic fields that remain are related to the mechanical stress. The residual magnetic field density H is characterized by the normal component *Hp(y)* and tangent component *Hp(x)* on the stress-concentration zones [8].

Hypothetically, the materialization of the magnetic area at the location of highest defect concentration together with the influence of an external magnetic field creates a full demagnetizing field of defects. The entire demagnetizing field that is liberated from the surface of the component is in the form of magnetic scattering fields. The regular component of the scattering field *H(y)* indicates the location of the metal structure's highest level of inhomogeneity [10]. A substantial number of studies have been performed in efforts to link the MMM signals to fatigue crack growth parameters. Xing et al. [11] investigated the association between the crack growth rate *da/dN* and the MMM regular component signal *H(y)*. It was observed that the *H(y)* increased in tandem with the rise of *da/dN* while the correlation curve between *H(y)* and *da/dN* exhibited a sigmoidal figure. Dong et al. [12] described the behaviour of the peak to peak value of *H(y)* in a tension-tension fatigue crack growth test. According to them, the progression of a fatigue crack coincides with an elevation in the magnetic peak to peak value. Huang et al. [9] conducted dynamic three-point bending fatigue tests to delve into the option of harnessing magnetic flux gradient signals *dH(y)/dx* for the prediction of fatigue crack propagation. They discovered that *dH(y)/dx* can be brought into play for determining the degree of stress concentration. They also forwarded that *dH(y)/dx* is exponential with the loading cycles and just about linear to the crack length.

Due to the shortage of physical models and quantitative defect criteria, the MMM technique is currently deemed a qualitative or semi-quantitative process [13]. As such, its capacity to establish a linkage between MMM signals and the fatigue growth parameter remains questionable. The intention of this investigation is to assess the possibility of harnessing the magnetic flux gradient signals $dH(y)/dx$ to ascertain the location of a crack, and to recognize the development of fatigue cracks in materials commonly used in the automotive industry. The authors also intend to determine if the $dH(y)/dx$ can be utilized as a substitute for the stress intensity factor range ΔK in the Paris equation, an equation frequently used to explain the occurrence of fatigue crack growth.

2 Experimental set up

The objective of this experiment is to link the magnetic signal parameter of ferromagnetic materials released during crack propagation to the fatigue crack growth parameters. The fatigue test was conceived in line with the ASTM E647, a typical testing procedure for the measurement of fatigue crack growth rates. This procedure includes the determination of fatigue crack growth rates and the crack driving force of fatigue crack growth, namely, the stress intensity range ΔK. In the opinion of Dong et al. [12], the successful acquisition of the MMM data signal necessitates the use of a wide specimen to ensure the availability of a satisfactory working space. With this in mind, and in compliance to the ASTM E647 standard, a single edge-cracked specimen with a width of 60 mm, a thickness of 3 mm and an initial crack length width of 22 mm was employed. The specimen was made of SAE 1045 carbon steel which is extensively used in automotive transmission parts [2]. The previously executed tensile test on the SAE 1045 carbon steel material disclosed that its ultimate tensile strength is 611.7 Mpa, its yield stress is 363.5 Mpa, and its elastic modulus is 206.7 GPa.

The constant amplitude fatigue test was performed on the SAE 1045 carbon steel specimen using the 100 kN servo-hydraulic machine with a loading of 7.0 kN. Our previous work in this area established that this testing circumstance is appropriate for the scrutiny of crack growth propagation as well as for realizing correlation between the magnetic flux leakage signals and the fatigue crack growth parameter. The crack opening displacement (COD) device was employed to measure the propagation of crack length on the specimen, and the MMM scanning device TSCM-2FM was utilized to capture the magnetic flux leakage signals. Upon the arrival of the crack in the specimen at the pre-set length, the fatigue test was halted for the measurement of the magnetic flux leakage signal. The sensor of the MMM device rolled on the determined scanning line with the gap between the scanning line and the initial crack set at 20 mm. All related data were documented up to the point where the specimen failed. Figure 1 exhibits the schematic geometry of the specimen in relation to the scanning line, the MMM device, and the installation of the fatigue test involving the COD device.

3 Equations and mathematics

The concepts of linear elastic fracture mechanics were applied to examine the fatigue crack growth process stretching from initial crack detection to ultimate failure. The fundamental supposition here is that during the progression of fatigue, the state of the material is, for the most part, linear elastic. This examination entails an establishment of the linkage between the crack growth rate da/dN and the stress intensity factor range ΔK. Generally, the stress intensity factor range is used for describing the behaviour of fatigue crack growth in metals and alloys under constant amplitude test conditions [14]. Fundamentally, the da/dN can be ascertained through the slope of the graph for crack length a against the number of cycles N. The standard log-log plot of da/dN versus ΔK reveals that the fatigue crack growth process can be separated into three areas. Among them, the second area frequently displays a linear linkage between log *da/dN* and log ΔK. This can be expressed through the Paris law equation as in Equation (1).

$$da/dN = C(\Delta K)^m \quad (1)$$

where coefficient C and exponent m are the material constants [14]. During this endeavour, we made an adjustment to Equation 1 to realize an innovative relationship between *dH(y)/dx* and *da/dN*.

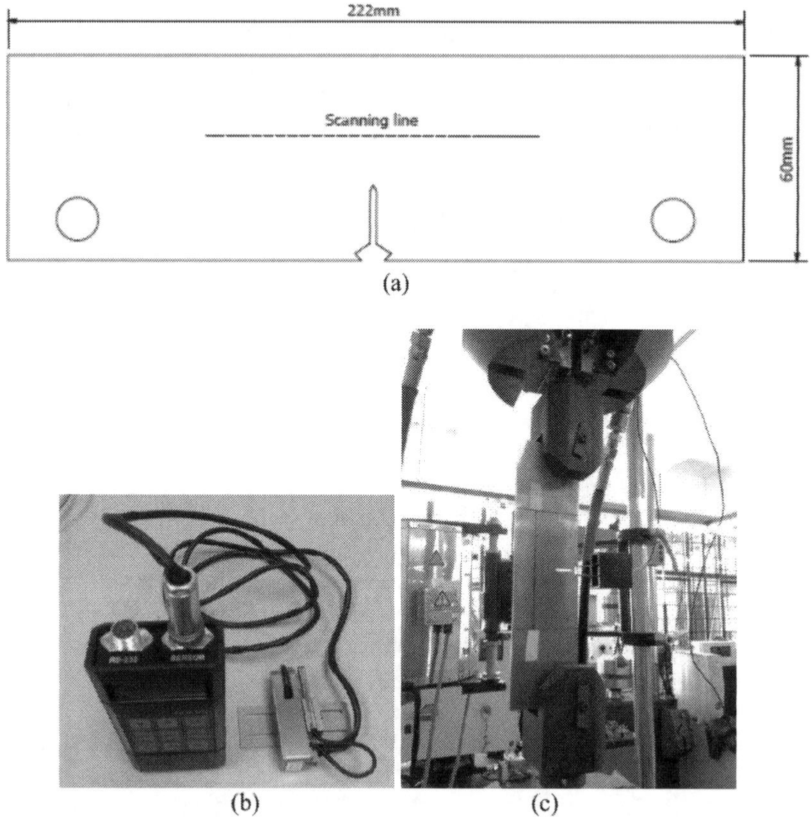

Fig. 1. Experimental setup (a) schematic geometry of specimen with scanning line; (b) MMM device; (c) installation of fatigue test with COD device.

4 Results and discussion

4.1 Fatigue Crack Growth Behaviour of SAE 1045 carbon steel

The fatigue test revealed that the log–log plots of *da/dN* versus ΔK assume a sigmoidal shape that can be separated into three regions. This situation is displayed in Figure 2. At Region I, the *da/dN* and ΔK values are extremely low and the early stage of the crack is apparent at the initial crack length. Region II is deemed the intermediate region of the entire fatigue crack growth process. Here, the propagation of the crack is stable. In line with the Paris equation, the relationship representing the straight line is $da/dN = 3E\text{-}9(\Delta K)^{3.0967}$.

According to this equation, the value of coefficient *C* is 3E-9 while the value of slope *m* is 3.0967. The R^2 value of 0.9796 indicates that the mathematical equation offers a plausible description of the association between log *da/dN* and log Δ*K*. As can be observed in Figure 2, crack propagation occurs within Region II from 2 mm to 18 mm. In Region III, the *da/dN* values rise briskly and the value of Δ*K* is at an exceptionally elevated level. Crack growth is unstable in this region and crack fracture in the specimen ultimately occurs at the crack length of 20 mm.

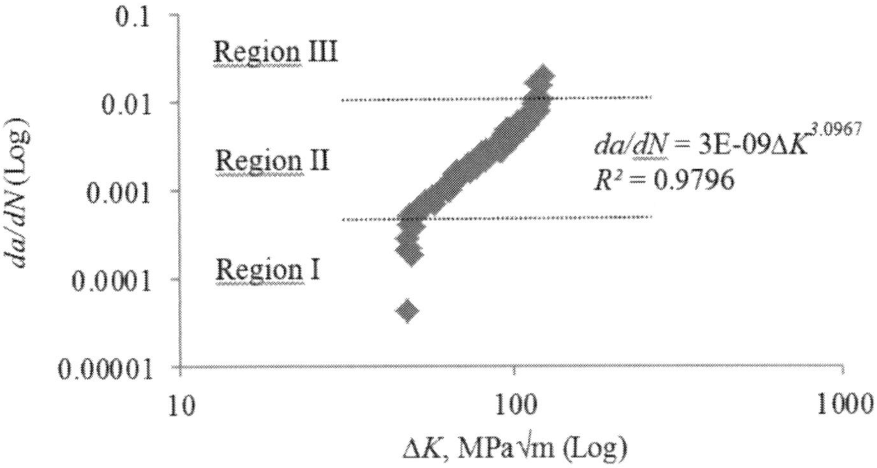

Fig. 2. The relationship between crack growth rate, *da/dN* and stress intensity factor range, Δ*K*

4.2 Observation of magnetic flux gradient signals, *dH(y)/dx*

The magnetic flux gradient signals *dH(y)/dx* were captured during the fatigue test. Figures 3(a) to 3(c) exhibit the magnetic flux gradient signals *dH(y)/dx* during the propagation stage in which the crack length stretches from 2 mm to 18 mm. The figures portray alterations in the *dH(y)/dx* value as the MMM device sensor rolled along the scanning distance of 100 mm.

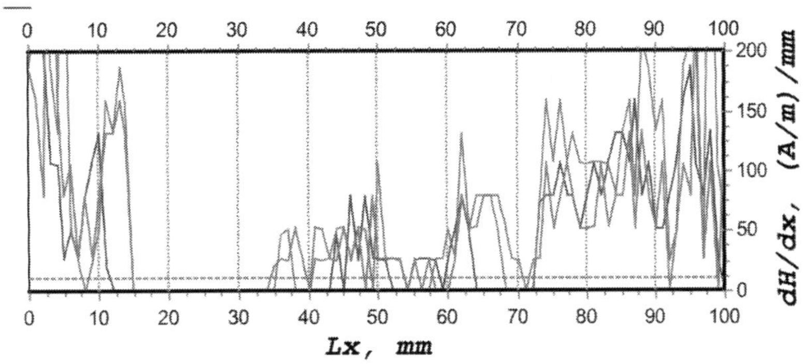

(a)

 DOI: 10.1051/matecconf/20179001035

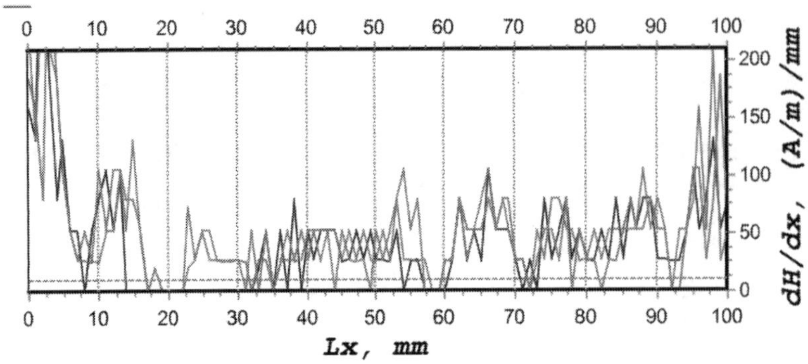

(b)

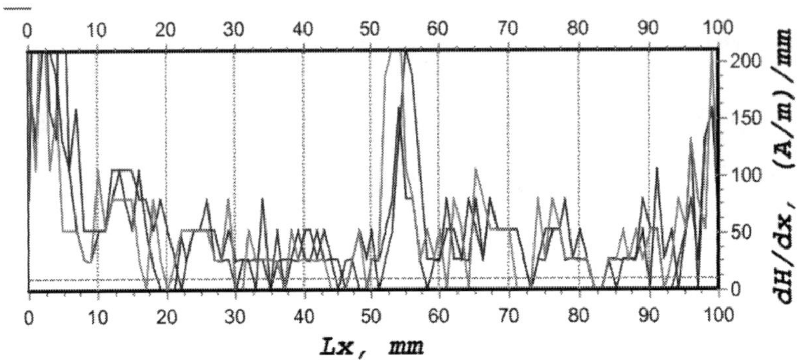

(c)

Fig. 3. The *dH(y)/dx* signals response at different length of crack propagation (a) 2–6 mm; (b) 8–12 mm; (c) 14–18 mm.

As illustrated in Figures 3(a) to 3(c), at the scanning distance of 50 mm to 60 mm, the value of *dH(y)/dx* rose in tandem with increases in the crack length. It is assumed that this represents the crack growth location. A previously conducted study disclosed that the *dH(y)/dx* can be utilized for describing the degree of stress concentration, as well as for determining the location of the crack [15, 16]. The notch initiates the occurrence of stress concentration which sets off the propagation of the fatigue crack. Eventually, the progression of the crack culminates in definitive material failure [4].

As portrayed in Figures 3(a) to 3(c), at a scanning distance of 0 to 50 mm, and 60 to 100 mm, the value of *dH(y)/dx* is fairly high. We presume that this is attributed to the influences of the original magnetic field from the fatigue testing machine clamps, as well as the environmental magnetic field [17, 18]. The *dH(y)/dx* emanating from the crack growth of the specimen become increasingly apparent with extensions in the length of the crack. This is illustrated in Figure 3(c).

Investigations were also conducted on regular alterations occurring at the scanning distance of 50 mm to 60 mm. Table 1 displays the *dH(y)/dx* values attained for different crack lengths at a scanning distance of 50 mm to 60 mm.

Table 1. *dH(y)/dx* value in crack propagation region at a scanning distance of 50–60 mm.

Crack length (mm)	2	4	6	8	10	12	14	16	18
dH(y)/dx value ((A/m)/mm)	27	27	27	54	80	107	160	213	267

The table reveals that the *dH(y)/dx* value rose in tandem with increases in the length of the crack. However, at the crack length of 2 mm to 4 mm, the *dH(y)/dx* values remained unaltered. This could be attributed to the distance of the set scanning line being slightly too remote from the location where the crack occurred. In such a circumstance, the sensor would fail to detect alterations in the values of the *dH(y)/dx*. Upon the arrival of the crack line at 8 mm, the crack location was drawn closer to the scanning line, and from then on, increases in the *dH(y)/dx* values were consistent. At the crack length of 18 mm the *dH(y)/dx* value is exceptionally elevated as the specimen is on the brink of experiencing an ultimate fracture. The experimental results disclosed that the MMM sensor device was unable to provide satisfactory *dH(y)/dx* values at the distance of 2 mm to 4 mm. In view of this, all following investigations were confined to the utilization of *dH(y)/dx* values at the distance of between 6 mm and 18 mm.

It is believed that *dH(y)/dx* related to the crack length and based on the previous study, Huang et al [9] and Chongchong et al [13] established that a linear correlation existed between the crack length and *dH(y)/dx*. Figure 4 portrays the linear relationship between *dH(y)/dx* and the crack length at the propagation stage of this study. With the R^2 value at 0.971, the relationship between the *dH(y)/dx* signals and the crack length is deemed adequate.

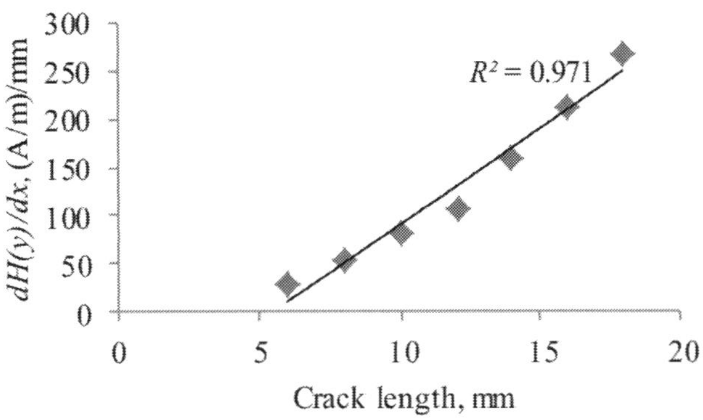

Fig. 4. The relationship between the magnetic flux gradient signals, *dH(y)/dx* and crack length.

4.3 Relationship between magnetic flux gradient signals, *dH/dx* and stress intensity factor range, Δ*K*

The MMM technique is a non-destructive procedure for ascertaining stress concentration zones. These zones are the major locations for the onset of failure growth [10]. The *dH(y)/dx* captured by means of the MMM technique can be used to express the fatigue crack growth by replacing ΔK in the Paris equation with *dH(y)/dx*. Equation 2 below represents the proposed modified Paris equation:

$$da/dN = C(dH(y)/dx)^m \quad (2)$$

The log-log plot of *da/dN* against the *dH(y)/dx* signals during the propagation stage is exhibited in Figure 5. As displayed, the pattern of this plot is similar to that of the second region in Figure 2. The *dH(y)/dx* value increased in tandem with the rise in da/dN. This indicates that the variation of *dH(y)/dx* signals reflects the alteration of ΔK during the crack growth process. The linear relationship between *da/dN* and *dH(y)/dx* is expressed as da/dN=5E-5$(dH(y)/dx)^{0.8893}$. According to the equation, the value of coefficient C is 5E-5 while the value of the slope m is 0.8893. The R^2 value attained from the relationship is

0.9554. This signifies that the correlation between *da/dN* and *dH(y)/dx* is acceptable. Hence, the fatigue crack growth rate can be assessed and foreseen through the correlation between *da/dN* and *dH(y)/dx*.

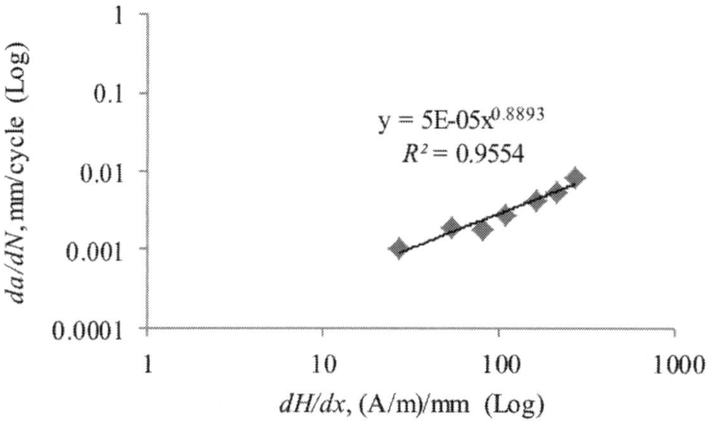

Fig. 5. The relationship between the crack growth rate, *da/dN* and magnetic flux gradient signals, *dH(y)/dx*

Hypothetically, *dH(y)/dx* and ΔK are similar in the sense that they both indicate the state of stress during fatigue crack growth. Xing et al. [11] and Ahmad et al. [19] investigate relationship between ΔK and *H(y)* and it was stated that *H(y)* variation reflects the change of stress intensity factor ΔK during the whole crack growth process and linear relation was represented between ΔK and *H(y)* graph. Figure 6 illustrates the investigation conducted on the relationship between *dH(y)/dx* and ΔK. The log-log plot of *dH(y)/dx* versus ΔK reveals that the relationship between *dH(y)/dx* and ΔK is linear. The acquired R^2 value of 0.9612 verifies that the relationship between the *dH(y)/dx* signals and ΔK is acceptable. As such, *dH(y)/dx* has the potential to replace ΔK during efforts to evaluate the fatigue crack growth process.

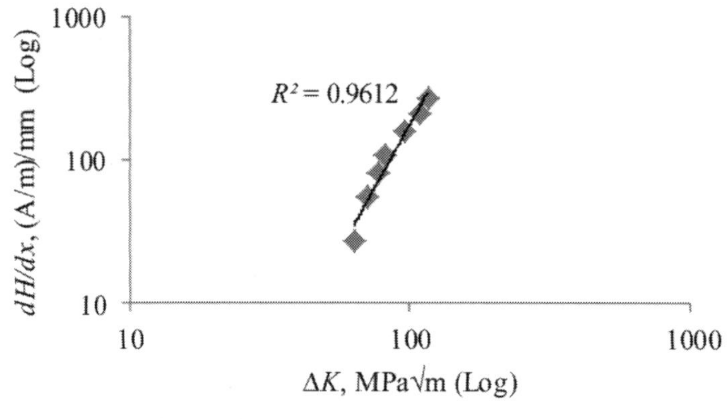

Fig. 6. The relationship between the crack growth rate, *da/dN* and magnetic flux gradient signals, *dH(y)/dx*

5 Conclusion

This investigation probes the likelihood of employing the MMM technique for the evaluation of fatigue crack growth in a ferromagnetic material. The fatigue crack growth behaviour of SAE 1045 carbon steel is expressed by way of the log-log plot of *da/dN* versus ΔK. The linear relationship obtained in Region II is represented by the Paris equation where the value of coefficient C is 3E-9 and the value of the slope m is 3.0967. The magnetic flux gradient signals *dH(y)/dx* captured by way of the MMM device sensor disclosed that at the scanning distance of 50 mm to 60 mm, the *dH(y)/dx* value increased along with extensions in the crack length. The *dH(y)/dx* value analysed at that range confirmed that the relationship between *dH(y)/dx* and the crack length is linear. The results from this investigation also suggest that it is possible to detect the location of a crack through the utilization of *dH(y)/dx*. A modified Paris equation entailing the replacement of the stress intensity factor range ΔK with *dH(y)/dx* was recommended by this study. The log-log plot of *da/dN* versus *dH(y)/dx* displays a similar curve trend to that of *da/dN* versus ΔK. According to the equation, the value of coefficient C is 5E-5 while the value of slope m is 0.8893. The outcomes from this investigation confirm that *dH(y)/dx* can be utilized for describing the fatigue crack behaviour of SAE 1045 carbon steel, a commonly used material in the manufacturing of automotive transmission components. Nonetheless, further investigations ought to be in the pipeline regarding the accuracy of the magnetic signals, the capacity of these signals to provide an early indication of crack development, and the applicability of the MMM technique in the automotive industry.

The authors would like to express their gratitude to Universiti Kebangsaan Malaysia and Ministry of Education Malaysia for funding under the grant of FRGS/2/2014/TK01/UKM/02/03 for supporting this project research.

References

1. Y. Verreman and H. Guo, Fatigue & Fracture of Engineering Materials & Structures, **30**, 932-946 (2007)
2. M. Y. Noordin, V. C. Venkatesh, S. Sharif, S. Elting, and A. Abdullah, Journal of Materials Processing Technology, **145**, 46-58 (2004)
3. A. Guimaraes, P. Brasileiro, G. Giovanni, L. Costa, and L. Araujo, Engineering Failure Analysis (2016)
4. T. Pals and R. Stephens, International journal of fatigue, **26**, 651-661 (2004)
5. M. Firat, Finite Elements in Analysis and Design, vol. **54**, 16-27 (2012)
6. A. Ince and G. Glinka, International Journal of Fatigue, **62**, 34-41(2014)
7. V. L. de Araújo Freitas, V. H. C. de Albuquerque, E. de Macedo Silva, A. A. Silva, and J. M. R. Tavares, Materials Science and Engineering, **527**, 4431-4437(2010)
8. H. Y. Xing, D. B. Wu, and M. Q. Xu, Advanced Materials Research, 408-411 (2011)
9. H. Huang, S. Jiang, R. Liu, and Z. Liu, Journal of Nondestructive Evaluation, **33**, 407-412 (2014)
10. A. Dubov, Metal Science and Heat Treatment, **39**, 401-405 (1997)
11. H. Y. Xing, R. X. Wang, M. Q. Xu, and J. Z. Zhang, Key Engineering Materials, 2293-2296 (2007)
12. L. H. Dong, B. S. Xu, H. P. Wang, and N. Xue, Applied Mechanics and Materials, **190**, 415-418 (2012)
13. L. Chongchong, D. Lihong, W. Haidou, L. Guolu, and X. Binshi, Journal of Magnetism and Magnetic Materials, **405**, 150-157 (2016)
14. C. A. R. P. Baptista, A. Adib, M. Torres, and V. Pastoukhov, Mechanics of Materials, **51**, 66-73 (2012)
15. S. Changliang, D. Shiyun, X. Binshi, and H. Peng, NDT & E International, **43**, 8-12, (2010)

16. Y. Zhang, R. Gou, J. Li, and G. Shen, Frontiers of Mechanical Engineering in China, **5**, 226-232 (2010)
17. J. Leng, M. Xu, S. Zhao, and J. Zhang, Measuring Technology and Mechatronics Automation, 784-786 (2009)
18. H. Xing, Y. Dang, B. Wang, and L. Jiancheng, Chinese Journal of Mechanical Engineering, 1-6 (2016)
19. M. Ahmad, A. Arifin, and S. Abdullah, Journal of Mechanical Engineering and Sciences, **9**, 1727-1733 (2015)

Investigation of emissions characteristics of secondary butyl alcohol-gasoline blends in a port fuel injection spark ignition engine

I.M. Yusri[1,*], *Rizalman* Mamat[1,2], *A.* Aziz[1,2], *A.F.* Yusop[1,2], *Omar* I. Awad[1] and *S.M.* Rosdi[1]

[1] Faculty of Mechanical Engineering, Universiti Malaysia Pahang (UMP), 26600 Pekan, Pahang, Malaysia
[2]Automotive Engineering Centre, Universiti Malaysia Pahang (UMP), 26600 Pekan, Pahang, Malaysia

Abstract. Exhaust emissions especially from light duty gasoline engine are a major contributor to air pollution due to the large number of vehicles on the road. The purpose of this study is to experimentally analyse the exhaust pollutant emissions of a four-stroke port fuel spark ignition engines operating using secondary butyl alcohol–gasoline blends by percentage volume of 5% (GBu5), 10% (GBu10) and 15% (GBu15) of secondary butyl- alcohol (2-butanol) additives in gasoline fuels at 50% of wide throttle open. The exhaust emissions characteristics of the engine using blended fuels was compared to the exhaust emissions of the engine with gasoline fuels (G100) as a reference fuels. Exhaust emissions analysis results show that all of the blended fuels produced lower CO by 8.6%, 11.6% and 24.8% for GBu5, GBu10 and GBu15 respectively from 2500 to 4000 RPM, while for HC, both GBu10 and GBu15 were lower than that G100 fuels at all engine speeds. In general, when the engine was operated using blended fuels, the engine produced lower CO and HC, but higher CO_2.

1 Introduction

Climatic change of the earth have triggered a global warning to each corner of this earth due to its adverse effects to each living creatures. Based on the estimation done by International Energy Agency (IEA), a rose by 53% in global energy consumption is foreseen by the year of 2030 [1]. Malaysia alone estimated to have an increment of gross domestic product (GDP) by 4.6% in between 2004 to 2030, which indicate that increased of GDP by 1% approximately resulted to growth of energy demand by 1% [2]. Transportation sector are one of the major contributor in rise of energy demand mainly from gasoline and diesel engine vehicles which consumed depleted fossilized fuels [3-5]. Perhaps one of the potential solution that could possibly bring back the balanced in energy consumption and the climatic change in this world is by introducing the biofuel in the transportation areas[6-7].

[*] Corresponding author: yusri890@yahoo.com

The use of alternative clean biofuels such as methanol, ethanol and butanol is one of the method to reduce the dependency on the energy demand for fossilized fuels in spark ignition engine [8-10]. However for the past few years the investigation of methanol and ethanol have received considerable critical attention with less attention paid to the butanol as a sustainable fuel substitutions alternative. Basically butanol is a four carbon chain of alcohol types. It exist in four types of isomers; 1-butanol, 2-butanol, tert-butanol and iso-butanol. Each types of isomer have different physicochemical properties. Butanol are considered as an advanced biofuel due to its superior characteristics compared to other alcohol family members [11-13]. Compared to methanol and ethanol butanol have the nearest fuel properties similarity to the gasoline fuel such as stoichiometric air fuel ratio, latent heating value, energy content, octane number and auto ignition temperature thus make it more suitable to be blended with gasoline fuel [14-16]. Furthermore butanol can be transported through the existing fuel pipeline as it is less corrosion. With all of the advantages offered by butanol without doubt it has been proposed as a next generation biofuel as an alternative to the conventional fuels [17-18].

Taking this into account, n-butanol undoubtedly to have a very remarkable prospective because its properties are almost similar to gasoline fuels. This can reduce the efforts that to be done to adapt their current range of vehicles to be able to run on butanol-gasoline blends. Various precious study has been done to investigate the butanol additive in a gasoline fuels. Among the recent study are from Feng et al. [19]. They used to investigate the effects of adding butanol additives by 30% and 35% of percentage volume in a gasoline fuels using single cylinder spark ignition (SI) engine. Based on his heat release analysis, butanol addition indicates higher knocking resistance. Szwaja and Naber [20] reported that the early combustion duration and length of combustion duration in a SI engine were shortened with increased of n-butanol volumes. In another study shorter early combustion duration stage, better combustion stability and faster combustion was stated by other researchers [21-22].

Galloni et al. [23] studied the effect of butanol and its blends by (20% and 40% of butanol volume) on engine performance and emissions by using port fuel injected turbocharged SI engine. The author found that the engine torque and thermal efficiency drop by approximately 4% for butanol-gasoline blends compared to the gasoline fuels. Singh et al. [24] conducted an experimental study on 5%, 10%, 20%, 50% and 75% of butanol volume percentage in a gasoline fuels with medium duty transportation SI engine. They found that reduction of brake thermal efficiency (BTE) and exhaust gas temperature, brake specific nitrogen oxides (BSNO), brake specific carbon monoxides (BSCO) and smoke emissions for butanol-gasoline blends compared to the pure gasoline fuels. Elfasakhany [25] analyzed the effects of performance and emissions of an engine fueled with low proportion of n-butanol by 0, 3, 7 and 10% volume n-butanol-gasoline blends. Experimental investigation have been done without any modifications on the SI engine systems. Based on the results show that engine in-cylinder pressure, torque, and exhaust gas temperature of the engine slightly decrease when n-butanol-gasoline blended fuels are used. Moreover blended fuel also produced lower CO, CO_2 and HC concentrations compared to those of neat gasoline. Yacoub et al. [26] examined butanol-gasoline blends with carbon numbers C1 to C5. The results showed that all n-butanol blends had lower CO and UHC emissions. Alasfour [27-28] evaluated the effect of using 30% n-butanol by volume blended with gasoline in a single-cylinder SI engine. He found that the lower engine efficiency by 7% compared to pure gasoline fuel.

This study aim to integrate the existing experimental investigation on combustion performance and emissions characteristics of a SI engine fueled with butanol-gasoline blends in a low proportion of butanol mixture by 5%, 10% and 15% at 50% of wide throttle open (WTO) varied from 1000 to 4000 RPM with interval of 500 RPM. Butanol used in this study are from the second butanol family namely secondary butyl alcohol (sec-butanol). This research can contribute to further the knowledge on the effects of butanol mixture in a four cylinder four strokes port fuel injection SI engines. In addition, as far as the author concern there are little attention has been paid to butanol-gasoline blends mixture from the secondary butyl alcohol family. The investigation of emissions characteristics analyses were performed; in particular CO, CO_2, and HC.

2 Experimental setup

2.1 Materials

In this research investigation, engine testing was done with gasoline fuels as a reference fuels (G100) and blends of 5%, 10% and 15% by volume of sec-butanol in a gasoline fuels indicated as GBu5, GBu10 and GBu15 respectively. Briefly, 2-butanol was added into gasoline fuels and mixed at low stirring rate using an electric magnetic. The mixture was stirred continuously for 15 minutes at room temperature to prepare the blended fuels. Gasoline fuels was bought from local petrol station and stored in the lab inside the proper container. The 2-butanol with percentage of purity of 99.5% were bought from Merck distributor in Malaysia as in Figure 1. The properties of G100 and 2-butanol fuels are specified in Table 1. The fuel blends were prepared just before the start of experiment to ensure that the fuel mixture was homogenous.

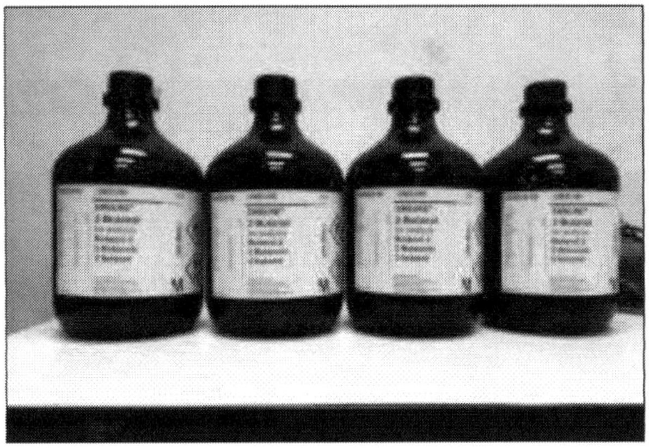

Fig 1. 2-butanol purchased through Merck distributor in Malaysia

2.2 Experimental procedure

In this experimental study, the experiments were performed on a Mitsubishi 1.8 single overhead camshaft (SOHC) engines with four cylinders, four stroke and spark ignition engines. The engine specifications are specified in Table 2. Figure. 2a and b present the actual engine and schematic diagram of the engine experimental test setup. A 100 kW of Dynalec Controls eddy current dynamometer was fixed to the engine in order to apply a consistent 50% of WTO conditions. The load exerted on the engine is measured by the load cell connected to the eddy current dynamometer. All the experiments are conducted and the results are recorded under steady state conditions. Fuel consumption was occupied using AIC fuel flow rate meter with an accuracy of 1% reading. Air consumption was recorded using Benetech GM8903 hot wire type anemometer with the air speeds resolution by 0.001 m/s. The relative air fuel ratio was measured using an accurate calibrated KANE gas

DOI: 10.1051/matecconf/20179001036

analyzer version autoplus 5-2. Sensitivity and measurements accuracy of the exhaust gas concentration have been described in Table 3.

Table 1. Properties of gasoline and 2-butanol [11,21,29-30].

Property	Gasoline	2-butanol
Molar C/H ratio	0.44 – 0.50	-
Density (kg/m^3)	736	806.3
Latent heating value (kJ/kg)	44, 300	33, 000
Stoichiometric air/fuel ratio	14.6	11.1
RON/MON	95/85	101/92~97
Auto – ignition temperature (°C)	228 – 470	406.1
Boiling point (°C)	27 – 225	99.5
Heat of vaporization (kJ/kg)	349	551
Flammable limits (%volume)	1.4 – 7.6	1.7 – 9.8
Laminar flame speeds [31]	~33	~48

Table 2. Engine specifications.

Engine descriptions	
Bore x Stroke	81.0mm x 89.0mm
Piston displacement	1834cc
Compression ratio	9.5:1
Fuel injection type	ECI-Multi (Electronically Controlled Multi-point Fuel Injection
Max power	86kW @ 5500rpm
Max torque	161Nm @ 4500rpm

Table 3. Sensitivity and measurements accuracy of instruments used for measuring the exhaust gas concentration.

Exhaust gas	Measurements domain	Measurement accuracy
CO	0 – 21%	+/- 5% or 0.06% volume^{-1}
CO_2	0 – 16%	+/- 5% or 0.5% volume^{-1}
HC	0 – 5000ppm	+/- 5% or 12ppm volume^{-1}

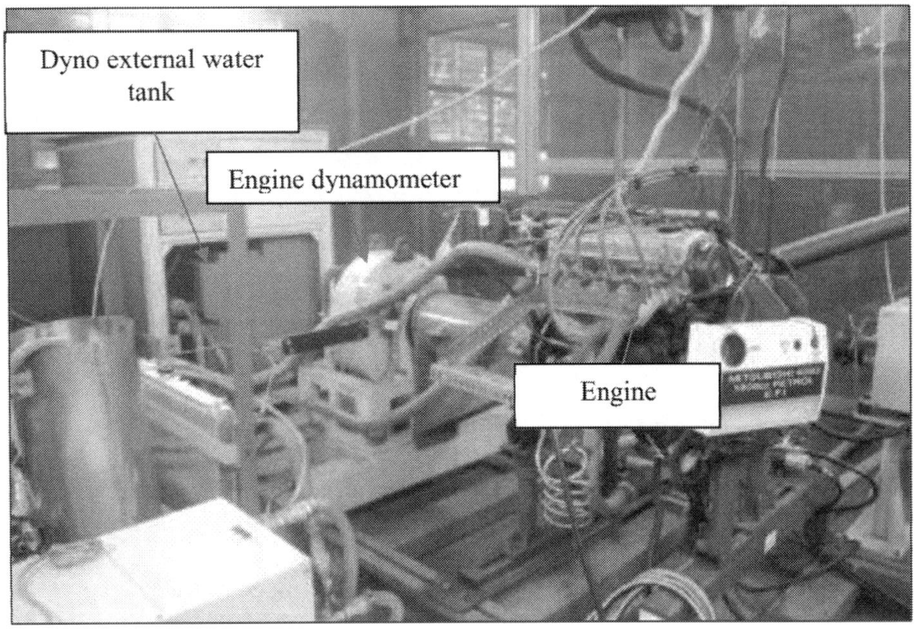

(a)

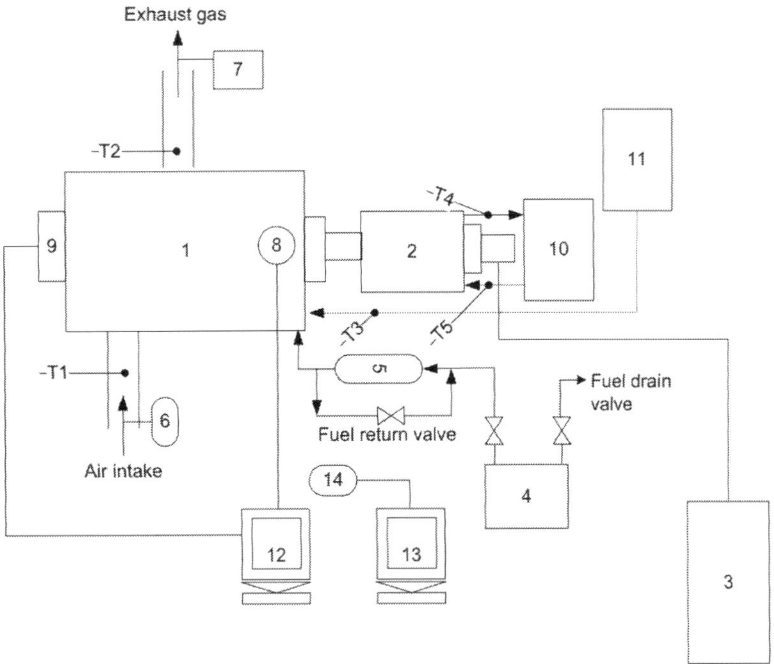

(b)

Fig 2. Engine test bed and test instruments (a) actual and (b) schematic

1. Engine test setup
2. Eddy current dynamometer
3. Dyno controller
4. Fuel tank
5. Fuel pump
6. Air flow rate
7. Exhaust gas analyzer
8. In-cylinder pressure sensor
9. Kistler crank encoder
10. Dyno external water tank
11. Engine external water tank
12. Dewe-5000 combustion analyzer
13. Computer
14. Data logger

2.3 Emissions index

The emissions data were reported using emission index basis to allow comparisons to be made between the different sizes of engines and fuel chemical compositions. According to Saxena and Jotshi [32], the emissions index (EI) can be calculated using the following equations:

$$EI_i = \left(\frac{X_i}{X_{CO} + X_{CO_2} + X_{HC}} \right) \left(\frac{\chi MV_i}{MW_f} \right) \tag{1}$$

Where it can be simplified as:

$$EI_{CO} = \left(\frac{CO}{CO + CO_2 + HC} \right) \times 100\% \tag{2}$$

$$EI_{CO_2} = \left(\frac{CO_2}{CO + CO_2 + HC} \right) \times 100\% \tag{3}$$

$$EI_{HC} = \left(\frac{HC}{CO + CO_2 + HC} \right) \times 100\% \tag{4}$$

Where CO, CO_2 and HC are in parts per million (PPM).

3 Results and discussions

In this study, the sec-butanol-gasoline blended fuels are examined in three different proportions (5%, 10% and 15%) and are compared to the reference fuels neat gasoline fuels in terms of emissions characteristics. The quantity of GBuX represents a blend consisting of X% of sec-butanol by percentage of volume, e.g., GBu5 indicates a blend consisting of 5% of 2-butanol in 95% of gasoline. Four test fuels were used in this study: gasoline (G100); 5% of 2-butanol (GBu5); 10% of butanol (GBu10); and 15% of (GBu15). Incomplete combustion and poor mixing of air and fuel are the major causes of CO productions [33].In Figure 3 presents effects of sec-butanol additions in gasoline fuels to the carbon monoxides (CO) emissions index (EI). From Figure 3, a slight increase was observed for blended fuels from engine speeds 1000 to 2500 RPM. However, as the engine speeds achieved engine speeds of 2500 to 4000 RPM, G100 fuels produced higher CO emissions as compared to blended fuels. The average reduction of CO emissions was calculated for blended fuels compared to G100 fuels in order to distinguish the effects of sec-butanol addition in G100 fuels. A significant of reduction by average of 8.6%, 11.6% and 24.8% for GBu5, GBu10 and GBu15 respectively throughout the speed range of 2500 to 4000 RPM. Hence, the blended fuels is more combustible than the G100 fuels. It appears that this result is in accordance with the studies which have already been reported such as in Ref. [21,34].

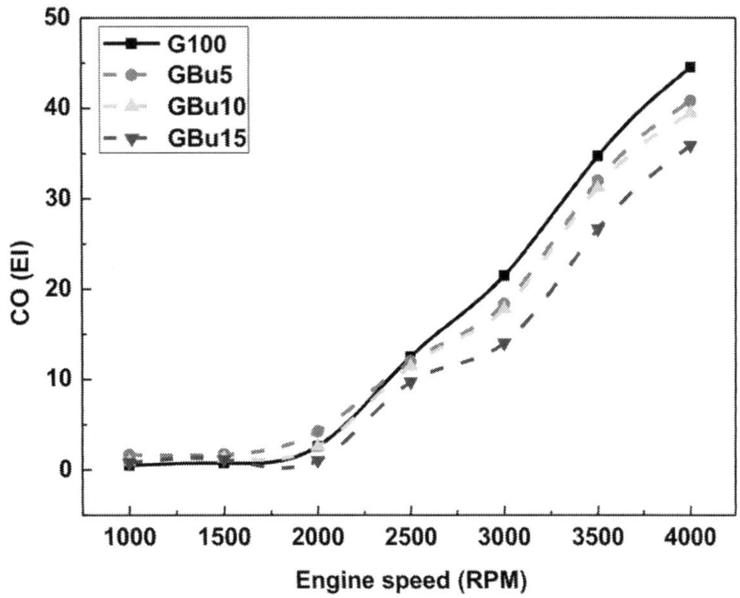

Fig 3. Emission index of carbon monoxide at 50% of WTO

Figure 4 shows effects of sec-butanol additions in a gasoline fuels to the carbon dioxides (CO_2). The CO_2 EI of the G100 was found to be lower than that blended fuels throughout the engine speed of 2500 4000 RPM and on the average it was around 3.7%, 4.7% and 9.1% lower than that of the GBu5, GBu10 and GBu15 respectively. Such increasing trends in the CO_2 EI for the blended fuels may refer to oxygen contents in the sec-butanol. Based on the chemical composition, gasoline only contains carbon and hydrogen atoms; however sec-butanol contains excess oxygen contents including carbon and hydrogen atom. Accordingly, it is realistic to get such CO_2 EI enlargement at using blended fuels.

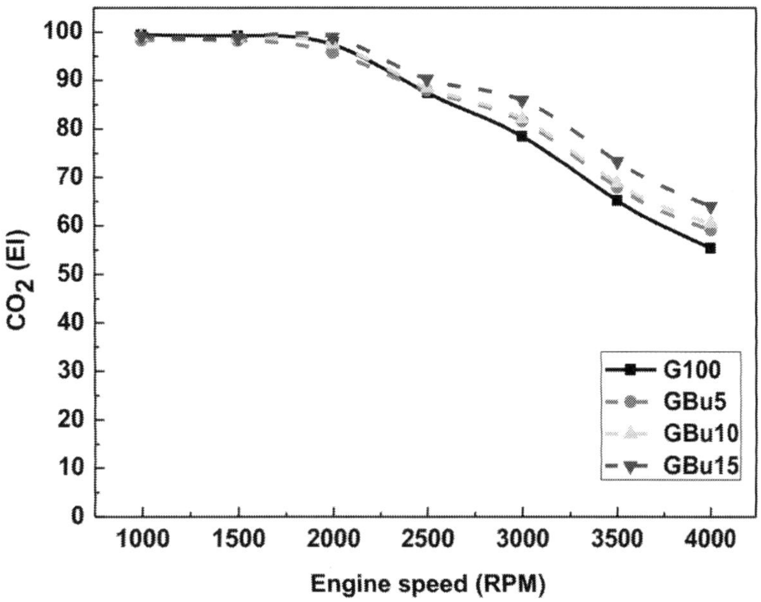

Fig 4. Emission index of carbon dioxide at 50% of WTO

Figure 5 indicates the experimental results on effects of sec-butanol additions in a gasoline fuels to the unburned hydrocarbon (HC). Strong evidence of HC reductions was found especially for 10% and 15% by volume of sec-butanol additions in gasoline fuels. However, comparing the two results between G100 and GBu5, it can be seen that the EI of GBu5 from engine speed 1000 to 2500 RPM was higher than G100 fuels. The rate of HC release is mainly influenced by the chemical compositions of carbon, hydrogen and oxygen of the respective fuels. In a four stroke engine process, particularly in expansion process, drop in cylinder draws compressed unburnt fuel from crevice volume to create unreacted fuel particle that remains in the exhaust. The unreacted unburned fuel continues to increase as the combustion process of the engine continues. The molecular weight of gasoline (114) is much higher than sec-butanol (74.1). Mixture of sec-butanol-gasoline blends produce lower molecular weight of blended fuel. Being a much more light weight fuel sec-butanol-gasoline blends capable to form much better homogenous air-fuel mixture. In addition, presence of oxygen further improve the combustion of the blended fuels. From the calculation, the HC EI of GBu10 and GBu15 was lower than that of the G100 throughout the engine speed range, and on average of 13.4% and 27.1% for GBu10 and GBu15 respectively.

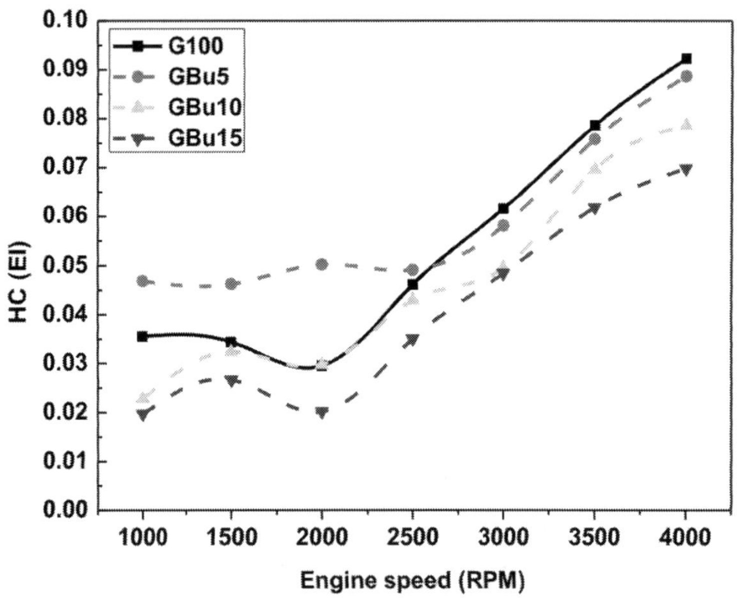

Fig 5. Emission index of unburned hydrocarbon at 50% of WTO

4 Conclusions

From this research investigation, it can be concluded that 2-butanol gasoline blends of GBu5, GBu10 and GBu15 can be directly used in gasoline engines without modifications. The evidence from this experimental study suggests that throughout the engine speeds of 1000 to 2500 RPM GBu5 produces higher CO and HC, but lower CO2 EI compared to the G100 fuels. On the other hand, following the increase of engine speeds from 2500 to 4000 RPM, all of the blended fuels significantly reduced CO by percentage average of 8.6%, 11.6% and 24.8% for GBu5, GBu10 and GBu15 respectively. Meanwhile G100 fuels produced lower CO¬2 by averaged of 3.7%, 4.7% and 9.1% lower than that of the GBu5, GBu10 and GBu15 respectively. It is also worth noting that HC EI deteriorated by averaged of 13.4% and 27.1% for GBu10 and GBu15 respectively compared to the G100 fuels throughout all engine speeds.

Appreciation and acknowledgement to the Ministry of Higher Education (KPT) for providing author the scholarship under My Brain 15 schemes and financial support under Universiti Malaysia Pahang Grand RDU 1403156. Sincere thanks to Mr. Muhd Hafietz Bin Yusoff for bountiful assistant in term of technical supports during the engine testing. Finally, the authors thank the anonymous referees and the editor for carefully reading this paper and suggesting many helpful comments on improving the original manuscript.

References

1 H. C. Ong, T. M. I. Mahlia and H. H. Masjuki. (2012). A review on energy pattern and policy for transportation sector in Malaysia. *Renewable and Sustainable Energy Reviews*, 16(1), 532-542.

2 A. Fazeli, F. Bakhtvar, L. Jahanshaloo, N. A. Che Sidik and A. E. Bayat. (2016). Malaysia's stand on municipal solid waste conversion to energy: A review. *Renewable and Sustainable Energy Reviews*, 58, 1007-1016.

3 Y. F. Wang, K. P. Li, X. M. Xu and Y. R. Zhang. (2014). Transport energy consumption and saving in China. *Renewable and Sustainable Energy Reviews*, 29, 641-655.

4 Y. S. H. Najjar. (2013). Protection of the environment by using innovative greening technologies in land transport. *Renewable and Sustainable Energy Reviews*, 26, 480-491.

5 S. A. Osman, A. J. Alimin, M. Y. Ismail and K. W. Hui. (2013). Performance and Emission Characteristics of Direct Injection C.I Engine Retrofitted with Mono-CNG System. *Applied Mechanics and Materials*, 446-447, 443-447.

6 B. L. Salvi, K. A. Subramanian and N. L. Panwar. (2013). Alternative fuels for transportation vehicles: A technical review. *Renewable and Sustainable Energy Reviews*, 25, 404-419.

7 B. L. Salvi and N. L. Panwar. (2012). Biodiesel resources and production technologies – A review. *Renewable and Sustainable Energy Reviews*, 16(6), 3680-3689.

8 B. M. Masum, H. H. Masjuki, M. A. Kalam, S. M. Palash, M. A. Wakil and S. Imtenan. (2014). Tailoring the key fuel properties using different alcohols (C2–C6) and their evaluation in gasoline engine. *Energy Conversion and Management*, 88, 382-390.

9 B. M. Masum, H. H. Masjuki, M. A. Kalam, S. M. Palash and M. Habibullah. (2015). Effect of alcohol–gasoline blends optimization on fuel properties, performance and emissions of a SI engine. *Journal of Cleaner Production*, 86, 230-237.

10 A. Elfasakhany. (2015). Investigations on the effects of ethanol–methanol–gasoline blends in a spark-ignition engine: Performance and emissions analysis. *Engineering Science and Technology, an International Journal*, 18(4), 713-719.

11 C. Jin, M. Yao, H. Liu, C.-f. F. Lee and J. Ji. (2011). Progress in the production and application of n-butanol as a biofuel. *Renewable and Sustainable Energy Reviews*, 15(8), 4080-4106.

12 I. M. Yusri, M. K. Akasyah, R. Mamat and O. M. Ali. (2015). Combustion and emissions characteristics of a compression ignition engine fuelled with n-butanol blends. *Jurnal Teknologi (Science and Engineering)*, 77(8), 69-73.

13 I. M. Yusri, R. Mamat, O. M. Ali, A. Aziz, M. Akasyah, M. K. Kamarulzaman, et al. (2016). The combustion of n-butanol-diesel fuel blends and its cycle to cycle variability in a modern common-rail diesel engine. *Journal of Engineering and Applied Sciences*, 11(4), 2297-2301.

14 A. Irimescu. (2012). Performance and fuel conversion efficiency of a spark ignition engine fueled with iso-butanol. *Applied Energy*, 96, 477-483.
15 Z. Chen, Z. Wu, J. Liu and C. Lee. (2014). Combustion and emissions characteristics of high n-butanol/diesel ratio blend in a heavy-duty diesel engine and EGR impact. *Energy Conversion and Management*, 78, 787-795.
16 A. Uyumaz. (2015). An experimental investigation into combustion and performance characteristics of an HCCI gasoline engine fueled with n-heptane, isopropanol and n-butanol fuel blends at different inlet air temperatures. *Energy Conversion and Management*, 98, 199-207.
17 E. G. Giakoumis, C. D. Rakopoulos, A. M. Dimaratos and D. C. Rakopoulos. (2013). Exhaust emissions with ethanol or n-butanol diesel fuel blends during transient operation: A review. *Renewable and Sustainable Energy Reviews*, 17, 170-190.
18 A. Demirbas. (2008). Biofuels sources, biofuel policy, biofuel economy and global biofuel projections. *Energy Conversion and Management*, 49(8), 2106-2116.
19 R. Feng, J. Fu, J. Yang, Y. Wang, Y. Li, B. Deng, et al. (2015). Combustion and emissions study on motorcycle engine fueled with butanol-gasoline blend. *Renewable Energy*, 81, 113-122.
20 S. Szwaja and J. D. Naber. (2010). Combustion of n-butanol in a spark-ignition IC engine. *Fuel*, 89(7), 1573-1582.
21 X. Gu, Z. Huang, J. Cai, J. Gong, X. Wu and C.-f. Lee. (2012). Emission characteristics of a spark-ignition engine fuelled with gasoline-n-butanol blends in combination with EGR. *Fuel*, 93, 611-617.
22 J. Dernotte, C. Mounaim-Rousselle, F. Halter and P. Seers. (2009). Evaluation of butanol–gasoline blends in a port fuel-injection, spark-ignition engine. *Oil & Gas Science and Technology – Revue de l'Institut Français du Pétrole*, 65(2), 345-351.
23 E. Galloni, G. Fontana, S. Staccone and F. Scala. (2016). Performance analyses of a spark-ignition engine firing with gasoline–butanol blends at partial load operation. *Energy Conversion and Management*, 110, 319-326.
24 S. B. Singh, A. Dhar and A. K. Agarwal. (2015). Technical feasibility study of butanol–gasoline blends for powering medium-duty transportation spark ignition engine. *Renewable Energy*, 76, 706-716.
25 A. Elfasakhany. (2014). Experimental study on emissions and performance of an internal combustion engine fueled with gasoline and gasoline/n-butanol blends. *Energy Conversion and Management*, 88, 277-283.
26 Y. Yacoub, R. Bata and M. Gautam. (1998). The performance and emissions characteristics of C1-C5 alcohol-gasoline blends with matched oxygen content in a single-cylinder spark igntion engine. *Proceedings of the Institution of Mechanical Engineers, Part A: Journal of Power and Energy*, 212.
27 F. N. Alasfour. (1997). Butanol - A single-engine study: Availability analysis. *Applied Thermal Engineering*, 17(6), 537-549.
28 F. Alasfour. (1999). The effect of using 30% iso-butanol-gasoline blend on hydrocarbon emissions from a spark-ignition engine. *Energy Sources*, 21(5), 379-394.
29 M. K. Balki and C. Sayin. (2014). The effect of compression ratio on the performance, emissions and combustion of an SI (spark ignition) engine fueled with pure ethanol, methanol and unleaded gasoline. *Energy*, 71, 194-201.
30 E. Christensen, J. Yanowitz, M. Ratcliff and R. L. McCormick. (2011). Renewable oxygenate blending effects on gasoline properties. *Energy & Fuels*, 25(10), 4723-4733.
31 K. Nithyanandan, H. Wu, M. Huo and C.-F. Lee. (; 2014). A preliminary investigation of the performance and emissions of a port-fuel Injected SI engine fueled with acetone-butanol-ethanol (ABE) and gasoline. *SAE technical paper 2014-04-01*.

32 S. Saxena and C. Jotshi. (1994). Fluidized-bed incineration of waste materials. *Progress in Energy and combustion Science*, 20(4), 281-324.
33 M. I. Jahirul, H. Masjuki, R. Saidur, M. Kalam, M. Jayed and M. Wazed. (2010). Comparative engine performance and emission analysis of CNG and gasoline in a retrofitted car engine. *Applied Thermal Engineering*, 30(14), 2219-2226.
34 G. Broustail, F. Halter, P. Seers, G. Moréac and C. Mounaim-Rousselle. (2012). Comparison of regulated and non-regulated pollutants with iso-octane/butanol and iso-octane/ethanol blends in a port-fuel injection spark-ignition engine. *Fuel*, 94, 251-261.

MATEC Web of Conferences **90**, 01037 (2017) DOI: 10.1051/matecconf/20179001037
AiGEV 2016

Investigating bending strength of spur gear: A review

Ismail Ali Abdul Aziz[1,*], *Daing Mohamad Nafiz* Daing Idris[1], *and Wafiuddin* Mohd Ghazali[1]

[1]Faculty of Mechanical Engineering, Universiti Malaysia Pahang, 26600 Pekan, Pahang, Malaysia,

Abstract. Gear tends to play a very vital role in all industries. As other mechanical systems, it is subject to design parameter, installation errors, method of manufacture and load uncertainties arising from randomness. Before engineers could design an efficient and safe gear, it is importance to understand how gear can fail is much needed. This article reviews the methodology used to investigate bending strength of spur gear; Finite Element Method (FEM), Numerical Calculation and Investigational Techniques were usually carried out in order to understand the bending strength of thin-rimmed spur gear. Works and experiment from the literature were studied, and their findings were extracted to understand the methods used. The most common method used to investigate bending strength is the numerical calculation. This method was used with several types of established equations and standards to predict gear failures. Next stage is to simulate the gear using Finite Element Method (FEM) in order to get the analysis of gear strength. The most important stage is to put the gears to physical experiment or testing facilities to determine and validate all data from the numerical calculation and FEM method.

1 Introduction

Nowadays, there are huge demands to have lightweight and efficient gear in the industry especially in aerospace and transportation applications[3]. Since the creation of gearing system to transmit water from river thousand years ago, the importance of gearing system become more crucial. With growing demands for a sustainable environment, it is an utmost importance for industries to save natural resources and cut the global carbon footprint [4]. Aircraft manufacturers and automakers are showing more interest in developing innovative solutions that reduce emissions with improved fuel economy by reducing weight. In gearing applications such as gearbox for vehicles and aircraft, thin-rimmed gears can help to reduce the weight of gearbox. Therefore, the usage of thin-rimmed gear became more important and often used in an application where lightweight and compact designs were demand.

Gears could fail in many unlike ways, and except for an increase noise level and vibration, there is no early indication of difficulty until total failure occurs[5]. The general types of failure modes (in decreasing order of frequency) include fatigue[6], fracture[7],

[*] Corresponding author: asmaelali82@gmail.com

wear[8], and stress rupture. Gear tooth failures occur in two distinct regions, the tooth flank and the root fillet[9]. For thin-rimmed gears, bending stress can be different because of rim and web thickness factor [10-12]. In thin-rimmed case, the rim of a gear usually fails between two adjacent teeth. Cracks propagate through the rim and into the web[12]. Sometimes cracks immerge in the web near the rim and web joint without disturbing the rim itself. Damaged gear can cause the entire operation of a machine to come to a standstill. Therefore fracture of gear tooth is the cause to be most concerned about, and worst can lead to the destruction of the machine. A crack patenting in the root section of gear tooth mostly end up with gear tooth failure from bending fatigue. There was usually evidence of a fatigue focal point of the break that leads to fracture under cyclic loading. Tooth fracture also occurs under overload loading. Tooth breakage caused by an overload, which exceeds the tensile in a short-cycle break, has a finer stringy appearance but still displays evidence of being pulled apart abruptly [1].

The importance of understanding on how gear can fail is much needed before engineers could design efficient and safe gears. Fig. **1** shows approaches done via various research in understanding gear bending strength. Gear testing can help designers to understand various kinds of gear failures, as well as estimating the gear life in service[13]. The aid to the designer is the ability to recognize the exact type of initial failure and come out with a method to avoid those failures. There are several numerical standards used to design gear as published by American Gear Manufacturers Association (AGMA) and International Organization for Standardization (ISO) with complex equations and calculations. In addition, Finite Elements Method (FEM) used to predict gear failure, estimates gears life in service[14] and later being verified by experimental test on bending strength. On top of numerical calculation and FEM, experimental gear testing is the best approach to find out gear bending strength. Gear design calculations and experimental tests are necessary to develop and confirm a hypothesis for the probable cause of failure [2].

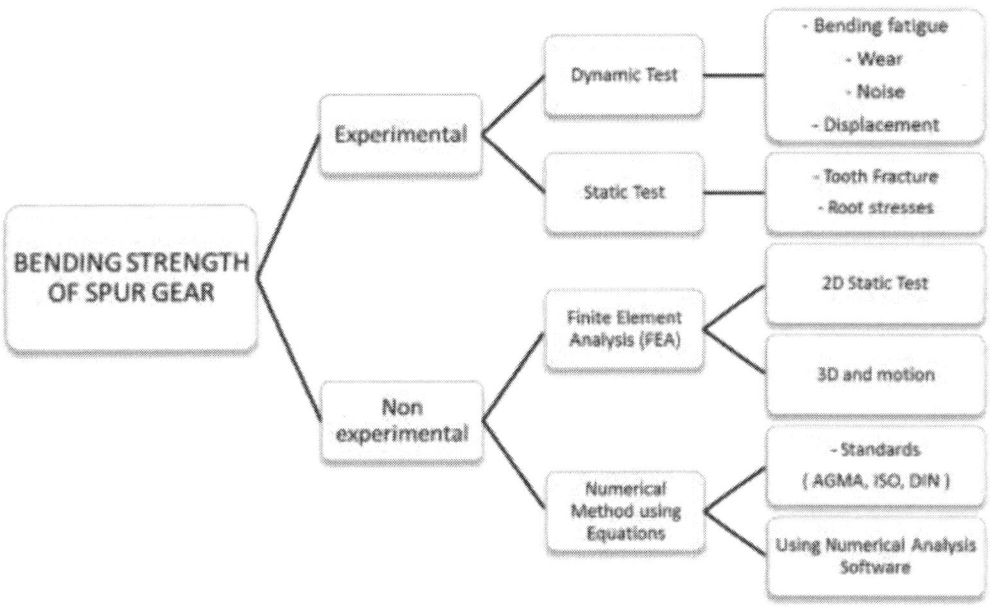

Fig. 1. Approach in understanding gear bending strength.

DOI: 10.1051/matecconf/20179001037

1.1 Gear failures

A pair of gear tooth in action generally subjected to cyclic stress failure that includes tooth-bending fatigue, tooth-bending impact, and tooth wear[1]. Gear tooth failures occur in two distinct regions, the tooth flank, and the root fillet. Gear can fail in several modes as shown in Fig. **2**.

A fracture of a tooth is the cause to be most concerned about as damaged gear can cause the entire operation of a machine to come to a standstill and worst can lead to the destruction of the machine. Tooth breakage caused by an overload, which exceeds the tensile in a short-cycle break, has a finer stringy presence but still shows evidence of being pulled apart abruptly [1]. In thin-rimmed case, sometimes cracks appear in the web near the rim and web junction without disturbing the rim itself. The rim of gear might fail between two adjacent teeth. Cracks spread through the rim and into the web. Gear tooth failure from bending fatigue generally appears from a crack initiation at the root section of the gear tooth[6]. There are other causes lead to bending fatigue failures, too much load, and cyclic loading that stress at the gear tooth beyond the endurance limits of the material may be one of major concern.

Gear tooth failure from bending fatigue mostly resulted from a crack patenting in the root section of the gear tooth. Based on gears rotational motion and the amount of load they carry, it's subjected to fatigue[15]. Failures that occur under applied load less than 10, 000 cycles considered as low cycle fatigue. Whereas high cycle fatigue defines as a failure that occurred after more than 10,000 cycles of applied load. The stress levels involved in low cycle fatigue are normally higher than the yield strength of the material.

As the nature of gearing system to have contact with each other, it tends to wear out. The most common wear failure is moderate wear. This type of wear happens for a huge number of cycles and might involve lack of lubricant or fine dirt in the system. Abrasive wear occurs due to the presence of metal particles from the gears and bearings, weld spatters, scale, rust, sand and dirt in the lubricant of the gears [1]. Meanwhile, corrosive wear happens due to corrosion action on the teeth of the gears. This type of wear failure is visually looked at as spotty dots appearing scattered on the tooth of the gear.

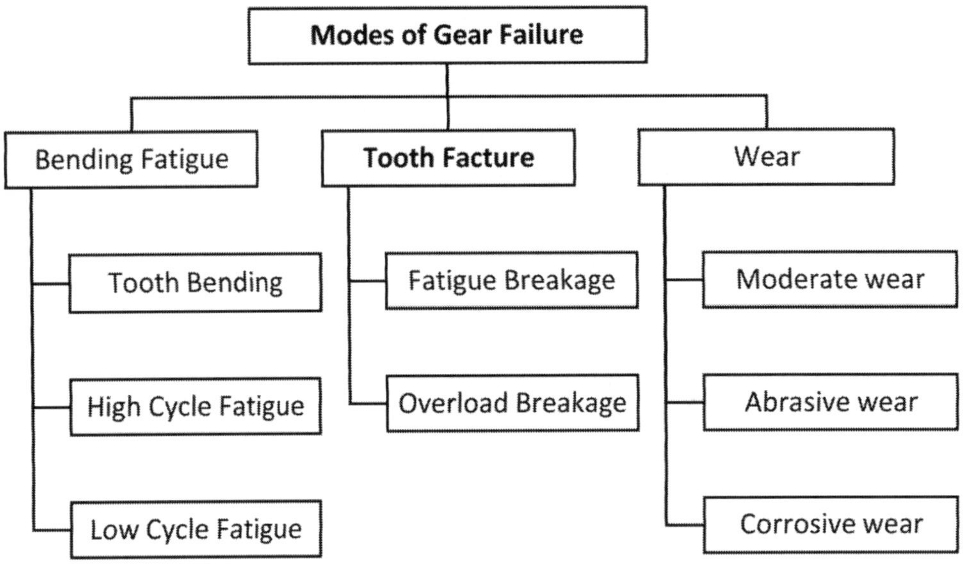

Fig. 2. Modes of Gear Failure[1, 2].

2 Finite element method (FEM)

Before actual experiment was conducted on a specimen, usually researchers will conduct simulations of the specimen. Works that use the finite element method could help the researchers to predict and understand the bending strength of gear before conducting a physical experiment or testing.

Toni Jabbour et. al[16] verified their calculation of the root and contact stresses with finite element calculations. Another gear associated to enable the simultaneous rotation of the two gears with consistent gear ratio. In order to choose the most effective mesh refinement, the maximum contact pressure values, obtained from several analyses were compared because the precision of the contact pressure value depends on the number of elements. From Finite element calculation, the critical tooth-root stress which leads from the position of the point of contact depends on the contact ratio of the pair of gears the increases number of gear teeth.

Jiaxing Zhan et. al[17] suggested that attention should be given to the tip relief of gear when conducting testing on gears. From their simulation, contact stress was decreased constantly when teeth pair starts to contact. Tooth tip with no modification indicates reading up to 476 MPa in contact stress value compare to the tip radius of 1.00mm indicates 190 MPa. This five teeth gear model were fine meshed with up to 31 740 elements and the total number of nodes was 143 938.Tthe time-varying load capacity of gears could be demonstrated in the quasi-static FEM in a complete mesh cycle.

A most recent study by Vincent Savaria [4] used FE model calibrated with representative axial and torsion laboratory specimens, based on the well-known Crossland criterion. The geometry used was a simplified 3D FE model including only one tooth and being compared to Von Mises strain maps obtained by digital image correlation and the FE model as shown in Fig. **3**. Results from 3D FE model shows the maximum root bending stress was up to 2083 MPa on model A while 1651 MPa for model D and baseline model with no induction shows 1049 MPa of maximum root bending stress. Results from FEA test were validated from data obtained by experiment on single tooth bending fatigue rig (STBF).

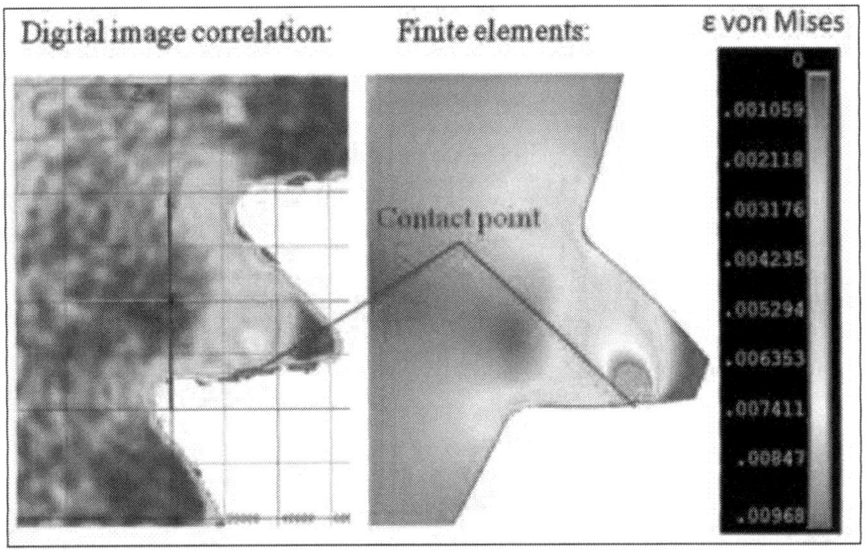

Fig. 3. Comparison of Von Mises strain maps obtained by digital image correlation and the FE model [4].

In order to reduce bending stress in external spur gears by redesign the standard cutting tools, Niels L. Pederson [18] simulated Finite Element Modelling of external spur gear with Poisson's ratio of $v = 0.3$. The load and supports, as well as, the geometry were earlier specified. Results from FEA simulated by F.Cura et. al [19] views that crack propagation path and crack initiation point were strongly influenced by centrifugal load. Spur gears with different web and thin-rimmed were modeled before boundary conditions consist in displacement loads to simulate the presence of the whole gear.

3 Analytical calculations method

Analytical Calculations were one of the most important stages in designing a gear. Through this method, design and analysis of the spur gears to resist bending failure of tooth surfaces can be implemented.

3.1 The Lewis bending equation

Wilfred Lewis[20] in 1892 introduced the bending equation that remains the most common or basic standard for gear design. Lewis calculated stress in the gear using a cantilever beam under an applied bending moment[21]. A rectangular cantilever beam of cross-sectional dimensions F and t, having a length l and a load W^t, uniformly distributed across the face width F. The section modulus I/C is $Ft^2/6$ end therefore, the bending stress is

$$\sigma = \frac{M}{I/c} = \frac{6W^t l}{Ft^2} \tag{1}$$

This basic formula being completed by including factor y that is called the Lewis form factor, so we have

$$\sigma = \frac{W^t P}{FY} \tag{2}$$

where P is the diametrical pitch. The used of this equation for Y means that only the bending of the tooth is considered and that the compression due to the radial component of the force is neglected.

3.2 AGMA stress equations

Based on Equation (2), ANSI/AGMA 2001-D04 and 2101-D04 contains dynamic factor K_v. Two fundamental stress equations are used in the AGMA methodology, one for bending stress and another for pitting resistance (contact stress). The fundamental equations are

$$\sigma = W^t K_o K_v K_s \frac{1}{bm_t}\frac{K_H K_B}{Y_J} \quad \text{(SI Unit)} \tag{3}$$

where

W^t is the tangential load, (N)	K_H is the load –distribution factor
K_o is the overload factor	K_B is the rim-thickness factor
K_v is the dynamic factor	b is the face width of the narrower member, (mm)
K_s is the size factor	
m_t is the transverse module	Y_J is the geometry factor for bending strength

3.3 Aida & Terauchi – On the bending stress of a spur gear

In 1962, Aida & Terauchi [22] arrived at the conclusion that was rational to take the sharing force into attention for the calculation of the bending stress of a gear tooth. They have calculated the bending stress of a standard gear tooth by defining the thickness and the position of the weakest section and the worst loaded point.

Therefore, the subsequent formula has been presented for the bending stress of a gear tooth:

$$\sigma_t = \left(1 + 0.08\,\frac{S}{\rho}\right)\left(0.66\,\sigma N_b + 0.40\,\sqrt{\sigma N_b{}^2 + 36\,\tau N^2} + 1.15\,\sigma\,N_c\right) \quad (4)$$

where

S	is tooth thickness on the weakest section
ρ	is the radius of curvature
σN_b	is nominal stress
$\sigma\,N_c$	is compressive stress
τN	is nominal shearing stress

the nominal stress σN_B, σN_c and τN can be calculated using of the following equations;

$$\sigma N_b = \frac{6P\,sin\theta\,(l_a + l_d)}{S^2 b} \quad (5\text{-}1)$$

$$\sigma N_c = -\frac{P\,cos\theta}{Sb} - \frac{6P\,cos\theta\,y}{S^2 b} \quad (5\text{-}2)$$

$$\tau N = \frac{P\,sin\theta}{Sb} \quad (5\text{-}3)$$

where P is load, θ is the angle between neutral line of the tooth, l_a is distance from the pitch point to the foot of the normal to the neutral line of tooth crossing the loading point, l_d is distance between the pitch point and the weakest section, b is tooth width and y is the distance between the loading point and the neutral line of tooth.

Many researchers since then have performed analytical studies to investigate gear tooth bending stress [23-26]. Chen et. al [27] came out with a method for the calculation of root and contact stresses for metal, spur and helical gears. They verified their finding by finite elements calculations. Finding shows the spur gears, the location of the point of contact that leads to the critical tooth-root stress depends on the contact ratio of the pair of gears that rises with the number of teeth.

Li Shunting, since 2002 [11, 28-30] has come out with several studies on bending strength of thin-rimmed spur gears. One of his recent studies on analyses of thin-rimmed spur gear with inclined webs founds that the web location and the web angle of the thin-rimmed gears have a substantial effect on the root bending stresses, the tooth contact stresses, and the joint stresses. His findings were aligned with the calculation from Miyachika [31], F Cura [19] and Opalic [32] that also take consideration on the effect of rim thickness and web arrangements on tooth bending strength of thin-rimmed spur gears.

DOI: 10.1051/matecconf/20179001037

4 Experimental techniques

Reviews also revealed several set up to study gear strength been made by researchers. Most of the researchers focusing the leading causes of failure appeared to be tooth-bending fatigue, tooth-bending impact and abrasive tooth wear. From literature, there are two different approaches being conducted that are single tooth static test and gear pair dynamic test.

David B. Stringer et al [33] set up a new high-speed test capability for determining the high-cycle bending-fatigue characteristics of gear teeth. These test various in load circumstances and cycle-rates. The cycle-rate varied from 50 to 1000 Hz. Li, Shuting [34] assembled a ''power-circulating form'' test rig to performed vibration tests of the thin-walled gear. The test was done at the speed range 500–3000 rpm and used strain gauge phase method to detect the resonance mode shapes of the thin-walled gears when they are running in a complete resonance state. M. J Handschuh et al. [35] established a test setup with equipment for the measurement of root stresses as shown in Fig. **4**(b). They used an open-architecture gearbox with drive and load capacity equipped with motor and brake to operate gears under high-load and low-speed conditions. Later, a study by Qi Zhang [36] used a setup of The FZG test rig that was a back to back gear test rig of the closed power loop type.

A most recent study done by Pawar, P. B et al [37] prepared hardness test of composite samples and conducted an experiment using on Automatic Optical Brinell Hardness Tester with range 250 kg to 3000 kg. A year before, Nizar Ahamed et al [38] develop a methodology which was tough for fault recognition of gears under fluctuating load and speed settings. A multiple-pulse independently rescaled-time synchronous averaging (MIR-TSA) technique in conjunction with conventional time synchronous averaging has been projected. It has been examined that the proposed method improves the fault detection under fluctuating speed conditions. A similar setup was used by Nick Bretl et al [39] to investigate tooth bending strength of case hardened gears in the range of high cycle fatigue. A pulsator test rig is shown in Fig. **4**(a) was consistently carried out the bending test.

(a) (b)

Fig. 4. (a) Illustration of a Pulsator Test Rig [39], (b) The gearbox setup for the measurement of root stress [35].

5 Conclusions

Failure of gear might occur in various type of modes from wear, bending fatigue, and tooth fracture. It is imperative for engineers and designers to perform an investigation on gear failure and most crucial is gear bending strength. From the reviewed articles, it could be concluded that there are several stages had been done in order to investigate gear bending strength.

1. Most researchers used the basic Lewis Bending Equation and AGMA Standards with modifications to get the first view of gear tooth strength. In fact, there are also some other equations such as Aida & Terauchi Equation, and many researchers also try to formulate their calculations to predict gear strength.
2. The second stage is to simulate gears using Finite Element Method (FEM) in order to have the analysis of gear strength. Technological advances in FEM provides an opportunity for researchers to conduct various tests based on various parameters.
3. The most important stage is to put the gears to physical experiment or testing facilities to determine and validate all data from numerical calculation and FEM method.

This work has been done under the grant no. RDU 1403132 and enormous support from Universiti Malaysia Pahang. The authors gratefully acknowledge the support given throughout this project.

References

1 E. E. Shipley, "Gear Failure," in *Machine Design*, ed Cleveland, Ohio: The Penton Publishing co., 1967.
2 J. R. Davis, "GEAR MATERIALS, PROPERTIES, AND MANUFACTURE," in *Mechanical Testing and Evaluation*, D. Associates, Ed., ed USA: ASM International, Materials Park, 2005.
3 S. K. R. G.D.Bibel, M Savage, "Effect of Rim Thickness on Spur Gear Bending Stress," p. 15, 1991.
4 V. Savaria, F. Bridier, and P. Bocher, "Predicting the effects of material properties gradient and residual stresses on the bending fatigue strength of induction hardened aeronautical gears," *International Journal of Fatigue,* vol. 85, pp. 70-84, 2016.
5 A. Belsak and J. Flasker, "Detecting cracks in the tooth root of gears," *Engineering Failure Analysis,* vol. 14, pp. 1466-1475, 2007.
6 P. J. L. Fernandes, "Tooth bending fatigue failures in gears," *Engineering Failure Analysis,* vol. 3, pp. 219-225, 1996.
7 S. Pehan, J. Kramberger, J. Flašker, and B. Zafošnik, "Investigation of crack propagation scatter in a gear tooth's root," *Engineering Fracture Mechanics,* vol. 75, pp. 1266-1283, 2008.
8 C. Hu, W. A. Smith, R. B. Randall, and Z. Peng, "Development of a gear vibration indicator and its application in gear wear monitoring," *Mechanical Systems and Signal Processing,* 2016.
9 H. J. Decker, J. M. Certo, R. F. Handschuh, and T. L. Krantz, "Development of a new single gear tooth bending fatigue test capability," in *62nd American Helicopter Society International Annual Forum* Phoenix, AZ, 2006, pp. 704-707.
10 K. M. Satoshi ODA, Takao KOIDE Toshiaki Fuji, "Dymanic Behavior of Thin-rimmed Spur Gear with Various Web Arrangemants," *Buletin of JSME,* vol. 29, pp. 247-235, 1986.
11 S. Li, "Effect of addendum on contact strength, bending strength and basic performance parameters of a pair of spur gears," *Mechanism and Machine Theory,* vol. 43, pp. 1557-1584, 2008.

12 G. Marunić, "Effects of rim and web thickness on gear tooth root, rim and web stresses," in *Key Engineering Materials* vol. 385-387, ed, 2008, pp. 117-120.
13 S. Glodež, S. Pehan, and J. Flašker, "Experimental results of the fatigue crack growth in a gear tooth root," *International Journal of Fatigue,* vol. 20, pp. 669-675, 1998.
14 J. Kramberger, M. Šraml, I. Potrč, and J. Flašker, "Numerical calculation of bending fatigue life of thin-rim spur gears," *Engineering Fracture Mechanics,* vol. 71, pp. 647-656, 2004.
15 Q. J. Yang, "Fatigue test and reliability design of gears," *International Journal of Fatigue,* vol. 18, pp. 171-177, 1996.
16 T. Jabbour and G. Asmar, "Tooth stress calculation of metal spur and helical gears," *Mechanism and Machine Theory,* vol. 92, pp. 375-390, 10// 2015.
17 J. Zhan, M. Fard, and R. Jazar, "A quasi-static FEM for estimating gear load capacity," *Measurement,* vol. 75, pp. 40-49, 11// 2015.
18 N. Pedersen, "Reducing bending stress in external spur gears by redesign of the standard cutting tool," *Structural and Multidisciplinary Optimization,* vol. 38, pp. 215-227, 2009.
19 F. Curà, A. Mura, and C. Rosso, "Effect of centrifugal load on crack path in thin-rimmed and webbed gears," *Frattura ed Integrita Strutturale,* vol. 9, pp. 447-455, 2015.
20 W. Lewis, "Investigation of the strength of gear teeth," *Proc. of the Engineers Club,* pp. 16-23, 1893.
21 E. J. Wellauer and A. Seireg, "Bending Strength of Gear Teeth by Cantilever-Plate Theory," *Journal of Engineering for Industry,* vol. 82, pp. 213-220, 1960.
22 Y. T. Toshio Aida, "On the Bending Stress of a Spur Gear," *The Japan Society of Mechanical Engineers,* vol. 5, pp. 161-183, 1962.
23 D. G. Lewicki and R. Ballarini, "Effect of Rim Thickness on Gear Crack Propagation Path," *Journal of Mechanical Design,* vol. 119, pp. 88-95, 1997.
24 M. A. Sahir Arikan and B. Kaftanoglu, "Dynamic Load and Root Stress Analysis of Spur Gears," *CIRP Annals - Manufacturing Technology,* vol. 38, pp. 171-174, 1989.
25 Y. Pandya and A. Parey, "Experimental investigation of spur gear tooth mesh stiffness in the presence of crack using photoelasticity technique," *Engineering Failure Analysis,* vol. 34, pp. 488-500, 2013.
26 M. Ristivojević, T. Lazović, and A. Vencl, "Studying the load carrying capacity of spur gear tooth flanks," *Mechanism and Machine Theory,* vol. 59, pp. 125-137, 2013.
27 M.-H. Chen and W.-J. Wu, "Bending stress in gear teeth for variable surface pressure distribution," *Journal of Mathematical Analysis and Applications,* vol. 167, pp. 182-202, 1992.
28 S. Li, "Gear contact model and loaded tooth contact analysis of a three-dimensional, thin-rimmed gear," *Journal of Mechanical Design, Transactions of the ASME,* vol. 124, pp. 511-517, 2002.
29 S. Li, "Contact Stress and Root Stress Analyses of Thin-Rimmed Spur Gears With Inclined Webs," *Journal of Mechanical Design,* vol. 134, pp. 051001-051001, 2012.
30 S. Li, "Effects of centrifugal load on tooth contact stresses and bending stresses of thin-rimmed spur gears with inclined webs," *Mechanism and Machine Theory,* vol. 59, pp. 34-47, 2013.
31 K. Miyachika, B. D. I. Daing Mohamad Nafiz, and T. Asano, "Root stresses and bending fatigue strength of thin-rimmed helical gears (helix angle β0 = 10, 20 and 30°)," in *International Gear Conference 2014: 26th–28th August 2014, Lyon*, P. Velex, Ed., ed Oxford: Chandos Publishing, 2014, pp. 836-845.

32 M. Opalić, K. Vučković, and D. Žeželj, "Effect of rotational speed on thin-rim gear bending fatigue crack initiation life," in *Key Engineering Materials* vol. 488-489, ed, 2012, pp. 456-459.
33 B. D. D. David B. Stringer, Kelsen E. LaBerge, Andrew J.Zakrajsek, Robert F.Handschuh, "A New High-Speed, High-Cycle, Gear-Tooth Bending Fatigue Test Capability," *67th Annual Forum and Technology Display (Forum 67),* 2011.
34 S. Li, "Experimental investigation and FEM analysis of resonance frequency behavior of three-dimensional, thin-walled spur gears with a power-circulating test rig," *Mechanism and Machine Theory,* vol. 43, pp. 934-963, 2008.
35 M. J. Handschuh, A. Kahraman, and M. R. Milliren, "Impact of Tooth Spacing Errors on the Root Stresses of Spur Gear Pairs," *Journal of Mechanical Design,* vol. 136, p. 10, 2014.
36 Q. Zhang, J. Zhang, C. Wu, Z. Xu, and S. Lyu, "The evaluation of contact fatigue strength for 20MnCr5 carburized gear," *International Journal of Precision Engineering and Manufacturing,* vol. 15, pp. 117-121, 2014.
37 P. B. Pawar and A. A. Utpat, "Analysis of Composite Material Spur Gear Under Static Loading Condition," *Materials Today: Proceedings,* vol. 2, pp. 2968-2974, 2015.
38 N. Ahamed, Y. Pandya, and A. Parey, "Spur gear tooth root crack detection using time synchronous averaging under fluctuating speed," *Measurement,* vol. 52, pp. 1-11, 2014.
39 S. S. Nick Bretl, Thomas Tobie, Karsten Stahl,Bernd-Robert Höhn, "Investigations on Tooth Root Bending Strength of Case Hardened Gears in the Range of High Cycle Fatigue," *Thermal Processing for Gear Solution,* 2012.

The future of end-of-life vehicles (elv) in Malaysia – A feasibility study among car users in Klang valley

Zulhaidi Mohd Jawi[1,*], *Mohd Hafzi* Md Isa[1], *Mohd Syazwan* Solah[1], *Aqbal Hafeez* Ariffin[1], *Akmalia* Shabadin[2] and *Mohd Rasid* Osman[1]

[1]Vehicle Safety & Biomechanics Research Centre (VSB), Malaysian Institute of Road Safety Research (MIROS), 43000 Kajang, Selangor, Malaysia
[2]Road Safety Engineering & Environment Research Centre (REER), Malaysian Institute of Road Safety Research (MIROS), 43000 Kajang, Selangor, Malaysia

Abstract. In its bid to become a developed nation in a few years' time, Malaysia has to consider various prevailing socio-economic and socio-technical issues in the country. In the transportation sector per se, the ELV policy and initiative is one of the lacking parts in the country's automotive ecosystem – in which a successful ELV program will not only cater the environmental concern but also help the safer car initiative for road users. This particular paper discusses what is regarded as the preliminary findings on the ELV policy from the Malaysia's automotive ecosystem study database. From a total of 484 respondents, 268 or 55.4% had agreed to the proposal to introduce an age limit for passenger vehicles in Malaysia. The majority of those who gave their nod to the policy choose 10 years of vehicle age as the limit (38.9%), and a staggering 79.8% of them supposed that the age limit should be between 5 to 10 years. Further analysis based on the Multiple Logistic Regression found out that from a total of nine important variables related to car users' profile and ownership status, the significant predictors to "the agreement to introduce vehicle age limit" were age, income and car status (new or used). Thus, this finding might be beneficial for the policymakers to strategize the ELV policy that sooner or later should be implemented in the Malaysia's "developed country" environment.

1 Introduction

In Malaysia's road transportation layout, private vehicles, in which the majority are cars and motorcycles, play an important role in everyone's daily routine. Since the public transport oriented mobility in the country is still not at satisfactory level according to the world's standard, both motorcycles and cars are the most convenient options for the daily commuting needs [1]. The motorcycles provide the economic option for mobility, while cars offer the comfort of a four-wheeler and technically more safety features than motorcycles [2]. Though there are many other reasons for someone to own car and/or

* Corresponding author: zulhaidi@miros.gov.my

motorcycle, the users in general regard the ownership of car as a symbol of success. In fact, according to the Nielsen Global Survey of Automotive Demand, Malaysians and users in other ASEAN countries have that particular believe, in which the rate is above the world's average of 52% [3].

For example, Malaysian users' obsession towards cars can be seen in the registration plate number popularity, in which some car owners have exceptional feeling and willingness to have the "stylish" and "meaningful" ones. It perhaps a big surprise to the non-Malaysians to know that prior to the introduction of "F" series, there were two highest bidding so far in the Road Transport Department (RTD) history [4]. They were "W 1 N" that symbolizes the word "WIN" (MYR 748,000 or about USD 200,000) and "WWW 1" that is linked to the internet's World Wide Web (MYR 520,000 or about USD 130,000). Also, it is expected that the recently launched "F" series registration number plate will contribute multi-million ringgits to the government revenue through RTD, in which the number "F1" will be the main target [4].

Furthermore, based on the abovementioned Nielsen's study, car ownership among Malaysians is not only the highest in the ASEAN but also has the highest incidence of multiple car ownership. It is reported that 54% of Malaysian households own more than one car. The abovementioned fact perhaps is adequate to explain the reason behind the high demand of both new and used cars in the country. According to the Total Industry Volume (TIV) in 2015, more than 660,000 new vehicles (including commercials) were sold and approximately 89% of that figure were new cars [5].

The sales of new cars that is stable and in ever-increasing trend seems contradict to the fact that Malaysia is among the countries that have high car prices [6]. Moreover, it is observed that the car users today have high willingness to own the so-called "non-national" cars that are essentially more expensive than the local marque, Proton and Perodua, due to the different taxing system. Though Perodua and Proton hold the biggest market share at around 47% in 2014 and 2015, the domination is actually in a downward trend from as high as 80% in the year 2000 [7]. In fact, the TIV shares among the non-national makes in 2014 had exceeded the 50% mark for the first time after decades i.e. since the establishment of Proton (1985) and Perodua (1993) [8].

Moreover, the Malaysia Automotive Institute (MAI) reckoned that Malaysia is the second cheapest country in the ASEAN region to own a vehicle after the Philippines despite the high car prices [9]. This is largely due to the relatively low fuel prices (previously subsidized), cheaper road tax and insurance premium – thus making the cost of vehicle ownership (CVO) at a very competitive figure. Even in a non-subsidized fuel scenario, the CVO would rank third as compared with its ASEAN counterparts and still cheaper than Thailand and Indonesia [9].

Nevertheless, it is quite unfortunate that the End-of-Life Vehicles (ELV) policy has never taken place in the system [2, 10]. By looking at the automotive world from a bigger picture, the current automotive ecosystem in Malaysia can be considered as incomplete or "unhealthy" (Figure 1) [2]. A car may only permanently or legitimately retire due to other reasons such as not roadworthy beyond economic repair or due to road accidents (premature ELV), rather than guided by a solid systemic intervention to produce a "healthy" system through ELV initiatives. Donella H. Meadows [11] mentioned about the dynamics of stocks and flows in order to understand the behaviour of complex systems using water in the bathtub as an example. If the bathtub is filled with water, its drain plugged up and its faucets turned off, it will produce an unchanging, not dynamic, and boring system (Figure 2). In the case of automotive ecosystem in Malaysia, the cars keep coming into the "tub" and there is no proper outflow from the system that raises two main concerns – environmental and safety issues.

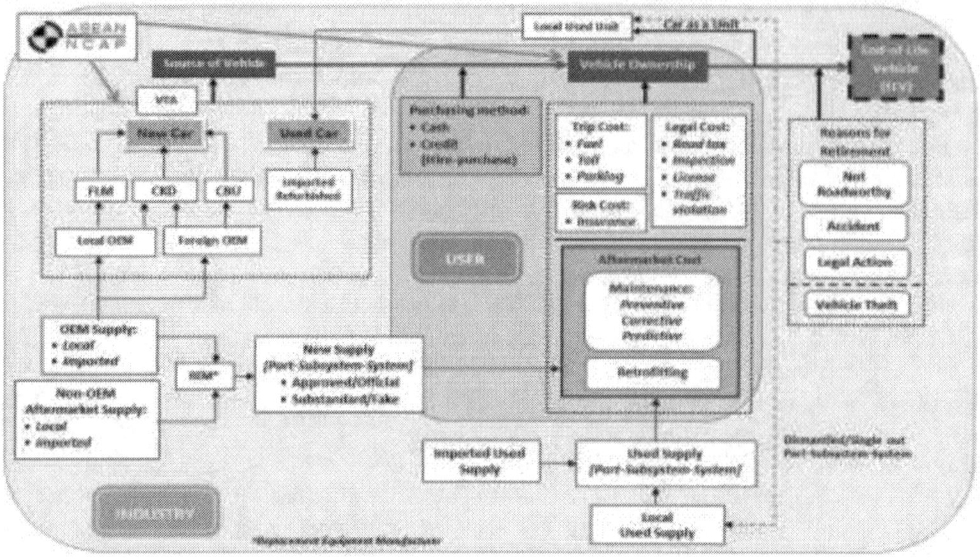

Fig. 1. The automotive ecosystem in Malaysia [2].

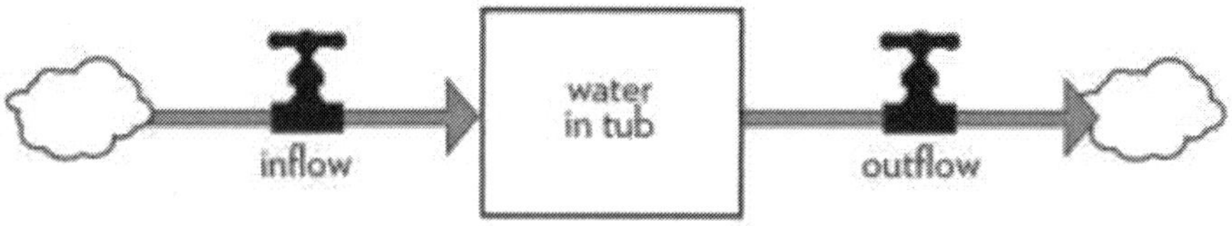

Fig. 2. Understanding complex systems using water in tub example [11].

A proper ELV plan can contribute to sustainable environmental control through the 3R's concept – Reduce, Reuse and Recycle [12]. The concept can be further elaborated into 6R's – Reduce, Reuse, Recycle, Recover, Remanufacture and Redesign – in which it signifies how extensive a proper ELV plan could achieve [12]. For example, certain countries had managed to reach more than 80% in recovery rate [13]. ELV could also help car maintenance by reusing parts or components that are obviously cheaper than new items [14].

Furthermore, aged car is not only deteriorating in its roadworthiness but also in the element of safety such as the body strength. Aged car also could not compete with newer cars that are equipped with latest technologies in both passive (better built and crashworthiness design element) and active safety (e.g. stability control). Thus, in an ideal environment, the old cars should be replaced by new cars by motivating the consumers to buy new cars and the ELV system forces aged or not roadworthy cars to retire.

This is how the safer vehicles initiative through New Car Assessment Program (NCAP) such as ASEAN NCAP and Vehicle Type Approval (VTA) by the Road Transport Department (RTD) can gradually "clean up" the system i.e. new cars that are generally safer than the old ones will replace the existing cars in the system [2]. Effective ELV policy also will be the catalyst to create more demand in new car market and the policy will efficiently "flush out" bad cars from the system. Consequently, the more the number of safer cars in the system may significantly reduce the probability of accidents to occur (mainly active safety) as well as the related injuries by virtue of the better protection in the more modern cars (passive safety).

The Malaysian government had once proposed the "gradual introduction of ELV policy" in the revised National Automotive Policy (NAP) in 2009 but immediately had taken it down due to public uproar upon the announcement [14]. Additionally, the latest version of NAP (2014) had changed its focus to "green vehicle" via the Energy Efficient Vehicle (EEV) scheme and complemented by the Voluntary Vehicle Inspection Policy (VVIP), which can be seen as a less superior move as compared to the controversial "full-scale ELV policy" [13]. Other than that, the government had once conducted the vehicle-scrappage subsidy program but that was merely under the economic stimulus exercise rather than a pure introduction to a permanent ELV structure [13].

It perhaps unique for Malaysia that any new changes to the so-called "routine" in the society will create an instant public uproar and negative perception, though the initiatives are for the better future of the country – economically and environmentally. This recently happened in the introduction of Goods and Services Tax (GST) to replace the old tax system [15], the introduction of Automated Enforcement System (AES) for speeding and red light running offences [16], as well as in the ELV issue [2].

Thus this paper discusses what is regarded as the preliminary findings on the ELV policy from the Malaysia's automotive ecosystem study database. The main aim of this paper is to explore the car users' opinion on the proposal to introduce an age limit to passenger vehicles in the country. Furthermore, it is important to understand the acceptance level of the most important stakeholders in the automotive ecosystem, i.e. the users, so that policymakers can better strategize the ELV policy that sooner or later should be implemented in the Malaysia's "developed country" environment in the year 2020.

2 Methodology

The data used in this paper was taken from the "automotive ecosystem" study's database, in which the effort started as a research cluster at MIROS in 2011. The most recent work in this subject was the cross-sectional self-administered survey done in Klang Valley under the theme of "automotive consumerism". In the abovementioned survey, the eligible respondents were those who owned cars and residing or commuting in the Klang Valley conurbation (or coined as Greater KL since 2010). It was a fairly lengthy questionnaire that consist of six sections and a total of 12 sub-sections.

The data was collected through five different approaches i.e. at public places, offices, industrial areas, schools (for teachers) and car clubs. A total of 500 responses were collected in a period of approximately a year (June 2015 to June 2016), and it is worth to mention that several sub-sections suffered a considerable amount of missing data due to a number of reasons such as the respondents' unfamiliarity with the formal or technical terms in automotive (e.g. NCD for No Claim Discount in car insurance), or they did not aware of certain items asked in the survey (e.g. all about cars are taken care by spouse).

Therefore, this paper would like to present the analysis on the final section of the survey, which is on the respondents' opinion on ELV initiative. The first part of the survey (ELV section) asked whether or not they will agree if there will be a law on vehicle age limit to be implemented in the country, and those who answered "Yes" was asked for the specific age that they think deemed suitable for the law. Simple Logistic Regression (SLR) analysis was conducted to determine the significance of individual predictors based on the car users' profile and ownership status towards "the agreement to vehicle age limit (ELV)" policy. Further analysis using the Multiple Logistic Regression (MLR) analysis was conducted to identify the significant variables using backward elimination method. The variable that has the p-value less than 0.05 was considered as significant variable(s).

3 Results and discussion

From a total of 484 valid responses with regard to the ELV section, 324 or 67% of the respondents were males and the majority of them were from the Malay ethnicity (85.5%) (Table 1). Also, most of the respondents were married (67.6%), had post-secondary level or tertiary education (combined percentage of 79%), worked at the private or Government-Linked Companies (GLC) sector (59.3%), had the monthly income below MYR 5,000 (approx. USD 1,250 at 4.00 exchange rate) (75.6%), owned only one car (74.4%), and using the cars that were bought as new (71.9%).

Furthermore, out of 484 valid responses, 268 (55.4%) of the respondents had agreed to the proposal of having a law on vehicle age limit in the country as part of the ELV solution. The majority of those who gave their nod to the policy choose 10 years of vehicle age as the limit (38.9%), and a staggering 79.8% of them supposed that the age limit should be between 5 to 10 years. Simple Logistic Regression analysis has been done to identify the important variables to predict the agreement to the vehicle age limit policy. Based on Table 1, the significant variables are: age, marital status, education level, income level (monthly) and car status.

The abovementioned five variables that are statistically significant in the Simple Logistic Regression were used in the Multiple Logistic Regression analysis. Out of the five variables, only three are significant, which are age, income and car status (Table 2). The model is a good fit since the significance value for Hosmer and Lemeshow Test is 0.305, which is greater than 0.05. There is also no multicollinearity among the independent variables.

The analysis demonstrated that the buying-in process for the implementation of ELV policy may very well accepted by those with higher income level. It is just a logical observation that users with higher income will prefer better and safer cars since they have the financial freedom to buy and "upgrading" cars. With the rising in income level among the users, Malaysians have the high tendency to upgrade their vehicle when they are financially able – at 93%, which is above the world's average at 89% [3]. In terms of car status, those who own the cars that they bought as "new" are having higher preference for ELV than the owners of used cars. Furthermore, the age factor is significantly decreasing towards older users – even though the impact might not be as big as the abovementioned factors.

The educational approach can also be considered as an additional catalyst in the buying-in process. Though the education level in not statistically significant in the Multiple Logistic Regression analysis, those with higher level of education (under- and post-graduate degree) had higher tendency to support the ELV initiative in the univariate analysis. According to D.A. Wahab and Z.F. Fadzil [17], the knowledge of the public community in Malaysia is at a fairly good level in terms of recycling and reuse of solid waste, but the knowledge on specific endeavors in environmental issues (specific ELV approach) is low. Thus, the educational approach must be done continuously with regard to the understanding of environmental-friendly automotive waste disposal via ELV framework.

According to Shameem et al. [13], the government must be ready to implement several strategies in order to build a solid ELV structure in Malaysia i.e. ELV recycling law, proper deregistration process and destruction permit, comprehensively aided remanufacturing activities, active research in policies and technologies, as well as collaborative effort among the government agencies and local authorities. Nevertheless, the public acceptance is another challenge to the government, especially when it is involved financial capacity. For example, certain ELV practice elsewhere need the user to pay for the disposal process either prior to car ownership (paid by first owner) or at the end of the car's life (paid by last owner) [18]. As mentioned earlier, any new changes in the society that incur additional cost to the users will not be a popular move, and yet the scrappage subsidy program is neither a sustainable option in the long run for the government. Therefore, it is important to monitor users' acceptance towards the Voluntary Vehicle Inspection Policy (VVIP) in order to indirectly gauge the growth in ELV awareness.

Table 1. Unadjusted odds ratio with 95% Confidence Interval (CI).

Factor	N (%)	Unadjusted OR (95% CI)	p-value
Gender			
Female	160 (33.1)		
Male	324 (66.9)	0.726 (0.494, 1.067)	0.103
Race			
Malay	414 (85.5)		
Non-Malay	70 (14.5)	1.435 (0.852, 2.419)	0.175
Age		0.978 (0.958, 0.999)	<0.05
Marital Status			
Single	151 (31.2)		
Married	327 (67.6)	0.654 (0.441, 0.971)	<0.05
Single Parent	6 (1.2)	0.118 (0.013, 1.035)	0.054
Education			
Primary, Secondary & Not Attended School	101 (20.9)		
Cert/Dip/STPM	143 (29.5)	1.293 (0.774, 2.160)	0.326
Degree	196 (40.5)	2.428 (1.486, 3.966)	<0.001
Post Graduate	44 (9.1)	2.608 (1.247, 5.453)	<0.05
Job			
Government	118 (24.4)		
Private/GLC	287 (59.3)	1.181 (0.767, 1.817)	0.450
Self-Employed	47 (9.7)	0.992 (0.504, 1.953)	0.982
Others	32 (6.6)	0.770 (0.352, 1.685)	0.514
Monthly Income			
< RM 3000	201 (41.5)		
RM 3001 - RM 5000	165 (34.1)	0.807 (0.534, 1.219)	0.308
RM 5001 - RM 8000	67 (13.8)	1.574 (0.889, 2.786)	0.120
> RM 8000	49 (10.1)	2.709 (1.335, 5.496)	<0.05
No. of Car(s)			
One car	360 (74.4)		
Two cars	105 (21.7)	0.860 (0.556, 1.330)	0.499
Three or more cars	19 (3.9)	1.075 (0.423, 2.737)	0.879
Car Status			
Used	136 (28.1)		
New	348 (71.9)	1.731 (1.161, 2.582)	<0.05

Note: Columns do not add to totals due to missing data

Table 2. Adjusted odds ratio with 95% Confidence Interval (CI)

Factor	Adjusted OR (95% CI)	p-value
Age	0.956 (0.932, 0.980)	<0.001
Income		
< RM 3000		
RM 3001 - RM 5000	0.877 (0.569, 1.350)	0.551
RM 5001 - RM 8000	1.998 (1.084, 3.684)	<0.05
> RM 8000	4.578 (2.096, 9.997)	<0.05
Car Status		
Used		
New	1.892(1.245, 2.874)	<0.05

DOI: 10.1051/matecconf/20179001038

4 Conclusions

A proper End-of-Life Vehicles (ELV) plan can contribute to sustainable environmental control as well as in promoting safety in the more modern cars. It is quite unfortunate that the ELV policy has never taken place in the Malaysia's automotive ecosystem and therefore the current ecosystem can be considered as incomplete or "unhealthy". The analysis of the data from the Malaysia's automotive ecosystem study database has shown that the government has to face certain challenges with regard to the users in implementing the ELV policy in the country. Though the slight majority of the respondents in this particular study had agreed to the idea of vehicle age limit law, the preference is largely driven by the users' economic stability (income level and car status) and age factor. Thus, the government has to consider these users' influences in developing the ELV policy that sooner or later is needed in the targeted developed country environment.

The authors would like to express their gratitude to those who had given support during the data collection, as well as during the analysis and report writing stage. Special mention goes to the Director-General of MIROS, Prof. Dr. Wong Shaw Voon and the Research Management and Compliance (RMC) Unit, for the opportunity and guidance in completing the main project, "Understanding Car User's Practices & Behaviour in a Proposed Conceptual Framework of Automotive Consumerism" under the grant no. VS102077.

References

1. A.N.N.A. Rahman, Z.M. Yusoff, D. Omar, I.S. Aziz, Travel to work patterns of low-income people in urban area, J. Tek., **71**, 63-66 (2014)
2. Z. Mohd Jawi, K.A. Abu Kassim and A.F. Sadullah, ASEAN NCAP – The best bet for vehicle safety in the region's safe system? Int. J. of Trade, Econ. & Fin., **4**, 358-363 (2013)
3. Nielsen, *Rising middle class will drive global automotive demand in the coming two years*, 16 April 2014, Accessed in January 2016, from http://www.nielsen.com/
4. S. Tamrin, *RM5m bid for 'F1' number plate,* 14 June 2016, Accessed in June 2016, from http://www.thesundaily.my/news/1838787
5. G. Chin, *2015 Malaysia vehicle sales market review, by brand*, 21 January 2016, Accessed in February 2016, from http://www.paultan.org/
6. L.H. Toh, *Straight Talk: Factors driving up motor vehicle ownership*, 5 October 2009, Accessed in January 2016, from http://www.theedgemalaysia.com/
7. F. Mashahadi, M.G. Mohayidin, Consumers' buying behaviour towards local and imported cars: An outlook after the implementation of AFTA in Malaysia, OUM Int. J. of Bus. & Mgmt., **1(1)**, 44-54 (2015)
8. D. Tan, *Malaysia vehicle sales data for March 2016 by brand*, 25 April 2016, Accessed in April 2016, from http://www.paultan.org/
9. CBT, *Malaysian car ownership cost is second lowest in South-East Asia – MAI*, 16 February 2012, Accessed in February 2016, from http://www.cbt.com.my/
10. L. Amelia, D.A. Wahab, C.H. Che Haron, N. Muhamad, C.H. Azhari, Initiating automotive component reuse in Malaysia, J. of Cleaner Prod., **17**, 1572-1579 (2009)
11. D.H. Meadows, *Thinking in Systems – A Primer* (Earthscan, London, 2009)
12. M.Z. Mat Saman, N. Zakuan, G. Blount, Design for end-of-life value framework for vehicles design and development process, J. of Sust. Dev., **5**, 95-111 (2012)
13. S. Ahmed, S. Ahmed, M.R.H. Shumon, M. Abdul Quader, End-of-Life Vehicles (ELVs) management and future transformation in Malaysia, J. of App. Sc. & Agri., **9 (18)**, 227-237 (2014)

14. Z. Mohd Jawi, F. Lamin, A.R. Abdul Manap, K.A. Abu Kassim, F. Abbas, S.V. Wong, *Review of the National Automotive Policy on Car Maintenance Issues: Malaysia's Automotive Ecosystem Explained* (MIROS, Kuala Lumpur, 2012)
15. S.C. Ling, A. Osman, A.B. Arman Hadi, A. Muhammad Safizal, S.M. Rana, Public acceptance and compliance on Goods and Services Tax (GST) implementation: A case study of Malaysia, Asian J. of Soc. Sc. & Hum., **5(1)**, 1-12 (2016)
16. Y. Ghani, M. Musa, Have you heard of 'AES' lately? in *Conference of ASEAN Road Safety (CARS)*, 220-225 (2015)
17. D.A. Wahab, Z.F. Fadzil, Public community knowledge on reuse of End-of-Life Vehicles: A case study in an automotive industrial city in Malaysia, J. of App. Sc., **14(3)**, 212-220 (2014)
18. N. Kanari, J.L. Pineau, S. Shallari, *End-of-Life Vehicle Recycling in the European Union*, Overview: Recycling (2003)

MATEC Web of Conferences **90**, 01039 (2017) DOI: 10.1051/matecconf/20179001039
AiGEV 2016

Finite element analysis of hybrid energy harvesting of piezoelectric and electromagnetic

Muhammad Ammar Faris Muhammad Yazid[1] , *Norlida* Jamil[1] , *Nik Nurul Husna* Muhmed Razali[1], and *Ahmad Razlan* Yusoff[1,*]

[1]Faculty of Manufacturing Engineering, Universiti Malaysia Pahang, 26600, Pekan, Pahang, Malaysia

Abstract. Harvesting energy from ambient vibrations is a highly required method because of the wide range of available sources that produce vibration energy application from industrial machinery to human motion application. In this paper, the implementation of harvesting energy from two technologies to form a hybrid energy harvester system was analyzed. These two technologies involve the piezoelectric harvesting energy and the electromagnetic harvesting energy. A finite element model was developed using the Ansys software with the harmonic analysis solver to analyze and examine hybrid harvesting energy system. Both power output generated from the magnet and the piezoelectric is then combined to form one unit of energy. Further, it was found that the result shows the system generate the maximum power output of 14.85 μW from 100 Hz, 4.905 m/s2, and 0.6 cm3 for resonance frequency, acceleration, and the volume respectively from the optimal energy harvester design. Normalized Power Density (NPD) result of 10.29 kgs/m3 comparable with other literature also can be used in energy harvesting system for vibration application.

1 Introduction

The term energy harvesting refers to the generation of energy from sources such as ambient temperature, vibration or air flow. A typical energy harvesting system starts with an energy collector or transducer device and depends on the type of energy one is trying to convert. These are typically solar or photovoltaic cells for light energy, piezoelectric for pressure, kinetic for movement, inductive for rotational or motion, thermoelectric for heat or temperature differential, and electromagnetic. The energy harvesting system is quiet significant as it has the potential to replace the batteries for small, low power electronic devices. Besides, the stored electrical energy can then be used by various sensor nodes for applications such as sensing, actuating, or sending wireless signals.

The most common types of energy harvesting are photonic, thermal, and vibration. Common photonic harvesters rely on solar energy drawn with the use of photovoltaic. However, according to Kwang [1], solar cells are not 100% efficient because some spectrums of the sunlight are being reflected, some are too weak to create electricity and some create heat energy instead of electricity. Thermoelectric devices used for energy

* Corresponding author: razlan@ump.edu.my

harvesting convert thermal activity into electricity. They do not require generators or pumps or fluids, and do not require abundant amounts of materials to build. But it has low efficiency as compared to other mechanical generators in which thermoelectric power generators generate less electric power for the same heat flow [2].

The method of the hybrid has become more popular for the development of the energy harvesting devices. Hybrid energy harvesting devices combine more than one of the energy scavenger systems to create a more efficient unit. El-Hami et al. [3] proposed design of a cantilever beam with two magnets attached to a C-shaped core was able to increase the natural frequency of the system up to 322 Hz. However, the frequency produce is very high from the goal to obtain a system with a low frequency as less than 30 Hz to capture vibrations from the ambient. Three years later, Duffy et al. [4] has developed an electromagnetic energy harvester that consists of two fixed magnets at each end of the device which act as the springs. Through this design, the research done was integrated in a shoe sole with the intention of collecting energy from human motion. Nevertheless, this device is relatively large in size and interferes with the human motion. Familiar mechanism of energy harvester is presented by Hassan et al. [5]. It is found that the induced voltage is proportional to the increment of the number of turns and the length of beams. Moreover, Soliman et al. [6] presented a mechanical stopper that is in the range of motion of a cantilever beam in order to increase the range of operation. When the beam is contact with the stopper, the stiffness of the beam would rises and causing the alteration on the resonance frequency of the system. In 2001, White et al. [7] reported that alternative way of scavenging the energy using the piezoelectric. By applying stainless steel tapered base with two layers of piezoelectric film printed on both sides, the deflection of the piezoelectric layers can be maximized and apparently increase the power output. Besides, Yu et al. [8] reported that more power output can be generated than a stand-alone piezoelectric or electromagnetic device, and the output voltage is increased by using the mems pzt cantilever array architecture. This configuration has also been implemented by Roundy [9]. However this configuration would only make the area of the beam near the clamped end receive the maximum strain and causing an uneven strain distribution along the cantilever beam. Therefore, Challa et al. [10] has proposed configuration consisted of a piezoelectric cantilever beam with a tip mass and magnets on both sides of the free end.

In this paper, a model of hybrid energy harvesting system made of piezoelectric and electromagnetic will be presented by using the harmonic analysis implemented in the Ansys software. The hybrid energy harvesting system will also be analyze several number of sample consist with different width and length of bimorph piezoelectric beam to produce power from the deflection of the beam.

2 Analytical design and simulation

2.1 Calculation of piezoelectric power output

The output power produce from the bimorph cantilever beam of piezoelectric with different geometrical dimension using Finite Element Analysis (FEA) can be obtained by deriving from the constitutive equations for a piezoelectric material as follows:

$$\delta = \frac{\sigma}{Y} + dE \tag{1}$$

$$D = \varepsilon E + d\sigma \tag{2}$$

Where δ is mechanical strain, σ is mechanical stress, Y is the modulus of elasticity, d is the piezoelectric strain coefficient, E is the electric field, D is electric displacement, and ε is the dielectric constant of the piezoelectric material. According to Roundy [9], by assuming the driving frequency is not corresponding with the natural frequency, he suggested that the voltage magnitude that is transmitted to the load for a piezoelectric beam can be given as follows:

$$V = \frac{-\frac{j\omega Y_c d t_c b^*}{\varepsilon}}{\left[\frac{1}{RC_p}\omega_n^2 - \left(\frac{1}{RC_p} + 2\zeta\omega_n\right)\omega^2\right] + j\omega\left[\omega_n^2(1+k^2) + \frac{2\zeta\omega_n}{RC_p} - \omega^2\right]} A_{in}^2 \tag{3}$$

Where V is the generated voltage from the piezoelectric material, ω is the driving frequency, Yc is Young's Modulus for a piezoelectric material, d is the piezoelectric strain coefficient, tc is the thickness of the piezoelectric material, b* is strain related to vertical displacement of the beam, ε is the dielectric constant of the piezoelectric material, R is the load resistance, Cp is the capacitance of the piezoelectric device, ωn is the natural frequency of the system, ζ is the mechanical damping ratio, k is a coupling coefficient, and Ain is the magnitude of the input acceleration. Meanwhile as for the power transferred to the load, the general equation of V2/R can simply be use. Equation 3 can be further simplified if the natural frequency ωn is assumed that it is equivalent to the driving frequency, ω. This is because the generated power is maximized when the vibration frequency is equal to the resonant frequency. Thus, the natural frequency and the environment frequency must be very similar. The power output, P of the beam can then be formulated as follows:

$$P = \frac{1}{2\omega_n^2}\frac{RC_p^2\left(\frac{dt_c\sigma_{ave}}{\varepsilon Z_t}\right)^2}{(4\zeta^2 + k^4)(RC_p\omega_n)^2 + 4\zeta k^2(RC_p\omega_n) + (2\zeta)^2} \tag{4}$$

2.2 Calculation of electromagnet output

The output power produce from the electromagnetic using the finite element analysis (FEA) can be obtained by using the manipulation of resonant technique. Electromagnetic coils such as solenoids, inductors, electromagnets, and electric motors often time required high-current and high-frequency operation. Figure 1 shows the example of the amplifier circuit that implements the resonant technique. First of all, the basic equation to get the power can be simply by using the Equation 5 as follows where P is the power output, I is the current and V, is the voltage. However, the magnitudes of the voltage are needed to be acquired from the simulation first. Meanwhile, the current flow on the circuit can be simply calculated by using the current equation at the specified frequency by using the Equation 6. It shows that capacitor reactance, C whereby at a given resonant frequency, f can be calculated by using the Equation 7 with the properties of the capacitance given in the Table 3 later.

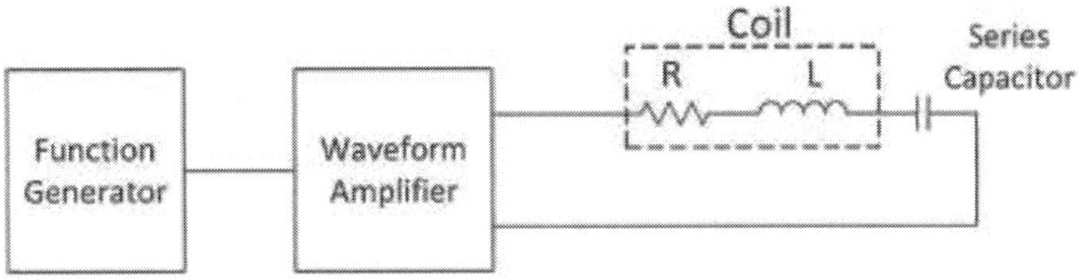

Fig. 1. Waveform amplifier drives current through the coil at resonance.

$$P = IV \tag{5}$$

$$I = V\omega C \tag{6}$$

$$C = \frac{1}{(2\pi f)^2} \tag{7}$$

2.3 Finite element analysis

Initially, the proposed design will be model via SolidWork software. Then the created model will be import into the Ansys software to perform the finite element simulation. A three dimensional harmonic analysis was run to determine both the deflection and the voltage output of the piezoelectric material. This study will focus on the design of hybrid harvester that able to scavenge the ambient energy and operate at low frequencies that are normally between 10-125 Hz. The proposed design comprised of the piezoelectric material that is mounted as bimorph cantilever beam. This design incorporates a piezoelectric cantilever beam with a magnet attached at the tip of the beam as shown in the Figure 2. The piezoelectric layer is fixed at the points of contact with the mount and as the system vibrates, it will experience the deflection. Therefore, the magnet that is attached at the tip of the piezoelectric beam will also move vertically inducing a voltage inside the copper coil through Faraday's law of induction. Hence, both piezoelectric and magnet will actively contribute to the energy harvesting process.

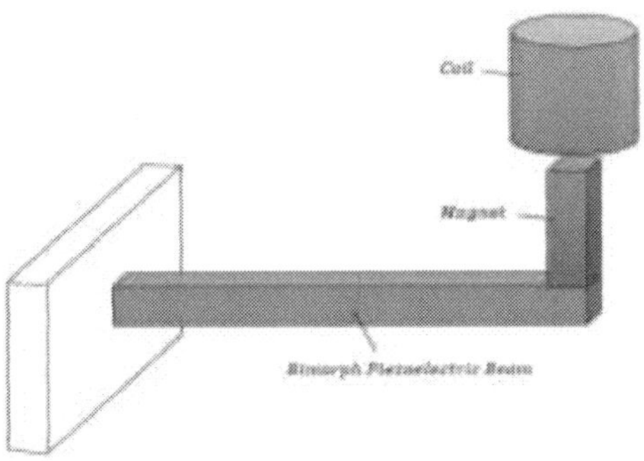

Fig. 2. Piezo-electromagnetic design concept.

The finite element analysis was executed with several numbers of samples that is made up with a variance of dimension for the cantilever beam. The variable parameter of dimension comprise of the length and the width of the beam. Whereas the thickness of the beam and load applied to the system were set to be constant. The sample and its dimension are listed in the Table 1.

Table 1. List of dimension for each cantilever beam sample.

Beam	Length (mm)	Width (mm)	Thickness (mm)
A	16.5	6.35	0.51
B	36.5	12.7	0.51
C	36.5	19.05	0.51
D	36.5	6.35	0.51
E	36.5	12.7	0.51
F	36.5	19.05	0.51

Two types of material used are magnet and piezoelectric materials. The designs of this energy harvesting consist of Neodymium Iron Boron (NdFeB) magnets. Neodymium magnets are commonly used in the harvesting application as it can be considered as one of the strongest permanent magnets that can be made. Based on the basic operation of magnet and coils, an oscillation movement of the coil through the magnet provides the change in magnetic field and induces voltage in the coil. Coil use for the design is Tinned Copper wire (MIL-W-3861 TYPE S) type. This tinned copper is electrolytic, soft drawn and annealed that allows to be easily formed. The piezoelectric material will be using the PZT-5A4E. There will be two layer of piezoelectric sandwich together with a layer of shim in between to form the bimorph beam. The shim layer will be made of the Brass material.

The full harmonic solver was used with element type Solid 226 with KEYOPT (1) = 1001 to activate the electromechanical coupling. It is a coupled field 20 nodes brick element. The structure is excited by a sinusoidal wave with an acceleration magnitude of 4.905 m/s2 and frequency range of 0 - 500 Hz. Because the tip mass is located of fcenter at the free end of the cantilever beam, it introduces a torque and force which cause the material to bend. Prior to the simulation, both material properties of piezoelectric and the brass are defined. These properties can be found in Table 2.

Table 2. Mechanical and electrical properties used in the FEA.

Parameters	Piezoelectric Material	Brass Shim
Young's modulus (GPa)	66	117
Yield Stress (MPa)	24	200
Maximum deflection (μm)	300	-
Poisson's ratio	0.31	0.324
Density (kg/m^3)	7800	7165
Relative dielectric constant	1800	-
d31 (m/V)	-190×10^{-12}	-

In general, meshing influences the accuracy of an analysis. Number of mesh element of the system is 11528 with default setting by system, as illustrated in Figure 3(a). Also, ANSYS also provides the user with interface tools to ease the process of assigning boundary conditions. In Figure 3(b), one end layer of the cantilever beam is fixed at three coordinates (x, y, and z). Meanwhile, a concentrated pressure is applied at the other free end of the cantilever beam. The load is defined by placing geometry with predefined mass properties of 4.1g. Because the tip mass is located off-center at the free end of the cantilever beam, it introduces a torque and force which cause the material to bend. The structure is excited by a sinusoidal wave with an acceleration of 4.905 m/s2 with the frequency of 0 - 500 Hz.

(a) Meshing the geometry

(b) The fixed support area is selected at the free end of the cantilever beam

(c) Magnetic model in Ansys Workbench

Fig. 3. Finite element analysis setting.

2.4 FEA: magnetostatic analysis

Figure 3(c) shows the geometry of a magnet model that was created and was analyses via magneto static analysis. There is a magnet that moves in the vertical direction and a stationary multi turn coil, surrounded by air. However, due to the limitation of this software, only the static analysis can only be performing to make the analysis. This geometry was then meshed with 89556 triangular elements. The values of the properties that where inputted are given in Table 3. To properly model the magnetic field around the magnet and coil, the air around the magnet and the coil must also be modeled and meshed in Ansys.

Table 3. Material property of magnet.

Property	Value
Multi Turn Coil	100 Turns
Coil Material	Copper
Wire Gauge Diameter	0.0897 mm

3 Results and discussion

3.1 Piezoelectric beam: analysis result

They are six sample of cantilever beam with the difference dimension were proposed to be analyzed. Therefore, it can be summarizes the difference between each beam according to the total volume and the total mass as follows. It clearly shows that cantilever beams with the high amount of volume are mostly come from the beam with the length of 36.5 mm. Beam F leading the volume amount with the 0.355 cm3. Meanwhile, Beam A has the least

 DOI: 10.1051/matecconf/20179001039

volume amount with the 0.0534 cm3. Analysis of this hybrid energy harvester can be described by analytical and finite element analysis. The results of this analysis are summarized in Table 4.

Table 4. Experimental result for hybrid energy harvester.

Beam	Max Stress of Beam (MPa)	Deflection of Beam (mm)	Resonance Frequency of Beam (Hz)	Voltage Output of Beam (V)	Power Output of Beam (μW)
A	0.49	0.029	150	8.55	14.9
B	0.72	0.042	200	9.04	16.9
C	0.24	0.047	250	2.42	1.50
D	0.11	0.058	300	3.04	2.00
E	0.26	0.059	100	8.24	14.0
F	0.94	0.141	450	2.89	1.90

The force applied on the piezoelectric beam would generate a stress. However, the stresses in the beam should be low enough so that it will not break or crack from the input acceleration. It is observed from the simulation that even the beam with the highest amount of stress applied (Beam F); will not break under their maximum stress as presented in the Table 5 which within the strength constraints of σ = 24 MPa. Moreover, from the data tabulated in the Table 5, it is shows that the beam D produces maximum deflection value of 0.11047 mm. In contrary, the beam C has the minimum deflection with the value of 0.0124 mm. The consequent results obtained from the deflection effect of each beam were also affecting the natural frequency range produce on each system. The minimum and maximum resonance frequency of the system occur at 50 Hz (beam D) and 450 Hz (beam F) respectively. Since the aim of analysis is scavenging the energy at low range of frequency as possible, thus, only beam A, D and E are satisfied the requirement with the frequency of 150 Hz, 50 Hz, and 100 Hz respectively.

Using Equation 3 and 4, the voltage and power generated of each beam are calculated. Beam B from length 16.5 mm conducted highest voltage output with 9.04mV. Meanwhile, the beams from width of 12.7 mm have generated the most voltage output to the system. Therefore, it can be deduced that the effective range of the width for this particular geometry with the constant material property and the load applied evolve around the range of 12.7 mm. Moreover, it shows the similar trend for power output determination. Highest magnitude of voltage comes from the width of 12.7mm. For the length of 16.5 mm, the maximum power output is 16.9 μW. While for the length of 36.5 mm generate slightly lower amount of the power output with 14.0 μW. However, considering the validity of the low range frequency as discussed before, only beam A and E (as illustrated in Figure 4); were approve to be implemented with the highest magnitude of the power output generated from the beam A, 14.9 μW.

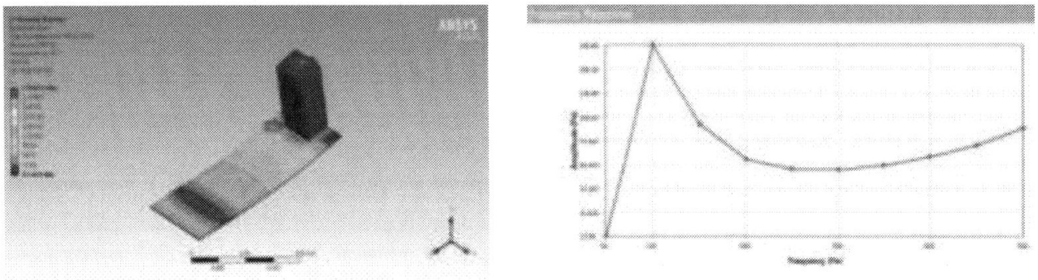

(a) Stress distribution of beam (b) Resonance frequency

Fig. 4. Beam E finite element analysis result.

3.2 Voltage and power output

Figure 5(a) shows the cross section geometry of the total magnetic flux density produced from the electromagnetic analysis. The value of the total magnetic flux density produced is 1.52 T. Meanwhile, the voltage output generated from the electromagnetic was showed in the Figure 5(b). The highest voltage output that can be produced is 7.25 mV is produced. Therefore, by using the Equation 5, the power output of the system performed at the frequency of 100Hz is calculated with the value of 0.85 μW.

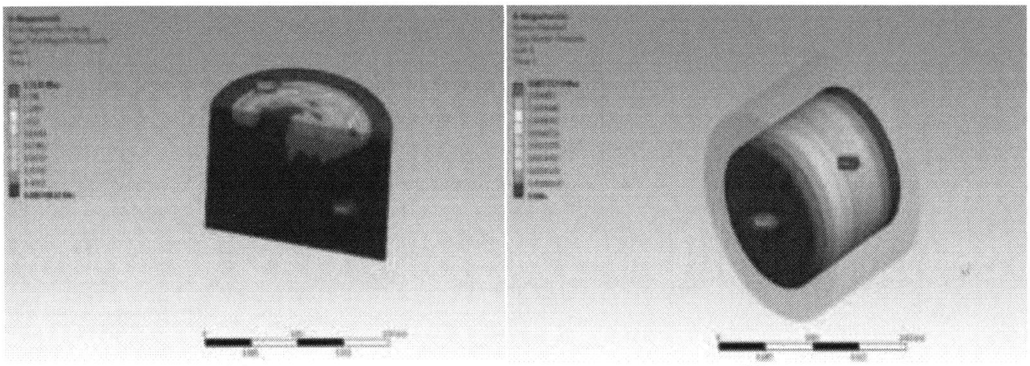

(a) (b)

Fig. 5. Electromagnetic analysis.

The highest amount of power output comes from the bimorph piezoelectric cantilever beam E with the value of 14.0 μW at the frequency range of 100 Hz. Meanwhile, the electromagnetic harvesting technology produced the output power of 0.85 μW. Generally, power output generated from both harvesting technologies was combined to form one power output for the system through the series circuit. By having both systems running under the same range of frequency which is 100 Hz, the hybrid energy harvesting system is able to produce the power output up to 14.85 μW.

3.3 Comparison of normalized power density

Currently in the field of energy harvesting there is not one single unified method for comparing the effectiveness of harvesters due to the large variations in the data that is presented in published works. This means some assumptions and calculations have to be made from the data provided. In 2007, Beeby et al. [11] has derived a measure for normalized power density (NPD) that was used only as an indication of relative performance. The measurement is described as in the Equation 8 where P is the peak power output, A is the base acceleration, and V is the volume of the harvester.

$$\mathrm{NPD} = \frac{P}{A^2V} \tag{8}$$

Table 5 compares some available harvesters with their respective NPD information. From presented data, it shows a few harvesters that have higher NPD values. In comparison, [12] has the least value of NPD with 0.09 while the study made by Beeby et al. [11] has the highest amount of value with 883. However, the calculation made was using only the volume of the magnets. If the full volume of the harvester which includes the bulk volume of the magnets, the tungsten mass used to tune the frequency, and the cantilever beam was to be considered, this value would drop to around 70.

Table 5. Comparison with other published energy harvester.

Energy Harvesters	Frequency (Hz)	Acceleration (m/s^2)	Volume (cm^3)	Power (μW)	NPD (kgs/m^3)
Current study	100	4.905	0.06	14.85	10.29
Beeby et al. [11]	52	0.589	0.15	46	883.97
Roundy [9]	157	2.5	1	375	60.00
Karami [13]	8	1.18	8.6	110	9.19
White et al. [7]	80	2.3	0.125	2.1	3.18
Ching et al. [12]	110	95.5	1	830	0.09

Besides, Roundy [9] has produced one of the highest amounts of NPD value and power output; however it compensate with the higher range of operating frequency. Although Karami [13] designed a harvester that operates at 8 Hz but it is large in size with the beam dimensions, 8.6 cm3 whereas the hybrid PE energy harvesting system has the least amount of volume of 0.06 cm3. In terms of power output, White et al. [7] has the least amount of power output produced with 2.1 μW while Ching et al. [12] obtained the highest amount of power output in compare to the others. Therefore, in comparison hybrid PE energy harvesting has the advantage in terms of the dimension which relatively small and can produce the acceptable amount of power output at a low range frequency.

4 Conclusions

The performance of the hybrid harvesting energy has been investigated. From the proposed design, all beam has passed the allowable stress as it well within the strength constraints of σ = 24MPa. The maximum power output is generated from bimorph piezoelectric beam comes from the beam A with the magnitude of 16.9 μW. However, due to the objective of this study in which to scavenge the energy at the lower range of frequency, therefore the

DOI: 10.1051/matecconf/20179001039

beam E has been selected as it has the least resonance frequency of 100 Hz and can produce up to is 14.0 μW. Meanwhile, the electromagnetic harvesting technology produced the output power of 0.85 μW. Combination of both piezoelectric and electromagnetic able to increase the power output of the system with 14.85 μW.

References

1. H. C. Kwang, Sun Power. Kuala Lumpur: Viz Urbana Sdn Bhd (2014)
2. D. Stuart, Thermoelectric Power Generators or Seebeck Power Generation. (online) http://www.electrical4u.com/thermoelectric-power-generators-or-seebeck - power-generation/ (31 November 2015) (2015)
3. M. El-Hami, P. Glynne-Jones, N.M. White, M. Hill, S. Beeby, E. James, A.D. Brown, J.N. Ross, Design and fabrication of a new vibration-based electromechanical power generator. *Sensors and Actuators A: Physical*, **92**(1), pp.335-342 (2001)
4. M. Duffy, D. Carroll, Electromagnetic generators for power harvesting. In *Power Electronics Specialists Conference, 2004. 2004 IEEE 35th Annual* , **3**, pp. 2075-2081. IEEE (2004)
5. M.F. Hassan, A.R. Yusoff, M.H. Mansor, M.T.C. Kar, Parametric Study of an Electromagnetic Energy Harvester. *Applied Mechanics and Materials*, *471*, p.113 (2013)
6. M.S. Soliman, E.M. Abdel-Rahman, E.F. El-Saadany, R.R. Mansour, A design procedure for wideband micropower generators. *Journal of Microelectromechanical Systems*, **18**(6), pp.1288-1299 (2009)
7. N.M White, P. Glynne-Jones, S.P. Beeby, A novel thick-film piezoelectric micro-generator. *Smart Materials and Structures*, **10**(4), p.850(2001)
8. S. Roundy, On the effectiveness of vibration-based energy harvesting.*Journal of intelligent material systems and structures*, **16**(10), pp.809-823 (2005)
9. H. Yu, J. Zhou, X. Yi, H. Wu, W. Wang, A hybrid micro vibration energy harvester with power management circuit. *Microelectronic Engineering*, *131*, pp.36-42 (2015)
10. V.R. Challa, M.G. Prasad, Y. Shi, F.T. Fisher, A vibration energy harvesting device with bidirectional resonance frequency tunability. *Smart Materials and Structures*, **17**(1), p.015035 (2008)
11. S.P. Beeby, R.N. Torah, M.J. Tudor, P. Glynne-Jones, T. O'Donnell, C.R. Saha, S. Roy, A micro electromagnetic generator for vibration energy harvesting. *Journal of Micromechanics and microengineering*, **17**(7), p.1257 (2007)
12. N.N. Ching, H.Y. Wong, W.J. Li, P.H. Leong, Z. Wen, A laser-micromachined vibrational to electrical power transducer for wireless sensing systems. In *Transducers' 01 Eurosensors XV* (pp. 38-41) (2001)
13. M.A. Karami, *Micro-scale and nonlinear vibrational energy harvesting* PhD Thesis, Virginia Polytechnic Institute and State University (2012)

Vehicle accelerated corrosion test procedures for automotive in Malaysia

Liza Anuar[1,*], *Astuty* Amrin[1], *Roslina* Mohammad[1], *Ali* Ourdjini[2]

[1]UTM Razak School of Engineering and Advanced Technology Universiti Teknologi Malaysia, Jalan Sultan Yahya Petra, 54100 Kuala Lumpur, Malaysia

[2]Department of Mechanical Engineering, University of Ottawa, 75 Laurier Ave E, Ottawa, ON K1N 6N5, Canada

Abstract. An accelerated corrosion test, known as proving ground accelerated test, is commonly performed by automotive manufacturers to evaluate the corrosion performance of a vehicle. The test combines corrosion and durability inputs to detect potential failures that may occur during in-service conditions. Currently, the test is conducted at an external test center overseas. Such test is aimed to simulate the effects of one year accelerated corrosion in severe corrosive environment of the north-east and south east of America. However, the test results obtained do not correlate with the actual corrosion conditions observed in the Malaysian market, which is likely attributed to the different test environment of the tropical climate of vehicles in service. Therefore, a vehicle accelerated corrosion test procedure that suits the Malaysian market is proposed and benchmarked with other global car manufacturers that have their own dedicated corrosion test procedure. In the present work, a test track is used as the corrosion test ground and consists of various types of roads for structural durability exposures. Corrosion related facilities like salt trough, mud trough and gravel road are constructed as addition to the existing facilities. The establishment of accelerated corrosion test facilities has contributed to the development of initial accelerated corrosion test procedure for the national car manufacturer. The corrosion exposure is monitored by fitting test coupons at the underbody of test vehicle using mass loss technique so that the desired corrosion rate capable of simulating the real time corrosion effects for its target market.

1 Introduction

Vehicle accelerated corrosion or proving ground test is a method used by many car manufacturers to evaluate the overall corrosion performance [1, 5]. It is able to combine corrosion and durability inputs as shown in Figure 1. Proving ground is a large scale laboratory that consists of a variety of road surfaces and facilities for automotive testing and validation. Examples of proving grounds by car manufacturers are listed in Table 1.

* Corresponding author: lizasri@yahoo.com

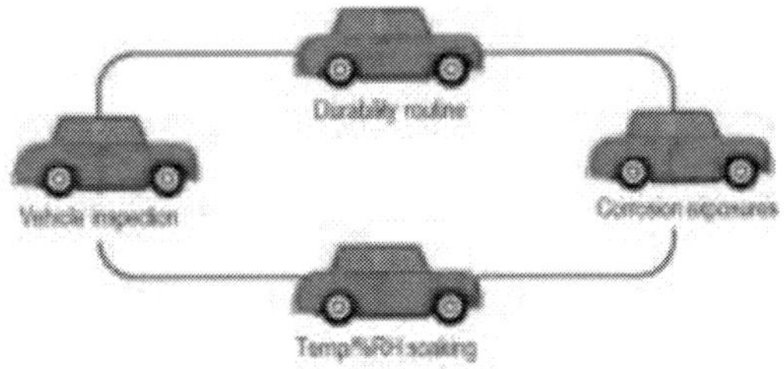

Fig. 1. General flow of proving ground accelerated corrosion test.

Table 1. Proving Grounds by car manufacturers.

Car Manufacturer	Proving Ground Location
General Motors	Milford, USA
Chrysler	Milford, USA
Volkswagen	Phoenix, USA
Ford	Kingman, USA
Navistar	Phoenix, USA
Mitsubishi	Japan

Corrosion test is important to ensure the lifespan of vehicles and relates with warranty periods [2]. It is used to predict potential failures that may occur during in-service conditions and which can be prevented. The national car manufacturer in Malaysia adopted the proving ground accelerated corrosion test, which is conducted at an external test centre overseas, to evaluate the corrosion performance of its products. This test simulates the effects of one year accelerated corrosion that mimics the severe corrosive environment of north-east and south east of America [3]. However, the test results obtained vary and do not correlate with the actual corrosion conditions observed in the Malaysia market. Because of this, the national car manufacturer is planning to establish corrosion facilities and design its own test procedure that suits the environment of the targeted market especially for tropical climate countries. Benchmarking the proposed test to other car manufacturers such as Volkswagen, General Motors and Ford is also one of the objectives in the development of this test [1].

2 Experimental setup

2.1 Test facilities

To replicate the proving ground test the required corrosion facilities should be available. A test track as shown in Figure 2 is used as the corrosion test ground [7] and consists of different types of roads for structural durability exposures. Corrosion related facilities like salt trough, mud trough and gravel road are constructed as additions to the existing facilities. A salt trough is designed to provide a corrosive environment to all vehicle surfaces – top, sides and bottom. The vehicle is driven through a shallow trough of salt water, which splashes the vehicle's underbody and sides. A mud trough is designed to introduce small particles into vehicle joints and crevices, primarily at the underbody, door lower hem flanges and engine compartment. The accumulated grit / poultice increases the time of wetness and may abrade coatings. A gravel road tests the physical integrity of protective coatings by introducing stone impact [4].

Fig. 2. Overall view of Test Track.

2.2 Test parameter

There are several important parameters that need to be considered in order to conduct the accelerated corrosion test. Table 2 describes these parameters [1-4].

Table 2. Corrosion test parameters.

No	Parameter	Details
1	Sodium Chloride (NaCl)	5% of NaCl solution as per ASTM B117 - Standard Practice for Operating Salt Spray (Fog) Apparatus. Purpose : Act as corrosive agent
2	Temperature and Humidity	Temperature of 50°C and 98% RH is widely used. Purpose : To create condition that will accelerate the natural corrosion
3	Gravel road	Aggregate from 25mm to dust is used. Purpose : To introduce stone chipping to vehicle body and components.
4	Mud	Water, sand and loam. Purpose : To build up deposits in related areas.

2.3 Test cycle

A test cycle is designed to suit the test track facilities lay-out and corrosion facilities. Figure 3 shows the corrosion driving cycle used in the present study. A vehicle is driven to achieve a target of 6 cycles per day to accumulate 140 km. the overall mileage targeted is 14000km that requires approximately 3 to 4 months of test duration.

2.4 Test Control and Monitoring

Control and monitoring of the test is conducted using test coupons made from recommended materials such as cold rolled mild steels ASTM C1008, AISI 1006-1010 [15] with the dimensions of 50 mm (H) x 25 mm (W) x 1.5 mm (T). The test coupons were installed on a coupon rack, which was then installed at under floor of the test vehicle. Each coupon rack contained 10 coupons. All materials used for the bolt, nut and washer were made from the plastic material of nylon. Figure 4 (a) and (b) show the installation of test coupon racks.

 DOI: 10.1051/matecconf/20179001040

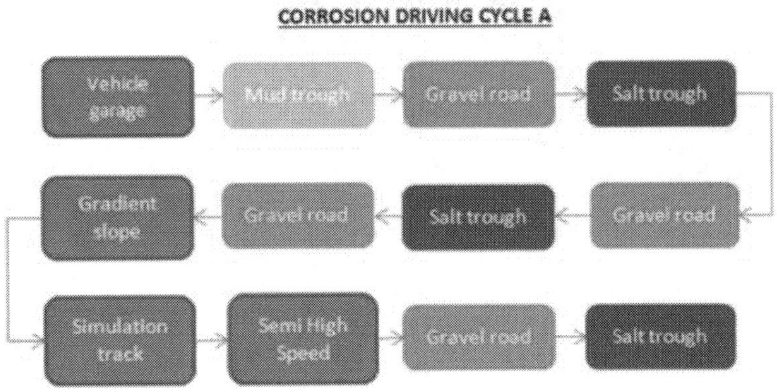

Fig. 3. Corrosion Driving cycle A.

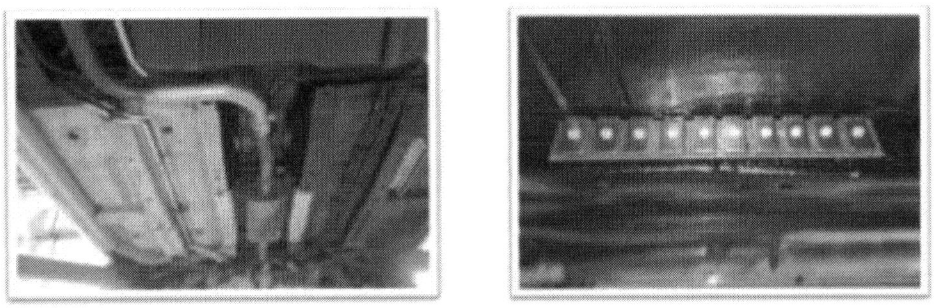

Fig. 4 (a) and (b) Test coupon racks at under floor of vehicle.

After each specified corrosion exposure interval, the test coupon was removed from the rack and analysed using the mass loss technique. The corrosion rate was calculated according to ASTM G1-03 standard using the formula in Equation 1.

$$\text{Corrosion Rate} = (K \times W)/(A \times T \times D) \quad (1)$$

where;
K = a constant
T = time of exposure in hours
A = area in cm^2
W = mass loss in grams
D = density in g/cm^3

2.5 Cosmetic Corrosion

Cosmetic corrosion inspection is conducted during the corrosion exposure interval. Cosmetic corrosion is characterized by blistering and/or rusting that is aesthetically displeasing, but may not result in catastrophic failure of the item [8]. The test driver will conduct inspection and provide cosmetic corrosion rating as shown in Table 3. The corrosion rating data is important and is used to correlate with the customer's acceptance towards corrosion [3].

Table 3. Cosmetic corrosion rating.

Scale	Rating	Customer's Perception
10	No visible corrosion	No Problem
9	Trace of corrosion	Not quite perfect
8	Slight corrosion	Readily visible but acceptable
7	Light corrosion	Almost acceptable : Critical customer would object
6	Moderate corrosion	Objectionable areas of corrosion
5	Medium corrosion	Very obvious corrosion : Everyone would object
4	Mostly corroded	Unacceptable but not totally imperfect
3	Totally corroded	Totally objectionable
2	Severe corrosion	Metal loss
1	Perforation	Hole in a panel/component

3 Results and discussion

3.1 Corrosion rate

The results of corrosion rate of the test vehicle calculated from the test coupon using the mass loss technique is plotted as shown in Figure 5. As can be seen in this figure the corrosion rate of the test vehicle is monitored against the target corrosion exposure line [10].

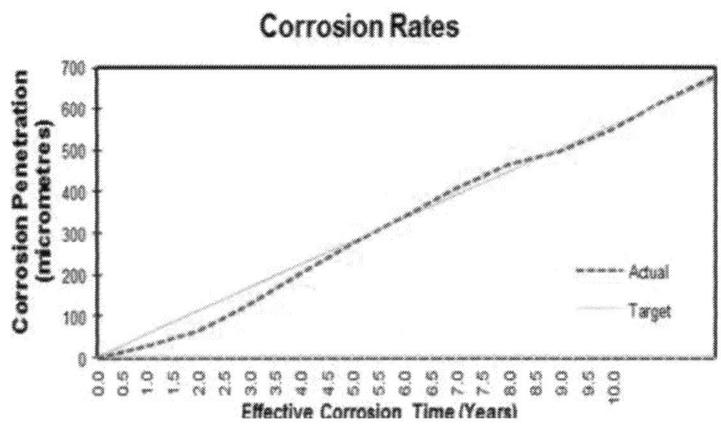

Fig. 5. Example of corrosion rate data

From Figure 5 it can be seen that the actual corrosion exposure rate has not reach the targeted exposure line during at the start of the effective corrosion years. It is during this period of exposure that the corrosion test cycles can be modified to provide additional corrosion exposure and durability inputs [9].

3.2 Corrosion cosmetic ratings

The overall cosmetic corrosion ratings are categorized into 6 observation areas as outlined in Table 4. It is a good indicator to evaluate the corrosion exposure given to the vehicle in relation with the corrosion rates obtained from the test coupons. Based on the ratings, it is shown that the test vehicles experienced medium to moderate corrosion behavior. Analysis on specific corroded components is also necessary so that proper assessment of the corrosion level is predicted. If too severe corrosion exposure is given to the test vehicle, it can be detrimental to the overall vehicle performance like breakdown and components malfunction.

Table 4. Cosmetic corrosion rating.

No	Area of vehicle	Requirement	Rating
1	Exterior	Rating 6 or better	6
2	Under hood		5
3	Underbody		6
4	Interior	Rating 7 or better	8
5	Perforation	No perforation	No perforation
6	Functional	No loss of function within 10 years exposure	Few functional issues

The combination of corrosion rate and cosmetic ratings obtained can be used as a guideline to improve the corrosion test procedures designed for Malaysia market [12].

4 Conclusions

The establishment of vehicle accelerated corrosion test in Malaysia is targeted to achieve the best correlation of the product performance with the local tropical climate. In vehicle and components development for the Malaysia's market, real time corrosion prediction can be done. Suitable material selection, coating application, product design and specification can be considered to meet the intended market. This can prevent an under-designed or over-designed component or vehicle and leads to a cost effective, durable and reliable products and higher overall performance. In addition, the test procedure can also be used in the test track for more product development and fast evaluation instead of outsourcing the test form overseas.

The authors are grateful to the national car company of Malaysia for providing support for PhD-Industry research study and to Universiti Teknologi Malaysia, RAZAK School for the technical support and supervision.

References

1. Society of Automotive Engineers (SAE), "SAE J1950 Proving Ground VehicleCorrosion Testing", Surface Vehicle Information Report, (1989-05)
2. Shinjong Oh, Byunghoon Min (Hyundai Motor Company) " Development of Accelerated Corrosion Test Mode Considering Environmental Condition," Society of Automotive Engineers, 2002-01-1231,(2002)
3. A W J Soper, "Accelerated Corrosion Testing," Millbrook Proving Ground, (2000)
4. Tie Wang, Dan Nymberg,Mats Strom, " A Comparative Study of Corrosion Test Environments at Three Proving Grounds," Society of Automotive Engineers, 2003-01-1240.
5. G.Hook, "History : A Treatise on the Development of a Proving Ground Corrosion Test," Society of Automotive Engineers, 770294,(1977)
6. Society of Automotive Engineers (SAE), "SAE J1293 Undervehicle Coupon Corrosion Tests", Surface Vehicle Information Report, Revised 1990-01 (1990)
7. Testing & Development Department of National Car Company, " Test Track Manual ," Rev 0, (2014).

 DOI: 10.1051/matecconf/20179001040

8. Society of Automotive Engineers (SAE), "SAE J1617 Body corrosion - A Comprehensive Introduction", Surface Vehicle Information Report, Issued 1993-11-02(1993).
9. Andrew Halfpenny and Mark Pompetzki, " Proving Ground Optimization and Damage Correlation with Customer Usage," SAE International, 2011-01-0484(2011).
10. W. A. McConnell, " How Good Is Testing? A correlation of customer, laboratory,and proving ground experience," Society of Automotive Engineers, **vol. 68**, pp.295, 1960.
11. Dennis D. Davidson, William A. Schumacher, " An Evaluation and Analysis of Commonly Used Accelerated Cosmetic Corrosion Tests Using Direct Comparisons with Actual Field Exposure," Society of Automotive Engineers, 912284.
12. N.LeBozec,N,Blandin,D,Theirry, "Accelerated corrosion tests in the automotive industry: A comparison of the performance towards cosmetic corrosion," Materials and corrosion, **vol. 59**,2008.
13. F.Zhu,D. Theirry,V.Kucera,S.Hedlund, " Perforation corrosion on Automotive Steel - Comparison of Accelerated Indoor Corrosion test and Field performance" SAE Technical Paper Series, 971001.
14. ASTM B117-11, "Standard Practice for Operating Salt Spray (Fog) Apparatus," ASTM International,(2011).
15. ASTM G4-95, "Standard Guide for Conducting Corrosion Coupon Tests in Field Applications," ASTM International,(1995)

MATEC Web of Conferences **90**, 01041 (2017) DOI: 10.1051/matecconf/20179001041
AiGEV 2016

Biodiesel unsaturation degree effects on diesel engine NOx emissions and cotton wick flame temperature

Mohd Fareez Edzuan Abdullah[1,*], *Sim Shu* Zhing[1], and *Clarence* Bilong Bugik[1]

[1]Faculty of Engineering, Universiti Malaysia Sarawak, 94300 Kota Samarahan, Sarawak, Malaysia

Abstract. As compared with conventional diesel fuel, biodiesel has better lubricity and lower particulate matter (PM) emissions however nitrogen oxides (NOx) emissions generally increase in biodiesel-fuelled diesel engine. Strict regulation on NOx emissions is being implemented in current Euro 6 standard and it is expected to be tighter in next standard, thus increase of NOx cannot be accepted. In this study, biodiesel unsaturation degree effects on NOx emissions are investigated. Canola, palm and coconut oils are selected as the feedstock based on their unsaturation degree. Biodiesel blends of B20 were used to fuel a single cylinder diesel engine and exhaust emissions were sampled directly at exhaust tailpipe with a flue gas analyser. Biodiesel flame temperature was measured from a cotton wick burned in simple atmospheric conditions using a thermocouple. Fourier transform infrared (FTIR) spectrometer was also used to identify the functional groups presence in the biodiesel blends. Oxygen content in biodiesel may promote complete combustion as the NOx emissions and flame temperatures were increased while the carbon monoxide (CO) emissions were decreased for all biodiesel blends. It is interesting to note that the NOx emissions and flame temperatures were directly proportional with biodiesel unsaturation degree. It might be suggested that apart from excess oxygen and free radical formation, higher NOx emissions can also be caused by the elevated flame temperatures due to the presence of double bonds in unsaturated biodiesel.

1 Introduction

Biodiesel is a mono-alkyl esters of fatty acids derived from vegetable oils or animal fats commonly produced by transesterification process where the triglycerides are reacted with alcohols usually methanol or ethanol in the presence of acid or base catalyst [1]. Biodiesel can reduce the dependence on petroleum-based diesel fuel where with current trend of oil consumption, half of the world total oil reserve would be depleted by 2025 [2]. Biodiesel is gaining attraction as alternative fuel because it can be used in diesel engine with only minor engine modification and the existing fuel supply infrastructure can be utilized directly [3]. Usage of biodiesel in diesel engine can reduce particulate matter (PM), carbon monoxide

* Corresponding author: amfedzuan@unimas.my

(CO), unburned hydrocarbon (HC) and improve life-cycle of carbon dioxide (CO2) emissions compare to petroleum diesel fuel [4]. Biodiesel production researches have been progressing actively where it can even be produced efficiently from waste oil, inedible oil and algae thus the "food/space vs fuel" can be avoided [5], [6].

Despite the promising aspects of biodiesel, it is widely reported that nitrogen oxides (NOx) emissions generally increased when diesel engine is fuelled with biodiesel [7]. Call for the use of biodiesel is facing some resistance from local authorities and private users due to the lack of fundamental knowledge of biodiesel combustion mechanisms [8]. Furthermore, automotive companies are obliged to reduce NOx emissions to 0.080 g/km in current Euro VI regulation standards, thus the increased of NOx in biodiesel case cannot be tolerated [9]. While most researchers agreed that the biodiesel increases NOx, some researchers observed that NOx emissions were unchanged or even decreased when biodiesel was used [10], [11]. Ozsecen et. al. suggested that the biodiesel NOx is a complex mechanism as it can be affected by many factors as such different physicochemical properties seem to present different combustion characteristics at different engine specifications and conditions [12].

1.1 Biodiesel NOx emissions

Before discussing several hypotheses of biodiesel NOx emissions, it is helpful to understand the fundamental of NOx formation. NOx is composed of mainly of nitrogen monoxide NO and nitrogen dioxide NO2. NOx is formed in high temperature conditions that mostly pronounce in burning of fossil fuel and naturally in a lightning flash. NOx formation is highly dependent on temperature where molecular nitrogen and oxygen in combustion air start to disassociate and form NOx at 1800 K and will exponentially increase as the temperature increase following reversible three steps Zeldovich mechanism are as follow [13].

$$N2 + O \leftrightarrow NO + N \quad (1)$$

$$N + O2 \leftrightarrow NO + O \quad (2)$$

$$N + OH \leftrightarrow NO + H \quad (3)$$

Noted that thermal NOx mechanism is governed by gas mixture temperature and it does not has a direct relationship with fuel properties. Although biodiesel NOx emissions cannot be quantitatively determined by a single factor, yet higher in-cylinder and/ or flame temperatures and longer combustion period would most likely increase NOx formation. Some common explanations on how biodiesel increase NOx emissions are discussed as follow [14]:

1. High bulk modulus of elasticity of biodiesel propagates wave faster in the fuel pump line thus the fuel is injected earlier and combustion occurs near to high pressure/ temperature top dead centre (TDC). NOx emissions are expected to increase significantly when longer injection period is coupled with high temperature in-cylinder temperatures.
2. Oxygen content in biodiesel is suspected to promote complete combustion in diesel engine. Fuel available is expected to improve air-fuel ratio that might increase flame temperature that leads to higher NOx formation.
3. Soot particles in diesel spray flame are heated up during combustion resulting in heat loss through radiation. Low soot biodiesel combustion is expected to have higher actual gas mixtures temperature that might causes higher NOx due to radiation heat loss.

Despite of having roughly 10% oxygen content, some authors argued that biodiesel should not produce higher actual or adiabatic flame temperature compare to the conventional diesel fuel due to it lower heating value [15]. Thus it might be considered that when biodiesel is used, there are differences in chemical kinetic pathway that form NOx other than the thermal NOx mechanism. Another NOx mechanism that is usually related to biodiesel properties (i.e. unsaturation degree) is prompt NOx or Fennimore mechanism where formation of free radicals in the flame front of hydrocarbon flames leads to rapid production of NO as below [16].

$$CH + N2 \leftrightarrow HCN + N \quad (4)$$

$$CH2 + N2 \leftrightarrow HCN + NH \quad (5)$$

$$N + O2 \leftrightarrow NO + O \quad (6)$$

$$HCN + OH \leftrightarrow CN + H2O \quad (7)$$

$$CN + O2 \leftrightarrow NO + CO \quad (8)$$

In order to elucidate the complex biodiesel combustion mechanism, the effects of the biodiesel unsaturation degree to NOx emissions will be investigated. Canola, palm and coconut biodiesels are produced to represent high, medium and low unsaturation degree biodiesel, respectively. NOx and CO emissions are measured directly from the exhaust tailpipe, flame temperatures of each biodiesels are measure in a cotton wick test and biodiesel functional groups are confirmed from their Fourier transform infrared (FTIR) spectroscopy.

2 Experimental setup

A Kubota single cylinder compression ignition (CI) engine was used and it specifications are provided in Table 1. The engine was operated under low-load condition at the speed of approximately 725 rpm throughout the tests. Although the engine was equipped with a hydraulic dynamometer as shown in Figure 1, it performance was not investigated in the present study. Exhaust gases were sampled directly from exhaust tailpipe. Since no dilution system was employ the effects of high particulate concentrations, high vapour concentrations, high temperature and pulsating flows on emissions measurement accuracy can be expected. Nevertheless, a relatively constant reading can be monitored throughout the tests.

The diesel engine NOx and CO emissions were measured with a Testo 350-XL flue gas analyser, where measurement probe was fixed at the tailpipe for about 3 minutes and 10 readings were taken for each fuel test. After new fuel was added into the fuel tank, the engine was operated for several minutes before any measurement was made to purge out previous residual fuel in the fuel supply system.

Biodiesel was produced from base-catalysed transesterification process of domestic use grade of coconut, palm and canola oil that were selected from their saturation degree as is summarized in Table 2. Methanol and potassium hydroxide KOH were mixed to produce potassium methoxide using heating orbital shaker at 60°C. Feedstock oil was mixed into potassium methoxide and stirred with a magnetic stirrer. After the mixture was settled for two days, biodiesel by-product glycerol was drained out of separatory funnel and the biodiesel was washed with warm distilled water for several times before it was dried using magnetic heater to avoid water contamination. Biodiesel was then seal tightly, stored in a dark container and used within 5 weeks to avoid oxidation.

To avoid fuel injector clogging, biodiesel blends were limited to 20 wt%; this biodiesel blend of canola, palm, and coconut as well as diesel fuel is abbreviated as Ca20, Pa20, Co20 and D100, respectively. It is assumed that generally biodiesel contained approximately 10 wt% of oxygen content [17].

Study of flame temperature in diesel combustion is necessary to validate thermal NOx emissions. However, to conduct combustion experiment with optically accessible diesel engine required high cost and extensive technical knowledge. Thus flame temperature of each pure biodiesels in a simple cotton wick at atmospheric conditions was measured. Figure 2 depicted the experiment where the flame temperatures were measured using a type K thermocouple inserted inside the hottest flame region just at tip of the wick and flame thermal images were recorded with a thermographic camera (ThermaCam P65). This experiment might yield useful insight in explaining biodiesel combustion phenomena despite of its simplicity compared to complex combustion mechanism in diesel engine [18]. In this study, due to unavailability of canola biodiesel, similarly unsaturated sesame biodiesel was used.

Finally a Shimadzu IRAffinity-1 FTIR spectrometer was used to measure the spectroscopy of canola, palm and coconut biodiesel blends. From the spectroscopy one can identify any functional groups present in the fuel and might be useful to identify fuel impurities such as water contamination that can be detected at wavelength region above 3400 cm-1.

Table 1. Diesel engine specifications.

Item	**Specification**
Manufacturer	Kubota Engine
Model	RK95-1-NB-RDK
Injection type	Direct injection
Air intake	Naturally aspirated
Bore × stroke (mm)	86 × 84
Displacement (cc)	487
Maximum output (kW)	7.09 (2400 rpm)
Continuos output (kW)	5.97 (2200 rpm)
Maximum torque (kgfm)	3.08 (1800 rpm)

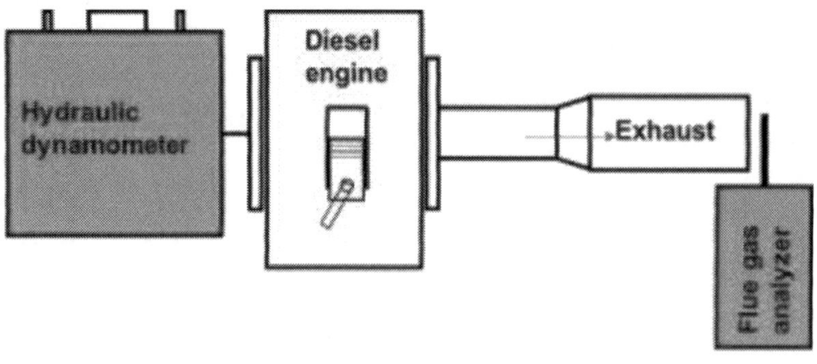

Fig. 1. Diesel engine experimental setup.

 DOI: 10.1051/matecconf/20179001041

Table 2. Properties of canola, palm and coconut biodiesel.

Biodiesel	Saturation degree (wt %)	Cetane number	Kinematic viscosity at 40°C (mm^2/s)
Canola	7.0	51.6	4.0
Palm	45.6	56.2	5.0
Coconut	81.5	68.0	2.7

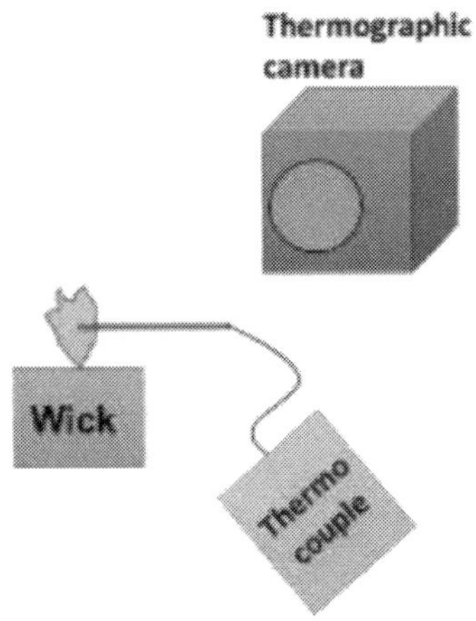

Fig. 2. Flame temperature experimental setup.

3 Results and discussion

3.1 NOx and CO emissions

Figure 3 shows NOx and CO emissions of diesel fuels and B20 blends sampled at the tailpipe where the error bars indicate standard deviation value. Figure 3(a) shows that the NOx emissions were increased in all biodiesel blends while Figure 3(b) shows that the CO emissions were decreased in all biodiesel blends. Although the oxygen content was not measured in each blend, B20 fuel (20% biodiesel blends) contains approximately 4% of oxygen assuming that biodiesel generally contains 10% oxygen. At first glance, it would be expected that excess oxygen in biodiesel blends will promote complete combustion thus will elevate formations of NOx and suppress formation CO. However, the results in Figure 3(a) shows that the NOx emissions appear to be directly proportional with unsaturation degree where highly unsaturation canola biodiesel produce highest NOx emissions and likewise. However, CO emissions shown in Figure 3(b) seem to have a weak relation with biodiesel unsaturation degree.

If we reason the biodiesel NOx emissions from thermal NOx mechanism only, the NOx emissions for all biodiesels in this study would be similar since oxygen contents in all blend is approximately 4%. It might be suggested that prompt NOx formation might contributed to significant portion of total NOx emission in biodiesel case considering that NOx emissions of canola biodiesel is 50% higher than coconut biodiesel. However further investigation needed to be conducted on the combustion kinetics of the biodiesel surrogate fuels with different unsaturation degree, i.e. stearic acid, oleic acid and linoleic acid.

DOI: 10.1051/matecconf/20179001041

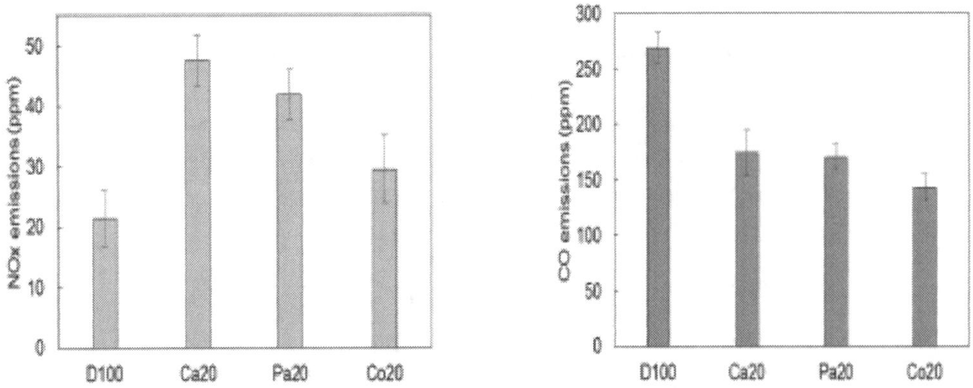

Fig. 3. NOx emissions (a) and CO emissions (b) of diesel, canola, palm and coconut biodiesel blends (20%).

3.2 Flame temperatures

Figure 4 shows the flame temperature of diesel fuel, B20 and B100 in a cotton wick flame test at atmospheric conditions using a K type thermocouple. Sesame biodiesel was chosen to represent canola biodiesel in this test since both feedstock oils are highly unsaturated. It can be seen that the flame temperatures were elevated as the biodiesel amounts were increased from B20 to B100; although biodiesel heating value is generally 10% lesser than that of diesel fuel, the flame temperatures for all biodiesels are higher than diesel fuel with the exception for coconut biodiesel. Equating from enthalpies, Jha et. al. concluded that flame temperature does not depends solely on heating value, but it is also a function of air-fuel ratio [18]. Additionally, high radiation heat loss might occur in diesel fuel case that will decrease its actual flame temperature. Hence oxygen content in biodiesel might promote leaner combustion, which leads to higher flame temperatures.

It is noticed that the flame temperatures is directly proportional to biodiesel unsaturation degree where highest unsaturation degree of sesame biodiesel yields highest flame temperature and likewise. These results seem to be in good agreement with the thermal NOx mechanism where higher flame temperatures will produce higher the NOx emissions as shown in Figure 3(a). Pierre-Alexandre et. al. THERGAS software calculation also showed that adiabatic flame temperatures increased when unsaturated bonds were present in the molecule [19]. However further study on how biodiesel unsaturation degree effects the temperature of actual diesel combustion can provide useful insight of the biodiesel NOx mechanism.

Figure 5 shows the variation of the flame thermal images of canola, palm and coconut biodiesel captured using a thermographic camera where the red colour and blue colour indicates maximum and minimum temperature, respectively. However the results of significantly low maximum flame temperatures and indistinctive flame shapes might be caused by the camera error.

 DOI: 10.1051/matecconf/20179001041

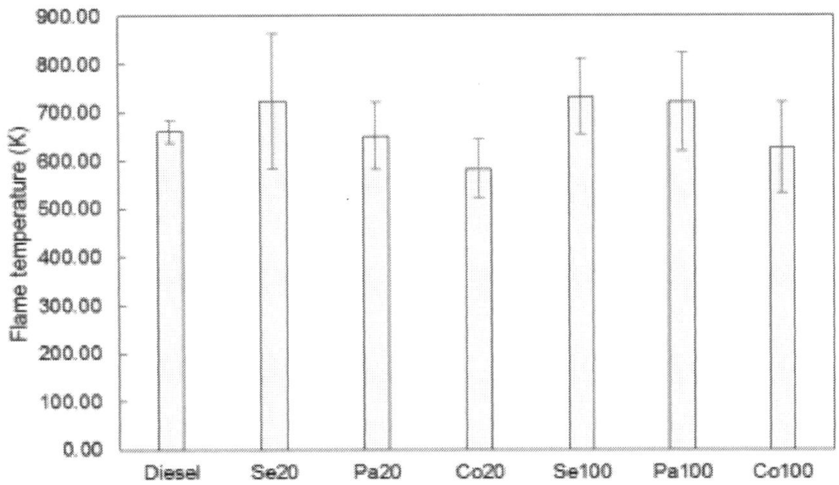

Fig. 4. Flame temperatures of diesel, sesame, palm and coconut biodiesel (canola biodiesel is substituted with similar saturation degree of sesame biodiesel).

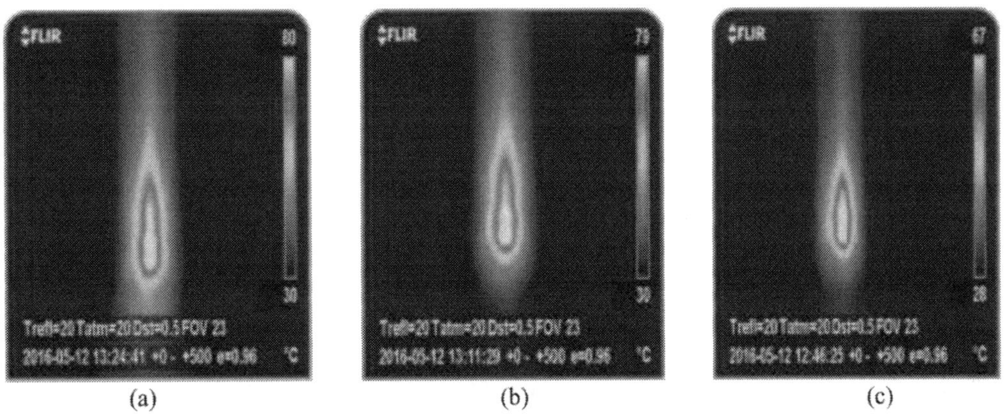

Fig. 5. Flame thermal images of canola (a), palm (b) and coconut (c) biodiesel.

3.3 FTIR spectroscopy

The spectrometer used in this study can only detects the presence of functional groups but cannot be used to measure quantitative property. Hence the results of all biodiesel blends used in this study are similar and can be represented by any blends. Figure 6 shows the FTIR spectroscopy of canola biodiesel blends and the peak wavelengths were indicated with red circles. Oxygen group presence was detected at 1745 cm-1 region and no water contamination was detected at region above 3400 cm-1 that confirm the quality of the biodiesels being produced to a certain extent.

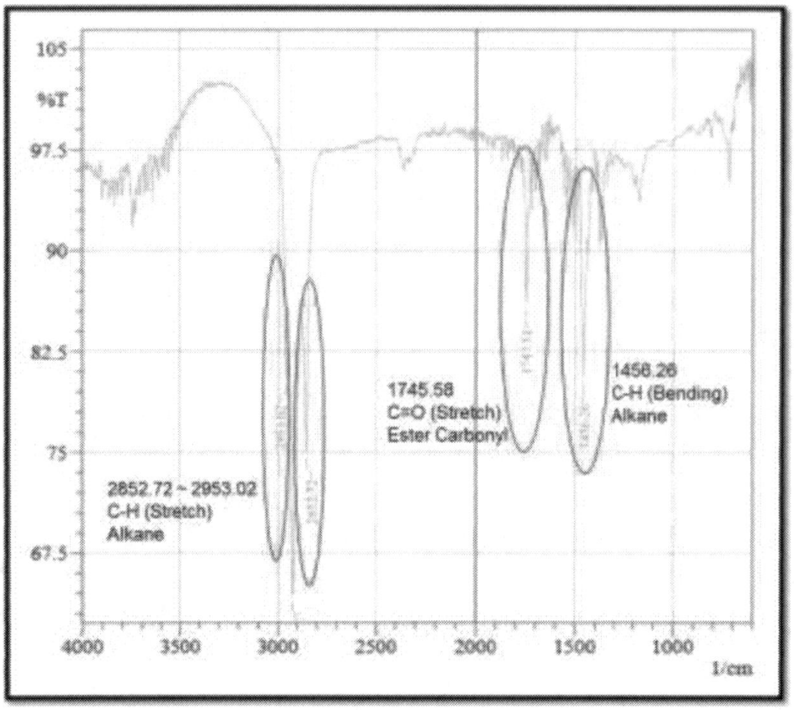

Fig. 6. FTIR spectroscopy of canola biodiesel (FTIR spectroscopy for palm and coconut biodiesel is similar).

4 Conclusions

In this study, it was attempted to understand the fundamental knowledge on biodiesel unsaturation degree effects on NOx emissions. Canola, palm and coconut oils were selected based on their unsaturation degree and converted to biodiesel by transesterification process. The tailpipe emissions and cotton wick flame temperature were measured and the results can be summarized as follow. NOx emissions and flame temperatures increased while CO emissions decreased in biodiesel cases compared to petroleum diesel fuel. Furthermore, the biodiesel unsaturation degree is directly proportional to NOx emissions and flame temperatures. It can be suggested that apart from the prompt NOx, the thermal NOx might also be significant for total NOx formation in highly unsaturated biodiesel case. In near future, the effects of biodiesel unsaturation degree on actual diesel combustion temperature and combustion kinetics will be investigated.

This work is supported by UNIMAS SGS grant, F02(S178)/1322/2016(2). The authors fully acknowledged Universiti Malaysia Sarawak for the approved fund which makes this research viable and the Department of Mechanical & Manufacturing technical staff.

References

1. S. K. Hoekman and C. Robbins, "Review of the effects of biodiesel on NOx emissions," Fuel Process. Technol., vol. **96**, pp. 237–249 (2012)
2. S. Fernando, C. Hall, and S. Jha, "NO x Reduction from Biodiesel Fuels," Energy & Fuels, vol. 20, no. **x**, pp. 376–382 (2006)

3. T. Aizawa, M. F. E. Bin Abdullah, A. Inoue, Y. Ishidzuka, N. Taki, and H. Kosaka, "Aromatic additive effects on soot formation in a fischer-tropsch diesel (FTD) spray flame via laser spectroscopy," Proc. 8th Int. Conf. Model. Diagnostics Adv. Engine Syst. COMODIA 2012, pp. 275–280 (2012)
4. S. M. Palash, M. A. Kalam, H. H. Masjuki, B. M. Masum, I. M. Rizwanul Fattah, and M. Mofijur, "Impacts of biodiesel combustion on NOx emissions and their reduction approaches," Renew. Sustain. Energy Rev., vol. **23**, no. 4, pp. 473–490, (2013)
5. S. Ismail, A. S. Ahmed, R. Anr, and S. Hamdan, "Biodiesel Production from Castor Oil by Using Calcium Oxide Derived from Mud Clam Shell," J. Renew. Energy, vol. **2016**, p. 8 (2016)
6. T. M. Mata, A. A. Martins, and N. S. Caetano, "Microalgae for biodiesel production and other applications: A review," Renew. Sustain. Energy Rev., vol. **14**, no. 1, pp. 217–232 (2010)
7. M. F. E. bin Abdullah, M. H. bin Sulaiman, and N. A. B. Abdul Majid, "NOx Emission of Diesel Fuel Blended with Different Saturation Degrees of Biofuel and with Oxygenator," Appl. Mech. Mater., vol. **660**, pp. 397–401 (2014)
8. M. Lapuerta, O. Armas, and J. Rodríguez-Fernández, "Effect of biodiesel fuels on diesel engine emissions," Prog. Energy Combust. Sci., vol. **34**, no. 2, pp. 198–223, (2008)
9. European Parliament and Council of the European Union, "REGULATION (EC) No 715/2007 OF THE EUROPEAN PARLIAMENT AND OF THE COUNCIL of 20 June 2007 on type approval of motor vehicles with respect to emissions from light passenger and commercial vehicles (Euro 5 and Euro 6) and on access to vehicle repair and mai," Off. J. Eur. Union, vol. **L171,** no. December 2006, pp. 1–16 (2007)
10. M. N. Nabi, "Theoretical investigation of engine thermal efficiency, adiabatic flame temperature, NOx emission and combustion-related parameters for different oxygenated fuels," Appl. Therm. Eng., vol. **30**, no. 8–9, pp. 839–844 (2010)
11. E. Sendzikiene, V. Makareviciene, and P. Janulis, "Influence of fuel oxygen content on diesel engine exhaust emissions," Renew. Energy, vol. **31**, no. 15, pp. 2505–2512 (2006)
12. N. Ozsezen, M. Canakci, A. Turkcan, and C. Sayin, "Performance and combustion characteristics of a DI diesel engine fueled with waste palm oil and canola oil methyl esters," Fuel, vol. **88**, no. 4, pp. 629–636 (2009)
13. J. Sun, J. A. Caton, and T. J. Jacobs, "Oxides of nitrogen emissions from biodiesel-fuelled diesel engines," Prog. Energy Combust. Sci., vol. **36**, no. 6, pp. 677–695 (2010)
14. J. Mueller, A. L. Boehman, and G. C. Martin, "An Experimental Investigation of the Origin of Increased NOx Emissions When Fueling a Heavy-Duty Compression-Ignition Engine with Soy Biodiesel," SAE Int. J. Fuels Lubr., vol. **2**, no. 1, pp. 789–816 (2009)
15. M. Lapuerta, O. Armas, R. Ballesteros, and J. Fernández, "Diesel emissions from biofuels derived from Spanish potential vegetable oils," Fuel, vol. **84**, no. 6 SPEC. ISS., pp. 773–780 (2005)
16. Fenimore, "Formation of nitric oxide in premixed hydrocarbon flames," Symp. Combust., vol. **13**, no. 1, pp. 373–380 (1971)
17. K. Varatharajan and M. Cheralathan, "Influence of fuel properties and composition on NO x emissions from biodiesel powered diesel engines: A review," Renew. Sustain. Energy Rev., vol. **16**, no. 6, pp. 3702–3710 (2012)
18. K. Jha, S. Fernando, and S. D. F. To, "Flame temperature analysis of biodiesel blends and components," Fuel, vol. **87**, no. x, pp. 1982–1988 (2008)
19. P. A. Glaude, R. Fournet, R. Bounaceur, and M. Molière, "Adiabatic flame temperature from biofuels and fossil fuels and derived effect on NOx emissions," Fuel Process. Technol., vol. **91**, no. 2, pp. 229–235 (2010)

Study on Differential Regenerative Braking Torque Control to Increase the Stability of the Small Electric Vehicle with Four In-Wheel Motors

N. M. Ali[1], *M. H.* Peeie[1,*], *M. A. H.* Rasid[1], *H.* Ogino[2]

[1]Faculty of Mechanical Engineering, Universiti Malaysia Pahang, 26600 Pekan, Pahang, Malaysia
[2]Faculty of Prime Mover Engineering, Tokai University, 4-1-1 Kitakaname, Hiratsuka-Shi, 259-1292 Hiratsuka Prefecture, Japan

Abstract. Based on the advantages of the electric motor such as fast and precise torque response, the performance of the electric vehicle (EV) can be improved. During braking or driving on the cornering, the vehicle will over steer or under steer if a car turns by more or less than the amount commanded by the driver. To improve the stability of the small EV with four in-wheel motors, the differential regenerative braking torque control is proposed. In this system, the regenerative braking torque at each wheel will be controlled individually based on the value of slip ratio. If the slip ratio is greater than the optimum value, the regenerative brake will turn off. In this situation, the electric motor will not produce the regenerative braking torque. Conversely, if the slip ratio lower than the optimum value, the regenerative brake will turn on and the electric motor will generate the regenerative braking torque. In the numerical analysis, to investigate the effectiveness of the proposed model, the road condition is set to an icy road on the left tire and dry asphalt on the right tire. From the simulation results, the differential regenerative braking torque control can prevent the tire from lock-up and avoid the vehicle from skidding.

1 Introduction

In recent years, due to the environmental problems and growing needs of energy, electric vehicles (EVs) are gradually gaining attention to the consumers and manufacturers [1]. The driving system of the EV is the electric motor and there are several advantages of the electric motor such as fast torque response, the generated torque can be detected accurately and motor can be detached on each wheel [2]. Based on these advantages, more effective anti-skid brake system and traction control system (TCS) can be developed on EV.

Over the past years, many studies on anti-lock braking system (ABS) and TCS focuses on to prevent the tire from lock up and vehicle from skidding by controlling the value of slip ratio [3–7]. The optimum slip ratio is important to ensure that the tire can develop a

* Corresponding author: mheerwan@ump.edu.my

sufficient high braking force to stop the vehicle [8]. Cai et al have designed two anti-skid controllers; a rule-based controller and a fuzzy logic controller to control the motor torque [9]. In that study, the slip level can be estimated by calculating the ratio of wheel acceleration to motor torque. S. Sakai et al have compared the electric motor with hydraulic brake system as an actuator of ABS [10]. They used a model following control (MFC) as a control algorithm in their comparison.

The research of skid control on small EV have been developed by S. Hasegawa et al, H. Ogino et al and M. H. Peeie et al [11], [12], [13]. S.Hasegawa have examined the braking system of the small EV with two in-wheel motors. From their study, due to the space limitation on the driven tire (rear tire), the hydraulic unit of ABS cannot be installed. ABS only on the front tire and during braking on the slippery road, the driven tire is locked up [11]. To improve the safety of the small EV with the hydraulic-mechanical hybrid brake system, H. Ogino et al have developed a regenerative braking system for a small EV with two in-wheel motors [12]. The ABS system was installed on the front tire and the regenerative brake on the rear tire. Their study shows that the regenerative brake can be controlled to prevent the tire from lock up during braking on the slippery road. M. H. Peeie et al. have developed the regenerative brake timing control to increase safety and stability of the small EV [13]. During braking on critical conditions (icy road and cornering), regenerative brake timing control can prevent the tire from lock up and vehicle from skidding. That model also can improve the steer performance of the vehicle.

In this study, the differential regenerative braking torque control is proposed to improve the stability of the small EV with four in-wheel motors. The control method of the differential regenerative braking torque control is based on the slip ratio on each tire. To examine the effectiveness of the proposed model, the road condition is set to an icy road on the left side and dry asphalt on the right side. The simulation result shows that the differential regenerative braking torque control can improve the stability of the small EV with four in-wheel motors.

2 Numerical analysis

2.1 Vehicle model

The analysis vehicle model in this study is Toyota COMS, model AK10E-PC. The driving system of this vehicle is two in-wheel motors at the both rear tires. In the numerical analysis, to design the small EV with four in-wheel motors, another two in-wheel motors have been added to the front tires. Figure 1 shows the vehicle model and the specifications of the vehicle as in Table 1.

2.2 Slip phenomenon

In this study, the control method of the differential regenerative braking torque control is based on the relation between slip ratio and friction coefficient. The equation of slip ratio during braking as below:

$$\rho = \frac{u - r\omega}{r\omega} \tag{1}$$

where u is the vehicle velocity, r is the radius of the tire and ω is the angular velocity of the tire.

The relations between slip ratio and friction coefficient on the dry asphalt and icy road are illustrated in Figure. 2 [13]. Skid phenomenon will occur when the large driving torque suddenly drops to 0. In this situation, the braking force and the side force generation will disappear and the slip ratio will rapidly increase towards 1.0. The vehicle will become unstable and skidding.

Fig. 1. Analysis vehicle model.

Table 1. Specifications of the analysis vehicle model.

Vehicle mass	361.9 (kg)
Maximum speed	60 (km/h)
Height of centre of gravity	0.105 (m)
Tread front	0.840 (m)
Tread rear	0.815 (m)
Wheelbase	1.280 (m)
Inertia of front and rear tire	2.530 (kgm^2)
Driving system	4 in-wheel motors
Lead acid batteries	6 set, 12 (V)

From Figure. 2, the optimum value of the slip ratio for an icy road is between 0.2 to 0.3. In this range, the tire can develop a sufficient high braking force to stop the vehicle and prevent the vehicle from skidding.

2.3 Regenerative braking torque control

Figure 3 shows the ideal braking force distribution used to determine the percentage of regenerative brake force and mechanical brake force [8]. From the actual brake curve, the maximum degree of rear brake force can be measured by taking the front brake force produced when the coefficient of friction is in maximum value. The equation of ideal braking force for the front and rear tyres are as below:

$$F_{mecha} + F_{reg} = F_{ideal} \tag{2}$$

where, F_{mecha} is the mechanical braking force, F_{reg} is the regenerative braking force and F_{ideal} is an ideal braking force.

During braking on an icy road, the regenerative braking torque control is proposed to improve the braking performance of the small EV. In this model, the regenerative braking torque control will function similarly as ABS, which is based on the slip ratio, ρ of the tire. From Figure 2, if ρ larger than 0.3, the in-wheel motor turns off and the vehicle in the neutral condition. In this situation, the in-wheel motors do not produce regenerative braking torque. The braking force on the tire is only mechanical braking force. However, if ρ becomes lower than 0.2, the in-wheel motor turns on and produce the regenerative braking torque to regain the ideal braking force.

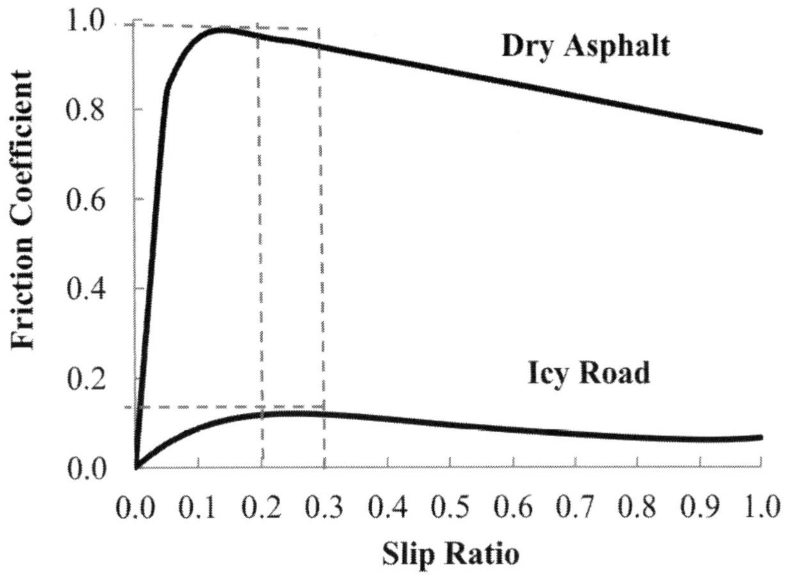

Fig. 2. Friction coefficient and slip ratio on the dry asphalt and icy road.

2.4 Dynamics equation of motion

Figure 4 shows the force vector at each wheel for the construction of equations of motion. The equation of motion in longitudinal and lateral direction and the equation of yaw motion are shown as below:

$$m\left(\frac{du}{dt} - vY\right) = \left(-X_{fr} - X_{fl}\right)\cos\theta + \left(-Y_{fr} - Y_{fl}\right)\sin\theta - X_{rr} - X_{rl} \quad (3)$$

$$m\left(\frac{dv}{dt} + uY\right) = \left(Y_{fr} + Y_{fl}\right)\cos\theta + \left(-X_{fr} - X_{fl}\right)\sin\theta + Y_{rr} + Y_{rl} \quad (4)$$

$$I\frac{dY}{dt} = l_f\left(Y_{fr}\cos\theta + Y_{fl}\cos\theta - X_{fr}\sin\theta - X_{fl}\sin\theta\right) - l_r(Y_{rr} + Y_{rl}) + \frac{d_f}{2}\left(-X_{fr}\cos\theta + X_{fl}\cos\theta - Y_{fr}\sin\theta + Y_{fl}\sin\theta\right) + \frac{d_r}{2}(-X_{rr} + X_{rl}) \quad (5)$$

where u and v is the velocity of the vehicle in longitudinal and lateral axis, Υ is the vehicle yaw rotational speed, X_{fr}, X_{fl}, X_{rr}, X_{rl} are the friction force while Y_{fr}, Y_{fl}, Y_{rr}, Y_{rl} are the cornering force.

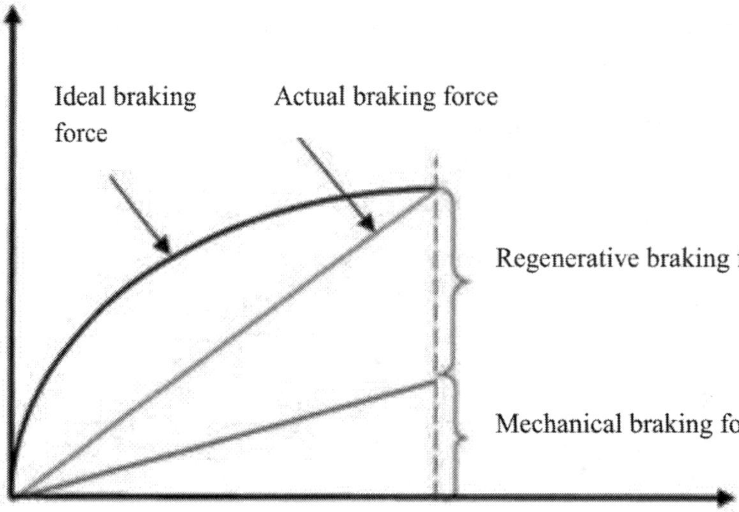

Fig. 3. Ideal braking force distribution.

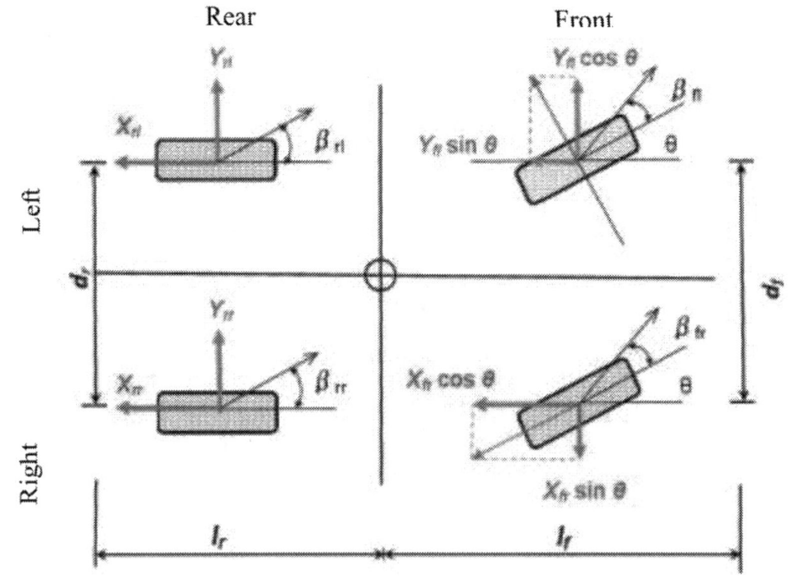

Fig. 4. Force diagram of the vehicle.

The brush tire model has been used in the simulation to measure the friction force, X_{fr}, X_{fl}, X_{rr}, X_{rl} and cornering force, Y_{fr}, Y_{fl}, Y_{rr}, Y_{rl}. This model allows elastic deformation in both the longitudinal and lateral directions. The equation of friction force and cornering force during braking as below:

$$\xi_p = 1 - \frac{K_s \lambda}{3\mu W_z (1-\rho)}$$

$$\lambda = \sqrt{\rho^2 + \left(\frac{K_\beta}{K_s}\right)^2 tan^2 \beta_T}$$

$$K_s = \frac{b l_T{}^2}{2} K_x \quad , \quad K_\beta = \frac{b l_T{}^2}{2} K_y$$

$$Xfr, Xfl, Xrr, Xrl \begin{cases} = -\frac{K_S \rho}{1-\rho} \xi_p{}^2 - 6\mu W_z \frac{\rho}{\lambda}\left(\frac{1}{6} - \frac{1}{2}\xi_p{}^2 + \frac{1}{3}\xi_p{}^3\right) & (\xi_p > 0) \\ = -\mu W_z \frac{\rho}{\lambda} & (\xi_p < 0) \end{cases} \tag{6}$$

$$Yfr, Yfl, Yrr, Yrl \begin{cases} = -\frac{K_\beta \tan \beta_T}{1-\rho} \xi_p{}^2 - 6\mu W_z \frac{K_\beta \tan \beta_T}{K_S \lambda}\left(\frac{1}{6} - \frac{1}{2}\xi_p{}^2 + \frac{1}{3}\xi_p{}^3\right) & (\xi_p > 0) \\ = -\mu W_z \frac{K_\beta \tan \beta_T}{K_S \lambda} & (\xi_p < 0) \end{cases} \tag{7}$$

Where, b, l is the width and length of the interacted tyre surface, K_x, K_y are the rigidness of tyre in longitudinal and lateral axis, and β_T is the side slip angle of the tyre. Based on the analysis vehicle model, the value of b, l, K_x and K_y are 10 *cm*, 15 *cm*, $3.3 \times 10^7 N/m^3$ and $3.3 \times 10^7 N/m^3$ respectively.

2.5 Calculation procedures

When a braking torque T_B is applied to the tyre, a corresponding torque friction of the tyre T_f is developed on the tyre ground contact patch, which acts in the opposite direction of the applied braking torque T_B. The difference between T_B and T_F causes an angular acceleration $\dot{\omega}$ of the tyre:

$$\begin{cases} T_{ffr} - T_{Bfr} = I_{fr} \frac{d\omega_{fr}}{dt} \\ T_{frr} - T_{Brr} = I_{rr} \frac{d\omega_{rr}}{dt} \end{cases} , \quad \begin{cases} T_{ffl} - T_{Bfl} = I_{fl} \frac{d\omega_{fl}}{dt} \\ T_{frr} - T_{Brr} = I_{rr} \frac{d\omega_{rr}}{dt} \end{cases} \tag{8}$$

where I is the inertia of the tyre. The inertia moment of the front tyre is 0.43 kgm^2 and the inertia moment of the rear tyre is 2.53 kgm^2 [1].

During braking, the in-wheel motor will produce the regenerative braking torque. Then, the total braking torque for on each tyre is the sum of the mechanical braking torque and the regenerative braking torque. Equation (9) shows the braking torque at each tyre.

$$\begin{cases} T_{Bfr} = BEF_{fr} \times r_{fr} B_{fr} \\ T_{Bfl} = BEF_{fl} \times r_{fl} B_{fl} \end{cases} \quad \begin{cases} T_{Brr} = BEF_{rr} \times r_{rr} B_{rr} + T_{Rr} \\ T_{Brl} = BEF_{rl} \times r_{rl} B_{rl} + T_{Rl} \end{cases} \tag{9}$$

Where *BEF* is braking efficiency factor and the value of *BEF* is 1.5.

If ρ is not in the optimum range (0.2 to 0.3), the ABS control unit will start to operate to control the braking pressure in the master cylinder. The braking pressure from the master cylinder will be directed to the front wheel cylinder and rear power cylinder. Without a time delay response from the master cylinder to the front wheel cylinder, and ABS is enough to maximise the slip ratio and cornering force of the front tyre. However, for the rear braking system, due to the rigidness of the mechanical braking system, a time delay response occurred during ABS operational. To compensate for the loss of friction force, the in-wheel motor will produce the regenerative braking torque. The regenerative braking control timing will be operational when the value of ρ is not in the optimum range. From the operation of the ABS and regenerative brake control timing, we can keep the optimum value of ρ and we also can maximise the braking force and cornering force, thus assuring the stability of the vehicle.

3 Results and discussion

3.1 Effect of the regenerative braking without electric motor control

In the first part of this study, the effect of the regenerative braking without electric motor control have been examined. The right side of the road is dry asphalt and the left side is an icy road. Figure 5 shows the tire rotational speed (uFL, uFR, uRL, uRR) and vehicle speed (u) when differential regenerative braking torque control does not apply to all tires. From this result, the rotational speed of the right tires (uFR, uRR) almost similar with the speed of the vehicle. However, due to the low friction between road and tire on the left tires (uFL, uRL), the rotational speed of uFL and uRL was decreased rapidly from 8.3 m/s to 1 m/s in 4 seconds. After 4 seconds, the rotational speed of uFL and uRL was similar with the speed of u, uFR and uRR. It means that in the low-speed condition, the regenerative braking force produced by the electric motor is every small and not affect the slip of the tire.

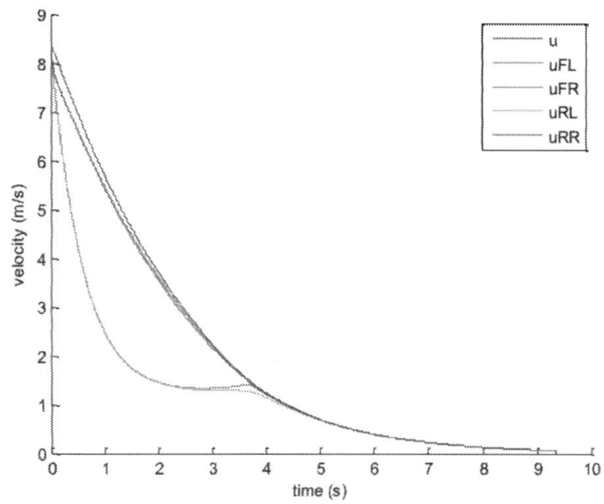

Fig. 5. Deceleration of the vehicle and tire without differential regenerative braking torque control.

The skidding condition of the vehicle can be measured from the slip ratio of the tire in Figure 6. From Figure 6, the slip ratio of the right tires (slip ratio FR, slip ratio RR), are very small. When braking on the dry asphalt, due to the high friction between the tire and road, the optimum braking force can be applied to stop the tire from rotating. However, on the icy road condition, the slip ratio of the tire (slip ratio FL, slip ratio RL) are increases rapidly from 0 to 0.6 in 1.5 seconds. Start from 1.5 seconds to 4 seconds, the slip ratio FL and RL were decreased rapidly to 0.5. The regenerative braking torque produced by the electric motor is proportional to the speed of the tire. Without regenerative braking torque control, in a high-speed situation, the tire will slip. On the other hand, in a low-speed condition, which is start from 4 seconds, the slip ratio is very small.

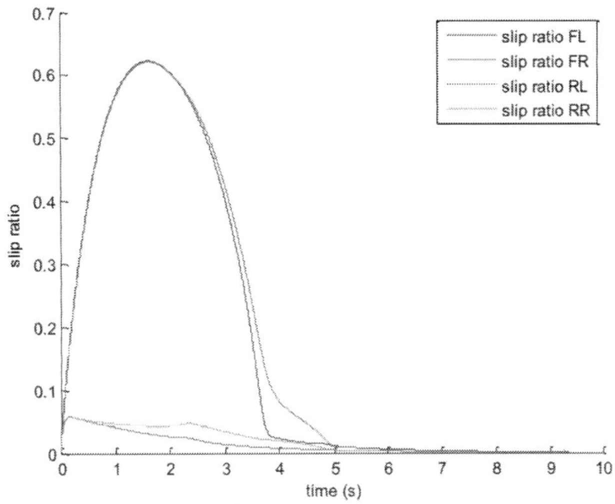

Fig. 6. Slip ratio of the tire without differential regenerative braking torque control.

3.2 Effect of the differential regenerative braking torque control

The second part of this study is to investigate the effect of the differential regenerative braking torque control on the dry asphalt and icy road. Figure 7 shows the deceleration speed of the vehicle (u) and tire (uFL, uFR, uRL, uRR). From this result, the rotational speed of the right tires (uFR, uRR) was decreased approximately to the speed of the vehicle. On the dry asphalt condition, the friction between contact patch of the tire and road are high, and the regenerative braking force produced by the electric motor can stop the tire without any slip. However, for the left tires (uFL, uRL), the deceleration of the tires are fluctuated start from the beginning until 3.5 seconds. In a high-speed situation, when braking on an icy road with differential regenerative braking torque control, the electric motor will turn off and turn on based on the slip ratio of the tire. From this mechanism, the rotational speed of the tire will fluctuate. From 3.5 seconds, the rotational speed of the uFL and uRL are similar with u, uFR and uRR. This situation occurs because, in a low-speed condition, the regenerative braking force is minimised and no slip on the tire.

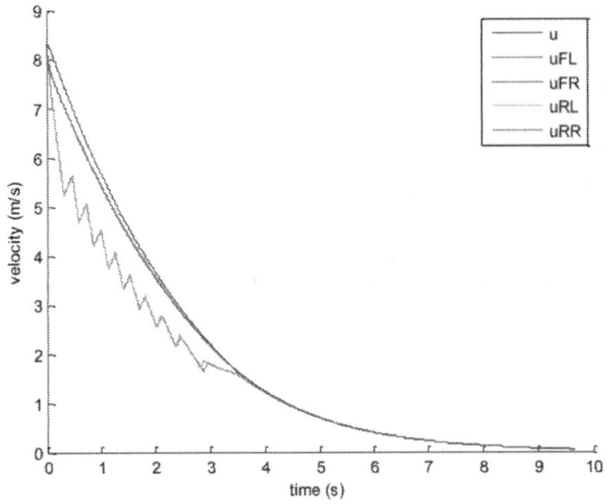

Fig. 7. Deceleration of the vehicle and tire when differential regenerative braking torque control is applied.

The slip ratio of the tire (slip ratio FL, FR, RL,RR) are shown in Figure 8. From this result, the slip ratio FR and RR are very small. On a dry asphalt condition, there is no slip on the tire. However, on an icy road, the slip ratio FL and RL have fluctuated in the 3.2 seconds. In this condition, the controller sends the signal to the electric motor to control the regenerative braking torque on each wheel to prevent the tire from lock up. After 3.2 seconds, there is no signal from the controller to control the regenerative braking torque because the rotational speed of the tire is very small.

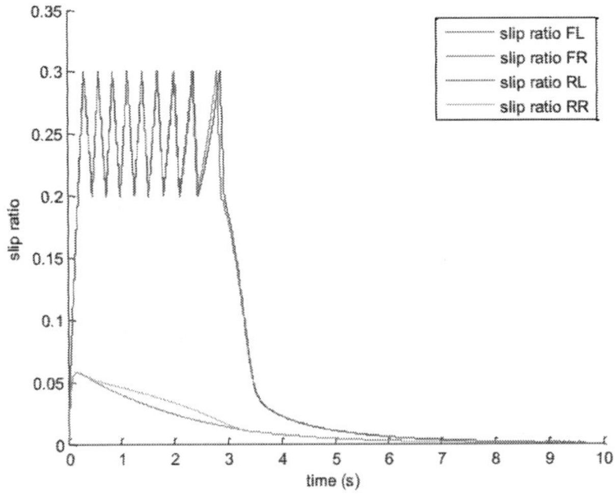

Fig. 8. Slip ratio when differential regenerative braking torque control is applied.

4 Conclusions

In this paper, the differential regenerative braking torque controller is developed to increase the stability of four small EV with four in-wheel motors. To investigate the effectiveness of the differential regenerative braking torque control, the simulation model has been developed in the Matlab-Simulink. In the simulation, the right side of the road is set to the dry asphalt and the left side of the road is set to an icy road. From the simulation result, when braking on the different road condition for the right and left side, the differential regenerative braking torque control can avoid the tire from lock up. Besides, it also can prevent the vehicle from skidding and increase the stability of the vehicle.

The authors would like to thank Universiti Malaysia Pahang (UMP) for providing laboratory facilities and financial assistance under project no. RDU160312. Special thanks to Ministry of Higher Education Malaysia for providing Research Acculturation Grant Scheme (RAGS) under project no RDU151403.

References

1. Y. O. Mohamad Heerwan Bin Peeie, Hirohiko Ogino, *Skid Control of Small Electric Vehicle*, Proc. 2013 IEEE Conf. Syst. Process Control, Dec, pp. 13–15. (2013).
2. H. F. Toru Suzuki, *Slip Ratio Estimation and Regenerative Brake Control without Detection of Vehicle Velocity at Urgent Brake Turning*, 11th IEEE International Workshop on Advanced Motion Control, (2010).
3. D. Peng, Y. Zhang, C. Yin, and J. Zhang, *Combined Control of a Regenerative Braking and Anti-lock Braking System for Hybrid Electric Vehicles*, vol. 9, no. 6, pp. 749–757. (2008).
4. S. Sakai and Y. Hori, *Road Condition Estimation for Traction Control in Electric Vehicle*, IEEE International Symposium on Industrial Electronics, pp. 973–978. (1999).
5. M. Wu and M. Shih, *Simulated and experimental study of hydraulic anti-lock braking system using sliding-mode PWM control*, Mechatronics, vol. 13, no. 4, pp. 331–351. (2003).
6. H. Fujimoto, T. Saito, a. Tsumasaka, and T. Noguchi, *Motion control and road condition estimation of electric vehicles with two in-wheel motors*, Proc. 2004 IEEE Int. Conf. Control Appl. 2004., vol. 2, pp. 1266–1271. (2004).
7. J. R. Cho, J. H. Choi, W. S. Yoo, G. J. Kim, and J. S. Woo, *Estimation of dry road braking distance considering frictional energy of patterned tires*, Finite Elem. Anal. Des., vol. 42, no. 14–15, pp. 1248–1257. (2006).
8. M. H. Peeie, H. Ogino, and Y. Oshinoya, *Skid control of small electric vehicle with hydraulic-mechanical hybrid brake system (Effect of ABS and regenerative brake control on an icy road)*, Mech. Eng. Journal, Japan Soc. Mech. Eng., vol. 1, no. 5, pp. 1–15. (2014).
9. Z. Cai, C. Ma, and Q. Zhao, *Acceleration-to-torque ratio based anti-skid control for electric vehicles*, Proc. 2010 IEEE/ASME Int. Conf. Mechatron. Embed. Syst. Appl., pp. 577–581. (2010).
10. S.-I. Sakai, H. Sado, and Y. Hori, *Anti-skid control with motor in electric vehicle*, 6th Int. Work. Adv. Motion Control Proc., pp. 317–322. (2000).
11. S. Hasegawa and H. Ogino, *Research on Skid Control of Small Electric Vehicle with Hydraulic-Mechanical Hybrid Brake System (1st Report; Hydraulic simulation of anti-lock brake system for two-wheel models)*, Proc. Sch. Eng. Tokai Univ., vol. 49, no. 2, pp. 89–94. (2009).

 DOI: 10.1051/matecconf/20179001042

12. H. Ogino, S. Kobayashi, and S. Hasegawa, *Research on Skid Control of Small Electric Vehicle (Effect of Regenerative Braking System in Skidding Condition)*, Trans. JAPAN Soc. Mech. Eng. Ser. C, vol. 77, no. 784, pp. 4605–4618. (2011).
13. M. H. Bin Peeie, H. Ogino, and Y. Oshinoya, *Skid control of a small electric vehicle with two in-wheel motors: simulation model of ABS and regenerative brake control*, Int. J. Crashworthiness, vol. 8265, no. April, pp. 1–1., (2016).

Structural dynamic investigation of frame structure with bolted joints

M.H.N. Izham[1], *N.A.Z.* Abdullah[1], *S.N.* Zahari[1] and *M.S.M.* Sani[1,2,*]

[1]Advanced Structural Integrity and Vibrational Research Group (ASIVR), Faculty of Mechanical Engineering, Universiti Malaysia Pahang
[2]Automotive Engineering Center, Universiti Malaysia Pahang

Abstract. Frame structure is widely used in many engineering structures. Besides, vibrational problem is one of the main challenges in industry. Bolted joints are commonly used in industry to connect two or more mechanical parts and it plays a significant role in the dynamics characteristic of the structure. This study aims to perform a model updating procedure on a portal frame structure which consists of bolted joints. Modal parameters such as the natural frequencies, mode shapes and damping ratios are gathered through finite element analysis (FEA) and experimental modal analysis (EMA). Frame structure is set to be fixed-free boundary condition and equivalence of nodes is performed at the area of bolted joints. Correlation between these two sets of data is carried out. With the selected parameters identified to perform model updating on the structure by using sensitivity analysis, the discrepancies in natural frequencies were reduced between FEA and EMA.

1 Introduction

Space frame structure is widely used as engineering structure. The application covers wide range of usage such as roofing, sports arena, transportation support, building supports, communication tower, defence tower and etc. Various application/function applies on the frame structure but how rigid/strong will the structure able to withstand the load. Structures are usually used to overcome the large moment due to the applied loading. In past few decades technology has driven frame structure to be more competitive to meet the current world requirement such as lightweight and economical, at the same time to maintain the stability of the structure [1–3].

Structural systems or space frame structure is build up or assemble of many components. One of the most important roles on frame structure is the joint. A mechanical joint is a section of a structure which was used to connect one component or more. Few types of connections such as weld, bolt and thread are used for frame structure. Joints will usually consume 15-30% of the steel consumption which lead to high cost [2].

Bolted connection is commonly used in the industry, easy to assemble/disassemble and low cost is what making it popular among user. Figure 1 shows the frame structure with bolted

* Corresponding author: mshahrir@ump.edu.my

joints with fixed boundary condition at the base of the structure. Frame structure with bolted joints definitely vibrates. In order to perform dynamic investigation of frame structure with bolted joints, finite element analysis was used in the early stage of the study.

Fig. 1. Frame structure with bolted joints.

Finite element method is common computer aided software used in industry and commonly used method to simulate the performance of the product/structure in early engineering design stage. Finite element analysis (FEA) famously used in strength and stress analysis, and lately dynamic investigation has brought interest to industry to simulate the product performance. FEA is used in industry to gain confidence level in early designing stage and to analyses the product performance and definitely to predict the dynamic characteristic of the structure. It's a common practice to use FEA to predict the product performance or dynamic characteristic and few studies have been using FEA method to gain data of the structures/products [4–7].

Experimental modal analysis (EMA) is a process to describe the vibration characteristic, vibration characteristics are divided into three which are natural frequency, damping ratio and mode shape which also called as modal parameters. Experimental modal analysis has been used since early 1970's [8]. Structure is excited by a force using tool called impact hammer and the output response will be measured by accelerometers with the fast fourier transform (FFT) analyzer, modal parameters will be extracted from the frequency response function (FRF) [8–10].

FEA method often have discrepancies between the experimental value [11] hence model updating is required to minimize the discrepancies between numerical studies and experimental. Before model updating fall in place, few most common errors in numerical studies need to be addressed such as model structure errors, model parameters errors and model order errors [12]. Model updating is a process adjusting selective parameters in finite element to reduce the results discrepancies between FEA and EMA [13].

Performing model updating by trial and error approach has been the technique classically [13], hence there are few exploration of model updating techniques has been studied by then [4,11–18]. Model updating applied on structural dynamic [19] explained general concept of FEA model updating and model updating method. Parameters are adjusted in model updating process [20-21] to reduce the discrepancies. Sensitivity analysis is one of the model updating technique develops algorithm to optimize the model updating technique, only sensitive parameters is selected to minimize the discrepancies [13, 22].

In this project, investigation on dynamic properties of space frame structure will be carried out to determine modal properties in numerical analysis and experimental. One of the approached to solve model updating is to use sensitivity method. Sensitivity method is focusing on the approach of model updating problem. Furthermore, model updating will be used to minimize the discrepancies between experimental modal analysis and finite element analysis.

2 Experimental modal analysis

Increasing development of data acquisition and processing capabilities has led to major advances in the experimental realm of the analysis, widely known as modal testing. Modal parameters such as natural frequency, mode shape and damping ratio were extracted using modal testing. The experiment was carried out in fixed-free boundary condition. The upper part of the structure is in free condition while the base structure is in fixed condition. The portal frame structure was divided into 48 grid points in order to achieve adequate spatial resolution of global structural mode shapes. The impact location and measurement points were carefully chosen in order to avoid nodal points. Figure 2 displays the location of grid points on the portal frame structure.

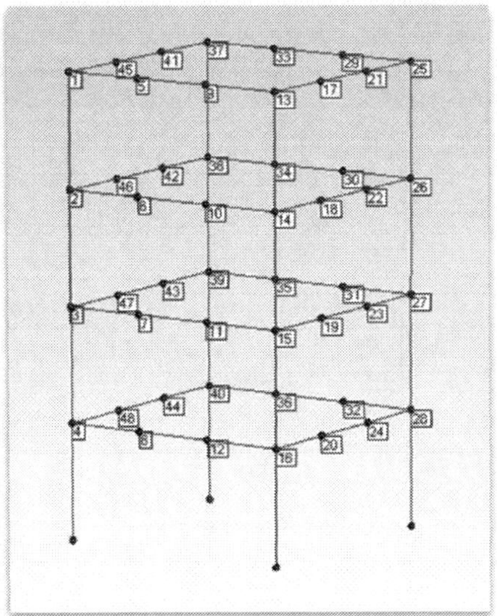

Fig. 2. Portal frame structure with lebeled grid points.

The excitation method used in this paper is impact hammer testing with roving accelerometer. The impact hammer with force transducer attached on its head used to excite the structure. The response was detected by tri-axial accelerometer attached on the structure and data analyser was used to convert the signal response which was in time domain to the frequency domain which the equipment shown in Figure 3.

 DOI: 10.1051/matecconf/20179001043

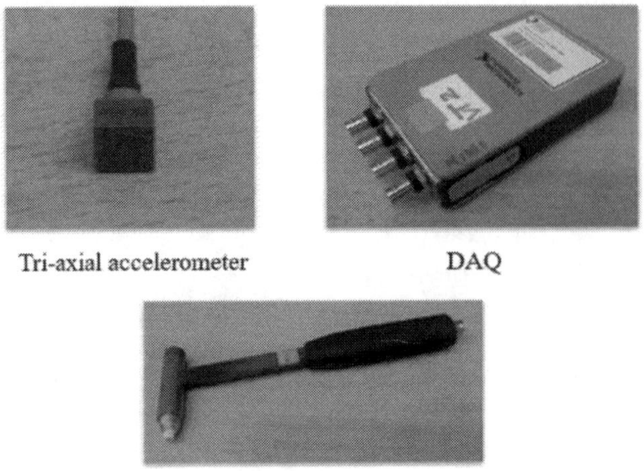

Tri-axial accelerometer DAQ

Impact hammer

Fig. 3. Equipment used for EMA.

Natural frequency and mode shapes were extracted using curve fitting method from ME'scopeVES software and being tabulated in Table 1

Table 1. Experimental natural frequencies results.

Mode	Natural Frequency (Hz)
1	34.5
2	40.0
3	57.9
4	65.8
5	81.4

3 Finite element analysis

Computer aided design (CAD) model of the frame structure was created by using CATIA software before the model was converted into FEA model. FEA was performed by using MSC Nastran/Patran software. The FEA model of the frame structure composed of 3476 nodes and 2760 shell elements (CQUAD4). Geometrically, the model consists of 4 angle-bars and 16 flat bars which their properties is stated in Table 2. Equivalence of nodes was performed at the area of bolted joints in the model in order to represent the rigidity of joints in the structure.

Boundary condition and external forces were assigned to the model as the model was let to exist in fixed-free boundary condition. Structure base is in fixed condition by wall plug. Calculation of modal properties in MSC.Natran/Patran was done by using SOL 103, which is the solution for normal modes analysis. By using the normal modes analysis, the natural frequencies and mode shapes of the structure can be computed. In normal modes analysis, the equation of motion is stated as Eq. (1).

$$\boldsymbol{M}u + \boldsymbol{K}u = 0 \tag{1}$$

where **K** and **M** are stiffness and mass matrices respectively. These system matrices are computed automatically by MSC Nastran/Patran, based on the geometry and properties of the FEA model. By assuming a harmonic solution, the equation from above can be reduced to an eigenvalue problem which is stated as below.

$$[\boldsymbol{K} - \lambda_i \boldsymbol{M}]\{\phi_i\} = 0 \qquad (2)$$

Where $\{\phi\}$is the eigenvector (mode shape) corresponding to the eigenvalue λ (natural frequency). Each eigenvalue is proportional to a natural frequency and is corresponding to eigenvector. The eigenvalues are related to the natural frequencies as

$$f = \frac{\sqrt{\lambda_i}}{2\pi} \qquad (3)$$

The constructed finite element model as viewed in graphic interface of MSC Nastran/Patran software is as shown in Figure 4 and results of FEA is obtained as shown in Table 3.

Table 2. Material properties of frame structure.

Structure	Properties		Value
Angle bar	Material properties (low carbon steel)	Young's modulus, E (GPa)	180
		Density (kg/m^3)	7860
		Poisson ratio	0.3
	Physical properties	Length (mm)	1000
		Width (mm)	37.5
		Thickness (mm)	4.0
Flat Bar	Material properties (low carbon steel)	Young's modulus, E (GPa)	180
		Density (kg/m^3)	7860
		Poisson ratio	0.3
	Physical properties	Length (mm)	500
		Width (mm)	25.0
		Thickness (mm)	6.0

4 Correlation of experimental and numerical results

The results of FEA often have discrepancies between the experimental [11], hence the results obtained from FEA and EMA will be correlated, where discrepancies between both result will be compared and the percentage error will be captured.

Table 4 shows the correlation of natural frequency from different modes between experimental vs numerical value of frame structure with bolted joints. Results observed the percentage of error which recorded highest of 16.29% on mode 1 and lowest of 0.23% on mode 4. Mode shape in between FEA and EMA is presented in table 5.

Fig. 4. Frame structure of FEA model.

Table 3. FEA natural frequency results.

Mode	Natural Frequencies (Hz)
1	40.12
2	43.77
3	54.65
4	65.95
5	86.33

Table 4. Correlation of natural frequencies between EMA and FEA.

Mode	Natural Frequency (Hz)		Percentage of error (%)
	EMA	FEA	
1	34.5	40.12	16.29
2	40.0	43.77	9.43
3	57.9	54.65	5.61
4	65.8	65.95	0.23
5	81.4	86.33	6.06
		Total Error	37.62

Model updating is a process adjusting selective parameters in finite element to reduce the results discrepancies between FEA and EMA [13]. The target of correlation is to minimize the percentage of error in between FEA and EMA. This can be completed by performing the model updating. The optimization algorithm called SOL 200 has been used in this study. The objective function is constructed as Eq. (4).

$$g(x) = \sum_{i=1}^{n} W\left(\frac{w_i^e}{w_i^a} - 1\right)^2 \tag{4}$$

where w_i^e and w_i^a are the experimental and computational natural frequencies respectively, with W as the real positive weighing factor.

Table 5. Correlation mode shape between FEA and EMA.

Mode	Mode shape	
	FEA	**EMA**
1	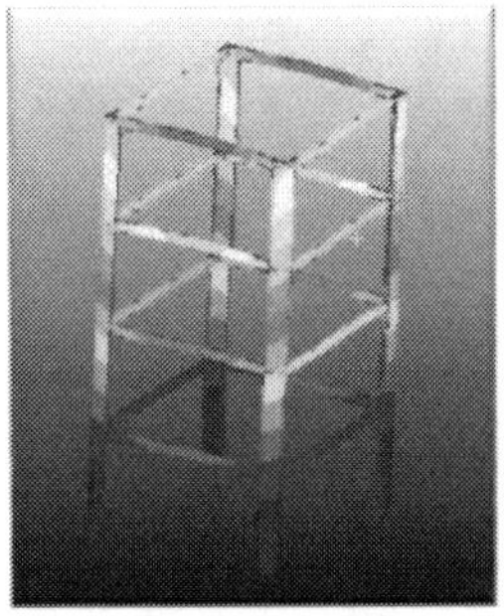	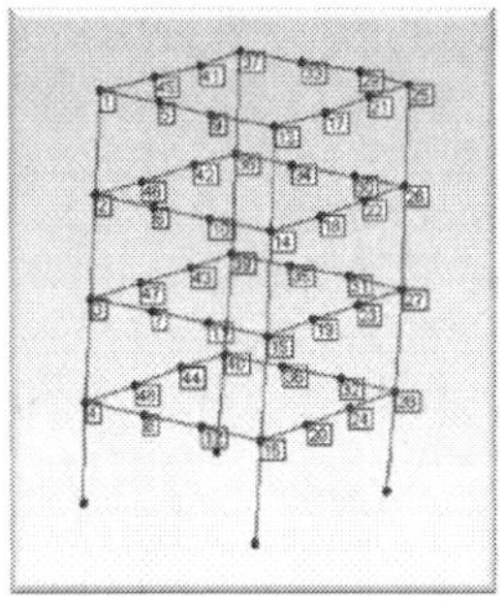
2	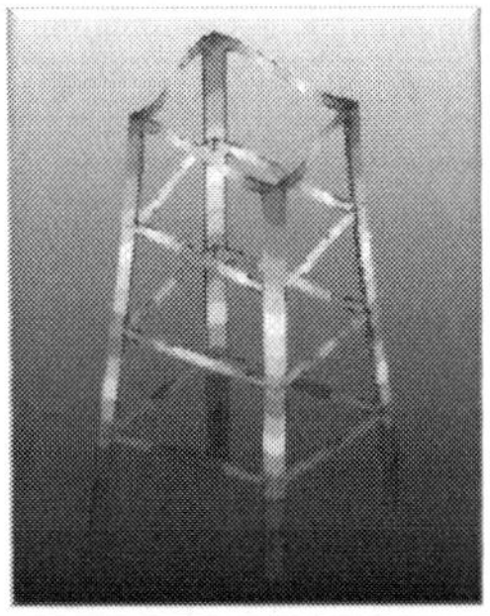	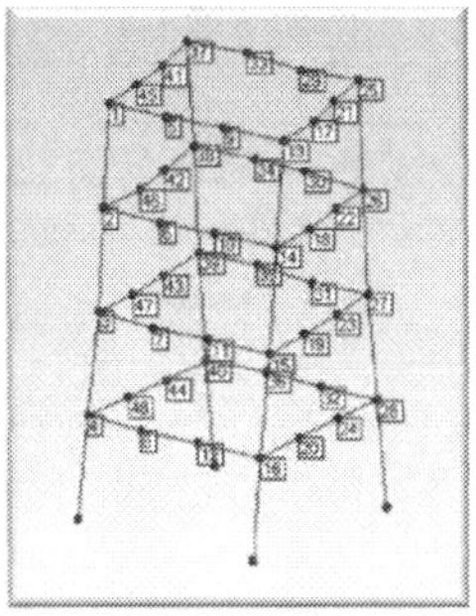
3	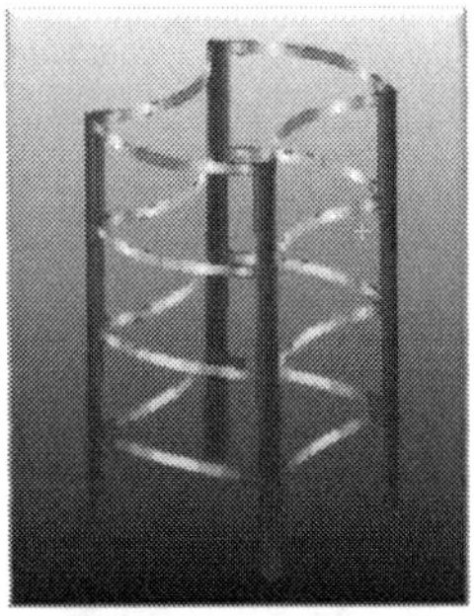	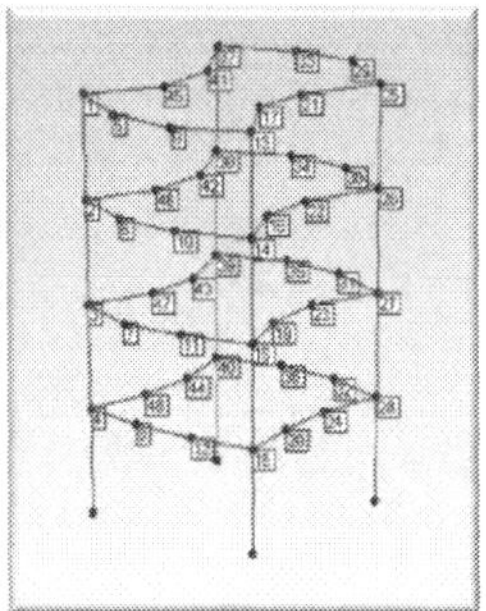
4	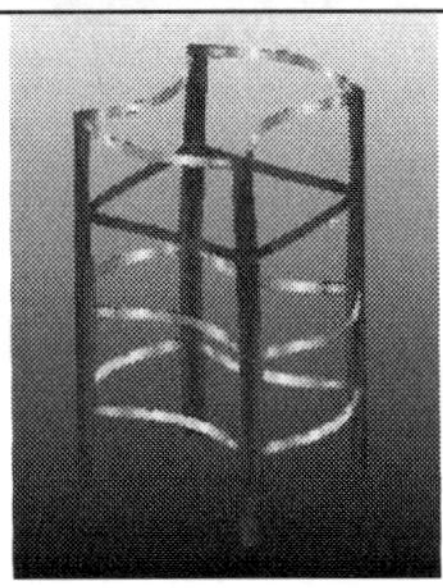	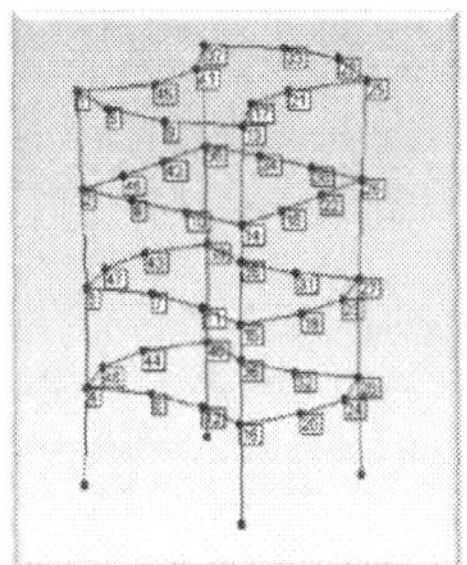
5	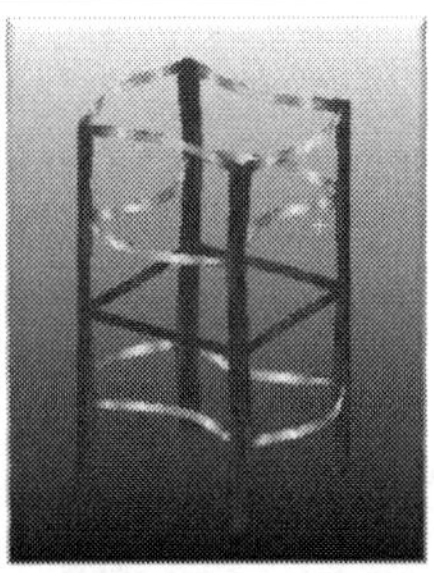	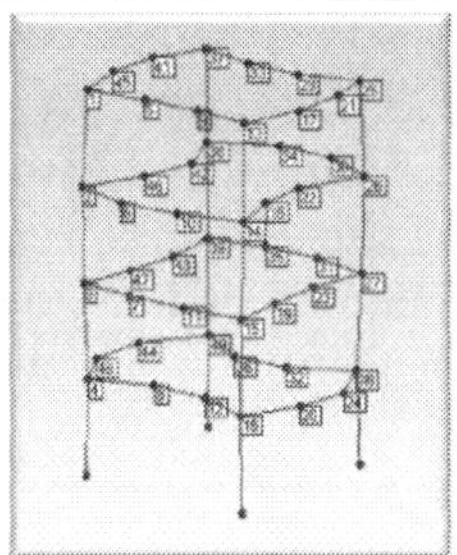

Performing model updating with a selection of sensitive parameters is very crucial in order to minimize the discrepancies. Which resulting to four parameters has been picked in this study, which are Modulus Young, poison ratio and thickness of the angle and flat bar.

Table 6 below is presenting the results prior to model updating of the frame structure, comparison of natural frequencies performed in FEA between before and after model updating. The significant errors were observed on mode 1, upon model updating the error rate reduces from 16.29% to 10.52%. Mode 4 showing an increasing percentage of error, but other modes shows a significant reduction in percentage of error.

Table 6. Comparison of natural frequencies values between initial and updated results.

Mode	EMA Natural Frequency (Hz)	Initial FEA results		Model updating FEA results	
		Natural Frequency (Hz)	Error (%)	Natural Frequency (Hz)	Error (%)
1	34.5	40.12	16.29	38.13	10.52
2	40.0	43.77	9.43	42.98	7.45
3	57.9	54.65	5.61	56.04	3.21
4	65.8	65.95	0.23	64.80	1.52
5	81.4	86.33	6.06	82.29	1.09
		Total Error	37.62	Total Error	23.79

Four parameters have been identified to reduce the discrepancies between EMA and FEA natural frequencies via model updating procedure. Table 7 shows the deviation of the updating parameter and it explains that young modulus, poison ratio and thickness play an important role to reduce the discrepancies. Flat bar thickness shows higher sensitivity value when sensitivity analysis was performed.

Table 7. Changes of updating parameters from the initial values.

Parameter	I	II	Changes (%) =
	Initial value	Updated value	|(II-I)/I| x 100
Young's modulus (GPa)	180	162	10
Poison Ratio	0.3	0.33	10
Angle Bar Thickness (m)	0.004	0.0039	2.5
Flat Bar Thickness (m)	0.006	0.0065	8.3

5 Conclusions

This study is to focus on reducing the discrepancies between EMA and FEA on frame structure. With the fixed-free boundary condition applied on the frame structure with bolted joints, fixed boundary was applied on the base of the structure and equivalence of nodes is performed at the area of bolted joints in FEA. Four parameters were chosen in this study for updating parameters. Parameters assumption has been made in earlier phase when performing FEA which resulting to discrepancies in natural frequencies, with updating procedure this gap is narrow down.

The discrepancies between EMA and FEA have been brought down by applying the model updating technique using sensitivity analysis optimization. This finding can be further improved by focusing updating on the joint, defining the joint element can probably reduce the discrepancies between EMA and FEA.

This work was supported by focus group of Advanced Structural Integrity of Vibration Research (ASIVR) Universiti Malaysia Pahang for providing all the equipment used for this project.

References

1. M. Ghonim, "11 . SPACE FRAMES CONSTRUCTION," pp. 1–11, 2015.
2. E. C. Wai-fah, "Lan, T.T. ' Space Frame Structures ,'" 1999.
3. M. Buijsen, "Dynamic Space Frame Structures," 2011.
4. K. He and W. D. Zhu, "Finite Element Modeling of Structures with L-shaped Beams and Bolted Joints : Model Updating and Predictive Modeling Approaches," *Asme*, vol. 133, no. February, pp. 1–18, 2009.
5. K. He and W. D. Zhu, "Modeling of Fillets in Thin-Walled Beams Using Shell/Plate and Beam Finite Elements," *J. Vib. Acoust.*, vol. 131, no. 5, p. 051002, 2009.
6. Andrew M. Brown and Richard M. Seugling, "Using Plate Finite Elements for Modeling Fillets in Global Response Analysis," *Finite Elem. Anal. Des.*, 2004.
7. R. M. Seugling and A. M. Brown, "Modeling of fillets in thin-walled structures for dynamic analysis," *Proc. Int. Modal Anal. Conf. - IMAC*, vol. 1, pp. 379–384, 2001.
8. B. J. Schwarz and M. H. Richardson, "EXPERIMENTAL MODAL ANALYSIS," 1999.
9. P. Avitabile, "Experimental Modal Analysis," *Sound Vib.*, vol. 35, no. 1, pp. 1–15, 2001.
10. N. LARBI and J. LARDIES, "Experimental Modal Analysis of a Structure Excited By a Random Force," *Mech. Syst. Signal Process.*, vol. 14, no. 2, pp. 181–192, 2000.
11. T. Marwala, L. Mdlazi, and S. Sibisi, "Finite Element Model Updating Using Bayesian Approach," *22th Int. Modal Anal. Conf.*, vol. 42, no. 1, p. 7, 2007.
12. J. E. Mottershead and M. I. Friswell, "Model Updating In Structural Dynamics: A Survey," *Journal of Sound and Vibration*, vol. 167, no. 2. pp. 347–375, 1993.
13. J. E. Mottershead, M. Link, and M. I. Friswell, "The sensitivity method in finite element model updating: A tutorial," *Mech. Syst. Signal Process.*, vol. 25, no. 7, pp. 2275–2296, 2011.
14. M. K. Sen, A. Datta-gupta, P. L. Stoffa, L. W. Lake, and G. A. Pope, "DYNAMIC FINITE ELEMENT MODEL UPDATING using Simulated Annealing and Genetic Algorithms," *SPE Form. Eval.*, vol. 4, pp. 49–55, 1995.
15. W. Wang, J. E. Mottershead, A. Ihle, T. Siebert, and H. Reinhard Schubach, "Finite element model updating from full-field vibration measurement using digital image correlation," *J. Sound Vib.*, vol. 330, no. 8, pp. 1599–1620, 2011.
16. A. Yan and J. Golinval, "Model Updating Based on Frequency Response Functions Using A General Finite Element Code," *Convergence*, vol. 48, pp. 149–162.
17. H. Grafe, "Model Updating of Large Structural Dynamics Models Using Measured Response Functions," no. October, p. 184, 1998.
18. N. Guo, Z. Yang, Y. Jia, and L. Wang, "Model updating using correlation analysis of strain frequency response function," *Mech. Syst. Signal Process.*, vol. 70–71, pp. 284–299, 2016.
19. N. A. Z. Abdullah, M. S. M. Sani, M. M. Rahman, and I. Zaman, "A review on model updating in structural dynamics," vol. 012015.
20. S. N. Zahari, M. S. M. Sani, N. A. Husain, M. Ishak, and I. Zaman, "Dynamic Analysis of Friction Stir Welding Joints in Dissimilar Material Plate Structure," *J. Teknol.*, vol. 78, no. 6–9, pp. 57–65, 2016.
21. N. A. Z. Abdullah, M. S. M. Sani, N. A. Husain, and M. M. Rahman, "Model Updating of Go-Kart Structure in Structural Dynamics," 2015.
22. N. A. Husain, H. H. Khodaparast, A. Snaylam, S. James, G. Dearden, and H. Ouyang, "Finite-element modelling and updating of laser spot weld joints in a top-hat structure for dynamic analysis," *J. Mech. Eng.*, vol. 224, pp. 851–860, 2010.

 DOI: 10.1051/matecconf/20179001044

Perodua Myvi engine fuel consumption map and fuel economy vehicle simulation on the drive cycles based on Malaysian roads

Muhammad Iftishah Ramdan[1,*]

[1]School of Mechanical Engineering, Universiti Sains Malaysia, 14000 Nibong Tebal, Penang, Malaysia

Abstract. This paper presents the fuel consumption engine map for a 1.3L Perodua Myvi passenger car. The engine dynamometer and the engine throttle are controlled, to create the operating conditions for the engine map. Interpolation work is done in MATLAB, to create a 3D fuel consumption engine map. The engine map is used in a fuel-economy estimation simulation, using the city and the highway drive cycles based on Malaysian roads. The fuel economy values generated from the simulations are similar to experimental fuel consumption results.

1 Introduction

Malaysia is a rapidly developing country in Asia, where the demand for passenger cars increases every year. The average car production in Malaysia is 24,403 units from 1984 until 2016, with maximum production of 56,745 units in April 2015 [1]. Perodua Myvi remained the best-selling car in Malaysia for eight consecutive years, between 2006 and 2013 [2]. Thus, it is important to study its fuel consumption performance as a measure of energy optimization.

A fuel consumption engine map refers to a 3-dimensional plot contour of fuel consumption on torque-speed maps, to show the engine fuel consumptions at given torque-speed combinations [3]. The engine's torque and speed can be measured using a load device called an engine dynamometer. Ramasamy [4] has previously built a brake specific fuel consumption (BSFC) map for K3-VE Perodua Myvi engine that presents the fuel consumption at wide-open throttle (WOT) condition. However, the map is not enough for vehicle simulations that estimate fuel economy. This is because any vehicle simulation would use broad range of speed and torque and not only at WOT condition.

In this study, a Perodua Myvi K3-VE engine is attached to an engine dynamometer. Its' throttle, along with the dynamometer load, are controlled simultaneously, to achieve the desired engine torque and speed. At the same time, the fuel consumed at the desired engine torque and speed, is recorded. The engine speed ranges from 1000 to 4500 RPM, and the torque is increased from 10Nm at increments of 10Nm until the engine stalls. The gathered fuel consumption data are then compiled and interpolated to create a 3D fuel consumption

[*] Corresponding author: shahramdan@usm.my

engine map. The engine map is very useful for any vehicle fuel economy simulation to represent its engine model.

This paper also presents the simulation work to estimate the fuel economy of a Perodua Myvi 1.3L automatic on Malaysian drive cycles. Although there is an effort to establish a Malaysian drive cycle [13], there is no official one published yet. Thus, paper uses two drive cycles data that are based on driving on roads in Alor Star, Kedah, Malaysia [5]. The simulation utilizes the fuel economy engine map generated from the dynamometer measurements to calculate the estimated fuel consumption. The fuel economy results from the simulations and the actual experiment [5] are compared and discussed.

2 Experimental setup

The experiment intends to obtain the fuel consumption values of the engine at different combinations of engine torque and speed. The engine has to be subjected to controlled loads in order to generate specific torque values. Thus, an eddy current engine dynamometer is connected to a K3-VE engine flywheel (fig. 1).

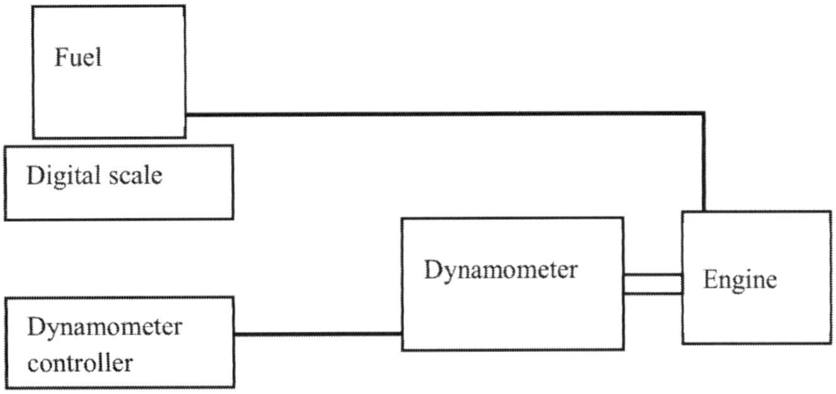

Fig. 1. Engine and dynamometer setup.

The experiment starts as the warmed-up engine reaches the desired torque and speed by controlling the engine throttle angle, and the dynamometer controller simultaneously. The engine speeds are varied from 1000 RPM to 4500 RPM, at increments of 500 RPM. At every speed, the torque values start at 10Nm, and are increased at 10Nm increments until the engine is stalled at a wide-open throttle (WOT).

Once the desired torque and speed are obtained, the fuel consumption value is collected gravimetrically, where the difference in the fuel mass is calculated. The fuel container is placed on top of a digital weight scale that is used to measure its initial and the final mass during 30 seconds of fuel consumption. The difference between the initial and the final weights is then divided by 30 seconds to obtain the mass of fuel consumed for every second (in g/sec) (equation 1). The experiment is run twice, and the average of the two fuel consumption values, is used. In this study, the fuel consumption data are gathered and plotted for 64 combinations of torque and speed in a 3D fuel consumption engine map (figure 3).

$$Fuel\ consumption = \frac{Mass_{initial} - Mass_{final}}{30\ sec} \tag{1}$$

3 Mathematical modelling and simulation

3.1 Vehicle mathematical modeling, control strategy and simulation algorithm

The simulation algorithm reads the vehicle speed from a drive cycle [5] and discretizes it into one-second time steps. At every time step, the average speed and acceleration of the vehicle from a drive cycle, are fed into the vehicle model to calculate the acceleration, the tractive force, and the torque needed at the wheels (equation 2 to equation 7) [12]. The simulation uses a four-speed, automatic transmission 1.3L Perodua Myvi's parameters (table 1) as its mathematical vehicle model.

Table 1. Automatic Perodua Myvi parameters [6,7].

Maximum engine torque	**Tengine**		116 Nm
Total gear ratio (gear ratio x final drive ratio)	**Gratio**	1st gear	11.01
		2nd gear	6.15
		3rd gear	4.03
		4th gear	2.81
Efficiency of transmission	**ηt**		0.85
Tire radius	**Rtire**		0.288 m
Frontal area	**A**		2306 m2
Air drag coefficient	**Cd**		0.32
Air density	**ρ**		1.225 kg/m3
Mass of vehicle + driver	**m**		1055 kg
Coefficient of rolling resistance	**Cr**		0.01†

*("Rolling friction and rolling resistance,")

At every time step, the vehicle wheels have to overcome four forces in order to move. They are the air drag force (F_{air}), the force of slope of the road (F_{slope}), the acceleration force ($F_{acceleration}$), and the rolling resistance ($F_{rolling}$). The sum of these forces is equal to the tractive force of the driving wheels (F_{wheel}) (Equation 2).

$$F_{wheel} = F_{air} + F_{slope} + F_{acceleration} + F_{rolling} \tag{2}$$

DOI: 10.1051/matecconf/20179001044

The air drag force is calculated using equation (3), where ρ is the density of air (kg/m³), v is velocity of the vehicle (m/s), C_d is the air drag coefficient, and A is the frontal area of the vehicle (m²).

$$F_{air} = \frac{1}{2}\rho v^2 C_d A \qquad (3)$$

Equation 4 shows the calculation for the force caused by the gradient, or slope, of the road, where m is the sum of the masses of the vehicle and the driver (kg), *g* is the gravity acceleration (m/s²), and α is angle of slope of the road (°). In this paper, the angle of slope is assumed to be zero, because most parts of the roads used in drive cycle are levelled.

$$F_{slope} = m\, g \sin\alpha \qquad (4)$$

The force caused by vehicle acceleration is shown in equation 5, where m is the sum of the masses of the vehicle and the driver (kg), and *a* is the vehicle's acceleration (m/s²).

$$F_{acceleration} = m\, a \qquad (5)$$

The rolling resistance is caused by the tires' contact patches that resist the movement of the tires (equation 6), where C_r is the rolling resistance coefficient [8], m is the sum of the masses of the vehicle and the driver (kg), and *g* is the gravity acceleration (m/s²).

$$F_{rolling} = C_r m\, g \qquad (6)$$

From the calculated tractive force at the wheels (F_{wheel}), the wheel torque can be calculated by using equation.

$$T_{wheel} = F_{wheel} R_{tire} \qquad (7)$$

Where T_{wheel} is the wheel torque (Nm), F_{wheel} is the total tractive force of driving wheels (N). The engine torque (T_{engine}) and the engine speed (ω_{engine}) are determined from equations 8 and equation 9 respectively, which involve the gear ratio (G_{ratio}). G_{ratio} is selected, by following a simple, rule-based control strategy algorithm, as shown in table 2. The engine torque (equation 8) is also influenced by the transmission efficiency (η_t) which models the losses of the drive train.

$$T_{engine} = T_{wheel} / G_{ratio} / \eta_t \qquad (8)$$

$$\omega_{engine} = \omega_{wheel} G_{ratio} \qquad (9)$$

Table 2. Gear selection rules.

Speed (m/s)	Gears
< 7.73	1
7.73 to 13.83	2
13.83 to 21.09	3
> 21.09	4

The combination of the engine torque and speed are used to estimate the fuel consumption from the fuel consumption engine map obtained experimentally (figure 3).

The total fuel consumption from all the time steps is obtained, to calculate the fuel economy of the vehicle operating on every drive cycle by using equation (10).

$$Fuel\ Economy\ (km/L) = \frac{DT}{FCT} \tag{10}$$

Where DT is the total distance travelled (km), and FCT is the total volume of the fuel consumed (L).

This paper presents the results from two simulations, with a different drive cycle in each simulation. One of the drive cycles represents city driving conditions, with a low-speed, frequent start-and-stop pattern. The other one represents a highway drive cycle, with a high-speed, in frequent start-and-stop driving pattern. The simulation fuel economy results are compared to actual fuel economy values obtained during the development of the drive cycles [5]. The overall simulation can be summarized in simulation diagram (figure 2).

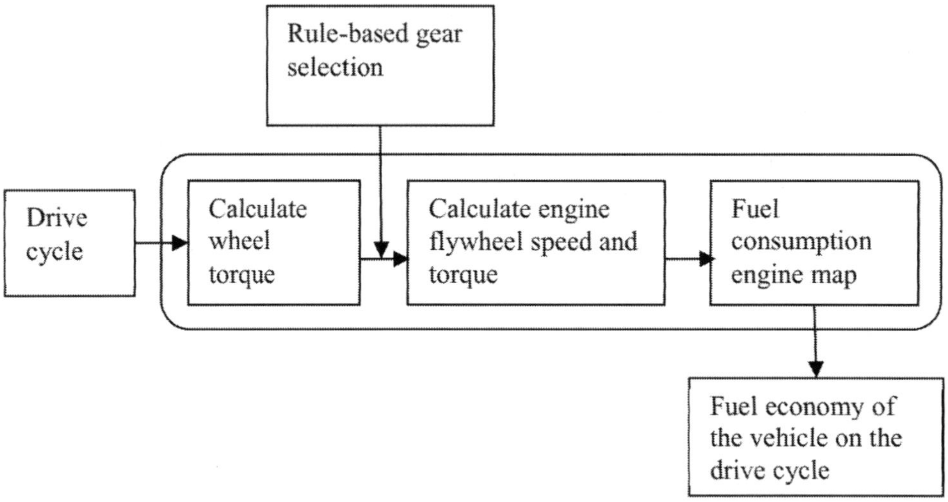

Fig. 2. Simulation algorithm.

4 Results and discussion

The 64 experimental fuel consumption data points are used to construct the 3D engine fuel consumption map using the 'nearest' interpolation function in MATLAB (figure 3). The 3D fuel consumption map presented here is used in the simulation to calculate the estimated fuel consumption.

DOI: 10.1051/matecconf/20179001044

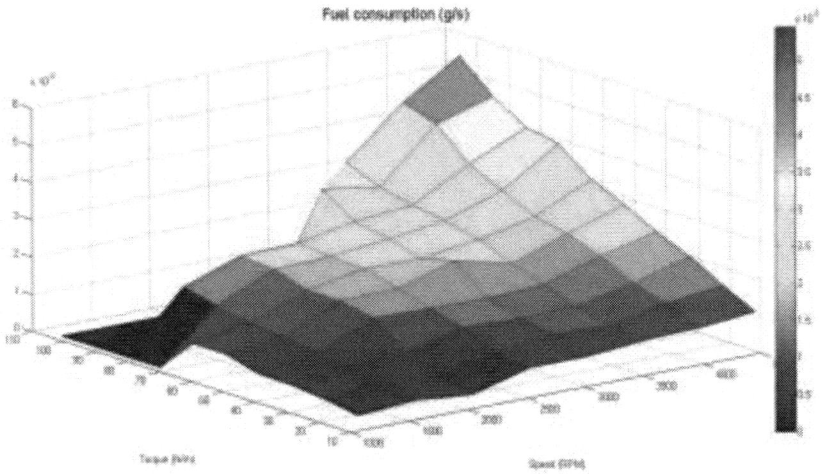

Fig. 3. 3D fuel consumption engine map.

4.1 Simulation and experimental fuel economy

The vehicle mathematical model is simulated on a city drive cycle and a highway drive cycle. The city drive cycle consists of 1,738 seconds of high-frequency stop-and-go driving conditions and low speeds, with the maximum speed of 19.5 m/s (70.2 km/h). The vehicle only utilizes the first, the second, and the third gears, with engine speeds of less than 3,000 RPM required to drive on this drive cycle (figure 4).

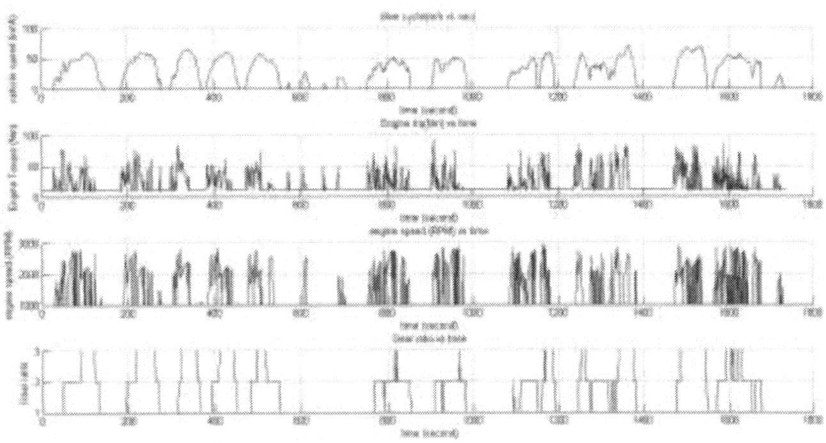

Fig. 4. Vehicle operating conditions for city drive cycle.

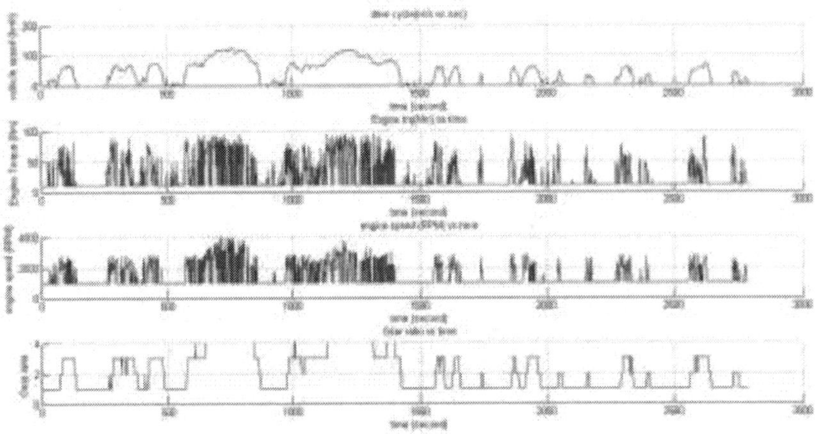

Fig. 5. Vehicle operating conditions for the highway drive cycle.

The highway drive cycle, on the other hand, consists of 2,878 seconds of low-frequency stop-and-go driving conditions and high speeds, with the maximum speed of 33.37 m/s (120.13 km/h). The vehicle utilizes all four gears, with the engine speeds of up to 4000 RPM required to drive on this drive cycle (figure 5).

The fuel economy values from the simulations are compared to the actual ones, as shown in table 3. The fuel economy of the vehicle driven on highway drive cycle is better compared to that when the car is driven on the city drive cycle. This is because the highway drive cycle requires less acceleration than the city drive cycle, which contributes to high torque (and power) demands from the wheel (equations 2 and 5).

All in all, the simulation and the actual fuel economy results from both drive cycles are similar, with percentage differences of less than 5%. The difference is due to the absence of transient elements from the simulation. The fuel consumption engine map used in the simulation records the fuel consumption when the engine reaches steady state torques and speeds. The difference of results between the steady-state and transient were mentioned by Z. Gao [9]. However, for estimation purpose, this simulation method is adequate as shown by the small difference in the simulation and the actual fuel economy results.

Table 3. Simulation and real (previous paper) fuel economy for city and highway drive cycles.

Drive cycles	**Simulation (Km/L)**	**Experiment (km/L)**	**Difference (km/L)**	**Percentage difference (%)**
City	10.13	9.77	0.37	3.76
Highway	12.15	12.77	0.62	4.89

5 Conclusions

This paper presents the development of a 1.3L Perodua Myvi fuel consumption engine map. The fuel consumption engine map is used in the simulations to calculate the estimated fuel economy of the vehicle driven on a city drive cycle and a highway drive cycle that are based on Malaysian roads. The estimated fuel economy results from the simulations are compared to the ones from the real world and they show similarity with a percentage difference of only 5%.

The authors would like to thank Universiti Sains Malaysia for providing laboratory facilities and financial assistance under short term grant #60313026.

References

1. Trading Economics. (2016, 28 May 2016) Available: http://www.tradingeconomics.com/malaysia/car-production
2. D. Tan. (2011, 30 May 2015). Perodua achieves record sales in 2010. Available: http://paultan.org/2011/01/12/peroduaachieves-record-sales-in-2010-5-year-old-myviis-still-malaysias-best-selling-car-alza-leads-thempv-charts/
3. C. E. Goering, and H. Cho, "Engine model for mapping BSFC contours," *Mathematical and Computer Modelling*, vol. **11**, pp. 514-518 (1988)
4. D. Ramasamy. "Improvement of in cylinder airflow burning rate for a four stroke spark ignition engine running on compressed natural gas (CNG) fuel" *(PHD thesis), Universiti Sains Malaysia*, 2014
5. M. I. Ramdan, and L.C. Piew, "Fuel economy comparison of Perodua Myvi passenger vehicle on Malaysian city and highway drive cycles," *Journal of Scientific Research and Development,* vol. **2**, 13, pp. 75-82 (2015)
6. PERODUA. (2013, 30 October 2014). Specification of Perodua Myvi. Available: http://www.perodua.com.my/ourcars/myvi/specifications
7. Toyota Peru. (15 March 2015). Daihatsu Data: 3SZ-VE/K3-VE. Available: http://intranet.toyotaperu.com.pe/tdp_sit/data/DAIHATSU/TERIOS/TERIOS_SHORT/pdf/manual/9514/D55B-2-E.pdf
8. Engineering Tool Box. (30 November 2014). Rolling friction and rolling resistance. Available: http://www.engineeringtoolbox.com/rollingfriction-resistance-d_1303.html
9. Z. Gao, J.C.Conklin, C.S. Daw, and V.K. Chakravathy. "A proposed methodology for estimating transient engine-out temperature and emissions from steady-state maps", *International Journal of Engine Research*, vol. **11**, 2 (2010)
10. Y. Liu, J. Li, and B. Shen. "Research on Driving Cycle of Long distance Passenger Vehicles Based on Principle Component Analysis and Cluster Algorithm". *International Journal of Control and Automation*, vol. **7**, 3, pp. 125-136 (2014)
11. Department of Energy, "Annual Energy Review 2003," *Energy Information Administration (EIA)*, Washington, D.C., USA, (2004)
12. L. Guzzella, A. Sciaretta, "Vehicle propulsion systems: Introduction to modeling and optimization," *Springer* (2007)
13. M. Abas, O. Salim, R. Martinez-Botas, and S. Rajoo, "Efforts to Establish Malaysian Urban Drive-Cycle for Fuel Economy Analysis," *SAE Technical Paper* (2014)

Simulation model for different bolted pre-combustion chamber (PCC) for spark ignition engine

Khairil Amri Muhamad Tajuddin[1], *Musthafah* Mohd Tahir[1,2,*], *Mohd Azli* Salim[1,2], *Muhd Ridzuan* Mansor[1,2], *Mohd Zaid* Akop[1,2] and *Adzni* Md. Saad[1,2]

[1]Faculty of Mechanical Engineering, Universiti Teknikal Malaysia Melaka, Hang Tuah Jaya, 76100, Durian Tunggal, Melaka, Malaysia.
[2]Centre for Advanced Research on Energy, Universiti Teknikal Malaysia Melaka, Hang Tuah Jaya, 76100, Durian Tunggal, Melaka, Malaysia.

Abstract. In an engine, two combustion chamber are present in the cylinder. First, the main combustion chamber and the other is called the auxiliary combustion chamber also known as pre-combustion chamber (PCC). The PCC size is small than the main combustion chamber size. In this research, four concept designs are developed for bolted PCC design. CATIA V5 R20 software is used for designing process. After that, Computational Fluid Dynamic (CFD) simulation is used to analyze each type of the bolted PCC powered with natural gas fuel by ANSYS Fluent 16.0 simulation software. Each of the bolted PCC simulation investigated by its flow behavior for the velocity vectors and turbulence kinetic energy (TKE). This PCC can give bigger effect to the engine performance by 10%-15% using natural gas as fuel source.

1 Introduction

Currently there are increasingly in number of internal combustion engines (ICE) that are converted from using liquid fuel such as gasoline or diesel to use alternative gas fuel. The alternative gas fuel such as Compressed Natural Gas (CNG). This is due to government demand for limited availability liquid fossil fuel, demand for improving fuel efficiency and the most important to reduce environmental pollution. CNG has emerged at most promising and better alternative fuels for engine uses. In worldwide, the use of natural gas in ICE has many advantages including their abundance, low price, and adaptability as engine fuels, higher octane number and less exhaust emission compared to liquid fossil fuels [1]. However, CNG suffer from the problems of very low energy density.

Nowadays, improvement of engine performance has become an important issues for automotive manufacturer's leaders. In order to increase engine performance, many studies are carried out on emission formation and also combustion process. Other than that, there are two common method use in research field. The method is the experimental study and

* Corresponding author: musthafah@utem.edu.my

mathematical simulations (thermo dynamical modelling, computational fluid dynamics (CFD) modelling, etc.) [2]. CFD is defined as the science of predicting fluid flow, heat transfer, mass transfer, chemical reactions, and related phenomena by solving the mathematical equations which govern these processes by means of computer based simulations [8]. It is easy to predict and analyze various details that are technically difficult such as cylinder process of gasoline combustion, temperature and pressure contours. The cylinder speed, the cylinder chamber geometry, fuel utilization, in-cylinder liquid dynamics and ignition devices utilized, speak to distinctive variables that influence the internal combustion [6]. For the most part, the principal air movements were filled into categories such as the rotary movement induced tangentially in the cylinder (swirl) and the rotary movement in the axial plane (tumble) [7].

Pre-combustion engine is an internal combustion engine in which carburetion is enhanced by method for a pre-combustion chamber. The fuel, or working blend, is initially sustained into the pre-combustion chamber for partially burned. At the point when the fuel mixture are ignited, the pressure in combustion chamber increases, as the result of heated fuel mixture and combustion products enters the combustion chamber over the piston cylinder. The steady rate of fuel combustion as it is removed from the pre-combustion chamber into the combustion chamber provide a uniform increment of pressure in cylinder for smoother operation of pre-combustion engine. For fuel pre-combustion chamber, the fuel utilization is about 6 to 8 percent lower by weight, and there are less toxic exhausts. In any case, stability and reliability of performance are diminished under few methods of operation. The pre-combustion chamber volume is generally from 25 to 30 percent of the main combustion chamber volume. Pre-combustion makes the turbulence required for the fuel and air to be properly mixed. Because of the differences in pre-combustion chamber designs, the way in which pre-combustion helps in making turbulence varies from one type of auxiliary combustion chamber to another. The disadvantages of pre-combustion chamber when the velocity of air and fuel mixture is too high during the entry from pre-chamber, so the heat loss is very high. This causes decrease in the thermal efficiency which can be counterbalanced by increasing the compression ratio. Cold starting will be troublesome as the air loses heat to chamber walls during compression process [9].

2 Simulation setup

2.1 Design geometry of pre-combustion chamber (PCC)

To begin the analysis, a solid geometry drawing of Pre-Combustion Chambers (PCC) must be created. A four geometry Pre-Combustion Chamber (PCC) has been designed by using CATIA V5R20.

2.2 Computational fluid dynamic (CFD) by ANSYS workbench (FLUENT)

Next, by using a Computational Fluid Dynamics (CFD), the fluent flow in these Pre-Combustion Chamber is calculated in order to choose which Pre-Combustion Chamber show higher performance. For this analysis, ANSYS Workbench is used. Simulation analysis consists of three elements which are pre-processing, solver and post-processing. Analysis on this simulation is at the volume inside a Pre-Combustion Chamber.

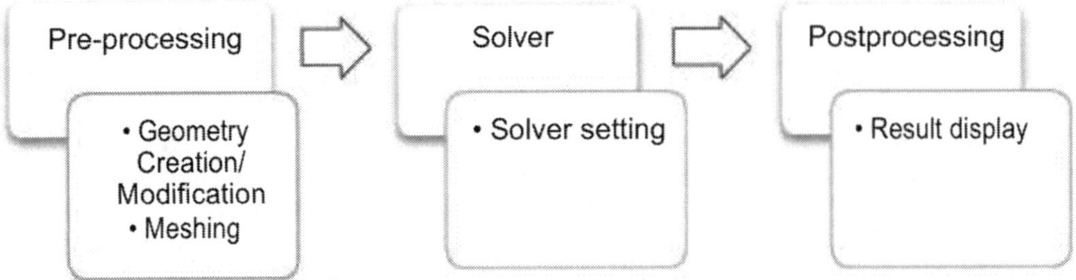

Fig. 1. Elements in CFD simulation.

2.3 Geometry creation / modification

The simulation was done at the volume inside the Pre-Combustion Chamber (PCC). The PCC geometry was imported into ANSYS. The modification is done by converting the caps inside the PCC into solid part body. Figure shows the PCC designs imported into ANSYS.

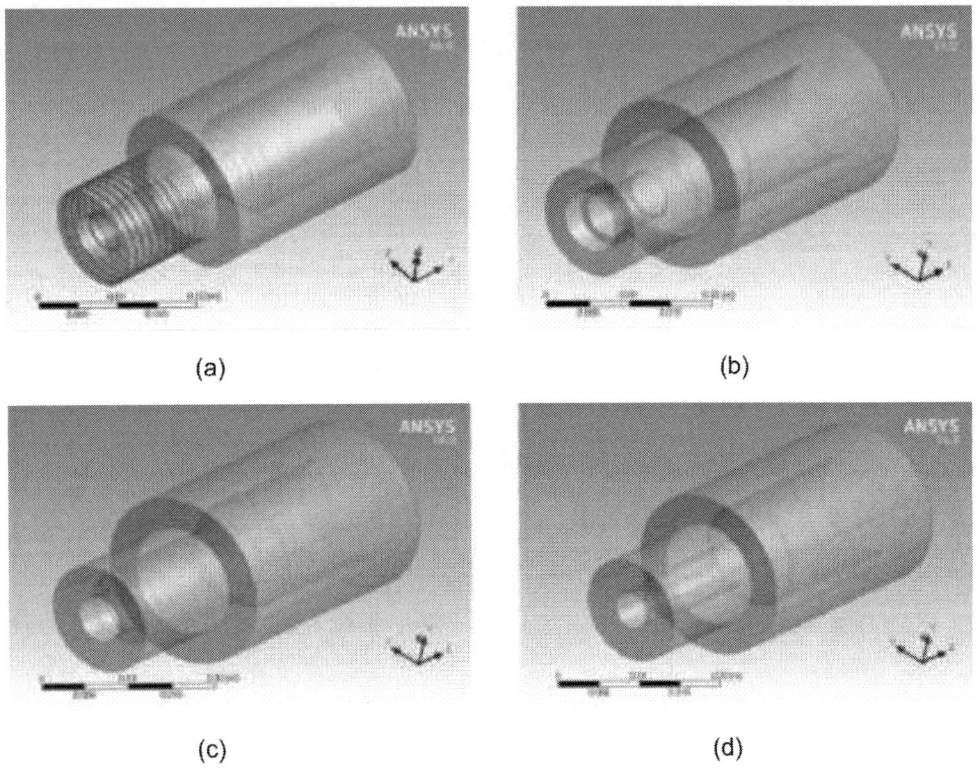

Fig. 2. Pre-combustion Chamber (PCC) geometry design.(a) Type 1 PCC design (b) Type 2 PCC design. (c) Type 3 PCC design. (d) Type 4 PCC design.

2.4 Meshing

Domain is required to be divided into discrete cells for a flow to be solved in order to determine the result's accuracy with the quality of the mesh. Boundary conditions is applied where the cells are grouped into boundary zones which are:

1. Inlet
2. Outlet
3. Wall solid

4. Interior solid

Mixture of air and fuel is stimulating in this project and the velocity of each inlet is to be determined. Stated that air-fuel mixture in Pre-Combustion Chamber (PCC) of any one of claims 7, 8, 9, 10, or 11, further comprising generating a flow filed velocity within about 1.0 meter per second and about 100.0 meter per second.

By using ANSYS-Mesh, the mesh was generated. From the mesh, the nodes, number of element and orthogonal quality of each PCC designs can be determined. Mesh quality are important before the analysis is run because they will affect the accuracy of the solution and the convergence time. The poor quality of meshing is defined when the orthogonal quality closer to 0 while the best quality of meshing is defined when the orthogonal quality closer to 1. Table below show the nodes, number of element and orthogonal quality of each PCC designs. Figure below show the mesh quality of the simulation geometry for each PCC.

Table 1. The nodes, number of element and orthogonal quality of each PCC designs.

Pre-Combustion Chamber (PCC)	**Nodes**	**Number of Element**	**Orthogonal Quality**		
			Minimum	**Maximum**	**Average**
Type 1 Design	12179	60952	0.30418	0.99394	0.85552
Type 2 Design	57247	305068	0.17495	0.99579	0.85868
Type 3 Design	2933	13328	0.30699	0.99858	0.85525
Type 4 Design	15087	75874	0.25746	0.99978	0.8557

2.5 Solver setting

Pressure based solver was used in this simulation which Energy model was defined. The time independent and the simulation in the solver is performed in steady state. K-epsilon model with near wall treatment of standard wall function was used as turbulent viscous model.

Next, the solution method parameter such as SIMPLE algorithm for Pressure-Velocity Coupling, Least Squares Cell Based for Gradient, Second Order for Pressure, Second Order Upwind for Momentum, First Order Upwind for Turbulent Kinetic Energy, First Order Upwind for Turbulent Dissipation Rate were used for all parameters.

3 Results and discussion

3.1 Fluid flow (FLUENT)

Pre-combustion chamber, and more particularly to the engine is designed to enhance fuel consumption and decontamination of exhaust gases. Whereby a stratification of the mixture gas and two stage combustion are performed. For this situation, the production of NOx of negative parts contained in exhaust gases may be suppressed. In any case, due of combustion of the rich mixture in the pre-combustion chamber, an unfavorable measure of hydrocarbons and carbon monoxide is produced [5]. In this situation, it is important that air should be supplied through the communicating passage into the pre-combustion chamber from the main combustion chamber without being adversely influenced by the throttling action because to the communicating passage and that the flame should be consistently diffused all through the main combustion chamber to effect the secondary or resulting combustion in that. In addition, the velocity of swirl flow of the mixture charge inside the pre-combustion chamber should be damped to this way lower the pressure increasing rate for the removal of the negative parts contained in exhaust gases and in addition the reduction in the level of noise incident to the combustion [5].

DOI: 10.1051/matecconf/20179001045

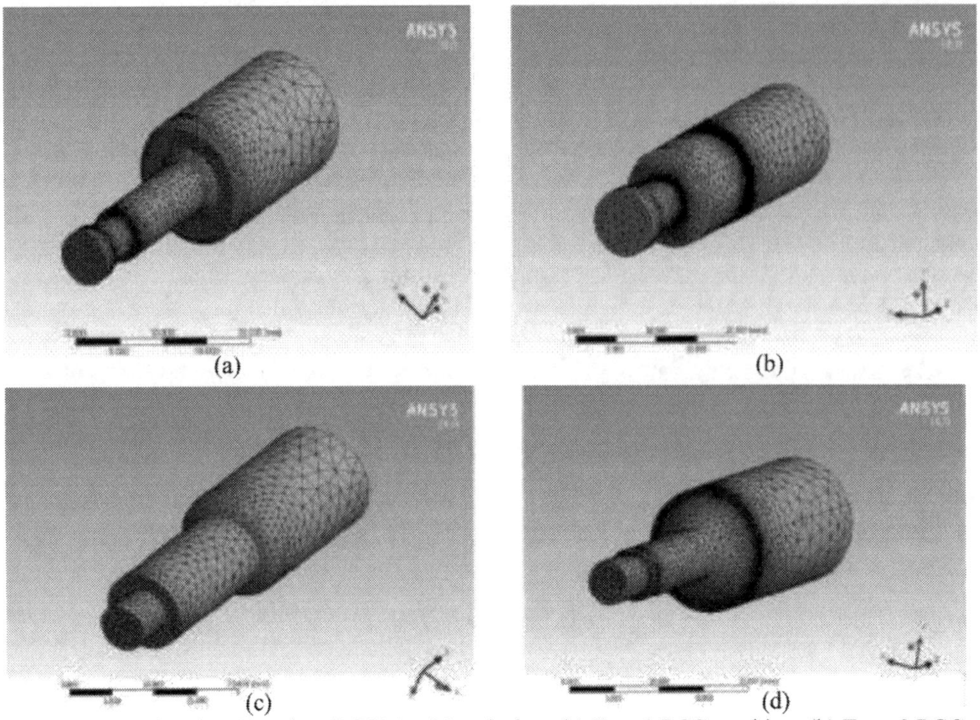

Fig. 3. Pre-combustion Chamber (PCC) meshing design. (a) Type1 PCC meshing. (b) Type 2 PCC meshing. (c) Type 3 PCC meshing. (d) Type 4 PCC meshing.

Likewise, the passage being considerably tangentially arranged relative to the inner surface of the pre-combustion chamber and being inclined relative to the axis of said cylinder whereby the swirl flows produced in the pre-combustion chamber interfere with each other as a result of their spaced centers to decrease the velocity of the swirl flows so that flame propagation of a fuel mixture charge is lowered and the amount of NOx in exhaust gas is decreased.

3.2 Velocity vectors

Results were analyzed and the simulation of the results were presented in graphical views. The velocity vector of the flow at the planes of each Pre-Combustion Chamber (PCC) designs were analyzed to determine the swirl and tumble pattern. To achieve the objectives of this project, the best of three of the PCC designs have been chose by comparing their velocity flow and the swirl pattern. Figure (a), (b), (c) and (d) show the plane where the results will be illustrated. From the figure below, determine the highest maximum velocity each of the PCC using ANSYS simulation result.

The velocity vector displayed on z-y axis plane indicates swirl flow in the each PCC. The vectors represent the velocity magnitudes. From figure above, it can be seen that the mixtures flows to the center of the PCC and the swirl is visible at the corner in each PCC designs. Fig.6 shows the highest velocity flow followed by the velocity flow in Fig. 7, Fig. 4 and Fig. 5. This is due to long inlet passage inside it. The swirl pattern for Type 3 design of PCC in Fig. 6 shows the best flow among the others design where the second best is Type 4 design of PCC in Fig. 7 and followed by Type 2 design of PCC in Fig.5 and Type 1 design of PCC in Fig. 4. From this comparison, it can concluded that the Type 2, Type 3 and Type 4 design of PCC are selected based on their best velocity and swirl pattern compared to the Type 1 designs. As the burning air and fuel and combustion products from the pre-combustion chamber involve a much bigger volume than the spark plug arc, the combustion process within the main combustion chamber is completed much more quickly, resulting in greater engine efficiency [3]. From the past study, a lean air/fuel ratio can increase the probability of irregularities in burning because there is a little margin of fuel in the combustion chamber in excess of that amount necessary for proper combustion process. Especially if the fuel is not mixed adequately, there can be regions of gas near the spark plug gap at the time of combustion that do not contain sufficient fuel for consistent burning. Accordingly, it is to give a combustion system including a main combustion chamber and a pre-combustion chamber which advances steady and consistent combustion, subsequently enhancing engine efficiency, reducing toxic exhaust emissions, and lessening fuel consumption [4].

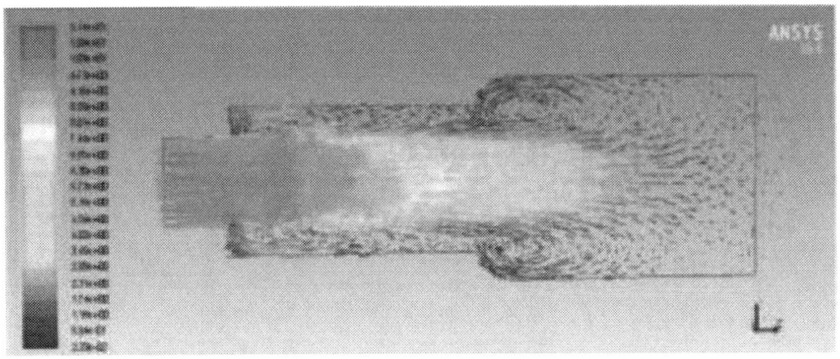

Fig. 4. Type 1 PCC velocity vectors (Maximum velocity = 1.6 m/s).

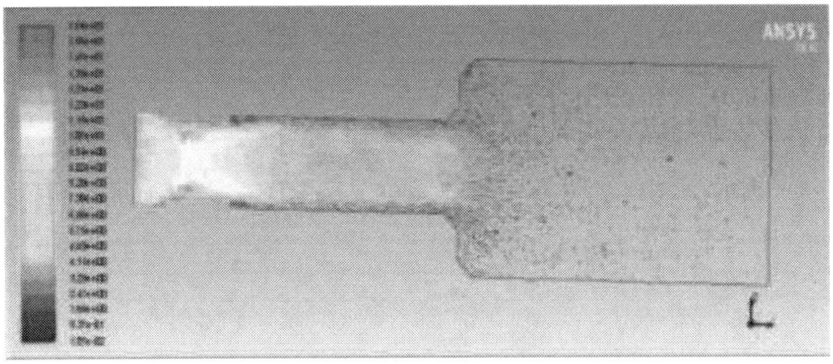

Fig. 5. Type 2 PCC velocity vectors (Maximum velocity = 2.5 m/s).

Fig. 6. Type 3 PCC velocity vectors (Maximum velocity = 1.14 m/s).

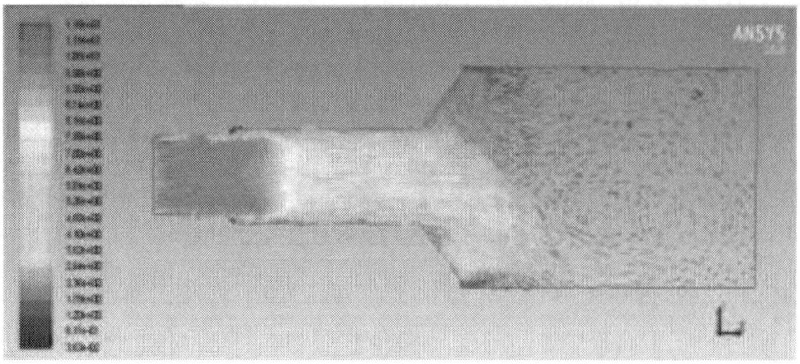

Fig. 7. Type 4 PCC velocity vectors (Maximum velocity = 1.16 m/s).

3.3 Turbulence kinetic energy (TKE)

The irregularity of flows is represented by turbulence and it is used to find the best air-fuel mixing preparation before combustion occurs and help to increase the combustion rates. By referring the Figure (e), (f), (g) and (h), the brightest color indicates higher value of Turbulence Kinetic Energy (TKE).

The brightest colors are more visible in Fig. 9, Fig. 10 and Fig. 11. From Fig. 9, the Turbulent Kinetic Energy show brightest at the center of Pre-Combustion Chamber (PCC). PCC in Fig. 10, its TKE show brightest at the inlet of the air-fuel mixture. Compare to PCC in Fig. 11, the turbulent kinetic energy show brighter than PCC in Fig. 8, even though PCC in Fig. 11, its turbulent kinetic energy are not at the best compare to PCC in Fig. 9 and Fig. 11. From this comparison, it can concluded that the Type 2 and Type 3 of PCC are selected based on their best turbulent kinetic energy compared to another designs.

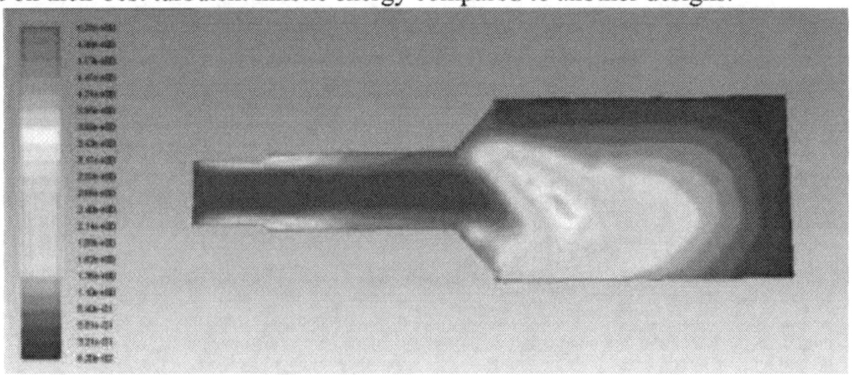

Fig. 8. Contours of Turbulence Kinetic Energy (TKE) for Type 1 PCC ($^2/$ ^{2}).

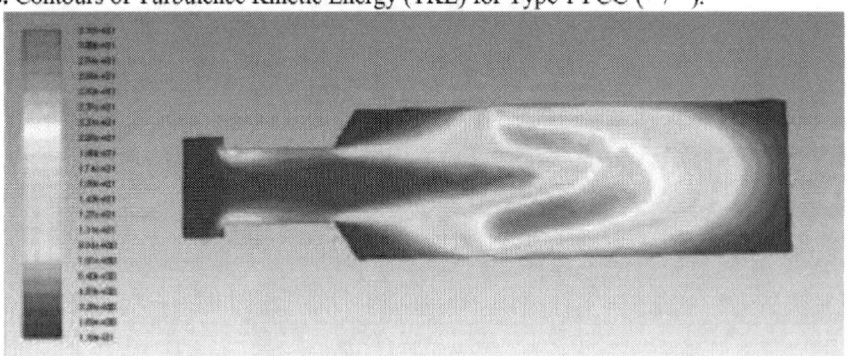

Fig. 9. Contours of Turbulence Kinetic Energy (TKE) for Type 2 PCC ($^2/$ ^{2}).

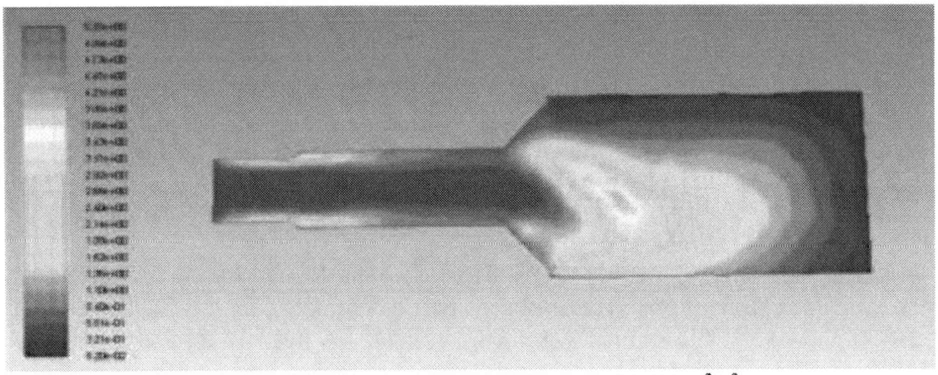

Fig. 10. Contours of Turbulence Kinetic Energy (TKE) for Type 3 PCC (^{2}/ ^{2}).

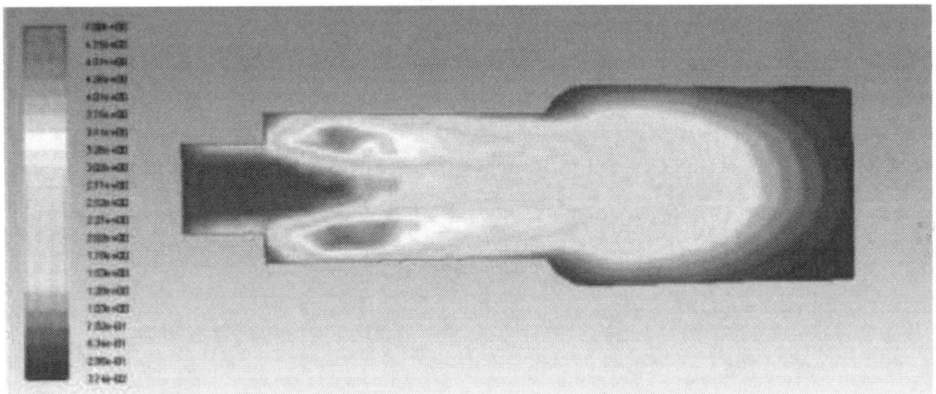

Fig. 11. Contours of Turbulence Kinetic Energy (TKE) for Type 4 PCC (^{2}/ ^{2}).

4 Conclusions

In this paper, the mixture formation, combustion process and flow field were simulated using computational fluid dynamic (CFD) by ANSYS simulation software. As for interpretation of the results:

1. Swirl is more visible in Fig. 6 in fluid fluent shows the highest velocity flow. When the burning fuel and air occupy much larger volume, the combustion process in PCC has completed much more rapidly, resulting in greater engine efficiency.
2. For the Turbulence Kinetic Energy (TKE), Fig. 9 shows the brightest colour at the centre of the PCC. Meanwhile, the Fig. 10 shows the brightest colour at the inlet of the PCC for the air-fuel mixture. Therefore, it can help to increase the combustion process rates.

The authors would like to acknowledge GTeV research group, CARE, UTeM for providing the technical support and facilities to perform this study. The author would like to also acknowledge UTeM and FKM under research grant PJP/2014/FKM(6A)/SA1323.

References

1. R.G. Papagiannakis C.D. Rakopoulos D.T. Hountalas D.C. Rakopoulos, "Emission characteristics of high speed, dual fuel, compression ignition engine operating in a wide range of natural gas/diesel fuel proportions", *Fuel* **89** (2010) 1397–1406.
2. R. Barzegar, S. Shafee, S. Khalilarya, : Computational Fluid Dynamic Simulation of

the Combustion Process, Emission Formation and the Flow Field in an In-Direct Injection Diesel Engine, *THERMAL SCIENCE: 2013*, Vol. **17**, No. 1, pp. 11-23.
3. D. F. Paul, "Swirl flow precombustion chamber," *U.S. Patent* 5555867, Sep. 17, (1996)
4. L. K. George, "Pre-combustion chamber spark plug and method of igniting lean fuel," *U.S. Patent* 4864989, Sept. 12, (1988)
5. T. Suzuki, "Internal combustion engine provided with precombustion chamber," *U.S. Patent* 4066045, Jan. 3, (1978)
6. Z. Barbouchi, and J. Bessrour, "Turbulence study in the internal combustion engine," *Journal of Engineering and Technology Research*, Vol. **1** (9), pp. 194-202, (2009)
7. J.B. Heywood, "Fluid motion within the cylinder of internal combustion engines," ASME, *J. Fluids Eng.*, **109**, pp. 3-35, (1987)
8. H.K. Versteeg, W. M., (1995). *An introduction to computational fluid dynamics: The finite volume method.* England: Longman Scientific & Technical

DOI: 10.1051/matecconf/20179001046

Nanostructure and oxidation properties investigation of engine using Jatropha biodiesel as engine fuel

Mohd Aznan Abdul Latif[1,*], *Ahmad Anas* Yusof[1], *Ahmad Zaki* Shukor[2], *Farah Aqilah* Habidi[1], *Nur Fathiah* Mohd Nor[1], *Mohd Zaid* Akop[1], *Mohd Hafidzal* mohd Hanafi[1], and *Aiman* Roslizar[1]

[1]Faculty of Mechanical Engineering, Universiti Teknikal Malaysia Melaka, Hang Tuah Jaya, 76100 Durian Tunggal, Melaka, Malaysia
[2]Faculty of Electrical Engineering, Universiti Teknikal Malaysia Melaka, Hang Tuah Jaya, 76100 Durian Tunggal, Melaka, Malaysia

Abstract. The purpose of this study is to investigate the soot nanostructure characteristic and soot oxidation behaviour that is emitted from a single cylinder diesel engine by using two different engine fuels that is Jatropha Biodiesel (JBD) with the presence of 3% of methanol as additive and Diesel fuel. In order to obtain the result for soot characteristic, Scanning Electron Microscope (SEM) is used, while Thermogravimetric Analysis (TGA) is used to analyze the oxidation of the soot itself. The parameter for the engine testing has been fixed by applying different loads of 25% and 50% load with constant speed of 2500 rpm. The result has been compared for each fuel in terms of their structure and oxidation behaviour. The soot was collected on a filter paper and the SEM grid from the exhaust stream within the time period set. The soot is then observed by using SEM and the result shows that diesel soot particulate matter is smaller with average size of 95 nm for 25% load and 63.5 nm for 50% load than those of B20, which has the average size of 122 nm for 25% load and 105 nm for 50%. It complies with the evidence that stated 11% of the oxygen content helps in lowering particulate emission by improving the combustion. The result also shows that the oxidation behaviour of B20 occurs faster than diesel either at 25% load or 50% load.

1 Introduction

Fuel is a basic need for vehicle to move around and its categories are three which is solid fuels, liquid fuels and gaseous fuel. As for diesel engine, biodiesel has become a good alternative fuel [1]. Generally, biodiesel and diesel are both in liquid fuels and as known, biodiesel is better than diesel. In order to reduce emission problems that cause from combustion engine, biodiesel is proposed as the solution [2]. This is because; biodiesel significantly reduces carbon emissions and other harmful particulate matter especially in producing the soot when engine is running. There is a lot of new technology that makes the engine become more powerful but less harmless to the environment [3]. As known, biodiesel is made from vegetable oil or animal fats and can be used as fuel for vehicle and

* Corresponding author: golden3744@yahoo.com

diesel engine either in their pure form or mixed with mineral diesel but it is usually used as diesel additive in order to reduce level of gas emission or soot, carbon-monoxide and hydrocarbons or called as renewable resources.

The increasing population and high demand in the use of vehicles with the transformation of climate may affect the environment, creates global warming and affect human health in Malaysia. With the rising of demand for vehicle, the collection of fuel that is non-toxic for environment, less emissions, as well as less soot is now being considered. In 1990, "Rudolf Diesel" has first demonstrate the use of biodiesel in diesel engine by using peanut oil at the world exhibition that was held in Paris [4].

Diesel in engine exhaust nowadays produces black soot where it consists of carbon compounds that were not combusted and this makes the soot chemically/biologically highly active and hazardous. In order to reduce it, biodiesel fuel is used where it is a green fuel which is non-toxic biodegradable and also reduces engine emissions. As for the soot characteristic, diesel soot is formally formed by short exposure times to high temperatures and has bigger soot size than biodiesel fuel.

The main purpose to use biodiesel as an alternative fuel is to reduce emission by taking into account their nanostructure characteristic and the oxidation behaviour. Biodiesel is a renewable source and it creates less pollution than diesel. Unlike other alternative sources which often need new equipment or modification, by using biodiesel, no expensive modifications are required to the engine, and a lot easier to integrate biodiesel into current system. Table 1 shows the comparison between commercial diesel and biodiesel [5].

Table 1. Comparison between diesel and biodiesel.

Fuel Property	**Diesel**	**Biodiesel**
	ASTM D975	**ASTM D6751**
Lower Heating Value, BTU/gal	129,050	118,170
Kinematic Viscosity @ 40^0 C., cSt	1.3-4.1	4.0-6.0
Specific Gravity @ 60^0 C., g/cm³	0.85	0.88
Carbon, wt %	87	77
Hydrogen, wt %	13	123
Oxygen, by dif. Wt %	0	11
Sulfur, ppm	500	0
Boiling Point, ^{0}C	180 to 340	315 to 350
Flash Point, ^{0}C	60 to 80	100 to 170
Cloud Point, ^{0}C	~15 to 5	~3 to 12
Pour Point, ^{0}C	~35 to ~15	~15 to 10
Cetane Number	40-50	48-65
Lubricity (HFRR), µm	300-600	<300

Biodiesel is a liquid that is eco friendly. It has a low processing temperature and lower oxidation stability. This is why biodiesel is accepted by the community. [6]. Different from diesel fuel and pollutants from diesel engine, it affected environment and human health such as lung cancer. From the previous study, by using biodiesel like Jatropha plant, it could reduce global warming and provide a clean air. Jatropha oil can be produce worldwide because of its potential [7]. Oil can be produced up to 40% from its seeds and also the properties of the oil is better compared to soybean oil and palm oil [8]. Table 2 shows that Jatropha biodiesel is considered as good as diesel.

Table 2. Properties of Jatropha biodiesel versus diesel.

Parameters	Jatropha bio-diesel	Diesel oil
Specific Gravity @ 15^0 C	0.860-0.933	0.82-0.86
Sulphur	0.13	1.2
Viscosity (cST)	37.00-54.80 at 30 ^{0}C	37.00-54.80 at 30 ^{0}C
Pour Point, ^{0}C	-3	33 to -15
Cloud Point, ^{0}C	2	15 to -5
Flash Point, ^{0}C	210-240	60-80
Cetane Number	38-51	40-55
Heating Value (Mj/kg)	37.83-42.05	42

In order to make the comparison for the nanostructure characteristics and oxidative behaviour of the soot from biodiesel and diesel fuel, an experiment is carried out to analyze the characteristic and oxidative behaviour of the soot by using Jatropha biodiesel and diesel as engine fuel. For this experiment, a new pipeline will be designed and fabricated where filter paper and SEM grid will be located at the centre of the pipeline to take samples of the soot. Then, the soot sample will be analysed and compared between diesel and Jatropha biodiesel fuel for the nanostructure and oxidative reaction of the soot by using thermogravimetric analysis (TGA) and scanning electron microscopy (SEM). The objective of this research is to determine the oxidative behavior of soot using Jatropha biodiesel as engine fuel with the presence of methanol as fuel additives. Other than that, the objective is also to investigate of the nanostructure characteristics of soot emitted from a single cylinder diesel engine and also to compare the nanostructure of soot produced by two different fuels that is emitted from Jatropha biodiesel and diesel.

2 Experimental setup

2.1 Design of experiment

The apparatus, specifications of the apparatus, procedure of the experiment, design pipe from exhaust engine and flowchart of the experiment is described in this section. The apparatus that is used is a single-diesel engine for producing soot, Scanning Electron Microscope (SEM) to investigate the nanostructure characteristics.During oxidation, the structural change can be revealed using electron microscopy[9]. Thermogravimetric Analysis (TGA)is being used to investigate the oxidation behaviour of soot that is created by two different test fuels that is *Jatropha* biodiesel with presence of 3% methanol and diesel. This experiment will be run at constant speed (2500rpm) with different load (25% and 50%) form hydraulic system that attached to the engine. Thus, for each different load for both different fuels, the nanostructure characteristic and oxidation behaviour will be investigated, analysed and compared. The main apparatus is the single-cylinder diesel engine which is use to produce emission and collect the soot sample. The specification or the parameters of the engine is shown in Table 3.

Table 3. Engine Specifications/Parameter of KIPOR model.

Content	Details
Model	KM186FB
Type	Vertical cooling load with water, 4 cycles
Combustion System	Direct Injection
No. Of Cylinder	1
Bore x Stroke	86x75
Shift	0.857 (0.226)
Rated Output	14.0/2400 (10.3)
Rated Output as per 1 hr hp/rpm (kW)	16.0/2400 (11.8)
Compression Ratio	19.6
Fuel	Diesel
Injection Time	BTDC 17.0
Injection Pressure (kg/cm2,psi)	200 (2845)
Fuel Tank Capacity (1)	14.3 (3.78)
Lubrication System	3.0 (3.78)
Lub Oil Capacity	3.0 (0.79)
Cooling system	Water-Radiator
Water Cooling Capacity	3.0 (0.79)

The scanning electron microscope (SEM) is a scientific equipment which uses a beam of high- energetic electrons to generate a variety of signals in order to examine an object at the surface of solid specimens on very small fine scales. Because of the signals, it could reveal several of information about the sample such as; external morphology (texture), crystalline structure and their chemical composite/element. SEM have magnification ranging from 20x to approximately 30,000x, spatial resolution of 50-100nm and also analyses selected point locations on the sample which the technique is useful to identify or determine chemical composition (using EDS), crystalline structure and crystal orientation. Thus, by using SEM, it is able to analyze at high level that have been specified at nano-level and used to measure nano particles size of the soot for this experiment.

Thermogravimetric Analysis (TGA) is a technique or method of thermal analysis which changes in physical or chemical properties of materials upon temperature with constant heating rate or upon time with constant temperature and/or constant mass loss. For measurement, TGA used to determine the composition of materials or chemical and predict their thermal stability at temperature up to 1000°C. This kind of technique can characterize materials or chemical that exhibit weight loss or gain due to decomposition, oxidation or dehydration. The samples weights for TGA are range from 1mg to 150 mg but sample weights of more than 25mg are preferred. For best result, sample weight of 1 mg is enough. This sample can be analyzed in form of powder or small pieces so the interior sample temperature remains close to the measured gas temperature. For this experiment, TGA is used to predict the thermal stability at certain temperature due to oxidation behaviour of the soot from different kind of load (low to high load) at constant speed.

2.2 Experimental procedure

Before conducting the experiment to determine the soot oxidation and soot nanostructure of both fuels, the mixture of the Jatropha biodiesel with diesel is needed. The mixture of the fuel is B20. 25 ml of Jahropa methyl-ester and 475 ml of diesel fuel is mixed and stirred until it reaches homogenous consistency. The procedure on how to analyze the soot sample were basically refered to the SEM standard procedure [10].

DOI: 10.1051/matecconf/20179001046

3 Results and discussion

For the nanostructure characteristic and oxidation behaviour, the result will be discussed based on the result from the SEM image and TGA graph which had been obtained from Scanning Electron Microscopy (SEM) and Thermogravimetric Analysis (TGA). The result is compared based on the different test fuels which Diesel (D2) and *Jatropha* Biodiesel (B20). Besides that, the differences of the result obtained also been compared with the differences of the load given during the experiment. For the soot sample, the soot particle are in fuel spray cores formed which it obtained from the exhaust pipe that has been attached to the single cylinder engine exhaust that operate in constant speed of 2500 rpm with 25 % and 50% load. For about 15-30 minutes, the emitted diesel and biodiesel soot particle were trapped at the filter paper and SEM grid that attached inside the pipe at the exhaust.

(a) (b) (c) (d)

Fig. 1. (a) Filter paper of trapped fine particle of Diesel (a) 25% load, (b) 50% load and *Jatropha* biodiesel (B20) (c) 25% load, (d) 50% load with trapping duration for 30 minutes each.

As in Figure 1, all the sub-figures show the particulate concentration of soot on filter papers. The result were reported as the particle that is emitted from diesel combustion were more than Jatropha biodiesel (B20) combustion because both have same duration of trapping that is 30 minutes and particle concentration of Jatropha biodiesel are slightly lower than diesel as shown in Figure 1. By referring to the particle concentration, it can be observed that soot particle in diesel combustion are higher than soot particulate in Jatropha biodiesel combustion and it can be observed clearly due to the increasing load for both fuel. Due to Jatropha biodiesel is made from vegetation plant, the oil contain oxygen within the molecule fuel which it is readily been oxidized with the available oxygen in the flame zone. Besides at 25% engine load, flaky particulates soot were collected as in Figure 1 on the left for image (a) diesel and (b) B20. At 50% engine load, particle collection is much darker to observed at the filter papers which significantly presents a higher emissions at higher engine loads.

3.1 Soot nanostructure characteristic using SEM

By using scanning electron microscope (SEM), sample preparation of the soot nanostructure characteristics can be elaborated and analyzed for both Diesel (D2) and Jatropha Biodiesel (B20) fuel. The particle diameter can be visualize using SEM method [11]. Figure 2 shows the image of the soot nanostructure for Diesel (D2) fuel and Jatropha Biodiesel (B20) fuel for load of 25% with constant speed 2500 rpm.

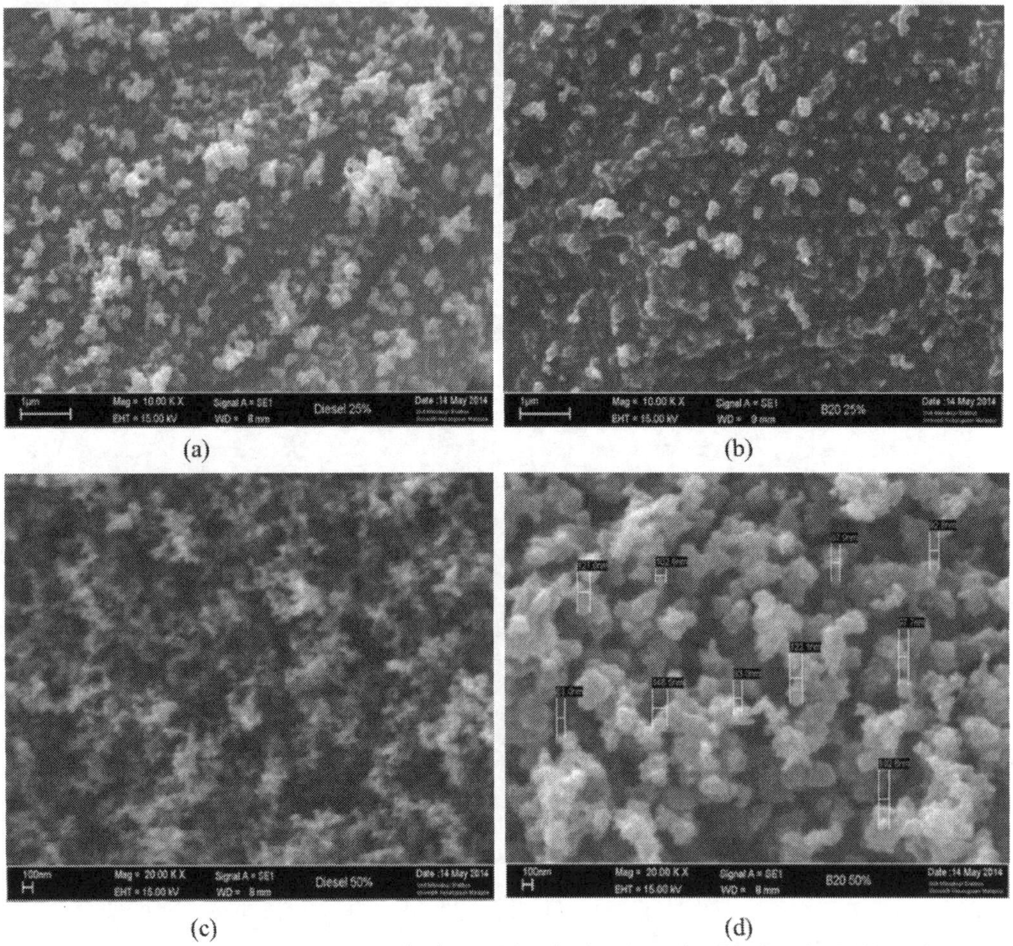

Fig. 2. Soot nanostructure at 25% load of (a) Diesel, (b) Jatropha Biodiesel (B20) and soot nanostructure at 50% load of (c) Diesel, (d) Jatropha Biodiesel (B20).

Based on Figure 2, the morphology features of both soot were analyzed and presented where the surface of morphology of the diesel soot (a) and (c) is seen to be non-uniform either at 25% or 50 % load and the diesel soot particles are slightly small than Jatropha biodiesel soot particles.In Figure 2, the image shows the micro-image of diesel and Jatropha biodiesel particulates by SEM which in fine ranges particle. In this experiment, the average size of the soot particle for Jatropha biodiesel is 122 nm for 25% load and 105 nm for 50%. As for soot particles size for diesel obtained are approximately about 95 nm for 25% load and 63.5 nm for 50% load. In this experiment, the particle image of soot is taken in order to verify and identify the particle size by micro- and nano- image. The image showed that diesel particle sizes are slightly smaller than of Jatropha biodiesel particle. The Jatropha biodiesel particulate size is approximately 80 – 160 nm whereas the diesel size is around 40 – 120 nm. Besides, at higher engine load, ultrafine particle are obtained which mainly composed of organic carbon (OC), where at higher engine loads, these particle are overall covered of elemental carbon (EC). Based from previous study, the particulate mass emission can be reduce on operating condition by using biodiesel blend and also give a better engine performance compared to diesel [12]. Biodiesel fuel have a higher oxygen content in it which help in lowering particulate emission by improving the combustion [13]. Besides that, with presence of 3% additive of methanol inside B20, it also improves biodiesel performance by reducing particulates, NO_x, fuel dilution and carbon dioxide.

3.2 Oxidation behaviour using TGA

There is method can be used to determine particulate reactivity that is quantified by the temperature and oxidation rate [14]. By using thermogravimetric analysis (TGA), sample preparation of the soot oxidation behaviour can be analyze for both Diesel (D2) and Jatropha Biodiesel (B20) fuel. Figure 3 (a), (b), (c) and (d) below, show the graph of soot oxidation and the thermal behaviour of both diesel and B20 fuel for load of 25% and 50% with constant speed of 2500 rpm.

These samples were examined under the following experimental conditions:

Instrument : TGA Q 500
Heating Rate : 10 K/min
Temperature range : 25°C to 750°C
Atmosphere : Nitrogen, 50 ml/min

As in Figure 3, the test condition for all cases start from 25°C to 750°C with nitrogen and from 50°C to 300°C is considered as evaporation moisture. Then, from 320°C to 400°C a significant major mass loss occurred with about above 80% and this is assumed to be due to hydrocarbon and carbon oxidation desorption. Finally, at above 400°C shows that biodiesel has more ashes than diesel which may conclude that, the oxidation rate of biodiesel is faster than diesel. The weight loss curves which are plotted in Figure 3 shows that soot from B20 starts to oxidize at slightly higher temperature than soot from diesel. Furthermore, soot from B20 oxidizes significantly at 390°C to 750°C while Diesel soot experiences weight loss at temperature of 400°C and above. therefore, it can be concluded that Diesel produces soot that is less reactive than soot from B20.

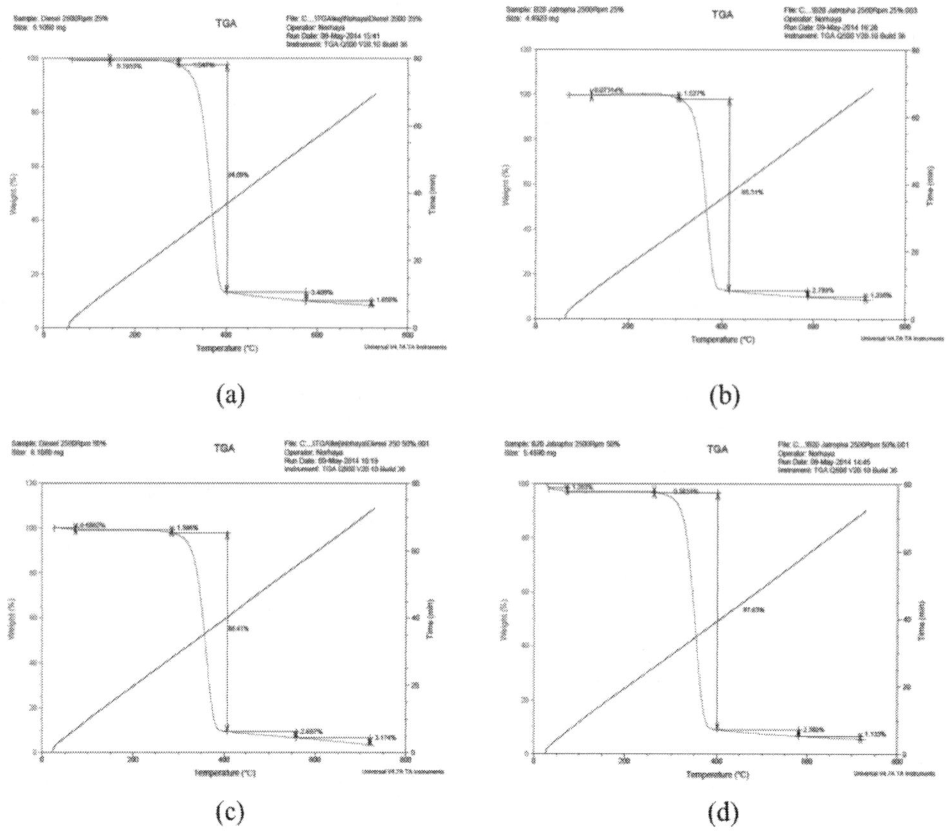

Fig. 3. Comparison of TGA curve and conversion of (a) diesel 25% load, (b) B20 25% load, (c) diesel 50% load and (d) B20 50% load with constant speed 2500 rpm for all cases.

4 Conclusions

Equations As a conclusion, it can be said that the objectives of the experiment has been achieved. The investigation and comparison of soot nanostructure characteristic of soot emitted from single cylinder diesel engine of diesel and B20 is successfully determined by using SEM. The result shows that, the soot nanostructure for diesel and B20 at two different loads where diesel soot particulate matter are smaller with average size of 95 nm for 25% load and 63.5 nm for 50% load than B20 with average size of 122 nm for 25% load and 105 nm for 50%. The surface of morphology of diesel soot is seen to be non-uniform and chain-like agglomerations than B20.

For the oxidation behaviour of soot, there are three possibilities for the oxidation behaviour reactions that are, at higher engine load, it produces higher in-cylinder temperature which leads to faster oxidation reaction. Secondly, oxidation behaviour could react when the sample is in low particle size and have high surface area and a last possibility is because of the presence of oxygen in the biodiesel. Besides that the presence of methanol as an additive improves the performance of biodiesel by reducing the particulates thus could also help on the oxidation behaviour in B20.

Through out this experiment, there were several limitations that becomes an obstacle to get a more in-depth result for both sample in order to obtain better results. First, the limitation that had been faced is the engine load. Due to the condition of the engine itself, the engine that can be imposed is below 75%. This is because, the engine uses a hydraulic system, if the load exceeds 75%, it will cause the engine to wear out faster than their long term life cyle.

Second is the cost limitation. In order to get an exact and detail image of the soot nanostructure, Transmission Electron Microscopy (TEM) were suggested, where by using TEM, details result of soot nanostructure can be observed and obtained for example their soot morphology, chemical compositions and crystalline structure. But, due to cost limitation, the analysis of soot nanostructure can only afforded by using Scanning Electron Microscope (SEM). Same goes for Thermogravimetric Analysis (TGA) where, it only can observe and determine the oxidation behaviour. To analyze deeper on the oxidation behaviour, the differential scanning calorimetry (DSC) can be used to investigate the thermal characteristic which is the thermal critical points like melting points, enthalpy heat temperature or substance, but it requires a higher cost.

As for recommendation, this experiment should provide with various types of fuel sample in the future. There should be at least three samples of fuel blended *Jatropha* Biodiesel (JBD) that is B15, B20 and B100 which then can be compared with conventional diesel. This is because; there will be more variable result that can be obtained during the experiment. The fuel test sample should also include fuel blended without the addition of 5% methanol as additives so that the results can differentiate between conventional diesel and blended fuel with presence of additives.

Another recommendation is to do an Energy dispersive spectroscopy (EDS) to analyse deeper on soot nanostructure or soot morphology and determining their chemical compositions and crystalline structure. For oxidation behaviour, the differential scanning calorimetry (DSC) can be used to investigate the thermal characteristic that is the thermal critical points like melting points, enthalpy heat temperature or substance.

The authors wish to acknowledge Ministry of Education, Malaysia (KPM) through research grant RAGS/2013/FKM/TK01/02/B00041 for supporting and funding these research activities. The authors also thanks to Universiti Teknikal Malaysia Melaka (UTeM) and Green & Energy Efficient Technology Research Group (GrEET) under Center for Advance Research on Energy (CARE) for the facilities provided.

References

1. P. Tan, Z. Hu, D. Lou, and Z. Li, Exhaust emissions from a light-duty diesel engine with Jatropha biodiesel fuel, *Energy*, **39**, no. 1, pp. 356–362, (2012)
2. M. Happonen, T. Lähde, M. E. Messing, T. Sarjovaara, M. Larmi, L. R. Wallenberg, A. Virtanen, and J. Keskinen, The comparison of particle oxidation and surface structure of diesel soot particles between fossil fuel and novel renewable diesel fuel, *Fuel*, **89**, no. 12, pp. 4008–4013, (2010)
3. M. Balat, H. Balat, C. Öz, Progress in bioethanol processing, *ScienceDirect,* **34**, pp. 551–573, (2008)
4. P. K. Sahoo and L. M. Das, Combustion analysis of Jatropha , Karanja and Polanga based biodiesel as fuel in a diesel engine, *Fuel*, **88**, no. 6, pp. 994–999, (2009)
5. M Herskowitz, M. Landau, I. Reizner and kaliya, Production of diesel fuel from vegetable and animal oils. U.S. Patent 2006 0207166 A1, Sep. 21, (2006)
6. S. Jain and M. P. Sharma, Oxidation and thermal behavior of Jatropha curcas biodiesel influenced by antioxidants and metal contaminants, *International Journal of Engineering, Science and Technology,* **3**, no. 4, pp. 65–75, (2011)
7. A. B. Chhetri, M. S. Tango, S. M. Budge, K. C. Watts and M. R. Islam, Non-Edible Plant Oils as New Sources for Biodiesel Production, *International Journal of Molecular Science,* pp. 169–180, (2008)

8. V. C. Pandey, K. Singh, J. S. Singh, A. Kumar, B. Singh, and R. P. Singh, Jatropha curcas: A potential biofuel plant for sustainable environmental development, *Renewable Sustainable Energy Reviews*, **16**, no. 5, pp. 2870–2883, (2012)
9. R. L. V. Wal, A. Yezerets, N. W. Currier, D. H. Kim, and C. M. Wang, HRTEM Study of diesel soot collected from diesel particulate filters, *Sciencedirect*, **45**, pp. 70–77, (2007)
10. D. Kim and C. M. Feng. SEM Standard operating procedure. Internet http://www.sci.ccny.cuny.edu, [June 14,2016]
11. Y. Songsaengchan, C. Chareonphonphanich, P. Karin, N. Chollacoop, M. Tongroon, and K. Hanamura, Physical Characterization of Biodiesel Particulate Matter by SEM, *The Second TSME International Conference on Mechanical Engineering,* (2011)
12. J. N. Gangwar, T. Gupta, and A. K. Agarwal, Composition and comparative toxicity of particulate matter emitted from a diesel and biodiesel fuelled CRDI engine, *Atmospheric Environment,* **46**, pp. 472-481, January, (2012)
13. H. Jung, D. B. Kittelson and M. R. Zachariah, Characteristics of SME Biodiesel-Fueled Diesel Particle Emissions and the Kinetics of Oxidation, *Environmental Science and Technology,* **40**, pp. 4949–4955, (2006)
14. J. Song, M. Alam, A. L. Boehman, and U. Kim, Examination of the oxidation behavior of biodiesel soot, *sciencedirect,* **146**, pp. 589–604, (2006)

Analysis of reconfigurable assembly system framing systems in automotive industry

Mohamad Zamri Md Zain[1], *Ahmad Majdi* Abdul Rani[2,*], *Mohd Amin* Abd Majid[1]

[1]Mechanical Engineering Department, Universiti Teknologi PETRONAS, 31750 Tronoh, Perak, Malaysia
[2]Centre for Intelligent Signal & Imaging Research (CISIR), Universiti Teknologi PETRONAS

Abstract. Current trend in automotive industry shows increasing demand for multiple models with lean production. Prior to that, automotive manufacturing systems evolved from mass production to flexible automation. Material handling systems and equipment in a single assembly line with multiple models require high investment but with low throughput thus making production cost relatively high. Current assembly process of side structure and undercarriage with downtime occurrence during assembly process affecting production performance (quality, cost and delivery). Manufacturing facilities should allow more flexibility and increase intelligence evolving toward novel reconfigurable assembly systems (RAS). RAS is envisaged capable of increasing factor flexibility and responsiveness by incorporating assembly jig, robot and framing, which could be next generation of world class automotive assembly systems. This project research proposes a new methodology of framework reconfigurable assembly systems principles in automotive framing systems i.e. enhance assembly process between side structure assembly and undercarriage assembly which a new RAS is capable to reconfigure the assembly processes of multiple model on a single assembly line. Simulation software (Witness) will be used to simulate and validate current and proposed assembly process. RAS is expected to be a solution for rapid change in structure and for a responsively adjustable production capacity. Quality, cost and delivery are production key parameters that can be achieved by implementing RAS.

1 Introduction

Current global economy scenario shows new enablers are needed in both product development and manufacturing systems due to open and unpredictable market changes. Automotive manufacturers explore economies of scale and scope through globally shared features of their products or models, for example: (1) sharing common automotive development, (2) collaborative engineering processes, (3) incorporated manufacturing processes. In small automotive industry, particularly in Association of Southeast Asian Nations (ASEAN) region; the primary issues with the current systems are capability to

* Corresponding author: mohzam11@yahoo.com

 DOI: 10.1051/matecconf/20179001047

produce multiple models in a single line and an integration between material handling and equipment. Thus to overcome these problems new trends evolution of automotive framing systems should be considered.

Based on Organisation Internationale des Constructuers d' Automobiles or The International Organization of Motor Vehicle Manufacturers (OICA), automotive manufacturer which Total Industry Volume (TIV) for passenger cars production annually less than 200,000 units can be categorized as small automotive industry [1]. Manufacturing enterprises are forced to re-evaluate their production paradigms, so that their products can be designed to maximize potential achievable variants, while the manufacturing systems to make them can be designed to operate efficiently by robustly accommodating future product changes, minimizing time to market and providing a reliable production base. Manufacturing facilities have to possess a high degree of flexibility, enabling mass customization of production. Mass customization is the capability of a firm to offer high productivity (product variety and overall volume), low manufacturing cost and also fast delivery [2].

Multiple models can now be assembled on a single production line. Normally, more than 70% of the tooling and equipment in a body shop are not specific to an individual model but can be used for all models produced. The production line can be planned to assemble various models, such as sedans, hatchbacks or SUVs (sport utility vehicles). While in a paint shop, all equipments are programmable to cover all styles efficiently and cost effectively. Also in the final assembly, most of major automakers are building more models, gained from global platforms, on the production lines that can simultaneously handle multiple products allowing for efficient utilization of people and equipment. Fig. 1 shows the automotive framing system assembling process through combination of a common under-body platform and side structure framing to become body in white (BIW). The automotive structure with the outer metal panels (skin) define the vehicle styling, which at the end become special appearance characteristics.

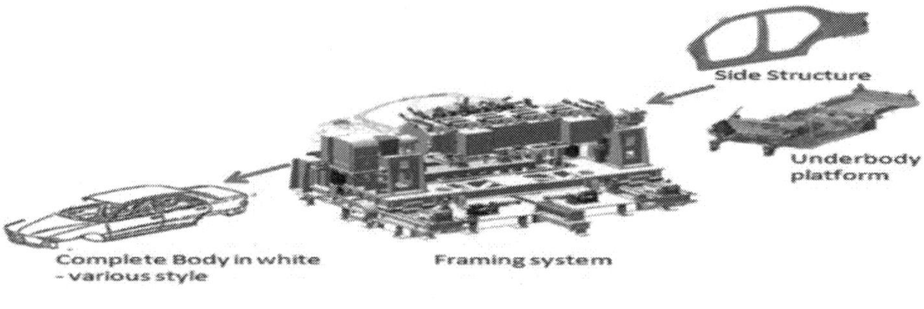

Fig. 1. Automotive framing system.

This paper focus on the manufacturing systems by identifying various potential flexible elements for product platform design and also identify their physical interaction with manufacturing mainbody framing systems. The hypothesis is that if the correct subsets of automotive body segments and production capabilities are designed considering for future flexibility based on the reconfigurability principles, then the manufacturing system can better adapt body styling changes, variants of family production without the need for tooling changeover with significant increase in throughputs.

2 Optimization of assembly process

Assembly includes all assembly or sub assembly processes and equipment which required to:

a) Bring together, configure, orient, align and adjust components and materials to form the end product.

b) Physically attach parts, materials and components such as stapling, nailing, screwing, riveting, gluing, interlocking, wrapping, tying, fusing, welding, sewing, bonding, pegging, soldering, laminating, insertion, sealing, coupling and similar activities.

Assembly is an essential part of total manufacturing process which assembly cost typically 25% to 50% of total manufacturing cost. The total manufacturing cost consists of fixed cost and variable cost. When the costs are added, it has the following total cost in Eq. (1):

$$TC = FC + VC(Q) \quad (1)$$

where TC = total annual cost (RM/yr), FC = fixed annual cost (RM/yr), VC = variable cost (RM/pc), and Q = annual quantity produced (pc/yr)

Assembly system is a portion of manufacturing system, which carries out the assembly processes. Assembly systems can be classified in terms of different criteria. Based on system reconfigurability, assembly systems are classified into dedicated, flexible and reconfigurable. A dedicated manufacturing system is planned for manufacturing specific or customize product at a fixed line. While a flexible manufacturing system is designed for production of a part family, which consists of fixed hardware and fixed, but programmable, software to handle changes in work orders, production schedules, part programmes and tooling for several types of parts in the family. A reconfigurable assembly system (RAS) is designed at the outset for rapid change in structure, as well as in hardware and software components, in order to quickly adjust product capacity and functionality within a part family in response to sudden changes in market or in regulatory requirements.

It should be noted that the terminology 'reconfigurable' has many alternatives such as 'flexible', 'modular', 'agile', 'recomposable', 'retransformable', 'recomponentable' [4]; although these terminologies have a little different meanings, their creators have had the same motivation for them to deal with changes and uncertainties of the system. Among a challenge of current manufacturing systems is to deal with the changes and uncertainties in area of product functions, variations, product volume, delivery time and quality. RAS can be a solution to overcome the said challenges.

From academic perspective, flexibility as a metric of the system's behavior can be expressed in many forms, depending on the system's desired characteristics. So far several types of flexibility have been proposed such as product, operation, process, volume, expansion and labor flexibility [5]. G. Michalos et al. also highlighted that in general, four approaches could be distinguished in the design of an assembly system: (1) Manual assembly, (2) Fixed assembly, (3) Semi-automated assembly, (4) Flexible assembly

Platforms can be defined as a common infrastructure that serves as a backbone for multiple product or model variants. It cover common components, processes, and interfaces that allow end products to achieve unique variety in their product attributes by adding unique elements to product platforms. Unique elements are only found in individual variants, but not in the common platform [4]. A platform provides a common base for the assembling of a great number of vehicle variations, involving completely different models (e.g. Audi TT and Volkswagen Golf) and different model versions (i.e. 3 door, 5 door, coupe and sedan). Based on that, its required a factory to be flexible enough to handle this product variety.

 DOI: 10.1051/matecconf/20179001047

3 Overview of automotive framing system

'Framing system' is a process and the related basic structure for a precise positioning and securing under-body platform with the upper body components during assembling process. The body framing systems is focused because it is a critical stage in terms of assuring proper geometry and structural integrity of the automotive vehicle body. Nowadays, most passenger vehicles made comprises of 100–150 stamped metal panels, assembled into 18–22 modules (sub-assemblies) in order to become a structural body. These modules are assembled into 3 systems to create the vehicle body formed (prior to painting, is referred to as body-in-white) in a multistep manufacturing process, during which the modules are joined together by welding. In order to assure the final body geometry according to design specifications, the panels to be welded are positioned (by locating pins and NC blocks), constrained by clamps and presented to the process by highly complex automated fixture devices.

Overall, a vehicle framing system can be divided into two subsystems, as shown in Fig. 2:

a) Under-body platform (consisting of up to 10–12 subassemblies).

b) BIW framing (3–5 sub-assemblies added prior to framing), grouped into up to 4 assembly cells.

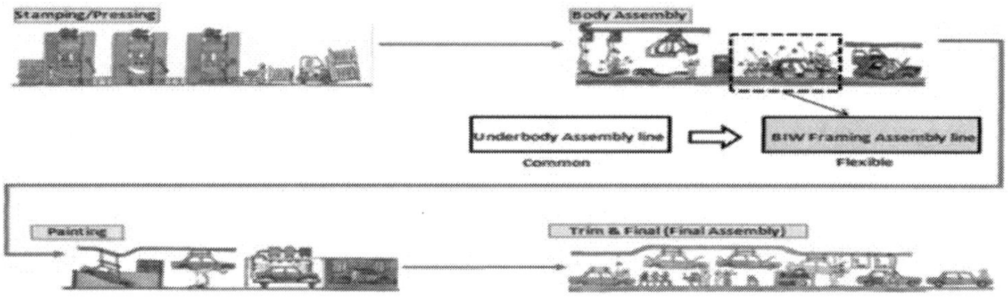

Fig. 2. Assembly process of BIW framing systems.

Main body jig or framing station has been dedicated and complex for a model of body. Over the years, different concepts of framing jigs have been developed with objective of high productivity and assured quality. However, with changing manufacturing strategy to attain lean inventory and more variety, flexibility of framing jig has become a more demanding requirement. A change-over or introduction small automotive of new models, manufacturing industry has three alternatives: (1) Change the old line with new one by stopping production for a considerable period, (2) Allocate the new line in different location, (3) Change manufacturing line for production of forthcoming models while production of the current model/s go on.

The best option for the manufacturers is the third alternative and flexibility is a necessity for reasons i.e. to improve mixed production, enormous investment for change over when the model is changed, stoppage of manufacturing lines prior to change over, surplus tied up production capacity during phasing out of model and huge manpower to change the layout and installation of new equipment [6].

Historically, all current framing systems have evolved from the Open-Gate and Robo-Gate system, initially developed in the 1980s. There are several types of framing systems i.e. (1) 2 slider framer, (2) Robot change framer, (3) Separated type robot framer, (4) One-body type robot framer, (5) Manifold cylinder framer, (6) Roof robot framer, (7) Sliding framer system (Open-gate system), (8) Index framer

Among the concern items with the current framing systems are currently the gates structure is integral and strongly coupled design; changeover times have to be accommodated for in the production plan, as the line has to stop running; gates storage systems require significant amount of floor space; overall high cost, high lead-times for engineering and build time; very expensive to manufacture and maintain; high cost of lost production due to downtime (breakdown or retooling) and high risk in capital investment.

4 Evolution of automotive framing systems

The aim in the future production systems is to quickly be able to adjust to product changes. A new trend come out in the manufacturing systems is called knowledge capture. Evolution/history of developments of vehicle framing systems are shown in Fig. 3. Due to markets increasingly require more customized products with shorter life cycles in the last few year, so manufacturing systems have evolved from mass production techniques through flexible automation and mass customization, to produce at mass production costs. Manufacturing facilities must incorporate more flexibility and intelligence, evolving toward reconfigurable assembly systems (RAS) prior to support effectiveness of manufacturing operations.

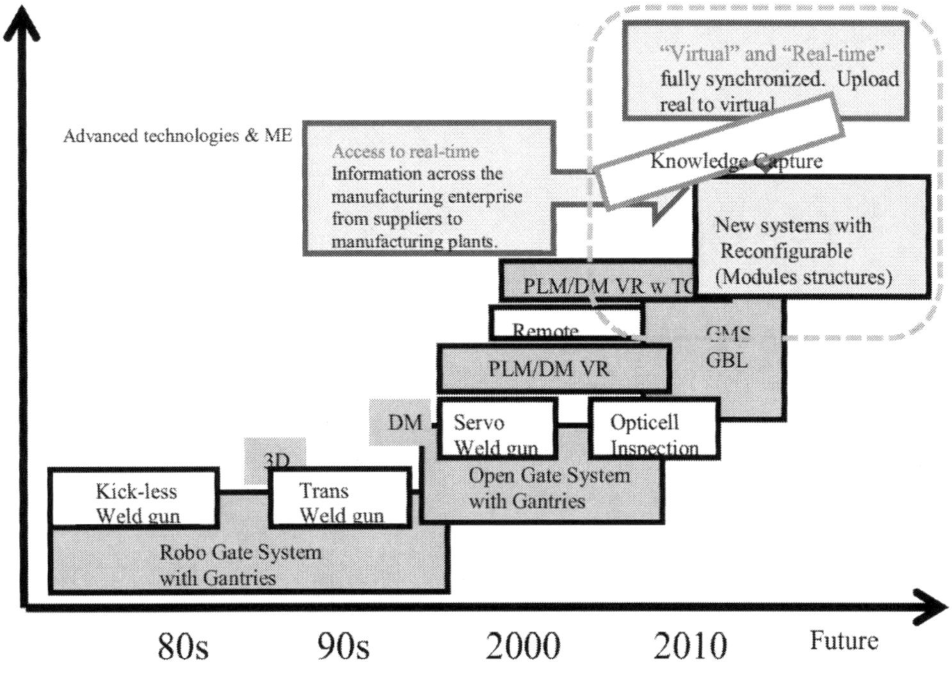

Fig. 3. Evolution of Automotive Framing Systems [4].

5 Problem statement

One of an element in the most of automotive manufacturer including small manufacturer to remain competitive is manufacturing cost. Therefore minimize manufacturing expenses always have been put priority in their operations . With limited flexibility that they have in order to achieve a good productivity i.e. increase of production volume, small car industry also have intention to produce multi model in their plant at current area. Sedan, Hatchback

5-doors, Hatchback 3-doors, Coupe, MPV (Multi Purpose Van) and SUV (Sport Utility Vehicle) are among the models categorized under passenger car/vehicle.

Rigidity and complexity are characteristics of current production systems of BIW vehicle framing which it different with product structure. Experience, expertise and knowledge among the factors in automotive industry prior to enhance a systematic methodology because manufacturing process design and integration are subjective. The methodology can be a guideline for systems designers and developers in order to react quickly and more economically to unpredictable of market change by utilizing flexibility in both product and production systems. The hypothesis is a requirement to have a potential to assemble side structure and underbody at the single assembly line can be materialized if the right subset of vehicle body elements and assembly processes are designed with proper care for future flexibility, then the proposed production assembly system compatible with multiple models i.e. variants of family production.

6 Research goal

The goal is to reconfigure a production assembly process, to identify and incorporate flexibility in the assembly process of production, to verify the proposed reconfigurable assembly system through simulation and validation and respond to a requirement for implementing multiple model manufacturing system at Mainbody line and minimize production downtime.

7 Research flow

In this research, some kinds of methodology are being conducted as shown in the Fig. 4 below:

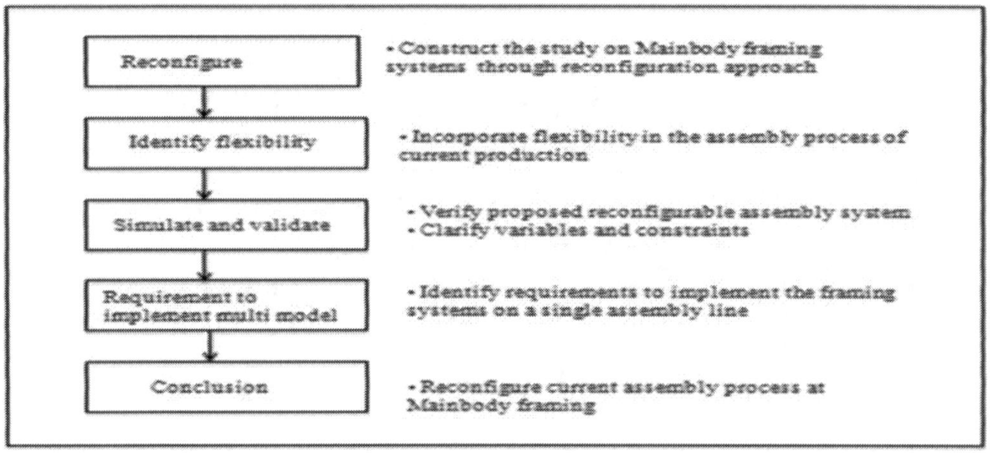

Fig. 4. The research methodology flow.

In the methodology, simulation software is used i.e. WITNESS 12. Simulation model development is developed based on information from the automotive manufacturing plant data.

8 Results and discussion

By changing Mainbody assembling process from Mainbody Presetting method to enhance Mainbody framing i.e. Robot gripper, some positive result are expected to achieve based on Quality, Cost and Delivery perspective. In term of quality, Mainbody accuracy percentage will be increased and become consistent. Cost of down time and loss of production can be reduced or eliminated and last but not least, delivery of production volume can be improved and achieve as per planned. Optimal number of models for a Mainbody framing also can be determined.

In manufacturing or assembling process, the results of product output are measured based on quality, cost and delivery. The manufacturer or assembler will enhance their current process and engineering know-how in order to achieve better result of quality, cost and delivery target indicators.

Body in White (BIW) is a jig for fitment of components parts in making an automotive CBU (Complete Build up Unit). Therefore, factor accuracy of the mainbody plays a major role in achieving good quality of the CBU. By achieving quality target accuracy of 85% for mainbody, it can help a lot to achieve quality of final product later. Gap, parts fitting and flushness can be improved through good accuracy of mainbody.

Operating in minimum cost is a requirement from the management. Profit is one of the key performance indicator of most of company at the end of financial year. Through reconfigurable assembly systems i.e. by changing assembly process from Mainbody Presetting to Mainbody Robot Gripper type, cost of downtime can be reduced. Cost of an opportunity loss also need to take in account once production output is not achieve target. For example, if daily production plan is to produce 400 units BIW but in actual only produce 300 units, meaning minus 100 units which is considered as opportunity loss.

Among the main key task in production is to deliver product or finish good as per delivery plan. Reconfigurable assembly systems can be a medium in order to achieve the delivery plan. Value analysis of mainbody sub systems can be validated through RAS approach.

9 Conclusions

Reconfigurable assembly systems (RAS) can be a solution for rapid change in automotive framing structure and for a responsively adjustable production capacity. It also contribute in improving manufacturing or production target indicators i.e. quality, cost and delivery. Mainbody accuracy percentage also be able to improve by implementing RAS. Future study also can be explored for Mainbody framing systems by using Aluminium material for side gate.

The author would like to be obliged to Ministry of Higher Education Malaysia (MOHE) for providing financial assistance.

References

1. des Constructeurs d'Automobiles, Organisation Internationale, World motor vehicle production by manufacturer (2012)
2. A. Al-Zaher, W. El Maraghy, Z.J Pasek, J. Mfg. Sys., **32**, 436-448 (2013)
3. M.P. Groover, Automation, Production Systems and Computer-Integrated Manufacturing, 2nd Edition, Prentice Hall (2001)
4. Z.M. Bi, L. Wang, S.Y.T. Lang, Int. J. Mfg. Research, **2**, 303-328 (2007)
5. G. Michalos, S. Makris, N. Papakostas, D. Mourtziz, G. Chryssolouris, CIRP J. Mfg. Sc. & Tech. **2**, 81- 91 (2010)
6. Drishtikona, Body Welding – Can it be flexible, http://www.drishtikona.com/books/automobile-manufacturing/ch6, 1-27 (2010)

Lean manufacturing implementation in reducing waste for electronic assembly line

Nurul Husna Zakaria[1,*], *Nik Mohd Zuki* Nik Mohamed[1,2], *Mohd Fadzil Faisae* Ab Rahid[1], and *Ahmad Nasser* Mohd Rose[1]

[1]Faculty of Mechanical Engineering, Universiti Malaysia Pahang, 26600 Pekan, Pahang, Malaysia.
[2]Automotive Engineering Centre, Universiti Malaysia Pahang, 26600 Pekan, Pahang, Malaysia.

Abstract. Lean manufacturing is the most convenient way to eliminate unnecessary waste and can provide what customers demand. This paper presents possibilities and sustainability of application of lean manufacturing method by using a virtual simulation of the workers performance in a line production of small and medium industry. Actual case study and Witness simulation were used in this study to find the waste that exists in the production and identified the performance of workers in the production line. Lean manufacturing concept has identified and rectified problems related to low productivity in the assembly line. The case study is involved a line production for electronic part assembly. The result of this preliminary study should illustrate the relationship of worker's performance by lean manufacturing method as well as the productivity improvements which help to reduce cost for manufacturer. Lean manufacturing method has been used during the study to reduce the cost when waste is eliminated by reducing the workstation without reducing the performance of the production. The performance of the production is increased when allocating the labor in a needed working area. Lastly, the study also proves that the new layout has improved the process to be used for future production process.

1 Introduction

Lean manufacturing is a thinking that aims to remove and reduce wastes in each range of production counting the customer relations, supplier systems, product design, and plant management [1]. It can be implemented to remain competitive by aiding its clients in a better way and continually reducingcosts needed [2,3].

Lean manufacturing concepts can be organized into three levels consist of lean manufacturing objectives and fundamental principles, prime management and manufacturing approaches and also the implementation techniques that are the actions for applying and maintaining the stratagems [2,3]. The orderly removal of waste will ease the cost of working the prolonged enterprises and satisfies the customer's need for supreme value at a lower cost [13]. In lean production or process, mistakes are continually made but

* Corresponding author: nhusnazakaria@gmail.com

these faults are not usually repetitive since this is one of waste that the lean aim to exclude [1,9].

1.1 Benefits of lean manufacturing

Some typical benefits that lean offers to the industries are financial saving, less rework, reduced lead-time, increased process understanding, less process waste and reduced inventory [5,10]. Based on a research that had been done in 200 SME companies, their perceptions toward lean benefits are cost saving, continuous improvement, waste reduction, tools and technique to improve production operation, fully integrated management philosophy, a way of life, comprehensive system of organizing and managing the supply chain and lastly is headcount reduction, which are shown in figure 1 [16,14].

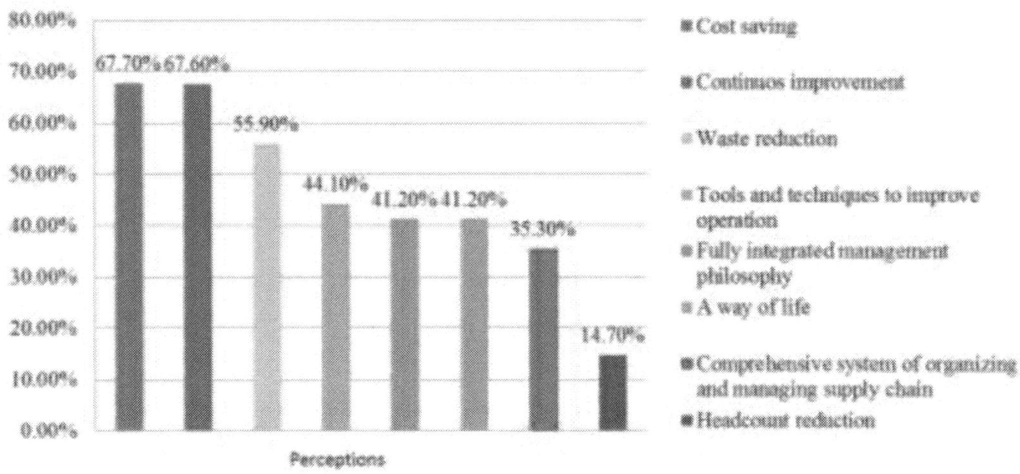

Fig. 1. Graph of perceptions towards lean practices.

1.2 Methods in lean manufacturing

A cluster of lean production methods is grouped into five categories. According to [4,6,8,12,15], methods in these categories in Table 1.

1.3 Type of waste

In Lean Manufacturing, there have seven types of waste targeted which is product defects, waiting time, processing waste, overproduction, motion waste, inventory, and transportation. In general, a basic understanding about these wastes is, a defect is a manufacture product that out of specification, and service components that cause in rework, scrap, checkup, flawed materials or replacement production. Waiting time means postponements allied with stock-outs, equipment downtime, machine breakdown, capacity bottlenecks and lot processing delays. Processing waste is processed phases that are not essential to manufacturing the product. Overproduction is more products being produced than demand and before it is needed. Motion waste is defined as a human motion that are pointless and work-in-process (WIP) moving extensive distances. Then, inventory is additional raw material, finished goods, or WIP, which contributes towards pending further processing, while transportation, waste means work done to transport unnecessary stock to a warehouse [1]. The types of waste are simplified in Table 2.

Table 1. Cluster of lean production methods.

Type	**Lean Method**
Machinery and equipment	Low Cost Automation
	Overall Equipment Effectiveness (OEE)
	Preventive Maintenance
	Single Minutes Exchange of Dies (SMED)
	Total Productive Maintenance
Material flow and layout	Cellular Manufacturing
	First in First Out (FIFO)
	One-piece-flow
	Simulation software
	Optimization of the supply chain
	Value Stream Mapping (VSM)
	Workstation design
Organization and staff	5S
	Autonomous work groups
	Benchmarking
	Ideas management
	Job rotation
	Lean office (Administration)
	Kaizen
	Standardization
Quality	FMEA
	Poka Yoke
	Quality Circles

This study will focused on material flow and layout type which is using simulation software method and workstation design in order to provide improvement for current production that met the objectives for the research which is to study the existing process in the company and to analyze the data by using the lean method, that lastly will provide a comparison of improved process to be a basis for future set up.

2 Methodology

In conducting this study, some methods are scheduled in details. This phase of activity is following Plan-Do-Check-Act (PDCA) method. After getting enough information in data collections, a gap analysis is proceeded to know the difference from the standard where steps needed to change from the current state to desire state or output is determined. Some visit is planned to complete things needed to do this project.

2.1 Observation and data collection

The observation and data collection was made by having an interview session with the company supervisor and also their workers. Then another approach to get the data collections is by 'Genchi Genbutsu', go and understand the actual situation. Literally, by doing this approach, it helps to have a better understanding of current condition in the factory. Other than 'Genchi Genbutsu', Gemba is the other approach. It is a method of discovery deformity over observation and it aids to recognize main processes in production. Based on this, a process mapping is created by using material and information flow chart. According to the process mapping, by using material and information flow chart, any unnecessary waste or non-value added which hiding in the production can be detected. Other than that, problems that linked and related to material flow can be recognized and can produce a new improve layout. In creating a new layout, the simulation of it is designed by using Witness software.

Table 2. Type of waste.

Types of Waste	Descriptions
Product Defects	Faults during the process that each demanding re-works or supplementary work.
Waiting Time	Tools or product pauses are a non-added value to the buyer.
Processing Waste	The situation of a particular process step that does not contribute any benefit to the product.
Over Production	Product prepared without a specific client. The progress of a product, a procedure or a production competence for no added value.
Motion Waste	The unnecessary motion of the people who control the manufacturing equipment is wasteful. Where they are in a movement that did not support the manufacturing of the product. An unnecessary motion of data, information, and decisions.
Inventory	Storage of products, intermediates, raw materials that will lead to cost money.
Transportation	Transporting the product to numerous places. While the product is in a movement, it is in a situation that will not be processed and thus did not count values to the client.

2.2 Applying lean method

Once observation and identification are done, data that have been collected will be used to identify which method of lean is suitable to be implemented in the production area. This will be justified according to the seven elements of waste which occur in lean manufacturing.

2.3 Simulation by using software

The Witness simulation makes it possible to creating a variety of discrete and continuous elements. Subject on the type of component, each can be in any quantity of 'states' which means either idle, busy, blocked, in-setup, broken down, and waiting labor [7]. The end of the simulation done by using Witness software, the result from the data will be analyzed whether it brings any improvement towards the company.

2.4 Data analysis

The result obtained from simulation in Witness will be analyzed whether it brings any improvement towards the company and can be used for future setup.

3 Results and discussion

The first step in collecting the data in the focused cell is observed the layout in the production by detailing the sequences of workstation involved. Figure 2 below is the layout and process in the production.

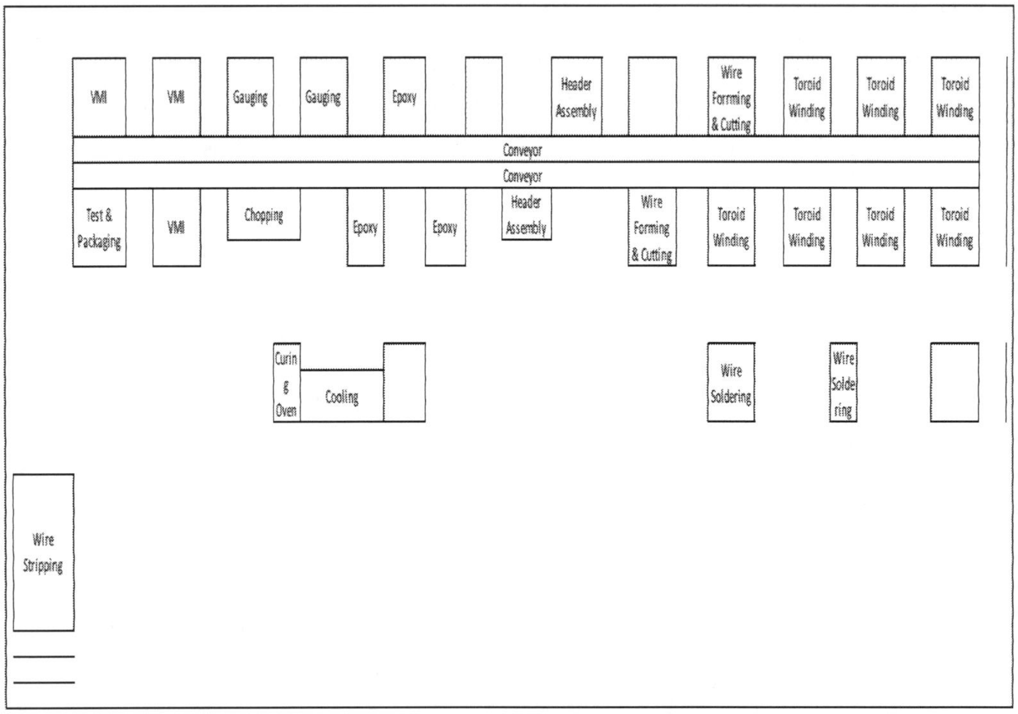

Fig. 2. Current layout in focused cell.

Other than observing the layout of the focused cell, the data involved in the production are being collected. Below is the type of data that has been tabulated in a Table 3 and 4. The cycle time for each worker is taken three different session; morning, afternoon and evening, in order to get the average cycle time.

The performance for current production is run in software and the result of performance is shown in Table 5. According to the results of performance of each labor, it shows that Gauging Process has workers with a high percent of idleness. By referring to this result, an improved process can be made.

Table 3. Data collections in existing production.

Type of data	Time
Demand per day	4000
Working Hours per day	10 hours
Break per day	0.75 hours
Available Working Hours per day	9.25 hours
Takt Time	8.3 seconds
Target per Hour	433
Conveyor Speed (m/s)	0.03

Table 4. Cycle time for each worker.

Workers	Process	Cycle time 1 (s)	Cycle time 2 (s)	Cycle time 3 (s)	Avg. Cycle Time (s)
1	Wire Soldering	59	58	46	54.3
2	Wire Soldering	60	58	48	55.3
3	Toroid Winding	99	99	86	95
4	Toroid Winding	56	55	58	56.3
5	Toroid Winding	67	64	68	66
6	Toroid Winding	92	91	89	91
7	Toroid Winding	62	65	64	63.7
8	Toroid Winding	34	29	31	31.3
9	Toroid Winding	40	38	39	39
10	Wire Forming & Cutting	12	10	11	11
11	Wire Forming & Cutting	13	12	13	12.7
12	Header Assembly	13	15	14	14
13	Header Assembly	7	8	7	7.3
14	Epoxy	40	37	37	38
15	Epoxy	40	30	30	33.3
16	Epoxy	24	22	22	22.7
17	Chopping	3	3	3	3
18	Gauging	4	5	4	4.7
19	Gauging	2	2	2	2
20	VMI	5	15	7	9
21	VMI	13	15	14	14
22	VMI	31	27	28	28.7
23	Test & Packaging	6	5	5	5.3

3.1 Data analysis for first improvement

Based on existing data, an improved process is produced to be one of the future setups for this company. The method used to create an improved process is by referring to the percentage of the idleness of the labors in existing production where it is found that two labors that are in the gauging process have a high percent of idleness compared to others. According to that, one workstation is reduced, and the performance of it is being analyzed by using Witness. The two labor of Gauging Process in current production has been reduced to one labor, which means the workstation is reduced from 23 to 22. Lady18 is chosen to be removed because she has the highest percent of idleness that is 93.84%. After

DOI: 10.1051/matecconf/20179001048

the workstation has been removed, based on the performance in Table 6, the labor's performance did not have any different except for Lady17, which is 85.53% of idleness reduce to 71.05 since she handles the process that had been removed from Lady18. The performance calculation of this improvement is shown by Eq. (1).

$$Performance = \frac{(1041piece/288min)}{7piece\,per\,min} x100 = 51.64\% \quad (1)$$

Table 5. Result of performance by using Witness.

Job Element	Name	% Busy	% Idle	No. Jobs Started	No. Jobs Ended	Avg Job Time (s)
Wire Soldering	Lady01	99.87	0.13	334	333	54.3
	Lady02	99.78	0.22	327	326	55.3
Toroid Winding	Guy01	99.57	0.43	238	237	75.53
	Guy02	99.57	0.43	380	379	47.53
	Guy03	99.57	0.43	406	405	44.39
	Lady03	99.48	0.52	351	350	51.5
	Guy04	99.48	0.52	374	373	48.22
	Guy05	99.48	0.52	704	703	25.61
	Lady12	99.48	0.52	820	819	21.97
Wire Forming & Cutting	Lady04	78.03	21.97	684	683	20.7
	Lady13	99.24	0.76	621	620	29
Header Assembly	Lady05	99.05	0.95	2565	2564	7
	Lady14	51.61	48.39	1282	1281	7.3
Epoxy	Lady06	89.45	10.55	427	426	38
	Lady16	78.33	21.67	427	426	33.3
	Lady07	53.49	46.51	427	427	22.7
Chopping	Lady09	18.48	81.52	1116	1116	3
Gauging	Lady17	14.47	85.53	558	558	4.7
	Lady18	6.16	93.84	558	558	2
VMI	Lady19	55.43	44.57	1116	1116	9
	Lady10	85.7	14.3	501	500	31
	Lady20	85.75	14.25	542	541	28.7
Test & Packaging	Lady11	13.21	86.79	1041	1041	2.3

Statistics for the number of job ended in Table 6 did not have any changes even the workstation had been reduced and it shows that the output is same as current production. This shows that it is significant to reduce the workstation since the performance of it is equal to the performance of the current production.

Table 6. Data on labors performance for first improvement.

Job Element	Name	% Busy	% Idle	No. Of Jobs Started	No. Of Jobs Ended	Avg Job Time
Wire Soldering	Lady01	99.87	0.13	334	333	54.3
	Lady02	99.78	0.22	327	326	55.3
Toroidal Winding	Guy01	99.57	0.43	238	237	75.53
	Guy02	99.57	0.43	380	379	47.47
	Guy03	99.57	0.43	406	405	44.39
	Lady03	99.48	0.52	351	350	51.5
	Guy04	99.48	0.52	374	373	48.22
	Guy05	99.48	0.52	704	703	25.61
	Lady12	99.48	0.52	820	819	21.97
Wire Forming and Cutting	Lady04	78.03	21.97	684	683	20.7
	Lady13	99.24	0.76	621	620	29
Header Assembly	Lady05	99.05	0.95	2565	2564	7
	Lady14	51.61	48.39	1282	1281	7.3
Epoxy	Lady06	89.45	10.55	427	426	38
	Lady16	78.32	21.68	427	426	33.3
	Lady07	53.49	46.51	427	427	22.7
Chopping	Lady09	18.48	81.52	1116	1116	3
Gauging	Lady17	28.95	71.05	1116	1116	4.7
VMI	Lady19	55.43	44.57	1116	1116	9
	Lady10	85.7	14.3	501	500	31
	Lady20	85.75	14.25	542	541	28.7
Test & Packaging	Lady11	13.21	86.79	1041	1041	2.3

3.2 Data analysis of second improvement

A reallocating method is used in this second improvement by reallocating the free labor to a workstation that has a bottleneck which has a low capacity of buffer. The final improvement that gives increment in productivity when the labor is put into that process is by reallocating the labor to VMI process, since this process can be considered as final process before test and packaging stage. The analysis for this second improvement gives increment in productivity of the production, which is from 1041 total output it becomes 1109. The productivity for this improvement is calculated below.

$$Performance = \frac{(1109 pieces/288 min)}{7 piece\ permin} x100 = 55.01\% \quad (2)$$

Below is the labors performance for second improvement where it shows that the output gained in this improvement is increased a bit. This output can be seen in number of job ended in the final process in Table 7, which in test and packaging stage.

Table 7. Data on labors performance for second improved process

Job Element	Name	% Busy	% Idle	No. Of Jobs Started	No. Of Jobs Ended	Avg Job Time
Wire Soldering	Lady01	99.87	0.13	334	333	54.3
	Lady02	99.78	0.22	327	326	55.3
Toroidal Winding	Guy01	99.57	0.43	238	237	75.53
	Guy02	99.57	0.43	380	379	47.53
	Guy03	99.57	0.43	406	405	44.39
	Lady03	99.48	0.52	351	350	51.5
	Guy04	99.48	0.52	374	373	48.22
	Guy05	99.48	0.52	704	703	25.61
	Lady12	99.48	0.52	820	819	21.97
Wire Forming and Cutting	Lady04	78.03	21.97	684	683	20.7
	Lady13	99.24	0.76	621	620	29
Header Assembly	Lady05	99.05	0.95	2565	2564	7
	Lady14	51.61	48.39	1282	1281	7.3
Epoxy	Lady06	89.45	10.55	427	426	38
	Lady16	78.33	21.67	427	426	33.3
	Lady07	53.49	46.51	427	427	22.7
Chopping	Lady09	18.48	81.52	1116	1116	3
Gauging	Lady17	28.95	71.05	1116	1116	4.7
VMI	Lady18	55.43	44.57	1116	1116	9
	Lady19	68.66	31.34	434	433	28.7
	Lady10	75.35	24.65	370	369	37
	Lady20	70.61	29.39	310	309	41.4
Test & Packaging	Lady11	14.08	85.92	1110	1109	2.3

4 Conclusions

In this paper, it was aimed to reduce the waste exist in line production that related by the performance of workers in line. The virtual simulation was simulated based on data collections gained during observation process. The significant of each improvement was verified according to simulation result in the software used which the performance of production was calculated to support the result. The performance was reduced from 56.04% to 55.01% with total production from 1041 to 1109 pieces. According to the results obtained, the waste that had been addressed by this simulation is processing waste; a situation of a particular process step that does not contribute any benefit to the product where the two labors in one process are having a small percent of busy. In addition, with this improvement, the productivity of it is increased, since one of the bottlenecks of the production is solved.

References

1. Hicks, B. J. Lean information management: Understanding and eliminating waste. *International Journal of Information Management.* **27(4)**, 233-249, (2007)
2. Lewis, M. A., Lean Production and Sustainable Competitive Advantage, *International Journal of Operation and Production Management.* **20(0)**,121-125, (2000)
3. Matt, D. T., and Rauch, E. Implementation of Lean Production in Small Sized Enterprises. *Procedia CIRP.* **12(0)**, 420-425, (2013)
4. Mehta, R. K, Mehta, D., Mehta, and N. K.,. Lean Manufacturing Practices: Problems and Prospects. *ANNALS.* **3(0)**, 119-124, (2013)
5. Melton, T. The Benefits of Lean Manufacturing: What Lean Thinking has to Offer the Process Industries. *Chemical Engineering Research and Design,* **83(6)**, 662-673, (2005)
6. Ojaghi, Y., Khademi, A., Yusof, N. M., Renani, N. G., and Hassan, S. A. H. b. S, Production Layout Optimization for Small and Medium Scale Food Industry. *Procedia CIRP.* **26(0)**, 247-251, (2015)
7. Pam, L. M., and Michael, H. M.,, Witness Simulation Software A Flexible Suite of Simulation Tools. *Proceedings of the 1997 Winter Simulation Conference.* **17(0)**, 711-717, (1997)
8. Raja, M. I. Lean Manufacturing-an Integrated Socio-Technical Systems Approach to Work Design. *Clemson University.* **5(0)** 35-43, (2011)
9. Relkar, A. S., and Nandurkar, K. N. Optimizing & Analysing Overall Equipment Effectiveness (OEE) Through Design of Experiments (DOE). *Procedia Engineering.* **38(0)** 2973-2980, (2012)
10. Rose, A.M.N., Deros, B.Md., Rahman, M.N.Ab. and Nordin, N. Lean manufacturing best practices in SMEs, *Proceedings of the 2011 International Conference on Industrial Engineering and Operations Management.* **2(5)** 872-877, (2011)
11. Ulutas, B., and Islier, A. A. Dynamic facility layout problem in footwear industry. *Journal of Manufacturing Systems.* **36(0)** 55-61, (2015)
12. Villa, A., and Taurino, T. From JIT to Seru, for a Production as Lean as Possible. *Procedia Engineering.* **63(0)** 956-965, (2013)
13. Wahab, A. N. A., Mukhtar, M., and Sulaiman, R.. A Conceptual Model of Lean Manufacturing Dimensions. *Procedia Technology.* **11(0)** 1292-1298, (2013)
14. Womack, JamesP., Jones, and DanielT, Lean Thinking: Banish Waste and Create Wealth In Your Corporation, Simon & Schuster, New York. **45(0)** 457-470, (1996)
15. Yu, C. W., Kuan, Y. W., and.Anwar A., A study on lean manufacturing implementation in the Malaysia electrical and electronics industry. *European Journal of Scientific Research.* **38(4)** 521-535, (2009)
16. Zhou, B., Lean principle, practices, and impacts: a study on small and medium-sized enterprises (SMEs). *Springer Science Business Media LLC.* **10(0)** 112-117, (2012)

Preparation and stability of silicone dioxide dispersed in polyalkylene glycol based nanolubricants

M.Z. Sharif[1,*], *W.H.* Azmi[1,2], *A.A.M.* Redhwan[1,3], *N.M.M.* Zawawi[1]

[1]Automotive Engineering Research Group (AERG), Faculty of Mechanical Engineering, Universiti Malaysia Pahang (UMP), 26600 Pekan, Pahang, Malaysia,
[2]Automotive Engineering Centre, Universiti Malaysia Pahang (UMP), 26600 Pekan, Pahang, Malaysia.
[3]Faculty Manufacturing Engineering Technology, TATI Universiti College (TATIUC), 24000 Kemaman, Terengganu, Malaysia.

Abstract. Nanolubricant is one of the efforts introduced by researchers to increase efficiency in many mechanical application, especially in refrigeration. Two-step method is the most common method used in the process of adding nanoparticles dispersed in base lubricant because of the simplicity of the process to prepare a stable solution. In this work, the SiO_2/PAG nanolubricants were prepared using two-steps method without the use of surfactant. The stability of SiO_2/PAG nanolubricant was observed trough sedimentation photograph capturing technique and UV-Vis spectrophotometric test. The results shown that there are minimum sedimentation observed over one month. This result also confirmed via the UV-Vis test; the nanolubricant relative concentration was maintained at over 70 % compared to the initial concentration.

1 Introduction

Different lubricants with various properties are created nowadays to be utilized in many types of application involving mechanical parts, especially in refrigeration compressor. In a mechanical system, friction is a primary cause for energy loss. Hence, lubricant is the best solution to reduce friction consequently decreasing energy losses and heat production [1]. Thus, to improve the energy-efficiency systems involving mechanical system, it is important to enhance the lubricant properties. Application of nanotechnology dispersed in lubricant is a new way to improve the lubricating properties and interested many scholars to explore further in this area. For example, the researcher has proven that SiO_2 not only have the advantages of better heat transfer effect, but also give improvement to the tribulogical properties due to reduce friction coefficient and better anti wear properties [2].

The nanotechnology utilization as property's enhancement effectiveness in numerous applications, such as, motor and transmission oil cooling, refrigeration (domestic and chillers), evaporator fumes warming and cooling of buildings; cooling of electronics

* Corresponding author: sharif5865@yahoo.com

 DOI: 10.1051/matecconf/20179001049

component, oils, biomedical application and nanofluids in transformer oil has been studied by the different researchers [3-5]. According to Alawi et al. [6], there are few advantages of applying nanoparticles as additives in a lubricant for application in refrigerant. The advantage is by enhancing the lubricant-refrigerant solubility and also significantly increases the thermo-physical properties of lubricant. It is also improves the wear rate and friction coefficient. Currently, there are two available methods for preparing the nanolubricant which is a one-step method and two-step method. Two-step method is the most common method used in producing nanolubricant because of the simplicity of the process in order to prepare the stable solution. However, this method has a drawback on agglomeration due to large and active in surface area [7]. Hence, a suitable method for stabilize the solution is required by adding pH and obtaining an optimum ultrasonic sonification time. Previously, the effect of sonification time on stability was being done [8]. The sonification time may vary according to the types of base fluid.

Therefore, the purpose of this paper is to establish the preparation of the SiO_2 dispersed in Polyalkylene Glycol nanolubricant. Then, the study on stability of the nanolubricant evaluated using visual analysis and UV-Vis Spectrometer absorbance analysis.

2 Materials

Nanoparticles of SiO_2 are obtained from Sigma-Aldrich Corporation. The SiO_2 nanoparticle average sizes are 30 nm and displayed in Fig. 1. The properties of the nanoparticles are shown in Table 1.

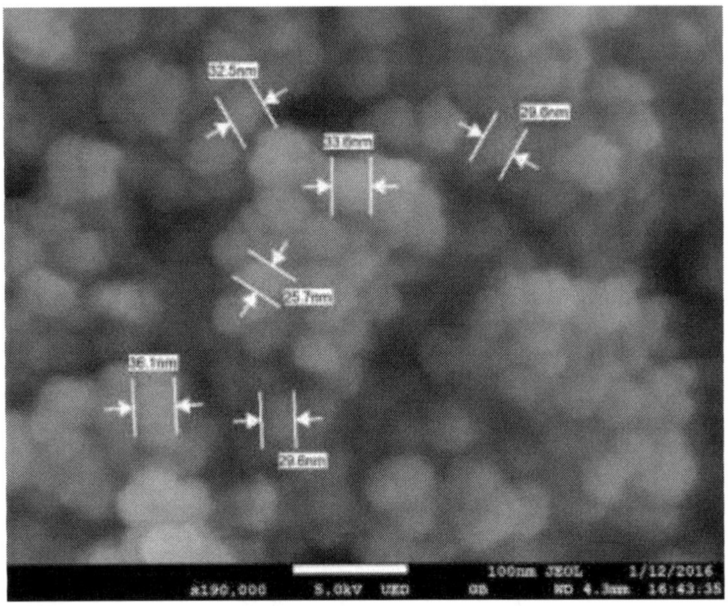

Fig. 1. SiO_2 nanoparticles with average size of 30 nm.

Polyalkylene Glycol (PAG) lubricant, is specially designed to be used in compressor for air conditioning in automobiles and lorry obtained from Denso. The properties of the PAG lubricant are shown in Table 2.

Table 1. Properties of SiO_2 nanoparticles used in this experiment [9, 10].

Property	SiO_2
Molecular mass, g mol^{-1}	60.08
Average Particle diameter, nm	30
Density, kg m^{-3}	2220
Thermal Conductivity, W m^{-1} k^{-1}	1.4
Specific heat, J kg^{-1} K^{-1}	745

Table 2. Properties of PAG lubricant [11, 12].

Specifications	Value
Kinematic viscosity, cSt @ 40 °C	41.4-50
Kinematic viscosity, cSt @ 100 °C	11
Density, kg/cm^3@ 15 °C	0.9954
Viscosity index	184
Pour point, °C	-51

3 Methodology

3.1 Preparation of nanolubricant

Nanolubricant of SiO_2 nanoparticles dispersed in Polyalkylene Glycol (PAG) lubricant was prepared using two-step methods.In order to prepare the nanolubricant, proper mixing and the stabilization of the particles are required. In the present study, SiO_2nanoparticles will mix with PAG lubricant using a magnetic stirrer for 1 hour, and then dispersed continuously using ultrasonic bath vibrator Fisherbrand (model: FB15051) generating ultrasonic pulses of 230V at 50 ± 3kHz for 2 hour. The 2 hour sonification time was determined using UV-Vis spectrophotometer analysis as an optimum time for the sonification to ensure that the nanoparticles are mixed well with less sedimentation potential and better mean particle size. The detailed results will be explained in the next sub-chapter 4.1.The sonicator will break down the agglomeration and ensure the nanoparticle well dispersed into the base solution. The detailed step of preparation was shown by Azmi et al.[13]. The volume concentrations between 0.2 to 1.0%of nanolubricant was prepared in this experiment. The nanolubricant volume concentration was calculated using the formula in Eq. (1).

$$\phi = \frac{\dfrac{m_p}{\rho_p}}{\dfrac{m_p}{\rho_p} + \dfrac{m_L}{\rho_L}} \times 100 \qquad (1)$$

3.2 Stability evaluation of the nanolubricant

Stability of nanolubricant can be determined in different ways. It this experiment, the relative stability of nanosupension was observed using sedimentation photograph capturing methods and UV- Vis Spectrophotometer analysis method. Therefore, the rate of sedimentation will be identified by analyzing the data. The visual sedimentation photograph method was the main method to find out the visual sedimentation of nanolubricant [14]. The sample was observed for a month, and the image of nano suspension sedimentation apparentness will be compared.Another method for stability is using UV-Vis Spectrophotometer (model: Genesys 10). The UV-Vis can measure the colloidal stability of the dispersions of nanolubricant. The picture of the instrument was shown in Fig.2.

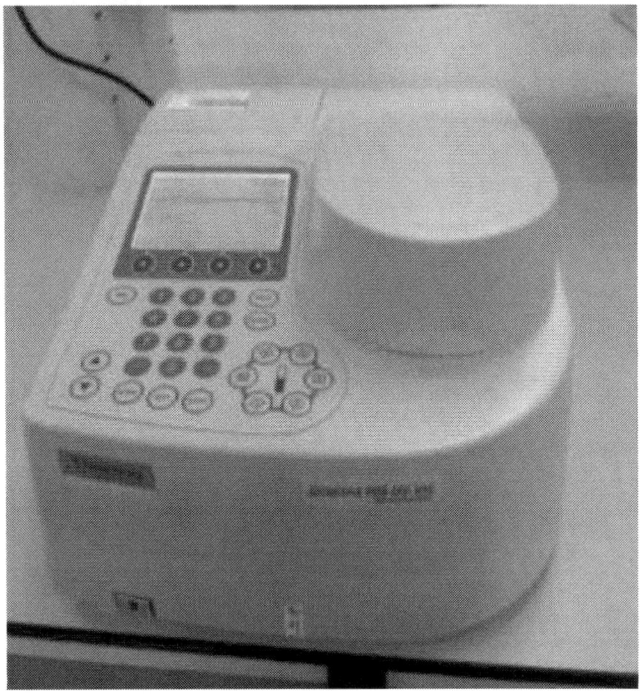

Fig. 2. Genesys 10 UV-Vis Spectrometer.

UV–vis spectrophotometer measures the absorption and the scattering of light by comparing the intensity of the light of the nanolubricant with the reference, the base lubricant [14]. In addition, the absorption and dispersions in the nanolubricant or nanofluid generally measured in the range of 200-900 nm wavelength [15]. Normally, nanosuspension stability is analyzed by measuring the sediment volume and absorbance versus the sediment time [16]. The absorbance for three volume concentration will be observed in this experiment for over a month.

4. Results and discussion

4.1 Sedimentation photograph capturing methods

SiO_2 nanoparticles were added to PAG lubricant used in the refrigeration system.The samples were left within a month. Sample pictures were taken for each hour, day, and week to see the sedimentation and deposition of the nanoparticle in the liquid.The images of the nanoparticle suspension sedimentation were compared. Fig.3 shows the sedimentation photograph of SiO_2/PAG nanolubricant after preparation and a month of preparation. In the figures., it can be seen that the sedimentation was minimum even after a month. There is not much change in terms of color form observed the experiment even after a month of preparation. Furthermore, only a small layer of two-layer layer form signifying that fewer sedimentation occurred in that particular liquid.

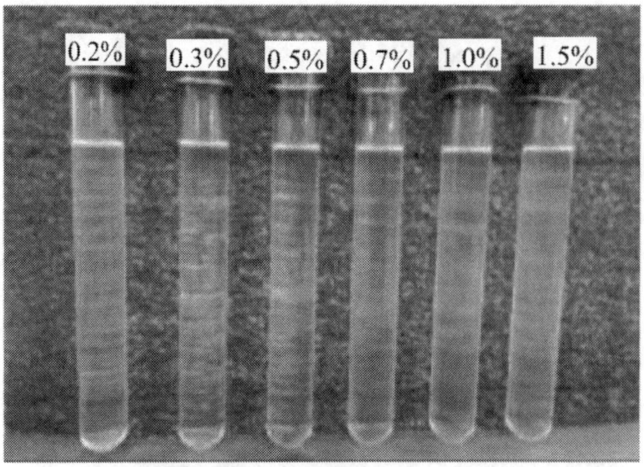

(a) Visual sedimentation after preparation

(b) Visual sedimentation after one month

Fig. 3. SiO_2/PAG nanolubricant sedimentation photograph.

4.2 UV-vis spectrophotometer analysis methods

After the preparation, the particle morphology was observed for different concentrations. Fig. 4 shows the UV–Vis spectra of SiO_2 with their respective concentrations. The absorbance of all nanolubricants demonstrates the characteristic absorption in the range of 200 to 400 nm wavelengths. Besides, the absorbance of nanolubricant increases from lower concentration to higher concentration. From the observation, it can be concluded that lower concentration of nanolubricant have more potential for agglomeration and faster sedimentation time.In the UV-vis spectrometer sedimentation test, firstly, the peak absorbance wavelength of the SiO_2/PAG nanolubricant at dilute concentration, which is 1.5% was tested. Fig. 5 shows the result of the scanning of the UV-vis spectrometer in the range of 200 nm to 400 nm. The scanning results show that the peak absorbance for SiO_2/PAG lies at 317 nm wavelength.

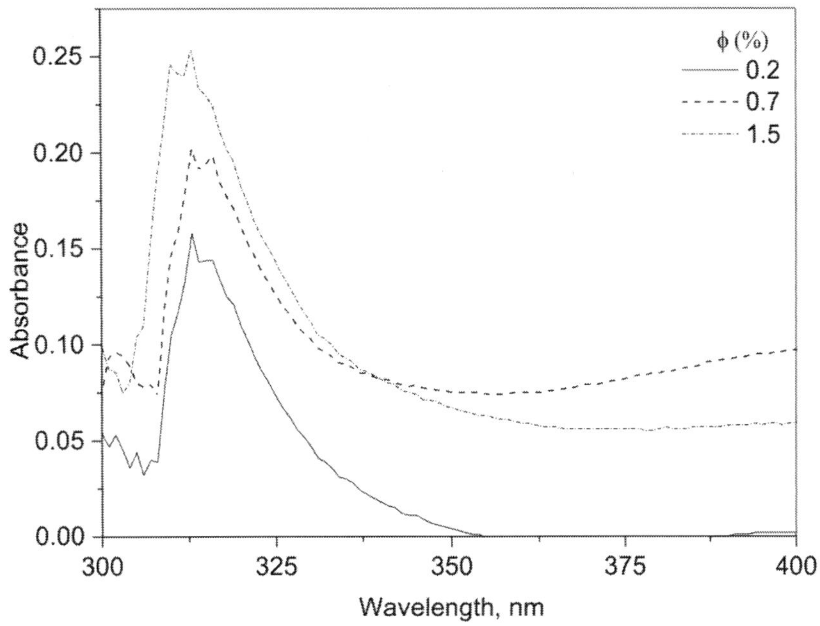

Fig. 4. UV-Vis spectrometer of different concentrations.

The UV-Vis spectrometer analysis was also done to confirm what is the best time needed for the sonification time to obtain great stability. The same concentration of SiO_2/PAG nanolubricant was prepared for different duration of sonification process time to achieve this purpose. The absorbance ratio over sedimentation time in hours for five samples graph was shown in Fig. 6. The data plotted in Fig. 6 clearly show that the absorbance ratio is decreasing with sedimentation time and strongly dependent to sonication time. Finally, the best or optimum sonication time for SiO_2/PAG nanolubricant is attained at 2.0 hours' sonication time. The sample remained stable during the sedimentation time and absorbance ratio was above approximately 80% even after 336 hours (two weeks).

After that, the relative stability measurement will be done on three concentrations of nanolubricant (0.5 %, 1.0 %, and 1.5 % volume concentration) for one month. The nanolubricants was prepared using the optimum time for sonification process which is 2 hours. The supernatant particle absorbance for three concentrations against the sedimentation graph was shown in Fig.7.

From the graph and the visual sedimentation photograph result, it has clearly shown that SiO_2/PAG nanolubricant was stable. After a month, the absorbance was maintained within the range of 0.24 to 0.35 values, showing that the relative concentration of these nanolubricant was maintained at over 70 % compared to the initial concentration.

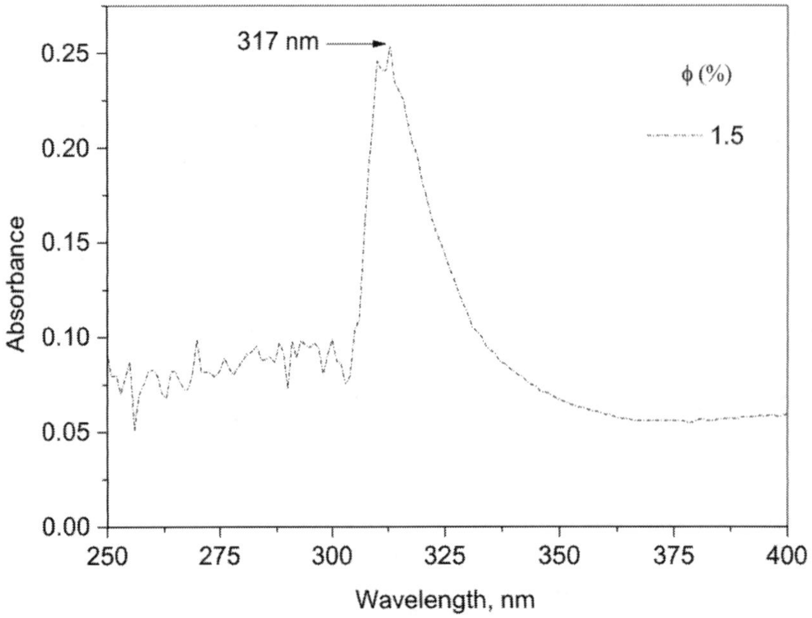

Fig. 5. UV-Vis spectrometer scanning graph at 200 nm to 400 nm.

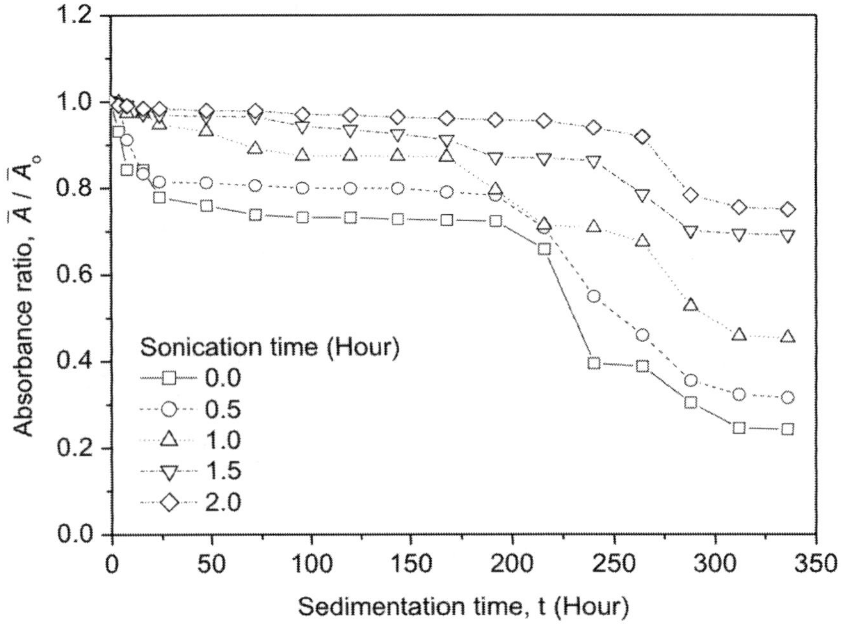

Fig. 6. Absorbance ratio of SiO_2/PAG nanolubricant as a function of sedimentation time.

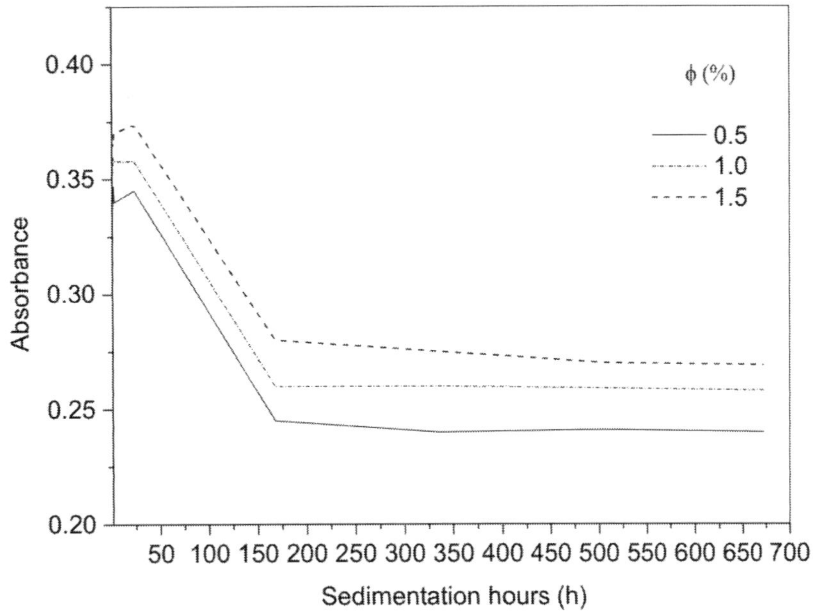

Fig. 7. UV-Vis spectrometer scanning graph at 200 nm to 400 nm.

5 Conclusions

SiO_2 was dispersed into the PAG lubricant which normally utilize in the refrigeration system using two-steps method without any surface stabilizer or surfactants. The best or optimum sonication time for SiO_2/PAG nanolubricant is attained at 2.0 hours' sonication process duration which was confirmed by using the UV-Vis spectrophotometer analysis. The stability properties of SiO_2/PAG nanolubricant were evaluated at few concentrations by visual sedimentation photograph and UV-Vis spectrophotometer method. From the sedimentation photograph, it can be seen that the sedimentation was minimum even after a month. There are no apparent changes on the samples after one month signifying that fewer sedimentation occurred in that particular liquid In the UV-Vis spectrometer test, it was shown that the peak absorbance of the SiO_2/PAG nanolubricant lies at 317nm wavelengths. After one month, the result from the UV-Vis showing that the relative concentration of these nanolubricant was maintained at over 70 % compared to the initial concentration. It can be concluded from the graph and the visual sedimentation photograph result; it has clearly shown that SiO_2/PAG nanolubricant was in a stable condition over a month of observation.

The authors are grateful to the Universiti Malaysia Pahang (UMP) and the Automotive Engineering Centre (AEC) for financial supports given under RDU1403153 and RDU151411 (RAGS/1/2015/TK0/UMP/03/2).

References

1. C. R. Ferguson and A. T. Kirkpatrick, *Internal combustion engines: applied thermosciences*: John Wiley & Sons, (2015).
2. X. Li, Z. Cao, Z. Zhang, and H. Dang, "Surface-modification in situ of nano-SiO_2 and its structure and tribological properties," *Applied Surface Science,* vol. 252, pp. 7856-7861, (2006).

3. J. Eastman, U. Choi, S. Li, L. Thompson, and S. Lee, "Enhanced thermal conductivity through the development of nanofluids," in *MRS proceedings*, p. 3,(1996).
4. E. Serrano, G. Rus, and J. Garcia-Martinez, "Nanotechnology for sustainable energy," *Renewable and Sustainable Energy Reviews,* vol. 13, pp. 2373-2384, (2009).
5. R. Saidur, K. Leong, and H. Mohammad, "A review on applications and challenges of nanofluids," *Renewable and Sustainable Energy Reviews,* vol. 15, pp. 1646-1668, (2011).
6. O. A. Alawi, N. A. C. Sidik, and H. Mohammed, "A comprehensive review of fundamentals, preparation and applications of nanorefrigerants," *International Communications in Heat and Mass Transfer,* vol. 54, pp. 81-95, (2014).
7. M. Drzazga, M. Lemanowicz, G. Dzido, and A. Gierczycki, "Preparation of metal oxide-water nanofluids by two-step method," *Inż. Ap. Chem,* vol. 51, pp. 213-215, (2012).
8. S. J. Chung, J. P. Leonard, I. Nettleship, J.-K. Lee, Y. Soong, D. V. Martello*, et al.*, "Characterization of ZnO nanoparticle suspension in water: effectiveness of ultrasonic dispersion," *Powder Technology,* vol. 194, pp. 75-80, (2009).
9. B. C. Pak and Y. I. Cho, "Hydrodynamic and heat transfer study of dispersed fluids with submicron metallic oxide particles," *Experimental Heat Transfer an International Journal,* vol. 11, pp. 151-170, (1998).
10. R. S. Vajjha, D. K. Das, and D. P. Kulkarni, "Development of new correlations for convective heat transfer and friction factor in turbulent regime for nanofluids," *International Journal of Heat and Mass Transfer,* vol. 53, pp. 4607-4618, (2010).
11. W. L. Brown, "Polyalkylene glycols," *CRC Handbook of Lubrication and Tribology,* vol. 3, pp. 253-267, (1993).
12. Dow, "Material Safety Data Sheet," *Ucon Refrigerant Lubricant 213,* (2013).
13. W. H. Azmi, K. Sharma, P. Sarma, R. Mamat, S. Anuar, and V. D. Rao, "Experimental determination of turbulent forced convection heat transfer and friction factor with SiO_2 nanofluid," *Experimental Thermal and Fluid Science,* vol. 51, pp. 103-111, (2013).
14. A. Ghadimi, R. Saidur, and H. Metselaar, "A review of nanofluid stability properties and characterization in stationary conditions," *International Journal of Heat and Mass Transfer,* vol. 54, pp. 4051-4068, (2011).
15. K. Lee, Y. Hwang, S. Cheong, L. Kwon, S. Kim, and J. Lee, "Performance evaluation of nano-lubricants of fullerene nanoparticles in refrigeration mineral oil," *Current Applied Physics,* vol. 9, pp. e128-e131, (2009).
16. Y. Hwang, J.-K. Lee, J.-K. Lee, Y.-M. Jeong, S.-i. Cheong, Y.-C. Ahn, *et al.*, "Production and dispersion stability of nanoparticles in nanofluids," *Powder Technology,* vol. 186, pp. 145-153, (2008).

Development of nanolubricant automotive air conditioning (AAC) test rig

A.A.M. Redhwan[1,3,*], *W.H.* Azmi[1,2],*M.Z.* Sharif [1], *Ftwi Y.* Hagos[1]

[1]Faculty of Mechanical Engineering, Universiti Malaysia Pahang (UMP), 26600 Pekan, Pahang, Malaysia.
[2]Automotive Engineering Centre, Universiti Malaysia Pahang, 26600 Pekan, Pahang, Malaysia.
[3]Faculty of Manufacturing Engineering Technology, TATI University College, 24000 Kemaman, Terengganu, Malaysia.

Abstract. Nanolubricant been introduced in compressor might improve the performance of automotive air conditioning system. Prior testing of the nanolubricant enhancement performance, an automotive air conditioning (AAC) system test rig base on compact car has to be developed; therefore this paper presented the development process of AAC test rig. There are 15 thermocouples, 2 pressure gauges and power analyzer were assembled on the system in order to analyse its performance. The experiment was conducted with four different charged of refrigerant. The charging was based on initial weight charged. At each quantity of refrigerant charge, performance of the AAC system was evaluated by determining three important parameters which is cooling capacity, compressor work and coefficient of performance (COP). The maximum average COP is achieved at 900 RPM is 7.07. The average and maximum COP enhancement of 7.07 % and 13.34 % were achieved by applying SiO_2 nanolubricant inside the compressor.

1 Introduction

The automotive air-conditioning (AAC) is becoming a need for thermal comfort in passenger's cabin especially in equatorial countries experiencing hot and humid climate throughout the year. The AAC is expected able to remove the heat produced in the passenger compartments of the cars as quickly as possible under any given environmental condition. In automotive sectors, air conditioning compressor is the single largest auxiliary load on an automotive engine [1]. Extra load used by the AAC system implies less efficiency, causing more fuel consumption and more greenhouse gas discharges. A study in 2001 shows ACC utilization decreases mileage around 20% and builds discharged of nitrogen oxides (NOx) by 80% and carbon dioxide (CO) by 70% [1] [2].

One of the best ways to improve the efficiency of AAC is by implementing nanofluids in the refrigeration system; by replacing conventional lubricant with nanolubricant. Adding nanoparticles to the base liquids can altogether increase their transport properties and the efficiency of the system, regardless of the possibility that the impact on pressure must be

* Corresponding author: redhwan323@gmail.com

deliberately evaluated. In addition, nanolubricants can enhance their tribological properties (lubricity, against wear properties, high pressure condition) with clear advantages for the compressors.

The uses of nanoparticle in refrigerant/lubricant mixture in improving coefficient performance/cooling capacity in refrigeration system were done by few researchers [4-10]. For example, Bi, et al. [3] conducted experiment of mineral oil with TiO_2 nanoparticles mixtures as lubricant in domestic refrigerator. They concluded that 26.1% less energy consumption was achievable. The same authors also conducted experimental work for performance of a domestic refrigerator using TiO_2/R600a nanorefrigerant as working fluid. The performance of refrigerator showed 9.6% less energy used [4]. Wang, et al. [5] mixed $NiFe_2O_4$ in mineral-based nanorefrigeration oil (MNRO). The outcome of the test showed that cooling/heating EER of the RAC were increased by 6 %. Jwo, et al. [6] dispersed Al_2O_3 nanoparticles in polyester lubricants. They found that the power consumption decreased by 2.4% and the COP was enhanced up to 4.4%. Subramani and Prakash [7] dispersed Al_2O_3 in MO. It is observed that by using Al_2O_3 /R134a/MO nanorefrigerant, the power consumption was reduced by 25 % and up to 1.53 enhancement factor in the evaporator was attained. Sabareesh, et al. [8] were experimenting the effect of dispersing low concentration of TiO_2 nanoparticles in the mineral oil based lubricant. This experimental results showed that the enhancement in COP and average heat transfer rate were 17% and 3.6% respectively. Kumar and Elansezhian [9] dispersed Al_2O_3 nanoparticle in polyalkylene glycol (PAG) oil used in R134a vapor compression refrigeration system. When compared to pure R134a/PAG mixture, 0.2 vol. % concentration Al_2O_3 /R134a/PAG nanorefrigerant showing better performance of the refrigeration system with 10.32% less energy used. Even though many research works were done on the refrigeration system, none of them are using automotive air conditioning (AAC) as the test rig.

Hence in this work, an AAC system test bench was developed and built in according to SAE International standard, J2765 OCT 2008. AAC system of a compact car consists of basics refrigeration system components was selected. The Coefficient of Performance (COP) test was conducted to compare the performance of the AAC with and without nanolubricant providing the basic data for the application of the nanoparticles in the AAC system.

2 Methodology

The experimental set-up was made up of original components from the air conditioning system of a compact car, is arranged in such a way to imitate those in the actual vehicle. The experimental bench setup was made of AAC system from a compact car, electrical motor and inverter frequency controller, water bath for evaporator, piping system and various measurement instruments as shown in Figure 1. The components used are listed in Table 1.

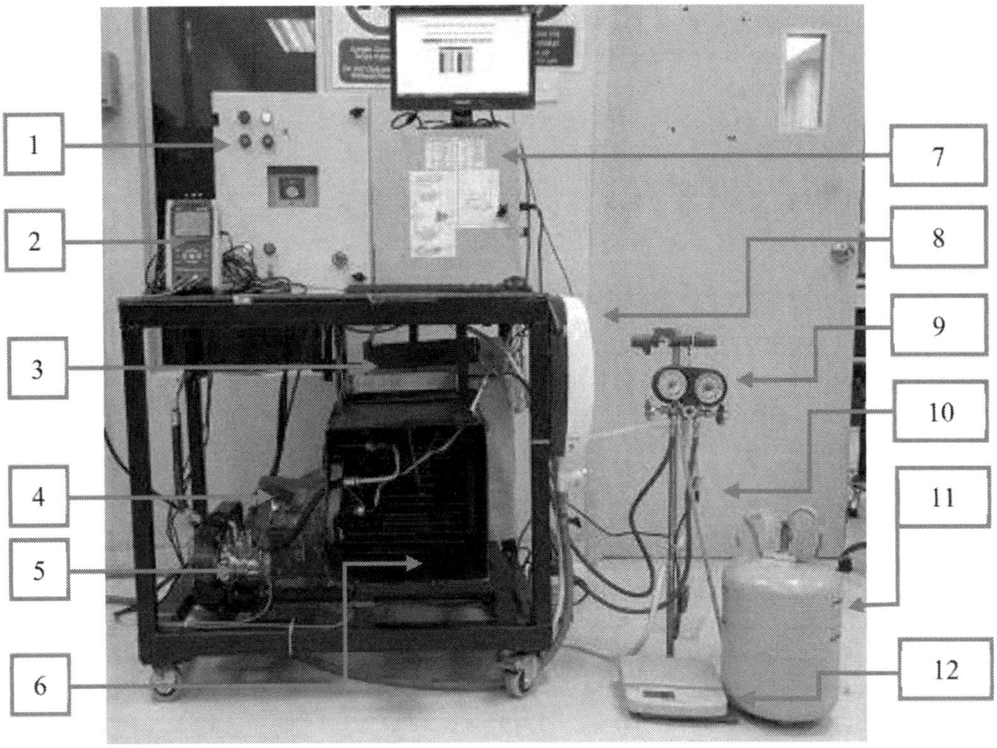

Fig. 1. The setup of AAC Test rig.

Table 1. List of component used in AAC Test rig

No.	Description	No.	Description
1	Frequency Inverter Controller	7	Data Logger Module
2	Power Analyzer	8	Water Heater
3	Evaporator (inside Water bath)	9	Pressure Gauge
4	Three phase induction motor	10	Flow Transducer
5	Compressor	11	R134a gas
6	Condenser	12	Weight Scale

The AAC system used in the present study was from a Kancil SD specification. The compressor used is a SWJ -7B08 Sanden compressor and ran with R134a as refrigerant. The compressor and its AAC system is been lubricated by Polyalkylene Glycol (PAG). The main components of AAC system were compressor, condenser, evaporator and expansion valve. The compressor pulley is driven by a 2.2 kW, 3-phase electrical motor acted upon by a frequency inverter in order to control and vary the rotational speed. The speed used in this study is varies between 900 and 2100 revolution per minute (RPM) by varying the motor's frequency. For ease and consistent evaporator temperature measurement reading, the evaporator is been immersed in water by implementing water bath system. The calculation of Coefficient of performance (COP) by immersing the evaporator in water bath is done according to ASHRAE 41.9-2000. The water bath system was equip with a fully thermal insulated tank, a flow transducer, 3-ways inlet and outlet pipes for achieving homogenous water temperature in the water bath and a water heater that regulate the water inlet temperature and flow rate. The AAC system was appropriately equip with temperature sensors, digital power analyzer and pressure gauges. The temperatures at 15 points were assessed using calibrated K-type thermocouples following to ANSI/ASHRAE Standard,

1986. Out of fifteen (15) thermocouples installed, eight (8) were installed at along of the AAC system pipeline at specific location at inlet and outlet of each main component. The calibration of thermocouple were done in contra with the Standard Platinum Resistance Thermometer (SPRT) as performed by [10] and [11]. Table 2 shows the two sets of data with different conditions of water used during calibration. The thermocouple used was of K-type, 0.3 mm diameter and aimed for temperature span between -40 to 375 °C with ±1.5 °C tolerance.

Table 2. Calibration data for thermocouple points and its location.

Thermocouple Type/ No.	Temperature Measurement (°C)		Locations/ Remarks
	Freezing	Boiling	
Platinum	0.0	100.0	Calibration set
1	0.1	100.1	$Comp_{in}$
2	0.2	100.2	$Comp_{out}$
3	0.0	100.1	$Cond_{in}$
4	0.0	100.0	$Cond_{out}$
5	-0.1	100.0	Exp. $valve_{in}$
6	0.0	100.2	Exp $valve_{out}$
7	0.1	100.2	$Evap_{in}$
8	-0.1	100.0	$Evap_{out}$
9	-0.1	99.8	Water bath $_{in}$
10	0.1	100.0	Water bath $_{out1}$
11	0.1	100.1	Water bath $_{out2}$
12	0.0	100.0	Water bath $_{out3}$
13	0.0	100.1	Air $cond_{in}$
14	0.1	100.2	Air $cond_{out}$
15	-0.1	99.7	Amb. Temp

The temperatures and water flow rate measurement were recorded and observed via a data logger. Pressure gauges used in this experiment was of bourdon tube type gauges and was calibrated. Evaporator and condenser pressure were assessed through calibrated pressure gauges. The pressure data were tracked manually. The power consumption of the AAC system is measured by using power analyzer. As stated earlier, all measurement devices were calibrated and the measurement devices were found to be working accordingly. The initial refrigerant charge is been weighted in grams. The initial charged is ranged between 100 to 130 grams. Last but not least, the temperature and humidity of the room where the test rig is allocated is been monitored and controlled between 24.5 to 25.5 °C and at 45 to 65 % respectively. A schematic diagram of the AAC system test rig is depicted in Figure 2.

Temperatures and pressures assessed at the specific points are vital associated to cooling capacity, compressor work and COP. The measured temperature and pressure were used to ascertain the specific point of enthalpy value. In this experiment, five sets of temperatures and pressures data were collected for five different refrigerant initial charges. Each set of data was repeated 3 times on different days to ensure its reliability. The temperature and pressure were assessed for 10 minutes, after the AAC was in steady-state condition as suggested by SAE International standard, J2765 OCT 2008. The enthalphy at point of interest was determined by assessing the temperature and pressure at that point. The final average enthalpy value then was acquired from the 3 repeated experiments. The calculated enthalpy value then been used in determining the cooling capacity (Q_L), the compressor work (W_{in}) and Coefficient of performance (COP) by using Eqs. (1) – (3) as shown respectively:-

$$Q_L = h_1 - h_4 \quad (1)$$

$$W_{in} = h_2 - h_1 \quad (2)$$

$$COP = \frac{Q_L}{W_{in}} = \frac{h_1 - h_4}{h_2 - h_1} \quad (3)$$

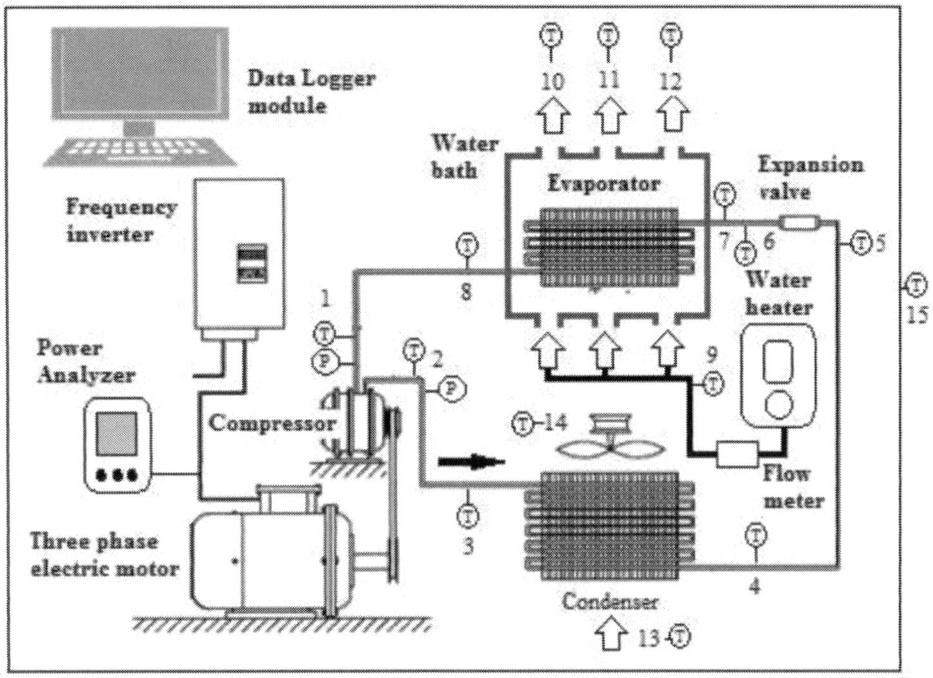

Fig. 2. The schematic diagram of AAC test rig.

3 Results and discussion

Figure 3 depicts the cooling capacity per unit mass for the test rig in the function of initial refrigerant charged for three different RPM. The cooling capacity is increased by increasing the initial charge. The cooling capacity also increases by increasing the speed of the compressor. The result data is agreed with [12]. It is observed that the cooling capacity result is increasing almost linearly with refrigerant charged. Kim et al. [13] observed a similar pattern and concluded that specific cooling capacity increased linearly with the amount charged. At lower refrigerant charge, the cooling capacity is low and it may due to the two-phase refrigerant flow from condenser to the expansion valve. Figure 4 depicts the compressor work per unit mass in the function of initial refrigerant charged. The compressor work is increasing by increasing the compressor speed. The higher the speed of the compressor, the harder work needs to be done by the compressor, reducing its efficiency.

DOI: 10.1051/matecconf/20179001050

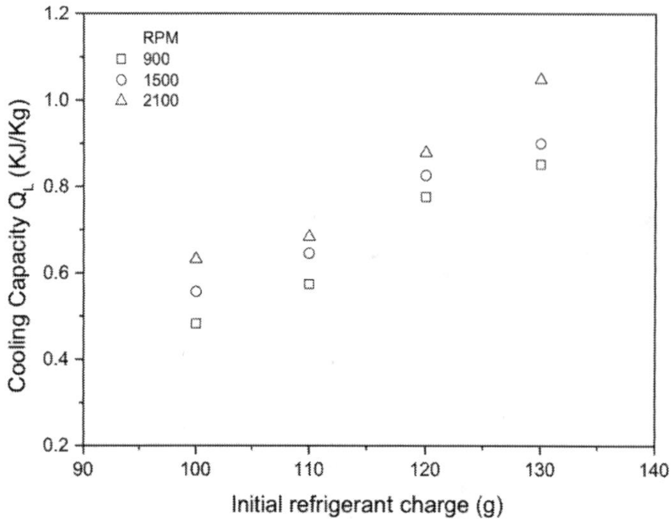

Fig. 3. Cooling capacity per unit mass in the function of initial refrigerant charge.

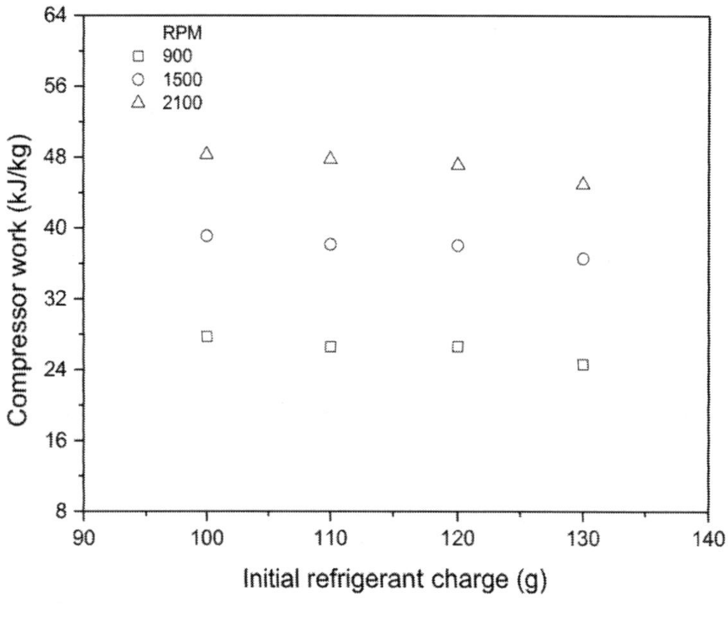

Fig. 4. Compressor work per unit mass in the function of initial refrigerant charge.

While Figure 5 shows the coefficient of performance (COP) of the test rig in the function of initial refrigerant charge. The COP is determined by using equation 3, where cooling capacity and compressor work play important roles in deciding the efficiency. From the graph, it shows that the COP is increased with the increase of refrigerant initial charge. This is due to the increasing of the cooling capacity (QL) and decreasing in the compressor work with the increasing of initial charge, hence the COP is increasing. On the other hand, the COP is decreased with the increasing of the compressor speed. The experimental results here reflect that the compressor work plays more dominant roles in deciding the efficiency of the test rig. As mention earlier, the higher the compressor speed, the higher work has to be done by the compressor and it gives negative effect to the compressor efficiency. The highest average COP is recorded at 900 RPM with the average value of 7.17. COP of the SiO_2 nanolubricant also has been determined and plotted in Figure 5. The value is then compared with the pure lubricant as a preliminary discussion. When 0.3% volume concentration PAG/SiO_2 is placed inside the compressor, the compressor work is slightly lower than an AAC with pure PAG. This is a positive result because less compressor work leads to a higher coefficient of performance (COP). The average enhancement of the COP is 7.07 % and maximum COP achieved is 13.34 %.

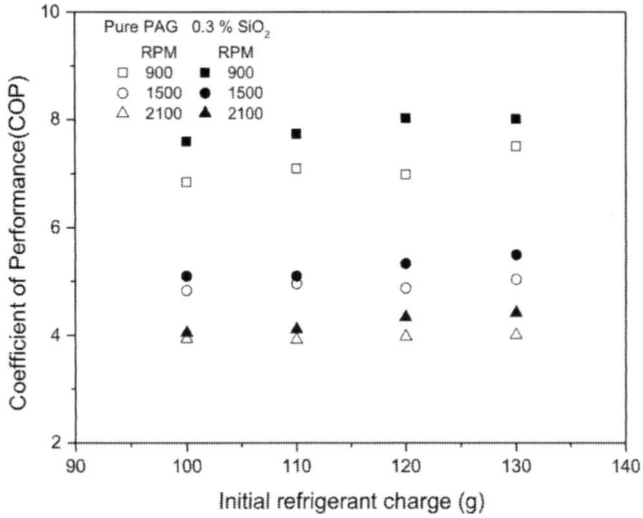

Fig. 5. Coefficient performance of the test rig in the function of initial charge.

4 Conclusions

Development of the AAC test rig has been explained thoroughly. The location and calibration of the thermocouple also have been discussed. Cooling capacity and compressor work are plotted against refrigerant initial charge before the coefficient of the performance was plotted. From experimental findings, maximum average COP of the test rig is 7.07 is recorded. The average and maximum COP enhancement of 7.07 % and 13.34 % are achieved by implementing SiO_2 nanolubricant.

The authors are grateful to the Universiti Malaysia Pahang (UMP) and Automotive Engineering Centre (AEC) for financial supports given under RDU1403153 and RDU151411 (RAGS/1/2015/TK0/UMP/03/2).

References

1. J. P. Rugh, T. J. Hendricks, and K. Koram, "Effect of solar reflective glazing on Ford Explorer climate control, fuel economy, and emissions," in *Proceedings of the International Body Engineering Conference*, 2001.

2. O. Bevilacqua, "Effect of air conditioning on regulated emissions for in-use vehicles," *Clean Air Vehicle Technology Center, Oakland, CA, Phase I Final Report Prepared for Coordinating Research Council, Inc., Atlanta, GA, CRC Project E-37,* 1999.
3. S.-s. Bi, L. Shi, and L.-l. Zhang, "Application of nanoparticles in domestic refrigerators," *Applied Thermal Engineering,* vol. 28, pp. 1834-1843, 2008.
4. S. Bi, K. Guo, Z. Liu, and J. Wu, "Performance of a domestic refrigerator using TiO_2-R600a nano-refrigerant as working fluid," *Energy Conversion and Management,* vol. 52, pp. 733-737, 2011.
5. R. Wang, Q. Wu, and Y. Wu, "Use of nanoparticles to make mineral oil lubricants feasible for use in a residential air conditioner employing hydro-fluorocarbons refrigerants," *Energy and Buildings,* vol. 42, pp. 2111-2117, 2010.
6. C. S. Jwo, L. Y. Jeng, T. P. Teng, and H. Chang, "Effects of nanolubricant on performance of hydrocarbon refrigerant system," *Journal of Vacuum Science & Technology B,* vol. 27, pp. 1473-1477, 2009.
7. N. Subramani and M. Prakash, "Experimental studies on a vapour compression system using nanorefrigerants," *International Journal of Engineering, Science and Technology,* vol. 3, pp. 95-102, 2011.
8. R. K. Sabareesh, N. Gobinath, V. Sajith, S. Das, and C. Sobhan, "Application of TiO_2 nanoparticles as a lubricant-additive for vapor compression refrigeration systems–An experimental investigation," *International Journal of Refrigeration,* vol. 35, pp. 1989-1996, 2012.
9. D. S. Kumar and R. Elansezhian, "Experimental study on Al_2O_3-R134a nano refrigerant in refrigeration system," *International Journal of Modern Engineering Research,* vol. 2, pp. 3927-3929, 2012.
10. K. Sreejith, "Experimental Investigation of A Domestic Refrigerator Having Water-Cooled Condenser Using Various Compressor Oils," *International Journal of Engineering and Science, Vols,* vol. 2, pp. 27-31, 2013.
11. T. Mohd Yusof and A. A. Azizuddin, "Performance Analysis Of A Domestic Refrigerator," 2010.
12. K. Atik and A. Aktas, "An experimental investigation of the effect of refrigerant charge level on an automotive air conditioning system," *Journal of Thermal Science and Technology,* vol. 31, pp. 11-17, 2011.
13. J. H. Kim, J. M. Cho, I. H. Lee, J. S. Lee, and M. S. Kim, "Circulation concentration of CO2/propane mixtures and the effect of their charge on the cooling performance in an air-conditioning system," *International Journal of Refrigeration,* vol. 30, pp. 43-49, 1// 2007.

Thermal conductivity enhancement of Al_2O_3 and SiO_2 nanolubricants for application in automotive air conditioning (AAC) system

A.A.M. Redhwan[1,3,*], *W.H.* Azmi[1,2], *M.Z.* Sharif[1], *N.N.M.* Zawawi[1]

[1]Faculty of Mechanical Engineering, Universiti Malaysia Pahang (UMP), 26600 Pekan, Pahang, Malaysia
[2]Automotive Engineering Centre, Universiti Malaysia Pahang, 26600 Pekan, Pahang, Malaysia.
[3]Faculty of Manufacturing Engineering Technology, TATI University College, 24000 Kemaman, Terengganu, Malaysia.

Abstract. Nanolubricant been introduced in compressor might improve the performance of automotive air conditioning system. Prior testing of the nanolubricant enhancement performance, thermal conductivity of Al_2O_3/PAG and SiO_2/PAG nanolubricants has to be investigated and compared. Al_2O_3 and SiO_2 nanoparticles first been dispersed in Polyalkylene Glycol (PAG) for different volume concentrations. KD2 Pro was used in determining the thermal conductivity of the nanolubricant. The experimental results showed that the thermal conductivity of the Al_2O_3/PAG and SiO_2/PAG nanolubricants increased by volume concentration but decreased by temperature. The highest thermal conductivity was observed to be 0.153 W.(m·K)-1 and enhancement of 1.04 times higher than the base lubricant for Al_2O_3 with 1.0 volume concentration. Finally regression equations were developed in order to estimate the thermal conductivity for these nanolubricants.

1 Introduction

Nanolubricant was produced when nanoparticles were been dispersed in refrigerant/lubricant based. Nanolubricant intended to be put in compressor might improve the performance of automotive air conditioning system. Bi et al. [1] concluded three main bonus obtained when nanoparticles were emlpoyed in refrigerant/lubricant; (1) adding nanoparticle could enhance the solvability between the lubricant and refrigerant. (2) Thermal conductivity and heat transfer characteristics of the refrigerant could be increased. (3) Nanoparticles dispersion into lubricant might reduce the friction coefficient and wear rate. The studies on mixture of nanolubricants and nanorefrigerants have shown enhanced performance compared to their based fluid. For example, Kedzierski [2] studied the influence of Al_2O_3 nanoparticles and R134a/polyolester mixtures on the pool boiling performance on a rectangular finned surface. He found that the boiling performance enhanced up to 113% on a rectangular finned surface.

* Corresponding author: redhwan323@gmail.com

Judging by the potential of nanolubricants in improving the efficiency of air conditioning and refrigeration systems, thermal conductivity measurement of potential nanolubricants is necessary not only on the fundamental research but also to the design consideration. But to the best of authors' knowledge, there are only few literatures which are available for the experimental works on thermal conductivity of nanolubricant. For example, Kedzeisky et al [3] investigated the thermal conductivity of Al_2O_3 and ZnO dispersed in polyolester based lubricant. They also summarized that thermal conductivities of nanolubricants were increased with the increased of nanoparticle mass fraction.

Contradict to the research on Al_2O_3 nanolubricant, not much research work was done on SiO_2 nanoparticle. Through literature review, it was found out only one research was done on this particular nanoparticle. R134a refrigerant was used by Henderson, et al. [4] in intention to study the effect of SiO_2 and CuO nanoparticles on the flow-boiling of R-134a and R-134a/polyolester mixtures. It is observed that the heat transfer coefficient decreases by 55 % when SiO_2 nanoparticles were directly dispersed in R134a refrigerant. The decrement of the enhancement is due to difficulties in attaining a stable dispersion. It is noticeable that conclusion made by this research make SiO_2 nanoparticle is less desirable to be explored by other scholars. If further investigation was done in showing or producing more stable SiO_2 nanolubricant/nanorefrigerant, it might open a new sub area of nanorefrigerant research.

As different base fluids have different thermal conductivity properties, the model implemented may not suit for PAG base nanolubricants purposes. Therefore if experimental data of the thermal conductivity properties of the nanolubricants are obtained, it would be used for better understanding on the enhancement of heat transfer, coefficient of performance (COP), energy saving and others.

2 Methodology

2.1 Material preparation

Nanoparticles used are Al_2O_3 and SiO_2. Al_2O_3 with 99.8 % purity and 13 nm in size while SiO_2 nanoparticles with 99.5 % purity and 30 nm in size. Both nanoparticles are procured from Sigma-Aldrich. The characterizations of these nanoparticles are obtained by the field emission scanning electron microscopy (FESEM) technique. The images of FESEM at magnification of 300,000 and 200,000 are shown in Figure 1. From the FESEM image it has been observed that both nanoparticles are in nearly spherical shape and the sizes are about 13 nm and 30 nm respectively. The properties of Al_2O_3 and SiO_2 are shown in Table 1. The properties of these particles are important to be known in order to estimate the model of physical properties such as thermal conductivity. Polyalkylene gycol (PAG) lubricant was intended to be tested in automotive air conditioning system. PAG have been used mainly in automotive air-conditioning systems due to the compatibility characteristic with most of elastomers [5]. Table 2 shows the properties of the PAG 46 lubricant at atmospheric pressure [6, 7].

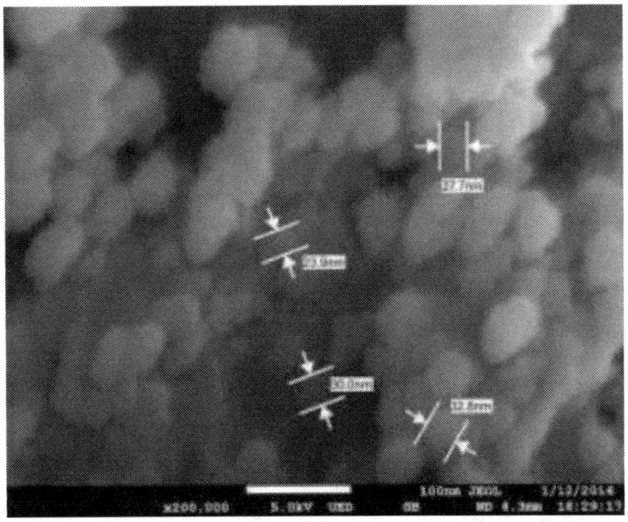

(a)

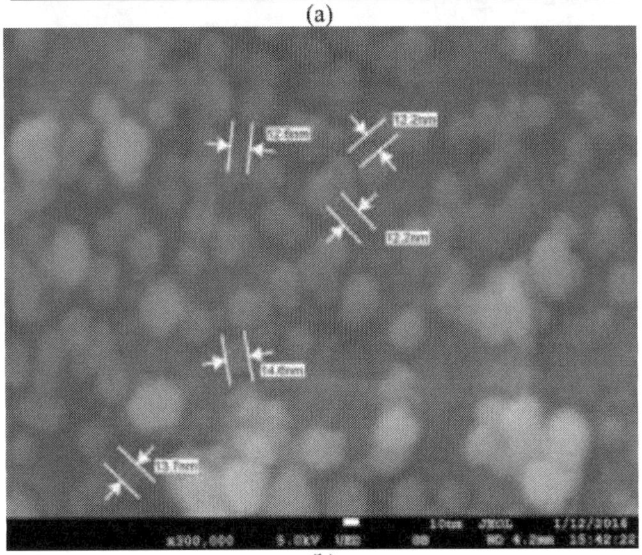

(b)

Fig. 1. Fesem image of dry nanoparticles (a) SiO_2 at 200,000 (b) Al_2O_3 and 300,000 magnification respectively.

Table 1. Properties of Al_2O_3 and SiO_2 nanoparticles.

Property	Al_2O_3	SiO_2
Molecular mass (g mol^{-1})	101.96	60.08
Average Particle diameter (nm)	13	30
Density (kg m^{-3})	4000	2220
Thermal Conductivity (W m^{-1} K^{-1})	36	1.4
Specific heat (J kg^{-1} K^{-1})	773	745

Table 2. Properties of Polyalkylene Glycol (PAG).

Property	**PAG**
Density, g/cm^3@ 20 °C	0.9954
Flash Point, °C	174
Kinematic viscosity, cSt @ 40°C	41.4-50.6
Pour Point, °C	-51

Two step method recommended by Yu et al. [8] is used in preparation of nanolubricants. The preparation of nanolubricant was according to Sharif et al. [9]. The mixing started with magnetic stirrer process for half an hour. Prior mixing, the required mass of the Al_2O_3 and SiO_2 nanoparticles to be dispersed in lubricant was measured using high accuracy electronic balance. The mixtures then are subjected to ultrasonic homogenization for specific hour(s) according to nanoparticles to ensure good dispersions of nanolubricants. Dispersion stability is observed visually after a month of preparation and found that no sedimentations were occurred in the samples as shown in Figure 2. No surfactant is been added in this experiment.

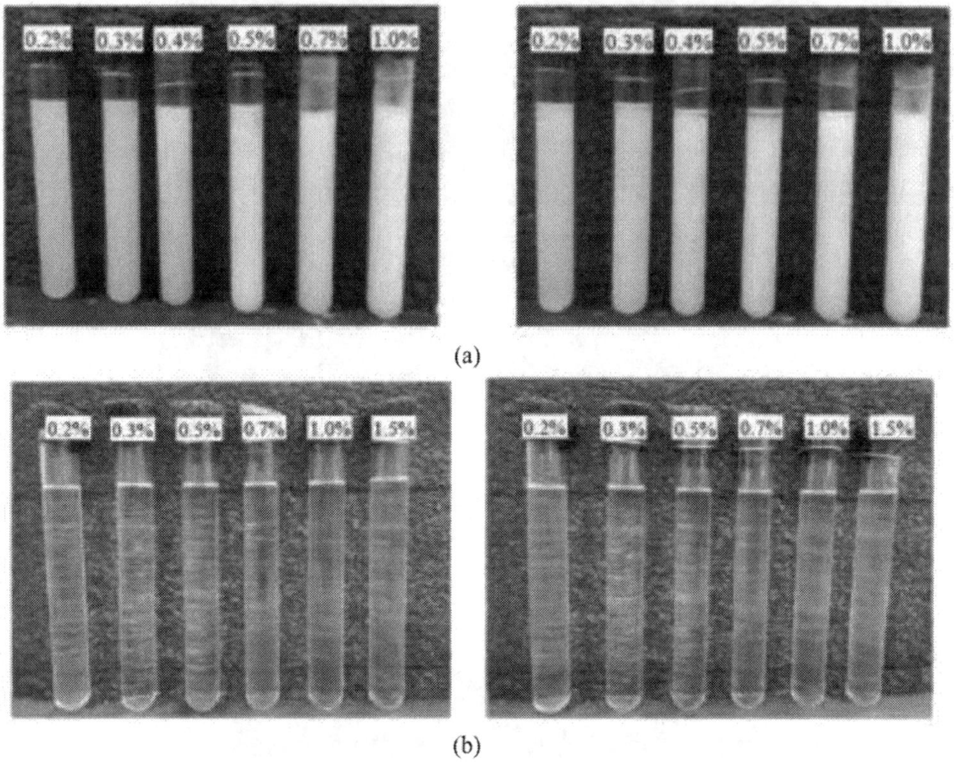

Fig. 2. (a) Al_2O_3/PAG (b) SiO_2/PAG nanolubricants at different concentrations after a month of preparation

2.2 Thermal conductivity measurement

As shown in Figure 3, KD2 Pro thermal property analyzer was used in measuring thermal conductivity. This apparatus uses the transient line heat source to establish the thermal properties of liquids and solids. The apparatus meets the standards of both ASTM D5334 and IEEE 442-1981. Single needle sensor (KS-1) in the range of 0.002 to 2.00 $W.(m \cdot K)^{-1}$ is used. A water bath of WNB7L1 model, is been used to maintain a constant temperature of the sample with accuracy of 0.1 °C [10]. The thermal conductivity of 0.2 to 1.0% volume concentrations of nanolubricants were measured for temperature range of 30 to 80 °C. The sensor was validated by measuring the thermal conductivity of the verification liquid (glycerin) given by the supplier. The measured value of glycerin at 25 °C is 0.286 $W(m \cdot K)^{-1}$, which is in accord with the calibrated data of 0.285 $W(m \cdot K)^{-1}$ and within ± 0.35 % accuracy. The validation process was done each time before the thermal conductivity measurement was taken. In order to ensure the consistency of data measurement, minimum five data were taken for every concentration at a specific

DOI: 10.1051/matecconf/20179001051

temperature. Two thermal conductivity models, Maxwell [11] and Yu and Choi [12] in equations 1 and 2 were used to verify the results of nanolubricants thermal conductivity.

$$k_r = \frac{k_{eff}}{k_{bf}} = \left[\frac{k_p + 2\,k_{bf} + 2\left(k_p - k_{bf}\right)\varphi}{k_p + 2\,k_{bf} - 2\left(k_p - k_{bf}\right)\varphi} \right] \quad (1)$$

$$k_r = \frac{k_{eff}}{k_{bf}} = \left[\frac{k_p + 2k_{bf} + 2\left(k_p - k_{bf}\right)(1+\beta)^3 \varphi}{k_p + 2k_{bf} - \left(k_p - k_{bf}\right)(1+\beta)^3 \varphi} \right] \quad (2)$$

Fig. 1. KD2 Pro Thermal properties analyzer with water bath.

3 Results and discussion

Figure 4(a) and (b) show the thermal conductivity of the Al_2O_3/PAG and SiO_2/PAG nanolubricants at 30 °C for 0.2 to 1.0% volume concentrations. The figure shows that the thermal conductivity of the nanolubricant increases with volume concentration. The experimental data were also been compared to the estimated values obtained from earlier published model in literature. The experimental values for these study were found to be slightly higher than the Maxwell [11] model. However, the model by Yu and Choi [12] seem agreed with the experimental value in some extent. The mean and maximum deviation of the experimental values of Al_2O_3/PAG nanolubricant compared to Yu and Choi [12] as depicted in Figure 4(a) is 0.36 and 0.80 %, respectively. While for SiO_2/PAG nanolubricant, the mean and maximum deviation compared to the same model as shown in Figure 4(b) is 0.34 and 0.63 %, respectively. Mahbubul et al. [13] on the hand have compared their thermal conductivity experimental value of Al_2O_3-R141b nanorefrigerant with Maxwell [11]. The results showed that their experimental data also is much higher compared to Maxwell [11] by 34% deviation. By comparing with other researcher result, the thermal conductivity result in this study is found to be in a good agreement with these two models.

DOI: 10.1051/matecconf/20179001051

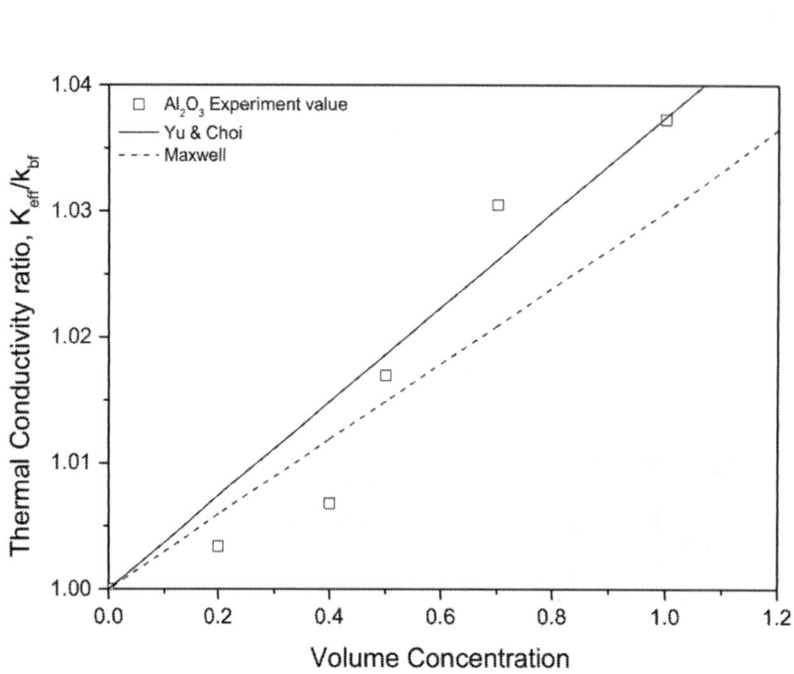

(a)

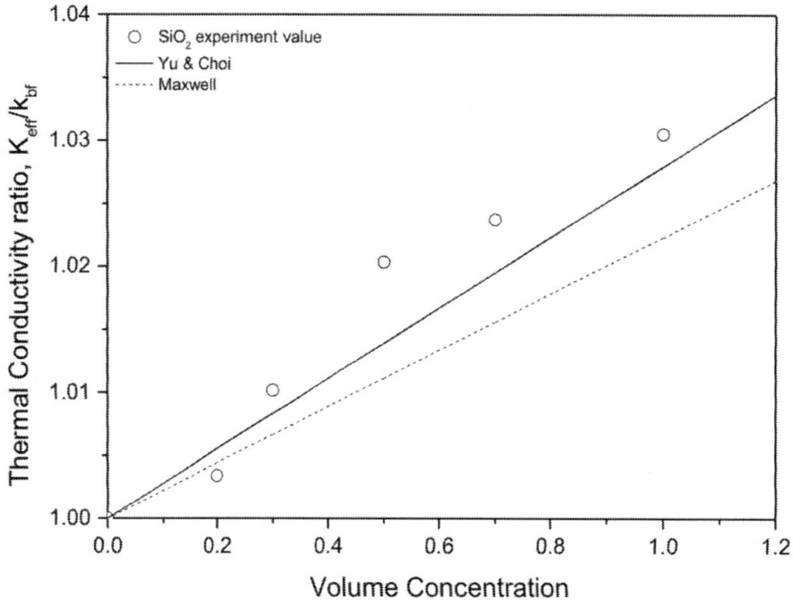

(b)

Fig. 4. Variations of thermal conductivity ratio as a function of volume concentration at 30°C (a) Al_2O_3 nanolubricant (b) SiO_2 nanolubricant.

Figure 5 shows the thermal conductivity of the nanolubricants as a function of temperature. The highest thermal conductivity is 0.153 W.$(m \cdot K)^{-1}$ achieved by Al_2O_3/PAG at 1.0% volume concentration and temperature of 30 °C. In addition, the enhancement ratio of Al_2O_3/PAG nanolubricant is 1.04 higher when compared to pure PAG under the same volume concentration and temperature. While for SiO_2/PAG nanolubricant, the maximum thermal conductivity achieved is 0.152 W.(m · K)$^{-1}$, slightly lower compared to Al_2O_3/PAG under the same condition. This is reasonable because the thermal conductivity of SiO_2 nanoparticle is 1.4 W.(m · K)$^{-1}$ lower compared to 36 W.(m · K)$^{-1}$ by Al_2O_3 nanoparticle. The enhancement ratio is 1.03, just slightly lower to Al_2O_3/PAG. This result is supported by Jiang et al. [14] when they concluded that thermal conductivities of nanorefrigerants with various types of nanoparticles are quite similar to one another if the nanoparticle volume fractions are similar.

Figure 5 also depicted that the measured thermal conductivity for all volume concentrations decreased with the increasing of temperature. The pattern is well agreed with the pure PAG behaviour as plotted using the data from Booser [15] which presented by the solid straight line in Figure 5. This behavior could be explained when liquid is heated, the molecules of the liquid move apart, hence increase the mean path. Consequently the probability of collision of molecules will be reduced. As the result, thermal conductivity decrease with increase in temperature.

Consequently, Eq. (3) is developed to estimate the thermal conductivity of nanolubricants for different volume concentrations and wide range of temperature. The correlation has an average deviation of 0.34% and standard deviation of 0.26%. The equation is in good agreement within ± 1.5% deviation compared to the experimental data and applicable for $0 \leq \phi \leq 1.0\ \%$ and $30 \leq T \leq 80\,^{\circ}\mathrm{C}$.

$$k_r = \frac{k_{NL}}{k_L} = 0.15\left(1+\frac{\phi}{100}\right)^4\left(1+\frac{T}{80}\right)^{-0.05} \tag{3}$$

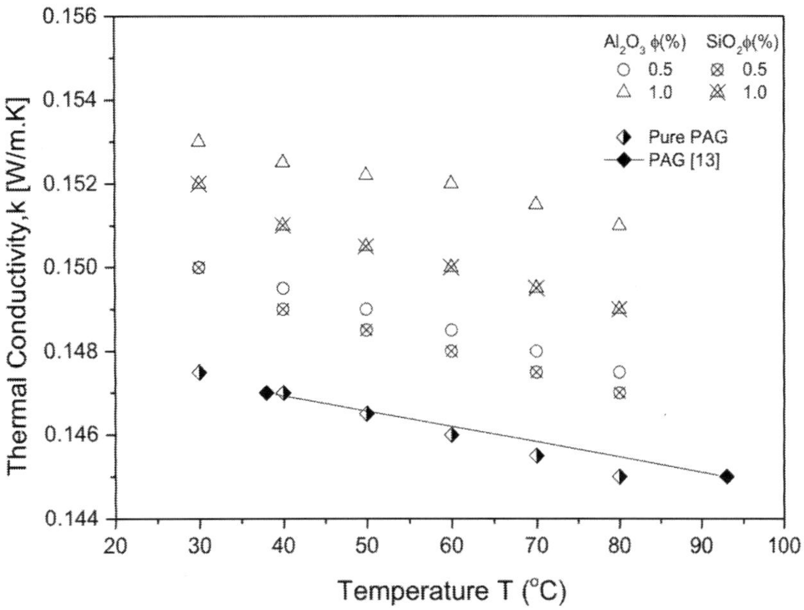

Fig. 5. Thermal conductivity of nanolubricants as a function of temperature at different volume concentration.

4 Conclusions

Thermal conductivity of nanolubricants has been studied. The experimental investigation found that the thermal conductivity of the nanolubricants increases with volume concentration, but decreases with the increase of the temperature.

Thermal conductivity of Al_2O_3/PAG and SiO_2/PAG nanolubricants also has been compared. It is found that the thermal conductivity of Al_2O_3/PAG nanolubricant is slightly higher but not significant compared to SiO_2/PAG nanolubricant. The highest value of experimental reading is recorded by Al_2O_3/PAG nanolubricant with value of 0.153 W.(m·K)$^{-1}$ with enhancement of 1.04. Referring to thermal conductivity result, the Al_2O_3/PAG nanolubricant is suggested to be used as nanolubricant in compressor because its thermal conductivity is superior compared to SiO_2/PAG nanolubricant. Further thermo-physical investigation is needed to be done in order to achieve concrete conclusion of better enhancement between these two nanolubricant. Therefore, it is recommended to further investigate the thermo-physical of these nanolubricants especially their viscosity behavior towards the increment of volume concentrations and temperature.

The authors are grateful to the Universiti Malaysia Pahang (UMP) and Automotive Engineering Centre (AEC) for financial supports given under RDU1403153 and RDU151411 (RAGS/1/2015/TK0/UMP/03/2).

References

1. S. Bi, K. Guo, Z. Liu, and J. Wu, "Performance of a domestic refrigerator using TiO_2-R600a nano-refrigerant as working fluid," *Energy Conversion and Management,* vol. 52, pp. 733-737, 2011.
2. M. A. Kedzierski, "R134a/Al_2O_3 nanolubricant mixture pool boiling on a rectangular finned surface," *Journal of Heat Transfer,* vol. 134, p. 121501, 2012.
3. M. A. Kedzierski, R. Brignoli, K. T. Quine, and J. S. Brown, "Viscosity, Density, And Thermal Conductivity Of Aluminum Oxide And Zinc Oxide Nanolubricants," *International Journal of Refrigeration.*
4. K. Henderson, Y.-G. Park, L. Liu, and A. M. Jacobi, "Flow-boiling heat transfer of R-134a-based nanofluids in a horizontal tube," *International Journal of Heat and Mass Transfer,* vol. 53, pp. 944-951, 2010.
5. P. L. Matlock, W. L. Brown, and N. A. Clinton, "Polyalkylene glycols," *Chemical Industries-New York-Marcel Dekker,* pp. 159-194, 1999.
6. W. L. Brown, "Polyalkylene glycols," *CRC Handbook of Lubrication and Tribology,* vol. 3, pp. 253-267, 1993.
7. Dow, "Material Safety Data Sheet," *Ucon Refrigerant Lubricant 213,* 2013.
8. W. Yu and H. Xie, "A review on nanofluids: preparation, stability mechanisms, and applications," *Journal of Nanomaterials,* vol. 2012, p. 1, 2012.
9. M. Z. Sharif, W. H. Azmi, A. A. M. Redhwan, and R. Mamat, "Investigation of thermal conductivity and viscosity of Al2O3/PAG nanolubricant for application in automotive air conditioning system," *International Journal of Refrigeration,* vol. 70, pp. 93-102, 10// 2016.
10. I. Zakaria, W. H. Azmi, W. A. N. W. Mohamed, R. Mamat, and G. Najafi, "Experimental investigation of thermal conductivity and electrical conductivity of Al_2O_3 nanofluid in water-ethylene glycol mixture for proton exchange membrane fuel cell application," *International Communications in Heat and Mass Transfer,* vol. 61, pp. 61-68, 2015.

11. J. C. Maxwell, *A Treatise on Electricity and Magnetism*, 2nd ed. Cambridge, U.K.: Oxford University Press, 1904.
12. W. Yu and U. S. Choi, "The role of interfacial layers in the enhanced thermal conductivity of nanofluids: A renovated Maxwell model," *Journal of Nanoparticle Research,* vol. 5, pp. 167-171, 2003.
13. I. M. Mahbubul, R. Saidur, and M. A. Amalina, "Influence of particle concentration and temperature on thermal conductivity and viscosity of Al_2O_3/R141b nanorefrigerant," *International Communications in Heat and Mass Transfer,* vol. 43, pp. 100-104, 2013.
14. W. Jiang, G. Ding, H. Peng, Y. Gao, and K. Wang, "Experimental and model research on nanorefrigerant thermal conductivity," *HVAC&R Research,* vol. 15, pp. 651-669, 2009.
15. E. R. Booser, *CRC Handbook of Lubrication and Tribology, Volume III: Monitoring, Materials, Synthetic Lubricants and Applications* vol. 3. USA: CRC Press LLC, 1994.

Simulation of storage performance on hydro-pneumatic driveline in dual hybrid hydraulic passenger car

Faizil Wasbari[1,2,*], *Rosli* Abu Bakar[1], *Leong Ming* Gan[1], *Ahmad Anas* Yusof[2], and *Zarina* Amat Jafar[1]

[1]Faculty of Mechanical Engineering, Universiti Malaysia Pahang (UMP), 26600 Pekan, Pahang, Malaysia
[2]Faculty of Mechanical Engineering, Universiti Teknikal Malaysia Melaka (UTeM), Hang Tuah Jaya, 76100 Durian Tunggal, Melaka, Malaysia

Abstract. The charging process is one of the critical processes in the hydro-pneumatic driveline storage system. It converts the kinetic energy of the vehicle braking and coasting to the compression energy. This energy is stored in the storage device called the accumulator. The system is planned to be used on the dual hydro-pneumatic hybrid driveline and applied to a hydraulic hybrid passenger car. The aim of this paper is to find the effect of charging parameters on the storage performance through simulation. Through the storage behaviour, the desirable and optimal sizing of the accumulator can be selected. The paper emphasized on the effect of pressure elevation, pre-charge pressure, effective volume, thermal reaction and required time of the accumulator's charging process. The circuit of charging process has been designed and simulated by using the hydraulic tool in the Automation Studio software. The simulation results were corroborated through the component specification for data rationality. Through the simulation, it was found that pre-charge pressure had a significant effect on the charging process. It determined the efficiency of the effective volume. The higher the pressure elevation, the higher the effective volume. Nevertheless, the more energy required to compress the nitrogen gas in the bladder. Besides, in term of volume displacement, higher volume displacement reduced charging time and lower the fluid temperature. The simulation had been positively highlighted the critical point in charging process which later on, benefited the sizing process in the component selection specification.

1 Introduction

Lately, hybrid technology has become so popular in the field of automotive due to the factor of energy-saving, environmentally friendly and also more efficient compared to conventional vehicles [1]. In the field of automotive, hybrid means the vehicle has more than one propulsion system to move the vehicle. There are several hybrid technology has

* Corresponding author: faizilwasbari13@gmail.com

been developed such as hybrid electric, hybrid flywheel, hybrid hydraulic and pneumatic. However, hybrid electric is more dominant as it has been in the phase of commercialization. Based on hybridcars.com and US department of energy report on 2015, a total of 3.5 million number of the hybrid electric vehicle were sold by several manufacturers such as Toyota, Honda, Audi, Volkswagen, BMW and much more in the United States from 1999 until December 2014 [2][3]. Hybrid pneumatic and flywheel are still in the phase of research and development. Hybrid hydraulic turns popular among heavy vehicles such as buses and garbage trucks [4][5][6]. This statement is entirely realistic because it is a power density type of energy. It can drive almost all kinds of vehicles. This technology has been in the market and becomes increasingly popular in its categories. So far, innovation of hydraulic hybrid technology is focused on applying the technology to a passenger car [7]. Many research were done to ensure the success of this technology. However, it is still in the phase of research and development [8][9].

The hydraulic hybrid technology uses ICE and hydraulic motor as a propulsion while the hydro-pneumatic accumulator acts as power storage. Many types of research have been conducted related to this technology and it can be classified into four categories. There are internal combustion engine (ICE), hydro-pneumatic driveline, transmission system, and control system as illustrated in Fig. 1 [10][11][12]. The aim of this paper is to focus on the study of the hydro-pneumatic driveline. The driveline consist of five subsystems called propulsion, a regenerative system, storage, transmission, and control subsystem. The propulsion is a mechanism to move the vehicle. The regenerative is utilizing braking and coasting energy and changes them into the compression energy. The energy is then stored as potential energy in the hydro-pneumatic accumulator. The stored energy will be channelled to the second propulsion unit when the energy is required. The last subsystem is the main control system. It controls all activities in the hybrid system.

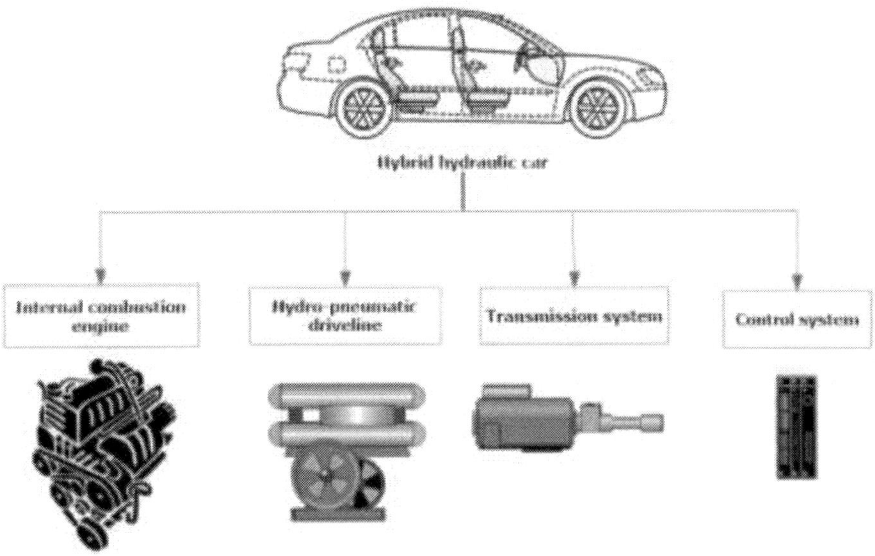

Fig. 1. Research regarding hydraulic hybrid car mostly focuses on the sub-systems breakdown.

One of the key components for hydro-pneumatic driveline is a storage system. It consists of the accumulator, pressure valve and safety block. More detail, the accumulator comprises a casing, bladder with 99.9% dry nitrogen, pressure valve and safety block. Casing design is based on the pressure vessel standard because it involves security matters. Bladder made out of an elastomer material to withstand high-pressure hydraulic oil. Nitrogen gas used as compression medium because it has a high heat absorption resistance to pressure changes. Air do not have endurance in such a way, and it can cause an explosion. The pressure gauge is used as an indicator of the pressure level in the accumulator. Last but not least is safety block. It acts as a pressure relief valve to protect the accumulator from the over-pressure. With this component, charging operation is safer and more reliable. Storage system carries out two main functions called charging and discharging process [13]. The charging process involves compression of hydraulic oil to the nitrogen bladder. This process creates compression energy which was then referred as potential energy. Along the compression process, the temperature changes are imperative for review. In comparison with pneumatic compression, hydro-pneumatic compression produces less heat because hydraulic oil is incompressible and also, act as a coolant to absorb the heat generated by nitrogen compression. It can be said that the system is like built-in with a heat exchanger [14].

Lammert et al. (2014) conducted lab scale experiment to parcel delivery truck. The truck used an 83.3 L accumulator that operated at 241 to 276 bar. He found that the configuration was able to increase 19% to 52% of fuel consumption in the diesel engine while 30% to 56% saving in the gasoline engine [15]. Kepner (2002) had been using the 54.5 L gas bladder accumulator on the 5.4 L V8 sports utility vehicle. The accumulator operated at 172 to 345 bar. He found that the arrangement was likewise able to cut the emission that will affect pollutions [16]. Boretti & Zanforlin (2014) took their first step to simulate the hybrid system in passenger car application. The high-pressure accumulator operates between 135 to 485 bars and the low-pressure tank 3.5 to 13.5 bars. The accumulators have a low specific energy (1.8-4 Wh/Kg) but a wide range of specific power (7-4,000 W/kg). They found that the system was able to achieve 30% better fuel economy [8]. The study has contributed significantly to the research for developing hydro-pneumatic hybrid driveline, but none of the studies related specifically to the subsystem. References also do not state clearly why the related pressure was used.

To answer the above statement, the Automation Studio software has been used to simulate the effect of fluid power parameters on the performance of charging process. Automation Studio is a tool for design, simulation of complex automation, documentation, and training. The software includes hydraulic, pneumatic and electrical operative devices as well as a command part diagram. It also provides technical and commercial data for simulation. Undoubtedly that the software such as Matlab has high flexibility to run a simulation that is related to the case study, but the Automation Studio has the advantage in terms of circuit design, functional simulation, fluid power component sizing, system design, validation and virtual simulation [17]. These advantages had resulted the Automation Studio more suited for this project. Currently, Automation Studio has improved one step ahead by connecting some applications with Simulink, Matlab [18]. It is hoped that more efforts will be made to improve the scope of the software application.

At the moment, many types of research focus on the simulation of the fuel consumption in the hybrid passenger car [19][20][8][21]. Nonetheless, a few only carrying out the detailed research related to the subsystem itself. Therefore, this research emphasizes on the effects of fluid power parameters to the performance of charging process on the hydro-pneumatic driveline. By knowing the effects of each parameter, perhaps, the process of selecting the right storage size and capacity can be conducted with more accurate. The accuracy is important because it ensures that the storage system always safe and operate at optimum levels. In the future, the hydro-pneumatic driveline is planned to be adopted on the dual hybrid hydraulic system as shown in Fig. 2.

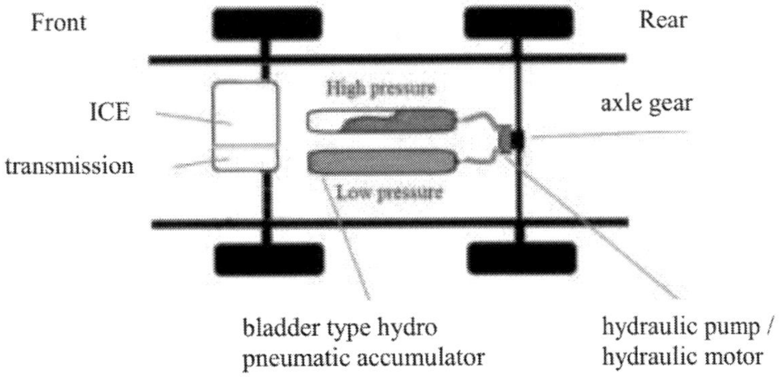

Fig. 2. Configuration of dual hybrid hydraulic car.

2 Methodology

The task started by designing a schematic diagram by using Automation Studio software. There are two types of simulation involve in the design which is functional and analysis simulation. Functional simulation is to ensure operation of the charging function works as desired while analysis is to get output data when all specifications are included. The analysis is divided into two. First, the internal effect of nitrogen bladder on compression and second is the effects of the schematic system to charging performance. For the internal compression effects on nitrogen bladder, the analysis made by using equations derived from Boyle-Mariotte's law for gasses [22][14], which takes into account the compressibility of the gas as illustrated in Fig. 3.

$$\Delta V_{ideal} \leq V_1 - V_2 \tag{1}$$

where ΔV_{ideal} is the effective volume, V_1 is volume before compression and V_2 is volume after compression.

$$p_0 \times V_0 = p_1 \times V_1 = p_2 \times V_2 \tag{2}$$

where p_0, p_1, p_2 is pressure at a certain stage of compression and V_0, V_1, V_2 is the volume at a certain stage of compression. Refer to **Error! Reference source not found.** for detail stage of charging. However, if the condition is not in isothermal then, the equation is as follows,

$$V_{effective} = \Delta V_{ideal} \cdot K_i \tag{3}$$

where $V_{effective}$ is effective volume, ΔV_{ideal} is ideal volume under isothermal and K_i is correction factor.

This is the simplified equation to calculate the average time required for the filling process. It is assumed that the flow entering the accumulator is constant and transient effect is neglected.

$$V_{effective} = Q_i \cdot t_i \tag{4}$$

where $V_{effective}$ is effective volume, Q_i is the flow rate and t_i is the filling time. The equation (4) is then rearranged,

$$t_i = \frac{\Delta V_{effective}}{Q_i} \quad (5)$$

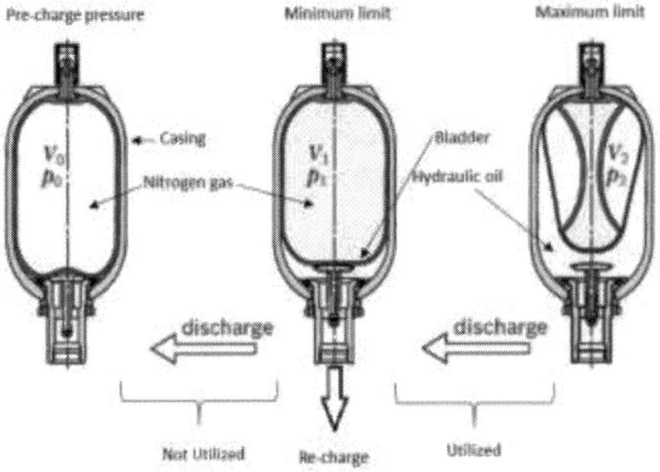

Edited source: https://www.boschrexroth.com/[14]

Fig. 3. Bladder condition during charging and discharging.

Fig. 4 shows the schematic that has been used to simulate the accumulator's process of charging. The diagram considers the flow from the pump is constant at all points of the system. Charging process starts with pumping fluid from pump to accumulator unit. In this process, the pressure change occurred. The pressure different through the pump can be expressed as follows [22],

$$\Delta p = p_{out} - p_{in} \quad (6)$$

where Δp is the pressure different (N/m^2), p_{out} is pump output pressure (N/m^2) and p_{in} is pump input pressure (N/m^2). Since the main function of the pump is to create flow, the actual output flows produced by the pump can be expressed as

$$Q_{out} = C \cdot \omega \cdot \eta_V \quad (7)$$

where Q_{out} is pump output flow rate (m^3/s), C is displacement (m^3/rad), ω is nominal speed (rad/s) and η_V is volumetric efficiency. Then, equation (8) is used to calculate the shaft torque required by the pump.

$$T = \frac{C \times \Delta p}{\eta_M} \quad (8)$$

where T is torque (Nm), C is displacement (m^3/rad), Δp is the pressure different and η_M is mechanical efficiency. Shaft power for the pump can be calculated by inserting equation (8) into equation (9).

$$P_S = T_A \cdot \omega \cdot \eta_{Overall} \quad (9)$$

where P_S is shaft power (Watt), T_A is torque (Nm), ω (rad/s) and $\eta_{Overall}$ is overall efficiency. Theoretical fluid power produced by the pump can be expressed as

$$P_P = Q_{out} \cdot \Delta p \cdot \eta_{Overall} \tag{10}$$

where P_P is pump power (Watt), Q_{out} is pump output flow rate (m^3/s), Δp is the pressure different and $\eta_{Overall}$ is overall efficiency. The pressure relief valve (PRV) works as a safety valve. The pre-set pressure is the maximum pressure that is allowed to be working in the system. If the charging pressure were exceeded, the PRV would be opened to stabilize the pressure. So, the overpressure never occur.

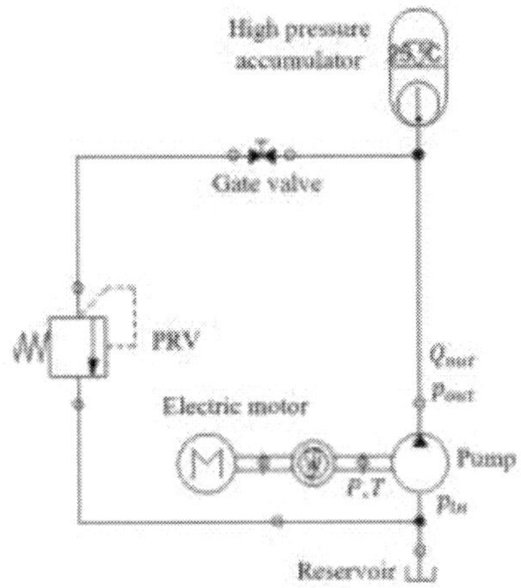

Fig. 4. Schematic diagram for charging process.

Based on the literature, the prescribed pressure is set from 0 to 400 bar while pre-charge pressure is from 10, 30 and 50 bar. If the value of pre-charge is too high, it produces the greater resistive force in the filling or charging process. This will give impact to the power to be supplied. Flow rate affects directly to the charging process, however in operation, the volume displacement is more suitable because it is associated with the specification of the pump. The combination of volume displacement and rotational speed produce flow rate. The volume displacement facilitate determination of the pump specifications including the speed and the appropriate type of pump to be used. The output is expected to present the effects of pressure elevation, pre-charge pressure and volume displacement to the storage performance which is effective volume, charging time and temperature.

3 Simulation

Simulation was done by using Automation Studio 6.1 software. Simulation is based on a Graphical User Interface (GUI) which is integrated with the code and equation. The equation is flexible which can be programmed by changing the root equation setting. All specifications for components, fitting, and measurement tool can be inserted into the tool's setting. Since the study involves bulk modulus and heat transfer, so the two functions were enabled. Effect of bulk modulus was considered because the present of compression process in the bladder. By enabling the bulk modulus setting, the simulation considers the effects of pressure increase respect to the volume in charging and discharging process. The thermal setting takes into account the fluid heating in contact with the pipe component inner surface. This simulation assumed that the heat transformation is isothermal where the fluid temperature remains constant. The heat exchange between gas and atmosphere is instantaneous. Pressure elevation is set as independent variable, temperature, effective volume and charging time as dependence variable. As for pre-charge pressure and volume displacement, the parameters are set as a control variable. For this project, the specifications used are as Table 1.

Table 1. Simulation parameters.

Motor rotational speed	*2000 RPM*	Initial liquid volume	*0 liter*
Pump thermal efficiency	*90%*	Accumulator int. diameter	*225 mm*
Pump heat transfer coeff.	*10 W/m2K*	Type of process	*Isothermal*
Hydraulic oil	*Hyd. AW-32*	Gas type	*Nitrogen*
Ambient temperature	*25°C*	Gas temperature	*25°C*
Cracking pressure	*400 bar*	Hydraulic line	*Steel*
Port 1 (C_v)	*12*	Line type	*NPS ¼-DN8*

4 Results and discussion

4.1 Effective volume

Fig. 5, Fig. 6 and Fig. 7 have shown the effects of effective volume with regards to the pressure change for V_d= 10 cm^3/rev and V_d= 50 cm^3/rev at a pre-charge pressure of 10 bar, 30 bar and 50 bar. The graphs show that the effective volume has proportional relation to the pressure elevation. At 10 bar pre-charge pressure, it was found that the filling process will only start after preceding pressure of 10 bar. Initially, hydraulic oil pressure is at the level of negative (-ve) due to its level of pressure is lower than pre-charge value [14]. While the pressure increased, the hydraulic oil pressure will be on the same footing with the value of pre-charge. At this stage, the pressure difference is zero then; there is nothing happened even little compression. After the pressure exceeds the pre-charge value, the pressure difference is positive (+ve), and the filling process begins. The selfsame concept also happens to pre-charge 30 bar and 50 bar. This is proved through the manual obtained from manufacturer Bosch and Parker [23][14]. Regarding bulk modulus, the nitrogen gas bulk modulus is low, so it has high compressibility. That means it has a higher level of compression maneuverability than hydraulic oil. Thus, the hydraulic oil is a medium to compress nitrogen gas in the bladder.

The effect of compression can be seen on two conditions; first, on the level below 100 bar and second, the level of effective maximum volume. At the level below 100 bar, the graph with pre-charge 10 bar experienced sharp increases from 0 to 40 liters compared to pre-charge 30 to 50 bar. This happens because of the value of the compression resistance of 10 bar is lower than 30 and 50 bar. Therefore, it is faster and more easily compressed. It can be sorted as compression resistance value of pre-charge 10bar < 30 bar < 50 bar. There is research found 30% losses in stored energy [24]. The maximum effective volume level does not change when the volume displacement increase but it is becoming less when the pre-charge is increased. Manual for accumulator also supports this statement. In Fig. 6 and Fig. 7, there had been a small bump at the negative pressure different. It happens because the higher the value of pre-charge led to the higher resistance to the compression. The scenario creates a condition less-refuse between both medium and results in fluctuation. This value forms the small bump at the beginning of the graph. With 10 bar pre-charge, the average effective volume efficiency about 96%. When the pre-charge was increased to 30 bar, the efficiency was reduced to 89% and at 50 bar pre-charge, the efficiency reduced to 86%.

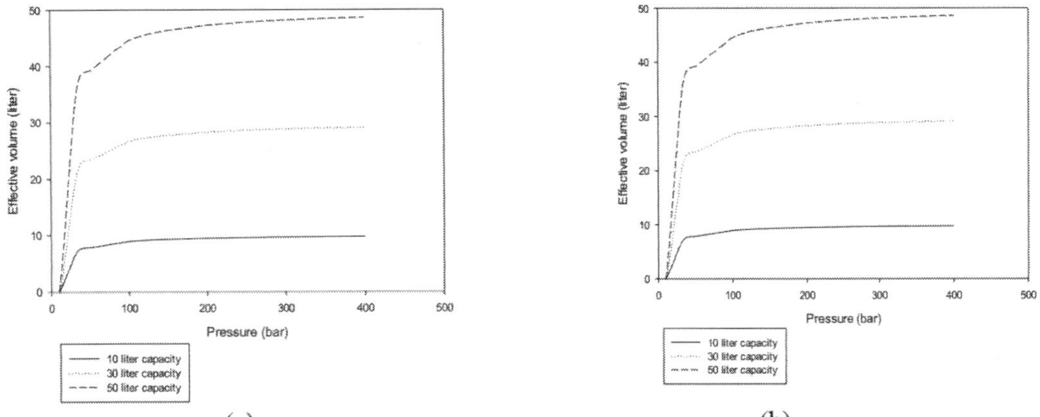

(a) (b)

Fig. 5. Effects of effective volume (liter) with regards to pressure change for V_d= 10 cm^3/rev (a) and V_d= 50 cm^3/rev (b) at pre-charge pressure of 10 bar.

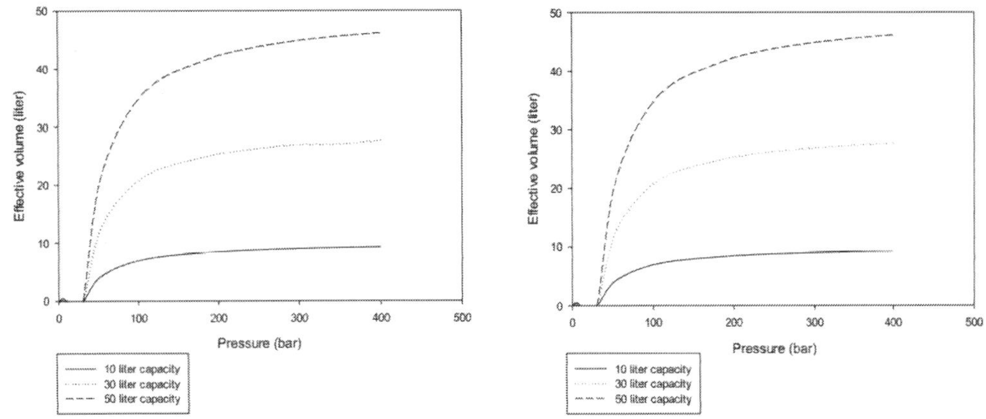

(a) (b)

Fig. 6. Effects of effective volume (liter) with regards to pressure change for V_d= 10 cm^3/rev (a) and V_d= 50 cm^3/rev (b) at pre-charge pressure of 30 bar.

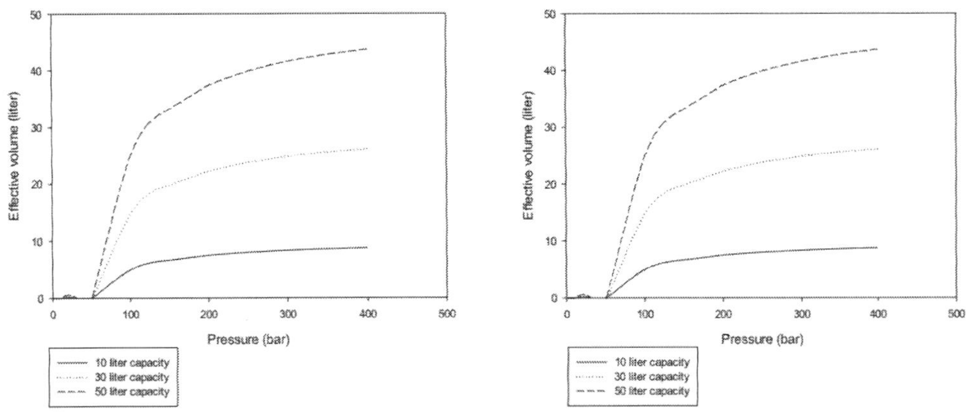

(a) (b)

Fig. 7. Effects of effective volume (liter) with regards to pressure change for V_d= 10 cm^3/rev (a) and V_d= 50 cm^3/rev (b) at pre-charge pressure of 50 bar.

4.2 Temperature effect

The effects of pressure elevation, pre-charge pressure and volume displacement on temperature elevation were shown in Fig. 8, Fig. 9 and Fig. 10. Overall, the temperature of volume displacement 50 cm^3/rev is lower compared to volume displacement 10 cm^3/rev for all pre-charge pressure 10, 30 and 50 bar. The lowest temperature is at negative (-ve) pressure different which means lower than pre-charge value and before compression happened. The supreme temperature for all case is 55°C at pre-charge 10, 30 and 50 bar, volume displacement 10 cm^3/rev and capacity 50 liters. It was about 30°C different from the ambient temperature. The statement that the compression process causes the rise in temperature can be proved here. Before pre-charge pressure, the temperature is at ambient and unchanged. However, when the pressure increased exceeded the pre-charge value, this is the start of the rise in temperature. Thus, the assertion that the compression process leads to the gas temperature to rise is correct and proven [24]. However, there is one situation where the occurrence of a sudden spike of temperature happened. It was at the volume displacement 50 cm^3/rev, pre-charge 10 bar and 10 liters capacity. Significant elevation occurred at 300 bar to 400 bar. The rise happened because of the occurrence of over-compression on the bladder. Since the bladder is usually made of elastomer, thus, it is elastic. However, it has a certain limit of elasticity which if the pressure exceeds the limits then it can damage the bladder or perhaps reduced the service life. No wonder some manufacturers limits the pressure difference to keep the bladder from over stressed, leading to excessive compression and strong gas heating.

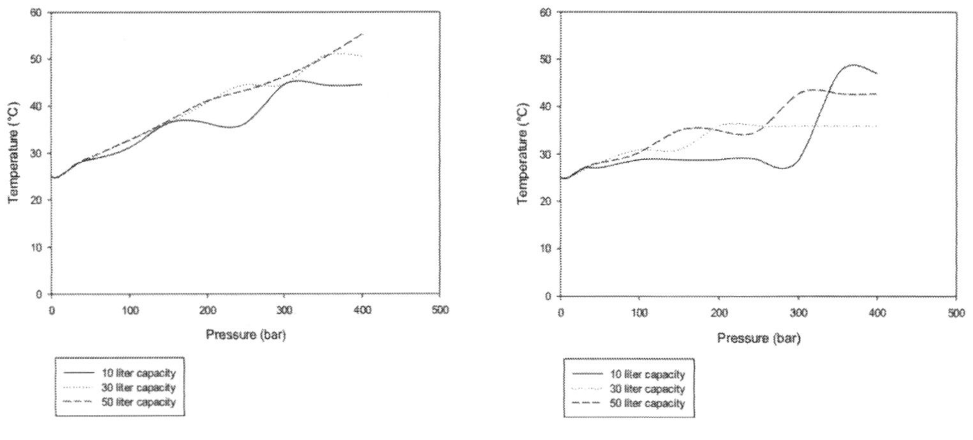

(a) (b)

Fig. 8. Temperature elevation (°C) for V_d= 10 cm^3/rev (a) and V_d= 50 cm^3/rev (b) at pre-charge pressure of 10 bar.

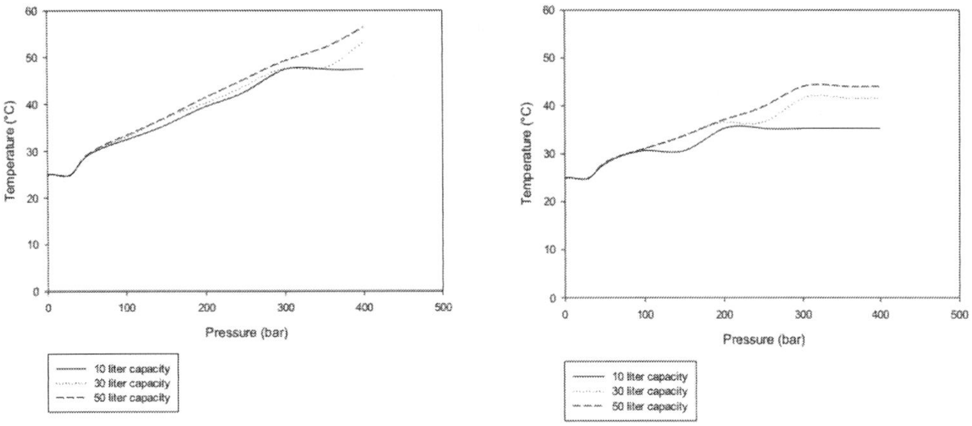

(a) (b)

Fig. 9. Temperature elevation (°C) for V_d= 10 cm^3/rev (a) and V_d= 50 cm^3/rev (b) at pre-charge pressure of 30 bar.

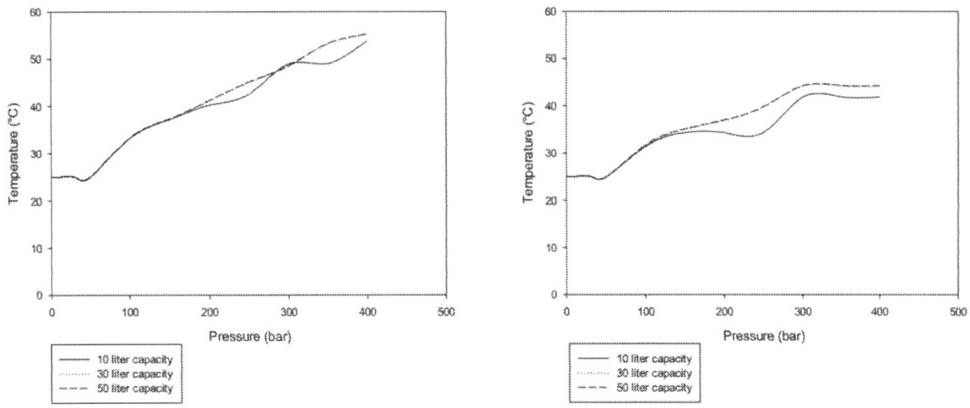

(a) (b)

Fig. 10. Temperature elevation (°C) for V_d= 10 cm^3/rev (a) and V_d= 50 cm^3/rev (b) at pre-charge pressure of 50 bar.

4.3 Charging time

Fig. 11, Fig. 12 and Fig. 13 have shown the effects of pressure elevation, pre-charge pressure and volume displacement on charging time. Overall, the volume displacement V_d= 50 cm^3/rev for each 10, 30 and 50 bar pre-charge produces shortest charging time compared to V_d= 10 cm^3/rev. This difference is due to the flow rate factor. Pump that has a high displacement volume can fill the storage at a faster rate. Therefore, the volume displacement is imperative and needs to be considered during the selection of the pump type and capacity while performing the process of sizing. The greater the capacity of the storage, it will take longer to be filled. The graph shows the existence of negative pressure differences in each level. It happens because of the higher the value of pre-charge then the higher resistance to the compression. The scenario creates a condition less-refuse between both medium and results in fluctuation. This value forms the small bump at the beginning of the graph. The graph also shows that the charging process takes quite long to compress 0

DOI: 10.1051/matecconf/20179001052

to 100 bar relative to 100 bar to 400 bar. This possibility is related to the characteristics of bladder elasticity and also nitrogen gas properties. The important thing that should be avoided for the charging process is the use of an unsuitable pump. If the volume displacement is small but storage capacity and pre-charge pressure are high, then the filling process took a very long time to complete as shown in Fig. 13(a).

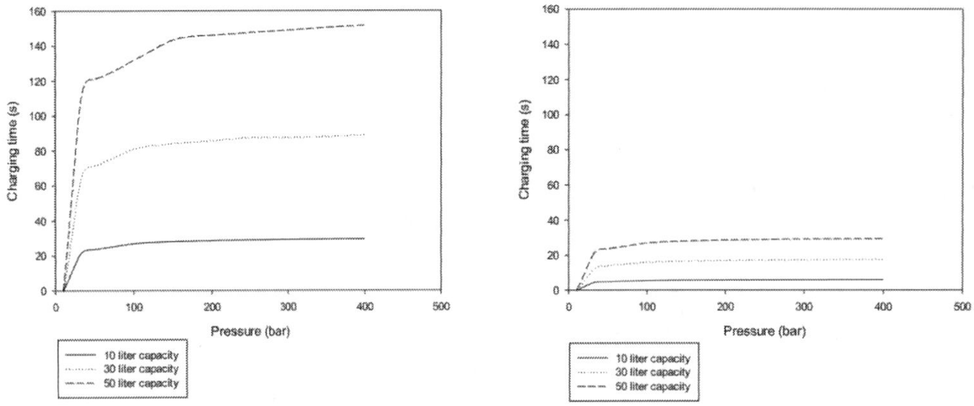

(a) (b)

Fig. 11. Charging time for V_d= 10 cm^3/rev (a) and V_d= 50 cm^3/rev (b) at pre-charge pressure of 10 bar.

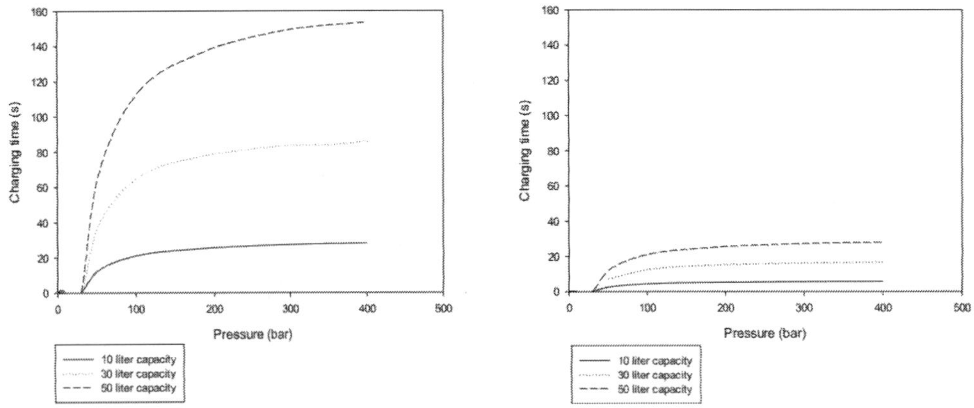

(a) (b)

Fig. 12. Charging time for V_d= 10 cm^3/rev (a) and V_d= 50 cm^3/rev (b) at pre-charge pressure of 30 bar.

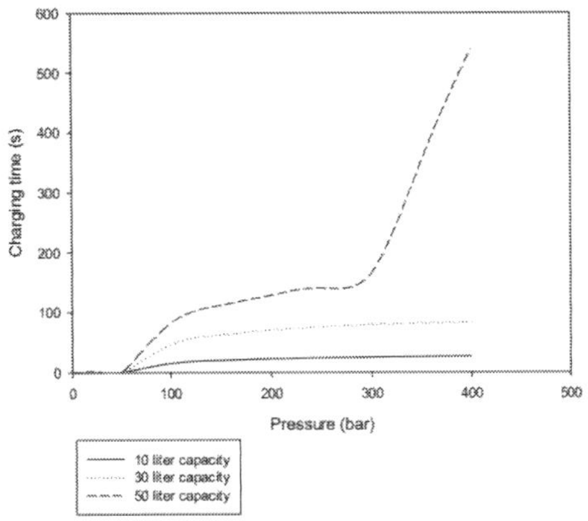

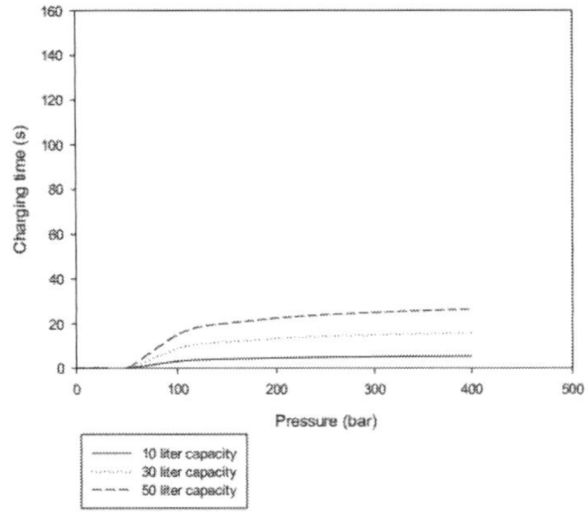

(a) (b)

Fig. 13. Charging time for V_d= 10 cm^3/rev (a) and V_d= 50 cm^3/rev (b) at pre-charge pressure of 50 bar.

5 Conclusions

In conclusion, the simulation result has managed to show some of the critical parameters which affect the performance of charging process. Effective volume relationship has shown that volume displacement does not influence the level of volume in accumulator instead pre-charge value do. The increase in pre-charge value gives adverse effect to the effective volume. Effect of temperature witnessed that the higher the volume displacement, the lower the heat in the system. In most cases, the larger the storage capacity, the higher the resulting temperature. However, this effect can be reduced by raising the volume displacement. In this context, pre-charge doesn't give significant effect to the temperature. The effect of charging time has shown that the higher the volume displacement causes the increasing of time required for charging. However, the size of the storage capacity has an adverse effect on the charging time. One thing to avoid in this process is the combination of low volume displacement, high storage capacity, and high pre-charge pressure because it gave severe impact to the charging time in pressure elevation around 300 to 400 bar. Charging time increased from 7 to 21 times greater than the 10-litre storage, 10 cm^3/rev displacement, and 50 bar pre-charge setting.

This paper was made possible by a scholarship from the Ministry of Higher Education and Universiti Teknikal Malaysia Melaka. The authors also would like to be obliged to Universiti Malaysia Pahang for giving the opportunities and providing facilities to complete this study.

References

1. K. D. Huang, S. C. Tzeng, Appl. Energy, **80**, 47 (2005)
2. J. Cobb, March 2016 Dashboard, http://www.hybridcars.com/, 11-May-2016
3. *US HEV sales by model*, http://www.afdc.energy.gov/data/10301. 11-May-2016
4. A. Boretti, J. Stecki, SAE Commer. Veh. Eng. Congr., (2012)
5. T. Lin, Q. Wang, B. Hu, W. Gong, Automation in Construction, **19**, 11 (2010)
6. P. Mrdja, N. Miljic, S. J. Popovic, M. Kitanovic, V. Petrovic, *Green Design Conference*, 129 (2012)
7. W. Backé, *Proc. Inst. Mech. Eng.*, **207**, 193 (1993)

8. A. Boretti, S. Zanforlin, SAE Int., (2014)
9. H. Holloway, *Peugeot 2008 Hybrid Air prototype first drive review*, http://www.autocar.co.uk/car-review/peugeot/2008/first-drives/peugeot-2008-hybrid-air-prototype-first-drive-review, 30-Jan-2015
10. U. Diego-ayala, *An investigation into hybrid power trains for vehicles with regenerative braking*, (2007)
11. Z. Dimitrova, P. Lourdais, F. Mar, Energy, **86**, 574 (2015)
12. F. Tavares, R. Johri, Z. Filipi, J. of Eng. Gas Turbines Power, **133**, no. 12, (2011)
13. F. Tavares, *Thermally Boosted Concept for Improved Energy Storage Capacity of a Hydro - Pneumatic Accumulator*, (2011)
14. *Accumulators*, https://www.boschrexroth.com/en/us, 11-May-2016
15. M. P. Lammert, J. Burton, P. Sindler, A. Duran, SAE Int. J., **4**, no. 1, (2014)
16. R. P. Kepner, SAE Tech. Pap., (2002)
17. *Automation Studio*, http://www.automationstudio.com/, 11-May-2016
18. *Automation Studio Target for Simulink*, http://www.br-automation.com/en/products/software/automation-studio-target-for-simulink/, 11-May-2016
19. J. Ma, H. Schock, U. Carlson, A. Hoglund, M. Hedman, SAE Tech. Pap. (2006)
20. J. Zhang, C. Lv, J. Gou, D. Kong, *Proc. Inst. Mech. Eng.*, **226**, 1289 (2012)
21. P. Achten, G. Vael, M. I. Sokar, T. Kohmäscher, *Proc. JFPS*, 47 (2008)
22. A. Esposito, *Fluid Power with Applications*, (2013)
23. *Bladder Accumulators*, http://www.parker.com/portal/site/, 11-May-2016
24. P. Puddu, M. Paderi, Energy, **57**, 326 (2013)

Diesel engine performance and exhaust emission analysis using diesel-organic germanium fuel blend

Zulkifli Syafiq[1], *Othman* Fahmi[1] , *Nurul* Syuhaida[1] , *Ang* F. Chen[1] and *Abdullah* Adam[1,2,*]

[1]Faculty of Mechanical Engineering, Universiti Malaysia Pahang, 26600 Pekan, Pahang, Malaysia
[2]Automotive Engineering Center, Universiti Malaysia Pahang, 26600 Pekan, Pahang, Malaysia

Abstract. Alternative fuels such as biodiesel, bio-alcohol and other biomass sources have been extensively research to find its potential as an alternative sources to fossil fuels. This experiment compared the performance of diesel (D), biodiesel (BD) and diesel-organic germanium blend (BG5) at five different speeds ranging from 1200-2400 rpm. BG5 shows significant combustion performance compared to BD. No significant changes of power observed between BG5 and BD at a low speed (1200 rpm). On the contrary, at higher speeds (1800 rpm and 2400 rpm), BG5 blend fuel shows increased engine power of 12.2 % and 9.2 %, respectively. Similarly, torque shows similar findings as engine power, whereby the improvement could be seen at higher speeds (1800 rpm and 2400 rpm) when torque increased by 7.3 % and 2.3 %, respectively. In addition, the emission results indicated that for all speeds, CO_2, and NO had reduced at an average of 2.1 % and 177 %, respectively. Meanwhile, CO emission had slightly increased compared to BD at low speeds by 0.04 %. However, the amount of CO released had decreased at an average of 0.03 % as the engine speed increased. Finally, measurement of O_2 shows an increment at 16.4 % at all speed range.

1 Introduction

Nowadays, the use of diesel engines in motor vehicles and electric generators has tremendously increased due to its higher efficiency and better performance compared to gasoline engines. Nevertheless, the diesel engine produces major problems associated with combustion characteristics and exhaust emissions such as a high NO_x and smoke emission in spite of yielding lower HC and CO[1, 2]. In order to improve diesel fuel properties and obtain lower exhaust emissions, various investigations on alternative liquid and gas fuels were performed.

Researchers reported that there are several ways of improving the self-ignitability of fuel such as increasing the compress ratio, adding an extra spark plug and redesigning the combustion chamber. On the other hand, another simple and efficient way is by adding a certain amount of cetane improvers to the fuel [3-5]. The function of cetane improvers is to

* Corresponding author: adam@ump.edu.my

shorten the ignition delay period and accelerate the burning rate of the fuel. Normally, the maximum amount of improvers added to diesel fuel is not more than 5% [6]. Meanwhile, the utilization of metal additives is an elective decision aimed towards better engine performance and decreasing pollutants in diesel engines. Metal additives have reactant impact with respect to enhancing the combustion process. These additives function as a combustion catalyst to promote combustion and decrease fuel consumption by reducing pollutant emissions. The metal work used reacts with water vapor to process hydroxyl radicals or specifically responds with carbon atoms as a catalyst, thereby discharging oxidation temperature [7-9]. Consequently, remarkable reductions in smoke and CO emissions were obtained. In order to determine the impact of metal-additive fuel properties and engine performance-emission characteristics, different metal additives such as titanium, manganese, iron, platinum, barium and cerium were investigated in numerous studies [8, 10-13]. The results indicated that metal additives helped fuel consumption in aspects such as viscosity, streak point, pour point, calorific value and cetane value. Furthermore, the characteristics of engine performance and emission improved significantly.

In this experiment, bis-carboxyethyl germanium sesquioxide also known as organic germanium (Ge-132) was blended with pure diesel and biodiesel fuel. Germanium is a chemical with physical properties similar to silicon and has a jewel similar to a crystalline structure. When organic germanium blends into diesel fuel, the blend shows an increase of oxygen content due to the organic germanium function is by attach to the oxygen molecules. Germanium is one of the element inside the carbon group in periodic table with carbon (C), silicon (Si), tin (Sn), lead (Pb) and flerovium (Fl) and it can easily reacts to form complexes with oxygen in nature. In addition, germanium is stable in air and water and also unaffected by alkalis and acids.

The purpose of this research is to evaluate diesel engine performance and exhaust emissions by using organic germanium (Ge-132) blend with pure diesel and biodiesel. The engine performance parameters included power and torque with data collected from experiments on the test engine. The emission parameters measured during the engine performance test were CO_2, CO, O_2 and NO.

2 Experimental setup

Three fuel types were used in the experiments. Organic germanium was blended with biodiesel using a mechanical stirrer at speed of 2000 rpm. The biodiesel–germanium blend was prepared with 5 mg organic germanium concentration by volume basis (BG5) while other test fuels are pure diesel (D) and biodiesel (BD). The data extracted for the blend was then compared with pure diesel and biodiesel. The measurement of the fuel properties was collected which all the biodiesel complied with standard ASTM. The chemical properties for all types of fuel are shown in Table 1:

Table 1. Physicochemical properties of test fuels.

Properties	Diesel	Biodiesel	BG5
Density (g/cm^2)	0.843	0.85	0.83
Cetane number	46.6	52.2	46.4
Kinematic viscosity at 40° C (cst)	3.718	4.57	4.29
Calorific value (MJ/kg)	42.32	46.1	NA
Cloud point (^{o}C)	-3	-8.3	NA

 DOI: 10.1051/matecconf/20179001053

The experimental test was performed using YANMAR TF120M single–cylinder, direct injection, four-stroke diesel engine. The specifications of the test engine are listed in Table 2. Figure 1 shows the schematic diagram that represents the experimental setup. The fuel control unit consists of two tanks; the diesel fuel tank and the alternative fuel tank. A hydraulic type dynamometer employed to apply load to the engine. All the data were collected and processed using the DAQ by TFX Engineering. The parameters measured and analyzed in this project were engine power, torque and exhaust emissions. Engine speed was in the range of 1200 rpm, 1500 rpm, 1800 rpm, 2100 rpm and 2400 rpm and at 30% load.

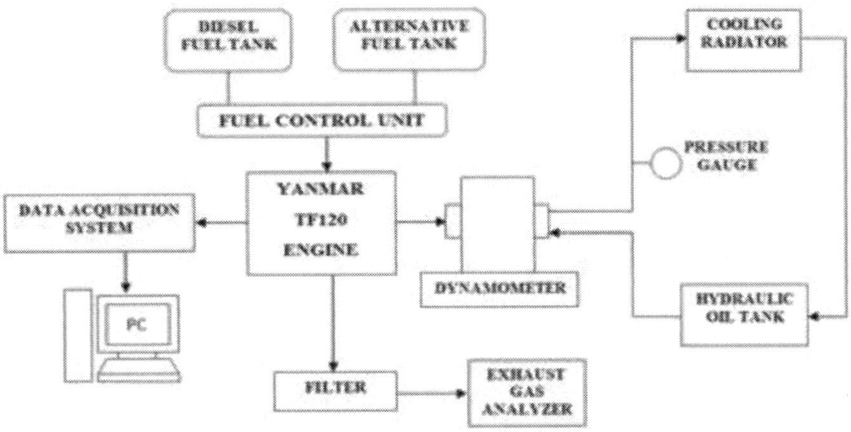

Fig. 1. Schematic diagram of an experiment setup.

Table 2. Engine specification.

Description	**Specification**
Engine model	YANMAR TF120M
Engine type	Single cylinder, horizontal, diesel 4 stroke cycle
Combustion system	direct injection
Number of cylinders	1
Bore x stroke (mm)	92 x 96
Displacement (L)	0.638
Injection timing	18° BTDC
Compression ratio	17.7
Continuous output (HP)	10.5 HP at 2400 rpm
Rated output (HP)	12 HP at 2400 rpm

3 Results and discussion

3.1 Power performance

Figure 2(a) shows the graph for engine power at all engine speed for D, BD and BG5. From the observations, BG5 overall power was larger than BD at all speeds. In addition, at low

DOI: 10.1051/matecconf/20179001053

speeds of 1500 rpm, the increase of BG5 compared to D and BD was at 12.1 %. At a medium speed of 1800 rpm, the power increment of BG5 was 15 % and 12.2 % compared to D and BD, respectively. At a high speed of 2400 rpm, BG5 showed a power increase of 9.2 % compared to BD, while a reduction of 6.5 % compared to diesel.

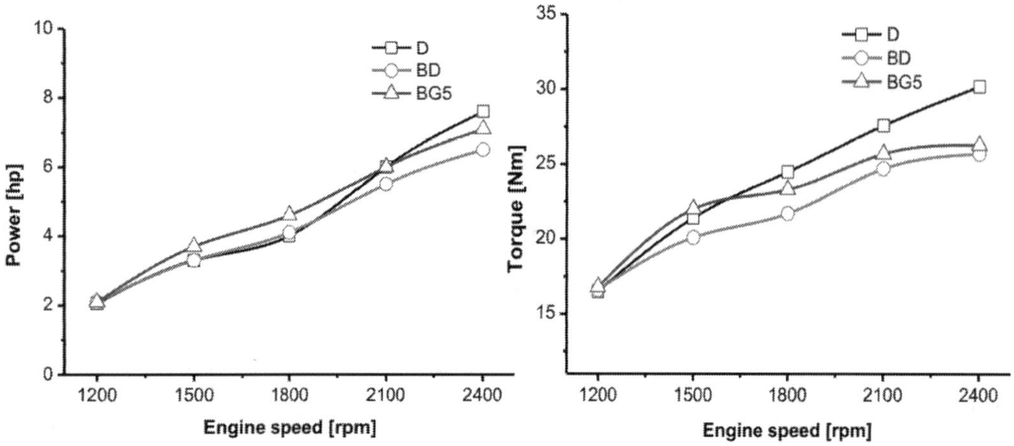

a) b)

Fig. 2. Comparison of (a) engine power and (b) engine torque for all test fuel.

Overall, the engine power is increase proportionally with the increment of engine speed for all test fuels. Additionally, BD high density and viscosity resulted in poor atomization and uneven combustion, therefore producing low levels of power [14]. The high density and kinematic viscosity of BD produce large size droplets during fuel injection into the combustion chamber, affecting the vaporization process and leads to incomplete combustion. Comparatively, the power produced by BG5 was higher than BD, which leads to the conclusion that BG5 is a more effective fuel than BD in terms of the combustion process.

3.2 Torque performance

Referring to Figure 2(b), at low to medium engine speeds of 1200-1500 rpm, BG5 torque was the highest compared to other fuels. The highest torque was measured by BG5 (22 Nm) followed by D (21.4 Nm) and BD (20.1 Nm). However, from 1600-2400 rpm, D torque had overtaken that of other fuels. The largest difference in percentage of D compared to BG5 was 12.9 % at a speed of 2400 rpm. Torque for BG5 was higher than BD at all ranges of speed. BG5 average percentage of BG5 was 4.8% higher than BD. Nevertheless, at higher speeds of 2400 rpm the percentage difference between BG5 and BD got smaller at 2.3 %. D torque showed an increase from a medium speed of 1800 rpm to a high speed of 2400 rpm.

Overall observation shows the torque graph shows that it is directly proportional, which is similar to the power. Pure diesel (D) marked the highest compared to BD and BG5 due to its low kinematic viscosity value among all test fuels. Torque increases because of complete combustion due to the good atomization of fuel, which then produces droplets in low density, calorific value and viscosity that is easier to vaporize. BD showed the lowest values in torque compared to all other fuel types. Again, the high density and high kinematic viscosity of BD resulted in lower velocity of air fuel mixtures, lack of turbulence, and the lack of mixing air and fuel particles in the combustion chamber, which affected combustion efficiency when compared to diesel [15]. On the other hand, BG5 produced good results when compared to BD. Adding germanium into fuel will increase higher combustion temperature thus improve combustion characteristics, leading to the efficient combustion process.

3.3 NO emission

Figure 3(a) shows the variation of NO emissions for the test engine with BG5, in reference to diesel fuel. NO emission decreased as engine speed increased for fuel type. At a low speed of 1200rpm, D shows the lowest emissions compared to other fuels. At low and medium engine speeds, the percentage difference between BG5 compared to D increased by 47.8% and 22.69%, respectively. Meanwhile, NO exhaust from BG5 compared to BD reduced by 8.5% and 25.13%, respectively. Moreover, at high speeds the pattern was similar when compared to low and medium speeds. Trend for NO emissions for all test fuel were declined gradually with the increment of engine speed. Among all the fuels, the highest NO was emitted by BD followed by BG5 and D. The results can be correlated with the value of kinematic viscosity of the fuel. BD has the highest kinematic viscosity at 4.57 cst compared to D at 3.718 cst and BD5 at 4.29 cst. In addition, high density of BD at 0.85 g/cm^2 contributes to the trend. The overall results show that addition of Ge-132 into BD could reduce the NO emission.

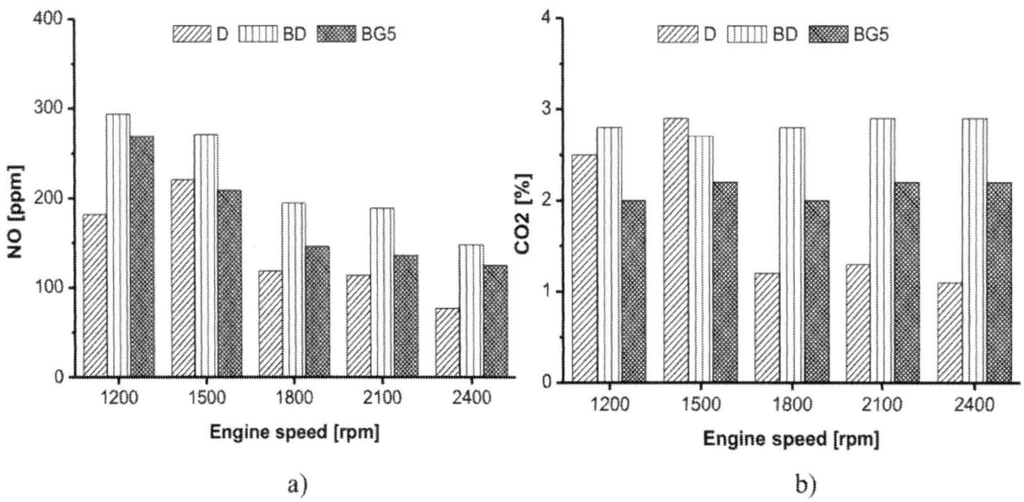

Fig. 3. Effect of test fuel on (a) NO emission and (b) CO_2 emission.

3.4 CO_2 emission

CO_2 emission is generated by the complete burning of fuel and it is a critical part in worldwide warming. Research has shown that carbon discharged during combustion of biodiesel is absorbed by plants through the process of photosynthesis [16-18]. As shown in Figure 3(b), BD has the highest CO emission among all types of fuels. Referring to Table 1, BD had the highest calorific value of 46.1 MJ/kg compared to BG5 and D. Figure 3(b) shows that the CO_2 emission of BD on average was 2.8 %, which was higher than that of diesel at all engine speeds. It is clear that BD marked the highest CO_2 emissions at all engine speeds. Overall, when comparing BG5 to BD for CO_2 emission, there was a decrease of 24.8 %. As compared to D, the overall average reduction was 1.8 % when BD was compared to BG5; there was a reduction of 2.1 %. Similar to BD, BG5 showed constant CO_2 emissions for all engine speeds. On the contrary, D showed an increasing trend of emissions at low engine speeds until 1500 rpm. At higher engine speeds from 1800-2400 rpm CO_2 emissions released by D was reduced by 0.028 %. At a speed of 1800 rpm the CO_2 emissions by BG5 increased to 66.6 % when compared to D. From a medium engine speed to a high engine speed, the percentage of increase by BG5, when compared to D, was by 66.7 % and up to 100 % at 2400 rpm.

3.5 CO emission

For overall CO emissions, Figure 4(a) shows that the trend is similar to CO_2 emissions, where BG5 at low engine speeds had decreased CO emissions while at medium to high engine speeds the trend for BG5 was constant. For D at low engine speeds, the graph showed that there was an increase of CO emission but at medium speed the CO emission for D dropped by 0.04 %. When at high speeds, CO emissions for D became constant with small changes. At low engine speeds, CO emission for BD decreased, while at medium to high engine speeds, the trend of the emissions had increased. When comparing BG5 with D at low engine speeds, the trend graph decreased, while at medium engine speed and the percentage increase by 50 %. At a high speed, there were small changes between BG5 and D. When comparing BG5 with BD, BG5 show an increment by 33.3 % at 1200 rpm but there was no change in value at 1500 rpm. However, at speeds of 1800 rpm, 2100 rpm and 2400 rpm, BG5 when compared to BD had decreased by 25 %, 50 % and 62.5 %, respectively.

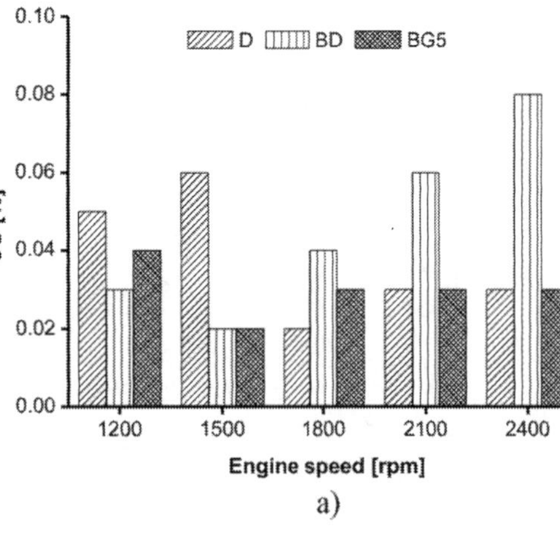

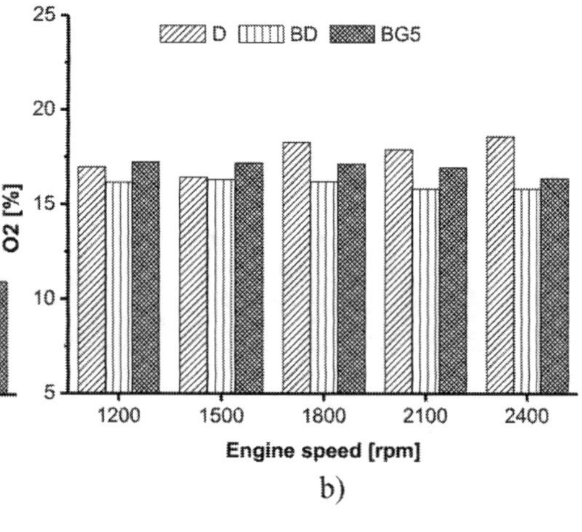

Fig. 4. Effect of test fuel on (a) CO emission and (b) O_2 emission.

3.6 O_2 emission

Figure 4(b) shows O_2 emission emitted by all test fuels. Among all test fuels, D record highest O_2 emission at all engine speed. Meanwhile, BD shows that the trend of O_2 emission from a low engine speed to a high engine speed is constant, although there were small changes. Similar, BG5 also showed a constant according to the graph. Moreover, at low engine speeds, emissions by BG5 increased by 1.6 % compared to D. However, at medium and high engine speeds, emissions by BG5 had reduced by 6.4 % and 11.9 %, respectively. Besides that, when BG5 was compared to BD, the graph showed that emissions by BG5 were the highest from 1800 rpm until 2400 rpm. At low engine speeds, emissions by BG5 increased by 6.6 % while at medium and high engine speeds BG5 showed that emissions increased by 5.7 % and 3.47 % when compared to BD.

4 Conclusions

The research had compared the engine performance in terms of engine power, engine torque and exhaust emission characteristics of diesel (D), biodiesel (BD) and biodiesel (BG5). The findings show that the power and torque characteristics of BG5 blends were almost similar in trend with D and better than BD. BG5 shows highest engine power among all fuel while D shows highest engine torque at all engine speed. BD marked lowest at engine power and torque among all fuel at all engine speed. Moreover, BG5 also showed reduced NO, CO_2 and CO emissions when compared to biodiesel. However, addition of germanium in biodiesel BG5 resulting an increment of O_2 emission compared to BD at all engine speeds. This is due to the addition of organic germanium results an increment of oxygen content inside the fuel blends. In general, the overall results indicate that BG5, as a new additive fuel, could be used as a diesel substitute without any modification to the engine and for long-term use.

Malaysian Ministry of Higher Education and Automotive Engineering Centre, Universiti Malaysia Pahang is greatly acknowledged for the technical and financial supports under Fundamental Research Grant Scheme FRGS (RDU130134).

References

1. S. H. Yoon, J. P. Cha, C. S. Lee, Fuel Prop. Tech., **91**, 1364 (2010)
2. L. Li, J. Wang, Z. Wang, J. Xiao, Fuel, **156**, 211 (2015)
3. Y. Ren, Z. Huang, H. Miao, Y. Di, D. Jiang, K. Zeng, B. Liu, X. Wang, Fuel, **87**, 2691 (2008)
4. A. Atmanli, Fuel, **172**, 209 (2016)
5. F. Frusteri, L. Spadaro, C. Beatrice, C. Guido, Chem. Eng. Jrnl., **134**, 239 (2007)
6. R. Li, Z. Wang, P. Ni, Y. Zhao, M. Li, L. Li, Fuel, **128**, 180 (2014)
7. G. R. Kannan, R. Karvembu, R. Anand, Applied Energy, **88**, 3694 (2011)
8. A. Keskin, M. Gürü, D. Altıparmak, Enrg. Conv. & Mgmt., **52**, 60 (2011)
9. M. A. Lenin, M. R. Swaminathan, G. Kumaresan, Fuel, **109**, 362 (2013)
10. A. Keskin, M. Gürü, D. Altıparmak, Fuel, **86**, 1139 (2007)
11. M. Gürü, B. D. Artukoğlu, A. Keskin, A. Koca, Enrg. Conv. & Mgmt., **50**, 498 (2009)
12. G. Kannan, R. Karvembu, R. Anand, Appl. Enrg., **88**, 3694 (2011)
13. M. Jones, C. H. Li, A. Afjeh, G. Peterson, Nanoscale research letters, **6**, 1 (2011)
14. H. Hazar, Rnwble. Enrg., **34**, 1533 (2009)
15. R. Behçet, R. Yumrutaş, H. Oktay, Enrg., **71**, 645 (2014)
16. C. L. Peterson, T. Hustrulid, Biomass and Bioenergy, **14**, 91 (1998)
17. W. Körbitz, Renewable Energy, **16**, 1078 (1999)
18. A. K. A. a. L. M. Das, Engineering for Gas Turbines and Power, 440 (2000)

The effect of tap water emulsified fuel on exhaust emission of single cylinder compression ignition engine

Nurul Aiyshah Mazlan[1], *Wira Jazair* Yahya[1,*], *Ahmad Muhsin* Ithnin[1], and *Mohamad Azrin* Ahmad[1]

[1]Vehicle System Engineering, Malaysia-Japan International Institute Of Technology (MJIIT), Universiti Teknologi Malaysia, 54100, Kuala Lumpur, Malaysia.

Abstract. An experimental investigation on exhaust emissions with emulsion fuel was conducted in a diesel engine that equipped with a "Real Time Non-Surfactant Emulsion Fuel Supply System, RTES" under four different loads operations (1, 2, 3 and 4 kW). RTES is a mixing device that able to produces non-surfactant emulsion fuel which is one of the alternative ways to improve the exhaust emissions of diesel engine, especially Nitrogen Oxides (NOx) and Particulate Matter (PM). As for the test fuel, neat diesel (D2) and tap water-in-diesel emulsion fuel (W/D) are tested as a comparison. Based on the experimental results, emulsion fuel decrease NOx radically compare to D2 in all load conditions with an average reduction of 18.99% respectively. As for the PM, emulsion fuel is lower compare to D2 at all load conditions and lowest at high load. However, tap water emulsion fuel shows high formation of Carbon Monoxide (CO) at all load conditions which due to lower combustion temperature. This significant increment is aligned with the reduction of NOx emissions.

1 Introduction

Nowadays, greenhouse gasses in our atmosphere have been increasing. Over 90% of the causes of the climate change come from human activities [1] of which the biggest cause of this catastrophe is contributed from tmospheric emission. The largest growth in this emission has come from industry and transportation [2]. Diesel engine, due to better power economy, power performances and durability are widely applied in mass transportation, heavy industries machinery and agricultural sectors. In spite of their preferable superiority, diesel engines are suffering from harmful exhaust emissions especially particulate matters, PM and nitrogen oxides, NOx [3]. These emissions are not only harmful to the environment but also hazardous to human's health, but they also lead to a serious damage to the health like lung cancer, asthma, cardiovascular issues and other fatal illness that would cause

* Corresponding author: wira@utm.my

death. However, introduction of water into diesel engine has been figured out by the researcher to simultaneously reduce the formation of NOx and PM emissions.

Water-in-Diesel (W/D) emulsion fuel is the promising alternative fuel that meets the requirement of emission regulations hence become a potential alternative fuel that could contribute the world's needs. There are several emulsification methods that have been introduced by researchers all over the world. For example, A. Kulchitskiy and A. Attia [4] used membrane emulsification and investigated the effect of emulsion fuel structure to the diesel engine. Moreover, Yilmaz et. al [5] also has prepared emulsion fuel with different method by using auxiliary emulsifier mono ethylene glycol, tested it in a turbocharged diesel engine and analyzed the emissions characteristics. However recently, there is a mechanical approach to produce emulsion fuel and the main objective of this concept is to continuously emulsified diesel and water without the presence of surfactant before instantaneously supplied to the engine. This new emulsification concept has been introduced by A.M. Ithnin et. al [6] which known as Real-Time Non-Surfactant Emulsion Fuel Supply System (RTES). As for this research, the purpose is to investigate the exhaust emissions which are NOx, PM and Carbon Monoxide (CO) of diesel engine fuelled with different types of fuel which one of that is W/D emulsion fuel that emulsified by using RTES. Tap water has been selected as water source to produce the W/D emulsion fuel.

1.1 Emulsion fuel

W/D emulsion fuel is the promising alternative fuel that meets the requirement of emission regulations hence become a potential alternative fuel that could contribute the world's needs. Moreover, it does not require any modification of the engine. The water-in-diesel emulsion fuel type is preferable to be the alternative fuel compared to the other emulsion fuel is because of high difference in boiling point between water and diesel fuel. Term emulsion is defined as a mixture of two or more immiscible liquids [7] which are unblended in nature. It is formed with the help of mechanical agitation together with the chemical additives so called surfactant that attract both the unblended liquids in forming one stable solution. The presence of surfactant is to form a stable emulsion that important in order to ensure this alternative fuel run accordingly in the engine. In W/D, there is a special phenomenon that only happened during emulsion fuel combustion called as micro-explosion phenomenon (figure 1). Ivanov and Nevedov reported that W/D experienced a spontaneous explosion during the combustion [8], hence caused the smaller fuel particles to be in contact easily with the air for complete combustion and reduces the generation of PM, NOx, CO and other pollution materials without decrease the combustion efficiency [9].

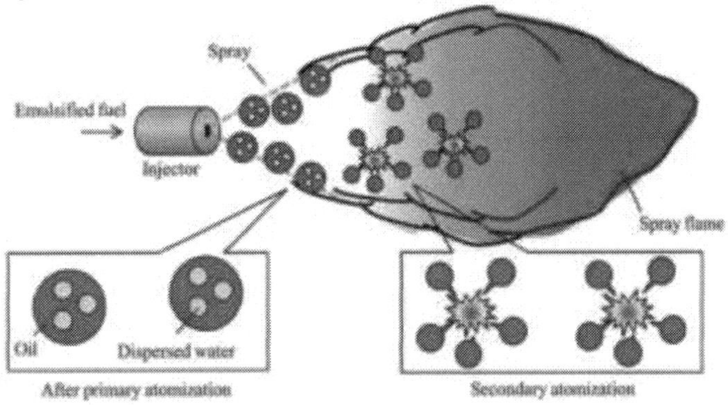

Fig. 1. Diagram of the occurrence of micro-explosion process [10].

1.2 Diesel fuel

Vehicle emission standard (Euro standard) is emission standard constructed specifically for vehicles and it is categorized based on the formation of Nitrogen Oxides (NO_x), Hydrocarbon (HC), Particulate Matter (PM) and Carbon Monoxide (CO) in vehicle exhaust emissions. Fuel quality is also required alongside Euro stages. In order to achieve air quality requirements that significantly reduce Sulfur (S), it is necessary for vehicles to use Euro 3 and above within other parameters such as Benzene (C_6H_6) and olefins or aromatics, which able to significantly reduce [11]. Low grade diesel fuel was widely used in non-developing countries and also in some developing nations [12]. Fuel quality contains very high Sulfur (S), Lead (Pb), Benzene (C_6H_6) and olefins or aromatics causing severe damage to the environment and also to the diesel engine. Table 1 shows the details of low grade diesel fuel specifications, which is Euro 2 standard (D2).

Table 1. D2 diesel fuel specification [13].

Specifications	**D2**
Calorific value, MJ/kg	45.280
Cloud point, °C	18
Density, kg/L, 15 °C	0.8538
Total Sulfur, mass %	0.28
Viscosity, cSt, 40 °C	4.642
Distillation temperature, 90% recovery, °C	367.9
Flash Point, °C	93.0
Pour Point, °C	12
Cetane number	54.6
Carbon (C), wt %	84.1
Hydrogen (H_2), wt %	12.8
Sulfur (S), wt %	0.2
Nitrogen (N), wt %	0.1
Oxygen (O_2), wt %	3.9

1.3 Tap water

In Malaysia, water supply management and development is not centralized. It is managed by state governments. The State Water Authorities or private companies will supply tap water from treatment plant to residential areas. This tap water comes from surface water

DOI: 10.1051/matecconf/20179001054

like lakes, reservoirs and streams and groundwater. The latest tap water quality criteria and frequency of monitoring were given by Engineering Service Division (ESD), Ministry of Health Malaysia. Quality of tap water is monitored by ESD at least four times in a year collected in shops, school and university and housing area. Tap water samples were analyzed by NovAA 400 Flame Atomic Spectrophotometer (AAS) system to evaluate chemical element [14]. Chemical element percentage in tap water is shown in Figure 2. Main chemical elements are Chloride (Cl), Sulphate (SO_4) and Natrium (Na). Metal element in chemical characteristics such as Magnesium (Mg), Iron (Fe) and Zinc (Zn) present in tap water due to the corrosion of plumbing material. Once tap water is treated, it must travel along pipes and become contaminated with metal elements when it flows from treatment plant to consumer.

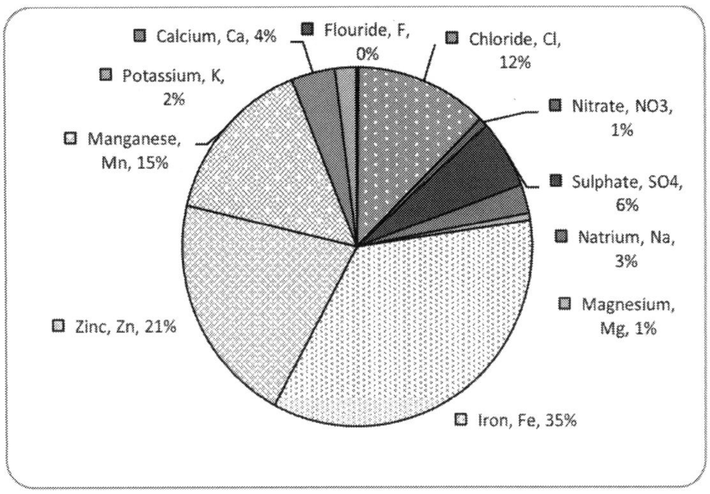

Fig. 2. Distribution of chemical element percentage bonding in tap water [14].

1.4 Real Time Non-Surfactant Emulsion Fuel Supply System (RTES)

RTES is a mechanical approach to produce emulsion fuel and the main objective of this concept is to continuously emulsified diesel and water without the presence of surfactant before instantaneously supplied to the engine [6]. RTES is a high potential concept because of its ability to remove the dependency of surfactant without affecting the benefits of emulsion fuel. On the market, there are variance types of surfactant available and the price is quite expensive. Due to that, the alternative fuels are much more expensive and not cost effective compare to the neat diesel fuel since they require a high amount of emulsifier and other chemical additives. Figure 4 shows the process flow of RTES.

2 Experimental setup

For the purpose of this experiment, two types of fuel have been tested which are neat diesel Euro 2 (D2) and tap water emulsion fuel (W/D tap water). W/D tap water emulsion fuel has been emulsified by using RTES. The experiment was performed under four different loads which are 1kW, 2kW, 3kW and 4kW by using an electric generator which has a direct injection diesel engine, single cylinder, 4-stroke and air-cooled. The detailed specifications of the engine are listed in table 2.

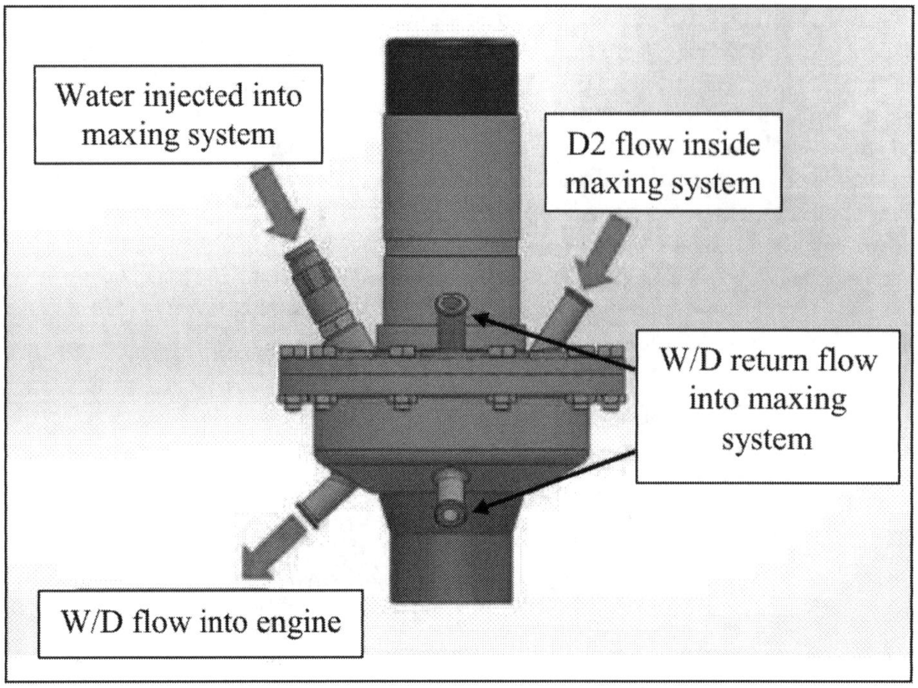

Fig. 4. Process flow of RTES [6].

Table 2. Diesel engine specification.

Parameter	**Specification**
Engine Type	4-stroke, single cylinder, direct injection
Cooling System	air cooling
Bore x Stroke (mm)	86 x 70
Displacement Volume (l)	0.406
Compression Ratio	19.3
Maximum Power (kW)	5
Fuel Injection Timing (bTDC)	13
Fuel Injection Pressure (MPa)	19.6
Rated Revolution (rpm)	3750

Diesel engine testing has been carried out in Low Carbon, Vehicle System Engineering laboratory, MJIIT UTM. Figure 5 shows the schematic diagram of engine test setup.

 DOI: 10.1051/matecconf/20179001054

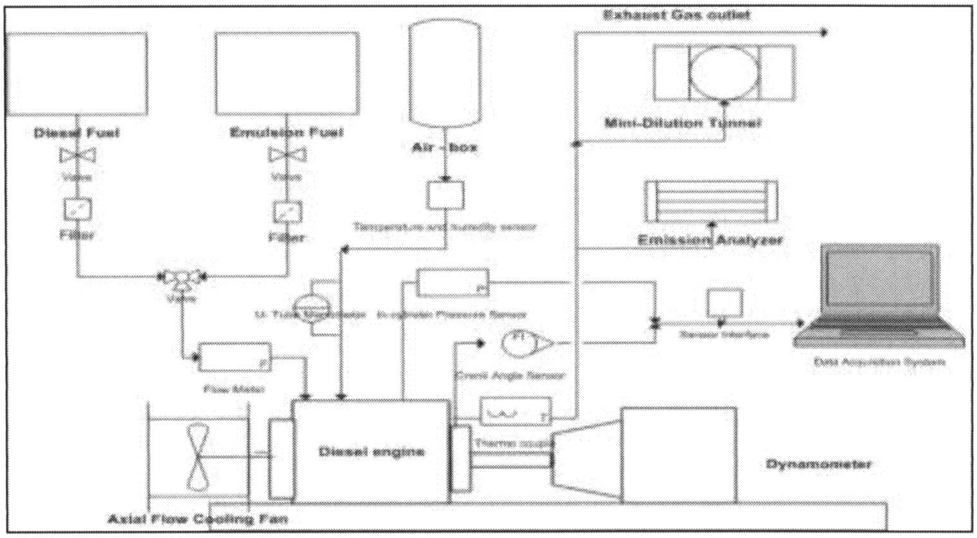

Fig. 5. Schematic diagram of diesel engine test bed setup.

As for emission measurement, two measurement instruments have been used in this experiment which is emission analyzer to measurement the emission of NO_x and CO and a mini dilution tunnel to measure the PM. Emission analyzer used in this test is Testo 350. During testing, probe of the analyzer was placed inside the exhaust of the engine and data has been transferred into the data logger. Figure 6 shows Testo 350 Emission Analyzer. Measurement is recorded every 5 seconds interval during 2 minute duration of engine running.

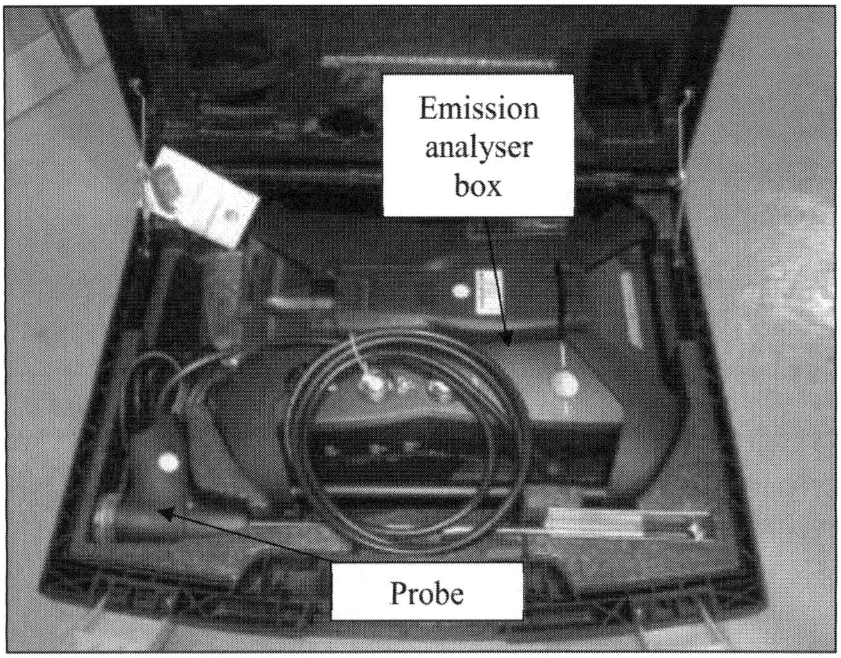

Fig. 6. Testo 350 Emission Analyser.

 DOI: 10.1051/matecconf/20179001054

Schematic diagram and photo of mini dilution tunnel to measure PM is shown in Figure 7. Exhaust smoke from engine flow through main tunnel. Distribution of exhaust smoke is sucked into dilution tunnel by dilution pump, together with heated clean air. Temperature of clean air is maintained to 50 °C that heated by heater. Air filter is used to filter impurities in air before it enters tunnel. Ratio of exhaust gasses that is diluted into dilution tunnel is set to 10:1. This can be done by measuring CO_2 concentration from main tunnel and also in dilution tunnel. If the ratio of CO_2 does not reach 10, speed of dilution pump is increased until it matches required value. The device used to measure CO_2 is Cambridge Sensotec Rrapidox 3100 dual gas analyser. It uses infrared (IR) sensor technology to detect the CO_2 measurement and has high accuracy with sensor range of 0 to 5000 ppm. When dilution ratio matches with value of 10, diaphragm pump is switched on to transfer diluted exhaust smoke through PM filter.

Filter used for trapping the PM is Teflon filter (Millipore FHLP04700) with diameter and orifice of 47 mm and 0.45 μm respectively. Gas meter is used to measure the volume of smoke through PM filter. The gas meter used is Shinagawa W-NK wet gas meter. Volume of absorbed gasses is set to 100 L in every filter. PM formation is determined by weighing PM filter before and after experiment. Filter is stored into a closed container equipped with silica gel for 12 hours. This is to absorb the moisture on the filter that may affect the measurement. The weighting device used is Quartz Crystal Microbalance Mettler AE240 with accuracy of 0.00001 g.

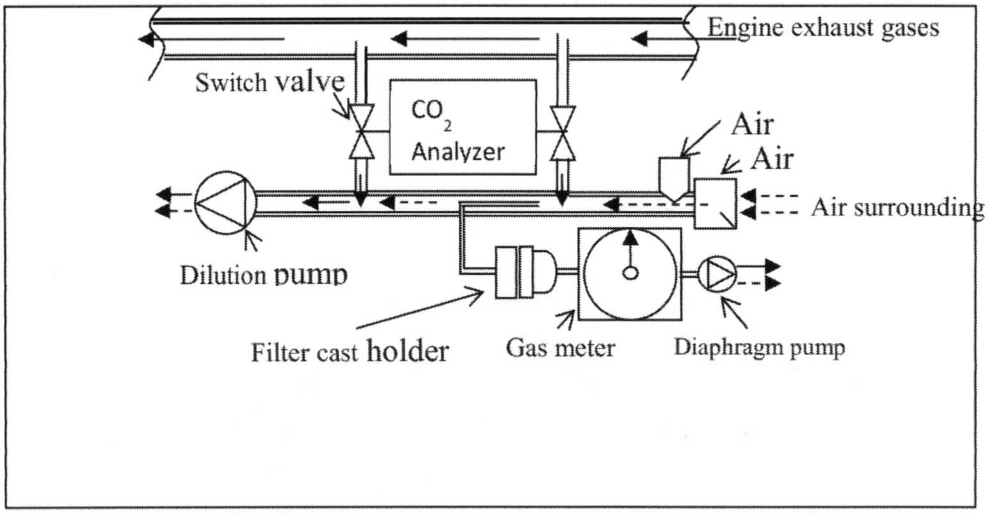

Fig. 7. Schematic diagram of mini-dilution tunnel.

3 Results and discussion

3.1 Nitrogen Oxides (NO_x)

Nitrogen Oxides (NO_x) emission for emulsion fuel under different engine load conditions compared with D2 is shown in Figure 8. It shows that W/D tap water emulsion fuel reduce NO_x significantly compared with D2 in all load condition. Average reduction NO_x per load for W/D tap water is 18.99%. In diesel engine, the creation of NOx is related to the accessibility of oxygen and high combustion chamber temperature. Consequently, this explained the reduction of NOx when using emulsion fuel. With emulsion fuel, the water existence reduced the peak temperature because of high latent heat from water vaporization that absorbs the heat during combustion. Besides, the increment in OH (hydroxyl) radicals which would react with excess oxygen also affects the reduction of NOx formation [15].

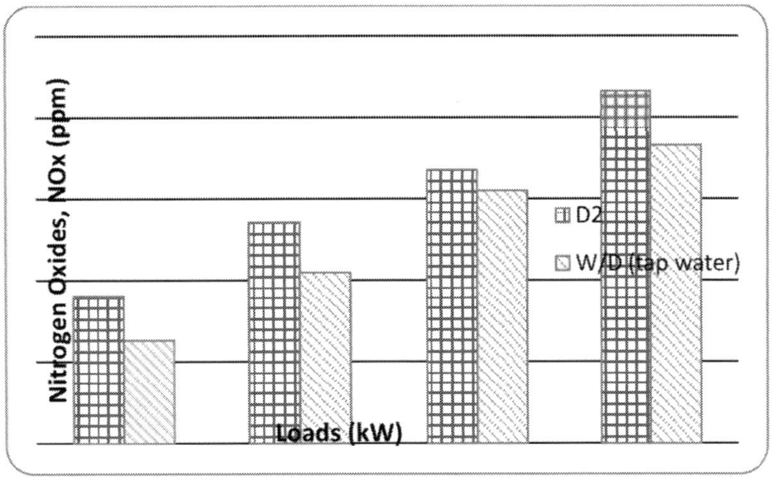

Fig. 8. NO_x emissions for W/D tap water emulsion fuel compared with D2 under different loads condition.

3.2 Particulate Matter (PM)

Particulate Matter (PM) for W/D tap water emulsion fuel under different engine load conditions compared with D2 is shown in Figure 9. Comparing with D2, emulsion fuel shows lower PM at all loads condition. The reduction of PM with the usage of emulsion fuel is due to the increase of OH radical with the addition of water during the combustion in the combustion chamber which leads to the oxidation of soot precursors. Besides, J.M.Ballester relates that PM reduction with improved combustion efficiency due to better air-fuel mixing process that produces micro-explosion phenomena. Micro-explosion occurs at secondary atomization of initial spray with rapid water evaporation process that is contained in droplet, tearing up droplet into very fine particles, thus helping to improve combustion [16].

3.3 Carbon Monoxide (CO)

Carbon Monoxide (CO) emissions for W/D tap water emulsion fuel under different engine load conditions compared with D2 is shown in Figure 10. W/D tap water emulsion fuel shows high formation of CO at initial load compare to D2. In general, lower combustion temperature of emulsion fuel yields higher CO emissions than conventional diesel fuel. Because of the presence of water, there is no sufficient temperature to convert CO to CO_2 [17]. K.A.Subramanian explained when temperature is lower than 1400 K, the oxidation process of CO will freeze [18]. However at high load, CO formation starts to reduce and becomes closer with D2 due to micro-explosion process at secondary atomization that explodes droplet into very fine droplets, thus improving air-fuel mixing process.

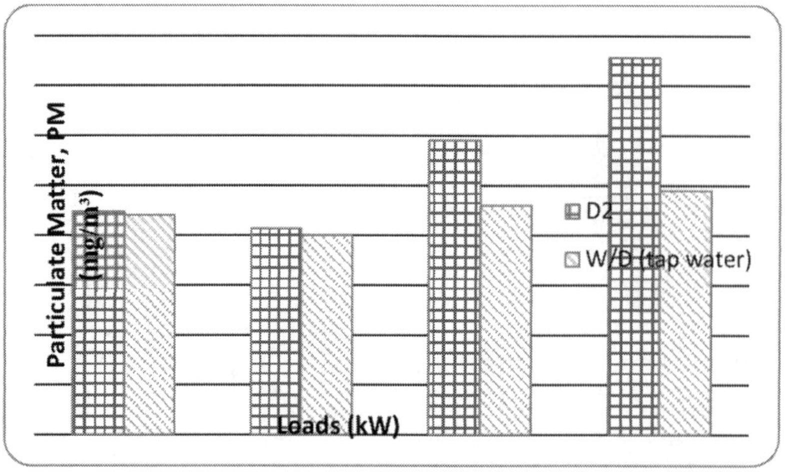

Fig. 9. PM for W/D tap water emulsion fuel compared with D2 under different loads conditions.

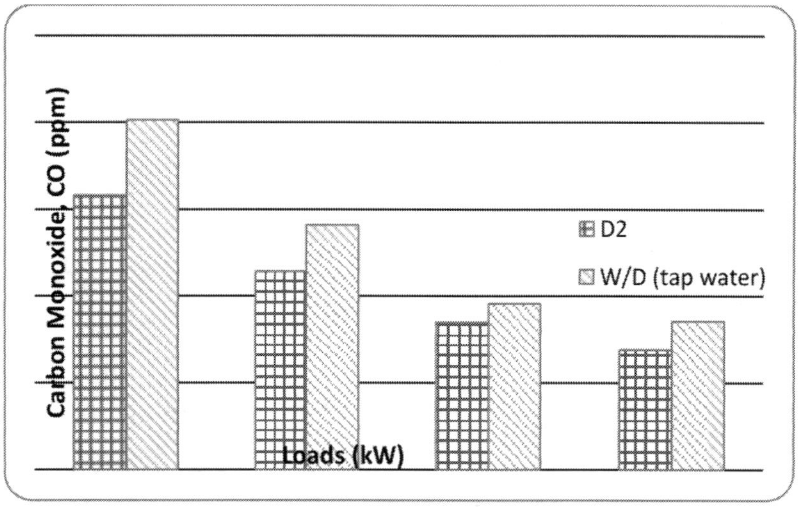

Fig. 10. CO emission for W/D tap water emulsion fuel compared with D2 under different loads conditions.

4 Conclusions

From this study, it is concluded that W/D tap water emulsion fuel gives better emissions compare to neat diesel fuel. As for NOx, average reduction when compared to D2 for W/D tap water emulsion fuel is 18.99%. Moreover, at all load conditions, W/D tap water emulsion fuel produce lower PM and the highest reduction was recorded at 3 kW and 4 kW. Lastly, for CO, due to lower combustion temperature, W/D tap water emulsion fuel shows high formation of CO at all load conditions compare to D2.

The authors would like to express their gratitude to the Malaysia–Japan International Institute of Technology (MJIIT) and UTM Centre for Low Carbon Transport, Universiti Teknologi Malaysia (UTM) for the financial support

References

1. S. Solomon, D. Qin, M. Manning, Z. Chen, M. Marquis, K. B. Averyt, M. Tignor, and H. L. Miller, IPCC, *Summary for Policymakers. In: Climate Change 2007: The Physical Science Basis. Contribution of Working Group I to the Fourth Assessment Report of the Intergovernmental Panel on Climate Change,* (2007)
2. T. Barker, *Climate Change 2007 : An Assessment of the Intergovernmental Panel on Climate Change,* 12–17, (2007)
3. A. Srivastava and R. Prasad, Renew. Sustain. Energy Rev. *Triglycerides-based diesel fuels,* **4**, 111–133, (2000).
4. AM Attia and R Kulchitskiy, Fuel, *Influence of the Structure of Water-In-Fuel Emulsion on Diesel Engine Performance*, **116**, 703 (2014)
5. T. Tan, J. Lu, K. Nie, L. Deng, and F. Wang, Biotechnol. Adv. *Biodiesel production with immobilized lipase: A review*, **28**, 628–34 (2010)
6. A.M. Ithnin *Real-time Water-in-Diesel Emulsion Fuel Supply System (RTES) for Diesel Electric Generator*, No PI 2014702681. (2014)
7. M. S. Kumar, J. Bellettre, and M. Tazerout, Power Energy, *The use of biofuel emulsions as fuel for diesel engines: A review,* **223**, 729–742 (2009)
8. VM Ivanov and PI Nefedov, NASA Tech Transla, *Experimental Investigation of the Combustion Process in Natural and Emulsified Fuels,* TIF-258
9. Dryer FL. Symposium International Combustion, *Water addition to practical combustion system concepts and applications.* **16,** 279-295 (1977)
10. H. Watanabe, Y. Suzuki, T. Harada, Y. Matsushita, H. Aoki, and T. Miura, Energy, *An experimental investigation of the breakup characteristics of secondary atomization of emulsified fuel droplet*, **35**, 806–813 (2010)
11. P. Sulphur, *Status of Fuel Quality and Vehicle Emission Standards in Asia-Pacific Status of Fuel Quality and Vehicle Emission Standards in Asia-Pacific*, 1–3 (2013)
12. S. Henningsen, SAE Pap, *Influence of the Fuel Injection Equipment on NOx Emissions and Particulates on a Large Heavy-Duty Two-Stroke Diesel Engine Operating on Water-in- Fuel Emulsion* (1994)
13. U. Szalkowska, Veh. Emiss. Terminol, *IFQC. Glob Euro*, 1–4 (2010)
14. A. Azlan, H.E. Khoo, M.A. Idris, A. Ismail, M.R. Razman, Scientific World Journal, *Evaluation of minerals content of drinking water in Malaysia*, (2011)
15. S Niko, B Kegl, RW Dibble, Fuel, *Numerical and experimental study of water/oil emulsified fuel combustion in a diesel engine*, **81**, 2035-2044 (2002)
16. WB Fu, LY Hou, L Wang, FH Mang, Fuel Process Technol, *A unified model for the micro-explosion of emulsified droplets of oil and water*, **79**, 107-119 (2002)
17. B John, Heywood, McGraw Hill Book Company New York, *Internal combustion engine fundementals* (1998)
18. M Abu Zaid, Energy Conversation Management, *Performance of single cylinder, direct injection diesel engine using water fuel emulsion*s, **45**, 697-705 (2004)

MATEC Web of Conferences **90**, 01055 (2017) DOI: 10.1051/matecconf/20179001055
AiGEV 2016

Methods of preparing internal combustion engine cylinder bore surfaces for frictional improvement

Kong Chung Hwa[1,2], *D.*Ramasamy[1,2,*], *K.* Kadirgama[1,2], *M.* M.Noor[1,2], *M.* M.Rahman[1,2], *M.* Samykano[1,2]

[1]Faculty of Mechanical Engineering, Universiti Malaysia Pahang, 26600 Pekan, Pahang, Malaysia
[2]Automotive Engineering Centre, Universiti Malaysia Pahang, 26600 Pekan, Pahang, Malaysia

Abstract. Frictional losses piston to cylinder bore contact is a major sources of mechanical losses in an internal combustion engine (ICE). Traditional plateau honing produces a relatively rough cylinder bore surface with many valleys for oil retention and plateau surfaces that are usually has micro roughness's that causes mechanical friction to act as a bearing surface. A smooth polished dimpled surface is more ideal to achieve low friction and wear in an ICE. Alternative methods to create a smooth dimpled surface on a hypereutectic aluminum ADC12 substrate for frictional improvements are evaluated in this study using an oscillating wear tester (OWT). The methods include casting in the dimples in the aluminum matrix, sandblasting as well as embossing the pits. The texture samples are evaluated by examining the surface properties, measuring frictional coefficient as well as wear characteristics. It was found that the samples embossed with #320 grit sandpaper and sandblasted with #240 sieve sand samples had a reduced coefficient of friction (μ) of 23% at low sliding speeds before hydrodynamic lubrication mode and 6.9% in the fully hydrodynamic lubrication region.

1 Introduction

The wheel revolutionized human civilization as it enabled heavy transportation of goods and commodities, fostered trade, made construction of larger structures possible and revolutionized human transport in the form of wheel barrows, horse carriages and ox carts. It wasn't until the accidental discovery by an engineer by the name of Beauchamp Tower whom in his experiment observed that the sliding surfaces can be separated by a pressurized oil film that the idea of two sliding surfaces can be separated by a lubricant achieving minimal friction and wear. Reynolds furthered his findings and established the theories of hydrodynamic lubrication and in 1922, Sir William Bate Hardy coined the term boundary lubrication [1, 2].

Since the turn of the 20th century, burning hydrocarbon based fuels in Internal Combustion Engines (ICE) has become one of the primary energy generation methods.

[*] Corresponding author: deva@ump.edu.my

Globally, the burning of fossil fuels provides 86.4% of energy demands [3]It saw applications far and wide from power generation to transport and even to operating a myriad of machines which together has powered mankind through the industrialization era and brought about the modern age as we know it today. The widespread application success of the ICE has hence contributed a major role in Carbon Dioxide (CO2) emissions to the atmosphere. According to the Intergovernmental Panel On Climate Change (IPCC), 57% of the total greenhouse emissions in 2007 came from the burning of fossil fuels alone which amounts to 32 billion metric tonnes of CO2 [4]. The average CO2 levels in the atmosphere has risen from 328 ppm in 1970 to 399 ppm in 2013[5]. As ICEs still are an integral to modern day living and its fuel source is not renewable, it is wise to invest some effort to make it cleaner and more fuel efficient to stretch oil reserve lifespans; allowing for more time for alternative renewable energy sources such as solar power to mature. An ICE that is cleaner and more efficient not only reduce running costs, but also helps reduce the its environmental footprint, preserving the environment for the future generations to come.

As governments worldwide begin to impose emission regulations to do their part to reduce greenhouse gasses in the atmosphere, there has been ever increasing importance for engine manufacturers to improve the emissions and
efficiency of their smaller engines as well. The first commitment period of the Kyoto Protocol (2008-2012) has just been passed where participating countries are required to reduce overall CO2 emissions from their 1990 levels over the next decade as stated in the agreement. The second commitment period (2013-2020) will see further CO2 cuts [6] which will push green engine designers to find extra efficiency from power plants and vehicles.

This in turn has pushed vehicle manufacturers to consider the use of aluminum as an engine block material to further reduce weight, saving fuel and reducing CO2 emissions in the process. Other improvements implemented include better fueling systems, improved engine head designs and changes aimed at reducing the mechanical friction of an ICE such as the use of smaller sized bearings as well as low friction cylinder bore surfaces.

Reducing the mechanical losses of an ICE is a key part in improving its efficiency. Besides the energy wasted heating up the engine as well as spent hot exhaust gasses, the second largest energy loss is from mechanical friction in the engine especially in the piston assembly [7-9]. In Figure 1, most of the power from combustion is converted in to heat. The remaining energy is turned in to work done and out of the work done by the ICE, a significant portion of it goes to overcoming mechanical and pumping losses [10, 11]. Figure 1 shows that the frictional losses from an ICE is mostly contributed by the piston assembly. Therefore, improvements in this area would yield the most effect to improving the overall fuel efficiency of an ICE. Aluminum especially in the form of its various silicate alloys is a lightweight metal that has good strength to weight ratios and has seen applications in many fields of engineering [12]. In the field of internal combustion engines, there has been an increasing trend where aluminium is used as the main material to produce engine blocks [13]. Aluminium is either cast around a cast iron liner or used as it is to from the cylinder block. The eutectic silicate alloy of aluminium, A390 has good strength to weight ratio and exhibits good wear properties due to its high silicate content preventing wear after the aluminium has been etched away exposing the silicate as the bearing surface.

This coupled with surface honing to create a crossed hatch pattern in the cylinder bore surface is the method currently deployed in making small ICEs for applications such as grass trimmers, chainsaws, small generators and other general purpose engines. To achieve minimal friction between the sliding surfaces in the engine bore, the ideal mating surfaces are polished and lubricated but this cannot be used as piston rings scrape off lubrication and the following reciprocations will result in unlubricated surfaces sliding past each other hence causing catastrophic wear and tear after short operation periods.[14, 15].

 DOI: 10.1051/matecconf/20179001055

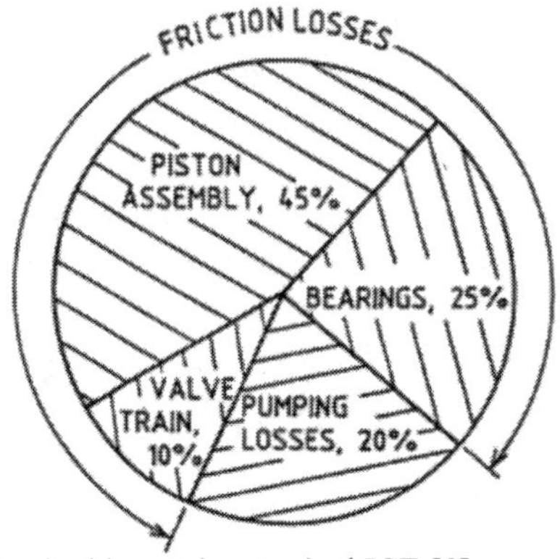

Fig. 1. The distribution of mechanical losses in a typical ICE [8].

Surface honing makes plateaued scratches on the surface of the cylinder but increases surface friction as it reduces surface conformity and creates more instances where the piston ring pack comes in to contact with the cylinder bore surface, which is known as boundary lubrication. Ideally, the cylinder bore has to be a surface that is polished to minimize inconformity between sliding surfaces which is the main cause of boundary lubrication mode [16]. Having said that however, an overly smooth conforming surface is detrimental when it comes to reciprocating wear systems as the reciprocating action wipes off the oil in one cycle and the following cycles would suffer from lubrication starvation and subsequently, catastrophic wear [17, 18]. To achieve reduction in friction and wear, the sliding surfaces ideally need to have a smooth surface and a uniformly distributed array of recesses or dimples to allow oil retention but at the same time ensuring good surface conformity.

2 Measurement of piston ring friction

Piston ring tension of the compression rings used in this study is measured using a method developed by Tan and Ripin where the piston assembly is placed on a V-block and the tensile machine is set in compression mode. In this test, an Instron 3360 universal testing machine is used. The compression speed of the machine is set at 0.005 mm/s and the compression is stopped when the diameter of the of the piston ring reaches the diameter of the cylinder bore. A V-block is used to locate the piston assembly during testing. The forces are resolved using a free body diagram as shown in Figure 2. The measurement is repeated three times and the average piston ring tension is calculated [19].The averaged result from the compression test shows that the piston ring tension of the compression ring is 17.9 N. The elastic tension of the piston ring is given by Eq. (1).

$$\Sigma F = F_1 + F_2 + F_3 \tag{1}$$

Since, $F_2 = F_3$

The surface area of contact between the piston ring and the cylinder bore is calculated via Eq. (2)

$$\text{Area} = \pi \times \text{r} \times \text{t} \tag{2}$$

where r is the radius of the cylinder bore and t is the thickness of the piston ring measured to be 1.5 mm giving an area of 169.6 mm2. Hence the nominal contact pressure of the

DOI: 10.1051/matecconf/20179001055

piston ring to the cylinder bore for each given piston ring is calculated to be 254.7 kPa using Eq. (3)

$$P = \frac{\text{Force}}{\text{Area}} \tag{3}$$

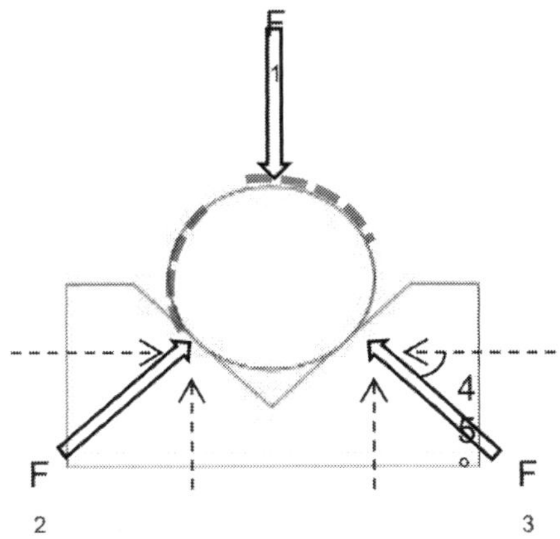

Fig. 2. Free body diagram of the piston assembly on the V-block during piston ring tension measurement. Dotted line indicate initial position of piston ring.

3 Initial oscillating wear tester fabrication

Initially an OWT design was conceptualized design criterions such as ease of sample fabrication to eliminate the need for complicated machining and enable higher repeatability in surface preparation and the ability to operate at a wide range of speeds that would be able to produce a broader representation of the friction curve behavior of an ICE as in Figure 3. To achieve these criterions that had been set, the OWT needed to have a small stroke that is similar to small ICEs and a low reciprocating weight to reduce the vibrations during testing. The next step was to make oil retention in the materials. The material was casted from aluminum and graphite was added to make peaks and valleys. The peaks and valleys undergo wear and due to their oil retention the friction coefficient is estimated to be reduced and a smoother engine operation is targeted.

Fig. 3. Oscillating Wear Tester.

4 Results and discussion

Oil retention volume, VO in $\mu m^3/\mu m^2$ is calculated via the Eq. (4) as described by [20] using the surface roughness values: Figure 4 shows the calculated oil retention volume for each of the samples in comparison with plateau honed surface. Conventional plateau honing would produce a surface with an oil retention volume of 0.83 μm3/μm2. Samples that were casted with graphite powder has even lower oil retention volume in comparison with the highest being only 0.61 $\mu m^3/\mu m^2$ and lowest being 0.49 $\mu m^3/\mu m^2$. The low oil retention volumes are attributed to the smoothness of the surface of the samples that were casted with graphite powder, having only small pits to retain oil after the piston rings sweep through the surface.

$$V_O = \frac{(100-MR2)}{200} \cdot R_{vk} \quad (4)$$

FigIn contrast to the low amounts of oil retention from the smooth surfaces of the samples casted with powdered graphite, surfaces that were sandblasted had much higher oil retention volumes which are attributed by their deep valleys that are formed when the impact of the sand particles on the surface of the substrate. Amongst the sandblasted samples, #180 sieve sandblasted sample has the highest oil retention volume with 3.68 $\mu m^3/\mu m^2$ followed by #240 sieve sandblasted sample with 2.41 $\mu m^3/\mu m^2$. #320 sieve sandblasted sample has an oil retention volume that is slightly higher than that of a conventional plateau honed surface with 1.18 $\mu m^3/\mu m^2$.

Embossing also creates surfaces that have equally high oil retention volumes as sandblasted surfaces with #320 grit sandpaper embossed surface having up to 4.46 μm3/μm2 of oil retention due to the very deep valleys created from the embossing process because of the higher contact pressure between the surface asperities to the surface. The oil retention sharply falls off to a more modest 1.85 μm3/μm2 and 1.40 μm3/μm2 when #480 grit and #640 grit sandpaper were used to emboss the surface as the valley depths of these two subsequent embossing are much less deep in comparison. Overall the sandblasted samples and embossed samples exhibited improved oil retention volumes over the conventional plateau honed sample which improves the oil film strength as well as their resistance to seizing during heavy operations. Their ability to maintain the oil retention even after excessive wear will also be good as the indentations created are deep enough to maintain the surface's ability to retain oil even after a significant amount of wear depth. Bearing area curve (BAC) or also known as the Abbott-Firestone curve or areal material ratio curve is cumulative probability density function of a surface profile's height. Typically a BAC is divided in to three sections; profile peaks, profile surface land and profile valleys. The cumulative material ratio of the peaks is the curve starting from the 100% point that falls rapidly in to a straight line which is the profile surface land. The surface profile land ends with a curve that steeply drops down to zero which represent the cumulative percentage of material height of the valleys on the surface profile. The beginning and the end of the bearing area land is denoted by the MR1 and MR2 points and the larger the bearing area land, generally the more tolerance it has for normal wear and tear. The BAC curve of the conventional plateau honed sample has a MR1 that starts at 9.8% and MR2 at 88.4%. Samples casted with graphite powder however has a MR1 value that occurs at an earlier between 4.6% to 5.3% and their MR2 points occur even later between 93.8% to 96.4%. This gives the graphite powder casted samples a slightly larger bearing area surface as seen in Figure 5 where the BAC of all three graphite powder casted samples are plotted along with the BAC of plateau honed surface for comparison. The bearing land area of the powder casted samples has a steeper gradient compared to plateau honed surface which indicates that the plateau honed surface flatter plateau region or peaks compared to the casted samples.

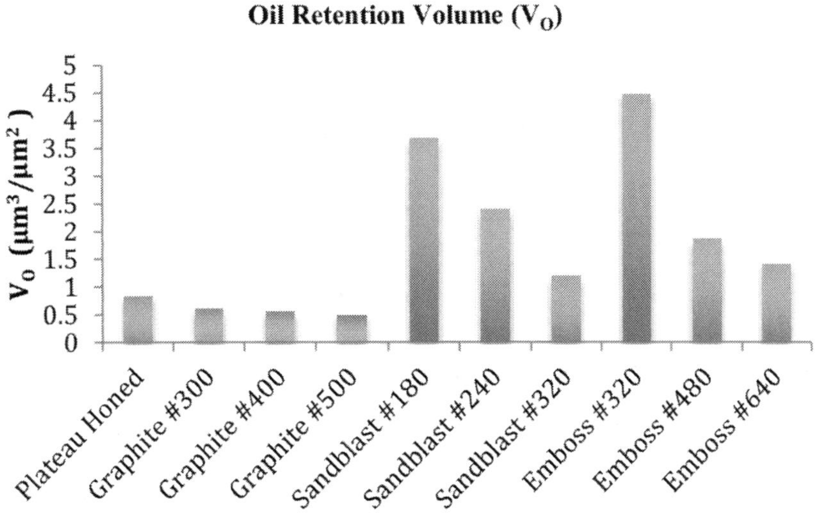

Fig. 4. Calculated oil retention volume of the samples.

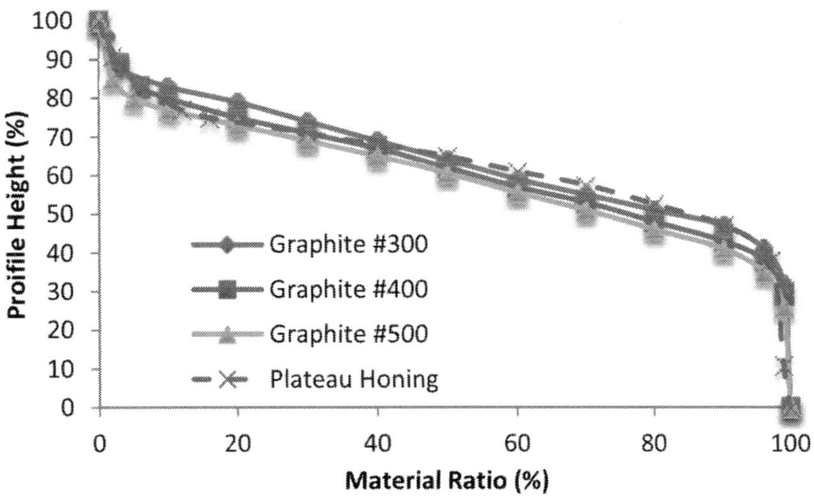

Fig. 5. Bearing area curve of Al-Si casted with sieved graphite powder in comparison to plateau honed bearing area curve.

The graphite addition seems to be beneficial in term peak and valley creation as in Figure 6. When the BAC of plateau honed sample was compared to sandblasted samples, the trend that can be seen is that the bearing area straight lines are above the bearing area lines of the conventional plateau honed sample even though the MR1 figures occur later between 8.7% to 12.8%. This is due to the nature of the surface roughness profiles that has valleys deeper than that of plateau honed surfaces. The deep valleys are shown at the end of the BAC of the sandblasted samples that fall off steeply from about 50% of the profile height as in Figure 7. The MR2 also occurs at about the same range as conventional plateau honed which is from 84.2% to 89.8% which would give it an advantage in oil retention volumes over conventional plateau honed surfaces.

DOI: 10.1051/matecconf/20179001055

The surface embossed with #320 grit sandpaper has BAC similar to that of a sandblasted sample with a low profile height % before MR1 at 11.7% and MR2 occurs at 83.5% with more than 50% of the profile height in valleys. For the sample embossed with #480 grit sandpaper, the BAC closely resembles that of a plateau honed surface with the MR1 and MR2 occurring slightly earlier at 8.5% and 84.4%. When #640 grit sandpaper was used, the resulting surface has a BAC that has higher peaks than a pleateau honed sample where the beginning part of the BAC is below the BAC of the plateau honed sample but the MR2 occurs at about the same range of profile height %. Though the profile maybe a bit peaky but the bearing landing is flatterer than the other samples.

Fig. 6. Graphite 3 Carbon Deposits on the SEM.

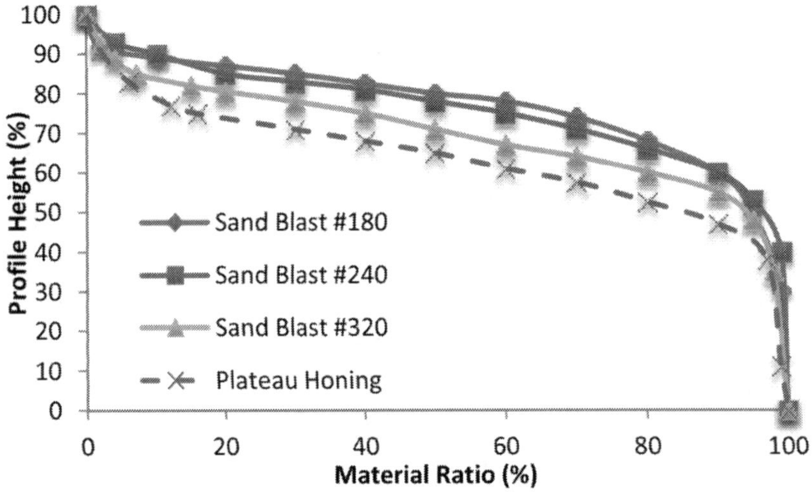

Fig. 7. Bearing area curve of Al-Si sandblasted with grated sand in comparison to plateau honed bearing area curve.

Peaks and valleys are created in sandblasted samples as in Figure 8 indicates that the samples are indeed creating oil retaining area which can aide the process of lubrication. For the embossed samples in comparison with plateau honed surface, there is a surprising amount of difference in the shape and the height of the BAC. Figure 9 is a plot of the BAC of the embossed samples in comparison to the plateau honed surface.

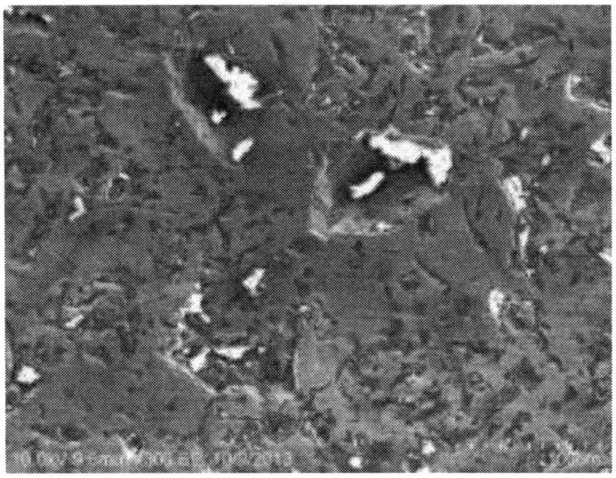

Fig. 8. Sand blast 3 porosity micro pits that increases oil retention.

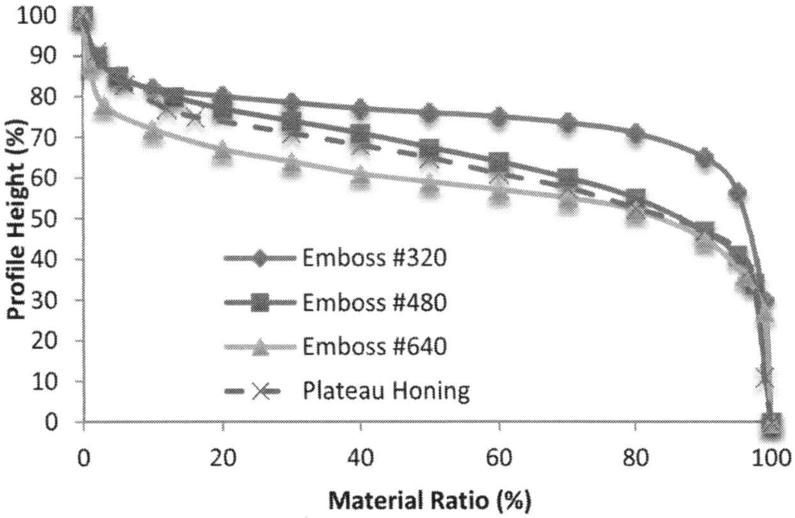

Fig. 9. Bearing area curve of Al-Si embossed with sandpaper in comparison to plateau honed bearing area curve.

Figure 10 shows that carbon deposit is similar to the EDX, with peak in Carbon value. The carbon addition shows good distribution of the valleys in Figure 11. The attrition of Aluminum in the sample increases as the wear is increased. The OWT running is make this possible. The Aluminum in the sample confirm that the material property is maintained. Based on Figure 11 the peaks and valleys occur as per BAC for the embosses sample. It is very similar to the plateau honed and can be described as a new method to increase the oil retention.

 DOI: 10.1051/matecconf/20179001055

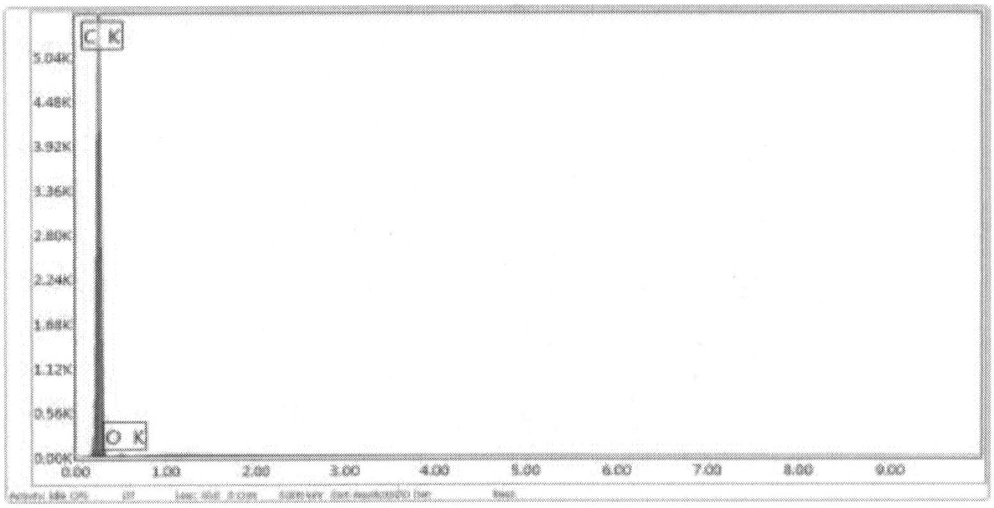

Fig. 10. EDX Showing the attrition of the sample.

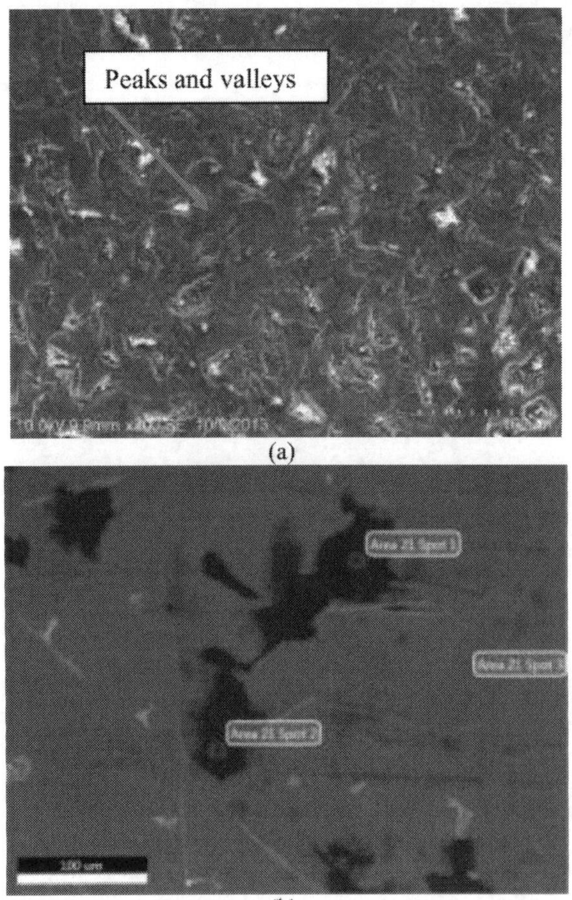

(a)

(b)

Fig. 11. Emboss 1 Surface is more distributed compared to plateau honing.

5 Conclusions

From the surface measurements, frictional measurements and wear studies of hypereutectic aluminium-silicon (ADC12) alloy, the following conclusions can be made:

1. The samples that have undergone polishing to flatten out the peaks and has lower Ra than plateau honing have on average, 24% lower coefficient of friction than plateau honed surface which shows especially significantly at speeds of 300 rpm and below and on average, 6% lower coefficient of friction at sliding speeds above 1200 rpm.
2. Hypereutectic aluminium-silicon alloy samples casted with graphite powder has significantly lower friction and wear rate attributed by the smooth polished surface that resembled a well-worn cylinder bore surface however suffer from low oil retention volumes which eventually led to lowered resistance to catastrophic wear.
3. Surfaces created from #240 sieve sandblasting and #320 grit embossing showed improvement over conventional plateau honed surface due to their lower mean surface roughness and similar skewedness values coupled with improved oil retention.

The authors would like to acknowledge Universiti Malaysia Pahang in the vote RDU 1403124 internal grants and Fakulti Kejuruteraan Mekanikal together with Automotive Engineering Research Group (AERG) for funding the research.

References

1. D. Dowson, A. Goldsmith, C. McNie, and S. Simith, *A Tribological Study of Metal-on-metal Total Replacement Hip Joints*: London, Professional Engineering Publishing, 2003.
2. G. Stachowiak and A. W. Batchelor, *Experimental methods in tribology* vol. 44: Elsevier, 2004.
3. A. E. Outlook, "Energy information administration," *Department of Energy,* 2010.
4. I. P. O. C. Change, "IPCC," *Aspectos Regionais e Setoriais da Contribuição do Grupo de Trabalho II ao 4º Relatório de Avaliação "Mudança Climática 2007" do IPCC,* 2007.
5. S. M. Miller, S. C. Wofsy, A. M. Michalak, E. A. Kort, A. E. Andrews, S. C. Biraud, *et al.*, "Anthropogenic emissions of methane in the United States," *Proceedings of the National Academy of Sciences,* vol. 110, pp. 20018-20022, 2013.
6. U. Nations, "Kyoto protocol to the united nations framework convention on climate change," 1998.
7. J. B. Heywood, *Internal combustion engine fundamentals* vol. 930: Mcgraw-hill New York, 1988.
8. C. Taylor, "Automobile engine tribology—design considerations for efficiency and durability," *Wear,* vol. 221, pp. 1-8, 1998.
9. D. Ramasamy, Z. Zainal, K. Kadirgama, and H. W.-G. Briggs, "Effect of dissimilar valve lift on a bi-fuel CNG engine operation," *Energy,* vol. 112, pp. 509-519, 2016.
10. P. Mishra, S. Balakrishnan, and H. Rahnejat, "Tribology of compression ring-to-cylinder contact at reversal," *Proceedings of the Institution of Mechanical Engineers, Part J: Journal of Engineering Tribology,* vol. 222, pp. 815-826, 2008.
11. A. Spencer, A. Almqvist, and R. Larsson, "A numerical model to investigate the effect of honing angle on the hydrodynamic lubrication between a combustion engine piston ring and cylinder liner," *Proceedings of the Institution of Mechanical Engineers, Part J: Journal of Engineering Tribology,* vol. 225, pp. 683-689, 2011.
12. J. G. Kaufman, *Introduction to aluminum alloys and tempers*: ASM International, 2000.

13. J.-m. Chen and J.-h. Jia, "Investigation of the tribological behavior of an aluminum alloy with embedded materials," *Materials Science and Engineering: A,* vol. 302, pp. 222-226, 2001.
14. Z. Dimkovski, N. Allard, C. Anderberg, F. Strömstedt, S. Johansson, B.-G. Rosen, *et al.*, "Cylinder Liner Honed Surface Optimisation-a Manufacture-Characterisation-Function Study," in *1st Swedish Production Symposium, 28-30 August, Gothenburg, Sweden*, 2007, p. 9.
15. E. Tomanik, "Friction and wear bench tests of different engine liner surface finishes," *Tribology International,* vol. 41, pp. 1032-1038, 2008.
16. L. Mourier, D. Mazuyer, F. Ninove, and A. Lubrecht, "Lubrication mechanisms with laser-surface-textured surfaces in elastohydrodynamic regime," *Proceedings of the Institution of Mechanical Engineers, Part J: Journal of Engineering Tribology,* vol. 224, pp. 697-711, 2010.
17. M. Organisciak, G. Cavallaro, and A. Lubrecht, "Variable lubricant supply of a starved hydrodynamic linear contact: lubricant lateral flow for smooth and laser textured surfaces," *Proceedings of the Institution of Mechanical Engineers, Part J: Journal of Engineering Tribology,* vol. 221, pp. 247-258, 2007.
18. D. Ramasamy, Z. Zainal, R. Bakar, and K. Kadirgama, "Mass fraction burn comparison of compressed natural gas and gasoline," in *Applied Mechanics and Materials*, 2014, pp. 442-446.
19. Y.-C. Tan and Z. M. Ripin, "Frictional behavior of piston rings of small utility two-stroke engine under secondary motion of piston," *Tribology International,* vol. 44, pp. 592-602, 2011.
20. M. Priest and C. Taylor, "Automobile engine tribology—approaching the surface," *Wear,* vol. 241, pp. 193-203, 2000.

Exploring cycle crash characteristics in Malaysia

A. Hamzah[1,*], *A. H.* Ariffin[1], *M. S.* Solah[1], *N. F.* Paiman[1], *M. R.* Osman[1]

[1]Vehicle Safety & Biomechanics Research Centre, Malaysian Institute of Road Safety Research (MIROS), Lot 125-135, Jalan TKS 1, Taman Kajang Sentral, 43000 Selangor, Malaysia.

Abstract. This paper explains the cycle road crash trend, characteristics and injuries in Malaysia. It analyses the in-depth road crash investigation data collected by the Royal Malaysian Police which was made available to MIROS crash database. Fatality data was utilized due to its consistency. Cycle casualties reflected a continual downward pattern for year 2009 to 2014 in which the number of cycle crash involvement reduced by 49% and fatalities dropped by 42%. Among the prevalent factors of cycle fatalities are >60 age group, federal and state roads and straight road sections, rural and small towns, evening peak hours, and mainly involving cars and motorcycles. It is hoped that all these information would spark interests to improve cycle safety in Malaysia.

1 Introduction

Bicycling or cycling in general, has been showing a resurgence in popularity on Malaysian roads recently. Cycling for sport and leisure are prevalently more visible than for commuting for work and utilities, though cycling is practically favorable for health and also the environment [1, 2]. Factors such as hot and humid weather were among the concerns for the low propensity to cycle. Connectivity, accessibility and traffic safety also influenced the cycling motivation [3]. With respect to road traffic injuries, cycling fatalities is ranked fourth in statistical records after motorcycle, car and pedestrian [4]. However, number of active cycles and cyclists are very challenging to be estimated due to fairly loose cycle registration policy. The requirement to register falls under the purview of local government authorities [5]. Nevertheless, observational studies of cycle use were able to provide preliminary indications of the users' population, though in fragments [6,7]. Additionally, cycle safety research seems yet to be a subject of high interest to the research community in Malaysia and it is hoped that this introductory information would spur the motives.

This report attempts to describe cyclists' road traffic injuries risks relative to fatality records and analyse the cycle crash characteristics and at the same provide a basis for future cycle-centric research.

* Corresponding author: azhar@miros.gov.my

2 Methodology

Road traffic casualties' data from the year 2009 to 2014 was acquired from the publication of Royal Malaysian Police and was used to reflect the overall injury trend [11]. For in-depth analyses, data for 2009 to 2013 which were documented in MIROS database system were utilized due to information extensiveness and accessibility [12]. Fatality records for all five years were aggregated together for ease of analyses and to provide clear visibility. The fatality data were considered in the analyses due to its high reliability and accuracy [8,9]. The severe and slight injuries records were disregarded due to concerns of incomplete and possibly massive underreporting, especially for cycle crashes [9,10]. This paper forms part of preliminary study on cycle safety in Malaysia.

It is to be noted that quite a substantial amount of unknown data was observable especially in Figure 2, 5 and 6 which concern cycle fatalities over age groups, time of day and also collision type, respectively. Such issue could be possibly attributed to incomplete information and data entry inaccuracies during processing, which should be looked into to improve data reliability.

3 Results

3.1 Cycle casualties, 2009-2014

Cycle injuries in Malaysia has been showing a consistent reduction for the last few years, as shown in Figure 1. Though the serious and minor injuries accident data were doubted by some studies, the trend reflected a similar downward pattern to fatal injuries. From 215 fatalities in 2009, the absolute fatal number has dropped to 124 in 2014, indicating a 42% reduction in the 6-year span and is shown by the trend line.

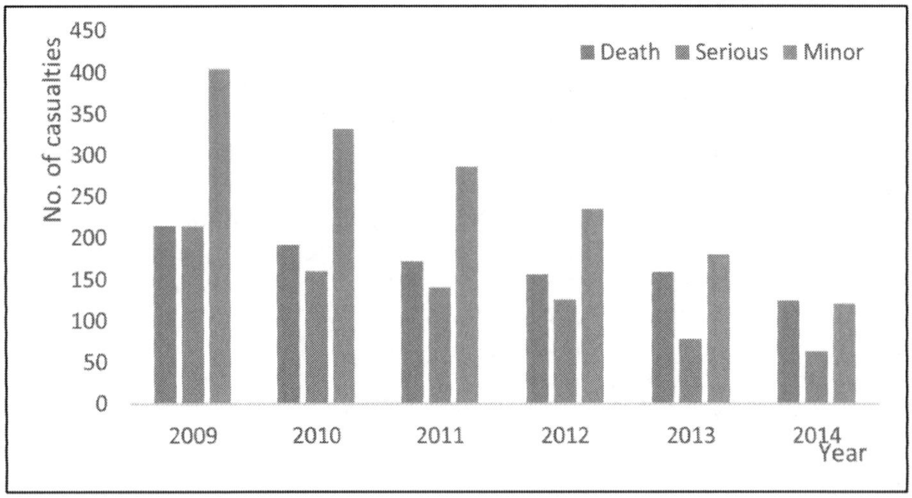

Fig. 1. Cycle casualties in road traffic [12].

Correspondingly, the number of cycles involved in road crashes has also reflected a continual downward pattern throughout the six years, with year 2013 recorded a considerable 31% drop. Comparing year 2009 to 2014, the cycle involvement in crashes has dropped significantly by 49%.

Table 1. Number of cycles involved in crashes [12].

Year	No. of cycles	Year to year reduction, %
2009	2486	-
2010	2178	12
2011	2033	7
2012	2000	2
2013	1370	31
2014	1275	7

3.2 Cycle fatalities by age group

The dominant age group was >60 year-old which represents about 26% (234 fatalities with average of 46 per year) of the total cycle fatalities (Figure 2) and displayed a substantial surge in comparison to other age groups. Fatalities among other age groups were scattered between 20 to 49 fatalities, excluding the age group of 1 to 5 year-old.

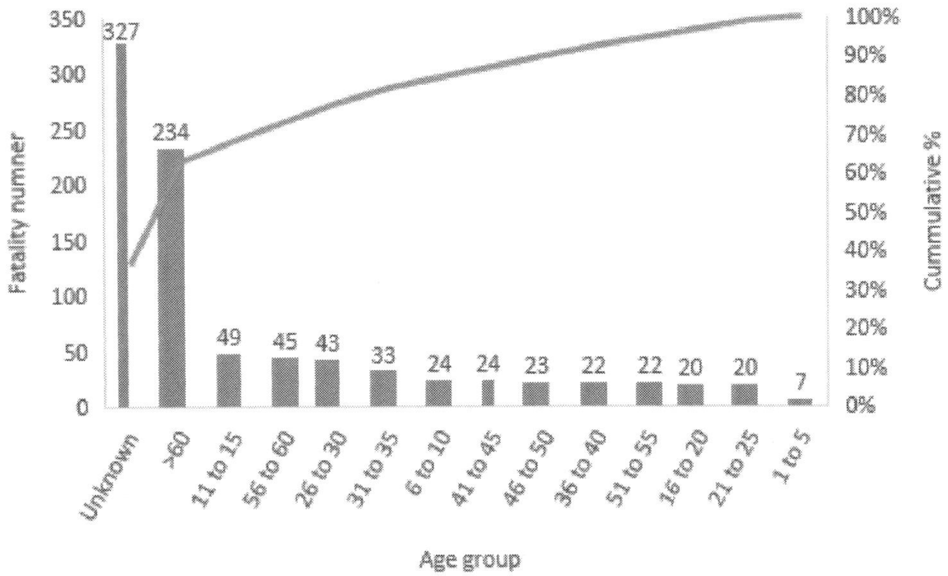

Fig. 2. Cycle fatalities by age group [12].

3.3 Cycle fatalities by road category and locality

In terms of road type, more than half of cycle fatalities occurred in the federal (28%) and state (25%) roads, as shown in Figure 3 (I). About 19% of fatalities occurred in municipal roads. For locality, rural (31%) and small town (14%) areas dominated the fatality record. Fatal cases in town and city were rather comparatively low (9% and 3%, respectively).

3.4 Cycle fatalities by road geometry

Straight road section was the most dangerous for cyclists (63%) and was followed by T/Y junctions (6.9%) and cross junction (4.5) (Figure 4). Curve sections and roundabout accounted for 3.7% and 1.7%, respectively. Cycle fatalities at staggered junction was low at 0.2%.

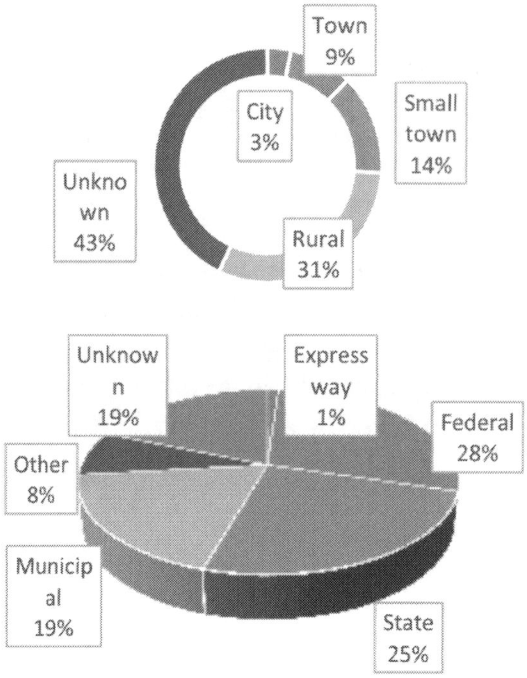

Fig. 3. Cycle fatalities by road category (l) & locality (r) [12].

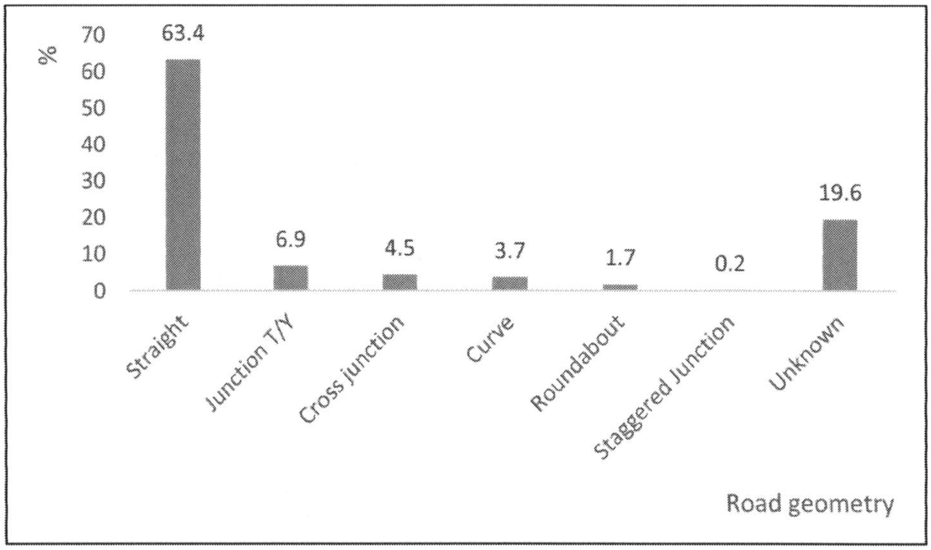

Fig. 4. Cycle fatalities by road geometry [12].

3.5 Cycle fatalities by time of day

More than 200 fatalities occurred around the time of 1801-2200 hours and were closely followed by 0601-1000 hours (>160 fatalities), as shown in Figure 5. These time intervals

DOI: 10.1051/matecconf/20179001056

coincide with the common peak commuting time. Moreover, cycle fatalities spread in all time intervals including midnight and early wee hours which was a startling fact.

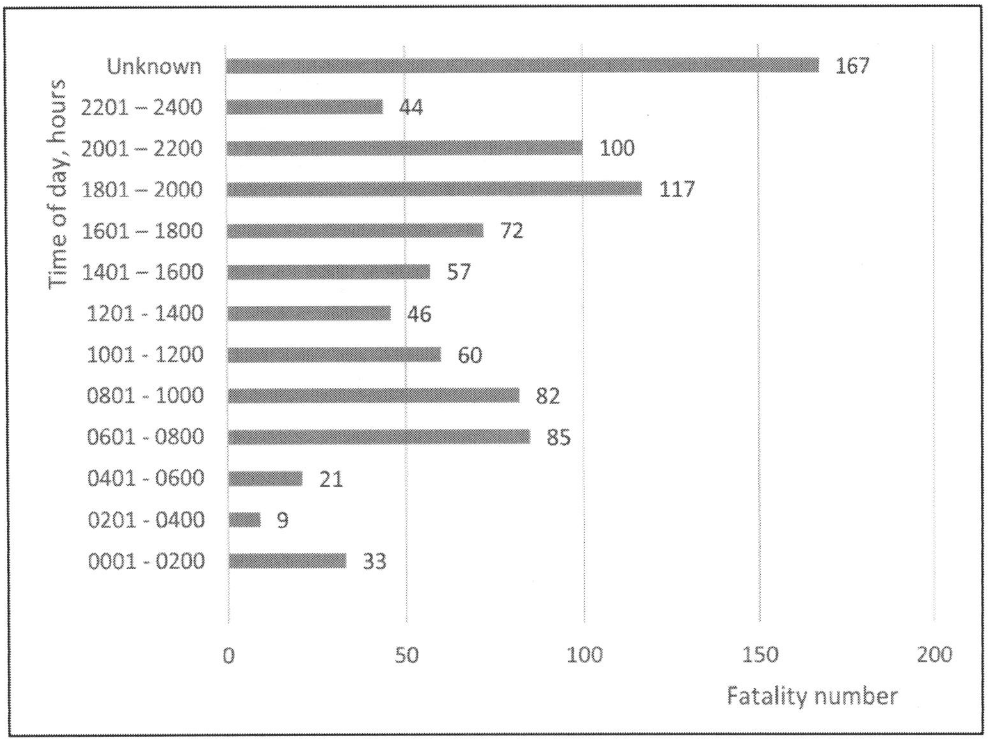

Fig. 5. Cycle fatalities by time of day [12].

3.6 Cycle fatalities by collision type

Based on Figure 6, angular/side collision was the most prevalent collision type (17.6%) with rear end (11.2%) and side swipe (10.3%) collisions were the next serious types. Other collision types include 90-degree (6.6%), head-on (5.7%) and out-of-control (1.1%). There were a few collision types which accounted less than 1% each, namely involving pedestrian, object on ground and off-road, squeezed and overturned.

3.7 Cycle fatalities by vehicle type

Car - cycle and motorcycle-cycle were the most common collisions (35.6% and 21.6% respectively) (Figure 7). There were extensive involvements of other mode of transports, ranging from few types of buses and lorries.

3.8 Cycle lane and facilities

In general, cycle lanes and cycle facilities are still lacking in cities and towns except in certain public parks and recreational areas. A brief survey of cycle facilities was conducted in few selected cities and areas, namely Putrajaya, Petaling Jaya and Pulau Pinang. The main aims were to have an overview of cycle lanes and cycle-related policies consideration in town planning and development. In short, cycle lanes and cycle-related facilities are indeed given considerable attention in existing town expansion plan and in new development projects where sustainable development and green initiative has been set as the key elements for growth. A number of cycle lanes and facilities have been constructed to encourage safe cycling and proper point to point connectivity (Figure 8). Nevertheless, a lot more are needed.

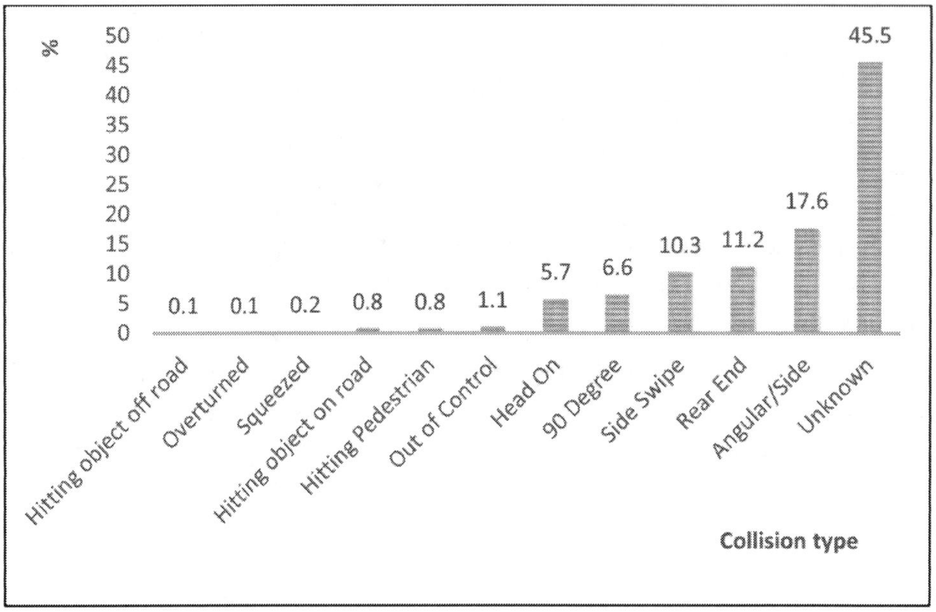

Fig. 6. Cycle fatalities by collision type [12].

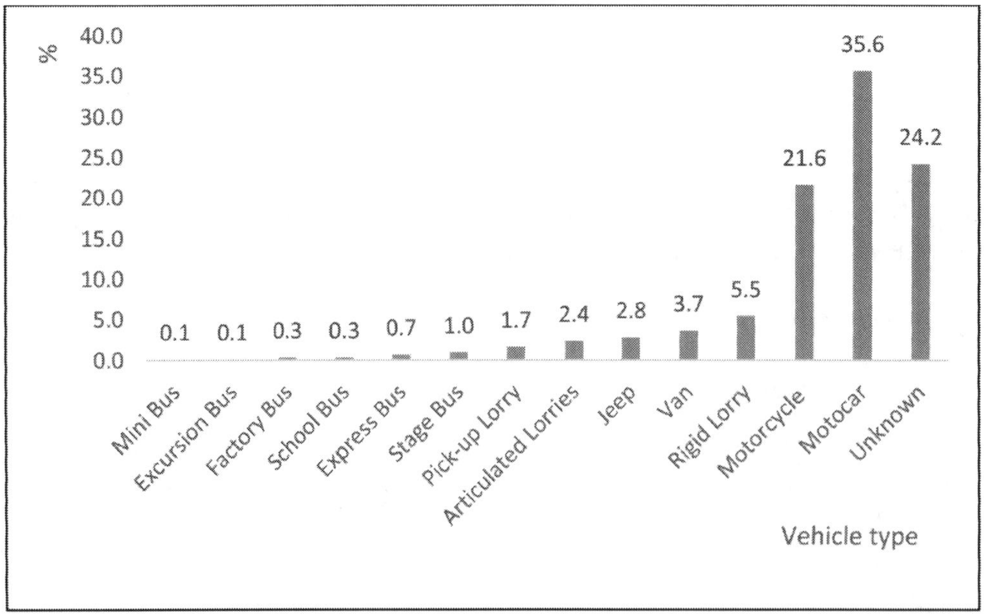

Fig. 7. Cycle fatalities by vehicle involvement [12].

Fig. 8. Cycle lanes in Petaling Jaya (l) and Pulau Pinang (r).

4 Discussion

Overall cycle fatalities reflected a downward trend from the year 2009 to 2014 and this favorable reduction was mainly originated from the age groups listed in Table 2. Main improvements were from the age group of 11 to 15 (57%) and 21 to 25 year-old (50%) while other age group displayed encouraging improvements. In term of absolute number, the unknown group was the biggest contributor to the decrease. 11 to 15 age group are school-going children however it could not be established whether the cases occurred during commuting to or from school.

Table 2. Reduction in cycle fatalities [12].

Age group, years	Mean	2009	2013	% Reduction
11 to 15	9.8	14	6	57
21 to 25	4	4	2	50
26 to 30	8.6	10	6	40
31 to 35	6.6	7	4	43
36 to 40	4.4	5	4	20
Unknown	65.6	69	35	49

Cyclists with age group of >60 year-old were the most dominant (26%) in cycle fatalities with an average of 46 numbers for the year 2009 to 2013. It is a known fact that elderly cyclists have the highest crash and injury risks among vulnerable road users and their overrepresentation in cycle deaths could go as high as 70% [13]. Cyclists in this age group involved in crashes most probably cycle for utility or leisure purposes, possibly at non-peak hours. A big portion (72%) of cycle fatalities took place in federal, state and municipal roads and were more prominent in small towns and rural areas (45%). With regards to road geometry, big majority of cycle deaths happened at straight road sections (63%). Correspondingly, straight section was also reported as the most hazardous in motorcycle crashes [4].

The time of day could be further classified into day, night and others with regards to the availability of natural light, though the cut-off time is arguable (Table 3). About 35% of fatal cycle crashes occurred during daylight. Cases at dusk and dawn (23%) were similar to cases at night although the time intervals at night were relatively more than double. A surprising fact is that there was a substantial fatality at night (23%), possibly indicating the presence of considerable number of active cyclists on roads. Cycle naturally has low conspicuity on roads due to its narrow size and low light environment may exacerbate it crash risk [14].

Table 3. Cycle fatalities according time of day [12].

Time of day	Time interval	Fatalities	%
Day	0801-1800	317	35
Dusk	1801-2000	117	13
Night	2001-0600	207	23
Dawn	0601-0800	85	10
Unknown	-	167	19

Angular/side collision was the most prominent collision type (17.6%) and considering fatal injuries and the common cycle speed, the opposing vehicles were presumably the most likely striker. The most two common opponents for cycle crash were passenger cars (35%) and motorcycles (22%) and could be explained by their high registration in the country. One way to improve the situation is by segregating cycles from other transport mode by creating designated cycle lanes [15]. The initiative to create cycle lanes in existing road networks is strongly encouraged to promote safe cycling.

5 Conclusions

Cycle involvement in road crashes in Malaysia has been showing a considerable decrease for the past few years till 2014. This positive improvement is reflected by lower cycle involvement in road crashes which consequently lead to fewer cycle deaths and injuries. However, the current records showed that the elderly group of >60 years old was overrepresented with an average of 46 deaths occurred annually. Most cycle deaths happened in straight road sections, at the state and federal road class, and in rural and small towns. By time of day, peak commuting hours in early morning (0600-1000 hours) and late afternoon (1800-2200 hours) were the commonest period for fatal crashes. The most fatal crash configuration was angular/side impacts involving predominantly passenger cars (motorcar) and motorcycles. Having all these information, it is hoped that more researches, effort and initiatives would be put in place to further improve cycle safety in the country.

The authors would like to extend their greatest appreciation to MIROS for providing the grant, to MIROS Crash Data & Exposure Analysis Unit for the assistance in providing traffic crash details. Lastly, greatest gratitude to Royal Malaysian Police for sharing their precious accident data.

References

1. http://malaysiacycling.blogspot.my, accessed May 21, (2016).
2. A. S. I. Almselati, R. Rahmat, O. Jaafar O, "An overview of urban transport in Malaysia," Social Sci. **6(1)**:24–33, (2011).
3. B. Afsar, Y. M. Golriz Nikjooy, "The consequences of the physical-environmental factors on encouraging pedestrian and cycling activities in UPM Campus," Int. J. Sci. & Eng. Research, **v6 (1)**, (2015).
4. M. M. Abdul Manan, A. Várhelyi. "Motorcycle fatalities in Malaysia,". IATSS Research. Jul; **36(1)**:30–9, (2012).
5. Act 171, Local Government Act, 1976, Laws of Malaysia.
6. M. R. M. Yazid, R. Ismail, R. Rahmat, M. Nazri, "City residence prepare towards implementation local lane of non-motorized". Res. J. App. Sci., Eng. and Tech., Selangor, Malaysia. 2012; **4(5)**:481–485, (2012).
7. S. A. A. Dawood, O. K. Rahmat R. A, "Factors that affect cycling transportation mode

for postgraduate students at Universiti Kebangsaan Malaysia by logit method. Jurnal Kejuruteraan. **27**:1–7, (2015).
8. N. Haworth, "How valid are motorcycle safety data?" The Road Safety Research, Policing and Education Conference, Sydney, Australia, (2003).
9. D. L. Rosman, M. W. Knuiman, "A comparison of hospital and police road injury data". Accident Analysis & Prevention. **26(2):**215–222, (1994).
10. R. Elvik, A. Mysen, "Incomplete accident reporting: meta-analysis of studies made in 13 countries". Transportation Research Record: Journal of the Transportation Research Board. Jan 1(1665):133-40, (1999).
11. Laporan Perangkaan Kemalangan Jalan Raya Malaysia, Polis Diraja Malaysia, 2009-2014.
12. MRoads cycle crash report, Malaysian Institute of Road Safety Research, (2016).
13. J. Oxley, B. Corben, B. Fildes, M. O'Hare, T. Rothengatter. "Older vulnerable road users- measures to reduce crash and injury risk". Monash University Accident Research Centre Reports. (218):162, (2004).
14. S. J. Thornley, A. Woodward, J. D. Langley, S. N. Ameratunga, A. Rodgers. "Conspicuity and bicycle crashes: preliminary findings of the Taupo Bicycle Study". . Injury Prevention.**14(1):**11-8, (2008).
15. J. Parkin, C. Meyers. "The effect of cycle lanes on the proximity between motor traffic and cycle traffic". Accident Analysis & Prevention. **42(1):**159-65, (2010).

MATEC Web of Conferences **90**, 01057 (2017) DOI: 10.1051/matecconf/20179001057
AiGEV 2016

Study on waste heat recovery from exhaust gas spark ignition (S.I.) engine using steam turbine mechanism

Kamarulhelmy Talib[1,*], *Safarudin* G. Herawan[1,2], *Musthafah* M. Tahir[1,2], *Azma* Putra[1,2], and *Shamsul* A. Shamsudin[1,2]

[1]Faculty of Mechanical Engineering, Universiti Teknikal Malaysia Melaka, Hang Tuah Jaya, 76100 Durian Tunggal, Melaka, Malaysia.
[2]Centre for Advanced Research on Energy, Universiti Teknikal Malaysia Melaka, Hang Tuah Jaya, 76100 Durian Tunggal, Melaka, Malaysia.

Abstract. The issue of global warming has pushed the effort of researchers not only to find alternative renewable energy, but also to improve the machine's energy efficiency. This includes the utilization of waste energy into 'useful energy'. For a vehicle using internal combustion engine (ICE), the waste energy produce by exhaust gas can be utilize to 'useful energy' up to 34%. The energy from the automotive exhaust can be harness by implementing heat pipe heat exchanger in the automotive system. In order to maximize the amount of waste energy that can be turned to 'useful energy', the used of appropriate fluid in the heat exchanger is important. In this study, the fluid used is water, thus converting the fluid into steam and thus drive the turbine that coupling with generator. The paper will explore the performance of a naturally aspirated spark ignition (S.I.) engine equipped with waste heat recovery mechanism (WHRM) that used water as the heat absorption medium. The experimental and simulation test suggest that the concept is thermodynamically feasible and could significantly enhance the system performance depending on the load applied to the engine.

1 Introduction

Years by years, the number of motor vehicles in the world continues to grow globally and therefore increases the reliance on the petroleum and increases the release of carbon dioxide into atmosphere. Eventually, this will contribute to global warming and the increase of earth's temperature. To overcome this trend, new vehicle technologies must be introduced to achieve better fuel economy consumption without increasing harmful gaseous emissions. For internal combustion engine (ICE) in most typical gasoline fuelled vehicles, for a typical 2.0 L gasoline engine used in passenger cars, it was estimated that 21% of the fuel energy is wasted through the exhaust at the most common load and speed range [1]. The rest of the fuel energy is lost in the form of waste heat in the coolant, as well as friction and parasitic losses.

* Corresponding author: helmy2127@gmail.com

Nowadays, the requirement of electric loads in a vehicle are increasing due to improvements of comfort, driving performance, and power transmission. Therefore, it is of interest to utilize the wasted energy by developing the heat recovery mechanism of exhaust gas from internal combustion engine. It has been identified in [2] that the temperature of the exhaust gas varies depending on the engine load and engine speed. As the engine speed increases, the temperature of the exhaust gas will also increase. By implementing heat recovery concept in the vehicle, a significant amounts of energy that would normally, be lost via engine exhaust can thus be recovered into electrical energy. Theoretically, the energy that loss from the exhaust can be recovered to supply an extra power source for vehicles. Thus will result in lowering fuel consumption, greater efficiency, and overall reduction in greenhouse gas emission.

The technologies on waste heat recovery of IC engine is consist of low grade heat from cooling system and high grade heat from exhaust system. For low grade waste heat, the Organic Rankine Cycle is the favourite choice to recover waste heat energy [3], whereas high grade heat, several techniques of recover the energy are applied such as thermoelectric generator [4], turbocharger [5], turbo-compound [6], Rankine cycle system [7], heat pipe [8], air conditioning [8], heat exchanger for thermal energy storage [9], and heat exchanger for fuel conservation, emission reduction [10], and power turbine of waste heat recovery mechanism [11]. The approaches either by theoretical, simulation or experimental works it leads to improve the brake fuel consumption that generated better overall efficiency.

Based on the application of waste heat recovery that are more attractive to be explored are turbocharger, and turbo-compound due to the simple construction and less cost. This is because the potential to produce output power and improve BSFC and efficiency are promising. To improve the performance of turbocharger, several studies [5, 11] used electrical motor assisted or Integrated Starter Generator (ISG) to turbocharger or to parallel hybrid power system on a diesel engine [11]. The aim of this method is to improve the fuel economy of diesel Homogeneous Charge Compression Ignition (HCCI), and the results are the fuel economy, the soot emission and NOx emission are improved by 10.9%, 6% and 12.1% respectively [11]. The electrical motor assisted turbocharger is the integration of a high-speed 7.5 kW electric motor-generator within a standard turbocharger of a heavy-duty vehicle, which the purpose of this system is to improve the fuel economy and turbo-lag. As a result, this system on an urban bus, the fuel economy can be improved by up to 6% or above depending on the actual driving cycle and an electrical assist motor can reduce turbo-lag by typically 50% [12].

2 Experimental setup

2.1 Design of experiment

The experiment was performed on a Toyota vehicle having 1.6 litre in-line four-cylinder gasoline engines. Table 1 shows the specification of the test engine. A schematic diagram of the experimental setup is shown in Figure 1. A 75 Watt bulb were used as a load causing the DC generator to produce an output current and voltage which were recorded in a computer through USB digital multimeter. The air duct to the intake manifold of engine was equipped with a pitot tube digital anemometer to measure the volume flow rate of the intake air. The engine speed and the WHRM turbine speed were continuously monitored using an optical tachometer allowing the digital data to be recorded in a computer through USB data acquisition module. This was also applied for the data of the throttle position for intake air captured using the existing throttle sensor in the experimental vehicle. The test was conducted on the road with variable vehicle speed up to 70 km/h with normal driving and full throttle driving to measure the performance of engine with and without WHRM. Some features of the instrumentation are summarized in Table 2. To determine the performance of engine on the full throttle driving, a simulation program is needed. This simulation is created under Matlab Simulink environment. The program is using demo Simulink program [13] with some modification to suit the target. From the simulation, we can get the power and torque of the engine experimental vehicle. The results are needed to compare between before and after implementation of waste heat recovery mechanism, which then can see whether these mechanisms affect the performance of engine or not. Figure-2 shows the original demo program Simulink and some modification had been made.

Table 1. Specification of the test engine.

Type	Specification
Valve train	DOHC 16 valves
Fuel system	Multi point fuel injection
Displacement	1587 cc (in-line)
Compression ratio	9:4:1
Bore	81 mm
Stroke	77 mm
Power	112Hp @ 6600 rpm
Torque	131Nm @ 4800 rpm

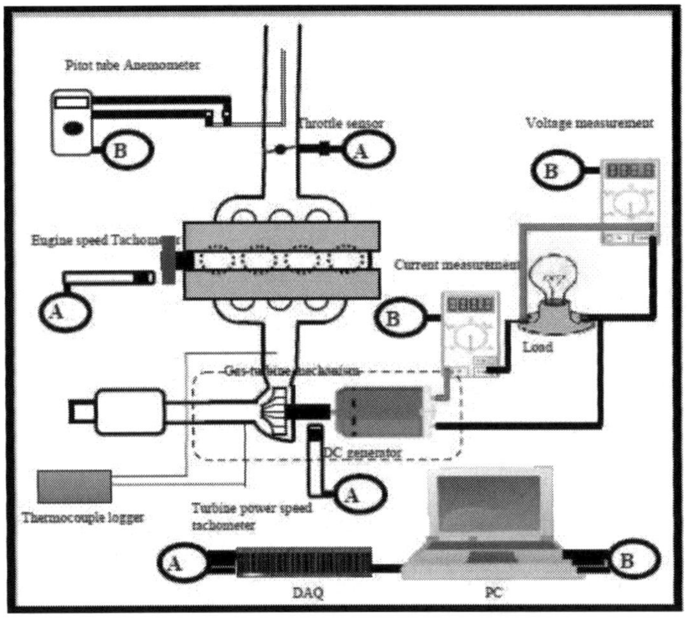

Fig. 1. Schematic layout of the experimental setup of waste heat recovery mechanism.

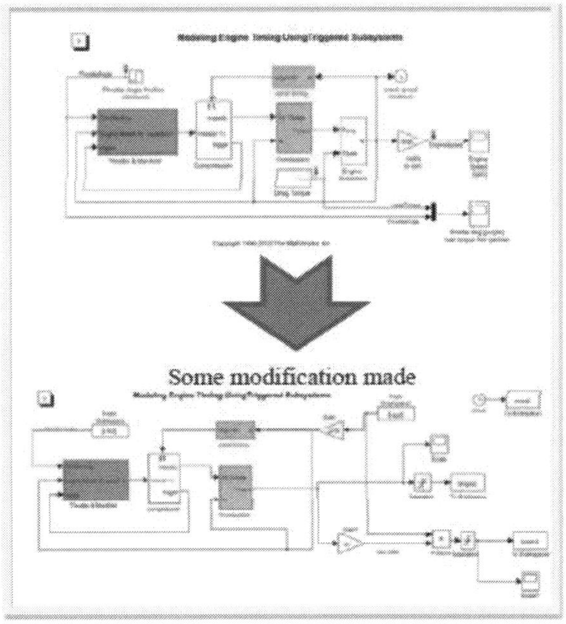

Fig. 2. Modified Demo Simulink Program.

2.1 Experimental procedure

One of the thing that make this project different from other waste heat recovery experimental is the data gather from the project is obtain from real condition, on the real road track without any controllable variable. The only constant variable is the track of the road test is same, that situated at Auto-City, MITC, Melaka as shown in Figure 3.

Fig. 3. The plan of the experiment track.

A 75 Watt bulb were used as a load causing the DC generator to produce an output current and voltage. The value of the current and voltage will then be recorded in a computer through USB digital multimeter. In this experiment, up to four 75 Watt bulbs will be used in order to generate bigger current and voltage values. As the number of the bulbs increases, the load given to the DC generator increases thus the value of current and voltage produces will also increase. The experiment is design to be run in ten (10) rounds as shown in Figure 4.

START

Round 1 and 2 – Free Load:
1) Round 1, the car will run on the track with normal speed average 60 – 70 km/h.
2) Round 2, there will be full-throttle acceleration (start point as shown in Figure 4) for about 8 seconds.
3) After that, the car will stop and added with one load (75 Watt bulb).

Round 3 and 4 – One Load (75 Watt bulb):
1) Round 3, the car will run on the track with normal speed average 60 – 70 km/h.
2) Round 4, there will be full-throttle acceleration (start point as shown in Figure 4) for about 8 seconds.
3) After that, the car will stop and added with one load (75 Watt bulb).

Round 5 and 6 – 2 Loads (75 Watt bulb):
1) Round 5, the car will run on the track with normal speed average 60 – 70 km/h.
2) Round 6, there will be full-throttle acceleration (start point as shown in Figure 4) for about 8 seconds.
3) After that, the car will stop and added with one load (75 Watt bulb).

Round 7 and 8 – 3 Loads (75 Watt bulb):
1) Round 7, the car will run on the track with normal speed average 60 – 70 km/h.
2) Round 8, there will be full-throttle acceleration (start point as shown in Figure 4) for about 8 seconds.
3) After that, the car will stop and added with one load (75 Watt bulb).

Round 9 and 10 – 4 Loads (75 Watt bulb):
1) Round 9, the car will run on the track with normal speed average 60 – 70 km/h.
2) Round 10, there will be full-throttle acceleration (start point as shown in Figure 4) for about 8 seconds.
3) After that, the car will stop and added with one load (75 Watt bulb).

END

Fig. 4. Flowchart of the road test.

 DOI: 10.1051/matecconf/20179001057

3 Results and discussion

Figure 5, Figure 6 and Figure 7 shows the heat energy come out from the exhaust gas on experimental vehicle in terms of engine speed, exhaust temperature, air flow rate and throttle angle.

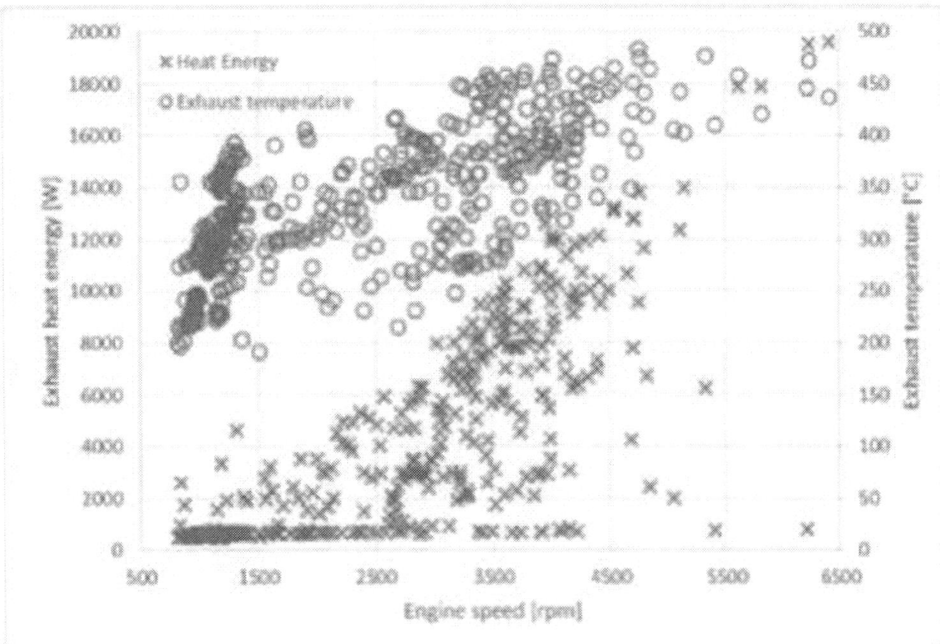

Fig. 5. Heat energy from exhaust gas of experimental vehicle on engine speed and exhaust temperature.

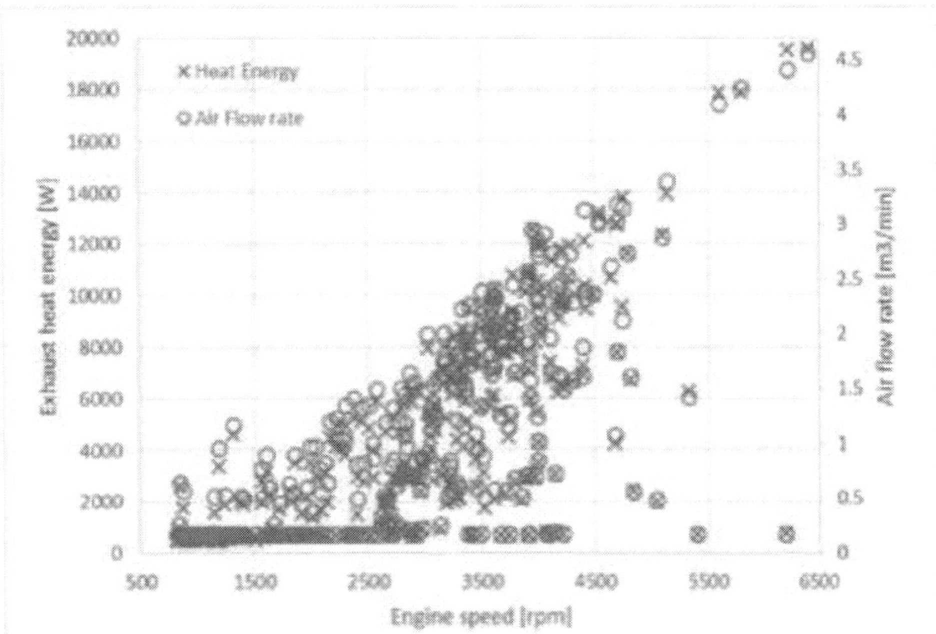

Fig. 6. Heat energy from exhaust gas of experimental vehicle on engine speed and air flow rate.

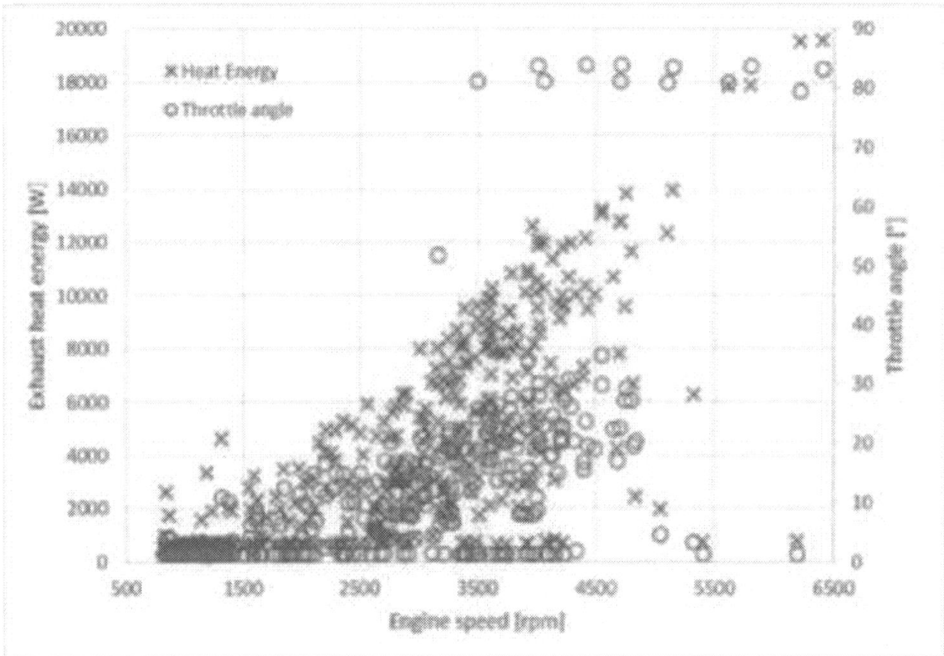

Fig. 7. Heat energy from exhaust gas of experimental vehicle on engine speed and throttle angle.

Theoretically, the heat energy produced from the exhaust is depending on the engine speed and exhaust temperature. As the engine speed increases, the exhaust temperature will increase. However, there are some behaviours that not accomplish with this argument. This is because the road test was conducted on-road test where no parameters were being controlled. The measurement is solely based on the normal driving and full throttle driving on the road, resulting both dynamic and transient conditions. The heat energy from exhaust is varies in the range of 500 W up to almost 23 kW, which is a good agreement with [14] where the heat energy is in the range of 5 kW and can reach up to 23 kW. Air flow rate give a significant effect on the heat energy as almost each single data of both parameters are match each other. Even though engine speed and exhaust temperature are high, but the air flow rate is low. The heat energy follows the air flow rate consistently.

In term of throttle angle, as the throttle angle increases, the heat energy will also increases as shown in Figure 7. By applying range of 10 to 30 degree on the normal driving, the heat energy produce is in the range of 2 kW to 12 kW. However, at the angle of almost 90 degrees (full throttle angle), not all the heat in the maximum value. This is once again due to the actual driving on the road as dynamic and transient conditions. Therefore, to optimize the heat energy from exhaust, the engine speed, exhaust temperature, throttle angle and air flow rate to be increases. As all these parameters be in optimum value, the heat energy from the exhaust can be generated at the optimum condition. It is obvious that the waste heat produced from the exhaust is high. Thus, it is significant and worthy to recover this wasted energy.

Figure 8 shows the measured data of the parameters which have already combined between the normal driving test and full throttle test. As seen in Figure 8, correlations between all the four parameters (engine speed, exhaust temperature, throttle angle, and air flow rate) can be clearly observed. The higher the value of throttle angle, the higher the air flow rate and thus increases the speed of the engine. The change of throttle angle also directly affects the WHRM turbine speed, which eventually changes the output current and voltage accordingly.

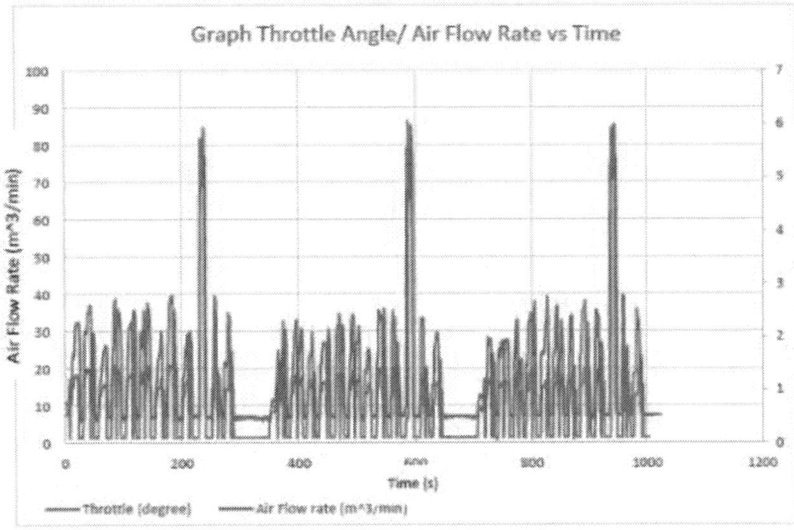

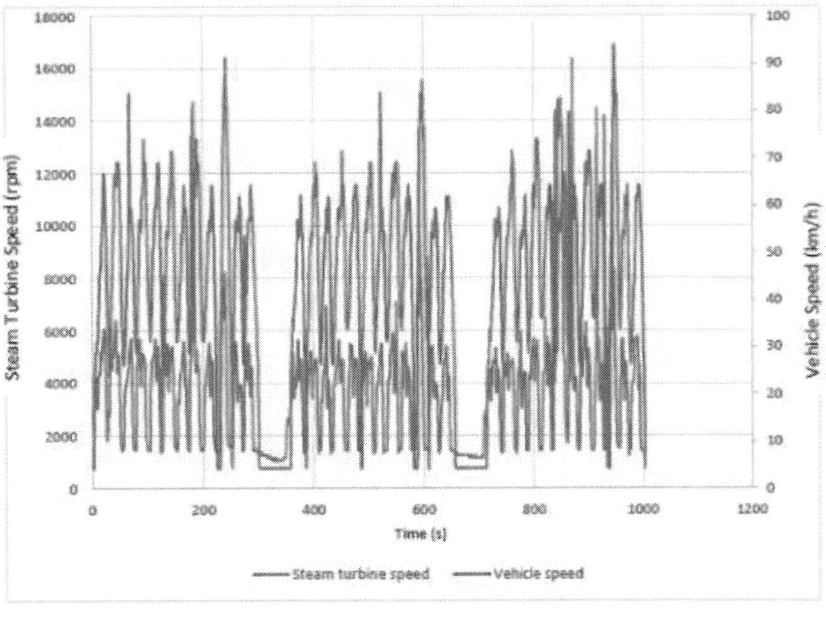

Fig. 8. The measured data from the experimental vehicle with WHRM.

Figure 9 shows the voltage produced from WHRM when applied two loads and three loads of 75 Watt bulbs. From the graph, the maximum voltage that produce by two loads and three loads are 21.04 V and 33.32 V respectively. This prove as the number of loads increases, the voltage produce will also increase and lead to producing higher power output.

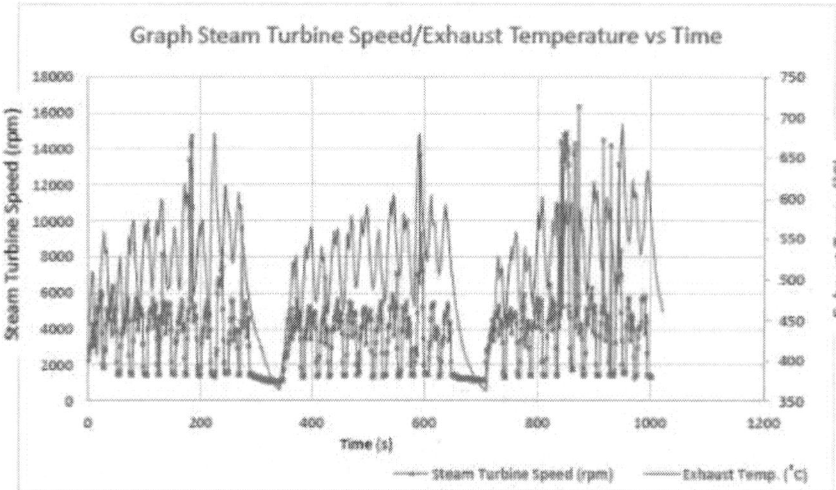

Fig. 9. Voltage produced from WHRM.

4 Conclusions

Utilization of waste heat energy from exhaust gas using WHRM system in a S.I. engine has been reported. The system is proved to produce voltage up to 33.32 V at normal driving conditions. The proposed system could become a potential energy recovery that can be stored in auxiliary battery and used for electrical appliances in automotive vehicle.

The authors would like to acknowledge the Universiti Teknikal Malaysia Melaka (UTeM) and Ministry of Higher Education Malaysia, for funding support and facilities through the Fundamental Research Grant Scheme (FRGS) no. FRGS/2/2014/TK06/FKM/02/F00236.

References

1. R.E. Chammas and D. Clodic, Combine cycle for hybrid vehicle, *SAE Paper 2005,* **01**, 1171 (2005)
2. Z. Peng, T. Wang, Y. He, X. Yang, L. Lu, Analysis of environmental and economic benefits of integrated Exhaust Energy Recovery (EER) for vehicles, *Applied Energy,* **105**, 238-243, (2013)
3. A. A. Boretti, Transient operation of internal combustion engines with Rankine waste heat recovery systems, *Applied Thermal Engineering,* **36**, 18-23, (2012)
4. J. Serrano, V. Dolz, R. Novella, A. Garcia, DH diesel engine equipped with a bottoming Rankine cycle as waste heat recovery system, Part 2: Evaluation of alternative solution, *Applied Thermal Engineering,* **36**, 279-287 (2012)
5. F. Yang, G. Gao, M. Ouyang, L. Chen, Y. Yang, Research on a diesel HCCI engine assisted by an ISG motor, *Applied Energy,* **101**, 718-79, (2013)
6. R. Zhao, W. Zhuge, Y. Zhang, Y. Yin, Z. Chen, Z. Li, Parametric study of power turbine for diesel engine waste heat recovery, *Applied Thermal Engineering*, **67(1)**, 308-319, (2014)
7. T. Wang, Y. Zhang, J, Zhang, G. Shu, Z. Peng, Analysis of recoverable exhaust energy from a light-duty gasoline engine, *Applied Thermal Energy*, **53(2)**, 414-419, (2013)
8. F. Yang, X. Yuan, G. Lin, Waste heat recovery using heat pipe heat exchanger for heating automobile using exhaust gas, *Applied Thermal Energy,* 23**(3)**, 367-372, (2003)
9. P. Kauranen, T. Elonen, L. Wikstrom, K. Heikkinen, J. Laurikko, Temperature optimisation of a diesel engine using exhaust gas heat recovery and thermal energy

storage (diesel engine with thermal energy storage), *Applied Thermal Energy,* **30(6)**, 631-638, (2010)
10. F. Will, Fuel conservation and emission reduction through novel waste heat recovery for internal combustion engines, *Fuel,* **102**, 247-255, (2012)
11. S. G. Herawan, A. H. Rohhaizan, A. Putra, A. F. Ismail, Prediction of Waste Heat Energy Recovery Performance in a Naturally Aspirated Engines Using Artificial Neural Network, *ISRN Mechanical Engineering,* (2014)
12. J. Bumby, E. Spooner, J. Carter, Electrically assisted turbochargers; their potential for energy recovery, 2016
13. T. MathWorks, Matlab, The MathWorks, Natick, MA, (2009)
14. N. Yamada, M.N. Mohamad, Efficiency of hydrogen internal combustion engine combined with open steam Rankine cycle recovering water and waste heat, *International Journal of Hydrogen Energy.* **35(3)**, 1430-1442, (2010)

 DOI: 10.1051/matecconf/20179001058

Dimensional accuracy of internal cooling channel made by selective laser melting (SLM) And direct metal laser sintering (DMLS) processes in fabrication of internally cooled cutting tools

S. A. C. Ghani [1,*], *M. H.* Zakaria[1], *W. S. W.* Harun [1], *Z.* Zaulkafilai [1]

[1]Human Engineering Group, Faculty of Mechanical Engineering,
Universiti Malaysia Pahang, 26600 Pekan Pahang, Malaysia.

Abstract. Selective laser melting(SLM) and direct metal laser sintering(DMLS) are preferred additive manufacturing processes in producing complex physical products directly from CAD computer data, nowadays. The advancement of additive manufacturing promotes the design of internally cooled cutting tool for effectively used in removing generated heat in metal machining. Despite the utilisation of SLM and DMLS in a fabrication of internally cooled cutting tool, the level of accuracy of the parts produced remains uncertain. This paper aims at comparing the dimensional accuracy of SLM and DMLS in machining internally cooled cutting tool with a special focus on geometrical dimensions such as hole diameter. The surface roughness produced by the two processes are measured with contact perthometer. To achieve the objectives, geometrical dimensions of identical tool holders for internally cooled cutting tools fabricated by SLM and DMLS have been determined by using digital vernier calliper and various magnification of a portable microscope. In the current study, comparing internally cooled cutting tools made of SLM and DMLS showed that generally the higher degree of accuracy could be obtained with DMLS process. However, the observed differences in surface roughness between SLM and DMLS in this study were not significant. The most obvious finding to emerge from this study is that the additive manufacturing processes selected for fabricating the tool holders for internally cooled cutting tool in this research are capable of producing the desired internal channel shape of internally cooled cutting tool.

1 Introduction

The temperature distribution at the tool-chip interface of a cutting tool has been a subject of investigation for many years due to resultant workpiece quality and prediction of tool wear [1, 2]. The conventional method, which is often called flood cooling, is using dousing

* Corresponding author: anwarcg@ump.edu.my

a large quantity of cutting fluid, usually a mixture of oil and water, to the area in the vicinity of the tool-chip contact region. However, cutting fluid or also addressed as metal working fluid (MWF) in several kinds of literature has been reported to have various adverse effects when in use abundantly [3]. Many cooling fluids contain harmful or damaging chemical constituents causing environment pollution and operator's health hazards, so strict environmental policies and health regulations have been introduced in connection with the increasing awareness of the environment and human health [4]. Based on the report by Canter [5] in 1999, Metalworking Fluids Standards Advisory Committee (MWFSAC) had recommended that the permissible exposure limit for metal working fluids be 0.5 mg/m^3. Therefore, Byrne, Dornfeld [4], Sreejith and Ngoi [6], and Wardle, Minton [7] have suggested that dry cutting, a machining process without the presence of external cutting fluid to be the desirable machining approach for the manufacturing industry for producing components and products, to satisfy both ecological and economical requirements.

A new method of cooling should be introduced though, as the excessive heat generated during machining may cause abrupt tool failure, short tool life or bad surface finish produced [8, 9]. Recently, there are a study on the new method of cooling which are Minimum Quantity Lubricant (MQL)[10, 11], internal coolant[12, 13], and cryogenic[14, 15]. In order to prevent from using excessive cutting fluid and dry machining, an experimental investigation of MQL as sustainable cooling technique was conduct by Rahim, Ibrahim [16]. The result of their experiment was proven that MQL technique performed better than dry machining.

Sun, Bateman [17] has modified a conventional tool to become an internally cooled smart cutting tool by drilling cooling channels in the tool shank and supply the cooling fluid beneath the cutting insert through a special insert support. Through experiment, this method has been demonstrated in that paper to offer great potential for effectively cooling the cutting insert because under the same thermal boundary conditions, the simulated cutting tip temperature and over- all temperature of internally cooled tools are much lower than the non-cooled cutting tools. Other than that, in real cutting situations, a tool with internal cooling can significantly decrease the cutting temperature and thus improve the cutting tool life.

The remaining issue in improving the design of internally cooled cutting tool is that the internally cooled cutting tool cannot be manufactured by using conventional machine due to the complex shape of the cooling channel. Nowadays, rapid prototyping (RP) techniques for metals, also known as metal additive manufacturing (MAM), have gained market share over the recent period, due to improvements in technologies and introduction of new metal powders that are adequate for metal end-products with applications in many fields such as aerospace, dental or medical sectors [18] [19]. Selective Laser Melting (SLM) and direct metal laser sintering (DMLS) are the most common MAM techniques [20]. To date, there are a limited number of works that has been identified which focus on SLM and DMLS part physical performance studies.

Mat Taib, Harun [21] have conducted an experimental investigation on open cellular structure CoCrMo Alloy to analyse the dimensional accuracy of the open cellular structure of CoCrMo samples which consist of unsupported struts (overhang region) fabricated by SLM process. Through the experiment, he found that with the increase in sample porosity leads to increase of shrinkage.

Pessard, Mognol [22] have conducted experimental investigation on complex cast part by using DMLS process to study and verify the geometrical quality of the obtained parts and capability of this process with respect to the required constraints of the die casting method. He found out that the geometrical accuracy of the rapid prototyping machine used in this work is approximately 0.2 mm. This corresponds to a workpiece tolerance of

 DOI: 10.1051/matecconf/20179001058

approximately ±0.1 mm, with 50% of this inaccuracy coming from the sintering machine and 50% coming from the shot-peening.

However, the main drawbacks of SLM and DMLS techniques are its inefficiency to produce high surface quality and dimensional accuracy. In order to have a better understanding on the matter, this study, therefore, aims at analysing the dimensional accuracy of the internally cooled cutting tool fabricated by using SLM and DMLS by comparing the produce to original dimension of CAD drawing.

2 Methodology

2.1 Part design of ICCT

The ICCT constitutes of three (3) different parts which are the shank, the support and internal flow channel. The shank is a part at where the tool holder is clamped on lathe machine. The role of the support is for allowing the cutting inserts to sit on it perfectly during machining and it is vital that the surface and dimensional geometries to have tight tolerances. The major part of ICCT is the internal flow channel where its inlet and outlet can be found on the inner side of the shank. The dimensions of internal channel are sensitive to defect since any deviation to the geometries will affect the flow regime. The modelling process of ICCT was carried out by using a CAD software which all the dimensions were defined clearly in the 3-dimensional (3D) model. Figure 1 depicts the 3D CAD model of ICCT without a cutting insert and a clamp bit that fixes the insert to the tool holder.

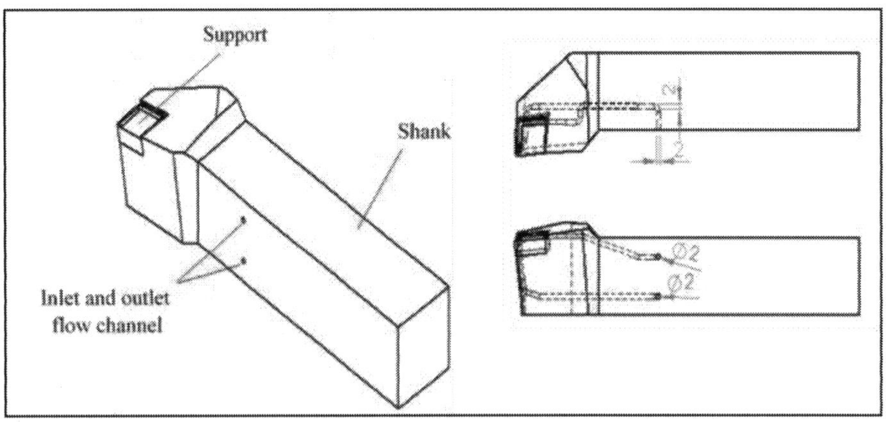

Fig. 1. CAD drawing of ICCT. The units displayed are in mm.

2.1 Manufacturing of ICCT using SLM process

The internally cooled cutting tool was fabricated by using SLM machine modelled SLM 280HL. The process started by slicing the 3D CAD file data into layers, usually from 20 to 100 micrometres thick to produce a 2D image of each layer and converted into a .stl file which can be read by most layer-based 3D printing technologies. This file was then loaded into a file preparation software package that assigns parameters, values and physical supports that allow the file to be interpreted and built by different types of additive manufacturing machines. Selective laser melting process commences with selectively melts thin layers of atomised fine metal powder are built onto a substrate plate that is fastened to an indexing table that moves in the vertical (Z) axis as illustrated in Figure 2. This takes place inside a chamber containing a tightly controlled atmosphere of inert gas (argon) to prevent oxidation and contamination. Once powder layer has been distributed, each 2D slice of the part geometry is fused by selectively melting the powder. The SLM process parameters in this study are listed in Table 1. The process is repeated layer after layer until the part is complete.

Table 1. Process parameters of SLM machines.

Parameters	Value
Power (W)	350
Hatching spacing (mm)	0.17
Scanning speed (mm/s)	930
Layer thickness (μm)	50
Chamber environment	Argon atmosphere

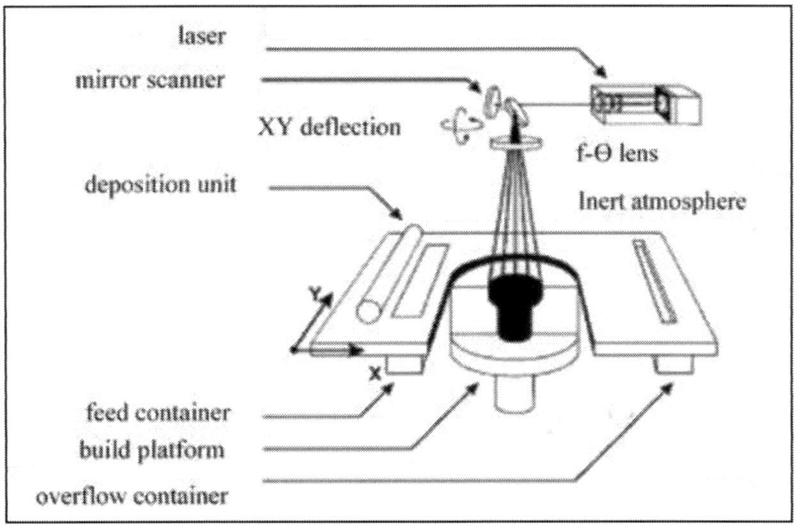

Fig. 2. Illustrated process of SLM.

2.2 Manufacturing of ICCT using DMLS process

The internally cooled cutting tool was fabricated by using DMLS machine modelled EOSINT M280. The DMLS process involves the use of a 3D CAD model whereby a .stl file is created and sent to the machine's software. The 3D model will be manually tuned to obtain the optimal orientation of the geometry for part building and supports structures will be added where necessitated. Inside the build chamber area, there is a material dispensing platform and a build platform along with a recoater blade used to move new powder over the build platform. The technology fuses metal powder into a solid part by melting it locally using the focused laser beam. Parts are built up additively layer by layer. Figure 3 show an illustrated working process of DMLS machine. This process allows for highly complex geometries to be created directly from the 3D CAD data, fully automatically, in a relatively short time and without any tooling. The DMLS process parameters in this study are listed in Table 2.

Table 2. Process parameters of DMLS machines.

Parameters	Value
Power (W)	195
Hatching spacing (μm)	0.1
Scanning speed (mm/s)	800
Layer thickness (μm)	20
Chamber environment	Nitrogen atmosphere

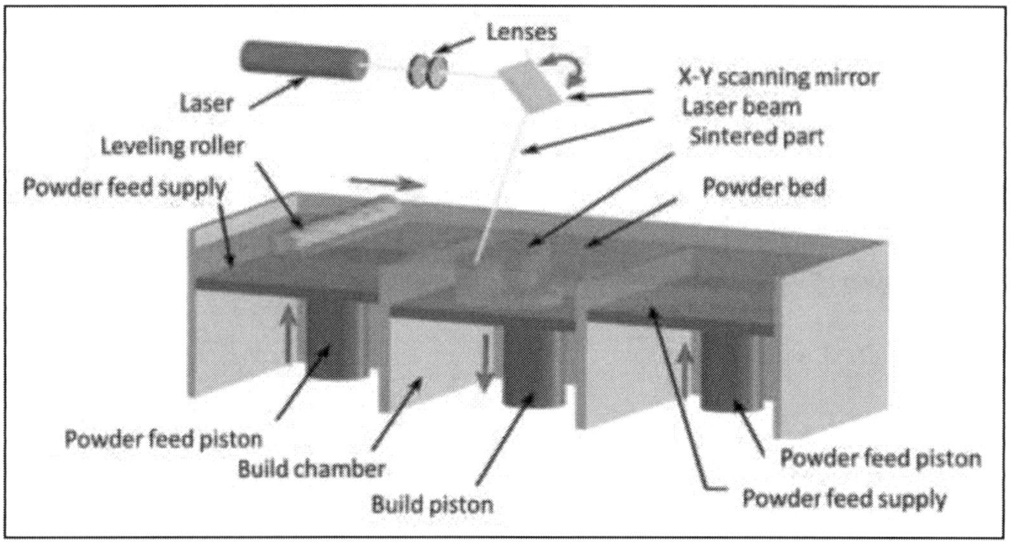

Fig. 3. Illustrated process of DMLS.

2.3 Measurements

The measurement was conducted on several sections as listed in Table 3, which is the diameter of inlet and outlet of the channel, diameter of the internal channel at various locations and the arithmetic mean of surface roughness, Ra. Measured dimensional data were then compared to the nominal value as per CAD design for obtaining build error data. In order to determine surface roughness, the MarSurf PS1 perthometer device was used. The geometrical measurement was carried out using Dino-lite edge digital microscope modelled AM4815T and a digital calliper. The features of digital microscope are 1.3MP sensor resolution, adjustable focus and magnification from 20x up to 220x and provided with software measurement and calibration.

Table 3. The investigated geometries of an internally cooled cutting tool.

Section	Symbol	Unit	Target Value
Inlet diameter	DI	mm	2
Outlet diameter	DO	mm	2
Channel diameter	DC	mm	2
Channel surface roughness	Ra	μm	The lowest

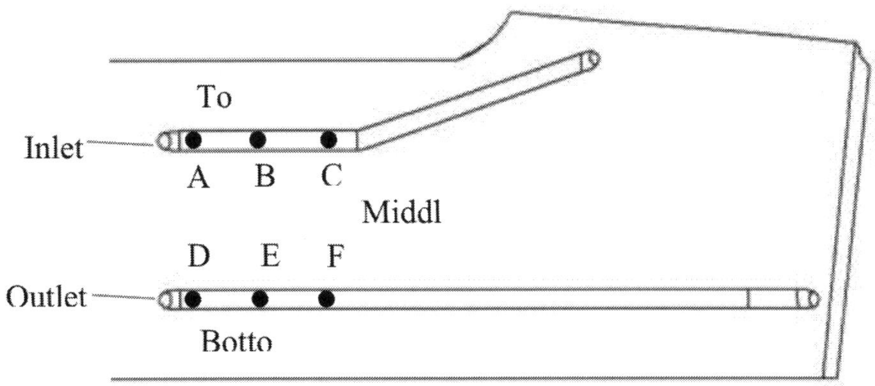

Fig. 4. Half cut CAD drawing ICCT. Black spots in the channel are the points at where the measurements along are taken.

3 Results and discussion

Table 4. Measurements of the channel dimension of ICCT.

		CAD	SLM			DMLS		
			$\bar{x}$	s	Error %	$\bar{x}$	s	Error %
	Inlet	2	1.79	0.006	10.5	1.91	0	4.5
	Outlet	2	1.92	0.006	4	1.92	0.006	4
Inlet channel	A	2	1.92	0.006	4	1.99	0.012	0.5
	B	2	1.9	0.056	5	1.96	0.015	2
	C	2	1.9	0.012	5	2.09	0.087	-4.5
Outlet channel	D	2	1.93	0.035	3.5	2.02	0.053	-1
	E	2	1.91	0.026	4.5	2.08	0.026	-4
	F	2	1.91	0.006	4.5	2.05	0.05	-2.5
	Mean Value :			0.019	5.12		0.031	2.88

The result of measured dimension of ICCT was shown in Table 4. The standard deviation of ICCT manufactured by using SLM process is 0.019 % which is lower than DMLS process 0.031. This means the precision of SLM process is better than DMLS. The calculated standard error of the process concealed that the accuracy of DMLS process is higher that SLM process with 2.88 % DMLS and 5.12 % SLM.

Figure 5 shows the graph of surface roughness of ICCT fabricated by using SLM and DMLS process. The surface roughness was measured on 3 different sections as labelled in figure 4. A standard deviation calculated from the data gives 0.383 and 2.01 for each process. The results were revealed that the process of SLM provided better roughness when compared with DMLS process.

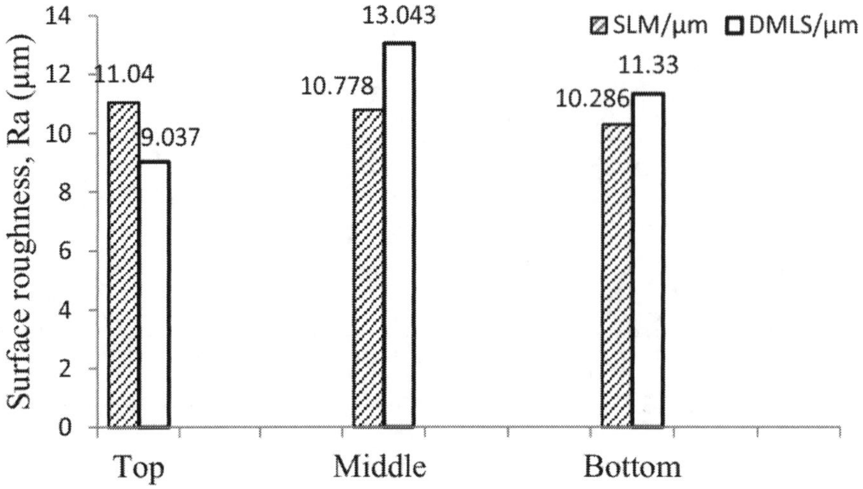

Fig. 5. Comparison surface roughness ICCT of different process.

4 Conclusion

This paper discussed a dimensional accuracy of internally cooled cutting tools manufactured by SLM and DMLS process. The following conclusions can be drawn from the present work:

- SLM process provided higher precision than DMLS process.
- DMLS process performed better in dimensional accuracy than SLM process.
- The surface roughness of SLM process is better than DMLS process.

The author would like to thanks to Universiti Malaysia Pahang for providing financial support under project no. RDU 140377 and RDU 150338. The author gratefully acknowledges the committed laboratory support from KKTM(Kuantan) and CONTRAVES SDN. BHD.

References

1. Usui, E., T. Shirakashi, and T. Kitagawa, *Analytical Prediction of Three Dimensional Cutting Process—Part 3: Cutting Temperature and Crater Wear of Carbide Tool.* Journal of Engineering for Industry, 1978. **100**(2): p. 236-243.
2. Davies, M.A., et al., *On The Measurement of Temperature in Material Removal Processes.* CIRP Annals - Manufacturing Technology, 2007. **56**(2): p. 581-604.
3. Ghani, B.C. and S. Anwar, *Design and analysis of the internally cooled smart cutting tools with the applications to adaptive machining.* 2013, Brunel University School of Engineering and Design PhD Theses.
4. Byrne, G., D. Dornfeld, and B. Denkena, *Advancing cutting technology.* CIRP Annals-Manufacturing Technology, 2003. **52**(2): p. 483-507.
5. Canter, N., *The possibilities and limitations of dry machining.* Tribology & Lubrication Technology, 2009. **65**(3): p. 40.
6. Sreejith, P. and B. Ngoi, *Dry machining: machining of the future.* Journal of materials processing technology, 2000. **101**(1): p. 287-291.
7. Wardle, F., et al., *Artificial neural networks for controlling the temperature of internally cooled turning tools.* 2013.

8. Umbrello, D., F. Micari, and I. Jawahir, *The effects of cryogenic cooling on surface integrity in hard machining: A comparison with dry machining.* CIRP Annals-Manufacturing Technology, 2012. **61**(1): p. 103-106.
9. Ferri, C., et al., *Efficiency in contamination-free machining using microfluidic structures.* CIRP Journal of Manufacturing Science and Technology, 2014. **7**(2): p. 97-105.
10. Dureja, J.S., et al., *Performance evaluation of coated carbide tool in machining of stainless steel (AISI 202) under minimum quantity lubrication (MQL).* International Journal of Precision Engineering and Manufacturing-Green Technology, 2015. **2**(2): p. 123-129.
11. Obikawa, T., *11.11 - Machining with Least Quantity Lubrication A2 - Hashmi, Saleem,* in *Comprehensive Materials Processing*, G.F. Batalha, C.J.V. Tyne, and B. Yilbas, Editors. 2014, Elsevier: Oxford. p. 255-281.
12. Minton, T., et al., *Temperature of internally-cooled diamond-coated tools for dry-cutting titanium.* International Journal of Machine Tools and Manufacture, 2013. **75**: p. 27-35.
13. Shu, S., et al., *An Innovative Method to Measure the Cutting Temperature in Process by Using an Internally Cooled Smart Cutting Tool.* Journal of Manufacturing Science and Engineering, 2013. **135**(6): p. 061018.
14. Bermingham, M., et al., *New observations on tool life, cutting forces and chip morphology in cryogenic machining Ti-6Al-4V.* International Journal of Machine Tools and Manufacture, 2011. **51**(6): p. 500-511.
15. Machai, C. and D. Biermann, *Machining of β-titanium-alloy Ti–10V–2Fe–3Al under cryogenic conditions: Cooling with carbon dioxide snow.* Journal of Materials Processing Technology, 2011. **211**(6): p. 1175-1183.
16. Rahim, E., et al., *Experimental investigation of minimum quantity lubrication (MQL) as a sustainable cooling technique.* Procedia CIRP, 2015. **26**: p. 351-354.
17. Sun, X., et al., *Design and analysis of an internally cooled smart cutting tool for dry cutting.* Proceedings of the Institution of Mechanical Engineers, Part B: Journal of Engineering Manufacture, 2012. **226**(4): p. 585-591.
18. Giannatsis, J. and V. Dedoussis, *Additive fabrication technologies applied to medicine and health care: a review.* The International Journal of Advanced Manufacturing Technology, 2009. **40**(1-2): p. 116-127.
19. Melchels, F.P.W., et al., *Additive manufacturing of tissues and organs.* Progress in Polymer Science, 2012. **37**(8): p. 1079-1104.
20. Murr, L.E., et al., *Metal fabrication by additive manufacturing using laser and electron beam melting technologies.* Journal of Materials Science & Technology, 2012. **28**(1): p. 1-14.
21. Mat Taib, Z.A., et al. *Dimensional Accuracy Study of Open Cellular Structure CoCrMo Alloy Fabricated by Selective Laser Melting Process.* in *Advanced Materials Research*. 2016. Trans Tech Publ.
22. Pessard, E., et al., *Complex cast parts with rapid tooling: rapid manufacturing point of view.* The International Journal of Advanced Manufacturing Technology, 2008. **39**(9-10): p. 898-904.

Stability behavior of non-surfactant water-in-diesel emulsion fuel using microscopic observation

Siti Amiliyana Norazni[1], *Wira Jazair* Yahya[1,*], *Ahmad Muhsin* Ithnin[1], *Hasanuddin* Abd Kadir[1], *Ili Fatimah* Abd Razak[1], *Nadia Dayana* Bahar[1], *Dhani Avianto* Sugeng[1], and *Nur Atiqah* Ramlan[1]

[1]Vehicle System Engineering, Malaysia-Japan International Institute of Technology, Universiti Teknologi Malaysia, Jalan Semarak, 54100 Kuala Lumpur, Malaysia

Abstract. Water-in-diesel emulsion fuel (W/D) is considered to be a potential alternative fuel that can reduces nitrogen oxides (NO_x) and particulate matter (PM). W/D is normally produced with addition of surfactant to prolong its stability. However, the dependency on surfactant leads to higher production cost. A concept that can eliminate dependency on the surfactant was introduced by strong mixing concept and direct supply the non-surfactant W/D to a diesel engine. Therefore, the objective of this study is to investigate the stability behavior of non-surfactant W/D using microscopic observation. Water and diesel fuel were mixed using a combination of high-shear mixer and an ultrasonic transducer. The amount of water injected into the system are 5% and 10% by volume, which were labelled as E5 and E10 respectively. The produced non-surfactant W/D was then placed into a petri dish for microscopic observation. The results showed that the average diameter of water droplets observed in E5 and E10 at sedimentation stage were 15.38 and 22.41 μm respectively. The stability period of E5 and E10 before it sediment were 25 and 67 seconds correspondingly. Overall, it is observed that microscopic observation is a reliable approach to determine the stability behavior of non-surfactant W/D.

1 Introduction

The interest in water-in-diesel emulsion fuel (W/D) for diesel engine application has been growing worldwide as it is regarded as an effective way to reduce the harmful pollutants such as nitrogen oxide (NO_x) and particulate matter (PM) simultaneously. Moreover, it also improves the combustion efficiency [1–4]. An emulsion is a mixture of two or more liquids that are immiscible in nature where one present as droplet, or the dispersed phase throughout another liquid known which acts as the continuous phase [3]. Emulsion is produced with the assistance of some external means, either by mechanical or ultrasonic agitation; and eventually, with an addition of surfactant to help bind both diesel and water together as one solution [5–8]. The mechanical homogenizing machine is the most used emulsifying device in emulsion preparation; it stirs up the mixture of continuous and

* Corresponding author: wira@utm.my

dispersed phase with the presence of surfactant [8]. Furthermore, there are also applications using ultrasonic agitation to produce emulsified fuel. The ultrasonic waves cause cavitation which creates alternating positive and negative pressure waves to form a strong mixing effect. It has been proven that the combination of mechanical and ultrasonic components in producing W/D yields small water droplet size [7].

The surfactant is required in the formation of W/D to lower the interfacial tension between the dispersed phase and the continuous phase while altogether to stabilize the water droplet phase within the diesel fuel phase to avoid coalescence [9]. Surfactant, also known as surface-active agent is a substance that contain both polar and non-polar groups in their molecules. Hydrophilic surfactants have more attraction to polar liquids such as water, in contrast to lipophilic surfactants which are more soluble in nonpolar liquids such as oil [8]. In W/D, the most commonly used surfactants by researchers are SPAN 80 and TWEEN 80, which are added to a certain percentage to achieve emulsion stability[10]. M.T. Ghannam *et. al* [9] reported that the separation time of water and diesel (i.e. destabilization of W/D) is dependent on water percentage. The increase of water percentage reduces the W/D stability, in other words, the separation time of water droplets phase from emulsion decreases. The stability of W/D relies on two factors which are the existence of surfactant around the surface droplets and coalescence mechanism [11]. In diesel engine, if the W/D is destabilized during storage or while the engine is in operation, it may cause engine failure and damage other components.

Utilizing the surfactant to a great large scale W/D production is not viable as the price of the surfactant is very high. This hinders the commercialization of W/D in the real-world applications. Therefore, with the motivation to utilize the W/D without using surfactant or so called non-surfactant W/D, a novel concept of a mixing system was introduced in previous study [12,13]. The idea of the system was to focus on the production of W/D through an in-line mixing system by continuously mixing diesel fuel and water, then directly supplies the produced W/D to the engine. This concept of mixing system is known as Real-Time Emulsion Fuel Supply System (RTES). Through this concept, the need of W/D with extended stability period is not relevant anymore, and the addition of the surfactant in W/D can be omitted. The W/D produced from the RTES is thermodynamically unstable, but it needs to stay stable before it is injected into the engine. In previous studies, there are several methods of determining emulsion stability have been proposed such as droplet size analysis, measurements of the physical properties of emulsion [14], accelerated test [15], turbidity ratio measurement [16] and bottle test method. With the latter, the prepared emulsion is placed into beaker/bottle and the time for the emulsion to destabilize is recorded (layer of diesel and water is observed) [9,15]. However, none of the researchers use the microscopic approach to explaining the stability behavior of non-surfactant W/D.

Therefore, the objective of this paper is to investigate the stability behavior of non-surfactant W/D produced by RTES in term of droplet size measurement and stability period by microscopic observation. In the experimental work, a detailed analysis of the stability behavior was conducted with a variety of 5 % and 10 % water content.

2 Experimental setup

The experimental procedures comprised of two stages where the first stage was the preparation of the non-surfactant W/D with 5 and 10 volumetric percent of water using a strong mixing system, RTES. The non-surfactant W/D with 5% and 10% of water were labeled as E5 and E10 respectively. The second stage was the stability behavior analysis of the non-surfactant W/D, carried out by using a microscope. In this analysis, observations of the water droplet size and the duration of W/D to sediment were performed. The produced non-surfactant W/D need to stay intact or in other words, to be stable for at least 3 seconds, which is the time needed for the fuel travel from the pump to reach the engine. Figure 1 below shows the non-surfactant W/D preparation process and the test setup.

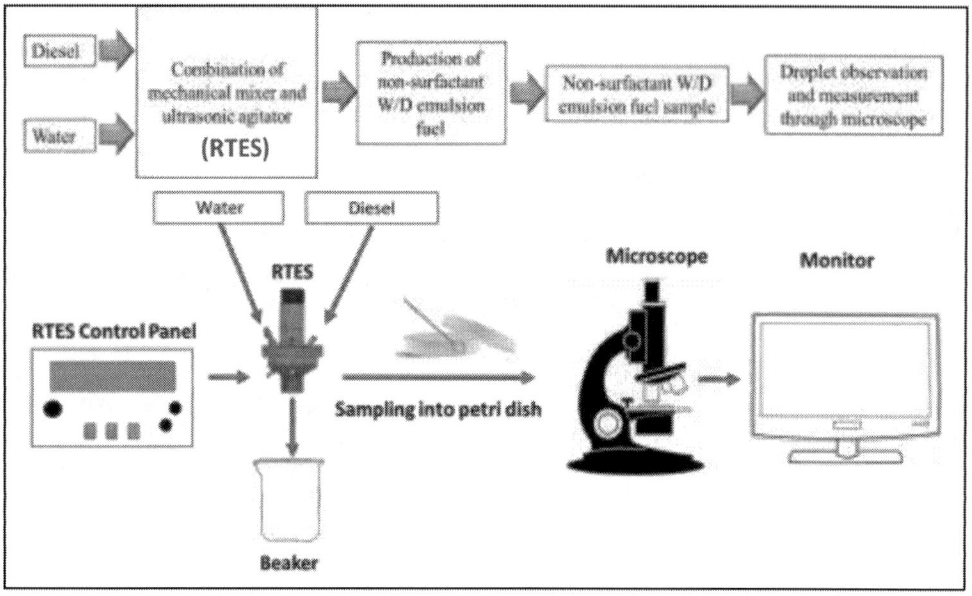

Fig. 1. Non-surfactant W/D preparation and test setup.

2.1 Preparation of water-in-diesel emulsion fuel (W/D)

For the preparation of E5 and E10, Malaysia grade 2 diesel fuel (D2) was mixed with water using RTES, which is a combination of high shear mixer with a constant motor speed of 1400 rpm and ultrasonic transducer with a frequency of 50 kHz. As shown in Figure 1, the diesel fuel flowed from the fuel tank and water was injected into the mixing chamber. Both liquids were mixed with high shear force and acoustic wave to produce W/D. Here, no surfactant was involved in the proses. The amount of water (5% and 10%) was regulated by a controller placed in the RTES control panel. RTES then supplied the non-surfactant W/D emulsion continuously and instantaneously. Table 1 below shows the properties of diesel fuel, used in the experiment.

Table 1. Specifications of Malaysia grade 2 diesel fuel [2].

Properties	
Cetane index	54.6
Viscosity at 40°C, cSt	4.642
Density at 15°C, kg/L	0.8538
Carbon, wt %	84.1
Calorific value, MJ/kg	45.280
Total sulphur, mass%	0.28
Flash point, °C	93

In the sampling process, the non-surfactant W/D was transferred into a graduated beaker. Then, a small portion of the fuel was placed into a petri dish for observation and measurement. While the behavior of the fuels were recorded by a microscope, the destabilization process of E5 and E10 emulsion process were also taken in video format with a duration of two minutes. For this purpose, a super resolution digital microscope with a magnification range of 100 to 1000 times was used. The specifications of the microscope are listed in Table 2.

Table 2. Specifications of the digital microscope.

Properties	
Model	KEYENCE VHX-200E digital microscope, equipped with VH-Z100R zoom lens
Magnification range	100 to 1000 times
Image size (pixels per image)	1600 x 1200
Video size (frames per second)	15

3 Results and discussion

As shown in Figure 2, the water droplet size gradually increased proportionally with time. At 10 seconds, the average diameter of water droplet size in E10 was 15.38 μm and the water droplets were moving freely at this stage. An emulsion is characterized as thermodynamically unstable; thus, the repulsive force between the droplets become weaker as time elapses. Ghannam *et al.* defined that the stability of the W/D by measuring the amount of water separated over time [9]. However, in this analysis, the stability of W/D ended when water layer began to appear at the bottom of the beaker. In other words, the W/D is described as destabilize when the sedimentation process occurs. As illustrated in Figure 2, after 25 seconds, the droplets became more attracted to each other, and the average droplet diameter reached 24.41 μm, which was bigger than previous droplet size, at 15.38 μm. The larger water droplet started to sink to the bottom of the beaker, which signified the sedimentation process to begin after the 25th second.

Figure 3 shows the sequence of sedimentation process of the same E10 emulsion, recorded after two minutes of mixing. It was evident that the size of water droplets become bigger with time. In the destabilization process of W/D emulsion, initially, the droplets underwent a flocculation process, where the water droplets move toward each other. This was due to the weaker repulsive bond between the droplets. They then formed three dimensional (3-D) clusters in which multiple droplets were bounded and separated only by a thin-film. The thickness of the thin-film was then slowly receding as it dictated by the Van der Waals force and when it reached its critical limit, the film broke, and the droplets formed larger water droplet (i.e. coalescence). Finally, followed by the sedimentation process, where large water droplets sank to the bottom of the beaker due to the density difference between water and diesel fuel. The process continued until the water was completely separated from diesel fuel and the W/D turned into two distinct layers.

The stability experiment for E5 was conducted under the same experimental condition as E10. Similar behavior and pattern with E10 was observed, where at the initial stage, the droplets were freely moving in the emulsion continuous phase. After 67 seconds, the sedimentation process occurred. Here, the average water droplet size observed in E5 is 9.54 μm which was smaller compared to the droplet size found in E10 (22.41 μm). This was because the amount of water in E5 injected into the system was less than in E10. In contrast, the sedimentation process in E5 takes 2.68 times longer than in E10. While in E5, it occurred at 67 seconds, in E10, it took place at 25 seconds. This implied that non-surfactant W/D with higher water percentage reached sedimentation stage faster compared

DOI: 10.1051/matecconf/20179001059

to one with lesser water percentage. Figure 4 shows the water droplets of E5 after 15 seconds of mixing, at a magnification of 1000x.

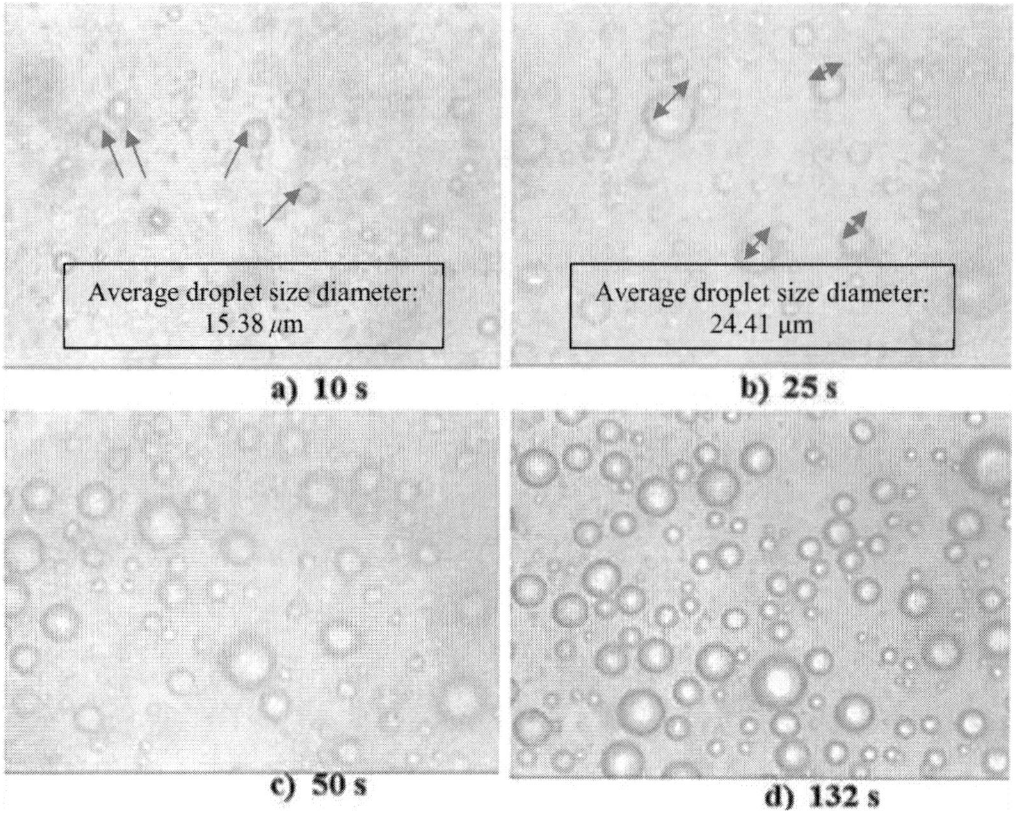

Fig. 2. The sequence of water droplet size in E10 emulsion.(1000x magnification).

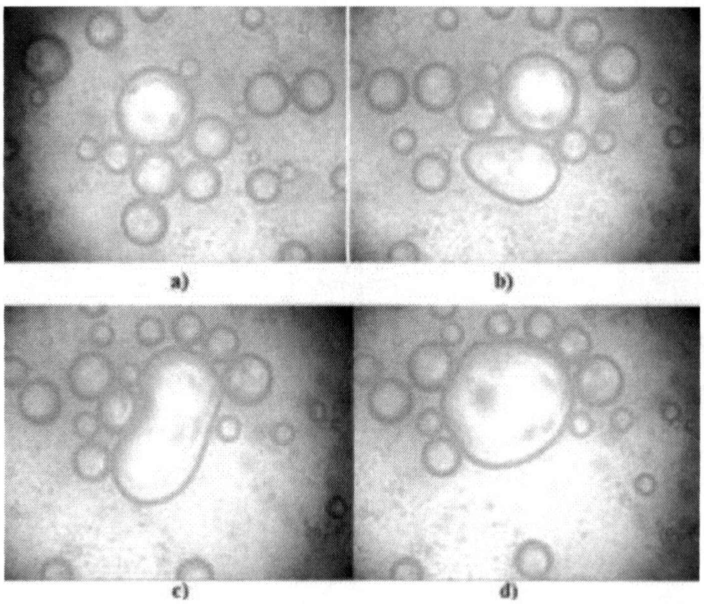

Fig. 3. The sequence of the water droplets destabilization behavior in E10 after 2 minutes of mixing. (100x magnification).

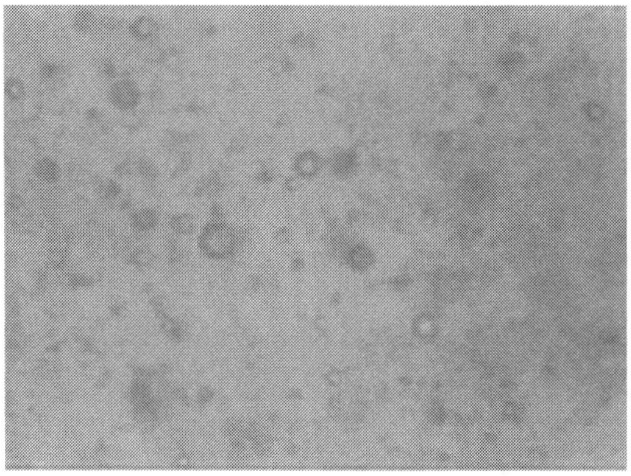

Fig. 4. Water droplet in E5 after 15 seconds of mixing.(1000x magnification).

4 Conclusions

This study highlights the stability behavior of non-surfactant W/D produced by a strong mixing system, RTES. Based on the experimental results, the average droplet size diameter of E5 and E10 at sedimentation stage was 9.54 μm and 22.41 μm respectively. Higher water percentage influence the droplet size and the stability period of non-surfactant W/D. From the observation, the stability period of E5 before reach the sedimentation stage was 67 seconds, while in E10, it was 25 seconds. This shows that both E5 and E10 are stable and exceeded the time it takes for the fuel from the pump to reach the engine.

Overall, it is observed that the evaluation of non-surfactant W/D stability behavior using microscopic observation technique is a reliable method. The characteristic of non-surfactant W/D that has been analyzed make it a potential alternative fuel which can be used in diesel engine.

The authors would like to express their appreciation to the Centre for Low Carbon Transport, (UTM) and Malaysia-Japan International Institute of Technology (MJIIT) for the financial and technical support.

References

1. A. M. A. Attia and A. R. Kulchitskiy, "Influence of the structure of water-in-fuel emulsion on diesel engine performance," *Fuel*, vol. **116**, pp. 703–708, Jan. (2014).
2. A. M. Ithnin, M. A. Ahmad, M. A. A. Bakar, S. Rajoo, and W. J. Yahya, "Combustion performance and emission analysis of diesel engine fuelled with water-in-diesel emulsion fuel made from low-grade diesel fuel," *Energy Convers. Manag.*, vol. **90**, pp. 375–382, (2015).
3. A. Alahmer, J. Yamin, A. Sakhrieh, and M. A. Hamdan, "Engine performance using emulsified diesel fuel," *Energy Convers. Manag.*, vol. **51**, no. 8, pp. 1708–1713, Aug. (2010).
4. A. Lif and K. Holmberg, "Water-in-diesel emulsions and related systems," *Adv. Colloid Interface Sci.*, vol. **123–126**, pp. 231–239, (2006).
5. C.-Y. Lin and K.-H. Wang, "Effects of an oxygenated additive on the emulsification characteristics of two- and three-phase diesel emulsions," *Fuel*, vol. **83**, no. 4–5, pp. 507–515, Mar. (2004).
6. C.-Y. Lin and L.-W. Chen, "Emulsification characteristics of three- and two-phase

emulsions prepared by the ultrasonic emulsification method," *Fuel Process. Technol.*, vol. **87**, no. 4, pp. 309–317, Apr. (2006).

7. C.-Y. Lin and L.-W. Chen, "Comparison of fuel properties and emission characteristics of two- and three-phase emulsions prepared by ultrasonically vibrating and mechanically homogenizing emulsification methods," *Fuel*, vol. **87**, no. 10–11, pp. 2154–2161, Aug. (2008).
8. C.-Y. Lin and K.-H. Wang, "The fuel properties of three-phase emulsions as an alternative fuel for diesel engines☆," *Fuel*, vol. **82**, no. 11, pp. 1367–1375, (2003).
9. M. T. Ghannam and M. Y. E. Selim, "Stability behavior of water-in-diesel fuel emulsion," *Pet. Sci. Technol.*, vol. **27**, no. 4, pp. 396–411, (2009).
10. B. K. Debnath, U. K. Saha, and N. Sahoo, "A comprehensive review on the application of emulsions as an alternative fuel for diesel engines," *Renew. Sustain. Energy Rev.*, vol. **42**, pp. 196–211, (2015).
11. P. Sherman and P. Becher, "*Encyclopedia of emulsion technology*," New York P. Becker ed, (1983).
12. N. A. binti Mazlan, H. A. Kadir, Z. Mohamad, and W. J. Yahya, "*Requirement for Real-Time Water Emulsion Fuel Supply System (RTES) Concept Design*," in A-DEWS 2015: Design Engineering in the Context of Asia, Asian Design Engineering workshop, 29th-30th October 2015, The Hong Kong Polytechnic University, (2015).
13. W. J. Yahya, A. M. Ithnin, H. Noge, and T. Koga, "*Determination of Characteristics for Real-Time Water Emulsion Fuel Supply System (RTES) Concept*," in A-DEWS 2015: Design Engineering in the Context of Asia, Asian Design Engineering workshop, 29th-30th October 2015, The Hong Kong Polytechnic University, (2015).
14. T. M. Dreher, J. Glass, A. J. O'Connor, and G. W. Stevens, "Effect of rheology on coalescence rates and emulsion stability," *AIChE J.*, vol. **45**, no. 6, pp. 1182–1190, (1999).
15. T. Förster, F. Schambil, and W. Von Rybinski, "Production of fine disperse and long-term stable oil-in-water emulsions by the phase inversion temperature method," *J. Dispers. Sci. Technol.*, vol. **13**, no. 2, pp. 183–193, (1992).
16. M.-G. Song, S.-H. Cho, J.-Y. Kim, and J.-D. Kim, "Novel evaluation method for the water-in-oil (W/O) emulsion stability by turbidity ratio measurements," *Korean J. Chem. Eng.*, vol. **19**, no. 3, pp. 425–430, (2002).
17. A. K. Hasannuddin, M. I. Ahmad, M. Zahari, S. S. Mohd, A. B. Aiman, S. A. Aizam, and J. Y. Wira, "Stability Studies of Water-in-Diesel Emulsion," in *Applied Mechanics and Materials*, (2014), vol. **663**, pp. 54–57.

Platooning strategy of mobile robot: simulation and experiment

K. Baarath[1], *Muhammad Aizzat* Zakaria[1,*], *M.V.* Suparmaniam[1], and *Mohd Yazid bin* Abu[1]

[1]Faculty of Manufacturing Engineering, Universiti Malaysia Pahang, 26600 Pekan, Pahang, Malaysia.

Abstract. Concurrent studies show vehicle platooning system as a promising approach for a new transportation system. The platooning strategy can be also applied to automated mobile robots. Including dynamic modelling in the simulation with kinematic model would yield a different result as the dynamic modelling would include the physical parameters of the mobile robot. The aim is to create a model that describes the motion of a robot that follows another robot based on predetermined distance. Dynamic model of the proposed mobile robot is simulated and the kinematic modelling was included in to simulate the motion of the mobile robot. PID controller will be used as a controller for robot's motion and platooning strategy. A reference distance is given as the input and the PID controller computes the error and sends input to the mobile robot in the form of voltage. The robot is able to follow the leader robot by maintaining a distance of one metre with a small deviation in the direction as the robot tends to move towards the left due to forces acting on the wheel. This method can be implemented in a human following mobile robot where the leader robot is replaced with a human user.

1 Introduction

Development of mobile robot has proven to be useful in many sectors. Factories, shopping malls, militaries are some of the target sectors for mobile robots. Often mobile robots are used to assist humans in their daily activities such as carrying certain objects. This task requires an important function which is to track a human [1]. For example, a mobile robot is developed for the elderly people to assist them as presented in [2], which could be helpful in taking care of the elderly. The navigation of mobile robots is a relevant and widely studied topic in the field of robotics [3 - 4]. The case of navigating robot in the presence of humans and other robots is still new to the field of robotics [5].

A logical approach to model the properties of mobile is through mathematical equations. The modelling of mobile robots can be divided into two section. First is the formulation of the plant (mobile robot) in mathematical equations. This plant formulation is known as dynamic equations that take the forces into consideration. The next part would be the derivation of the kinematic equations which describes the movement of the mobile robot. A subtle approach that is used in modelling for differential drive wheeled mobile

[*] Corresponding author: maizzat@ump.edu.my

robot (WMR) is, ignore the presence of either longitudinal or lateral wheel slippage or both to ensure the path planning is based purely on kinematics. The kinematic modelling is developed by taking into consideration of the wheel radius and angular velocity of the wheels. The angular velocity can be used to derive the moving velocity and acceleration of the WMR. The heading of the mobile robot can be determined by analysing the difference between the angular velocities of its wheels [6]. Ignoring dynamic forces in a purely kinematic approach has several known issues. The results were unrealistic for a WMR with elevated loads and speeds [7]. This imposes significant limitations to the applicability of purely kinematic system [8].

In order to overcome these limitations, many researchers have opted for the application of dynamic modelling in robotics [7 - 9]. A mathematical approach for dynamic modelling of WMR's can be classified into several methods. The first method will be the Newton-Euler which considers the forces applied to a system: the given forces and constraint forces [10]. Newton-Euler approach has few practical difficulties since these forces are not easily quantifiable. In Lagrange approach, the forces are expressed in terms of energies in the system which is the kinetic and potential energy. These energies are in scalar quantities and are easily expressible in terms of the system coordinates [10]. However, the Lagrangian approach usually provides a strong and versatile method in the formulation of the equation of motion for holonomic systems. Another approach would be Kane's method which uses generalised coordinated which is based on a vector approach [11 - 12].

Platooning is the particular formation of a moving object which has leader-follower formation. The follower tracks and follows the leader based on a predetermined distance. This system is commonly used in autonomous cars on a highway [13]. A follower requires information regarding the leader in order to achieve the platooning. There is two type of method used which is the global formation and local formation [14]. The global formation gathers the distance and orientation using an image sensor for example camera whereas the local formation uses distance and orientation from a motion sensor such as infrared and laser range finder. Communication is important in platooning. The conventional PID controller is being used in this project since the system is analysed based on a linear path while ignoring lateral path. PID is chosen due to its robustness and ease to use [15 - 16].

In this paper, the dynamic model of the proposed mobile robot and the kinematic modelling will be considered. The simulation results presented shows the behaviour of the mobile robot in following a moving object at a specified distance. The PID and error output is discussed to identify the behaviour of the robot. The output of the dynamic model gives the heading and velocity of the robot whereas the output of the kinematic model gives the motion of the robot with respect to the planar plane.

2 Simulation and experimental setup

2.1 Simulation setup

2.1.1 Simulation design and methods

The study aims at identifying the behaviour of the proposed mobile robot when following a specified target. The system is simulated in SIMULINK using block diagrams based on the proposed mathematical modelling. Both dynamic and kinematic model were used in determining the behaviour of the robot. The kinematic model is developed based on the differential drive system since the proposed robot uses two independent wheels and a free wheel.

Kinematic Model

The kinematic model is derived based on differential drive system where two independent wheels are considered. The speed of each wheel can be represented as rω, thus the average velocity of the two wheels is:

$$\dot{x}_R = r\frac{\omega_1 + \omega_2}{2} \tag{1}$$

The instantaneous rotation for one wheel:

$$\omega_1 = \frac{r\omega_1}{2l} \tag{2}$$

The total rotation:

$$\dot{\theta} = \frac{r}{2l}(\omega_1 - \omega_2) \tag{3}$$

The relation between the references frame is through the standard orthogonal rotation transformation:

$$R(\theta) = \begin{bmatrix} \cos(\theta) & \sin(\theta) & 0 \\ -\sin(\theta) & \cos(\theta) & 0 \\ 0 & 0 & 1 \end{bmatrix} \tag{4}$$

Considering a simple 90° point turn where $\theta = \frac{\pi}{2}$ and substitute into Eq (4):

$$\dot{\xi}_R = R\left(\frac{\pi}{2}\right)\dot{\xi}_I = \begin{bmatrix} 0 & 1 & 0 \\ -1 & 0 & 0 \\ 0 & 0 & 1 \end{bmatrix}\begin{bmatrix} \dot{x} \\ \dot{y} \\ \dot{\theta} \end{bmatrix} = \begin{bmatrix} \dot{y} \\ -\dot{x} \\ \dot{\theta} \end{bmatrix} \tag{5}$$

$$\dot{\xi}_I = R(\theta)^{-1}\dot{\xi}_R \tag{6}$$

Inverting the standard orthogonal rotation transformation, Eq (4):

$$R(\theta)^{-1} = \begin{bmatrix} cos(\theta) & -sin(\theta) & 0 \\ sin(\theta) & cos(\theta) & 0 \\ 0 & 0 & 1 \end{bmatrix} \tag{7}$$

Thus the final form of the differential equation is as shown below where Eq (1) & (3),

$$\dot{\xi}_I = \begin{bmatrix} cos(\theta) & -sin(\theta) & 0 \\ sin(\theta) & cos(\theta) & 0 \\ 0 & 0 & 1 \end{bmatrix}\frac{r}{2}\begin{bmatrix} \omega_1 + \omega_2 \\ 0 \\ \frac{\dot{\omega}_1 - \dot{\omega}_2}{l} \end{bmatrix} = \begin{bmatrix} \dot{x} \\ \dot{y} \\ \dot{\theta} \end{bmatrix} \tag{8}$$

DOI: 10.1051/matecconf/20179001060

Dynamic Model

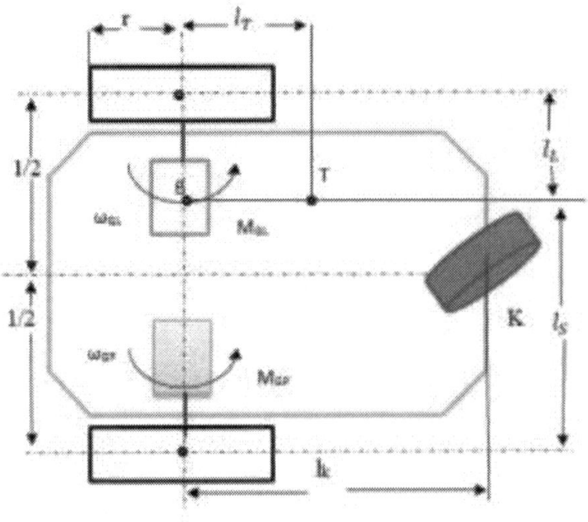

Fig. 1. The mobile robot chassis parameter.

The dynamic model is developed based on the configuration in figure.1. The reduced linear part is represented in standard state-space model in matrix form as shown in Equation 9.

$$\frac{dx}{dt} = Ax + Bu \quad (9)$$

$$y = Cx$$

Where

$$x = \begin{bmatrix} i_L \\ i_P \\ \omega_L \\ \omega_P \end{bmatrix} \; u = \begin{bmatrix} u_L \\ u_P \end{bmatrix} \; y = \begin{bmatrix} v_B \\ \omega_B \end{bmatrix}$$

The constant A, B and C are matrices as shown below.

$$A = \begin{bmatrix} -\frac{R+R_z}{L} & -\frac{R_z}{L} & -\frac{K}{L} & 0 \\ -\frac{R_z}{L} & -\frac{R+R_z}{L} & 0 & -\frac{K}{L} \\ \frac{K(d_P + b_P l_L)}{b_L d_P + b_P d_L} & \frac{K(d_P - b_P l_P)}{b_L d_P + b_P d_L} & -\frac{d_P a_L + b_P c_L}{b_L d_P + b_P d_L} & -\frac{d_P a_P + b_P c_P}{b_L d_P + b_P d_L} \\ \frac{K(d_L - b_L l_L)}{b_L d_P + b_P d_L} & \frac{K(d_L + b_L l_P)}{b_L d_P + b_P d_L} & -\frac{d_L a_L - b_L c_L}{b_L d_P + b_P d_L} & -\frac{d_L a_P + b_L c_P}{b_L d_P + b_P d_L} \end{bmatrix}$$

$$B = \begin{bmatrix} \frac{U_0}{L} & 0 \\ 0 & \frac{U_0}{L} \\ 0 & 0 \\ 0 & 0 \end{bmatrix}$$

$$C = \begin{bmatrix} 0 & 0 & \frac{l_P r_G}{l_L + l_P} & \frac{l_L r_G}{l_L + l_P} \\ 0 & 0 & \frac{r_G}{l_L + l_P} & \frac{r_G}{l_L + l_P} \end{bmatrix}$$

Table 1. The mobile robot chassis parameters.

Notation	Value	Dimension	Definition
l_L	0.1125	m	The distance of the left wheel from point B
l_P	0.3855	m	The distance of the right wheel from point B
l_T	0.1815	m	The distance of the centre of gravity from join between wheels
l_K	0.040	m	The distance of the caster wheel from the join between wheels
r	0.0635	m	Semi-diameter of driving wheel
M	6.0	kg	Total weight of the robot
k_v	0.100	$kg.s^{-1}$	The coefficient of the resistance against robot linear motion
JT	0.550	$kg.m^2$	Moment of inertia of robot with respect to the centre of gravity
k_ω	1.350	$kg.m^2.s^{-1}$	The coefficient of the resistance against the robot.

Table 2. DC motor parameters.

Notation	Value	Dimension	Definition
R	2.000	Ω	Motor winding resistivity
L	0.050	H	Motor inductance
K	0.100	$kg.m^2.s^{-2}.A^{-1}$	Electro motoric constant
R_z	0.200	Ω	Source resistance
U_0	12.00	V	Source voltage
J	0.025	$kg.m^2$.	Total moment of inertia of rotor and gearbox
k_r	0.002	$kg.m^2.s^{-1}$	The coefficient of the resistance against rotating of rotor and gearbox
p_G	25	-	Gearbox transmission ratio

SIMULINK Model

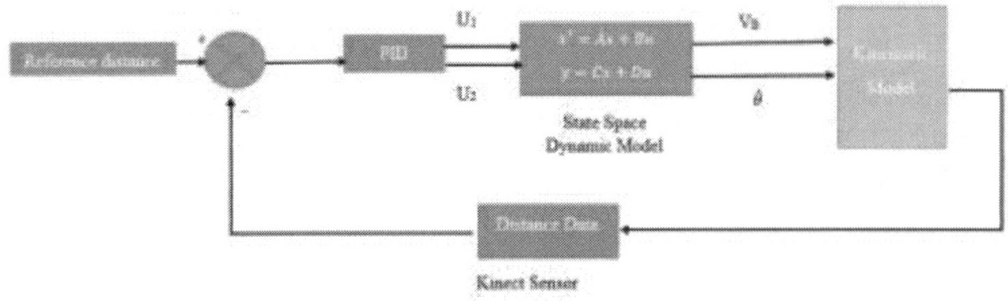

Fig. 2. Overall simulation model.

The input signal is given in the form of time variance multiplied with a gain defined gain. The input signal given is only in the form of X-axis since only linear motion is considered in this project. The simulation of the mobile robot is conducted using SIMULINK in MATLAB. The MATLAB editor is used to set the parameter of the dynamic model (i.e. state-space). The block diagrams are arranged as shown in figure 2. The kinematic model derived is used in the modelling to compute the new position of the mobile robot after it follows the user by maintaining pre-defined distance. The output of the kinematic model is the velocity in the x-direction ($\dot{x}$)and velocity in the y-direction ($\dot{y}$) which is then integrated to get the coordinate of the robot with respect to x direction. The output is then fed into the error computation to calculate the distance required to travel by the robot in order to maintain the distance.

2.2 Experimental setup

The experiment of the system developed is conducted on a self-developed mobile robot. The sensor used in the experiment is the Kinect sensor. Kinect sensor is used because it has depth sensor and also a camera to detect a specific object to be tracked. In this experiment, the robot is made to follow a human in linear a direction. The myRIO is used as the controller in the robot. The setup of the mobile robot is as shown in figure 3 and figure 4 shows the flow of the experimental system.

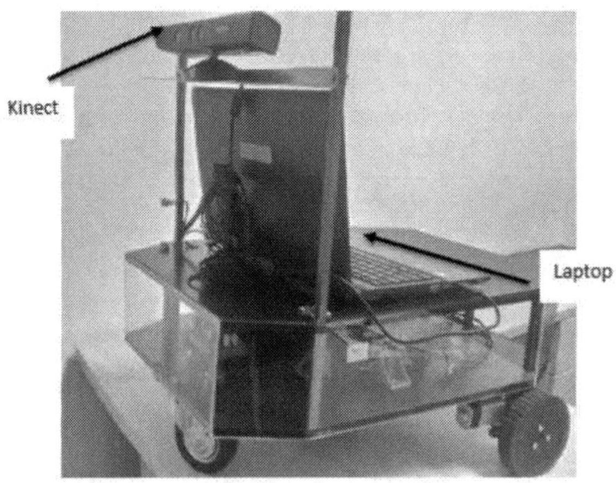

Fig. 3. The setup of the mobile robot.

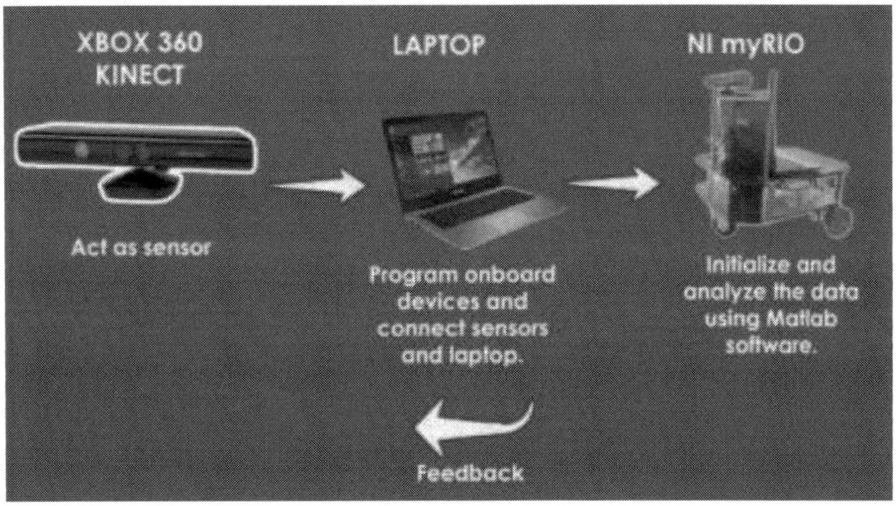

Fig. 4. The flow of the system.

The Kinect sensor detect the human and its distance from the human. MATLAB is used as interfacing software to send the distance data from Kinect to myRIO using User Datagram Protocol (UDP). The distance data received is computed in the myRIO and signal is transmitted to the motor to control the distance of the robot from the human.

3 Results and discussion

3.1 Modelling results

The simulation of the platooning system of the mobile robot is conducted. The result developed is shown in figure 3 below. A tracking reference of one metre is set as the input. The target is originally set at the position (1, -2) and the robot to be at (-5, -2). When the system is initiated the robot detects a huge difference in between the target and the robot. This is shown in figure 4. The PID calculates the error and gives maximum output to the motors. Based on figure 3 the saturation is at 20V which is the maximum voltage supplied to the robot. The robot detects the huge distance and tries to maintain the distance of one metre. Thus maximum voltage is supplied and gradually decreases as the robot gets closer to the target. The voltage supplied is negative which indicates the robot is moving backwards due to overshoot. The robot undergoes series of negative and positive voltage input to the robot until the distance is one metre from the target.

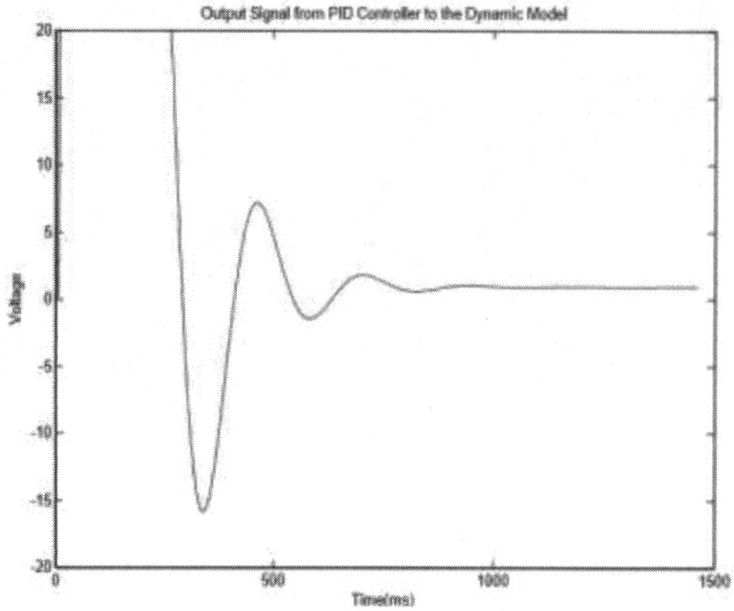

Fig. 5. The output of PID controller for dynamic model.

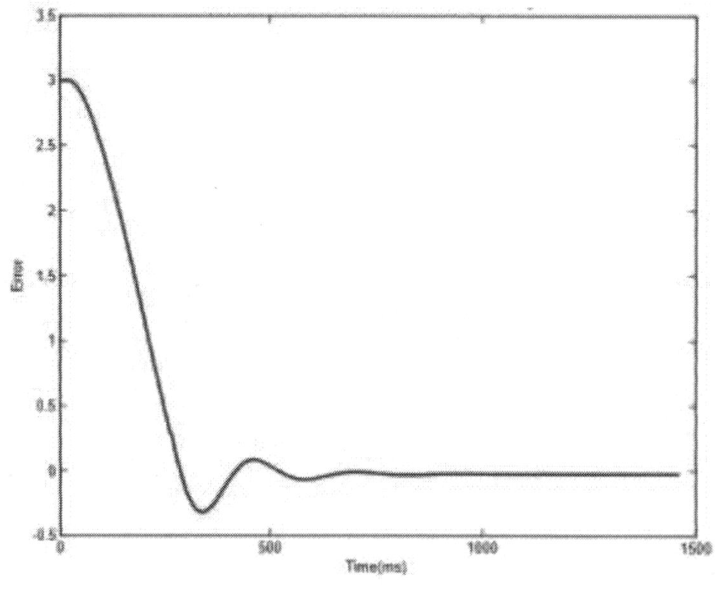

Fig. 6. The difference in distance between the robot and target.

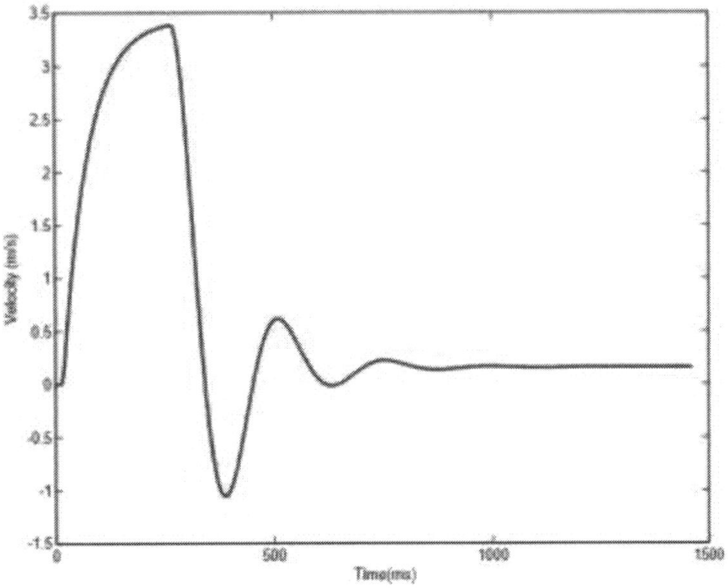

Fig. 7. The average velocity of both wheels.

The velocity of the wheels is at the maximum when the robot first initiates since a maximum voltage is supplied to both motors. The maximum speed achieved is approximately 3.4ms^{-1} as shown in figure 5. The negative speed shows the motor moves in opposite direction to correct its position and stabilises at 0ms^{-1}since the distance between the target and the robot is one metre.

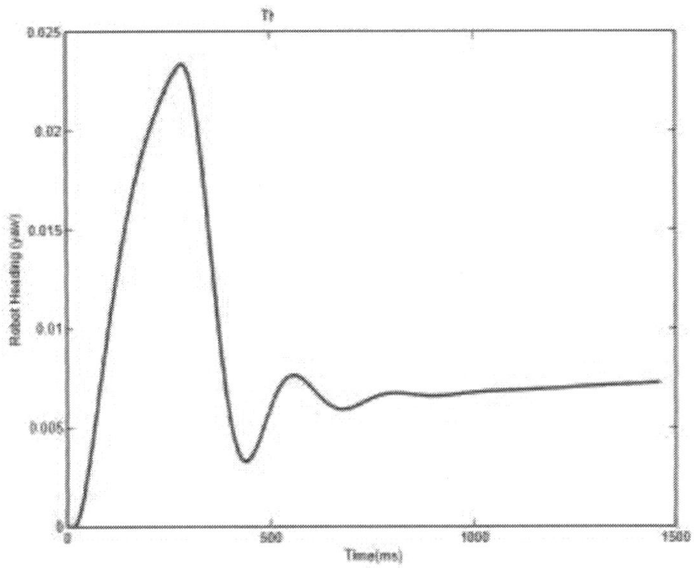

Fig. 8. Heading of the robot.

 DOI: 10.1051/matecconf/20179001060

The robot had a small heading due to the slip forces acting on it. The dynamic model of the robot includes slip forces acting on the robot which causes the robot to head left when the robot initiates. High speed and heavy chassis cause the robot to have a small amount of heading towards left. This can be observed in figure 6 where the robot has the highest heading which is 0.024 degree and gradually decreases to 0.0048 degrees.

3.2 Experimental results

The experiment of the developed system is conducted by setting up the experiment as mentioned earlier in methodology. The image processed by MATLAB is shown in figure 9. A point in the image is taken which is respective to X and Y axis of the Kinect (321, 117). A human is allowed to stand in front of the camera. Then Kinect measures the distance of the human from the camera and stores it in array form. The array of data is transmitted to myRIO using UDP. myRIO receives the depth data and process it to send a signal to the motors. In experimental setup, the PID controller was not used. Only the proportional from the PID is used in the experimental setup as an early testing. The result of the error output (difference in distance) is shown in myRIO. myRIO uses LABVIEW as programming software which enables monitoring of the output at each node similar to SIMULINK in MATLAB. The figure 10 and 11 shows the myRIO output.

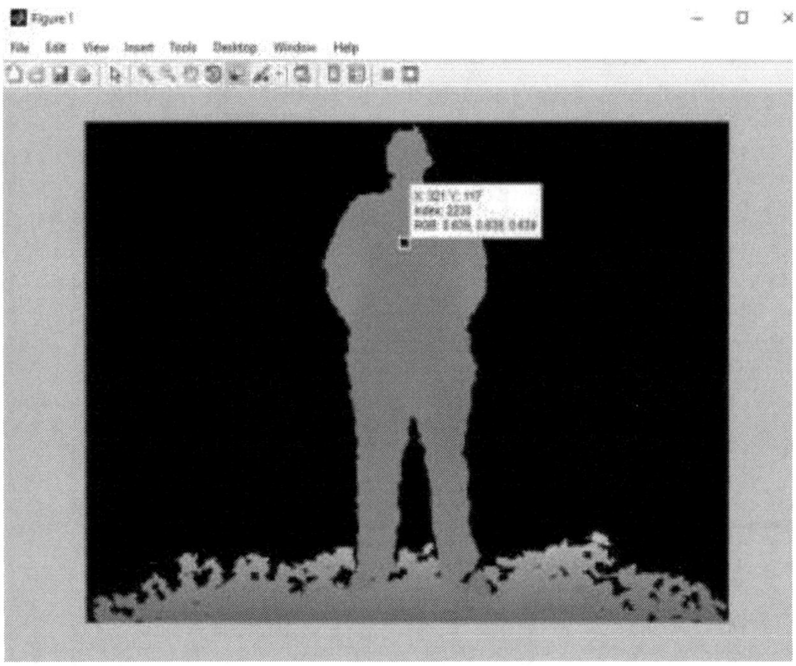

Fig. 9. Depth Image seen by Kinect.

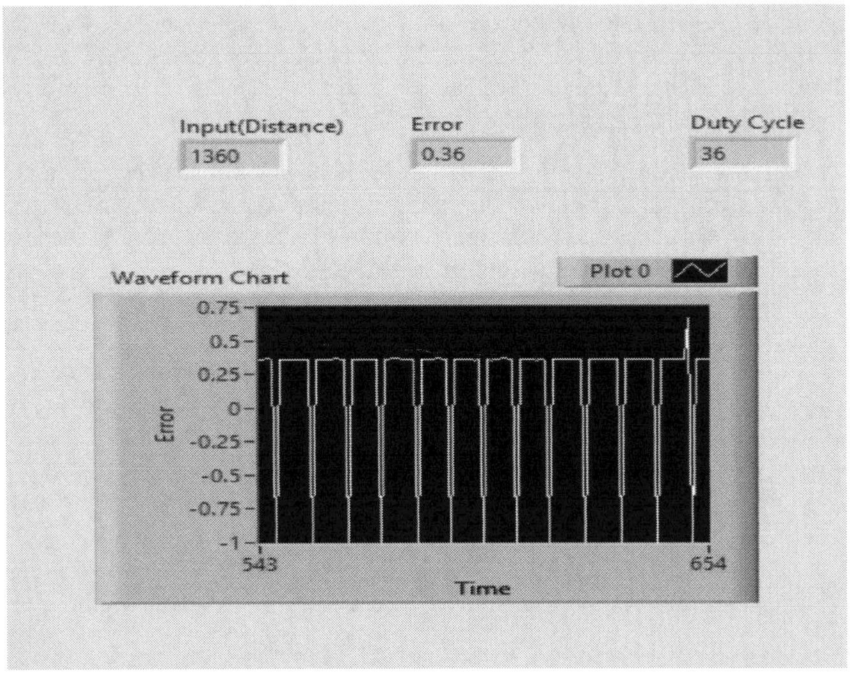

Fig. 10. The error output for distance more than 1 metre.

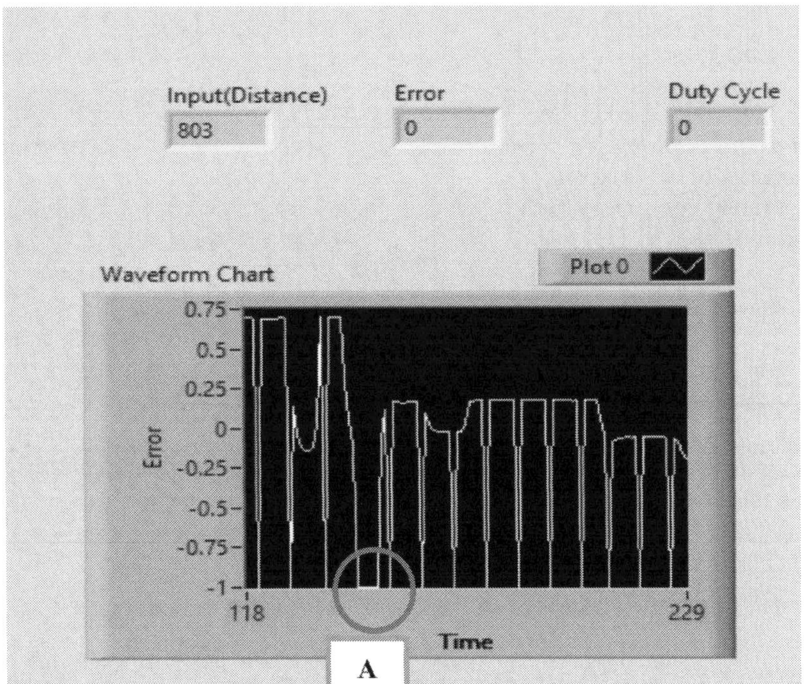

Fig. 11. The distance changes to maintain 1metre distance from the target.

The graph in figure 10 represents the error between the reference distance and the original distance of the target from the robot. The robot detects the human at 1.36 metre which is 0.36 metre more than the reference distance. When the motor is turned on, the robot moves closer to the target. This is shown in figure 11 where an initial 0.7m error is detected. The robot quickly moves forward to maintain the reference distance of 1 metre.

However, as the target moves the error varies as the distance enlarges. Thus the robot moves accordingly to maintain the distance of 1 metre. At point A, the error is at -1 which means the robot is closer to the target. The robot can be seen to correct the distance by moving further from the target to maintain the reference distance.

Based on the experimental result and the simulation result, the error graph can be compared. The error graph in figure 6 shows the error reduces to 0 as the robot gets closer to the target. This is same as shown in figure 11 of the experimental result where the error reduces to zero as the robot gets closer to the target.

4 Conclusions

In this paper, the behaviour of the mobile robot under platooning system was investigated. The mathematical model developed was used in the simulation profile along with the predetermined chassis parameters. PID controller was used as feedback controller to give input to the robot to move. The dynamic model presented shows the behaviour of the robot at given conditions Furthermore, the kinematic model is used to identify the position of the robot in the planar plane. The system developed can be used by researchers to develop a human following mobile robot that moves along with the human within a specified distance.

The authors are obliged to thank the Universiti Malaysia Pahang for providing a platform for the project and providing financial assistance under research grant, RDU16043.

References

1. Hu, J.S., J.J. Wang, and D.M. Ho, *Design of Sensing System and Anticipative Behavior for Human Following of Mobile Robots.* IEEE Transactions on Industrial Electronics, **61**(4): p. 1916-1927.(2014)
2. Mast, M., et al., *Design of the Human-Robot Interaction for a Semi-Autonomous Service Robot to Assist Elderly People*, in *Ambient Assisted Living: 7. AAL-Kongress 2014 Berlin, Germany, January 21-22, 2014*, R. Wichert and H. Klausing, Editors. 2015, Springer International Publishing: Cham. p. 15-29.
3. Vedova, M.L.D., et al. *Platooning control of autonomous nonholonomic mobile robots in a human-robot coexisting environment.* in *2012 American Control Conference (ACC),Montreal,QC,pp.6569-6574.* (2012).
4. Zakaria, M.A., et al., *A path tracking algorithm using future prediction control with spike detection for an autonomous vehicle robot.* International Journal of Advanced Robotic Systems, **10**.(2013.)
5. Lam, C.P., et al., *Human-Centered Robot Navigation—Towards a Harmoniously Human–Robot Coexisting Environment.* IEEE Transactions on Robotics, **27**(1): p. 99-112,(2011).
6. Kelly, A. and N. Seegmiller, *A Vector Algebra Formulation of Mobile Robot Velocity Kinematics*, in *Field and Service Robotics: Results of the 8th International Conference*, K. Yoshida and S. Tadokoro, Editors, Springer Berlin Heidelberg: Berlin, Heidelberg. p. 613-627,(2014)
7. Yoon, J., et al. *Autonomous dynamic driving control of wheeled mobile robots.* in *2014 IEEE International Conference on Robotics and Automation (ICRA)*,pp. 5274-5279,(2014).
8. Koubaa, Y., M. Boukattaya, and T. Dammak, *Adaptive Sliding-Mode Dynamic Control For Path Tracking of Nonholonomic Wheeled Mobile Robot.* 2015.

9. Köker, R., *A genetic algorithm approach to a neural-network-based inverse kinematics solution of robotic manipulators based on error minimization.* Information Sciences, **222**: p. 528-543,(2013).
10. Dhaouadi, R. and A.A. Hatab, *Dynamic modelling of differential-drive mobile robots using lagrange and newton-euler methodologies: A unified framework.* Advances in Robotics & Automation, (2013).
11. Azizi, M.R. and D. Naderi, *Dynamic modeling and trajectory planning for a mobile spherical robot with a 3Dof inner mechanism.* Mechanism and Machine Theory, **64**: p. 251-261,(2013).
12. Sandino, L., M. Bejar, and A. Ollero. *Tutorial for the application of Kane's method to model a small-size helicopter.* in *Proceedings of the 1st Workshop on Research, Development and Education on Unmanned Aerial Systems*. 2011.
13. Sipahi, R., et al., *Stability and Stabilization of Systems with Time Delay.* IEEE Control Systems,. **31**(1): p. 38-65,(2011).
14. Chevalier, A., et al. *Emulation of a highway bottleneck using leader-follower formation control.* in *Applied Computational Intelligence and Informatics (SACI), IEEE 8th International Symposium on*. 2013,(*2013).*
15. Fateh, M.M. and A. Arab, *Robust control of a wheeled mobile robot by voltage control strategy.* Nonlinear Dynamics, 2015. **79**(1): p. 335-348.
16. Shen, D., W. Sun, and Z. Sun, *Adaptive PID formation control of nonholonomic robots without leader's velocity information.* ISA Transactions, **53**(2): p. 474-480, (2014).

Magnetic field simulation of a thermal conductivity measurement instrument for magnetorheological fluid

M.S.A. Rahim[1,*], *I.* Ismail[1,2], *S.A.* Wahid[1], *S.* Aid[1], and *S.N.* Aqida[2,3]

[1]Faculty of Manufacturing Engineering, Universiti Malaysia Pahang, 26600 Pekan, Pahang, Malaysia
[2]Automotive Engineering Centre, Universiti Malaysia Pahang, 26600 Pekan, Pahang, Malaysia
[3]Faculty of Mechanical Engineering, Universiti Malaysia Pahang, 26600 Pekan, Pahang, Malaysia

Abstract. Technological advancements in thermal systems demand an innovative heat dissipation technology. Magnetorheological (MR) fluid has a huge potential to solve the problem. However, characterising thermal conductivity of the materials in magnetic fields required tailored instruments. This paper presents a concept design of the MR fluids thermal conductivity measurement instrument. The developed instrument was designed to be able to measure thermal conductivity in both parallel and perpendicular orientations with magnetic field. Magnetic fields distribution of the proposed concept design was analysed using finite element method for magnetics. Design modification then conducted to improve the magnetic fields strength. Findings of this study showed that gap thickness played a significant factor in determining the optimal design. Simulated magnetic fields strength at both parallel and perpendicular orientations were found identical, yet varied in distributions.

1 Introduction

Magnetorheological (MR) fluid is a suspension that possesses alterable rheological properties in magnetic field presence. Due to this advantage, MR fluid is beneficial to be applied in active and semi-active devices [1]. At high temperature environment, shear thinning and decreasing shear stress of MR fluid devices could occur that eventually resulted in loss of efficiency. One of the ways to anticipate this is to improve on thermal conductivity property of MR fluid.

Thermal conductivity is greatly enhanced when MR fluid is subjected to magnetic field [2]. The finding was experimentally done by Yildirim and Genc [3] in which the thermal conductivity of MR fluid increased along with increasing magnetic field. This behaviour is attributed to the formation of chain-like structures by the magnetic particles when subjected to magnetic field, which provide high conductivity heat transfer paths [4]. Additionally, thermal conductivity of MR fluid is found to be anisotropic under the influence of magnetic field. Reinecke et al. [5] found that thermal conductivity of MR fluid increased by almost 100% when thermal gradient is parallel with magnetic field.

* Corresponding author: sallehrahim@icloud.com

There are several ways to measure thermal conductivity property of a suspension. The most widely used is transient hot-wire (THW) method. THW method has the capacity to eliminate error due to natural convection besides getting faster results compared to other methods [6]. Very recently, Khdher et al. [7] adopted this method to obtain thermal conductivity of bio glycol based Al2O3 nanofluids. Gavili et al. [8] have used THW to measure thermal conductivity of water based ferrofluids. Cheng et al. [9] used THW at different MR fluid concentrations in the absence of magnetic field.

However, the THW method is unable to measure anisotropic thermal conductivity of MR fluid. This limitation has pushed researchers to adopt different methods. Cha et al. [10] has developed a transient hot-square instrument that enabled direct measurements of thermal conductivity in the direction parallel to external magnetic field. Krichler and Odenbach [11] used hot-plate technique where specimen is positioned between two plates. The idea is for both parallel and perpendicular arrangements of thermal gradient and magnetic field are possible.

This paper is aimed to obtain the optimal design for a thermal conductivity measurement instrument. The instrument will be able to measure thermal conductivity of MR fluid with the presence of magnetic field. It is also capable to measure the anisotropic thermal conductivity of MR fluid at parallel and perpendicular directions to magnetic field. Geometry parameters of the instrument are being considered. The influence of each parameter for several designs will be measured by magnetic flux density. Analysis and design optimization are carried out by using Finite Element Method Magnetics (FEMM) software.

2 Design methodology

The method used to measure thermal conductivity is based on a guarded hot plate technique and followed ISO 8301 standard. Usually this technique is applied to solid body measurements, for which the specimen is positioned in between two plates. The upper plate is heated, and the lower plate maps the temperature increase. During each measurement, the heater located in the upper plate is subjected to a current pulse and the resulting change in its electrical resistance and hence temperature is monitored as a function of time. However, modification of the system has been done to accommodate the magnets for magnetic field induction. This modification is beyond requirement of the standard. This method has also been applied in previous research [3, 10, 11].

In this design, a MR fluid sample is filled inside a PMMA container. A plate heater (upper plate) is used as the heat source and is embedded inside the cap. Heat sink (lower plate) acted as the base for the instrument and to release heat from the top. A pocket is made each at the top of the cap and at the bottom of the heat sink as slots for the permanent magnets. The attraction of the magnets will then generate a vertically stable magnetic field through the PMMA. The slots at the side of the PMMA are also created for the magnets to produce horizontal magnetic field. Several thermocouples are installed around the PMMA wall to monitor the change of temperature. The schematic diagram of the thermal conductivity measurement instrument is shown in Figure 1.

The research began by identifying geometry parameters of the instrument. The parameters included were PMMA diameter, PMMA height and gap thickness between magnets and sample. They were then categorized to three different levels; low, medium and high. Geometry parameters of the proposed design were selected at medium level. Such design was chosen after considering that the design geometries are suitable for ease of manufacturing and assembly. However, consideration of low and high levels of geometry parameters are also significant as to observe the influence and magnitude of each parameter. The summary of geometry parameters is as shown in Table 1.

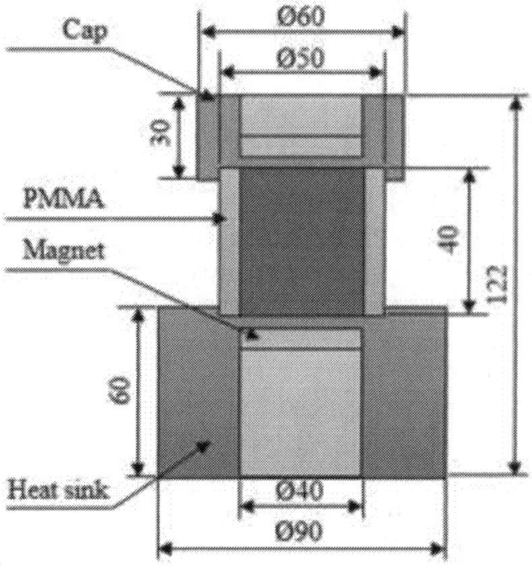

Fig. 1. Schematic diagram of thermal conductivity measurement instrument.

The diameter and height of the cylindrical PMMA were set to be equal in geometry regardless of their level. These were designed in such a way that the difference of thermal conductivity between parallel and perpendicular directions of magnetic field can be observed. The gap is what separates the magnet with the sample. On the other hand, the plate heater inside the cap indicates that thermal gradient is only unidirectional (vertical).

Table 1. Geometry parameters of thermal conductivity measurement instrument.

Geometry parameters	Low	Medium	High
PMMA diameter, mm	30	40	50
PMMA height, mm	30	40	50
Gap thickness, mm	5	10	15

Nevertheless, the main property to be focused on in this research was the magnetic flux density. Magnetic flux density was obtained from the two permanent magnets which were located and arranged either vertically or horizontally. The magnet thickness was 10 mm. Magnetic field from the magnets will create heat conduction paths across the instrument through the formation of chains in the MR fluid. Chain formation process in an MR fluid only acquires several milliseconds to complete after the magnetic field is applied [12]. The sample used in the simulation was MRF-132DG, a commercial MR fluid. The minimum magnetic flux density required to develop chain formation for MRF-132DG is approximately 0.0597 T [13].

2.1 Material of instrument

Due to many desirable properties, polymethyl methacrylate (PMMA) has been selected as the cylindrical wall material. It possesses high mechanical strength, high Young's modulus and low elongation at break. PMMA is also considered among the hardest thermoplastics and is highly scratch resistant. Meanwhile, thermal stability of standard PMMA is 65°C. Heat stabilised types can withstand temperatures of up to 100°C. The typical properties of PMMA material (Cyro Industries, USA) can be found in Table 2.

Table 2. Typical properties of polymethyl methacrylate (PMMA).

Properties	Value
Hardness, Rockwell	M80 – M100
Tensile strength	55 – 76 MPa
Tensile modulus	2413 – 3447 MPa
Max. operating temperature	65 - 93°C
Thermal conductivity	0.56 W/m.K
Melting point	130 – 140°C

Polytetrafluoroethylene (PTFE) offers high chemical resistance, low and high temperature capability, resistance to weathering, low friction, electrical and thermal insulation, and "slipperiness". PTFE's mechanical properties are low compared to other plastics, but its properties remain at a useful level over a wide temperature range of -73°C to 204°C. It has excellent thermal and electrical insulation properties and a low coefficient of friction. Hence, PTFE is very suitable as the cap material. The typical properties of PTFE material (DuPont, USA) are as depicted in Table 3.

Table 3. Typical properties of polytetrafluoroethylene (PTFE).

Properties	Value
Hardness, Shore D	D50
Tensile strength	26 – 27 MPa
Tensile modulus	550 MPa
Max. operating temperature	260°C
Thermal conductivity	0.25 W/m.K
Melting point	335°C

The material selected for heat sink was aluminium. Possessing high thermal conductivity, aluminium allows heat to pass through it swiftly. Meanwhile, the disk shaped permanent magnets used were NdFeB (Neodymium-Iron-Boron) N52 grade. N52 is the highest material grade with residual flux density of 1.43 T and coercivity of 891.3 kA/m. The maximum energy product for N52 is 52 MGOe.

2.2 Finite element method for magnetics

The simulation was a magnetostatic problem in which the fields are time-invariant. In this case, the magnetic field intensity, H and magnetic flux density, B must obey:

$$\nabla \times H = J \tag{1}$$

$$\nabla \cdot B = 0 \tag{2}$$

Finite Element Method Magnetics (FEMM) software goes about finding a field that satisfies Eqs. (1) and (2) via a magnetic vector potential approach [14]. Adopting the B and H relationship equation and flux density equation, (1) can be rewritten as:

$$\nabla \times \left(\frac{1}{\mu(B)} \nabla \times A\right) = J \tag{3}$$

∇ is the divergence of a vector field,
H is the magnetic field intensity,
J is the electric current density,
B is the magnetic flux density,
μ is the magnetic permeability,
A is the vector potential.

The thermal conductivity measurement instrument was designed in FEMM software and was analyzed as a 2-D planar model. The proposed design was with vertical magnet arrangement, PMMA geometries of Ø40 x 40 mm, and 10 mm gap thickness at each side. Materials were defined accordingly for each design parts. The gap material at the top was set to PTFE while at the bottom was set to Aluminum 6061-T6. This difference can be neglected as both are linear materials (relative magnetic permeability, $\mu_r = 1$) and have no coercivity. FEMM generated 8890 nodes and 17418 elements for the proposed design. The mesh shape was set to triangular while the mesh size had an average of 0.5 mm size. The boundary condition type was a mix of Robin or Neumann. For this simulation, the mixed boundary condition was treated as open boundary problem.

3 Magnetic field simulation

The magnetic field simulation results are given in Figure 2 and Figure 3. The low level of geometry parameters is represented by Figure 2 (a) and Figure 3 (a) while the high level is represented by Figure 2 (b) and Figure 3 (b). Warmer colours correspond to higher flux densities while solid black lines indicate magnetic field lines. The magnitude of flux density was at the maximum at the edge of each disk shaped magnets. It is also observed that the magnetic field lines were crossing through the PMMA. The closest point inside PMMA to the magnet had the highest magnitude of flux density while the centre point inside PMMA had the lowest magnitude of flux density. Hence, the MR fluid samples filled inside the PMMA containers for vertical and horizontal magnet arrangements were experiencing magnetization.

Since the magnets produced uniform magnetic fields, attractive or repulsive field-particle interactions do not exist due to the lack of a magnetic field gradient [15]. The particle-particle interactions in the MR fluid become dominant. The imposed field induces magnetic dipoles and as a result, mutual particle interactions occur. The magnetic particles in the MR fluid attract each other when aligned end to end. Due to the attractive forces, chain-like structures are developed .

The forming of chain-like structures from magnetic particles in the MR fluid created heat conduction paths. Since magnet arrangements in Figure 2 (a) and Figure 2 (b) were vertical with plate heater at the top, the conduction paths will be parallel to the thermal gradient. On the other hand, horizontal magnet arrangements in Figure 3 (a) and Figure 3 (b) will be perpendicular to the thermal gradient. Hence, the anisotropic thermal conductivity of MR fluid under the influence of magnetic field can be observed.

4 Design modifications

In vertical magnet arrangement, the magnitude of flux density at the centre point of PMMA in Figure 2 (a) and Figure 2 (b) were 0.166 T and 0.115 T, respectively. Comparing with the proposed design which has Ø40 x 40 mm of PMMA and 10 mm gap thickness, the magnitude was recorded at 0.143 T. Visually and mathematically, all designs produced higher flux densities than required (0.0597 T) to develop chain formations of magnetic particles [13].

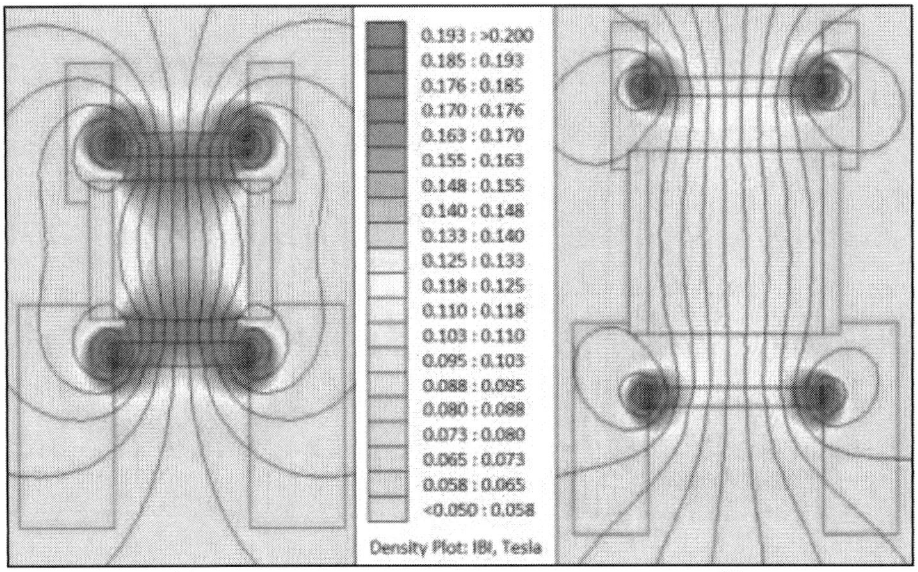

(a) (b)

Fig. 2. Magnetic flux density for vertical magnet arrangement of (a) Ø30 x 30 mm of PMMA and (b) Ø50 x 50 mm of PMMA at 5 mm and 15 mm magnet gap; respectively.

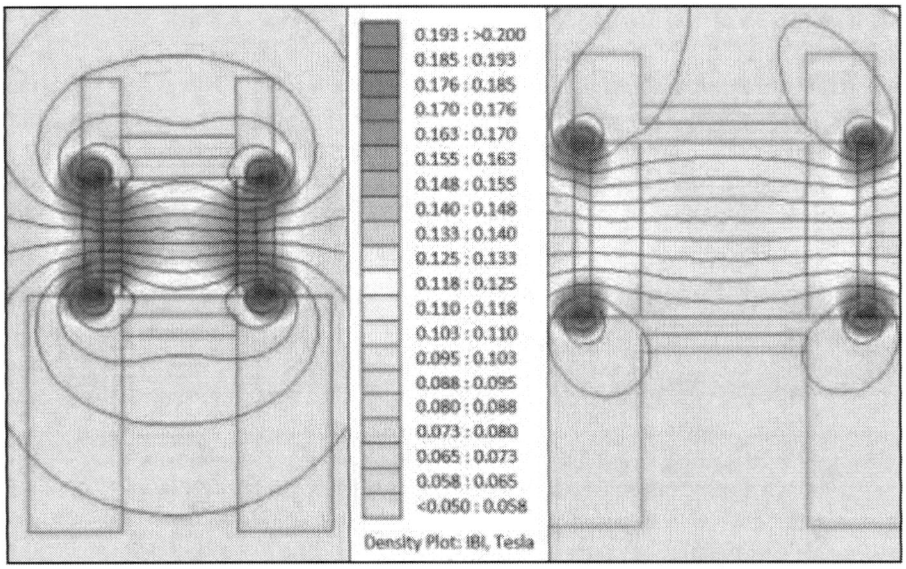

(a) (b)

Fig. 3. Magnetic flux density for horizontal magnet arrangement of (a) Ø30 x 30 mm of PMMA and (b) Ø50 x 50 mm of PMMA at 5 mm and 15 mm magnet gap; respectively.

Figure 4 highlights on the effect of PMMA geometries to magnetic flux density at 10 mm gap thickness. By varying PMMA heights, there is not much difference to the magnitude of flux density. A Ø30 x 30 mm of PMMA resulted in 0.159 T while a Ø30 x 50 mm of PMMA achieved at 0.155 T. At Ø40 mm and Ø50 mm, the difference is almost negligible where changes of flux density were only 0.001 T across all heights. The standard deviations for 30 mm, 40 mm and 50 mm of PMMA heights are 0.002 T, 0.001 T and 0.001 T, respectively.

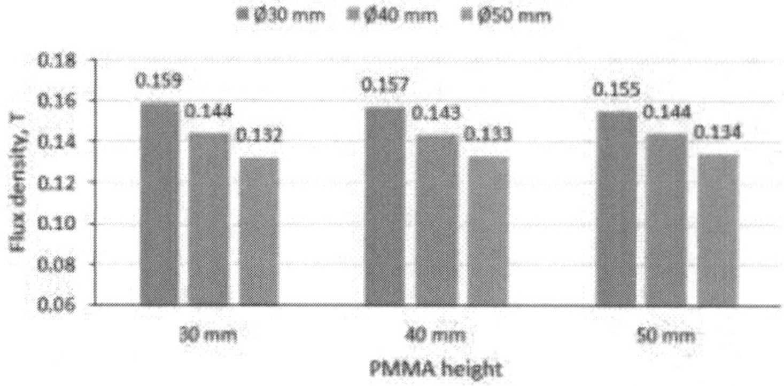

Fig. 4. Effect of different PMMA diameter and height (gap thickness = 10 mm).

A minor effect is also observed by changing the PMMA diameter. The trend shows that magnetic flux density decreases as PMMA diameter increases across any given height. The largest flux density difference can be seen at 30 mm PMMA height. Flux density was 0.159 T at Ø30 mm but decreased to 0.132 T at Ø50 mm. The standard deviations for Ø30 mm, Ø40 mm and Ø50 mm are 0.014 T, 0.012 T and 0.011 T, respectively. Hence, it is presumed that the distance between two magnets and diameter of the magnet (due to the increase of PMMA diameter) are insignificant due to oversupply of magnetic field strength from the strong magnets.

Meanwhile, the effect of gap thickness at constant PMMA geometry is as depicted in Figure 5. It is clearly found that 5 mm of gap thickness gave the highest magnitude of flux density. At Ø30 x 30 mm, the flux density was captured at 0.213 T for 5 mm gap thickness compared to 0.114 T for 15 mm gap. The impact is great as there is minimal boundary between magnets and sample. Then again, the declination also applies to Ø40 x 40 mm and Ø50 x 50 mm as gap thickness increases but has much smaller influence. At 15 mm gap thickness, there is hardly any change of flux density though PMMA geometries are different.

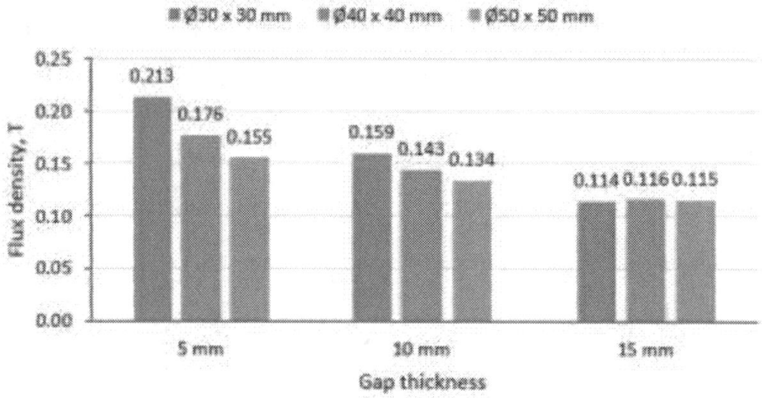

Fig. 5. Effect of gap thickness at constant PMMA diameter and height.

On the other hand, the magnitude of flux density for horizontal magnet arrangement was 0.215 T in Figure 3 (a) and 0.113 in Figure 3 (b). The proposed design was recorded at 0.147 T. As similar to vertical magnet arrangement, all designs are able to develop chain formations of magnetic particles.

 DOI: 10.1051/matecconf/20179001061

Figure 6 shows on the effect of PMMA geometry to flux density with 10 mm gap thickness. At different PMMA width, changes of flux density were not much. The flux density was 0.153 T at Ø30 x 30 mm and decreased to 0.151 T at Ø50 x 30 mm. This indicates that the PMMA width shows little influence on flux density. It also shows that the horizontal magnet arrangement is having similar effect with the vertical magnet arrangement. This is because the location of magnets is simply changed from vertical to horizontal and hence the height of PMMA at vertical becomes the width of PMMA at horizontal. The PMMA height also shows minimal impact but with greater influence than PMMA width. Flux density of 0.153 T was recorded at Ø30 x 30 mm and reduced to 0.136 T for Ø30 x 50 mm at the same width. The standard deviations for PMMA height of 30 mm, 40 mm and 50 mm are 0.009, 0.008 and 0.007; respectively.

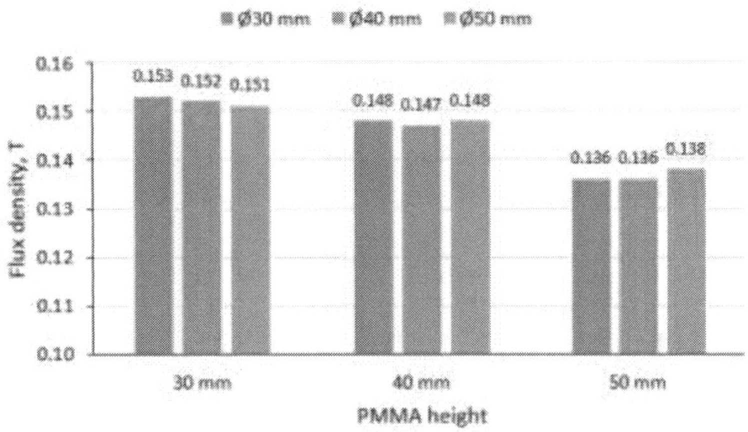

Fig. 6. Effect of different PMMA diameter and height (gap thickness = 10 mm).

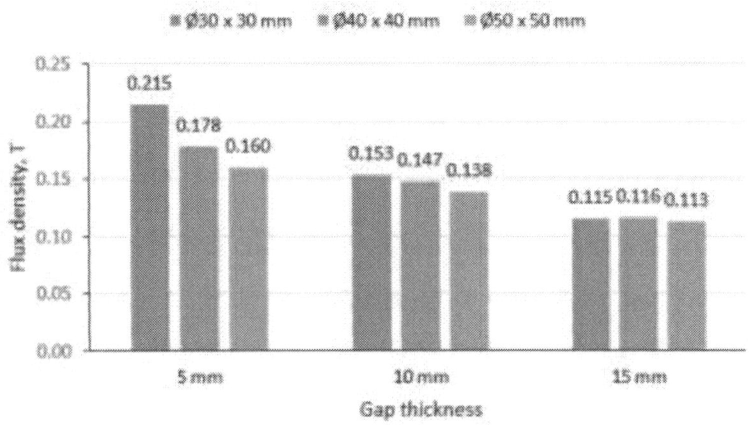

Fig. 7. Effect of gap thickness at constant PMMA diameter and height.

The effect of gap thickness with the same PMMA geometry is observed in Figure 7. As similar to vertical arrangement, the outcome is also larger at 5 mm gap thickness in horizontal magnet arrangement. Flux density was recorded at 0.215 T at 5 mm gap but reduced to 0.115 T at 15 mm gap. The changes are smaller as PMMA geometry is bigger due to reduced influence of magnetic field.

5 Conclusions

This work has suggested several design considerations for a thermal conductivity measurement instrument. Three parameters are being looked into namely PMMA diameter, PMMA height and gap thickness. By utilizing the FEMM software, simulation results were produced and magnetic flux density for all designs have been achieved. It is clearly found that gap thickness played a significant factor in determining the optimal design. For both vertical and horizontal magnet arrangements, the highest magnetic flux densities were produced with gap thickness at 5 mm. Meanwhile, the geometries of PMMA container indicated a minimal flux density change at all parameter levels. For vertical magnet arrangement, the low level (Ø30 mm) of PMMA diameter resulted in higher flux density regardless of the PMMA height. This is a complete opposite to horizontal magnet arrangement in which the PMMA height at low level (30 mm) is more dominant, irrespective of PMMA diameter. Hence, the optimal design for the thermal conductivity measurement instrument is by having PMMA geometries at medium level with 5 mm of gap thickness.

The authors are gratefully acknowledged Universiti Malaysia Pahang (UMP) for providing facilities and financial support.

References

1. I. Ismail, S.A. Mazlan, Z. Hairi, A.G. Olabi, Jpn. J. Appl. Phys. **51,** 1 (2012)
2. M.S.A. Rahim, I. Ismail, Mater. Sci. Eng. **100**, 1 (2015)
3. G. Yildirim, S. Genc, Smart Mater. Struct. **22**, 1 (2013)
4. S.I. Mistik, T. Shah, R.L. Hadimani, E. Siores, J. Intel. Mat. Syst. Str. **23**, 1277 (2012)
5. B.N. Reinecke, J.W. Shan, K.K. Suabedissen, A.S. Cherkasova, J. Appl. Phys. **104**, 1 (2008)
6. G. Paul, M. Chopkar, I. Manna, P.K. Das, Renew. Sust. Energ. Rev. **14**, 1913 (2010)
7. A.M. Khdher, N.A.C. Sidik, W.A.W. Hamzah, R. Mamat, Int. Commun. Heat. Mass. **73**, 75 (2016)
8. A. Gavili, F. Zabihi, T.D. Isfahani, J. Sabbaghzadeh, Exp. Therm. Fluid. Sci. **41**, 94 (2012)
9. H. Cheng, W. Hong, C. Liu, J. Chin. Adv. Mat. S. **2**, 130 (2014)
10. G. Cha, Y.S. Ju, L.A. Ahuré, N.M. Wereley, J. Appl. Phys. **107**, 1 (2010)
11. M. Krichler, S. Odenbach, J. Magn. Magn. Mater. **326**, 85 (2013)
12. J. de Vicente, D.J. Klingenberg, R.H. Alvarez, Soft Matter **7**, 3701 (2011)
13. N. Wiltsie, M. Lanzetta, K. Iagnemma, TePRA, 91 (2012)
14. D. Meeker, Finite Element Method Magnetics Version 4.2 User's Manual (2015)
15. M. Zrínyi, Colloid Polym Sci **278**, 98 (2000)

An investigation on the surface finish of sculptured surface utilizing reverse engineering data of crank case cover – (CATIA V5 & DELCAM)

Sundi Syahrul Azwan[1,*], *Jumali* Muhammad Syafik[1] , *Raffay* Mohamad Razly[2] and *R.Abdullah* R. Izamshah[2]

[1]Faculty of Engineering Technology, Universiti Teknikal Malaysia Melaka, Hang Tuah Jaya, 76100 Durian Tunggal, Melaka, Malaysia
[2]Faculty of Manufacturing Engineering, Universiti Teknikal Malaysia Melaka, Hang Tuah Jaya, 76100 Durian Tunggal, Melaka, Malaysia

Abstract. The ultimate aim of this study was to investigate the effect of surface finish for a machined part which was programmed by two popular Computer Aided Design / Computer Aided Manufacturing (CAD/CAM) software namely as Catia V5 and Delcam (Delcam for Solidworks - DFS) by using scanned data obtained from one of the Reverse Engineering methods namely Three-Dimensional (3D) scanning process. A crank case cover was chosen as the physical part to be scanned and machined because of its sculptured shape and complex geometry. In this study, simultaneous three-axis machining programs were created and machined using three-axis Computer Numerical Control (CNC) Milling machine; DMC 635 Ecoline. Furthermore, all machining parameters are remained the same for both programs. Initially, the physical crank case cover gone through the first process called scanning process using a 3D scanner; model 700 CX in order to capture the 3D CAD data in points cloud form. The raw scanned data then gone through editing process to obtain better surfaces using Geomagic Studio software. Moreover, the stable and edited CAD model then undergone CAD/CAM programming process for both mentioned software respectively. On the other hand, there were seven random points chosen to be analyzed for the surface roughness analysis. The arithmetic mean value (Ra) was used as the roughness value reference using Mitutoyo Surface Tester. Ultimately, it could be concluded that Catia V5 producing better surface finish compared to Delcam software concluded from the analysis carried out as presented in this paper.

1 Introduction

CAD/CAM system is commonly used in daily engineering tasks. CAM may be defined as the use of computer systems to plan, manage, and control the operations of a manufacturing

* Corresponding author: syahrul.azwan@utem.edu.my

plant through either direct or indirect computer interface with the production resources of the plant. The geometric model generated during the CAD process forms the basis for the CAM process. Various activities in CAM may require different types of information of the CAD process [1]. As it is developed through time and widely used, it needs the best decision to choose the best software to be owned. In the process of product design and manufacture, it is becoming increasingly important as competitive pressures call for improvements in product performance and quality, and for reductions in development time scale [2].

The process of discovering the technological principles of a device, object, or a system through analysis of its structure, function and operation is known as Reverse Engineering (RE). Reverse engineering has become a viable method to create a 3D virtual model of an existing physical part for use in 3D CAD, CAM, CAE or other software [3]. The measured data alone, usually represented as a point cloud, lacks topological information and is therefore often processed and modelled into a more usable format such as a triangular-faced mesh, a set of NURBS surfaces, or a CAD model. A digital 3D model must be constructed using a device that can analyse a real-world object or environment to collect data on its shape and possibly its appearance for example the colour. This device is known as 3D scanner. Nowadays, many different technologies can be used to build 3D-scanning devices, but each of the technology comes with its own limitations, advantages and costs. Some of the limitations of this technology that still present are optical technologies encounter many difficulties with shiny, mirroring or transparent objects [4].

The quality of the surface plays a very important role in the performance of milling as a good-quality milled surface significantly improves fatigue strength, corrosion resistance, or creep life. Surface roughness also affects several functional attributes of parts, such as contact causing surface friction, wearing, light reflection, heat transmission, ability of distributing and holding a lubricant, coating, or resisting fatigue. Therefore, the desired finish surface is usually specified and the appropriate processes are selected to reach the required quality. The machining quality was assessed through the comparison of surface roughness, surface texture and dimensional control parameters.

2 Methodology

Scanning process was the first step done in this research. A crank case cover is chosen as the physical part to be scanned and machined because of its sculptured and complex shape. Besides, this research initiated to be focused on application of reverse engineering in automotive industries. The scanning process was done by using 3D scanner model 700 CX. In 3D scanning, it was very hard to obtain fine surfaces from the raw data. Typical problems normally happened in the raw scanned data such as opened, overlapped and unfinished surfaces. Hence, it is really important to perform editing process in order to resolve all the problems mentioned. The editing process or smoothing process normally involves with sewing unattached surfaces, re-closing opened surfaces and re-creating overlapping surfaces. Geomagic Studio was the software used to perform the editing process. After editing process completed, the file must be saved in igs / IGES or stp / STEP file format so that it could be read in Catia V5 and Delcam software for CAM preparation. Figure 1 show the CAD model obtained after gone through editing process once the raw scanning data received from the 3D scanning process.

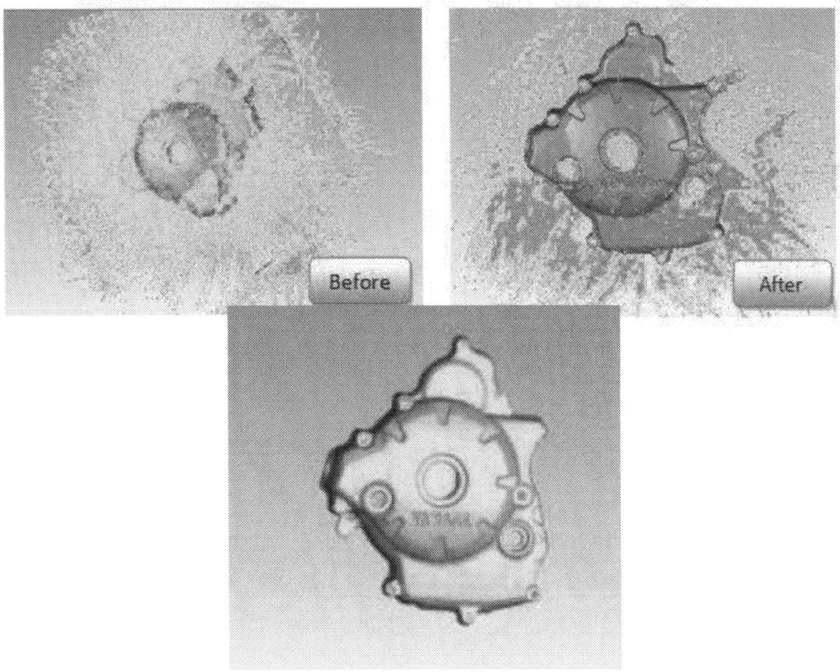

Fig. 1. The editing processes involved using Geomagic Studio.

For the physical machining, 3-Axis CNC milling machine model DMC 635 Ecoline DMG Mori machine was the machine selected. The size of raw materials or stocks prepared were 150 mm x 100 mm x 50 mm. The material was aluminium 3000 series. Cutting tool of End Mill D16 High Speed Steel (HSS) four flutes is used to perform the roughing process. Meanwhile, Ball Nose Mill R6 (HSS) cutting tools were chosen to perform the machining on the surfaces which were to be analyzed in surface roughness analysis. In Catia V5, Advanced Machining workbench is utilized to create the CAM program. Machining process offered by Catia V5 for surface machining called "Sweeping" process. On the other hand, in Delcam software it was automatically selected by the software once the specific geometries of the 3D CAD model are defined. There were setups or settings which need to be set for both CAMs software. It is realized that some different terms used carry the same meanings such as "depth of cut" used in Catia V5 whereas "Z increment" in Delcam. Table 1 indicates the settings have been applied for both CAMs software in preparing the CAM program. Meanwhile, Figure 2 shows the completed CAM programs for both CAMs software. Completed programs were then post processed using integrated post processor and finally transferred to the CNC machine for physical machining.

Surface Roughness is a measurable surface characteristic quantifying high frequency deviation from an ideal surface. It is usually described by the arithmetic mean value (Ra), based on the mean of the normal deviations from a nominal surface over a specified "cut off" length and normally measured in micron meters (µm). This process was done by using Mitutoyo surface roughness tester in metrology laboratory. Moreover, this analysis used a contact surface roughness tester to measure the surface finish. In other words, this method involves dragging a measurement stylus in traverse direction on the surface of the sample. The surface features were measured relatively to the drive unit reference surface. Hence, this accurately measures waviness and finally stepped features in addition to the surface roughness.

DOI: 10.1051/matecconf/20179001062

Table 1. Settings / parameters applied for both software in preparing the CAM program.

Software	Catia V5 software	Delcam software
Process	Roughing 1/ roughing 2	Rough Z-level
Spindle speed	7000 rpm	7000 rpm
Offset on part	1 mm / 0.25 mm	0.25 mm
Machining tolerance	0.1 mm	0.1 mm
Depth of cut	1.5 mm / 0.25 mm	0.25 mm

Software	Catia v5 software	Software	Delcam software
Process	sweeping	Process	Finish Z-level
Spindle speed	7000 rpm	Spindle speed	7000 rpm
Machining tolerance	0.1 mm	Machining tolerance	0.1 mm
Maximum distance between pass	0.25 mm	Step over	0.25 mm
Offset on part	0 mm	Finishing allowance	0 mm
Scallop height	0.002 mm	Z increment	0.002 mm

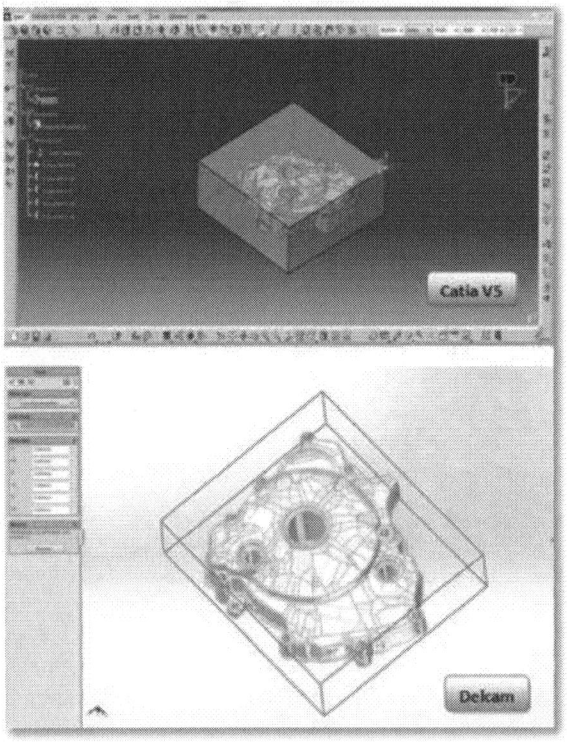

Fig. 2. Completed CAM program in Catia V5 (top) and Delcam (bottom).

3 Results and discussion

In measuring the surface finish, there were seven points randomly picked on the top surface of the machined parts; crank case cover. Various locations have been chosen to be analysed. Five readings taken on every each selected point and average value then calculated. All points taken for both machined parts were approximately on the same locations. Furthermore, three readings are taken for each identified point and the average value is calculated. The stylus traverse distance was set constantly at 5 mm at every single reading. Figure 3 indicates the location of the points taken for the surface analysis.

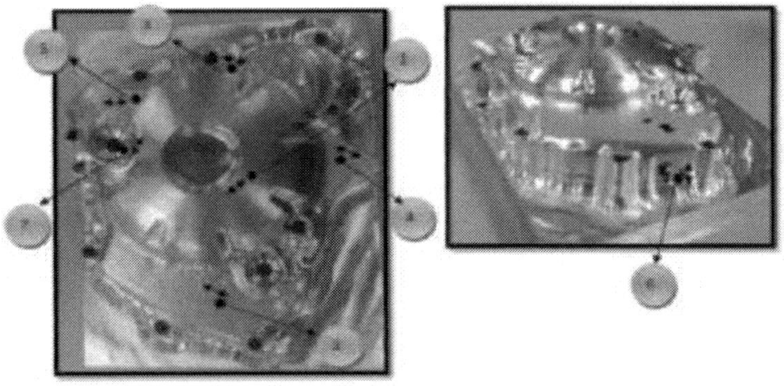

Fig. 3. Points taken in surface roughness analysis for both machined sample parts.

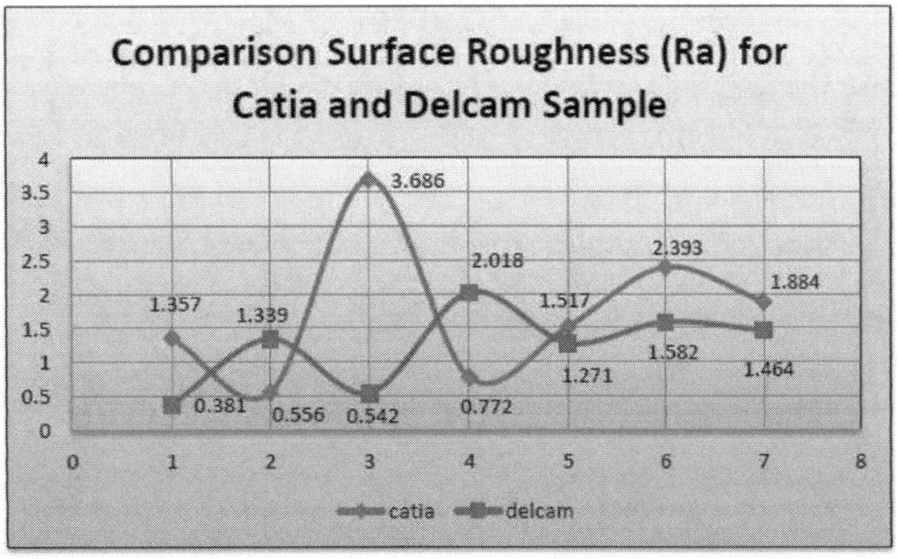

Fig. 4. Comparison of surface roughness average values for the 10 points taken.

From the result illustrates above, it is obviously shown that the highest average value for Catia V5 is 3.686 μm at the 3rd point whilst for Delcam's is 2.018 μm representing the 4th point. On the other hand, the best surface roughness value falls at the point marked as 2nd for Catia V5 with the value of 0.556 μm. Meanwhile, for Delcam it falls at the 1st point with 0.381 μm. This finding highlights that the same scanned 3D CAD model programmed by Delcam resulted slightly better surface finish in average than programmed by Catia V5. The main factor contributes to the result obtained was possibly caused by the different way of preparing the program. Although the same 3D CAD model were used and the same parameters applied, somehow the differences in determining the important settings whilst preparing the CAM program such as tool path direction and machining processes. In Catia V5 users were given options in determining the direction of tool path as this was not being offered in Delcam. In this study, the direction of the machining tool path selected was along the length of the part to be machined and in "Zigzag" motion. However, in Delcam the tool path was automatically calculated which was opposite direction from the one selected in Catia V5 and at certain areas the machining tool paths looked in "Spiral" movement. Furthermore, in Delcam there were too many shortcuts to simplify the machining program compared to Catia V5. Machining parameters were also automatically calculated from the library referring to the range set by the users as not in the practice of Catia V5. Catia V5 emphasizes details in each step and process compared to Delcam. Figure 5 shows the result of machining for both CAMs software.

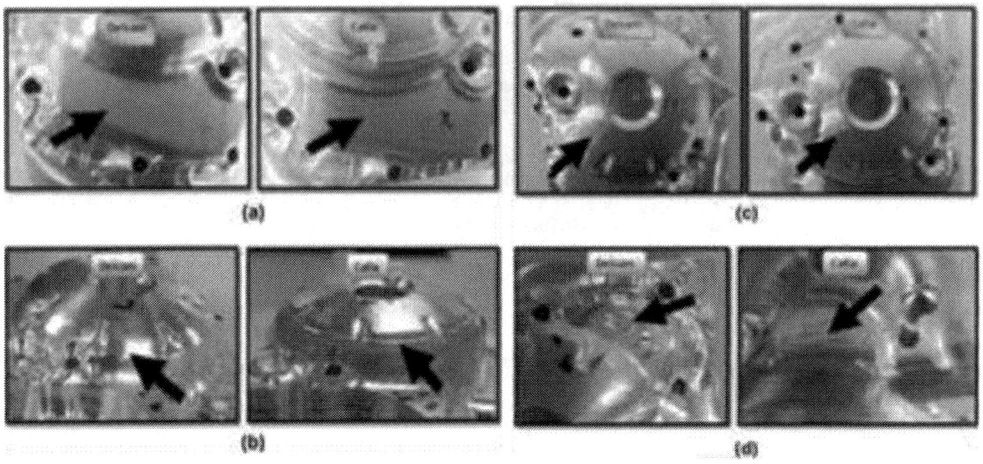

Fig. 5. Result of differences can be seen from machined sample parts - Delcam and Catia V5.

From Figure 5(a) both machined surfaces obviously indicate the differences. Surface machined by Delcam having linear lines all over the respected surface due to the effect of machining tool paths. Meanwhile, surface machined by Catia V5 has only macro lines to be compared to the Delcam's at certain specific area. This surface is marked as point 2 in the surface finish analysis. The average Ra value for Catia V5 sample is 0.556μm and 1.339μm for Delcam sample. Hence, from those two figures the surface of Catia V5 sample is definitely better than Delcam's. The main factor contributed to this result is believed due to the machining tool paths direction. In Catia V5 the direction of the machining tool path is set to along the length of the part to be machined and in "Zigzag" motion for finishing process. On the other hand, in Delcam it seems to be "Spiral" motion has been calculated for the finishing process.

Figure 5(b) illustrates the finest of profiles which were achieved by both mentioned software. Although the same machining parameters as well as the same CAM settings have been applied, the end machined results seems to be slightly different. In details, it could be seen that on certain particular areas surface machined by Delcam looks smoother than the Catia V5's. The size and type of cutting tool used for finishing process was ball end mill diameter 10mm. Again, "Spiral" direction machining tool path performed by Delcam's is believed was the main factor contributes to this result.

On the other hand, Figure 5 (c) presents the comparison of surface finish on the top area of the crank case cover. At a quick glance, both machined surfaces seems indicating approximately the same smoothness surfaces. However, the machined surface by Delcam obtained better result in Ra average value which stated 0.381μm. Meanwhile, the average Ra value for Catia V5 shows greater difference approximately by almost 1 μm or specifically 0.976 μm more than Delcam's which is 1.357μm. From the Ra value obtained, there is a significant conclusion can be drawn from this result. The differences in machining tool path direction or also called machining style / strategy really influence the result of final machined part. "Spiral" motion applied by Delcam software seems to be really effective in machining sculptured surfaces as carried out in this research. Nevertheless, "Zigzag" motion also able to give better surface finish on certain specified areas as reported.

Figure 5(d) exhibits very critical differences. The surface machined by Delcam unfortunately showed left out of unfinished materials. This is strongly believed due to the uncalculated areas determined by the automatic calculation machining tool paths method as default settings in Delcam software. On the other hand, the surface machined by Catia V5 seems to be completed and smoothly finished. Figure 6 below illustrates the machining tool paths directions for both meant software.

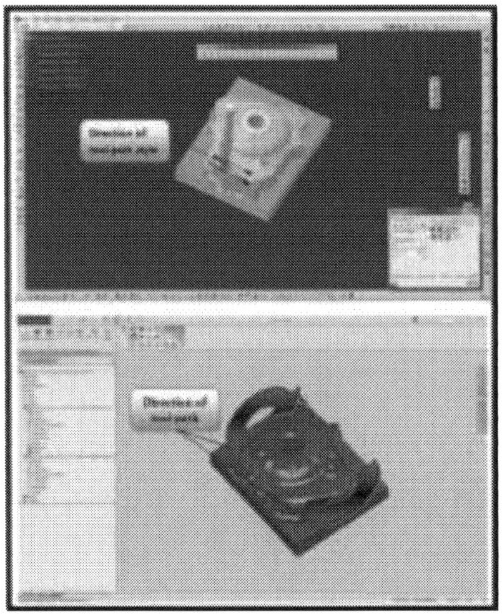

Fig. 6. Differences of tool path direction illustrated by both software.

4 Conclusions

In this study, the effect of surface finish of machined parts which were programmed by Catia V5 and Delcam applying one of Reverse Engineering methods namely 3D Scanned Data have been successfully investigated. The result reveals that the surface finish of machined part programmed by Delcam software is slightly better than Catia V5 by looking at the overall average Ra values 1.22 μm and 1.738 μm respectively. Important conclusion drawn from this work is that; differences in preparing the CAM program especially on the machining tool paths direction as well as the details of machining processes involve also somehow effect the surface finish of the machined part although the same 3D CAD model is used as the reference.

Ultimately, Catia V5 offers higher flexibility in determining the way of CAM program to be created than the Delcam software especially the version that been used in this research; Delcam for Solidworks (DFS). Importantly, the Delcam and the Catia V5 software are able to produce the desired machined part. In particular, further research should be conducted on study of dimensional accuracy in producing machined part from scanned data to be compared to the actual part for both meant software.

Author acknowledges, with gratitude, authors' debt of thanks to Universiti Teknikal Malaysia Melaka (UTeM) and to the funding body; Ministry of Higher Education Malaysia for the research grant (RAGS/1/2014/TK01/UTEM/B00079).

References

1. Abbott, Steven, SPE (Surface Profile Explorer), Abbott Apps. Steven Abbott TCNF Ltd. (Retrieved January 13, 2014)
2. Baptista R. and Antune Simoes J.F., Three and Five-Axis Milling of Sculptured Surfaces, Journal of Materials Processing Technology, **103**, Pages 398-403. (1999)
3. Bernardini, F., & Rushmeier, H. E., *The 3D Model Acquisition Pipeline*, Compute Graphics Forum. (2002)
4. Chris McMahon and Jimmie Brownie, *CAD/CAM Principle Practice and Manufacturing Management*, 2nd Edition Prentice-Hall. (1999)
5. Chennakesava R. Alavala, *CAD/CAM Concept and Application*, Prentice-Hall India. (2008)
6. Rozmarina Dubovska, Jaroslav Jambor and Jozef Majerik, Implementation of CAD/CAM system CATIA V5 in Simulation of CNC Machining Process, Procedia Engineering, **69**, pp. 638 – 645. (2014)
7. Sundi S.A., Muhamad M.R., Bakar B.A and Abdullah, R.I.R., Surface Roughness Analysis on Five-Axis Flank Milling for Curved Shape Part-Full Flute and Ground Shank End Mill, Science International (Lahore), **26**, pp. 1869-1973. (2014)
8. Sun Y.W., Guo D.M. and Jia Z.Y., Spiral Cutting Operation Strategy for Machining of Sculptured Surfaces by Conformal Map Approach, Journal of Material Processing Technology, **180**, pp. 74-82. (2006)
9. Várady, T., Martin, R. R., and Cox, J., Reverse Engineering of Geometric Models – Special issue: Computer-aided Design, pp. 253-254. (1997)

Ergonomic study on human-powered vehicles

Mohd Azman Abdullah[1,2,*], *Mohamad Haziq* Abdul Rahman[1], *Muhammad Zakwan* Azis[1], *Shamsul Anuar* Shamsudin[1,2], *Faiz Redza* Ramli[1,2] and *Mohd Nizam* Sudin[1,2]

[1]Faculty of Mechanical Engineering, Universiti Teknikal Malaysia Melaka, Hang Tuah Jaya 76100, Durian Tunggal, Melaka, Malaysia.
[2]Centre for Advanced Research on Energy, Universiti Teknikal Malaysia Melaka, Hang Tuah Jaya 76100, Durian Tunggal, Melaka, Malaysia.

Abstract. In this paper, new methods for ergonomic analysis of cyclist for 4-wheel recumbent seat human-powered vehicle (HPV) are performed. An ergonomic index with fundamental formulation is developed in order to determine the level of comfortness during handling of the HPV. Basic reference of sitting postures are produced from three HPVs for 20 different individuals. All the dimensions, angles and measurements are recorded. The same individuals are required to sit on the three HPV models to evaluate their comfortness and ergonomic by observing the same dimensions, angles and measurements of leg and hand postures. The data is compared with reference comfort sitting ergonomic. This study is limited to a number of individuals which are the students of a university in Malaysia with age range from 20 to 24 years old. However, the ergonomic index can be expanded for Asian people and with some improvement in the parameters, it can be used for other countries. Derivation of ergonomic index and formulation in determining the comfort level and ergonomic of HPV. Using the ergonomic index, a new improved HPV can be developed. The index is also applicable with modification on several parameters in the formulation for other countries.

1 Introduction

The ergonomic study for human posture always deal with workplace and workstation where people spend six to ten hours working and maintaining the same posture for long period of time [1]. There are also different studies for different regions of people in the world. There are not many studies of ergonomic involving driver and cyclist in recumbent seat. Many of the designs for driver's seat are following the design for human posture during sitting. In performing an ergonomics studies, human variability is used as a design parameter. The term success in ergonomics is measured by improved productivity, efficiency, safety, acceptance of the resultant system design and improved quality of human life [2]. The ergonomics plays the significance roles in human. Ergonomics not only important in establishing smooth interaction in using these appliances (comfortable and effective operation), but also contributes to the harmony and symbiosis of humans, machines and environments. Ergonomics is useful practical science which helps us to

* Corresponding author: mohdazman@utem.edu.my

achieve safety, security and comfort and to maintain and improve our health. Thus, the domains it covers vary significantly [3].

In ergonomics, there are 3 major goals. The main purpose goal is to generate tolerable working condition which neglects any danger to human life or in others word is safety. The second goal is productivity. At this condition, to enhance system performance, one can design a system which improves performance affordances. The final goal is to operator satisfaction. Satisfaction and dissatisfaction may be understood only in the both operator's and user's needs are clearly understood. Different type of people have different level of expectations, and these vary substantially between countries and cultures [4]. Furthermore, ergonomics provide working conditions which are good enough and above the minimum required to ensure health and safety of the workforce. This is not simple as that, as the musculoskeletal complaints are minimized. Then, a comfortable, productive and satisfying office environment can be achieved. To design such an environment, it is necessary to consider not only furniture and equipment, but also the job design, lighting, noise, air quality, office landscaping and personal space [5].

Based on research, there are several factors that lead to physical stress and injury on workers' bodies especially if coupled with poor machine design, tool, and workplace design or the use of improper tools. A dramatic increase in musculoskeletal disorder (MSDs) began in the 1970s when these disorders increasingly appeared on companies' injury and illness logs [5]. Appropriate general methodology for such an approach, which might be applicable to the following types of projects; selecting a commercially available product for purchase, improving an existing product or system, designing a new product system, adapting an individual workplace, refurbishing a business or workplace, for instance, after automation and designing a complete plant [1]. In this study, the ergonomic is focused on the cyclist for 4-wheel recumbent seat human-powered vehicle (HPV) [6] is performed. Propulsive force and cycling performance are affected by internal biomechanical variables which involve factors related to force or torque development and power production. These factors can be manipulated by modify and alter the effective muscle force or torque or power generated and transmitted to the vehicle.

These factors include position of initial and final muscle length, change in muscle length, muscle moment arm lengths, force arm length, load imposed on muscle, resistance arm length, joint angles, muscle angle of pull, single or multi-joint muscles, muscle fiber type and arrangement. Manipulations of external mechanical factors are affected the changes and interactions occurring in the internal biomechanical variables resulting in propulsive force [7]. Constraints imposed upon a cyclist are involving external mechanical factors which are the structure of the vehicle and how power is transmitted to the vehicle. These factors include the seat-to-pedal distance, seat tube angle, seat height, pedal crank arm length, handlebar height, length, and position, cycling body position, orientation, and joint configuration, foot-pedal position and losses in power transmission due to friction.

Manipulating these variables will alter joint angle position, joint angle ranges, muscle length, resistance load, muscle mechanical advantage, and the ability to produce force, torque and power. Thus the interaction between the internal biomechanical factors with the external mechanical factors in resulting propulsive force will be a function of developing force, torque, and power for propulsion. However, changes in the shape and structure of a human powered vehicle (from manipulation of the external mechanical factors) should be considered because it will further increased complexity. The consideration not only on how it affect propulsive forces, but also how it affects the type of resistive forces encountered from the environmental factors, and the resulting interaction between these forces [8]. HPV needs ergonomics analysis to increase their safety, comfortable and performance of HPV.

Safety is one of the important factors that people will consider. Thus, from ergonomics analysis the HPV can increase safety for the driver. For example, HPV use recumbent seat position which is more safety compared to normal bicycle. Based on ergonomics analysis and study, the comfort of HPV can be increased. Ergonomics study can increase HPV performance so that the driver can reach maximum speed with low energy consumption. The recumbent position has become popular due to its reduced frontal area thus reduced aerodynamic drag if compared to the normal upright position. It is suitable for high performance human powered vehicles. The center of mass of the vehicle is positioned relatively low to the ground making the vehicle more stable and safer because of less distance to fall if a crash were to occur. Recumbent position more protective because the head is protected as compared to the upright position the head leads the body if thrown forward over the handlebars. There are many advantages on recumbent position which are improved the visibility of driver by keeping the oncoming road in the field if vision, reduce neck strain, comfortable, reducing likelihood of crotch pain and injury, gives rider firm support to push against while pedaling, reduces the strain on the lower back and wrist as compared to the normal upright position. However, recumbent position still has it disadvantages. It may not allow for peak power production and sustained aerobic performances as high as normal upright position can produce [8]. There are several locations on sitting postures that have comfort angle.

In ergonomics study, angle of comfort is important to reduce stress and musculoskeletal disorder (MSDs) of HPV riders. Range of comfort angle for head location is 0° to 15°, upper arm is -15° to 45°, elbow comfort angle is from 80° to 120°, hip comfort angle is from 95° to 120°, knee is from 95° to 135°, ankle is from 85° to 110° and back is from 10° to 30°. Based on study, the optimum angle for knee is 120° while for elbow is 115° [8]. In the previous study of ergonomic via survey, the study was performed on a driving simulator cockpit where the cockpit represents the actual driver posture during driving and slightly in recumbent position [9]. The ergonomic scores were studied based on the opinions of the participants including seat position, steering wheel position, gear knob position, pedals positions and height of the seat. The results from the study is reliable only for statistical approach and require a lot of participants for the survey.

In this study, ergonomic study is performed on a human-powered vehicle (HPV) which was fabricated previously [6]. The HPVs were fabricated based on constraints of design such as seat type, track and wheel base dimensions and number of wheels. The previously successful design and analysis for manufacture (DAFM) [6, 9-12] was implemented in order to fabricate the HPVs. Nevertheless, the ergonomic aspect never been considered neither in design stage nor in fabrication stage. Three number of HPVs are selected for the study (Figure 1 2 & 3) and 2 approaches are performed which are actual human posture measurements and rapid upper limb assessment (RULA). Based on these three methods, the ergonomics of the HPVs are discussed and concluded.

2 Methodology

2.1 Cyclist posture measurement

There are several locations on sitting postures that have comfort angle. In ergonomics study, angle of comfort is important to reduce stress and musculoskeletal disorder (MSDs) of HPV riders. All respondent measurement and HPV measurement will be used to calculate leg and hand angle of respondent. Figs. 1-3 show the position and angle of leg with hand of the respondent sit on HPV 1, HPV 2 and HPV 3. The angles (Figure 4) are calculated by using the formula in Eq. (1), (2) and (3) for leg and hand angle of the respondent. The ergonomic index formula is shown by Eq. (4). The percentage of error is calculated to determine the best HPV for each of the respondent. Each HPV will have its one ergonomic index for each respondent. This formula is used to determine which HPV is

 DOI: 10.1051/matecconf/20179001063

suitable or more ergonomics for each respondent. The lower the percentage of error the higher percentage of ergonomic of the HPV. Therefore, this calculation is important to indicate the level of ergonomics of each HPV for each of the respondent. Each respondent will have their own calculation of angle, thus from this calculation will help to determine which HPV is the most ergonomics for each of the respondents.

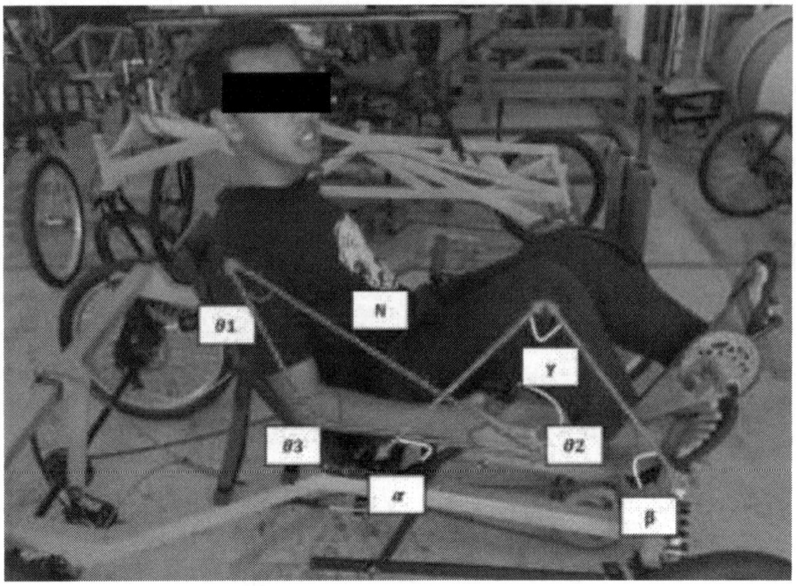

Fig. 1. Angles and measurements for respondent on HPV 1.

Fig. 2. Angles and measurements for respondent on HPV 2.

Fig. 3. Angles and measurements for respondent on HPV 3.

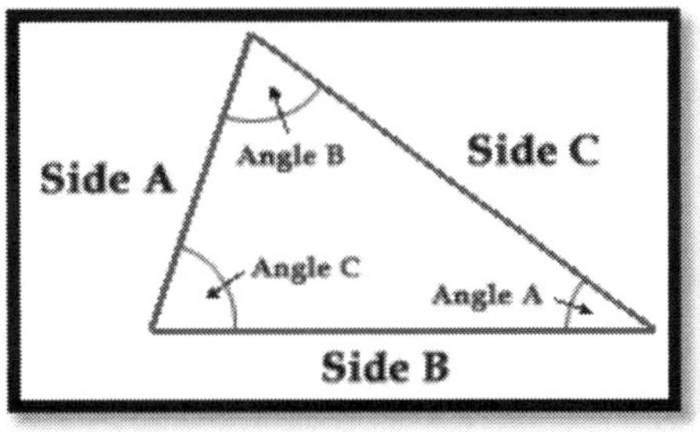

Fig. 4. Angles relation with dimensions.

$$\cos C = \frac{a^2 + b^2 - c^2}{2ab} \tag{1}$$

$$\cos A = \frac{c^2 + b^2 - a^2}{2cb} \tag{2}$$

$$\cos D = \frac{b^2 - c^2 - a^2}{2ac}\ , B = 180° - D \tag{3}$$

$$Percentage\ of\ error = \left|\frac{Theory\ value - Experimental\ value}{Theory\ value}\right| \times 100 \tag{4}$$

2.2 Rapid Upper Limb Assessment (RULA)

RULA (Rapid Upper Limb Assessment) is two quick postural targeting methods for evaluating workers postures and finding risk factor index. It was used to evaluate postural risk factors on pediatric population of two age groups where the purpose was to investigate the inter-rater and intra-rater reliability of RULA. From the results, it shows that RULA was more reliable and practical while evaluating the older age group of children, but its level of reliability necessitates caution for its lone use while evaluating younger children [13]. The RULA was designed for easy use without need for an advanced degree in ergonomics or expensive equipment. Using the RULA worksheet, the evaluator will assign a score for each of the following body regions: upper arm, lower arm, wrist, neck, trunk, and legs. After the data for each region is collected and scored, tables on the form are then used to compile the risk factor variables, generating a single score that represents the level of MSD risk as shown in Table 1 below.

Table 1. RULA score.

Score	**Level of MSD risk**
1-2	Negligible risk, no action required
3-4	Low risk, change may be needed
5-6	Medium risk, further investigation, change soon
6+	Very high risk, implement change now

3 Results and discussion

Table 2-5 show the ergonomic index (percentage) for all 20 respondents. Based on Table 2-5, it indicates that the most ergonomics HPV for respondent 1 is HPV 3. Only 9.75 percentage of error for elbow angle compared to the optimum elbow angle. Thus, it leads others HPV, which are 37.71 and 24.60 percentage of error for HPV 2 and HPV 1 respectively. HPV 2 is the least in percentage of error for knee angle (6.54%) compared to others which are 10.32 and 13.45 percentage of error for HPV 1 and HPV 3 respectively. Therefore the most ergonomic HPV for all respondents is HPV 3.

Table 2. Percentage of error for respondent 1 (R1) to 5 (R5).

HPV	**Part**	**Respondents Percentage of Error (%)**				
		R1	**R2**	**R3**	**R4**	**R5**
1	Knee	10.32	15.18	15.11	23.20	16.57
	Elbow	24.60	27.79	38.30	36.83	12.74
2	Knee	6.54	3.60	3.73	1.04	2.79
	Elbow	37.71	35.42	45.06	41.28	26.57
3	Knee	13.45	11.02	11.18	7.07	10.35
	Elbow	9.75	10.75	11.33	22.47	10.67

Table 3. Percentage of error for respondent 6 (R6) to 10 (R10).

HPV	**Part**	**Respondents Percentage of Error (%)**				
		R6	**R7**	**R8**	**R9**	**R10**
1	Knee	17.95	15.01	4.06	2.07	3.87
	Elbow	38.96	35.24	37.28	29.31	34.51
2	Knee	2.10	3.89	10.82	12.34	11.10
	Elbow	44.21	41.68	44.04	43.85	34.51
3	Knee	9.83	11.36	17.17	18.55	17.49
	Elbow	19.89	17.01	18.68	3.14	14.58

Table 4. Percentage of error for respondent 11 (R11) to 15 (R15).

HPV	Part	Respondents Percentage of Error (%)				
		R11	R12	R13	R14	R15
1	Knee	15.11	21.39	8.38	12.08	16.86
	Elbow	44.55	32.36	38.71	1.21	40.11
2	Knee	3.73	0.01	20.08	23.00	27.00
	Elbow	34.33	46.40	55.32	1.33	45.84
3	Knee	11.18	7.89	25.36	27.98	31.71
	Elbow	17.34	27.32	37.25	3.77	32.62

Table 5. Percentage of error for respondent 16 (R16) to 15 (R20).

HPV	Part	Respondents Percentage of Error (%)				
		R16	R17	R18	R19	R20
1	Knee	5.98	4.08	15.32	8.35	5.98
	Elbow	34.41	37.85	32.73	38.44	34.16
2	Knee	9.50	10.79	3.37	20.04	9.50
	Elbow	41.26	49.44	44.92	49.17	52.68
3	Knee	16.02	17.14	10.75	25.32	16.02
	Elbow	25.31	31.70	32.73	37.04	30.85

Based on Fig. 5, it shows that the differentiation percentage of comfort for knee and elbow for each of HPVs. Percentage of comfort for HPV 1 is 88.16% and 67.5% of knee and elbow respectively. This briefly describes that the HPV is comfort for respondents knees, but not really comfort in terms of respondents elbow part. Next is HPV 2 and from the bar chart, we can see that percentage of comfort for HPV 2 is 90.75% and 59.25% of knees and elbows respectively. This briefly describes that HPV 2 is really comforting for respondents' knee but not really comfort in terms of respondents' elbow part. Lastly is the percentage of comfort for HPV 3 and from the bar chart, we can know that percentage of comfort for HPV 3 is 84.16% and 79.29% of knee and elbow respectively. This briefly describes that HPV 3 is really comforting for respondent knee and also comfort in terms of respondents elbow part.

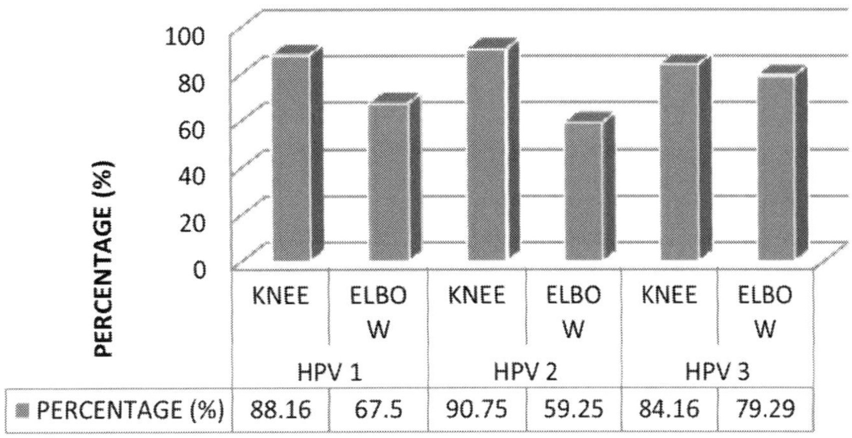

Fig. 5. Ergonomic index for each of the HPV for knee and elbow.

Results shown in Fig. 6-8 are from RULA analysis for ergonomic posture for all HPVs. The ergonomic score is 3 for all HPVs. From this ergonomic value, it shows that the design of all HPVs is acceptable in terms of ergonomic because nearly to the most ergonomic value in RULA analysis. As stated before, the most ergonomic value in a sitting position is two. RULA only evaluate the ergonomic factors based on the design of the HPVs not the actual HPVs. The ergonomic values for leg and upper arm are important. Forearm part ergonomic value is really closed to the most ergonomic value which is one. It is considered as good in terms of ergonomic since it gets two for ergonomic value for RULA analysis. Thus, all HPVs are ergonomic for cycling and handling based on their design.

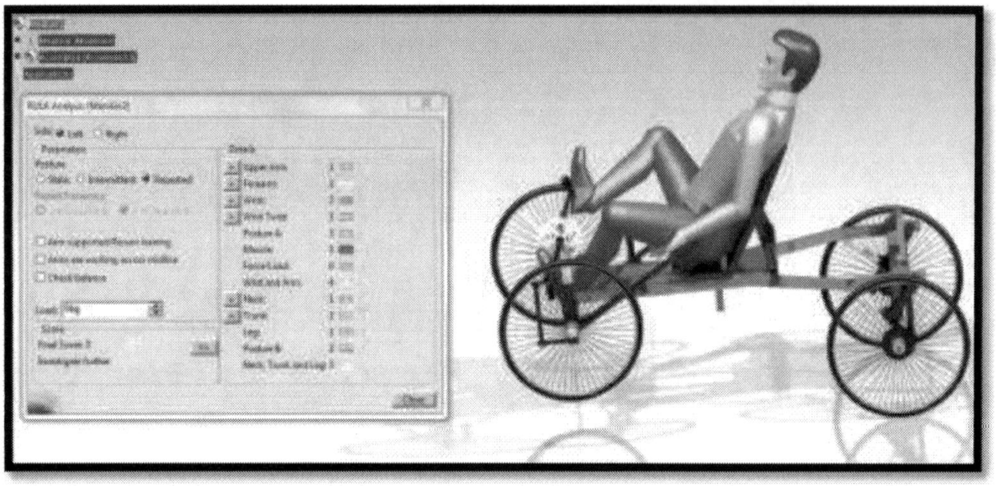

Fig. 6. RULA evaluation of HPV 1.

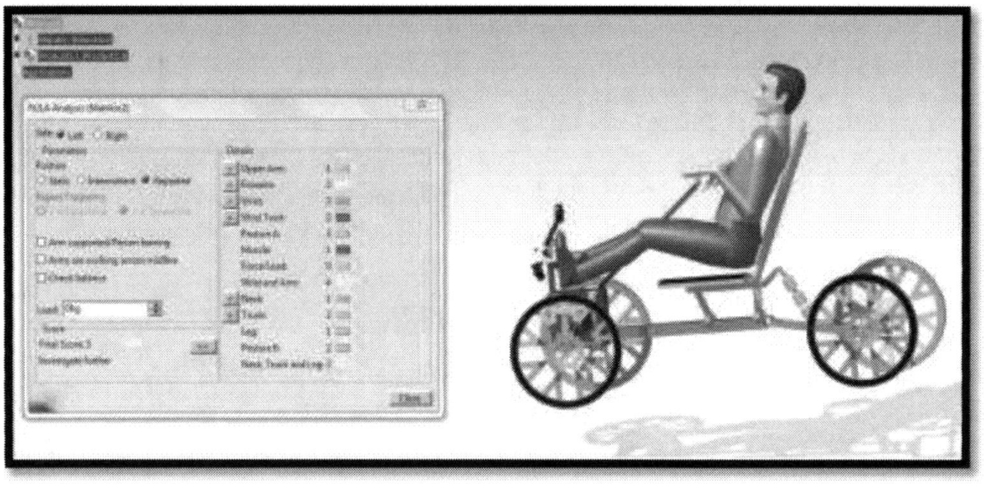

Fig. 7. RULA analysis of HPV 2.

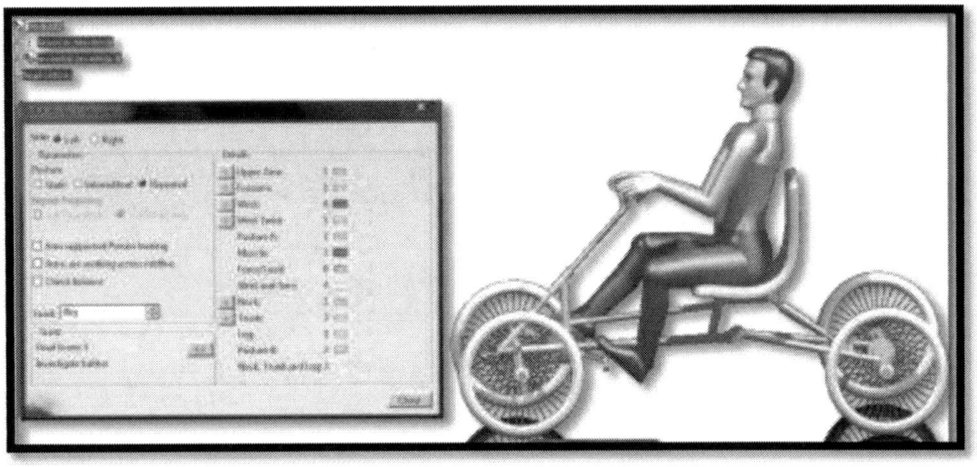

Fig. 8. RULA analysis of HPV 3.

4 Conclusions

Ergonomic index value analysis is used to determine the best HPV for all respondents. This was done through calculation of respondent angle and compared it with comfort angle in terms of ergonomic. High percentage of ergonomic index means high value in ergonomic aspect. RULA analysis is used to evaluate the HPVs in terms of ergonomic by the score. Low mark from RULA analysis indicate the excellent HPV design in terms of ergonomic. Based on the study, HPV 3 is the most ergonomic HPV among others. Even though HPV 1 and HPV 2 have high ergonomic index on knee, total score for ergonomic index (knee and elbow) goes to HPV 3. This method can be used for other design of recumbent type vehicle to evaluate its ergonomic index.

The authors gratefully acknowledged the Advanced Vehicle Technology (AcTiVe) research group of Centre for Advanced Research on Energy (CARe), the financial support from Universiti Teknikal Malaysia Melaka (UTeM) under Short Term Research Grant (PJP/2014/FKM (10A)/S01330).

References

1. B. W. Jan Dul, *Ergonomics for Beginners : a quick reference guide. (Third Edit), Boca Raton: CRC Press Taylor & Francis Group*, (2008).
2. K. Kroemer, H. Kroemer, and K. Kroemer-Elbert, *How to design for ease and efficiency, Ergonomics, Prentice-Hall, New Jersey*, (2001).
3. S. E. Burt, & L. J. Fine, *Musculoskeletal Disorders and Workplace Factors, National Institute for Occupational Safety and Health*, (1997).
4. O. Korhan, *Work-Related Musculoskeletal Discomfort in the Shoulder due to Computer Use, North Cyprus, Mersin, Turkey*, (2012).
5. Occupational Health and Safety Council of Ontario, *Musculoskeletal Disorders Prevention Series*, Occupational Health and safety Council of Ontario (OHSCO), (2007).
6. M. A. Abdullah, S. A. Shamsudin, F. R. Ramli, M. H. Harun and M. A. Yusuff, *Design and Fabrication of a Recreational Human-Powered Vehicle, International Journal of Engineering Science Invention*, ISSN (Online): 2319 – 6734, ISSN (Print): 2319 – 6726, www.ijesi.org, Volume 5 Issue 2, February 2016, PP.11-14, (2016).

7. D. Kee and W. Karwowski, *A comparison of three observational techniques for assessing postural loads in industry, International Journal of Occupational Safety and Ergonomics*, 13(1), 3–14, (2007).
8. A. R. Tilley, J. Anning, and R. Welles, *The Measure of Man and Woman: Human Factors in Design*, Revised Edition. *Wiley & Sons*, (2002).
9. M.A. Abdullah, A.H. Mohamad and F.R. Ramli, *Design Analysis and Fabrication of Fixed-Base Driving Simulator Frame, Journal of Engineering and Technology (JET)*, Penerbit Universiti, Universiti Teknikal Malaysia Melaka, Vol. 4, No. 2, July-December 2013, ISSN: 2180-3811, 85-101, (2013).
10. M. A. Abdullah, M. R.Mansor, M. Mohd Tahir, S. I. Abdul Kudus, M. Z. Hassan and M. N. Ngadiman, *Design, Analysis and Fabrication of Chassis Frame for UTeM Formula Varsity^{TM} Race Car, International Journal of Mining, Metallurgy & Mechanical Engineering (IJMMME)*, Volume 1, Issue 1, 75-77 ISSN 2320–4060 (Online), (2013).
11. M.A. Abdullah, M. R. Mansur, N. Tamaldin and K. Thanaraj, *Development of Formula Varsity Race Car Chassis, IOP Conference Series: Materials Science and Engineering*, Vols. 50, No. 1, doi:10.1088/1757-899X/50/1/012001, (2013).
12. M. A. Abdullah, N. Tamaldin, F. R. Ramli, M. N. Sudinand A. M. Mohamed Muslim, *Design and Development of Low Cost All-Terrain Vehicle (ATV), Applied Mechanics and Materials, Trans Tech Publications*, Vols. 663, pp 517-521, doi:10.4028/www.scientific.net/AMM.663.517, (2014).
13. S. Dockrell, E. O'Grady, K. Bennett, C. Mullarkey, R. Mc Connell, R. Ruddy, S. Twomey and C. Flannery, *An investigation of the reliability of Rapid Upper Limb Assessment (RULA) as a method of assessment of children's computing posture, Applied ergonomics*, *43*(3), pp.632-636, (2012).

An overview of the Oil Palm Empty Fruit Bunch (OPEFB) potential as reinforcing fibre in polymer composite for energy absorption applications

M.K. Faizi[1,2,*], *Shahriman* A.B.[1,2], *M.S.* Abdul Majid[1], *Shamsul* B.M.T.[3], *Ng* Y.G.[3], *S.N.* Basah[1], *E.M.* Cheng[1], *M.* Afendi[1], *Zuradzman* M.R.[1], *Khairunizam* WAN[1], *D.* Hazry[1]

[1]School of Mechatronic Engineering, Universiti Malaysia Perlis, Pauh Putra Campus, 02600 Arau, Perlis, Malaysia.
[2]Motorsport Technology Unit, Universiti Malaysia Perlis, Pauh Putra Campus, 02600 Arau, Perlis, Malaysia.
[3]Faculty of Medicine and Health Science,Universiti Putra Malaysia, 43400 UPM Serdang, Selangor Darul Ehsan.

Abstract. The oil palm empty fruit bunch (OPEFB) natural fibres were comprehensively reviewed to assess their potential as reinforcing materials in polymer composites for energy absorption during low-velocity impact. The typical oil palm wastes include trunks, fronds, kernel shells, and empty fruit bunches. This has a tendency to burden the industry players with disposal difficulties and escalates the operating cost. Thus, there are several initiatives have been employed to convert these wastes into value added products. The objective of this study is to review the potential of oil palm empty fruit bunch (OPEFB) as natural fibre polymer composite reinforcement to absorb the energy during low-velocity impact as another option for value added products. Initially, this paper reviewed the local oil palm waste issues. Previous research works on OPEFB polymer composite, and their mechanical characterization is appraised. Their potential for energy absorption in low-velocity impact application was also elaborated. The review suggests high potential applications of OPEFB as reinforcing materials in composite structures. Furthermore, it is wisely to utilize the oil palm biomass waste into a beneficial composite, hence, promotes the green environment.

1 Introduction

There has been a growing interest in recent years in finding fibre solutions for efficient and cheap energy absorbing composites. Among the numerous possibilities is using natural fibre to form polymer composite due to its eco-friendly and renewable nature. Natural

* Corresponding author: mkhairulfaizi@unimap.edu.my

fibres offer various advantages such as low density, low cost, biodegradability, acceptable specific properties, better thermal and insulating properties and low energy consumption during processing [1]. Currently, the natural fibres in polymer composite that extremely explored are jute, kenaf, cotton, flax, and hemp. These natural fibres commonly use with matrices to provide better particular tensile strength, flexural strength, stiffness, and elongation at break [2]. Examples of matrices that widely used are thermoplastic polymer: polypropylene (PP); polyvinyl chloride (PVC); polyethylene (PE); and polystyrene (PS), thermosets polymers: phenolics; epoxies; isocyanates, and unsaturated polyesters.

The oil palm empty fruit bunch (OPEFB) as shown in Fig. 1 is a potential natural fibre to be explored. Recent studies conducted investigate the OPEFB reinforced polymer composite found to yield even better in characteristics [1], [3]–[7]. OPEFB is categorized as a palm oil industry waste which is generated from the oil extraction mills process.

Fig. 1. The oil palm empty fruit bunch (OPEFB).

Malaysia is among the top most important oil palm producers in the world and experiencing a robust development in new plantations and palm oil mills through giant government companies (FELDA, FELCRA, and RISDA) and private estates (Guthrie, IOI Plantations, Genting Plantations, and Sime Darby). Fig. 2 shows the percentage of palm oil plantations ownership. Fig 3 on the other hand, demonstrate the expansion of palm oil cultivation area in Malaysia 1960 to 2014. The total area of oil palm cultivations by 2014 is 5.39 million hectares which private estates contribute a bigger percentage, 62%.

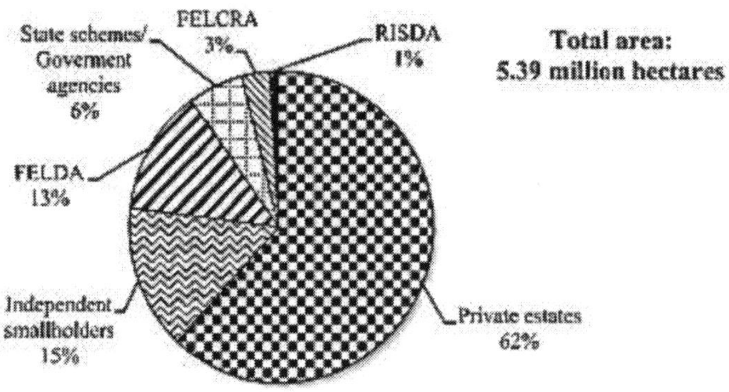

Fig. 2. Percentage of oil palm plantations ownership in 2014 [8].

The success of these oil palm plantation companies is strongly supported by responsible government agencies like the Malaysian Palm Oil Board (MPOB), Malaysian Palm Oil Council (MPOC), Palm Oil Research Institute of Malaysia (PORIM), and Forest Research Institute Malaysia (FRIM). This commodity plays a significant role in the Malaysia economic growth.

 DOI: 10.1051/matecconf/20179001064

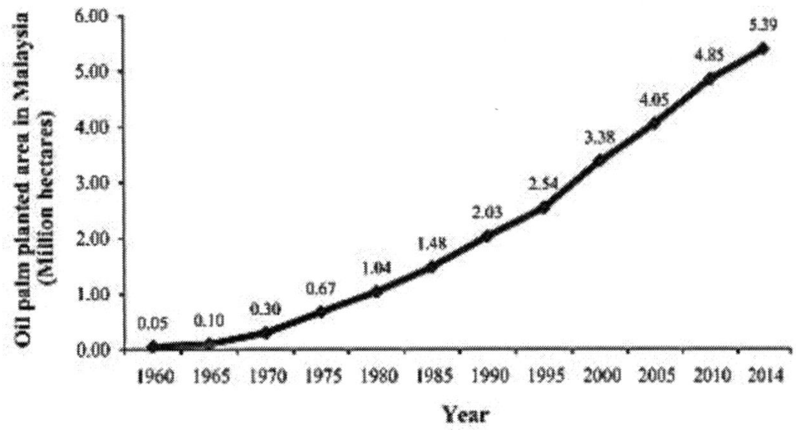

Fig. 3. Expansion of oil palm plantations cultivation in Malaysia area [8].

In palm oil industry, biomass wastes were generated by; [8] first, it comes from plantation routines, pruning and re-plantation work which produces wastes such as fronds and trunk. Second; wastes are also produced from oil extraction mills process such as oil palm empty fruit bunch (OPEFB) and mesocarp fibre. This abundance biomass wastes are a natural source of fibres and can be utilized as potential reinforcing materials in the polymer composite. However, this review will focus on OPEFB to investigate the capabilities of impact energy absorption applications due to its lack of downstream use. Unlike OPEFB, other oil palm wastes have established downstream use such as fronds and trunks were used for animal feedstock while kernel shells were used as fuel for the steam boiler in palm oil mills.

2 Oil palm wastes issues

The waste by-product from palm oil extraction process enormously could achieve 70–80 million tons per annum [4]. In the palm oil mill, palm oil consists only 10% of the total biomass, while the rest (90%) biomass are discarded as wastes [8]. This problem tends to burden the operators with disposal difficulties and escalates the operating cost.

To address the matter, these oil palm wastes were recycled in many applications. Among others; fronds and trunks were used as an animal feedstock [8], mesocarp turned into boiler fuel, kernel shells reused as road paving [9], an oil palm empty fruit bunch (OPEFB) transformed into soil conditioners [10], [11] and fiberboard [12]. However, the reused of OPEFB as soil conditioners still not of popular and there is still an enormous excessive waste yet to be handled. The steam from the sterilization process results in a high moisture content in the EFB; as high as 60%, make it unsuitable usage as boiler fuel [12]. Eventually, burn them without any benefits is the only option opted by the farmers. Even worse, the production of OPEFB shows increasing trend with 7 million tons of waste produced in 2007 [8]. Thus, this research purpose is to utilize the OPEFB waste fully into a beneficial polymer composite as an efficient and cheap energy absorbing the material in composite structures.

Apart from its eco-friendly, renewable nature, cheap, low density, better thermal and insulating properties, the processes of converting OPEFB's fibre into composite promotes low energy consumption [1].

DOI: 10.1051/matecconf/20179001064

3 The energy absorption in low-velocity impact

The energy absorption in mechanics can be exhibited as the power to absorb in various forms, conditions, and directions of impact energy. In the case of a projectile impacts hits into a structure, the absorbed energy can be estimated by energy loss after impact and mathematically can be calculated through the integration of the impact force with the displacement plot [13]. Usually, when performing the impact test to examine the energy absorption, the results were presented in load-displacement curves. The energy absorption during the impact test is obtained by calculating the area under the load- displacement curve with the unit Nm (Joule) as shown in Eq. 1:

$$E = \int_0^{\delta_1} P d\delta \tag{1}$$

where δ is the crushing distance, P is the applied load and E, the energy absorbed. However, the impact test result also can be presented in load-time curves. In fact, the load-time curves are easier to obtain from an experimental sample [14]. Thus, another way to calculate the energy absorption is to multiply the load signal by the average velocity and integrate over the impact time. The energy absorbed at any instant during the impact on the specimen depicted as shown in Eq. 2:

$$E = v_{ave} = \int_0^{t_1} P dt \tag{2}$$

where P is the applied load, v_{ave} the average velocity and E is the energy absorbed by the specimen at a specified time. The value of energy is normalized by the mass of the crushed portion or by the crushed volume. Furthermore, the energy loss is also could be calculated as a function of the difference between the incident kinetic energy and residual kinetic energy of the projectile [15]. Specifically, using the notation in Fig. 4, absorbed energy is expressed as shown by Eq. 3:

$$E_a = \frac{1}{2} m \left(v_i^2 - v_r^2 \right) \tag{3}$$

where Ea is the absorbed energy, m, Mass of the projectile,v_i^2, Incident velocity immediately before impact and v_r^2; Residual velocity immediately after impact.

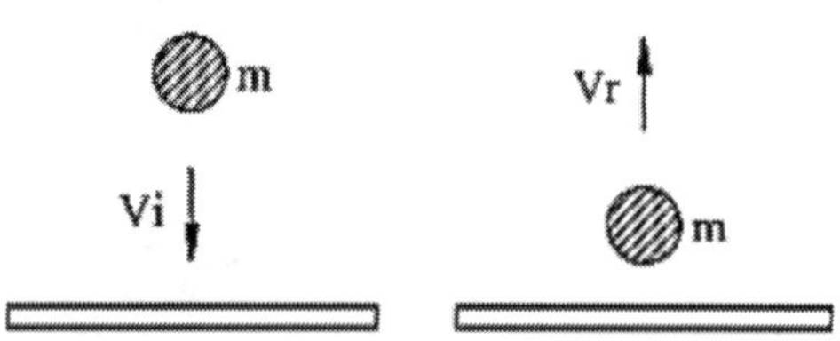

Fig. 4. Schematic view of a projectile impacting a structure [15].

Based on these formulae, the energy absorption is dependent on mass or load and projectile velocity. Usually, there are three types of velocity classifications in impact test: low velocity; high velocity; hyper velocity. However, this research is focusing on low-velocity impact condition. Low velocity impact occurs within 10 m/s [16]–[27].

4 The previous multidisciplinary OPEFB polymer reinforced composites studies

Recently, several investigations and efforts have been employed to utilize the OPEFB for value added products such as bio-adsorbent, bio-oil, pulp, and activated carbon [4]. The OPEFBs were also explored on electrical, bio-energy, bio-resource, and chemistry field [28]–[34]. Nevertheless, the OPEFB is still not been fully explored as biodegradable polymer composites. The OPEFB viewed as a potential to be highlighted in polymer composite due to its several advantages such as uncomplicated processing, low in density, real strength, relatively high in hardness, and the most crucial is the nature biodegradation.

A flexural and tensile properties of polypropylene (PP) composite which reinforced by OPEFB fiber and fiberglass agents were studied by Rozman et al. [7]. The incorporation of OPEFB fibre into the PP matrix has resulted in the reduction of flexural and tensile strength. The flexural and tensile modulus was found to improved with increasing level of overall fibre content loading. Along with the trend that used of PP's plastic packaging beverage increased, the huge waste of plastic waste generated also growing. Therefore, the wastes of PP's plastic (wPP) are derived with EFB to form a composite material [3]. However, some modifications had been made to lower the interfacial tension between the matrix and EFB fibre filler. So that, the maleic anhydride polypropylene-g-polypropylene (MAPP) was chosen to providing compatibilizer or treating the surface or both. The composite is then going through the impact strength and thermal degradation test. Results indicated that addition MAPP improved the impact resistance and increased degradation temperature.

The polyurethane (PU) oil palm empty fruit bunch (EFB) composites have been produced by [1]. EFB fibres were used in the form of mats where it was treated with hexamethylene diisocyanate (HMDI) and toluene diisocyanate (TDI). The found that the PU-EFB composite had acceptable properties which were thought to be mostly influenced by the type of bonding produced. The tensile strength, tensile modulus, elongation at break, tensile toughness, flexural strength, flexural modulus, flexural toughness, and impact strength optimally up to 23 MPa, 1.3 GPa, 4.4%, 4.5 Joule, 75 MPa, 1.7 GPa, 4.8 Joule, 19 j/sq.m respectively. The EFB fibres and its cellulose have also been matrixes with polypropylene (PP) to investigate the mechanical properties [6]. The PP-EFB and PP-cellulose gave a different performance in term of tensile strength, impact strength, and flexural modulus. The PP-cellulose performed better in all ways where the highest tensile strength is 35 Mpa, impact strength is 38 J/m, and flexural modulus is 4000 MPa while the PP-EFB is 20 MPa, 32 J/m, and 2700 MPa respectively.

The EFB fibre also explored as a hybrid composite for widening its potential. The hybrid of EFB/jute fibres reinforced epoxy had ever been studied to investigate the mechanical performance [5]. This hybrid composite with a different layer of the mat was prepared: EFB-jute-EFB; jute-EFB-jute. Through the bending test, the jute-EFB-jute exhibited the best performance with flexural strength and modulus in 49 MPa and 3.07 GPa respectively. As a comparison, the EFB-jute-EFB flexural strength and modulus in 44.6 MPa and 2.68 GPa respectively.

5 The OPEFB polymer reinforced composites potential for energy absorption application

Though the studies regarding energy absorption on the OPEFB polymer reinforced composites initiated, there are lots of parameters remaining unexplored. The wider experimental works through various impact tests require to be done in order to provide the precise energy absorption characteristics of OPEFB. However, the previous experimental results revealed its potential especially through low velocity impact. Rozman et al., (2004) improved the impact strength of OPEFB from 8 to 20 j/sq.m with the use of isocyanate treatment [1]. Khalid et al., (2008) increased the OPEFB impact strength from 32 to 38 J/m by deriving its cellulose [6]. In the meantime, Nasution et al., (2015) successfully characterized the impact strength based on maleic anhydride polypropylene-g-polypropylene (MAPP). As a result, the impact strength achieved 34000 J/m^2 for the content of 8% of the MAPP [3].

The most encouraging experimental result of OPEFB polymer reinforced composite was published by Jawaid et al., (2010) [5]. They compared the performance on energy absorption between different composite materials. The impact tests were performed on different sequence of fibre mate sandwich arrangement such as pure epoxy, pure OPEB, OPEFB/jute/OPEFB, jute/OPEFB/jute, and pure jute and were fabricated by hand lay-up method. Table 1 shows the impact strength experimental result. Referring to the Table 1, pure epoxy shows the weakest impact strength with 30.4 J/m and followed by pure jute in a bottom two with 32 J/m. Meanwhile, jute/OPEFB/jute and OPEFB/jute/OPEFB shows the better result with 57 J/m and 66.7 J/m respectively. Impressively, the pure OPEFB composite shows the tremendous result with 92.7 J/m of energy absorption capability. For this matter, Jawaid et al., claimed that the OPEFB fibre has higher fracture toughness compared to jute. It is because the OPEFB fibre is comprised of lignin, hemicellulose, and cellulose predominantly. Hence, the chemical composition during treatment and the architecture of cell wall gives the composite higher impact properties.

Table 1. The various composites fibre impact test result [5].

Composites	Impact strength (J/m)
Pure epoxy	30.4
Pure OPEFB	92.7
OPEFB/jute/OPEFB	66.7
Jute/OPEFB/jute	57
Pure jute	32

Based on the all findings from previous research in OPEFB polymer composite, there are huge potential for the energy absorption application on low velocity impact. These initial studies showed the positive experimental data in order to further investigation on OPEFB impact strength characteristic. Wider experimental parameters could be explored to discover the establish finding.

6 Conclusions

In this paper, the overview of OPEFB in multidisciplinary studies and the potential natural fibre polymer composite for energy absorption was elucidated. It reviewed the previous research regarding OPEFB composite development in the sense of methods, architectures, experiments, and results. However, this study was debated more on energy absorption of OPEFB polymer composite during low velocity impact. Based on the previous studies result, it's revealed the evidence that the OPEFB polymer composite had a potential in energy absorption applications. The local oil palm waste issue also had been discussed in this paper. With the oil palm empty fruit bunch (OPEFB) availability and disposal problem, it might boost the effort to convert them into a beneficial product. As a conclusion, the explorations of OPEFB reinforcing fibres in polymer composites suggest vast opportunity and encouraging towards promoting eco-products of the future.

The authors would like to be obliged to Universiti Malaysia Perlis (UniMAP) for providing laboratory facilities, Motorsport Technology Unit (MOTECH) UniMAP for providing financial assistance, and Universiti Putra Malaysia (UPM) as collaboration partnership.

References

1. H. D. Rozman, K. R. Ahmadhilmi, and A. Abubakar, "Polyurethane (PU) - Oil palm empty fruit bunch (EFB) composites: The effect of EFBG reinforcement in mat form and isocyanate treatment on the mechanical properties," *Polym. Test.*, vol. 23, no. 5, pp. 559–565, 2004.
2. M. Talimi, "Characterization of natural fibre reinforced biodegradable composites," 2011.
3. H. Nasution, S. Pandia, Maulida, and M. S. Sinaga, "Impact Strength and Thermal Degradation of Waste Polypropylene (wPP)/Oil Palm Empty Fruit Bunch (OPEFB) Composites: Effect of Maleic Anhydride -g-polypropylene (MAPP) Addition," *Procedia Chem.*, vol. 16, pp. 432–437, 2015.
4. R. Roslan, S. Zakaria, C. H. Chia, R. Boehm, and M. P. Laborie, "Physico-mechanical properties of resol phenolic adhesives derived from liquefaction of oil palm empty fruit bunch fibres," *Ind. Crops Prod.*, vol. 62, pp. 119–124, 2014.
5. M. Jawaid, H. P. S. Abdul Khalil, and A. Abu Bakar, "Mechanical performance of oil palm empty fruit bunches/jute fibres reinforced epoxy hybrid composites," *Mater. Sci. Eng. A*, vol. 527, no. 29–30, pp. 7944–7949, 2010.
6. M. Khalid, C. T. Ratnam, T. G. Chuah, S. Ali, and T. S. Y. Choong, "Comparative study of polypropylene composites reinforced with oil palm empty fruit bunch fiber and oil palm derived cellulose," *Mater. Des.*, vol. 29, no. 1, pp. 173–178, 2008.
7. H. D. Rozman, G. S. Tay, R. N. Kumar, A. Abusamah, H. Ismail, and Z. a M. Ishak, "Polypropylene ± oil palm empty fruit bunch ± glass ® bre hybrid composites : a preliminary study on the flexural and tensile properties," *Eur. Polym. J.*, vol. 37, no. 6, pp. 1283–1291, 2001.
8. M. F. Awalludin, O. Sulaiman, R. Hashim, and W. N. A. W. Nadhari, "An overview of the oil palm industry in Malaysia and its waste utilization through thermochemical conversion, specifically via liquefaction," *Renew. Sustain. Energy Rev.*, vol. 50, pp. 1469–1484, 2015.
9. F. Sulaiman, N. Abdullah, H. Gerhauser, and A. Shariff, "An outlook of Malaysian energy, oil palm industry and its utilization of wastes as useful resources," *Biomass and Bioenergy*, vol. 35, no. 9, pp. 3775–3786, 2011.
10. S. Yusoff, "Renewable energy from palm oil - Innovation on effective utilization of waste," *J. Clean. Prod.*, vol. 14, no. 1, pp. 87–93, 2006.
11. S. H. Shuit, K. T. Tan, K. T. Lee, and A. H. Kamaruddin, "Oil palm biomass as a sustainable energy source: A Malaysian case study," *Energy*, vol. 34, no. 9, pp. 1225–1235, 2009.
12. S. Prasertsan and P. Prasertsan, "Biomass residues from palm oil mills in Thailand: An overview on quantity and potential usage," *Biomass and Bioenergy*, vol. 11, no. 5, pp. 387–395, 1996.
13. L. Torre and J. M. Kenny, "Impact testing and simulation of composite sandwich structures for civil transportation," *Compos. Struct.*, vol. 50, no. 3, pp. 257–267, 2000.
14. E. E. Hong, "Energy absorption of composite sandwich structures during low-velocity impact" 1999.

15. H. Su, "Energy Absorption Capabilities of Composite Sandwich Panels under Blast Loads," 2011.
16. M. Akil Hazizan and W. J. Cantwell, "The low velocity impact response of foam-based sandwich structures," *Compos. Part B Eng.*, vol. 33, no. 3, pp. 193–204, 2002.
17. S. Heimbs, S. Heller, and P. Middendorf, "Simulation of Low Velocity Impact on Composite Plates with Compressive Preload," *LS-DYNA Anwenderforum*, pp. 11–24, 2008.
18. S. N. a Safri, M. T. H. Sultan, N. Yidris, and F. Mustapha, "Low Velocity and High Velocity Impact Test on Composite Materials – A review," *Int. J. Eng. Sci.*, pp. 50–60, 2014.
19. U. Farooq and P. Myler, "Ply level failure prediction of carbon fibre reinforced laminated composite panels subjected to low velocity drop-weight impact," *Acta Astronaut.*, vol. 102, pp. 169–177, 2014.
20. D. During, L. Weib, D. Stefaniak, N. Jordan, and C. Huhne, "Low-velocity impact response of composite laminates with steel and elastomer protective layer," *Compos. Struct.*, vol. 134, pp. 18–26, 2015.
21. B. Yang, Z. Wang, L. Zhou, J. Zhang, and W. Liang, "Experimental and numerical investigation of interply hybrid composites based on woven fabrics and PCBT resin subjected to low-velocity impact," *Compos. Struct.*, vol. 132, pp. 464–476, 2015.
22. M. Kathiresan and K. Manisekar, "Axial crush behaviours and energy absorption characteristics of aluminium and E-glass/epoxy over-wrapped aluminium conical frusta under low velocity impact loading," *Compos. Struct.*, vol. 136, pp. 86–100, 2016.
23. H. Zarei, M. Fallah, G. Minak, H. Bisadi, and A. Daneshmehr, "Low velocity impact analysis of Fiber Metal Laminates (FMLs) in thermal environments with various boundary conditions," *Compos. Struct.*, 2016.
24. V. Antonucci, F. Caputo, P. Ferraro, A. Langella, V. Lopresto, V. Pagliarulo, M. R. Ricciardi, A. Riccio, and C. Toscano, "Low velocity impact response of carbon fiber laminates fabricated by pulsed infusion: A review of damage investigation and semi-empirical models validation," *Prog. Aerosp. Sci.*, vol. 81, pp. 26–40, 2016.
25. P. Malekzadeh and M. Dehbozorgi, "Low velocity impact analysis of functionally graded carbon nanotubes reinforced composite skew plates," *Compos. Struct.*, vol. 140, pp. 728–748, 2016.
26. E. Panettieri, D. Fanteria, and F. Danzi, "A sensitivity study on cohesive elements parameters: Towards their effective use to predict delaminations in low-velocity impacts on composites," *Compos. Struct.*, vol. 137, pp. 130–139, 2016.
27. D. Nardi, L. Lampani, M. Pasquali, and P. Gaudenzi, "Detection of low-velocity impact-induced delaminations in composite laminates using Auto-Regressive models," *Compos. Struct.*, 2016.
28. L. Tan, M. Wang, X. Li, H. Li, J. Zhao, Y. Qu, Y. M. Choo, and S. K. Loh, "Fractionation of oil palm empty fruit bunch by bisulfite pretreatment for the production of bioethanol and high value products," *Bioresour. Technol.*, vol. 200, pp. 572–578, 2016.
29. N. Zulkiple, M. Y. Maskat, and O. Hassan, "Pretreatment of Oil Palm Empty Fruit Fiber (OPEFB) with Aquaeous Ammonia for High Production of Sugar," *Procedia Chem.*, vol. 18, no. Mcls 2015, pp. 155–161, 2016.
30. S. X. Chin, C. H. Chia, S. Zakaria, Z. Fang, and S. Ahmad, "Ball milling pretreatment and diluted acid hydrolysis of oil palm empty fruit bunch (EFB) fibres for the production of levulinic acid," *J. Taiwan Inst. Chem. Eng.*, vol. 52, pp. 85–92, 2015.
31. S. H. Chang, "An overview of empty fruit bunch from oil palm as feedstock for bio-oil production," *Biomass and Bioenergy*, vol. 62, pp. 174–181, 2014.

DOI: 10.1051/matecconf/20179001064

32. O. Hassan, T. P. Ling, M. Y. Maskat, R. M. Illias, K. Badri, J. Jahim, and N. M. Mahadi, "Optimization of pretreatments for the hydrolysis of oil palm empty fruit bunch fiber (EFBF) using enzyme mixtures," *Biomass and Bioenergy*, vol. 56, no. 0, pp. 137–146, 2013.
33. Y. L. Chiew and K. Y. Cheong, "Growth of SiC nanowires using oil palm empty fruit bunch fibres infiltrated with tetraethyl orthosilicate," *Phys. E Low-Dimensional Syst. Nanostructures*, vol. 44, no. 10, pp. 2041–2049, 2012.
34. M. H. Ahmad, M. F. Dolmat, N. Bashir, H. Ahmad, and a. a. a. Jamil, "Effects of Oil Palm Empty Fruit Bunch Filler on the Electrical Tree Propagation in Silicone Rubber," *APCBEE Procedia*, vol. 3, no. May, pp. 147–153, 2012.

Numerical study on dissimilar guide vane design with SCC piston for air and emulsified biofuel mixing improvement

Mohd Fadzli Hamid[1] , *Mohamad Yosof* Idroas[1,*], *Mohd Hafif* Basha[3], *Shukriwani* Sa'ad[4], *Sharzali* Che Mat[1,5], *Muhammad Khalil* Abdullah[2] and *Zainal Alimuddin* Zainal Alauddin[1]

[1]School of Mechanical Engineering, Universiti Sains Malaysia, Engineering Campus, Seri Ampangan, 14300 Nibong Tebal Penang, Malaysia.
[2]School of Material and Mineral Resources Engineering, Universiti Sains Malaysia, Engineering Campus, Seri Ampangan, 14300 Nibong Tebal Penang, Malaysia.
[3]Mechanical Engineering Programme, School of Mechatronic Engineering, Universiti Malaysia Perlis, Pauh Putra Campus, Arau, 02600 Perlis, Malaysia.
[4]Faculty of computer, Media and Technology Management, TATI University Collage, Jalan Panchor, Teluk kalong, 24000 Kemaman, Terenganu, Malaysia.
[5]Faculty of Mechanical Engineering, Universiti Teknologi Mara, 13500 Permatang Pauh, Penang, Malaysia.

Abstract. Crude palm oil (CPO) is one of the most potential biofuels that can be applied in the conventional diesel engines, where the chemical properties of CPO are comparable to diesel fuel. However, its higher viscosity and heavier molecules can contributes to several engine problems such as low atomization during injection, carbon deposit formation, injector clogging, low mixing with air and lower combustion efficiency. An emulsification of biofuel and modifications of few engine critical components have been identified to mitigate the issues. This paper presents the effects of dissimilar guide vane design (GVD) in terms of height variation of 0.25*R*, 0.3*R* and 0.35*R* at the intake manifold with shallow depth re-entrance combustion chamber (SCC) piston application to the in-cylinder air flow characteristics improvement. The simulation results show that the intake manifold with GVD improved the performance of the air flow characteristic particularly swirl, tumble and cross tumble ratios from the intake manifold to the engine. The GVD with the height of 0.3*R* was found to be the optimum design with respect to the overall improvement of the air flow characteristic. The improvement of the air flow characteristic with the application of GVD and SCC piston in the engine was expected to contribute to a better air fuel mixing, fuel atomization and combustion efficiency of the engine using emulsified biofuel as a source of fuel.

* Corresponding author: meyusof@usm.my

1 Introduction

Diesel engine becomes most important and reliable power source generation nowadays for its high thermal efficiency, high compression ratio and inherent lean burn which enable heat dissipation via excessive air. The diesel engine application is widely used in the industrial, transportation and agriculture application [1]. However, diesel engines have drawbacks of an enormous poison emission particularly soot, hydrocarbons (HC), nitrogen oxides (NO_x), sulfur dioxide (SO_x), carbon monoxide (CO) and particulate matter (PM). This emission contributes to a detrimental human health and risk to the ecological system. The stringent policies in many European countries have alerted both heavy and light engine manufacturers to introduce the engines that can fulfil clean and green technology. In order to pursue this matter particularly in combustion efficiency, several strategies and methodologies have been made such as the development of internal combustion engine designs; combustion chamber and intake manifold, injection strategies and elevated injection pressure pump and dual strategy [2]. Numerous researchers reported that the global energy demand will increase progressively, e.g. in 1980, the fuel demand was 6630 million tons of oil in equivalent (Mt), and by 2030, the fuel consumption demand will expected to increase to 53% [3-5]. For this reason, the searching for alternative fuels to offset the global demand of fuel consumption becomes more challenging in order to minimize the fossil fuel dependence and subsequently improve the quality of the ecological system.

It attracts many researchers to explore the potential application of crude palm oil (CPO) as an alternative fuel since it is widely available and can be mass produced. CPO uses less land energy to fertilizer than other resources and it is capable to produce ten times energy of it consumes [6]. However, a direct application of CPO in the conventional diesel engines will leads to several issues such as less efficient combustion process due to retardation of the atomization process and change of timing in ignition delay [7-8]. According to Maher et al [9], high viscosity and low volatility of CPO, the engine will confront problem i.e. carbon deposited formation. This is because of poor atomization and incomplete combustion due to large triglyceride molecule and their high molecule mass. To tackle this issue, the combustion efficiency has been highlighted by increasing the rate of evaporation and mixing biofuel into the air. One of the feasible methods is assisting the air flow such as swirl and tumble in the in-cylinder. Swirl is defined as a rotational flow about the vertical axis [10-11]. Tumble is defined as the rotational flow about an axis perpendicular to the vertical axis of the cylinder [12]. Cross tumble is defined as a rotational flow about an axis perpendicular to both swirl and tumble axes [13].

Hence, this paper proposed the dissimilar guide vane design (GVD) for enhancing the swirl and tumble air flow in the fuel injection region area (piston-bowl) and expected to aid the mixing homogeneously between emulsified biofuel and air. A computational fluid dynamic (CFD) solver is utilized to predict the best design of GVD that will improved the air flow characteristics in term of air fuel mixing, fuel atomization and combustion efficiency.

2 Methodology

In general, the curve vane assisted the air into combustion chamber by either swirl or tumble and their turbulence kinetic energy (TKE) inside it's really important. Saad et al. [13] demonstrated that the effect of various design of their devise namely guide vane swirl and tumble device (GVSTD). Their investigation shows that the GVSTD has been proven to improve air flow and eventually improve the combustion efficiency. To enhance the swirl flow, the large vane surface is required [14]. However, the large vane surface will obstacle the air flow and create a high resistance of flow [15]. Enlarge the vane size will benefit to the momentum of air flow but a loss in volumetric efficiency.

2.1 Design of GVD

The effectiveness of height vane configuration via air flow arrangement for suitability emulsified biofuel application is utterly focused in this research. Four vanes which equally space are equipped in the diameter of inlet manifold air inlet runner. The GVD specification is listed in Table 1 and their schematic diagram is shown in Fig. 1.

Table 1. Specification of the guide vane design (GVD).

No.	Parameter	Value
1	Number of Vane	4
2	Length of vane (*l*)	30mm
3	Width of vane	0.5mm
4	Height of vane (*Hv*)	0.25R 0.30R 0.35R
5	Vane twist angle (θ)	35°
6	Angle of incident	90°

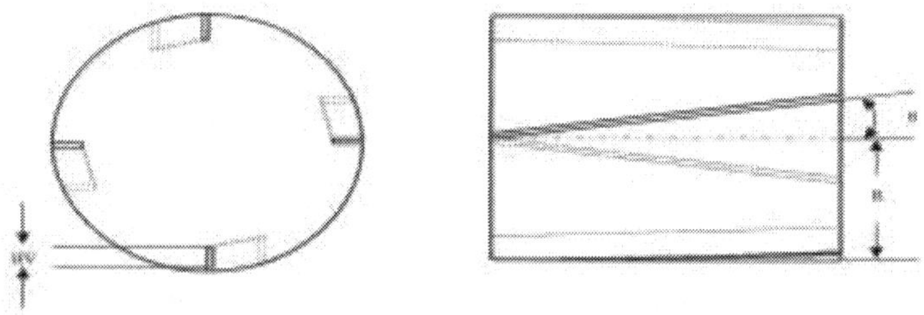

Fig. 1. Schematic diagram of GVD.

2.2 Governing equation

There are three fundamental physic principle involved, namely conservation of mass, conservation of momentum (Newton's second law) and energy.The equation for vector notation is express as below:

$$\frac{\partial \rho}{\partial t} + \nabla.[\rho \vec{u}] = 0 \quad (1)$$

where ρ is the fluid density and $\vec{u}$ is the three dimensional flow velocities in the *x, y, z* directions.

The equation for forces and surface force on the control volume:

$$\frac{\partial(\rho U)}{\partial t} + \nabla.(\rho U \times U) = -\nabla p + \nabla.\tau + S_M \quad (2)$$

where P, τ and S_M are the fluid pressure, strain rate and momentum source, respectively.
The energy equation is express as below:

DOI: 10.1051/matecconf/20179001065

$$\frac{\partial(\rho h_{tot})}{\partial t} - \frac{\partial p}{\partial t} + \nabla.(\lambda \nabla T) + \nabla.(U.\tau) + U.S_M \tag{3}$$

where h_{tot} and λ represent the total enthalpy and thermal conductivity of the fluid, respectively.

2.3 Grid independency test

The grid independent test (GIT) is a term usually used to describe the improvement via successively optimum cell sizes for analysis. Table 2 shows the case 3 was an appropriate meshing grid due to less nominal deviation.

Table 2. Summary of Grid Independency Test (GIT)

Case	1	2	3	4	5
Elements	103242	256242	302309	370061	447626
Average Cylinder Pressure	2.594×10^6	2.599×10^6	2.661×10^6	2.661×10^6	2.661×10^6

2.4 Numerical validation

Fig. 2. shows the in-cylinder pressure against crank angle (θ) diagram for a 3D in-cylinder air flow simulation at speed 2000 rpm. It was shows the good in agreement between experimental and numerical result and slightly 7% difference at peak pressure.

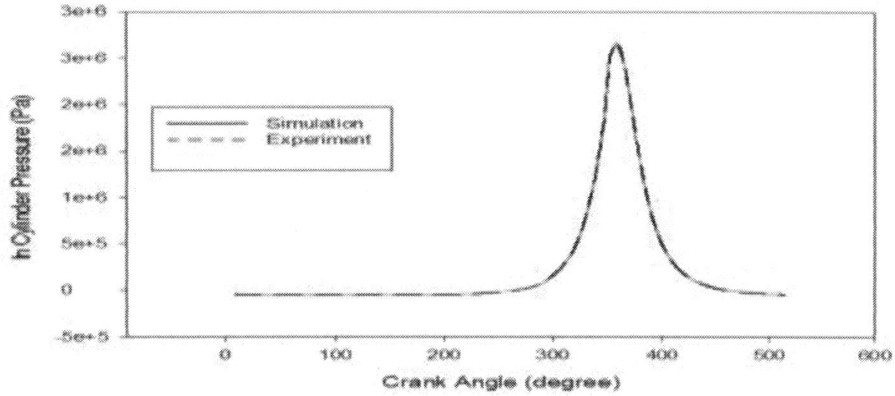

Fig. 2. In-cylinder pressure against crank angle (θ).

2.4 YANMAR L70AE diesel engine

The geometry of diesel engine was designed based on the Yanmar L70AE-D, engine speed in 2000 revolution per minutes (rpm) and their specification can be referred in Table 3. The diesel engine has been modelled using SOLIDWORK 2014, which considers the main six components of the model; intake runner, intake port, intake valve, cylinder, exhaust valve and exhaust port as can be seen in Figure 3. The components were assembled together to become one solid part before exporting to ANSYS- Design Modeller 15. The both valves; exhaust valve and intake valve were assigned as solid parts and the remaining were assigned as a fluid domain. Meanwhile, Figure 4 shows the GVD installed at the intake manifold.

Table 3. Specification of the engine

No.	Parameter	Value
1	Type	YANMAR L70AE-D
2	Bore	78mm
3	Stroke	62mm
4	Compression ratio	19.1
5	Intake Air	Natural Aspirated
6	Type of piston head	Bowl in-cylinder
7	Engine Speed	2000 rpm

No	Name
1	Intake Port
2	Intake Valve
3	Exhaust Port
4	Exhaust Valve
5	Bowl in SCC Piston
6	Clearance Volume

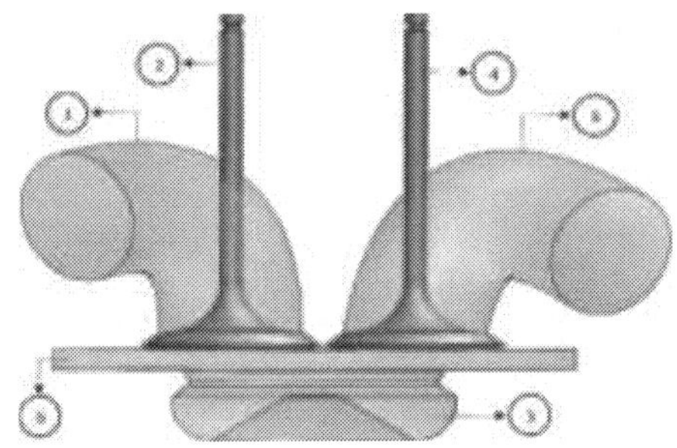

Fig. 3. Engine components.

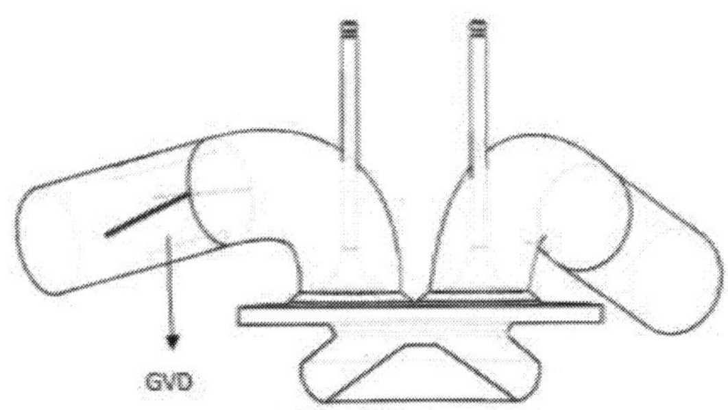

Fig. 4. GVD installation at the intake manifold.

The engine simulation was operated with a utilization of moving mesh to represent the motion of piston. The course mesh until the ultra-fine mesh sizes were verified and accomplished for better accuracy result and the optimization of meshing will save computational cost and time. The continuity, momentum and energy equations were used to compute this analysis and can be found at ANSYS FLUENT 15 solver theory guide. The initial pressure and temperature were set at 1atm and 300K respectively.

3 Results and discussion

3.1 Swirl Ratio, Tumble and Cross Tumble Ratio

Fig. 5 shows the swirl ratio against crank angle (θ) for three dissimilar models with GVD and no GVD installing at the air intake manifold. Swirl ratio with higher value promotes a better mixing of fuel and air homogeneously and aids the combustion efficiency. It is because of the swirl ratio is determined by their air flow pattern and their strength characteristics. The highest ratio of swirl will stimulate the better air mixing with low-grade and low volatility fuel. The ratio value either positive or negative value ratio is arbitrary. Thus, the important value to concern is a magnitude of the ratio. Based on Fig. 5. It shows that the air swirl ratio is produced higher once considering the GVD installation and consequently improves the air swirl ratio. The *0.3R* type of GVD produced higher of swirl ratio on the position start of injection (SOI) at 346° approximately 70% significant different compared to no GVD. The *0.25R* and *0.35R* also improved the swirl ratio approximately 14% and 22% respectively compared to no GVD. It clearly indicates that the result after considering GVD is a good agreement with Bari et al [17] whereby after consider the guide vane swirl tumble device (GVSTD), the air swirl flow characteristic is elevated instead of uninstalled GVSTD.

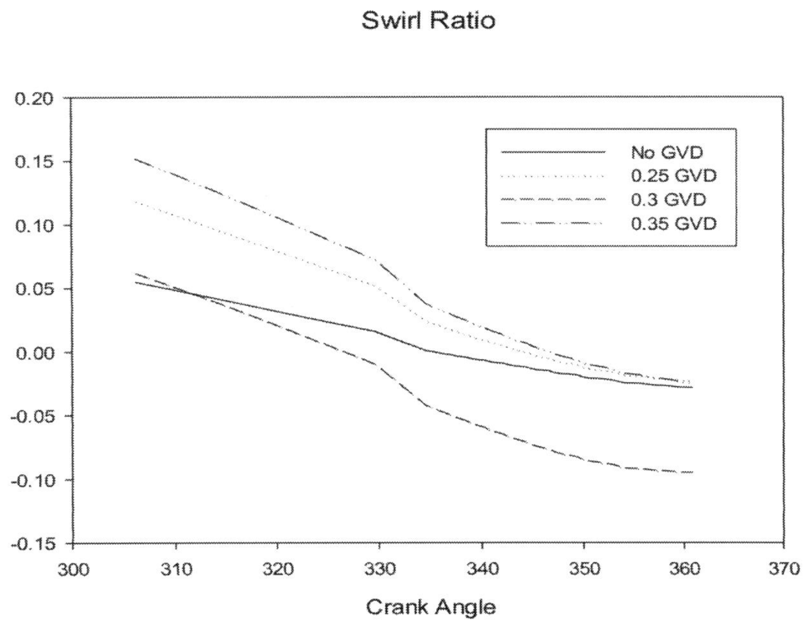

Fig. 5. Swirl ratio against crank angle (θ) with and without GVD.

Fig. 6. Shows the tumble and cross tumble ratio against crank angle (θ). In combustion theory, the effect of tumble and cross tumble are similar to swirl ratio which derive to the enhancement of turbulence flow and assist mixing fuel and air along the piston-bowl in a homogenous mixture. Fig. 6. (a) and (b) present the tumble and cross tumble ratio with installed GVD has significantly higher and strong magnitude ratio rather than the engine without GVD. Fig. 6. (a) Shows the tumble ratio which present the *0.3R* and *0.35R* type of GVD has higher tumble ratio and parallel both of them in term of magnitude ratio during SOI. Both of them has significantly difference approximately 75% compared to no GVD. Nevertheless, Fig. 6. (b) Shows the *0.35R* type of GVD has higher cross tumble ratio and approximately 60% significant difference with no GVD. This result implies that, once

 DOI: 10.1051/matecconf/20179001065

installing the GVD in the intake manifold substantially will give a stronger lateral flow of air within the cylinder. The lateral flow will assist the injected fuel spreading via consistent flow along the piston bowl.

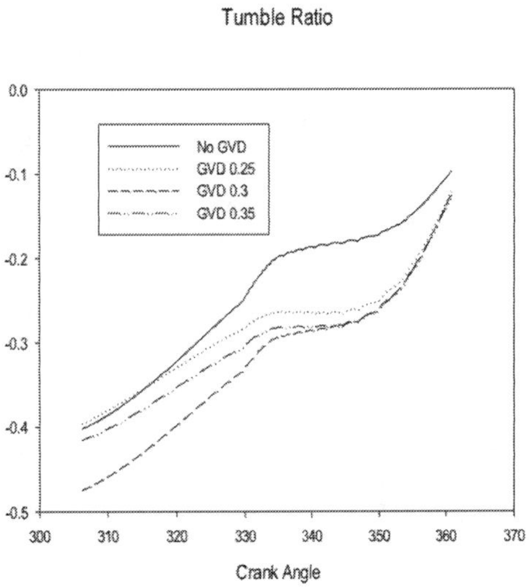

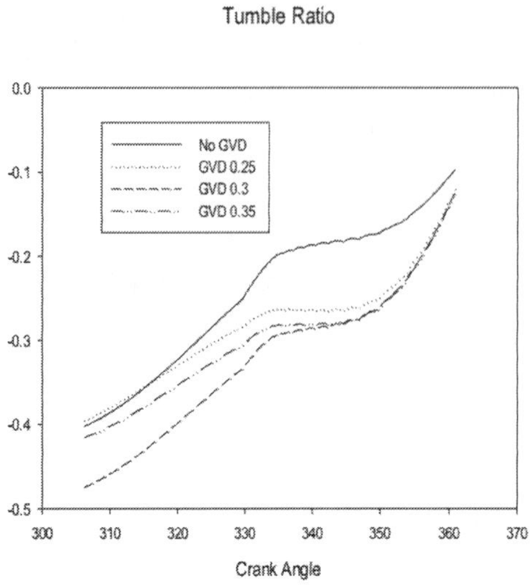

Fig. 6. (a) Tumble ratio against crank angle (θ) and (b) Cross tumble ratio against crank angle (θ).

3.2 Turbulence Kinetic Energy (TKE)

TKE is the mean kinetic energy per unit mass that is associated with eddies in turbulence flow [18]. TKE is mainly concern due to the capability of the energetic turbulence to assist the atomization molecule along the piston-bowl of the fuel flow. For the conventional diesel fuel, the atomization spraying profile injection is wide cone angle and short penetration during injection while the atomization profile for the emulsified biofuel is vice versa. The higher value of TKE implies that smoother and more efficient combustion, promotes the flame propagation, reduce carbon deposited and aid in break-up length penetration to mix with wide area inside the combustion chamber. From Fig. 7. It shows that the GVD has significant value of TKE at the start of combustion (SOC) at 352° compared to without GVD. The *0.3R* type of GVD has been overtake the no GVD TKE during SOC and leads higher until to top dead centre (TDC). However, during SOI at the crank angle 346°, the no GVD shows the good TKE rather than installing GVD and approximately 10% significant different compared to the *0.3R* type of GVD. Despite of installing GVD will detrimental TKE at SOI, nevertheless installing GVD will improve SOC and contribute in combustion efficiency, the detrimental of SOI is considering fewer. The GVD with 0.3*R* is the excellent design among the others of GVDs. This result indicates that, installing the GVD has been proven in the improvement of the TKE and consequently aids in the flame propagation along the piston bowl.

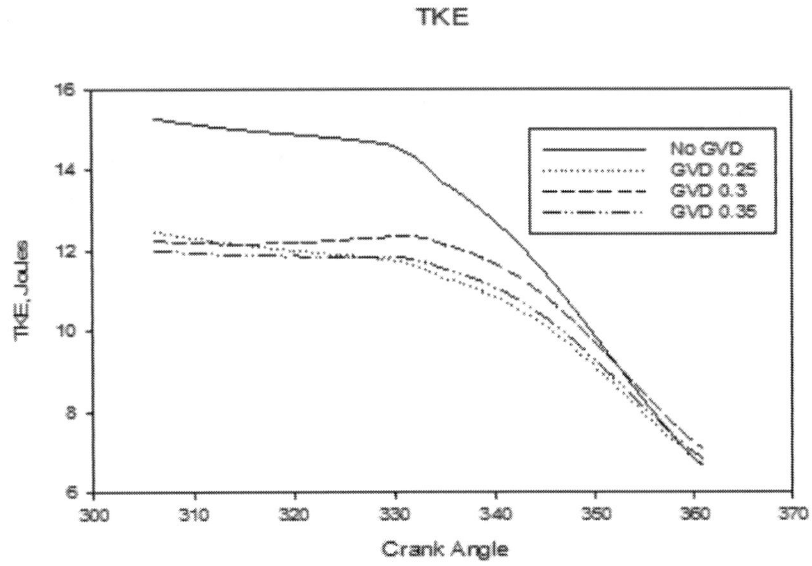

Fig. 7. In-cylinder TKE.

4 Conclusions

The effects of dissimilar GVD design in terms of height with SCC piston to the in-cylinder air flow characteristics have been studied. The in-cylinder air flow characteristics were measured and the profiles were developed via simulation in terms of swirl, tumble ratio, cross tumble ratio and turbulence kinetic energy. There was a significant improvement observed on the swirl, tumble and cross tumble ratios as a result of GVD installation in the intake manifold. The TKE result showed a higher turbulence producer which is essential in spreading atomization molecule along the piston bowl. The GVD with the height of $0.3R$ was found to be the optimum design that has contributed to the improvement of swirl, tumble and cross tumble ratios by 70%, 60% and 60%, respectively.

The GVD design was capable to increase the fuel dispersion in the cylinder and to break up the length penetration of the emulsified biofuel injection. The improved air flow characteristics can possibly mitigate the problems of carbon deposit formation on the piston bowl due to insufficient period of time to combust. Further investigation of GVD design parameters will be performed in order to determine the effects of vane length, vane angle, number of vane, thickness and diameter of the intake manifold to the overall improvement of the air flow characteristics.

This study was funded by the Research University Grant Scheme (project code of Mekanik/6071153) and Ministry of Higher Education under the Exploratory Research Grant Scheme (project code of Mekanik/6730085).

References

1. Qi, D. H., Chen, H., Matthews, R. D., & Bian, Y. Z. Combustion and emission characteristics of ethanol-biodiesel-water micro-emulsions used in a direct injection compression ignition engine. *Fuel*, **89**(5), 958–964, (2010)
2. Alahmer, a., Yamin, J., Sakhrieh, a., & Hamdan, M. a. Engine performance using emulsified diesel fuel. *Energy Conversion and Management*, **51**(8), 1708–1713, (2010)

3. Ashraful, A. M., Masjuki, H. H., Kalam, M. a., Rizwanul Fattah, I. M., Imtenan, S., Shahir, S. a., & Mobarak, H. M. Production and comparison of fuel properties, engine performance, and emission characteristics of biodiesel from various non-edible vegetable oils: A review. *Energy Conversion and Management*, **80**(June), 202–228, (2014)
4. Atabani, A. E., Silitonga, A. S., Ong, H. C., Mahlia, T. M. I., Masjuki, H. H., Badruddin, I. A., & Fayaz, H. Non-edible vegetable oils: A critical evaluation of oil extraction, fatty acid compositions, biodiesel production, characteristics, engine performance and emissions production. *Renewable and Sustainable Energy Reviews*, **18**, 211–245, (2013)
5. No, S.-Y. Inedible vegetable oils and their derivatives for alternative diesel fuels in CI engines: A review. *Renewable and Sustainable Energy Reviews*, **15**(1), 131–149, (2011)
6. Sumathi, S., Chai, S. P., & Mohamed, a. R. Utilization of oil palm as a source of renewable energy in Malaysia. *Renewable and Sustainable Energy Reviews*, **12**(9), 2404–2421, (2008)
7. Erazo, J. A., Parthasarathy, R., & Gollahalli, S. Atomization and combustion of canola methyl ester biofuel spray. *Fuel*, **89**(12), 3735–3741, (2010)
8. Chen, P. C., Wang, W. C., Roberts, W. L., & Fang, T. Spray and atomization of diesel fuel and its alternatives from a single-hole injector using a common rail fuel injection system. *Fuel*, **103**(x), 850–861, (2013)
9. Meher, L. C., VidyaSagar, D., & Naik, S. N. Technical aspects of biodiesel production by transesterification - A review. *Renewable and Sustainable Energy Reviews*, **10**, 248–268, (2006)
10. Heywood, J. B. *Internal combustion engine fundamentals. McGraw-Hill series in mechanical engineering* **Vol. 21**, (1988)
11. Pulkrabek, W. W. Engineering fundamentals of the internal combustion engine. *Journal of Engineering for Gas Turbines and Power*, **126**, 198, (2004)
12. Perini, F., Miles, P. C., & Reitz, R. D. A comprehensive modeling study of in-cylinder fluid flows in a high-swirl, light-duty optical diesel engine. *Computers & Fluids*, **105**, 113–124, (2014)
13. I. Saad, S. Bari, Effects of Guide Vane Swirl and Tumble Device (GVSTD) to the Air Flow of Naturally Aspirated CI Engine, in: International Conference on Mechanical Engineering 2011 (ICME2011), Progressive Printers Pvt. Ltd., Dhaka, Bangladesh, (2011)
14. J.S. Kim, Fluid Swirling Device for an Internal Combustion Engine, in: United States Patent, USA, (2003)
15. S.-S. Lin, J.-C. Yang, Intake swirl enhancing structure for internal combustion engine, in, Google Patents, (2000)
16. S. Bari, P. J. G. Johansen, and A. J. T. Alherz, "Simulation of Improvements to In-cylinder Mixing of Biodiesel with Air by Incorporating Guide Vanes into the Air Intake System," *Procedia Eng.*, **vol. 105**, pp. 480–487, (2015)
17. Bari, S., & Saad, I. CFD modelling of the effect of guide vane swirl and tumble device to generate better in-cylinder air flow in a CI engine fuelled by biodiesel. *Computers and Fluids*, **84**, 262–269, (2013)
18. F.M. White, *Fluid mechanics*, 1994, in, McGraw-Hill Inc., (1979)

Pedestrian-motorcycle collisions: associated risks and issues

Aqbal Hafeez Ariffin[1,*], *Azhar* Hamzah[1,2] , *Mohd Syazwan* Solah[1], *Noor Faradila* Paiman[1], *Siti Fairos* Mat Hussin[1] and *Mohd Rasid* Osman[1]

[1]Vehicle Safety and Biomechanics Research Centre, Malaysian Institute of Road Safety Research (MIROS), Lot 125-135, Jalan TKS 1, Taman Kajang Sentral, 43000 Kajang, Selangor, Malaysia
[2]MIROS PC3 Crash Laboratory, KM 15 Jalan Tiang Dua, 75450 Ayer Molek, Melaka, Malaysia

Abstract. From the statistics, there are serious concerns over the relatively high number of fatal motor vehicle crashes involving pedestrian-motorcycle in Malaysia. The high number of motorcycle registration on road, compounded by its popularity as the major mode of transportation in the nation, imposes safety risk to pedestrians, as well as to other road users. Data from 1,626 related road crashes of Royal Malaysia Police (RMP) for the 2009-2013 period were retrospectively collected via MIROS Road Accident Analysis and Database System (M-ROADS). The data were then analyzed via logistic regression method to determine associations between risks and injury severity in pedestrian-motorcycle collisions. The results indicate that five factors were significantly related to injury severity, which include age, location of body injury, as well as speed limit, road geometry and lighting condition of collision site. Subsequently, focus group discussions with stakeholders were also conducted to gather relevant data to identify related issues and suggestions on motorcycle safety technology with regards to collision with pedestrian.

1 Introduction

1.1 Overview of road crash statistics

According to the World Health Organization's (WHO) Global status report on road safety 2015, almost half of the estimated 1.25 million deaths of the world's roads per year are among those with least protection, i.e. vulnerable road users (VRU) which include motorcyclists (23%), pedestrians (22%) and cyclists (4%) (Fig. 1). The casualty statistics concerning VRUs are not just apparent in most low- and middle-income countries, but also in high-income countries such as European countries, the United States, the United Kingdom, China, Japan and Republic of Korea [1-4]. Comparable situation can be observed in Malaysia with motorcyclists (62.1%) and pedestrians (6.6%) constituted more than two-third of total road fatalities in 2013 and the fraction is almost the same each year [5-8].

* Corresponding author: aqbal@miros.gov.my

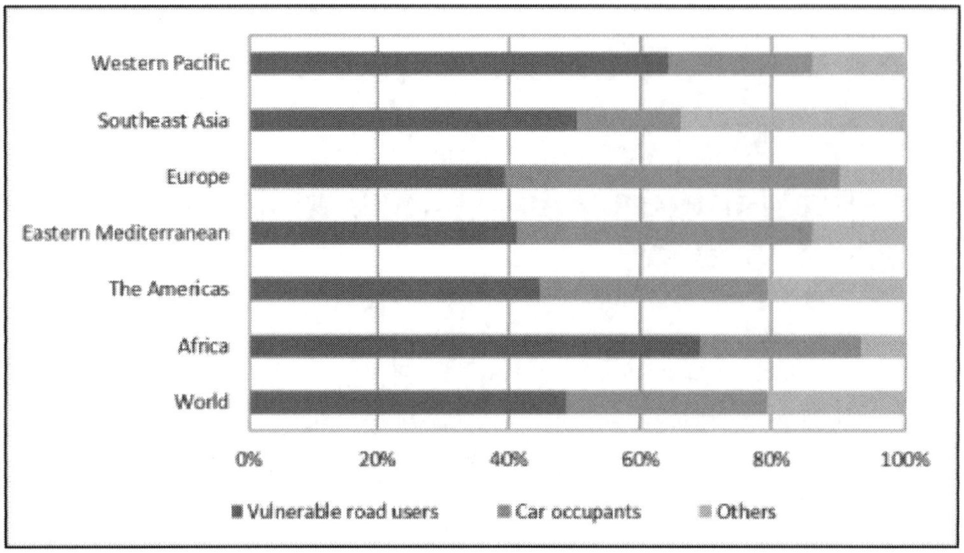

Fig. 1. Reported deaths by type of road user (%), by WHO region [3].

An analysis conducted for the past 10-year period (2001-2010) of local motorcycle crash data showed that motorcyclists were involved in 24.8% of killed and severely injured (KSI) crashes involving collision with pedestrian (by type of first collision) [9]. Aqbal Hafeez et al. (2010) in their study of pedestrian casualties in road crashes in Malaysia found motorcycles as the second-most striking crash opponents in single vehicle accident (SVA) with fatal single pedestrian after M1 and M2 category passenger vehicles (e.g. cars, vans, 4WDs). Nevertheless, for the same crash configuration, the study revealed most killed and severely injured pedestrians were struck by motorcycles (54%) followed by passenger vehicles (40%).

1.2 Road safety plan

In March 2010, the UN General Assembly proclaimed the Decade of Action (DOA) for Road Safety 2011-2020; it was subsequently launched globally on 11 May 2011, with the aim of reducing and stabilising road traffic fatalities worldwide by 2020 [10]. The DOA plan rests on five pillars in which one of them (Pillar 3) focusing on safer vehicles. Two related activities were set for elevating the safety of motorcycles and pedestrians by encouraging universal deployment of crash avoidance technologies such as Electronic Stability Control (ESC) and Antilock Braking System (ABS) in motor vehicles, and application of pedestrian protection regulations and increased research into safety technologies designed to reduce risks to VRUs. Similar programme has been adopted and outlined in the second Road Safety Plan Malaysia (RSPM) 2014-2020 to encourage implementations of UN Regulations through Vehicle Type Approval (VTA) which is by promotion of safety technology use for motorcycle (active safety e.g. ABS, ESC, lighting and signaling, passive safety e.g. crashworthiness) to reduce crash risk to motorcyclist as well as pedestrian and other road users [11].

1.3 Motivation of study and its importance

At present, to the best of knowledge of the author, limited study has been carried out concerning pedestrian-motorcycle collision safety either on worldwide or on Malaysia level, i.e. severity of the issue is unknown. From the statistics, there are serious concerns over the relatively high number of road accidents involving pedestrian-motorcycle in Malaysia. As reported by Road Transport Department, motorcycle constitutes almost half of total registered vehicles (46% of total 24.8 million registered vehicles) and its registration increases each year [12]. This, the high number of motorcycle registration on road, compounded by its popularity as the major mode of transportation in the nation, imposes risk to pedestrians, as well as to other road users. Based on the above point of views, a need exists to foster a great understanding of pedestrian-motorcycle crashes due to the worrying abovementioned crash statistics. Other than to investigate the characteristics of the said road crash, this study aims to identify via statistical analysis the associated risk factors that caused for road crashes involving pedestrian and motorcycle. Findings obtained from this study will provide vital information to the government and the motorcycle industry for improvement of related safety measures. Moreover, this study is expected to be a piece of useful reference for subsequent related researches.

2 Methodology

2.1 Data source, selection and description

The source of the road accident data was retrospectively obtained from the MIROS Road Accident Analysis and Database System (M-ROADS) which contains comprehensive accident data collected throughout the nation by the Royal Malaysia Police (RMP) Traffic Department. The form used by the police for the data collection is a road accident form known as POLIS 27 (POL 27) which contains a total of 91 variables comprising various information on the crash characteristics, the casualties and the vehicles involved. This study used a five-year data from 2009 through 2013 due to the data completeness at the time of the study was initiated. Dataset for this study includes fatal and non-fatal (severe and slight injuries) cases involving single injured pedestrians in single-vehicle accidents, i.e. one pedestrian struck by one motorcycle. The non-injury cases were omitted to maintain data accuracy and to avoid probable insufficient data due largely to underreporting. The final dataset contained a total of 1,626 related cases throughout Malaysia for the study period. Out of the figure, 18.0% (n = 293) contributed to fatal cases, 29.3% (n = 477) was severe injury cases and slight injury cases comprised of 52.6% (n = 856).

2.2 Statistical analysis

The current study focuses only on fatal and injury cases to pedestrian only (injury to motorcycle rider was not examined) to analyze factors contributing to the injuries. The data were filtered and analyzed using SPSS software version 17.0. Logistic regression technique was used to estimate the relative risks by approach of odds ratios (OR) at 95% confidence intervals (CI). An OR value that is greater than 1 indicates higher risk to the concerned attribute and vice versa. The pedestrian injury severity was divided into two injury levels which are fatal (attribute; coded as '1') and non-fatal (control; coded as '0') for the purpose of performing logistic regression that required for a binary typed outcome. Univariate logistic regression was performed for each selected predictor variable to identify potential risk factors which significantly influence the probability of injury to pedestrian. Multivariate logistic regression was then employed to develop injury risk model by taking consideration the effects of other predictor variables identified during the univariate analysis. The model was estimated manually to determine the variables that to be included in the final model. Entry and removal probabilities for the stepwise procedure were both set at 0.05 and 0.10, respectively. The dependent and independent variables considered in this study are summarized as follow:

Table 1. Description of variables selected for the analysis.

Factors	Categories	Variables and coding
Dependent variable	Injury severity	0: Non-fatal injuries 1: Fatal
Pedestrian	Age	0: ≤25 years old 1: 26-64 years old 2: >64 years old
	Body injury location	0: Others 1: Head injury
	Pedestrian location	0: Others 1: On roadways 2: At pedestrian crossings
	Pedestrian action	0: Others 1: Crossing road/junction
Environment	Speed limit	0: <50km/h 1: 50km/h 2: >50km/h
	Road geometry	0: Others (junctions, roundabouts) 1: Curved 2: Straight
	Lighting condition	0: Daylight 1: Dawn/Dusk 2: Dark-lit 3: Dark-unlit
Vehicle (Motorcycle)	Motorcycle's age	0: <5 years 1: 5-9 years 2: >9 years
	Engine capacity	0: <151 cc (underbone) 1: >150 cc

2.2.1 Logistic regression

As illustrated in numerous literatures, logistic regression is an imperative statistical method especially in road safety, in analyzing road accident data. It has been widely used by researchers to determine the factors associated with the cause of road accident and the resulted injury severities [e.g. 8, 13-15]. This regression method allows for prediction of a binary outcome from a set of explanatory dependent variables from recorded accident data which is normally categorical.

2.3 Focus group discussion

Interviews and technical discussions via FGD with experts from local motorcycle industry (assemblers and distributors) were conducted to explore their views and suggestions on motorcycle-related safety towards motorcycle-pedestrian collision. The selected assemblers and distributors are Boon Siew Honda, Suzuki Assemblers, Modenas, Hong Leong Yamaha Motor and Kawasaki Motors.

3 Results and discussion

3.1 Injury severity analysis

The following table describes the frequencies and ORs of risk factors to pedestrians.

Table 2. Univariate and multivariate logistic regression results with ORs (95% CI).

Factors	Frequency n (%)	Unadjusted ORs (95% CI)	Adjusted ORs (95% CI)
Age			
≤25 years old (*control*)	824 (50.7)		
26-64 years old	556 (34.2)	6.95 (4.72, 10.23)*	9.78 (5.59, 17.11)*
>64 years old	246 (15.1)	21.89 (14.42, 33.22)*	30.72 (17.01, 55.47)*
Body injury location			
Others (*control*)	1,135 (70.9)		
Head injury	465 (29.1)	6.94 (5.27, 9.13)*	6.22 (4.20, 9.20)*
Pedestrian location			
Others (*control*)	819 (51.2)		
On roadways	737 (46.0)	1.47 (1.13, 1.90)*	-
At pedestrian crossings	45 (2.8)	1.57 (0.76, 3.25)	-
Pedestrian action			
Others (*control*)	583 (36.4)		
Crossing road/junction	1,018 (63.6)	1.58 (1.20, 2.09)*	-
Speed limit			
<50km/h (*control*)	576 (35.4)		
50km/h	414 (25.5)	1.21 (0.85, 1.73)	1.49 (0.88, 2.54)
>50km/h	636 (39.1)	1.95 (1.45, 2.64)*	1.56 (1.00, 2.44)*
Road geometry			
Others (*control*)	106 (6.6)		
Curved	89 (5.5)	0.50 (0.24, 1.01)	0.53 (0.19, 1.50)
Straight	1,412 (87.9)	0.57 (0.37, 0.89)*	0.41 (0.20, 0.85)*
Lighting condition			
Daylight (*control*)	1,147 (70.7)		
Dawn/Dusk	88 (5.4)	1.71 (1.01, 2.90)*	1.46 (0.64, 3.33)
Dark-lit	229 (14.1)	2.07 (1.48, 2.90)*	1.10 (0.66, 1.85)
Dark-unlit	158 (9.7)	2.32 (1.58, 3.40)*	2.03 (1.16, 3.57)*
Motorcycle's age			
<5 years (*control*)	694 (53.4)		
5-9 years	276 (21.2)	1.05 (0.73, 1.50)	-
>9 years	330 (25.4)	1.06 (0.76, 1.49)	-
Engine capacity			
<151 cc (*control*)	1,270 (98.1)		
>150 cc	25 (1.9)	1.81 (0.75, 4.38)	-

* Statistically significant at the 0.05 level.

Univariate analysis revealed the variables of age, body injury location, pedestrian action, speed limit, road geometry, and lighting condition as significantly correlated with the dependent variable – possible important factors affecting injury severity. Nevertheless, motorcycle's age (calculated at time of accident based on year) and engine capacity was found insignificant despite being included in the analysis. Gender was also an unimportant factor.

Pedestrian age: Pedestrians 65 years old and above had the highest odds of death as compared to pedestrians between ages of 26 and 64. The former group of pedestrians were also involved in 50 percent of pedestrian fatalities, higher than the latter group's percentage (24.1%).

Body injury location: The odds of fatal for pedestrians with head injuries was nearly 7 times greater than those suffered other body injury locations.

Pedestrian location: Pedestrians at dedicated crossing areas and on roadways faced 1.57 and 1.47 times more risk of being fatal in road accident as compared to on other impact locations.

Pedestrian action: Pedestrians who cross the road/junction faced 1.58 times more risk of getting killed in road accident as compared to other actions.

Speed limit: The odds of death at the scene with speed limit of >50km/h and 50km/h were 1.95 and 1.21 times greater, respectively, than with speed limit of <50km/h.

Road geometry: Pedestrians who were involved in road accidents on straight roads were 0.57 times less likely to be fatal than those on other road geometries, such as such as junctions and roundabouts.

Lighting condition: Insufficient light especially during dawn/dusk (OR = 1.71), dark with (OR = 2.07) or without streetlight (OR = 2.32) posed higher risk of fatal outcome to pedestrians as opposed to road accidents occurred in daylight. Greater risk of fatal to pedestrians involved in road accidents that take place in ill-lit areas (dark without street lightings) (OR = 2.32) as opposed to road accidents occurred in daylight.

Risk factors which were statistically significant were included and their significance were assessed in the final model. Multivariate analysis showed that pedestrian location and action during collision with motorcycle were not significant variables and were further removed from the model. Similar risk factors were observed significant when comparing the results with the previous analysis. Pedestrian age remained significant and showed approximately 40% increase in ORs (from 6.95 to 9.78 and from 21.89 to 30.72 for 26-64 years old and >64 years old, respectively) in the final model. Several other factor levels even had close ORs although there were also relative differences between them. Head injury, speed limit >50km/h, straight road and ill-lit condition (dark-unlit) were also significant factors that increased the risk of fatal to pedestrians, but with slightly lower adjusted ORs as compared to those unadjusted.

3.2 Motorcycle-pedestrian safety from industry perspective

FGDs with local motorcycle industry revealed several invaluable information on motorcycle-pedestrian safety issues and suggestions as summarized in the following points.

- Pedestrian-motorcycle collision is a unique road safety issue especially in Malaysia where motorcycle registration accounted for almost half of the total vehicle registration.
- Crash avoidance technologies such as Antilock Brake System (ABS) and Automated Emergency Braking (AEB) pedestrian are some of the effective safety systems in passenger cars that can probably be incorporated into motorcycles to assist in preventing road collision between motorcycles and pedestrians.
- Potential design of "forgiving" structure (passive safety) through pedestrian protection regulations can be included for motorcycles to protect pedestrians and reduce the severity of injury in the event of collision. However, the design elements should not disregard the safety of motorcycle rider and pillion.
- There is no requirement in the local legislation for addressing pedestrian protection with respect to collision with motorcycle. With respect to ABS for motorcycle, some countries and region (European Union and Brazil) have started to mandate the related legislation for all new motorcycles over 125cc (Fig. 2). Whereas, manufacturers will have an option to provide either ABS of Combined Brake System (CBS) onto motorcycles with engine capacity under 125cc. There is a possibility for other countries

especially those with high number of accidents involving motorcycles to follow this example by making the legislation compulsory.

- Until recently, several motorcycle models of large engine capacity (>250cc) have been equipped with ABS. As for small engine capacity especially 'underbone' motorcycles (≤150cc), very limited models are possibly equipped with such safety system.

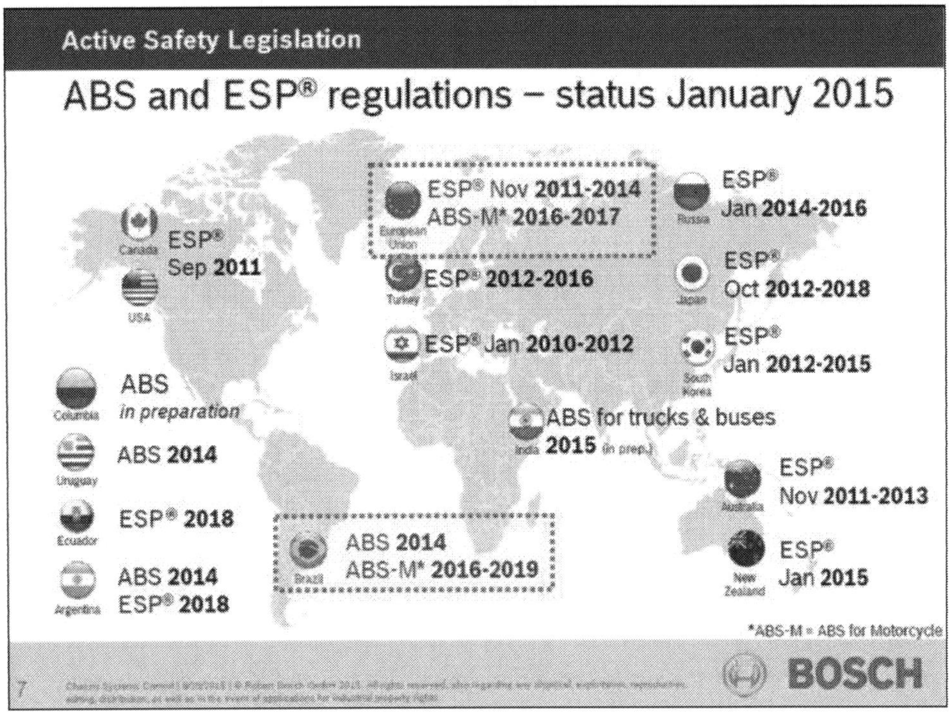

Fig. 2. ABS and ESP legislations status worldwide [16].

4 Conclusions

This study identifies risk factors that influence pedestrian injury severity in road collision with motorcycle via retrospective method by utilizing police-reported crash records. Results of the logistic regression analysis conducted on the dataset revealed age (>25 years old), body injury location (head), speed limit (>50km/h), road geometry (straight) and lighting condition (dark-unlit) as significant factors (in the final model) that led to a noticeably higher risk of pedestrian mortality. Discussions with motorcycle industry suggested inclusion of crash avoidance technologies (e.g. ABS and AEB) onto motorcycle as one of the possible vehicle engineering measures to reduce this type of road collision. Nevertheless, utilization of data from related in-depth accident investigations should be conducted to improve the overall analysis since this study is limited by the variables that are readily available in POL 27 - inadequate information on crash configuration (front, side, etc.) and detailed vehicle damage. Last but not least, the identification of these risk factors provides crucial information for interventions and safety control measures that help to prevent and mitigate casualties and injuries resulted from the pedestrian-motorcycle collisions, hence improve the road safety level in the country.

The authors would like to express their great appreciation to MIROS for granting this research work. Acknowledgements also go to RMP and the experts from motorcycle industry for the relevant road crash data and invaluable technical information shared in the FGDs, respectively. Last but not least, special thanks go to the research team members and those who have either directly or directly involved in supporting and assisting throughout the completion of the research and this paper.

References

1. http://www.swov.nl/rapport/Factsheets/UK/FS_Vulnerable_road_users.pdf
2. http://www.swov.nl/rapport/factsheets/uk/fs_pedestrians.pdf
3. http://apps.who.int/iris/bitstream/10665/189242/1/9789241565066_eng.pdf?ua=1
4. https://crashstats.nhtsa.dot.gov/Api/Public/ViewPublication/812124
5. *Statistical Report of Road Accidents 2013* (Royal Malaysian Police, Traffic Branch, Bukit Aman, Kuala Lumpur, 2014)
6. *Statistical Report of Road Accidents 2014* (Royal Malaysian Police, Traffic Branch, Bukit Aman, Kuala Lumpur, 2015)
7. A. H. Ariffin, Z. Mohd Jawi, M. H. Md Isa, K. A. Abu Kassim, S. V. Wong, *Proceedings of the MIROS Road Safety Conference*, 280-289 (MIROS, Kuala Lumpur, 2010)
8. https://www.jstage.jst.go.jp/article/eastpro/2011/0/2011_0_385/_pdf
9. https://www.miros.gov.my/1/dl.php?filename=Factsheet%202.pdf
10. http://www.who.int/roadsafety/decade_of_action/plan/plan_en.pdf
11. http://www.mot.gov.my/SiteCollectionDocuments/Darat/Pelan+KJRM+2014-2020.pdf
12. http://www.mot.gov.my/en/Statistik%20Tahunan%20Pengangkutan/Transport%20Statistics%20Malaysia%202014.pdf
13. A. S. Al-Ghamdi, Accid. Anal. Prev. **34**, 729-741 (2002)
14. H. L. Chang, T. H. Yeh, J. Transp. Eng. **132**, 227-236 (2006)
15. D. M. P. Wedagama, D. Dissayanake, Journal of Eastern Asia Society for Transportation Studies **8** (2009)
16. http://www.thaiauto.or.th/2012/Automotive-Summit/doc/ppt/2015/25-6-15/203_AM/05.Bosch_Automotive_Summit_25062015.pdf

Steady flow and dynamic analyses comparison of an air intake breathing capacity

Nur Adila Mohamad Shafie[1], *Mohd Farid* Muhamad Said[1,*], *Azhar* Abdul Aziz[1], *Zulkarnain* Abdul Latiff[1], *Ahmad Kamal* Mat Yamin[2], and *Noreffendy* Tamaldin[2]

[1]Automotive Development Centre (ADC), Faculty of Mechanical Engineering, Universiti Teknologi Malaysia (UTM), Johor, Malaysia.
[2]Automotive Department, Faculty of Mechanical Engineering, Universiti Teknikal Malaysia Melaka (UTeM)

Abstract. Analysis of intake breathing capacity is important in order to determine the performance output for any respective engine. Flow coefficient and discharge coefficient are the common parameters used to define the breathing capability and different analysis have been applied by researchers to determine these parameters which include experiments and Computational Fluid Dynamics (CFD) analyses. This study aims to investigate the difference in breathing parameters obtained from steady flow simulation and dynamics simulation. Proton CAMPRO 1.6-litre engine was selected as the reference engine in this study. The experiment involved was the flow bench test, while CFD simulation carried out were the port flow analysis (steady flow) and the cold flow analysis (dynamics). Results obtained indicate that flow coefficient and discharge coefficient from cold flow simulation are always lower than both parameter values obtained from port flow simulation and experiment with large deviations of minimum 15.6% and maximum 27%. Meanwhile, the breathing parameter values from port flow simulation were very close to the experimental data with the minimum deviation of 1.6%. This study concludes that port flow simulation is very accurate for the analysis of defining intake breathing capacity, meanwhile cold flow simulation can be used to predict the trend and lower limit of these parameters especially at low valve lift.

1 Introduction

As the most important aspect of the automobile system, engine performance is analysed in details in fulfilling the important parameters which include geometrical properties and also different engine efficiencies. Aside from thermal efficiency and mechanical efficiency, volumetric efficiency is another crucial parameter which represents the effectiveness of air induction in engine [1, 2]. Analysis of suction capability of the engine system is important as it influences the performances output of any respective engine [3]. In the effort of improving engine system, CFD simulation has emerged from years ago as an alternative

* Corresponding author: mdfarid@utm.my

and has been used alongside other methods which include Particle Image Velocimetry (PIV) and flow bench test. In analysing the intake air suction of engine, questions have surfaced on whether the steady flow test can accurately determine the engine breathing parameters of a moving car. Based on previous study, it has been defined that discharge coefficient and flow coefficient from steady flow analysis can represent real operation of car engine as long as it is under normal speed range [1]. CFD simulation is used for the analysis for optimization of the cost and timing [3]. There are many researchers that have adapted the use of CFD simulation and proved the significance of this method for engine study. Research by Balasankar *et. al* [3] had discussed on the optimization of intake port in order to improve the engine performance. This research had investigated on the computation of port flow at different valve lifts by CFD solver. Intake flow data was compared with the flow bench test, and the research proceeded into optimization of port geometry by DOE analysis. Balasankar *et. al* [3] had presented good correlation between port flow simulation and flow bench test, leading to a conclusion that CFD simulation is a reliable method to replace the conventional method of engine intake analysis.

Ismail and Abu Bakar [2] had discussed with great details about engine induction by experiment of Superflow bench test. It was proven that the steady flow bench test can provide high accuracy on the flow analysis. In the research, detail design of the port and the valve were described and the sources of flow losses were listed which include the bend at the valve guide, bend to exit valve and push rod expansion [2]. Analysis by Ismail and Abu Bakar [2] concluded that increasing the test pressure and the valve lift allowed the increment in sucked flow. This study also concluded that the increment of flow does not occur and stabilized after it reaches the ratio of maximum valve lift per diameter (L/D) of 0.25. Research by Kumar and Nagarajan [4] also investigated on those non-dimensional parameters with the inclusion of the swirl ratio analysis. Analysis showed the correlation of the obtained data from the flow bench test and also from the modelling by the commercial CFD code. This research discussed on the swirl improvement by modification of the valve design. Kumar and Nagarajan [4] defined that valve shrouding and twisted tape insert can improve the swirl in engine. From the research, it was also concluded that although flow increased along the valve lift, this took a penalty on the increased friction.

Port flow simulation in internal combustion engine was further analysed by Himanth and Jayashankar [5]. The geometry model created in ANSYS Fluent was constructed as close as possible to the proposed theoretical model. The main stages for problem solving in ANSYS Fluent involved decomposition, set up the engine case and performed the moving simulation. Himanth and Jayashankar [5] proposed on the use of growing prismatic cells at the surface of the valve to solve the turbulent boundary layer. Meanwhile, the computational model surface was meshed with triangular element and the remaining decomposed region was meshed with tetrahedral cells. This research concluded that as the losses in flow increased along the valve lift, the separation of flow becomes critical. It was also defined that eventhough valve lift influenced the flow, there was exception for the upstream of port bend [5].

The complex-problem solving capability of the CFD simulation is emphasised more in the analysis of engine by Chaudhari *et.al* [6], Morauszki *et.al* [7] and Schmidt *et.al* [8]. Injection of fuel and combustion were reflected close to reality in the research where it is proven that CFD simulation is superior in engine analysis with respect to time and cost optimization. In the research done by Chaudhari *et.al* [6] , analysis on a 2-dimensional spark ignition engine with the validation of experiment and zero-dimensional simulation had proven that deviation of data is larger at high engine speed compared to low one. Meanwhile, Morauszki *et.al* [7] had used CFD to provide greater details on the visualization of flow in engine, mixing, injection and combustion occurred in gasoline engine aided with fuel direct injection. This research had introduced the idea of reducing mesh number while countering the loss of data by refining the wall properties and its solver equation. On the other hand, Schmidt *et.al* [8] had also investigated on the operation in gasoline direct injection (GDI) engine using CFD. This study had visualized the engine phenomenon which include injection and combustion in the engine, and finding had defined the design parameter that influence the performance of engine.

Eventhough the solving capability of CFD simulation is undeniably excellent, the full analysis of engine is always separated into different stages; port flow or experiment for defining initial breathing parameters values, and CFD moving simulation for the other related operations involved in analysis of flow and combustion. Based on the study of previous researches, there is none of the study shows the use of moving simulation in finding breathing parameter. The study of engine breathing is always either by experiment or CFD port flow simulation where both are under the assumption of steady flow. Meanwhile, dynamics simulation by CFD has always been used to investigate other parameter aside from engine suction capability which include engine combustion (heat release, peak temperature, pressure) and internal flow properties (swirl, tumble). This has caused the questions to surface on whether moving simulation can be used to predict the ability of engine suction and what is the accuracy of dynamics simulation over steady flow simulation for the engine breathing study.

Based on the study of previous researches, steady flow analysis by CFD simulation has shown good correlation with experimental data. However, the tremendous improvement of CFD has make the dynamic analysis on engine much easier, and this has picked the interest in determining whether engine breathing analysis in dynamic operation also provides good correlation with the experiment. Thus, this research aims to investigate the difference of flow coefficient and discharge coefficient obtained from dynamic analysis (cold flow) and steady flow analysis (port flow) with the experiment (flow bench test). Research objectives are mainly to investigate the ability of moving simulation in engine breathing analysis, and to determine the accuracy of dynamics simulation compared to the steady flow simulation in study of engine suction. Proton CAMPRO 1.6-litre engine was selected as the analysed engine and commercial CFD software, Fluent is used to simulate the steady flow and dynamics analyses.

2 Literature review

2.1 Governing equations

In order to represent the fluid flow using the CFD codes, the governing equations are required for defining the flow cases. For internal combustion engine, the governing equation for conservation of mass is defined as [5]:

$$\frac{\partial \rho}{\partial t}+\frac{\partial(\rho.u)}{\partial x}+\frac{\partial(\rho.v)}{\partial y}+\frac{\partial(\rho.w)}{\partial z}=0 \tag{1}$$

The governing equation for the conservation of momentum can be written as [5]:

$$\rho\frac{\partial u}{\partial t}+\rho u\frac{\partial u}{\partial x}+\rho v\frac{\partial u}{\partial y}+\rho w\frac{\partial u}{\partial z}=\rho g_x-\frac{\partial p}{\partial x}+\mu\frac{\partial^2 u}{\partial x^2}++\mu\frac{\partial^2 u}{\partial y^2}+\mu\frac{\partial^2 u}{\partial z^2} \tag{2}$$

$$\rho\frac{\partial v}{\partial t}+\rho u\frac{\partial v}{\partial x}+\rho v\frac{\partial v}{\partial y}+\rho w\frac{\partial v}{\partial z}=\rho g_y-\frac{\partial p}{\partial y}+\mu\frac{\partial^2 v}{\partial x^2}++\mu\frac{\partial^2 v}{\partial y^2}+\mu\frac{\partial^2 v}{\partial z^2} \tag{3}$$

$$\rho\frac{\partial w}{\partial t}+\rho u\frac{\partial w}{\partial x}+\rho v\frac{\partial w}{\partial y}+\rho w\frac{\partial w}{\partial z}=\rho g_z-\frac{\partial p}{\partial z}+\mu\frac{\partial^2 w}{\partial x^2}++\mu\frac{\partial^2 w}{\partial y^2}+\mu\frac{\partial^2 w}{\partial z^2} \tag{4}$$

Equation for conservation of energy is defined as [5]:

$$\frac{\partial}{\partial t}\left(\rho e+\frac{1}{2}\rho v^2\right)+\frac{\partial}{\partial x}\left(\rho ue+\frac{1}{2}\rho uv^2\right)+\frac{\partial}{\partial y}\left(\rho ve+\frac{1}{2}\rho vv^2\right)+\frac{\partial}{\partial z}\left(\rho we+\frac{1}{2}\rho wv^2\right)=$$

$$k\left(\frac{\partial^2 T}{\partial x^2}+\frac{\partial^2 T}{\partial y^2}+\frac{\partial^2 T}{\partial xz^2}\right)-\left(u\frac{\partial p}{\partial x}+v\frac{\partial p}{\partial y}+w\frac{\partial p}{\partial z}\right)+$$

$$\mu\left[u\frac{\partial^2 u}{\partial x^2}+\frac{\partial}{\partial x}\left(v\frac{\partial v}{\partial x}+w\frac{\partial w}{\partial x}\right)+v\frac{\partial^2 u}{\partial y^2}+\frac{\partial}{\partial y}\left(u\frac{\partial u}{\partial y}+w\frac{\partial w}{\partial y}\right)+w\frac{\partial^2 u}{\partial z^2}+\frac{\partial}{\partial z}\left(u\frac{\partial u}{\partial z}+v\frac{\partial v}{\partial z}\right)\right]$$

$$+2\mu\left[\frac{\partial^2 u}{\partial x^2}+\frac{\partial u}{\partial y}\frac{\partial v}{\partial x}+\frac{\partial^2 v}{\partial y^2}+\frac{\partial v}{\partial z}\frac{\partial w}{\partial y}+\frac{\partial^2 w}{\partial z^2}+\frac{\partial w}{\partial x}\frac{\partial u}{\partial z}\right]+\rho ug_x+\rho vg_y+\rho wg_z \qquad (5)$$

2.2 Flow coefficient and discharge coefficient

In defining the capabilities of engine suction, this analysis represents the breathing capabilities in term of flow coefficient and discharge coefficient. Based on study of previous researches, there are different definitions that have been used for flow coefficient and discharge coefficient. This study however rely on the definition indicted by Xu [9] as it has been used more often in the intake flow analysis by car manufacturer. Flow coefficient, C_F is defined as [9-13]:

$$C_F=\frac{\dot{m}}{\rho A_p V_o} \qquad (6)$$

where $\dot{m}$ is air mass flow rate, ρ is air density at operating condition, V_o is the velocity of air and A_p is the area of valve inner seat. Valve inner seat area can be obtained by the following formula:

$$A_p=\pi D_v^2/4 \qquad (7)$$

where D_v is the valve diameter. Closely related to the flow coefficient, discharge coefficient is defined as [9-13]:

$$C_D=\frac{\dot{m}}{\rho A_c V_o} \qquad (8)$$

where A_c is the valve curtain area/gap area. Curtain area can be calculated by the formula below:

$$A_C=\pi D_V L_V \qquad (9)$$

where L_v is valve lifts and D_v is diameter of valve. Air velocity can be obtained by the following equation [4, 9, 11, 13]:

$$V_O=\sqrt{2\Delta P/\rho} \qquad (10)$$

where ΔP is the pressure difference.

3 Research methodology

This research focuses on the comparison between steady flow and dynamic analyses of CFD simulation with flow bench test. In CFD simulation for engine, it is divided into four different simulations which are port flow, cold flow, in-cylinder combustion and full-engine full-cycle simulation. Each simulation type differs from each other based on the operations and also the output. For steady flow analysis, it was computed by means of port flow simulation while cold flow simulation was computed for dynamics analysis of breathing parameter.

3.1 Experimental method (flow bench test)

Methodology:

1. For the actual flow bench test, test orifice plate was removed from the flow bench and the engine cylinder head, cylinder adapter and valve lifter were installed.
2. The dial indicator was set to zero with the valves completely closed.
3. At the FlowCom panel, the test pressure value of 25"H_2O was inserted.
4. As an initial step, leakage test must be performed with all the valves closed. All the parts that may lead to leakage were sealed with the clay.
5. The motor was then turned on to define the leakage value. The leakage value was then either added into the data sheet separately or to the FlowCom.
6. The valve in the test head was then lifted to 1mm opening.
7. The motor was turned on until desired test pressure is reached and the FlowCom automatically displayed the measured flow (cfm).
8. Step 6-7 were repeated with the increment of 1mm until it reach 8mm (maximum opening for the selected CAMPRO engine) and the reading of flow for each valve lift was recorded. Figure 1 shows the setup of flow bench test conducted in the analysis of intake air suction.

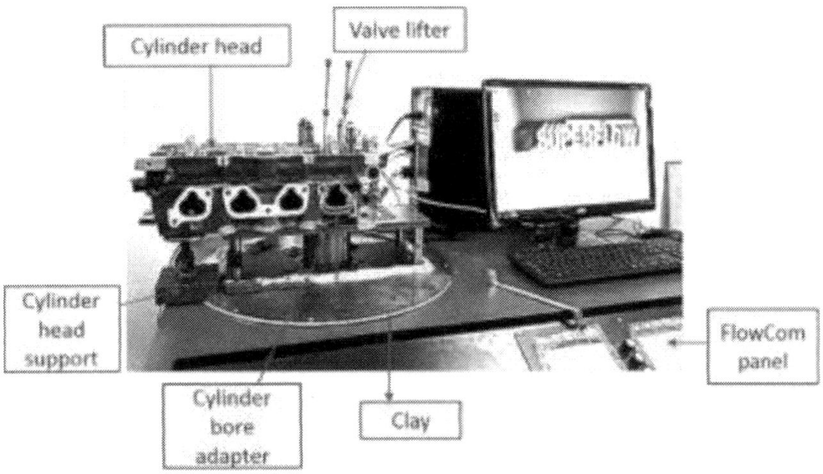

Fig. 1. Flow bench test.

3.2 Steady flow analysis (port flow simulation)

In the computation of the fluid flow in engine, one of the most important steps is the construction of the computational model. The computational model must be accurate in order to simulate the real operations of engine. Engine parameters for CAMPRO 1.6-litre engine is shown in Table 1 and the computational model is shown in Figure 2.

Table 1. Engine parameters [14].

Engine Parameters	Value
Bore (mm)	76
Stroke (mm)	88
Crank radius (mm)	44
Valve diameter (mm)	29
Number of cylinder	4
Connecting rod length (mm)	131
Total displacement (cc)	1,597
Compression ratio	10:1
Intake valve open (cad)	348.2
Intake valve close (cad)	555.3
Exhaust valve open (cad)	149.3
Exhaust valve close (cad)	372.2

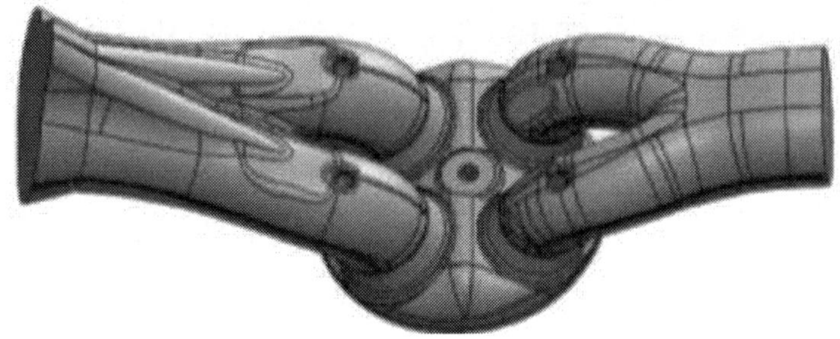

Fig. 2. Computational model of engine.

For port flow simulation, the original computational model was extended into "inplenum" and "outplenum" as shown in Figure 3. The inclusion of "inplenum" assisted the suction of air into the engine in the simulation. The box shape of "inplenum" was selected among other shapes which include spherical and hemisphere as the flow output for the CAMPRO engine is more accurate for the box "inplenum". The dimension for the box "inplenum" is 152.14mm (length) x 152.14mm (width) x 152.14mm (height) and the bending radius between "inplenum" and the intake port is 10 mm. As the simulation only involved analysis on intake breathing capacity, the exhaust port and exhaust valve were detached from the computational model.

DOI: 10.1051/matecconf/20179001067

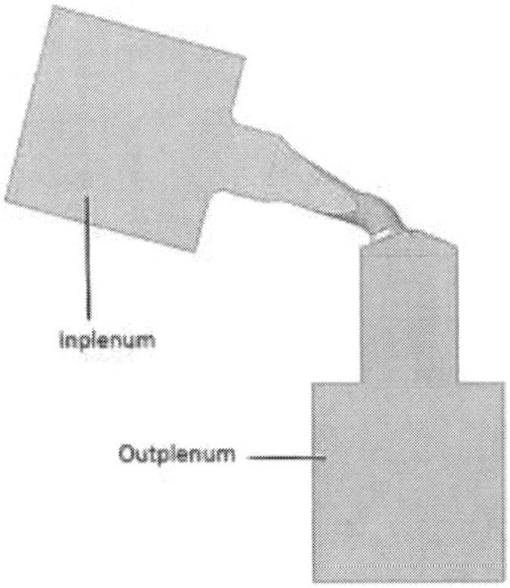

Fig. 3. Computational model for port flow.

The computational model was then decomposed mainly into "inplenum", port, valve, combustion chamber, and the "outplenum". At the intake port, three inflation layers were set for the intake wall and the rest of model parts were decomposed with tetrahedral cell. In port flow simulation, the model remains static thus the decomposition and meshing are much simpler than simulation involving dynamics process. The mesh quality was particularly refined near the valves in order to capture the properties of induced flow. In the setup of the engine case, the inlet pressure at "inplenum" was set to 0 and the outlet pressure at "outplenum" was set to -6225kPa.

3.3 Dynamic analysis (cold flow simulation)

Cold flow simulation is able to predict various phenomenons in engine without the reaction and combustion, for example the formation of swirl and tumble, wall impingement and turbulence under valves movement. For the analysis of only engine breathing parameters, cold flow is prior to in-cylinder combustion and full-engine full-cycle simulation which are much complex and time-consuming. Cold flow analysis is also adequate for this study because the simulation is only along the intake stroke which does not involve other critical process for example combustion and spark ignition. For cold flow simulation, as this analysis involves the transient process of the engine, the computational model is decomposed in different manner from the model in port flow simulation. Model was not extended into "inplenum" and "outplenum", and the original model was decomposed into different regions as shown in Figure 4.

Decomposition and meshing in cold flow analysis determine which parts remains static or moving during the simulation. As the piston and valves move along the engine stroke, another layers consist of hexahedral cells were added for piston and valve. Meanwhile, other regions of the computational model were meshed with tetrahedral cells.

 DOI: 10.1051/matecconf/20179001067

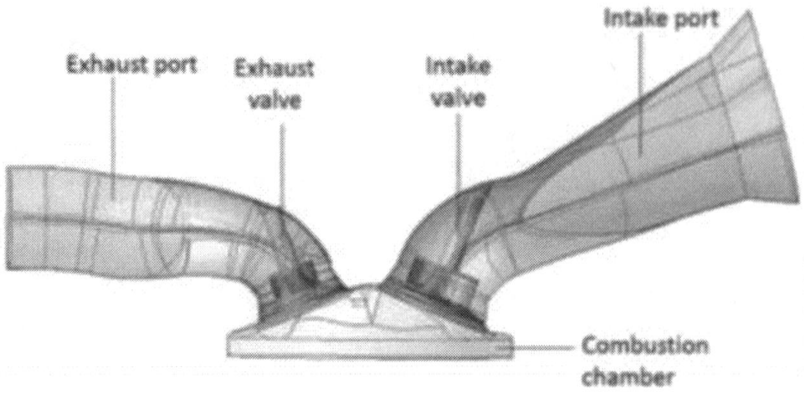

Fig. 4. Engine decomposition.

4 Results and discussion

In the conduction of flow bench testing, this experiment was repeated for a number of times in order to investigate the accuracy and the precision of the data obtained. Table 2 shows the volume flow rate (cfm) obtained for each valve lifts. From the data tabulated, it can be observed that flow data measured for each valve lifts are very close to each other. There is only a very small deviation of data for low valve opening (1mm) as the flow is more restricted at low valve lift [5].

Table 2. Flow (cfm) obtained from flow bench test.

Flow bench test	Valve lift (mm)							
	1	2	3	4	5	6	7	8
	Volume flow rate (cfm)							
1	12.3	26.6	39.7	52.6	65.4	75.3	82.3	83.4
2	12.2	26.6	39.7	52.6	65.4	75.3	82.3	83.4
3	12.4	26.6	39.7	52.6	65.4	75.3	82.3	83.4
Average	12.3	26.6	39.7	52.6	65.4	75.3	82.3	83.4

From the flow bench test, the flow (cfm) obtained were tabulated into the data sheet for the calculation of discharge coefficient and flow coefficient. Table 3 shows the calculation of both parameters with reference to other listed important parameters.

Flow coefficient obtained from flow bench test, cold flow and port flow simulation is shown in Figure 5 below. From 1mm valve lift until 8mm valve lift, flow coefficient increases as the valve lift increases. This is because the valve lift is critical in affecting the flow downstream as well as determining the separation of flow [5]. Based on previous research, the breathing parameter from flow bench test and port flow simulation are always in good correlation [5, 15]. This study has defined that for all of the methods; flow bench test, port flow and cold flow, the trend of the flow coefficient is the same. Comparing port flow data with flow bench test data, the maximum difference is as low as 5.1% which is very accurate. However, comparing cold flow data with experimental data, the deviation is quite large which is the maximum of 27% and the minimum of 15.6 %. This differences is expected as port flow closely resembled the flow bench in term of setup and boundary condition where engine is hold static at each critical point [3, 5, 12]. Meanwhile, cold flow simulation deals with the tremendous change of parameters during transition from one critical point to another, thus causing more losses in induced flow for moving simulation [5, 6].

Table 3. Calculation of flow coefficient and discharge coefficient.

		Valve lift (mm)							
Parameter	Unit	1	2	3	4	5	6	7	8
Test Pressure	"H2 O	25	25	25	25	25	25	25	25
Volume Flow Rate	cfm	12.3	26.6	39.7	52.6	65.4	75.3	82.3	83.4
Valve Curtain Area	m^2	0.000 18	0.000 36	0.000 55	0.000 73	0.000 91	0.001 1	0.001 3	0.001 5
Valve Seat Area	m^2	0.001 3	0.001 3	0.001 3	0.001 3	0.001 3	0.001 3	0.001 3	0.001 3
Flow Velocity	m/s	102.8 5	102.8 5	102.8 5	102.8 5	102.8 5	102.8 5	102.8 5	102.8 5
Air density	kg/ m^3	1.18	1.18	1.18	1.18	1.18	1.18	1.18	1.18
L/D		0.034	0.069	0.10	0.14	0.17	0.21	0.24	0.28
Discharge Coefficien t		0.310	0.335	0.334	0.332	0.330	0.317	0.297	0.263
Flow Coefficien t		0.042 8	0.092 5	0.138	0.183	0.227	0.262	0.286	0.29

Even though the deviation is big, the tabulated trend shows that cold flow simulation can be used to investigate the trend and predict the lower limit for the flow coefficient. After doing repeated data collection and analysis, it is defined that the data is more precise for lower valve lift which is around 1-4mm. Beyond 4 mm, repeated cold flow simulation shows large differences.

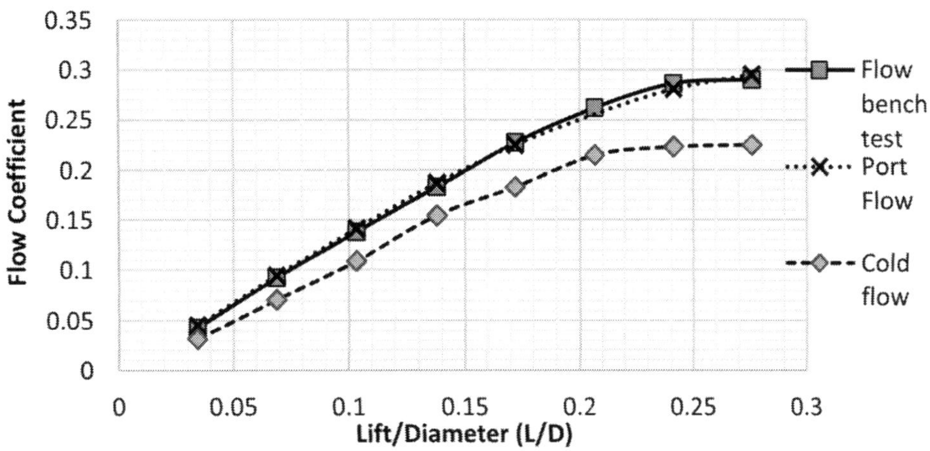

Fig. 5. Flow coefficient versus lift/ diameter (L/D).

The study on comparison between cold flow (dynamic analysis) and port flow (steady flow analysis) with flow bench test is further discussed in term of discharge coefficient as shown in Figure 6.

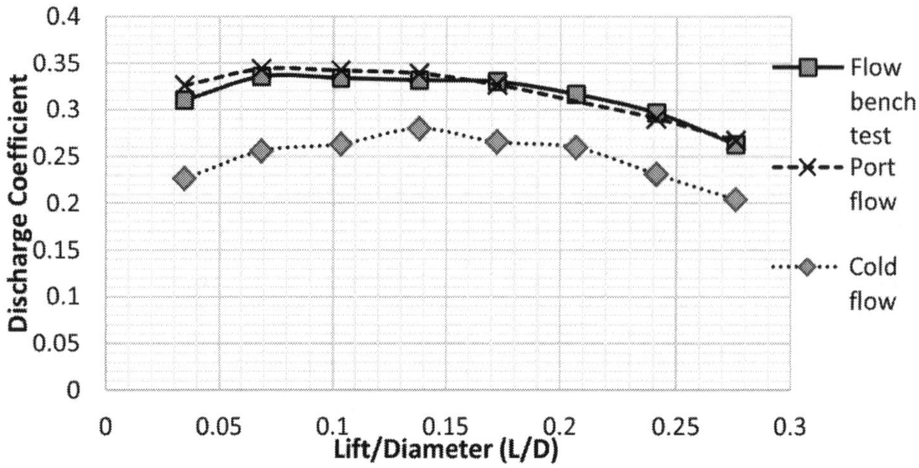

Fig. 6. Discharge coefficient versus lift/ diameter (L/D).

Similar to flow coefficient, the trend for discharge coefficient obtained from flow bench test, port flow and cold flow are all the same. Port flow simulation data is recorded with minimum and maximum deviation as small as 1.6% and 5.1% respectively which show that port flow simulation can accurately predict the breathing parameter of engine. For cold flow simulation, the maximum difference defined is 27% and minimum deviation is determined as 15.6%. Eventhough the large deviation causes the difficulties to accurately predict the real breathing parameter value, the similarities in trend shows that cold flow can be used to predict the trend of discharge coefficient and the lower limit of this breathing parameter. As discharge coefficient is closely related to flow coefficient and only differs in term of referred valve area, it is also deduced that the differences of data obtained is because there is more losses in the valve movement from lower valve lift to high valve lift in cold flow moving simulation [5, 6] compared to port flow engine simulation where losses in volume flow rate is measured at static critical point [3, 5, 12]. In a similar manner to flow coefficient data, repeated simulation shows that discharge coefficient is more precise at 1- 4mm valve lift, thus cold flow simulation is suitable for the prediction of trend and lower limit at lower valve lift value.

Both flow and discharge coefficients discuss on the mass flow rate that pass through the valve. These parameters only differ by reference area, where flow coefficient refers to inner seat area of valve, while discharge coefficient is dependent on gap of valve lip and valve seat. Results in Figure 5 and Figure 6 show that flow coefficient increase along valve lift, while discharge coefficient decreases along the opening of intake valve. This is because as the valve opens, greater flow is induced into the system. As valve inner seat area remains the same along valve lift, the flow coefficient remain increased [3, 5, 9]. Meanwhile, along valve lift, the valve curtain area also increase, thus causes the reduction of discharge coefficient value [5, 9, 12].

5 Conclusions

Based on the obtained result, it can be concluded that as valve lift increases, flow coefficient increases while the discharge coefficient decreases. As discharge coefficient refers on the valve curtain area, discharge coefficient should be used to perform analysis at low valve lift value while flow coefficient should be used for analysis at high valve lift. The port flow is proven to give good correlation with the experimental data of flow bench test. Meanwhile, cold flow simulation prediction on breathing parameter is not so close to the experiment and port flow simulation where even the smallest deviation is quite large, thus does not allow accurate prediction on the values of flow coefficient and discharge coefficient. Eventhough cold flow simulation give high deviation with the experiment, this research has defined that cold flow analysis has come out with similar trend of breathing parameter compared to port flow and flow bench test. This trend predicted by cold flow simulation is also more similar at lower valve lifts. Thus, it can be concluded that cold flow analysis can still be used to predict the trend and the lower limit of flow coefficient and discharge coefficient value.

This work acknowledges the financial support from Universiti Teknologi Malaysia (UTM) and Ministry of Higher Education Malaysia (MOHE) under the FRGS grant R.J130000.7809.4F174. Appreciation also goes to Automotive Laboratory, Universiti Teknikal Malaysia Melaka (UTeM) for assisting in the flow bench test.

References

1. J.B. Heywood, *Internal Combustion Engine Fundamentals*, (United States of America: McGraw-Hill, Inc., 1988)
2. A.R. Ismail and R. Abu Bakar, Amer. J. Appl. Sc., **5**, 963 (2008)
3. P. Balasankar and A. Gokhale, *HyperWork Technology Conference,* 1-5 (2012)
4. C.R. Kumar and G. Nagarajan, ARPN J. Eng. Appl. Sc, **7**, 180-186 (2012)
5. Himanth Kumar H.Y. and Jayashankar N., IJIERT, **2**, 1-9 (2015)
6. A.J. Chaudhari, N. Sahoo, and V.Kulkarni, *4th ICAER 2013,* 330-341, (Elsevier Ltd. 2014)
7. T. Morauszki, P. Mandli, Z. Horvath, and M.R Dreyer, Hung. J.Ind.Chem. **39,** 27-30 (2011)
8. S. Schmidt, M. Abart, O.Schogl, D. Jajcevid, A. Trattner, R.Kirchberger, and H.Eichlseder, *SETC2009,* 2009-32-0045 / 20097045 (2009)
9. H. Xu, *SAE Technical Paper Series* 2001-01-1308 (2001)
10. H. Bettes, *Engine Airflow: A Practical Guide to Airflow Theory, Parts Testing, Flow Bench Testing, and Analysing Data to Increase Performance for Any Street or Racing Engine,* (USA: HPBooks, 2010)
11. A. Mohammadebrahim, B. Shafiei, and S. K. Hannani, *The J.Engine.Research*, **26**, 11-19 (2012)
12. B. Soriano, and C. Rech, *Int. J. Mech. Eng. Autom,* **1**, 236-241(2014)
13. A. Algieri, Int.J.Eng.Tech, **2**, 140-148 (2013)
14. M. F. Muhamad Said, A. Abdul Aziz, Z. Abdul Latiff, and A.M. Andwari, *2014 SAE International.* 2014-01-2549 (2014)
15. H.Cui, M. Li, and Y. Han, *2010 WASE International Conference on Information Engineering*,255-259 (2010)

The effect of orifice diameter in constant volume chamber to spray characteristics in diesel engine using computational fluid dynamic

A. R. Andsaler[1,*], *A.* Khalid[1,2], *N. S. A.* Abdullah[1,2], *N.* Jaat[1,2]

[1]Centre of Energy and Industrial Environment, Universiti Tun Hussein Onn Malaysia, 86400 Parit Raja, Batu Pahat, Johor, Malaysia.
[2]Faculty of Mechanical Engineering, Universiti Tun Hseein Onn Malaysia, 86400 Parit Raja, Batu Pahat, Johor, Malaysia.

Abstract. Mixture formation of the ignition process is a crucial part in diesel combustion as it affects the combustion process and exhaust emission. The main focus of this study is to investigate the effects of spray characteristic to the fuel spray in a chamber using Computational Fluid Dynamics (CFD) simulation. However, simulations are done with the discrete phase of injection taken from the outlet of the nozzle hole to spray chamber. The details behavior of spray penetration and spray breakup length were visualized using Computational Fluid Dynamics. This simulation was done in different nozzle diameters 0.12 mm and 0.2 mm performed at a single ambient pressure of 4 MPa, the ambient temperature of 500 K and 700 K with different injection pressure 40 MPa, 70 MPa and 140 MPa. Results showed that the droplet diameter is influenced by high pressure to become smaller, and the penetration length also longer with high injection pressure applied. Besides, smaller orifice diameter gives a shorter length of the breakup. It is necessary to put afford on orifice diameter and ambient temperature conditions in order to improve the formation of spray.

1 Introduction

Internal combustion engine is the heat engines that convert properties of fuel into a mechanical system, and this application is on the rotating output shaft. The fuel injection equipment is one of the sub-systems that have been found to lend itself to development and leads to the improvement in engine performance and emission quality [1]. Fuel injected into the chamber is important to the diesel engine as it influences the performance and emission. Break-up and distribution of the spray are largely determined by the in-cylinder conditions, the injection pressure, and orifice geometry [1-2]. Design of the orifice for fuel injector is important on the performance and emission of diesel engine. All orifices must produce a

* Corresponding author: adiba_rhaodah@yahoo.com

fuel spray that encounters the necessary objectives of performance and emissions into the market for which the engine delivered regardless details of the fuel system design. This simulation study used computational fluid dynamics (CFD); ANSYS version 16.1 Fluent software to study the spray characteristics.

In the most basic sense, a spray is simply the introduction of liquid into a gaseous environment through an orifice such that the liquid, through its interaction with the surrounding gases and by its own unsteadiness, breaks-up into droplets [3]. The formation of a spray begins with the separation of droplets from the outer surface of a continuous liquid core extending from the orifice of injector. Spray characteristic is known to affect significantly the combustion and emission processes in diesel engines. By optimizing spray characteristics, the raw emissions from the diesel engine which are mainly NOx and PM can be minimized [2]. Therefore, investigations into Diesel spray characteristics had concentrated on the effect of spray characteristic on engine performance such as the spray tip penetration, break-up length and droplet size and velocity distributions [3].

More recent investigations by Hiroyasu and Arai [4], Naber and Siebers [5] had shown a strong dependency of spray penetration upon both in-cylinder pressure and fuel injection pressure. The injection pressure has significant effect on spray liquid penetration [5]. The effect of increasing pressure on spray tip penetration is depicted. The spray tip penetration gets longer as the injection pressure increases. This result is related to both higher quantity and higher velocity of the droplets at higher injection pressures [6]. When the spray lost its momentum related to the lower quantity or lower injection pressure, but difference for the penetration lengths which showed longer at the downstream region of spray. Proportional to injection pressure, the spray penetrates faster at higher injection pressures [7].

The break-up length characterizes a point of discontinuity, where the spray changes from a zone of liquid (bulk liquid, or interconnected ligaments and droplets), to a finely atomized regime of droplets [8]. After the disintegration of liquid column emerging from a nozzle, the generated droplets may further break-up into smaller ones as they moved into the surrounding gas [9]. The relative velocity between droplets and gas, and surrounding non-uniform pressure develops, causing the particles to deform. Development of this deformation leads to break-up into smaller droplets. The forces associated with dynamic pressure, surface tension and viscosity control the break-up of a drop [10].

In this research, the characteristics of spray diesel are investigated focusing on manipulate the ambient temperature and injection pressure with different orifice diameter. This study conducted by using Computational Fluid Dynamic simulation which can captured the spray penetration and breakup length.

2 Simulation setup

The modeling flow chart shows in Figure 1 about the flow of simulation. As shown in Figure 1, the first stage is pre-processing which consists of the flow problem in the simulation. Solver stage is complete the calculation of finite element model or solving of governing equation. Post-processing is a process of involving data collect and visualization.

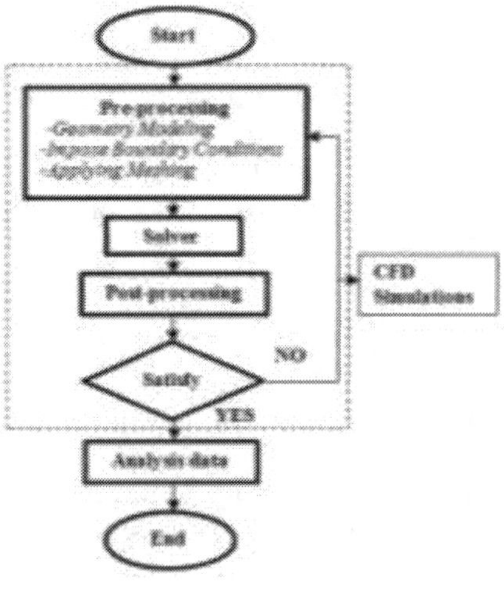

Fig. 1. Methodology and modeling flow chart.

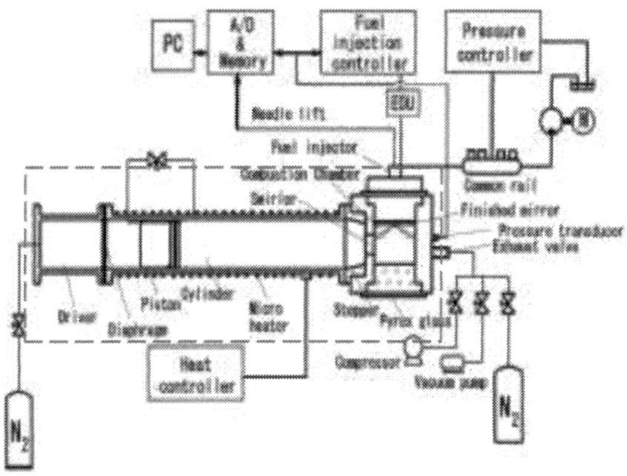

Fig. 2. Experimental setup of RCM.

A rapid compression machine (RCM) was used to generate the actual diesel combustion over a wide range of temperature, pressure and swirl velocities. Figure 2 shows a schematic diagram of RCM, together with an outline of the fuel injection system. Figure 3 and 4 shows the combustion chamber cross sectional drawing with the 3D. The RCM has a disc type combustion chamber with a diameter of 60 mm and width of 20 mm. The design was focused on injector nozzle and spray chamber. The fuel injector has six orifice holes and this design has been saved in IGES format and imported as geometry modeling in ANSYS Fluent 16.1.

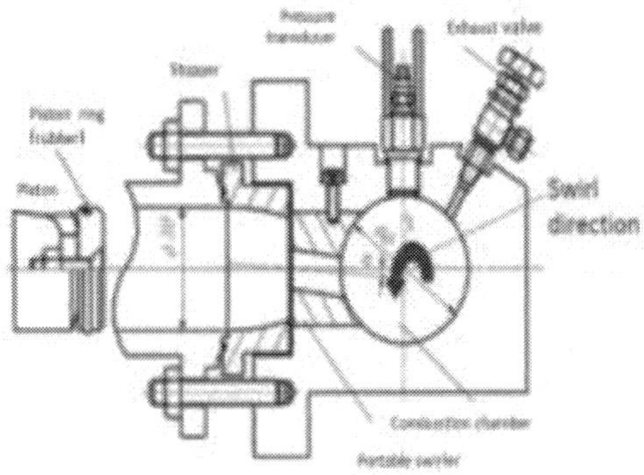

Fig 3. Combustion chamber.

Fig. 4. 3D section view of combustion chamber.

The geometry of injector was modeled by Solid Work software. In this study the combustion chamber is a space where the spray formation occurs. Moreover, model of the injector only considered the tip of nozzle. Figure 5 shows the 3D model of cross sectional area of the injector tip. Figure 6 shows the sections geometry of the combustion chamber. It is 1/6 section from the overall geometry because there are 6 orifices in the actual injector and only one orifice is adequate and considered sufficient for the simulation analysis.

The assigned boundary condition was served as the initiation and direct to motion of the flow. In addition, this simulation carried out with the multiphase model volume of fluid (VOF) and turbulence model of k-ε realizable in transient mode. Besides, a constant mass flow rate of air and diesel are applied as shown in Table 1 which indicates as solver setup of the simulation. Moreover, parameters to investigate are the droplet diameter of particle, penetration length and breakup length.

DOI: 10.1051/matecconf/20179001068

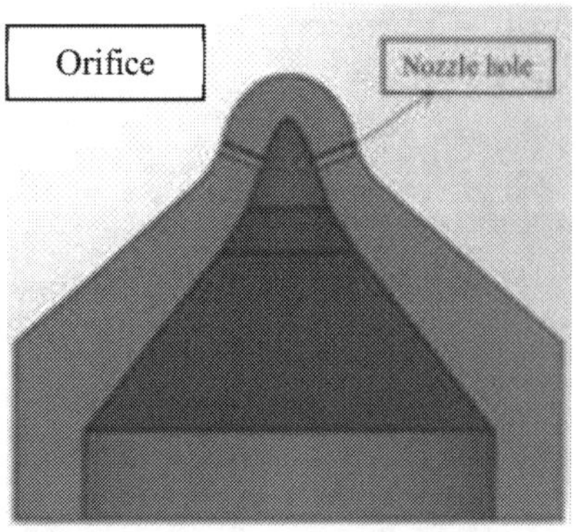

Fig. 5. Cross sectional model of spray injector.

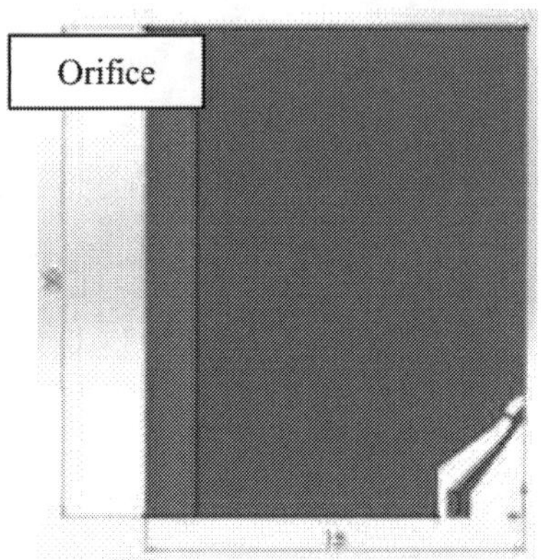

Fig. 6. Hollow model for the combustion chamber.

Table 1. Solver setup.

Options	Type
Multiphase Model	Volume of Fluid
Time	Transient
Model	Viscous K-Epsilon Model
K-Epsilon Model	Realizable
Material	Diesel Liquid
Boundary Condition	Injection Pressure = 40 MPa, 70 MPa, 140 MPa
	Temperature Inlet = 313 Kelvin
	Pressure Outlet = 4 MPa
	Temperature Outlet = 500 K, 700 K

3 Results and discussion

This section will explain the result obtained from the simulation. The consideration of this simulation is to make the predictions of spray characteristics located at downstream of injector by using CFD software. This simulation is done by three dimensional modelling in transient state with the multiphase model volume of fluid (VOF). Euler-Euler multiphase approach is applied in this study as it has two phases of materials; air and diesel fuel perform mixing inside the injector and chamber. Two different orifice diameters, 0.20 mm and 0.12 mm are investigated in this study. Meanwhile, this simulation also used several different ambient temperatures such as 500 K and 700 K for every injection pressures are 40 MPa, 70 MPa and 140 MPa. Additionally, 4 MPa is used as the boundary outlet for the ambient pressure in chamber.

Grid sensitivity was tested on the model with different grid sizes. The grid size divided into three levels which are coarse, medium and fine grids. Table 2 indicates the number of elements and nodes for each group of different grid sizes. Besides, the number of elements is ranging from 25888 to 118777 elements. The number of elements is obviously directly proportional to the numbers of nodes by ascending the level of grid sizes from coarse to fine with also level of smoothing from low to high.

Table 2. Grid sizes with number of elements and nodes.

Grid Sizes	Number of Elements	Number of Nodes
Coarse – Low	25888	5182
Coarse - Medium	25979	5234
Coarse – High	26015	5374
Medium – Low	47665	9683
Medium - Medium	47699	9762
Medium – High	48547	9896
Fine - Low	114562	22289
Fine- Medium	115998	22487
Fine - High	118777	22497

3.1 Effect of the ambient temperature, injection pressure and orifice diameter

Ambient temperature is influenced to the spray development which has been investigated throughout this simulation with the parameters of spray droplet diameter, penetration length and breakup length. Orifice diameter is known as one of factor that could influences to the flow inside orifice, while it also can affects to the spray development.

As illustrates in Figure 7, variant parameters influenced the droplet diameter from the qualitative development. At 40 MPa of injection, orifice diameter of 0.12mm with 700 K of ambient temperature shows more droplets with red in colour which represented as bigger droplet size compared to the 0.12 mm at 500 K. These results can be related to the prediction of droplet diameter could be influences by various parameters especially higher ambient temperature and injection pressure is predicted to generate smaller droplet diameter. However, smaller orifice diameter can also produce finer droplet diameter from the spray. Other than that, droplet diameter also depends on the aerodynamic drag in the chamber which can affects the smaller droplet diameter generates when the air collision with fuel. Nevertheless, little fuel injected is assumed to have lower combustion in the chamber as referred from the results prediction.

3.1.1 Penetration length

Ambient temperature has a strong effect on length of penetration, while injection pressure is predicted to have only slight effect on penetration length. However, the effects of most parameters and the mechanisms by which any of the parameters controlling penetration length are not completely understood. Apart from that, Figure 8 shows the visual penetration flow by single orifice simulation. The penetration length is considered from the outer of orifice until the tip of spray. The graphs plotted of penetration lengths against the time with two ambient temperatures of 500 K and 700 K for two different orifice diameters, 0.12 mm and 0.2 mm as shows in the Figure 9 and 10. The graphs showed penetration lengths for both ambient temperatures and also for both 0.12 mm and 0.2 mm orifice diameters increase constantly with the time taken. Besides, the penetration lengths for both

 DOI: 10.1051/matecconf/20179001068

ambient temperatures were not given much difference. This situation can be assumed as due to not much difference between the ranges of temperature.

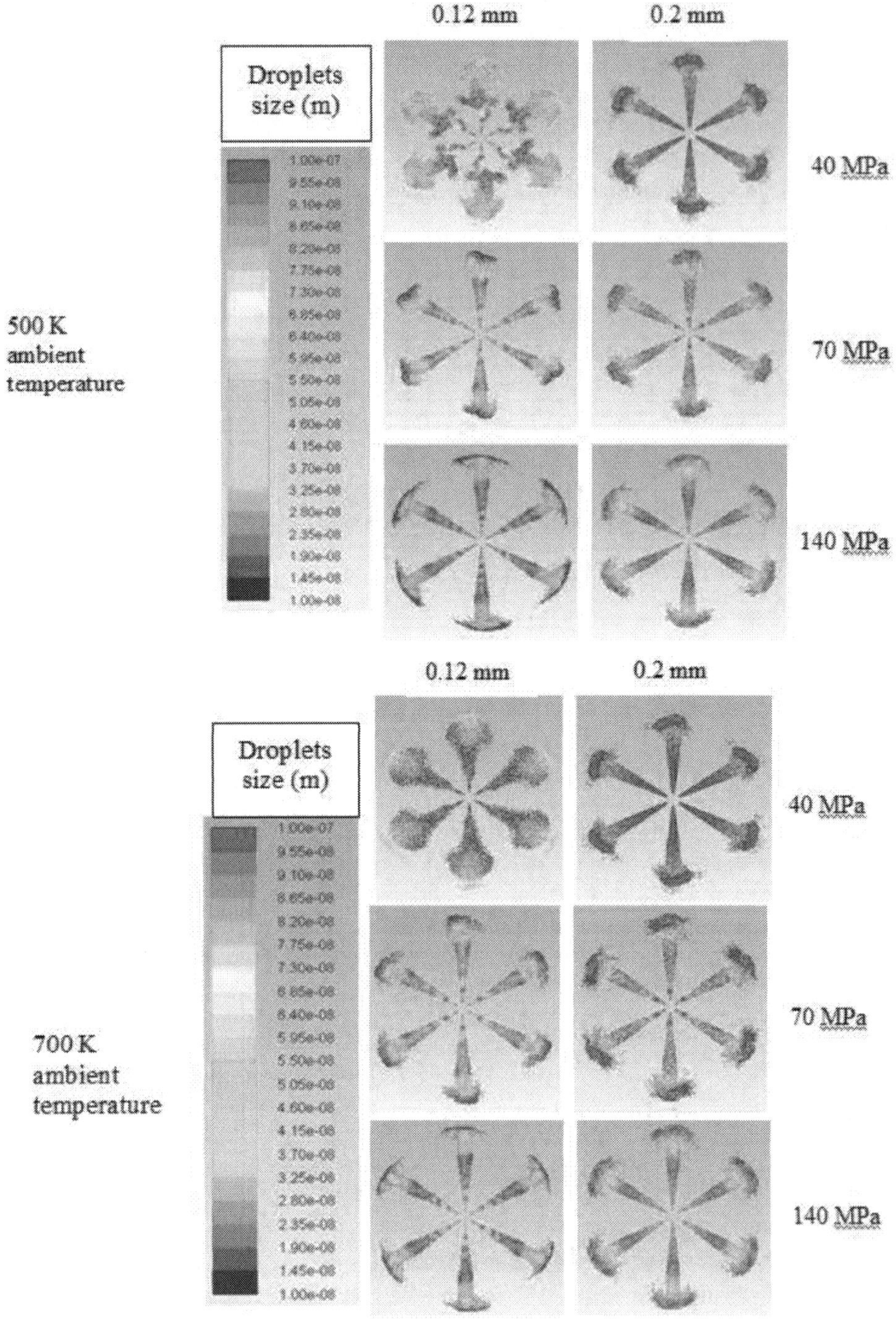

Fig. 7. Spray structure of droplet diameter produced by six nozzles, with colour scale represents the droplet diameter.

However, the injection pressure is also influenced the spray penetration length. The variant injection pressures applied which are 40 MPa, 70 MPa and 140 MPa with ambient pressure 4MPa in the chamber. As the higher injection pressure, the longer length of penetration due to the spray was resisted with the ambient pressure. Meanwhile, penetration length for the 0.12 mm orifice diameter performed a longer penetration compare to the 0.2 mm orifice diameter. This result can be explained by the bigger orifice diameter produces larger droplets and spray penetration travel become slower. Nonetheless, the spray will still hit the wall of chamber when it is at the end of injection.

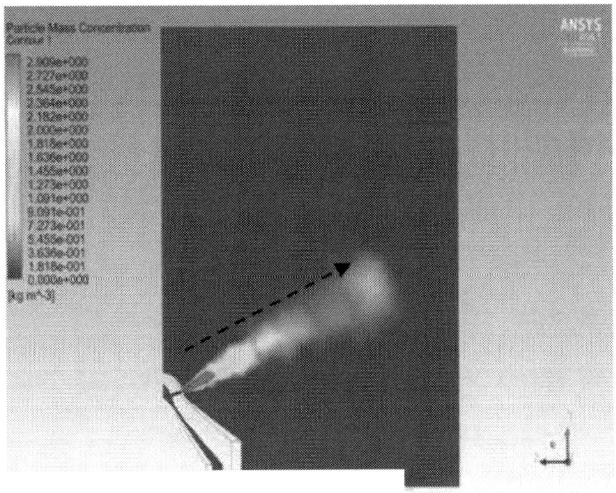

Fig. 8. Visual of penetration flow in simulation.

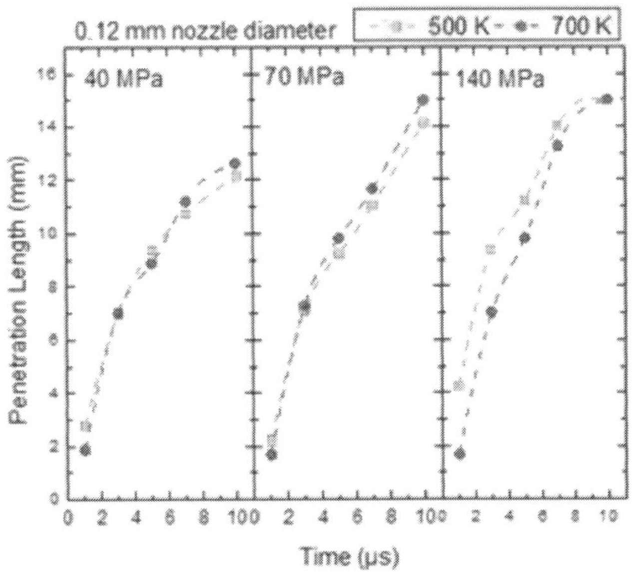

Fig. 9. Graph penetration length with orifice diameter of 0.12 mm.

 DOI: 10.1051/matecconf/20179001068

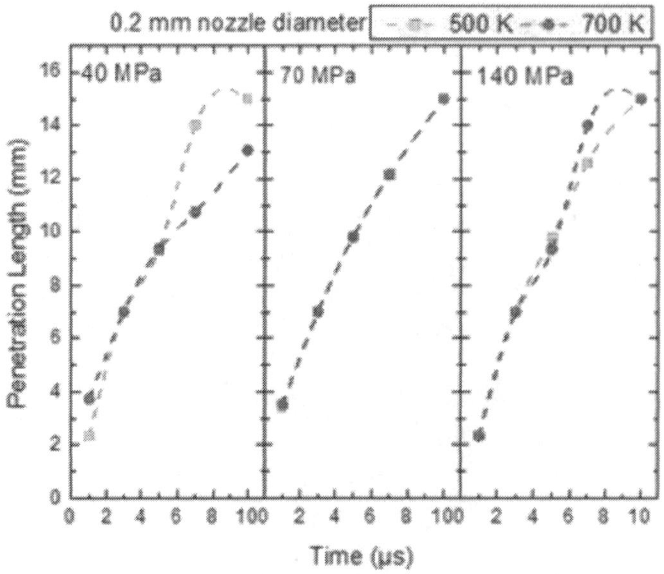

Fig. 10. Graph penetration length with orifice diameter of 0.2 mm.

3.1.2 Breakup length

The breakup length decreased with an increase of injection pressure. The decreased in breakup length could reduce the size of combustion in chamber. Moreover, Figure 11 shows the visualisation of breakup length from a single orifice simulation. The breakup lengths are measured by considering the tip of the spray diesel and the red contour as circuited in Figure 11. Besides, Figure 12 and 13 shows the graphs of breakup length based on the different orifice diameters.

The result of temperature 700 K shows slightly fluctuation, where injection pressure 140 MPa with orifice diameter of 0.12 mm implemented dramatically increases from 0.5 mm stood up to the 1.6 mm. On the contrary, the result of breakup length with temperature 500 K shows almost constant contribution. This phenomena is possibly due to the droplets did not breakup completely and they travelled further while caused the breakup length longer, which also called as the section of ligament. In addition, the breakup length for 140 MPa of injection pressure in orifice diameter of 0.2 mm shows flow steady with the time for both temperature 500 K and 700 K. Thus, it can be assumed that the breakup length could be influenced by ambient temperature used. Other than that the shorter breakup length happened at the core segment. The high injection pressure influenced to the breakup length, where the higher injection pressures caused shorter core of the breakup length which has been identified as shown in the Figure 12 and Figure 13. Nevertheless, diameter of orifice also plays an important role for the spray characteristic to being identified for the better combustion.

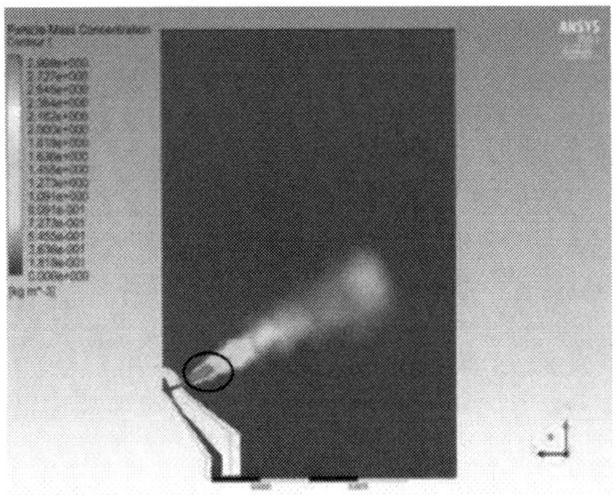

Fig. 11. Visualization of breakup length in simulation.

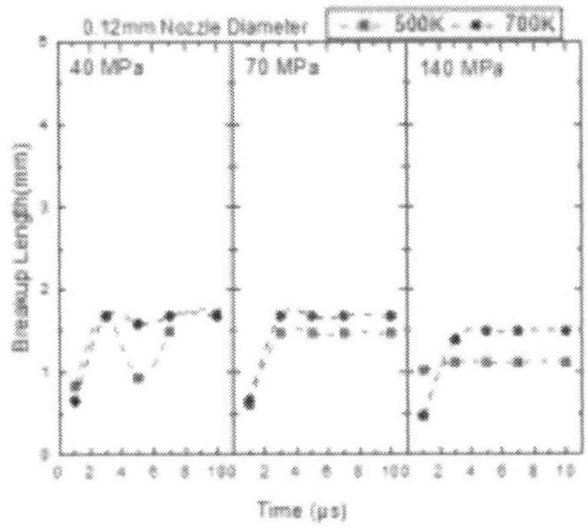

Fig. 12. Graph of breakup length with orifice diameter of 0.12 mm.

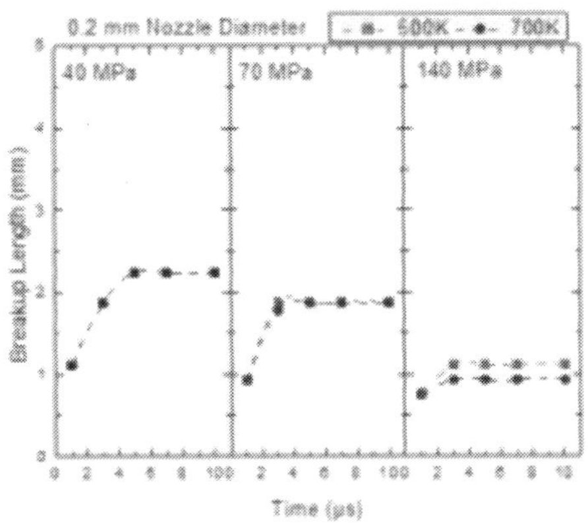

Fig. 13. Graph of breakup length with orifice diameter of 0.2 mm.

4 Conclusions

This study has shown a simulation of fuel-air flow through the orifice of injector to simulate a spray before the step of combustion. This study also found that the injection

 DOI: 10.1051/matecconf/20179001068

pressure and ambient temperature used influenced the formation of a spray in the chamber. This simulation was conducted with two different orifice diameters which are 0.12 mm and 0.2 mm. Ambient temperatures 500 K and 700 K are used in this study, while the ambient pressure applied in the chamber is 4 MPa. The results showed that the breakdown of fuel spray is been disturbed at low injection pressures in the chamber. Besides, very fine droplets are needed to start ignition at the fuel injection section. Moreover, temperature changed in the chamber also affected the spray penetration length. At the temperature of 700 K penetration length is predicted shorter than 500 K. For high ambient temperatures, fuel is assumed to evaporate more easily and faster ignition the fuel itself. In this simulation, the diameter of orifice is also predicted to give significant effect to the breakup length of the spray. The breakup length generated by 0.2 mm orifice diameter is longer than the 0.12 mm. Other than that, higher injection pressure also predicted to influence the breakup length of the spray, where the breakup length is shorted when high injection pressure applied. Last but not least, this simulation can be predicted that the injection pressure, ambient temperature and orifice diameter could affect the droplets diameter, penetration length and spray breakup length in the chamber.

The authors also would like to thank the Ministry of Higher Education, Malaysia for supporting this research under Fundamental Research Grant Scheme (FRGS) vot.1466 and partly supported by the Centre of Graduate Studies, UTHM.

References

1. D.A. Kennaird, C. Crua, J. Lacoste and M.R. Heikal, "In-Cylinder Penetration and Break-Up of Diesel Sprays Using a Common-Rail Injection System," *SAE Technical Paper,* 2002-01-1626, (2002).
2. Bergstrand, P., and Denbratt, I., "Diesel Combustion with Reduced Nozzle Diameter", *SAE paper,* 2001-01-2010, (2001).
3. Adam, A., Inukai, N., Kidoguchi, Y., Miwa, K. Et Al., 2007, "A Study on Droplets Evaporation at Diesel Spray Boundary during Ignition Delay Period," *SAE Technical Paper* 2007-01-1893, (2007).
4. Hiroyasu, H., and Arai, M. "Structure of Fuel Sprays in Diesel Engines", *Transactions of the SAE,* Vol **99**, Sect 3, pp 1050-1061, (1990).
5. Naber, J., and Siebers, D.L., "Effects of Gas Density and Vaporization on Penetration and Dispersion of Diesel Sprays". *SAE* 960034, (1996).
6. Khalid, A., Anuar, M.D., Ishak, Y., Manshoor, B., Sapit, A., Leman, M., Zaman, I., 2014" Emissions characteristics of small diesel engine fuelled by waste cooking oil", *MATEC Web of Conferences,* Volume **13**, Article number 06006,DOI:10.1051/matecconf/20141306006, (2006).
7. Khalid, A., Tamaldin, N., Jaat, M., Ali, M.F.M., Manshoor, B., Zaman, I.,"Impacts of Biodiesel Storage Duration on Fuel Properties and Emissions", *Procedia Engineering,* Volume **68**, Pages 225-230, Elsevier Ltd, DOI: 10.1016/j.proeng.2013.12.172, (2013).
8. Khalid, A. , Jaat, N., Sapit, A., Razali, A., Manshoor, B., Zaman, I., Abdullah, A.A., "Performance and emissions characteristics of crude jatropha oil biodiesel blends in a diesel engine" *International Journal of Automotive and Mechanical Engineering,* Volume **11**, Issue 1, Pages 2447-2457, DOI: 10.15282/ijame.11.2015.25.0206, (2015).
9. Khalid, A., Hayashi, K., Kidoguchi, Y., Yatsufusa, T., "Effect of air entrainment and oxygen concentration on endothermic and heat recovery process of diesel ignition", *SAE Technical Papers,* Society of Automotive Engineers of Japan, Inc. and SAE International, DOI: 10.4271/2011-01-1834, (2011).

10. Khalid, A., 2013, "Effect of Ambient Temperature and Oxygen Concentration on Ignition and Combustion Process of Diesel Spray", *Asian Journal of Scientific Research*, Volume **6**, Issue 3, Pages 434-444, Asian Network for Scientific Information, DOI: 10.3923/ajsr.2013.434-444, (2013).
11. Camille Hespel, Jean-Bernard Blaisot, Xandra Margot, S. Patouna, Armelle Cessou, Bertrand Lecordier, "Influence of Nozzle Geometry on Spray Shape, Particle Size, Spray Velocity and Air Entrainment of High Pressure Diesel Spray", *THIESEL*, (2010).
12. I. V. Roisman, L. Araneo, and C. Tropea,"Effect of Ambient Pressure on Penetration of a Diesel Spray," *International Journal of Multiphase Flow*, Vol. **33**, No. 8, Pp. 904–920, (2007).
13. Simon Martinez – Martinez, Fausto A. Sanchez – Cruz, Vicente R.Bermudez And Jose M.Riesco-Avila, "Liquids Spray Characteristic In Diesel Engine", *In: Daniela Siano (Ed.). Fuel Injection. Intech: Sciyo*. 20 – 48 Pages, (2010).
14. Ozg¨Ur O˘guz Tas¸Kiran,"Experimental Study on Diesel Spray Characteristics and Autoignition Process", *Hindawi Publishing Corporation*, (2011).
15. B. R. M. and B. R. Fukuda K., Ghasemi A., "An Open Cycle Simulation of DI Diesel Engine Flow Field Effect on Spray Processes," *SAE Paper*, pp. 2012–01–0696, (2012).
16. B. R. and B. R. M. Ghasemi A., Fukuda K., "Numerical Investigation of Spray Characteristics of Diesel Alternative Fuels," *SAE Paper*, pp. 2012–01–1265, (2012).

MATEC Web of Conferences **90**, 01069 (2017) DOI: 10.1051/matecconf/20179001069
AiGEV 2016

Autonomous mobile robot localization using Kalman filter

Nabil Zhafri Mohd Nasir[1], *Muhammad Aizzat* Zakaria[1,*], *Saifudin* Razali[1], *and Mohd Yazid* bin Abu[1]

[1]Faculty of Manufacturing Engineering, Universiti Malaysia Pahang, 26600 Pekan, Pahang, Malaysia

Abstract. Autonomous mobile robot field has gain interest among researchers in recent years. The ability of a mobile robot to locate its current position and surrounding environment is the fundamental in order for it to operate autonomously, which commonly known as localization. Localization of mobile robot are commonly affected by the inaccuracy of the sensors. These inaccuracies are caused by various factors which includes internal interferences of the sensor and external environment noises. In order to overcome these noises, a filtering method is required in order to improve the mobile robot's localization. In this research, a 2-wheeled-drive (2WD) mobile robot will be used as platform. The odometers, inertial measurement unit (IMU), and ultrasonic sensors are used for data collection. Data collected is processed using Kalman filter to predict and correct the error from these sensors reading. The differential drive model and measurement model which estimates the environmental noises and predict a correction are used in this research. Based on the simulation and experimental results, the x, y and heading was corrected by converging the error to10 mm, 10 mm and 0.06 rad respectively.

1 Introduction

A fully automated mobile robot will require the robot to be able to pinpoint its current poses and heading in a stated map of an environment. The process of determining its pose is named localization. Mobile robot localization often gets intact with accuracy and precision problem. Commonly known as *position tracking* or *position estimation*. Localization is a fundamental perceptual problem in robotics [1].

In a study [2] stated that the real challenge in localization of mobile robot is to perform correct error association between data collected from sensor and its environmental model. Localization is the process of establishing correspondence between robot's local coordinate system and the map coordinate system, which is described in a global coordinate system and entirely independent of a robot's pose. Knowing the facts that most sensors integrated to robot are not to be a noise-free sensor [3], pinpointing the pose of a robot requires to be inferred from a certain preset data. Depending on a single sensor is usually insufficient for defining the robot's pose. Several sensors are required in order to increase the precision of the integrated data.

[*] Corresponding author: maizzat@ump.edu.my

Acquired data from robot sensors is required to be combined in order to estimate the actual pose of the robot. The Kalman filter is considered to be the best possible, optimal, estimator for systems with uncertainty [4-5]. The Kalman filter is a data processing algorithm or filter, which is useful for the reason that only knowledge about system inputs and outputs is available for estimation purposes [6]. In short, Kalman filter is an iterative mathematical process which uses sets of equations and consecutive data inputs in estimating the true value in a system containing uncertainty. Kalman filter is implemented for combining the sensory data for estimating the state of the robot which is the location and current heading of the robot [7].

This paper presents the implementation of Kalman filter to improve the mobile robot localization by developing a sets of mathematical models correlates to the driving model, position prediction, and position estimation models. Furthermore, the fabrication of 2-wheeled mobile robot for the purpose of collecting measurements of its surrounding based on the integrated sensors. The study is bounded by only considering a static, known indoor environment.

2 Simulation and experimental setup

2.1 Kalman filter modelling

In general, mobile robots require a driving system which allows it to move about its specified surrounding environment. Mobile robot uses in this project are based on a two-wheeled mobile robot driving system. These two-wheeled mobile robot are controlled referencing differential drive system which makes it the least complex system and suitable for a common indoor environment navigation. Thus, less complexity in its position estimation.

2.1.1 Position prediction model

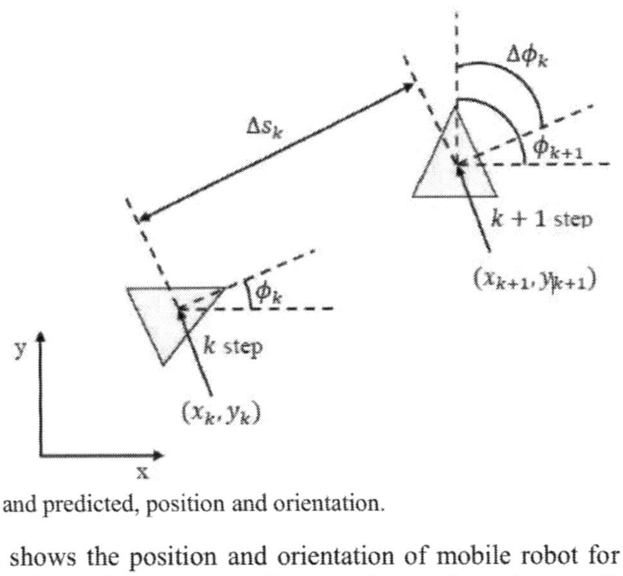

Fig. 1. Initial and predicted, position and orientation.

Figure 1 shows the position and orientation of mobile robot for certain different time steps. An initial coordinate denotes X_k are made up of x and y coordinate and heading angle θ referencing the centre of gravity of the mobile robot. The time step is denoted as k. The control input denotes as U_k are is based on robot relative displacement, s, and heading angle, ϕ. is first assumed that no noise is affecting the system for constructing the system model. This is to idealize the system to accurately estimate robots pose.

$$X_k = [x_k, y_k, \emptyset_k]^T \tag{1}$$

$$U_k = [\Delta s_k, \Delta\emptyset_k]^T \tag{2}$$

Updated position of a mobile robot X_k,,are estimated based on its previous position X_{k-1}, previous heading angle $\Delta\theta_{k-1}$, and the previous travels displacement s_{k-1}.

$$x_k = f_x(X_{k-1}, U_{k-1}) = x_{k-1} + \Delta s_k cos\left(\emptyset_{k-1} + \frac{\Delta\emptyset_k}{2}\right) \tag{3}$$

$$y_k = f_y(X_{k-1}, U_{k-1}) = y_{k-1} + \Delta s_k sin\left(\emptyset_{k-1} + \frac{\Delta\emptyset_k}{2}\right) \tag{4}$$

$$\emptyset_k = f_\theta(X_{k-1}, U_{k-1}) = \emptyset_{k-1} + \Delta\emptyset_k \tag{5}$$

The priori coordinate and priori error covariance can be measured as

$$\hat{X}_k^- = f\left(\hat{X}_{k-1}^+, \hat{U}_{k-1}\right) \tag{6}$$

$$P_k^- = \nabla f_x P_{k-1}^+ \nabla f_x^T + \nabla f_u U_{k-1} \nabla f_u^T + Q_{k-1} \tag{7}$$

Denoting U_k as controlled noise which is due respect to steer angle and velocity. ∇f_x and ∇f_u are Jacobian matrices describing the function of state transition referencing the state X and control input U.

2.1.2 Position estimation model

Reference walls are placed as landmarks with a ready set distance from square path sets for the mobile robot. Actual range between the robot and the reference wall are estimated using ultrasonic sensors. The centre of gravity of the mobile robot, (x_k, y_k) are set to be the origin on the local coordinate frame. Coordinate positioning the ultrasonic sensor are defined as (x_s', y_s') as shown in Figure 2.

DOI: 10.1051/matecconf/20179001069

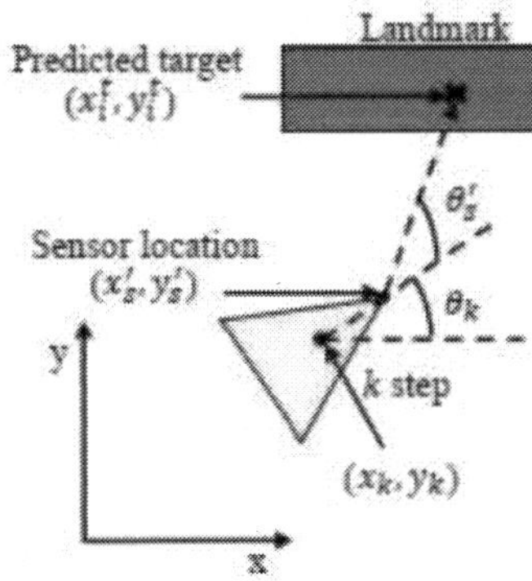

Fig. 2. Landmark prediction measurement.

Geometrical transformation of translation and rotation are used for converting the local coordinate frame to global coordinate frame.

$$\begin{bmatrix} X_{s,k} \\ Y_{s,k} \\ \theta_{s,k} \end{bmatrix} = \begin{bmatrix} x_k + x_s' cos(\theta_k) - y_k' \,.\, \sin(\theta_k) \\ y_k + x_s' sin(\theta_k) - y_k' \,.\, cos(\theta_k) \\ \theta_k + \theta_s' \end{bmatrix} \tag{8}$$

$$l_{i,k}^t = \begin{bmatrix} x_{i,k}^t \\ y_{i,k}^t \end{bmatrix} = \begin{bmatrix} X_{s,k} + h_{i,k} \cos(\theta_{s,k}) \\ Y_{s,k} + h_{i,k} \sin(\theta_{s,k}) \end{bmatrix} \tag{9}$$

denoting $l_{i,k}^t$ as the coordinate of the predicted landmark, and $h_{i,k}$ as the actual displacement measured from the ultrasonic sensor. The distance $h_{i,k}^t$ between ultrasonic sensor and landmark can be estimated one the predicted landmark coordinate $l_{i,k}^t$ is acquired, based on

$$h_{i,k} = \sqrt{(X_{s,k} - x_{i,k}^t)^2 + (Y_{s,k} - y_{i,k}^t)^2} \tag{10}$$

Difference between predicted heading and the actual heading measured by the encoder and yaw orientation $\emptyset_k$ get from IMU are determined by

$$Z_k = \begin{bmatrix} \sqrt{(X_{s,k} - x_{i,k}^t)^2 + (Y_{s,k} - y_{i,k}^t)^2} - h_{i,k}^t \\ \theta_{i,k} - \theta_{k,IMU} \end{bmatrix} \tag{11}$$

Kalman gain, K_k, the posterior state, $\hat{X}_k^+$, and the posterior error covariance, P_k^+ can be estimated and update by taking

$$K_k = P_k^- H^T (H P_k^- H^T + R_k)^{-1} \tag{12}$$

$$\hat{X}_k^+ = \hat{X}_k^- + K_k(Z_k) \tag{13}$$

$$P_k^+ = P_k^- - P_k^- K_k H \tag{14}$$

2.2 Experimental setup

2.2.1 Design of experiment

Figure 3 shows the robot and environmental setup being done to monitor the mobile robot poses in a known static environment. Four landmarks is set as B1, B2, B3, ad B4. The initial position of the robot is placed at the bottom right corner of the 1x1 meter square, and it is defined as the origin coordinate (x, y, θ). The robot is programmed to move along the four line creating a square path referencing the 1x1 meter square making counter-clockwise movements.

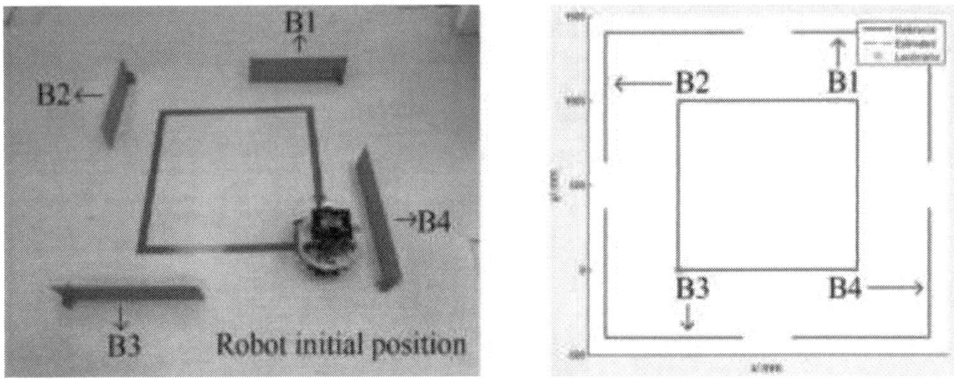

Fig. 3. Square path experimental setup and GUI representation.

Fig. 4. Ultrasonic actual distance measurement.

Error variance on the ultrasonic sensor are determined by comparing the raw data initially measured by the sensors, and the actual-measured distances between the sensors and the landmark placed. By means of performance, the sensors are capable of detecting range 0.04 to 30 meter, with resolution of 0.01 meter.

2.2.2 Robotics parameters

Based on the calibration done on sensors and several measurements on the mobile robot's base, robotics parameters are declared as presets for the system.

Table 1. Specified mobile robot parameters.

Parameter	Description	Value
w_d	Wheel diameter (*mm*)	145
w_b	Wheelbase diameter (*mm*)	285
n_p	Encoder pulse per revolution	2000
θ_s'	Ultrasonic angle (*deg*)	25°
$\sigma_{\Delta s}^2$	Odometry distance error (*mm*)	720
$\sigma_{\Delta \emptyset}^2$	Odometry orientation error (*deg*)	1°
σ_r^2	Dual ultrasonic error (*mm*)	72

3 Results and discussion

3.1 Error variance

To capture the motion data, mobile robot was programmed to navigate through 1x1 meter square path for 10 cycles to determine the encoder error. Odometry error variance, position displacement and robot's heading are closely noted.

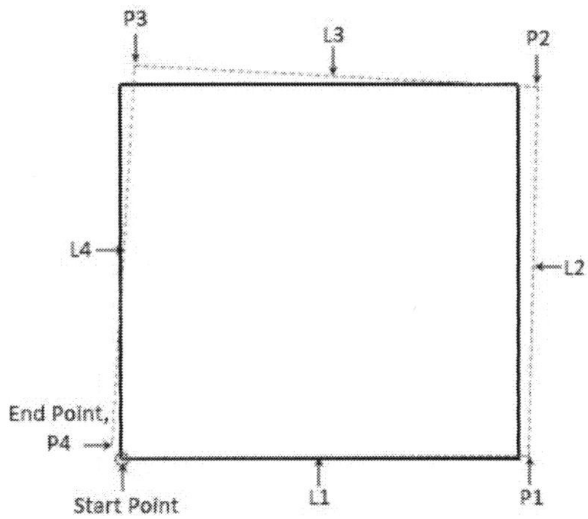

Fig. 5. Single run for programmed path.

Robotics trajectory for a single run is as shown in Figure 5 where points of measurement are denoted as labelled. From the figure, it can be seen that the reading is increasingly deviated from its original value due to error affecting its performance.

3.2 Kalman filter localization

Based on the previously stated mobile robot parameters, the error variance of the ultrasonic sensor, distances and orientation are uses as the initial state noise covariance and the measurement noise covariance to be post process in the Kalman filter for determining the mobile robot location in current environment. Figure 6 (a) shows the plotted estimated path of the mobile robot observing only on the odometry for localization. The estimated path is drifted away compared to the true trajectory of the robot. Since Kalman filter is done offline, thus the robot is not programmed with any measurement model which able to correct the position and heading of the mobile robot during its operation. This causes the deviation increases over time step.

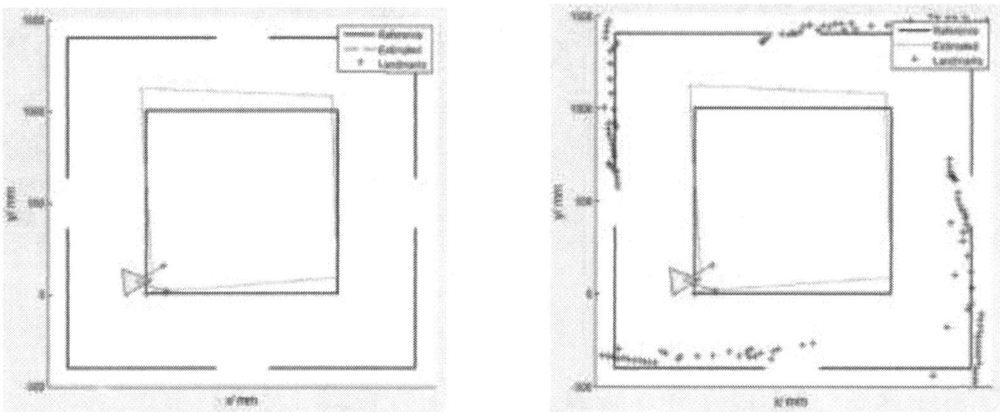

(a) (b)

Fig. 6. Estimated trajectory plotted data (a) Using odometry. (b) Includes ultrasonic sensor.

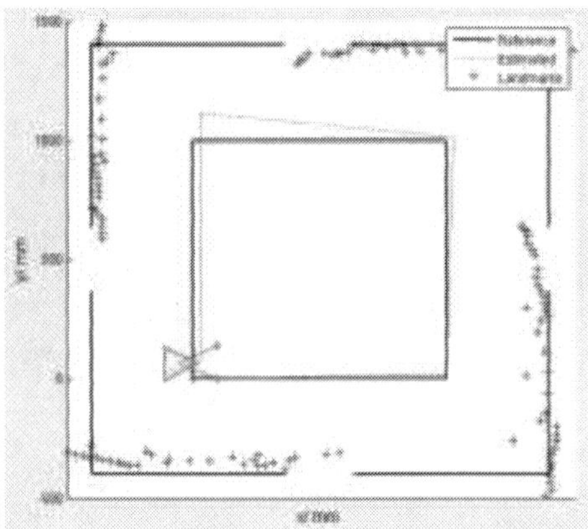

Fig. 7. Estimated trajectory plotted data using Kalman Filter.

Figure 7 shows the exact same data as in Figure 6 which plotted all the sensory data but using Kalman filter as inaccuracy filter. The path can be seen clearly to have improved in its accuracy and precision. It had improved the localization error through the runs. Based on the experiment done, it can be assumed that the average error affecting the robot's localization as shown in Table 2.

Table 2. Average position and orientation error.

Algorithm	Error		
	x, mm	y, mm	*ϕ, mm*
Odometry Localization	43	48	0.078
Kalman Filter Localization	10	10	0.060

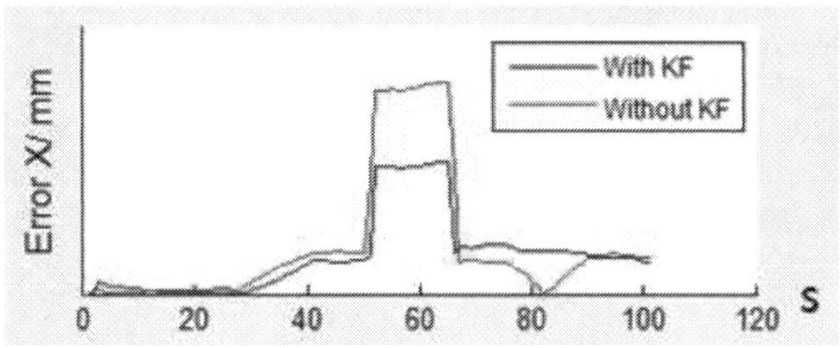

Fig. 8. Error in x-orientation.

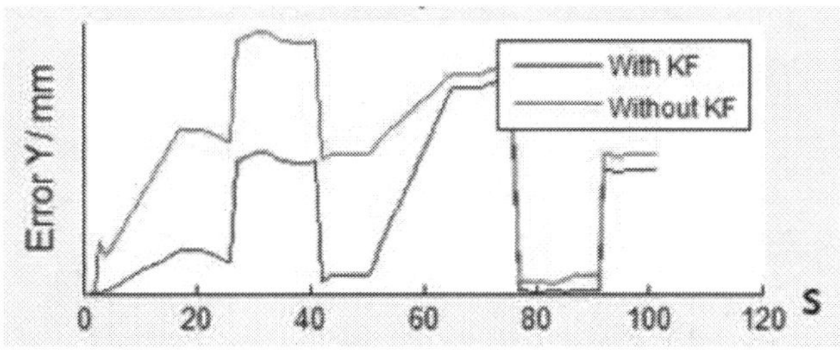

Fig. 9. Error in y-orientation.

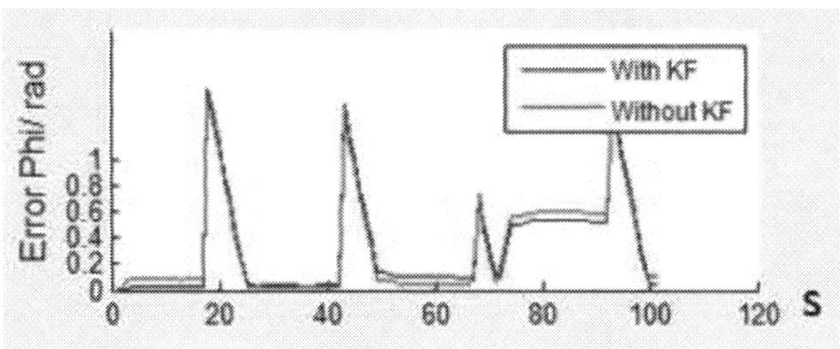

Fig. 10. Error in phi-orientation.

Figure 8, Figure 9, Figure 10 are showing the error comparisons between the estimated and the Kalman-filtered error for the x-coordinate, y-coordinate and the θ-orientation of the mobile robot for the above plotted mobile robot trajectory. Based on the figures, 0 on the x-axis represents the actual desired data which is comprised with zero error. Data plotted based on raw data are represented with the red lines, while the data implemented with Kalman filter are distinguished with the blue lines. From the results shown by the figure, it

DOI: 10.1051/matecconf/20179001069

can be clearly seen that Kalman filter implementation provides more accurate data which is much closer to the actual data. The Kalman filter implemented is proven to reduce an average of 23.25% of the error in mobile robot localization.

4 Conclusions

In this paper, the localization of the robot was improved by the implementation of Kalman filter. Based on the experimental and simulation results of the proposed system, the x, y, and heading or orientation errors is observed to be reduced to 0.1 meter, 0.1 meter and 0.06 rad angles respectively. The implementation of Kalman filter somehow introduces to delay in the systems processing. Kalman filter also depends greatly on the prediction model, an error in creating the prediction model will surely cause all the output data to be deviated from the required reference data.

The authors are obliged to thank the Universiti Malaysia Pahang for providing a platform for the project and providing financial assistance under research grant, RDU16043.

References

1. Siegwart and Nourbaksh, "Mobile Robot Localization," *Introd. to Auton. Mob. Robot.*, 2004.
2. F. Hardarson, "Stability analysis and synthesis of statically balanced walking for quadruped robots," *R. Inst. Technol.*, p. 154, 2002.
3. E. Elnahrawy and B. Nath, "Cleaning and querying noisy sensors," *Proc. 2nd ACM Int. Conf. Wirel. Sens. networks Appl. - WSNA '03*, p. 78, 2003.
4. R. E. Kalman, "A New Approach to Linear Filtering and Prediction Problems," *J. Basic Eng.*, vol. 82, no. 1, p. 35, 1960.
5. M. Grewal and A. Andrews, *Kalman filtering: theory and practice using MATLAB*, vol. 5. 2011.
6. R. Negenborn, "Robot localization and kalman filters," *Www. Negenborn Net/KalmanFilterImplementation (thesis). Pdf*, 2003.
7. J. Biswas and M. Veloso, "Multi-sensor mobile robot localization for diverse environments," *Lect. Notes Comput. Sci. (including Subser. Lect. Notes Artif. Intell. Lect. Notes Bioinformatics)*, vol. 8371 LNAI, pp. 468–479, 2014.

Effect on human metabolic rate of skin temperature in an office occupant

Rosli Abu Bakar[1,*], *Norfadzilah* Jusoh[1], *Ahmad* Rasdan Ismail[2], and *Tanti* Zanariah Shamshir Ali[1]

[1]Faculty of Mechanical Engineering, Universiti Malaysia Pahang (UMP), 26600 Pekan Pahang, Malaysia
[2]Faculty of Creative Technology and Heritage, Universiti Malaysia Kelantan (UMK), Beg Berkunci No. 01, 16300 Bachok, Kelantan, Malaysia

Abstract. The motivation of this study is to get a better understanding about the real thermal sensation of people in an office environment who undertakes moderate activities. The purpose of this study was to investigate occupant real thermal sensation at three tasks of activities and skin temperature to figure out its impact on thermal sensation in an office environment. All measurements were conducted in a climate chamber. In total, fifteen subjects were participated in three kinds of activities and skin temperatures, which is eleven locations were measured using thermocouple sensors. The results showed that the temperature is the one of the factors which affect metabolic rate. There is a strong linear relationship between these three tasks of activities and metabolic rate with the skin temperature. Furthermore, the results indicate that there was a greater increase thermal sensation when the metabolic rate 1.6. The conclusions of this study can provide some references to set a standard for air conditioning design and to set operating parameters of different functional architectures.

1 Introduction

Building environment has a great impact on people's health, comfort and productivity [1-3]. Thermal comfort studies are helpful to create a favorable indoor climate, maintain a productive thermal environment and keep reasonable energy consumption [4-6]. Most thermal comfort studies have focused on office building, classrooms or homes in which people usually sit or take low level activities. Actually, people may undertake various activities indoors such as those who work on assembling lines or those who exercise in building. With the increase of activity, the body heat production and thermal regulation process involved would be different from low activities [7]. Also, the thermal environment that makes people feel satisfied could not be similar.

Malaysia is a hot and humid tropical country that lies between 1° and 7° north and 100° and 120° east. Most locations have a relative humidity of 80-88% rising to nearly 90% in the highland areas and never fall below 60%. The mean maximum daytime temperature

* Corresponding author: rosli@ump.edu.my

recorded is 29-32 °C while the minimum temperature is 22-24 °C with a rainfall of 1000 mm per year [8,9].

Based on the literature review, few thermal comfort studies have been conducted in Malaysia before the 1990s. According to the Energy Efficiency Guideline by the Ministry of Energy, Telecommunications and Posts (1989) and the Malaysia Standard MS 1525 [10], the proposed temperature for the air-conditioned nonresidential buildings is 23-26 °C. These Figures are similar to the ASHRAE 55 [11] and ISO 7730 [12] standards. However, since 1990's with an increase of energy usage in the commercial sector, more research has been conducted in the area of thermal comfort to find appropriate means to provide comfortable indoor environments and to reduce energy consumption and costs. Studies have shown that an increase of 1.5 °C in indoor setting temperature give 15.8% energy saving [13,14].

The purpose of this study was to investigate occupant real thermal sensation at three tasks of activities and skin temperature to figure out its impact on thermal sensation in an office environment.

2 Experimental setup

The experiment was carried out in a climate chamber (L × W × H = 3.6 m × 2.4 m × 3.6 m), which was placed inside an air conditioned room. Other than that, the air temperature inside the chamber could be controlled within ±0.5 °C. Fig. 1 showed the arrangement of the study. This diagram illustrated the environment of equipment was placed behind the subject.

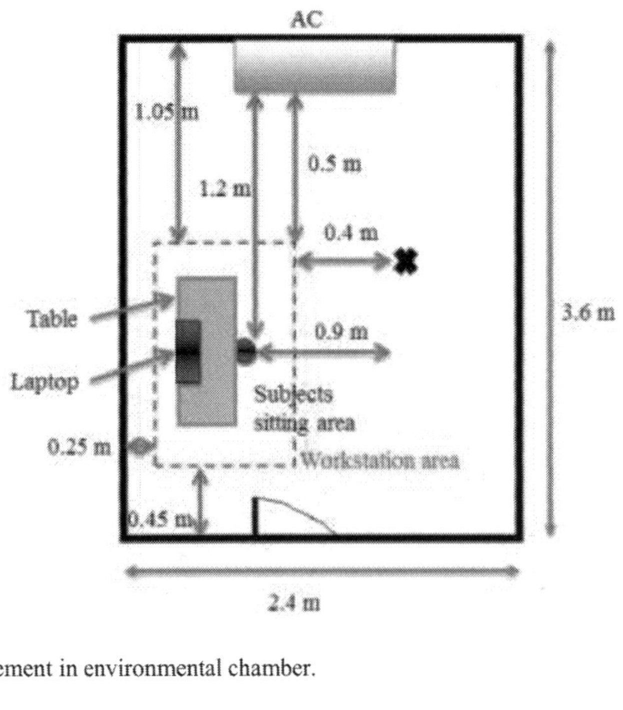

Fig. 1. The arrangement in environmental chamber.

2.1 Design of experiment

Thermal comfort tests with human subjects were conducted to test the thermal sensation of human for various thermal conditions. 15 volunteers (eight male and seven female) students were recruited in this study. The number of subjects in the study was agreed by Goldman [15]. He has 50 years' experience in the field of human thermal comfort and state that minimum subject for the study of human comfort is as many as six people.

All 15 subjects were using in this study to measure skin temperature while doing multitask in the chamber (represent an office room). There are three tasks the subjects must follow such as relaxing while sitting, typing or writing and printing paperwork as shown as in Table 1.

Table 1. Types of the task with the metabolic rate.

No. of Task	Types of the task	Metabolic rate (Met)
1	Thinking	1.0
2	Sitting while typing	1.2
3	Printing	1.6

Existing thermoregulation models have selected three to fifteen body segments for skin temperature measurement to estimate mean skin temperature [16]. The most frequently used body locations for assessing skin temperature in the 16 existing thermoregulation models are the chest, thigh, anterior-calf, wrist, posterior upper arm, forehead, abdomen, hand, foot and scapula. The body areas are ordered by the selection frequency in the existing models alongside the sum of the weight factors assigned in each model's formula.

These body locations for skin temperature sensors are important for both control effectiveness and for practical system application of wearable sensors. In this study, 11 body locations were selected from those most frequently used in the existing models to measure skin temperatures, which are forehead, posterior upper arm, wrist, back of the hand, chest, belly, thigh, anterior calf, posterior calf, and instep of a foot. The sensing interval was 10 min for all the measurements. As the posterior-calf has a larger weight factor than the scapula, and is a more convenient sensor location, the researchers selected the posterior-calf as the 10 skin area for the experiment.

Thermocouple sensors used to measure skin temperature during the experimental chamber is running. Usually, the thermocouple sensors were held to the skin of different sites on the subjects (see Fig. 2) with porous medical tape to measure the local skin temperatures.

2.2 Experimental procedure

The steps and duration of experiment were shown in Table 2. The subjects entered the chamber and spent 10 min at the chamber prior to the test acclimatize with the room temperature [17-19]. All the three tasks of activities during which the subjects were exposed to multiple thermal conditions.

Table 2. Steps and duration of experiment.

Step	Activity	Duration (min)
1	Measured demographic of subjects	30
2	Standby in a waiting room	10
3	Start the measurement with Task 1	10 x 5 = 50
4	Break	5
5	Task 2	10 x 5 = 50
6	Break	5
7	Task 3	10 x 5 = 50
	Total	**200**

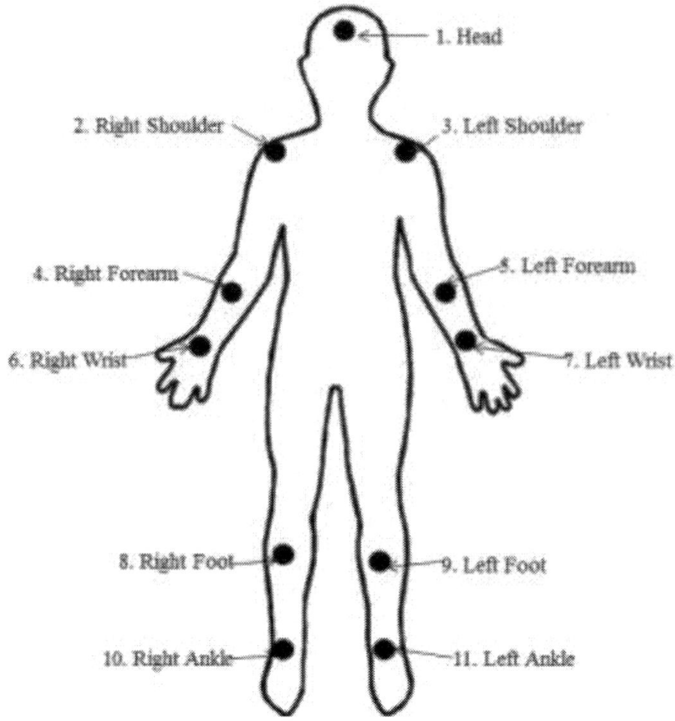

Fig. 2. Measurement points of skin temperature.

3 Theoretical models

This study is based on the thermal comfort of the human body. Therefore, it is necessary to understand what thermal comfort meaning and how to assess thermal comfort. Through many years' study on thermal comfort, Fanger [20] proposes conditions which is can satisfy thermal comfort; a) heat production and heat loss should satisfy the heat balance equation under the condition of steady state, b) skin temperature should be adapted to the thermal comfort level. Only satisfying these conditions at the same time can thermal comfort be realized. Thus, this paper based on these thermal comfort conditions to carry on this research.

3.1. The body heat balance equation

The suitable temperature is necessary to maintain the body's normal life people' activities. Thus, the body's heat production and heat dissipation should be in balanced. The body heat balance equation can be obtained [11] as Eq. (1):

$$M - W = (C + R + E_{sk}) + (C_{res} + E_{res}) + S \tag{1}$$

where, M [W/m^2] is metabolic rate, W [W/m^2] is mechanical work, C [W/m^2] is the convection heat loss of skin, R [W/m^2] is the radiation heat loss of skin, E_{sk} [W/m^2] the is skin evaporation heat loss, C_{res} and E_{res} [W/m^2] are sensible and latent respiratory heat loss, respectively, S [W/m^2] is heat storage of human body.

In general, W is very low, for most kinds of activities, $W \approx 0$ [21]. S is an important factor which affects the thermal comfort of dynamic environment [22]. However, the mainly consideration of this paper is steady state, so S=0. The detailed computational methods of variables in Eq. (1) can be expressed [11]:

a) Convection heat loss, C is the convection heat transfer between the body surface and the surrounding environment. When air temperature is higher than the surface skin temperature, the value of C is positive, otherwise, the value of C is negative. Convection heat loss can be obtained as follows:

$$C = f_{cl} h_c (t_{cl} - t_a) \tag{2}$$

$$f_{cl} = 1/(1 + 0.155 \times (h_c + h_r) \times I_{cl}) \tag{3}$$

$$t_{cl} = t_{sk} - 0.155 \times I_{cl} (M - W - (C_{res} + E_{res}) - E_{sk} - S) \tag{4}$$

where f_{cl} is the clothing area coefficient; h_c [W/(m^2°C)] is a convective heat transfer coefficient, h_c=8.3$V^{0.5}$; h_r [W/(m^2°C)] is radiation heat transfer coefficient; t_{cl} [°C] is the clothing surface temperature; t_a [°C] is air temperature; t_{sk}[°C] is skin temperature; I_{cl} [(m^{2}kPa)/W] is clothing insulation.

b) Radiation heat loss, R is the radiation heat transfer between the body surface and the surrounding environment. When the surface skin temperature is higher than the mean radiant temperature, the value of R is positive, otherwise, the value of R is negative, which can be seen in Eq. (5):

$$R = f_{cl} h_r (t_{cl} - t_r) \tag{5}$$

$$h_r = f_{rad} \varepsilon [(t_{cl} + 273.15)^2 + (t_r + 273.15)^2] \times$$
$$[(t_{cl} + 273.15) + (t_r + 273.15)] \times 5.67 \times 10^{-8}$$

where f_{rad} is the effective correction factor of human body surface, f_{rad} = 0.72; ε is the surface emissivity of the human body, ε = 0.97.

c) Skin evaporation heat loss, E_{sk} includes sweat evaporation heat loss and diffusion heat loss of moisture:

$$E_{sk} = \omega (p_{sk} - p_a) R_{ecl} h_e \tag{6}$$

$$R_{ecl} = 1/(1 + 0.143 \times h_c \times I_{cl} \tag{7}$$

$$LR = h_e / h_c \tag{8}$$

where ω is skin moisture index; p_{sk} [kPa] is a skin surface vapor pressure; p_a [kPa] is water vapor partial pressure of the surrounding air; R_{ecl} is clothing moisture permeability coefficient; LR = 16.5.

d) Respiratory heat loss includes sensible and latent respiratory heat loss, it can be obtained:

$$C_{res} + E_{res} = 0.0014M(34 - t_a) + 0.0173M(5.87 - p_a) \tag{9}$$

where f_{cl} and R_{cl} are determined by $I_c l$ and h_c; h_c and p_{sk} are related to air velocity (v) and t_{sk}, respectively; t_{cl} is determined by t_{sk} and t_a; p_a is determined by t_a and RH; t_{sk} is related to M and W. Thus, if t_a, t_r, RH, v and I_{cl} are known and the upper limit of ω determined, the biggest metabolic rate can be obtained under the condition of thermal comfort [23].

4 Results and discussion

4.1 Skin temperature

This section presents the relationship between the mean skin temperature and the thermal sensation which will also allow the prediction of thermal comfort. An analysis regarding the relationship between the mean skin temperature and the thermal sensation was conducted in order to determine the effect of metabolic rate on that relationship. The data were grouped based on the metabolic rate of the subjects.

Fig. 3 show effect of skin temperature on room temperature at different body segments. There are 11 body segments consists Head (1), Right Shoulder (2), Left Shoulder (3), Right Forearm (4), Left Forearm (5), Right Wrist (6), Left Wrist (7), Right Foot (8), Left Foot (9), Right Ankle (10) and Left Ankle (11) as illustrated in Fig. 2. Fig. 3 presented the highest body segment effect is left shoulder. This is because the left shoulder position is opposite with the door of the environmental chamber. All the body segment in left position is higher than right position. This also because, the right position faces the hanging wall of air conditioning. It also happens on the right forearm and right foot, both this body segments are placed on the table and on the floor. Furthermore, no barriers for temperature go through to these body segments.

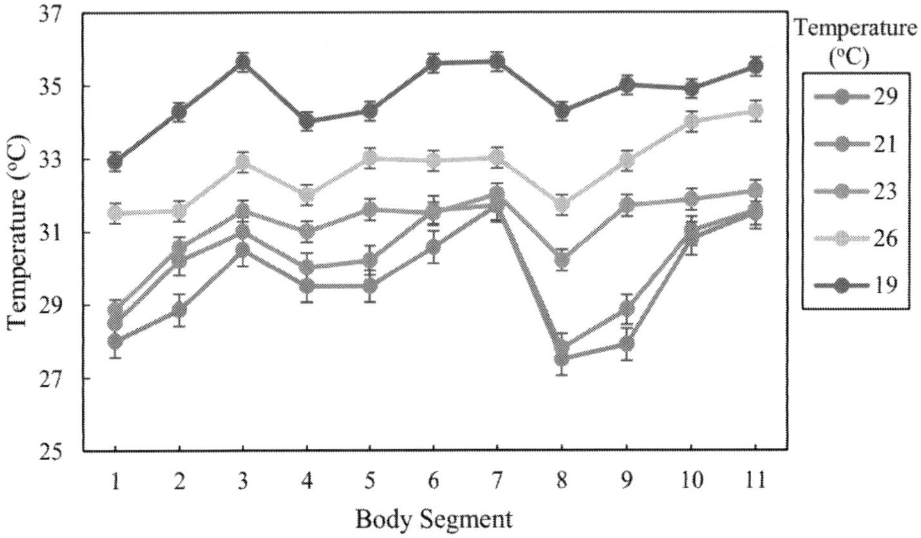

Fig. 3. Effect of skin temperature on room temperature at different body segments.

Fig. 4 shows that thermal sensation increased with the mean skin temperature in all three metabolic rate. For a given mean skin temperature, higher metabolic rate yielded a higher thermal sensation except around 34.89 °C mean skin temperature (p-value < 0.001). The results of the regression analysis for thermal sensation (see Figure 3) show a strong linear dependency for skin temperature of human. The change in mean skin temperature yielded similar changes in thermal sensation for low and medium metabolic rate conditions. However, the change in thermal sensation during the high metabolic conditions was less than the other two conditions. It can be seen that a straight line fits the data well for 1.0, 1.2 and 1.6 Met (R^2=0.7541, R^2=0.8679 and R^2=0.8467) respectively.

Under all conditions of activities, overall thermal sensation and skin temperature are correlated with each other closely. Skin temperature of 34.89 °C corresponds to thermal sensation of 0.79, that is to say, the subjects felt uncomfortable when their whole body is 34.89 °C which are more rigid than the definition proposed by Gagge et al. [24] and Fanger [25].

 DOI: 10.1051/matecconf/20179001070

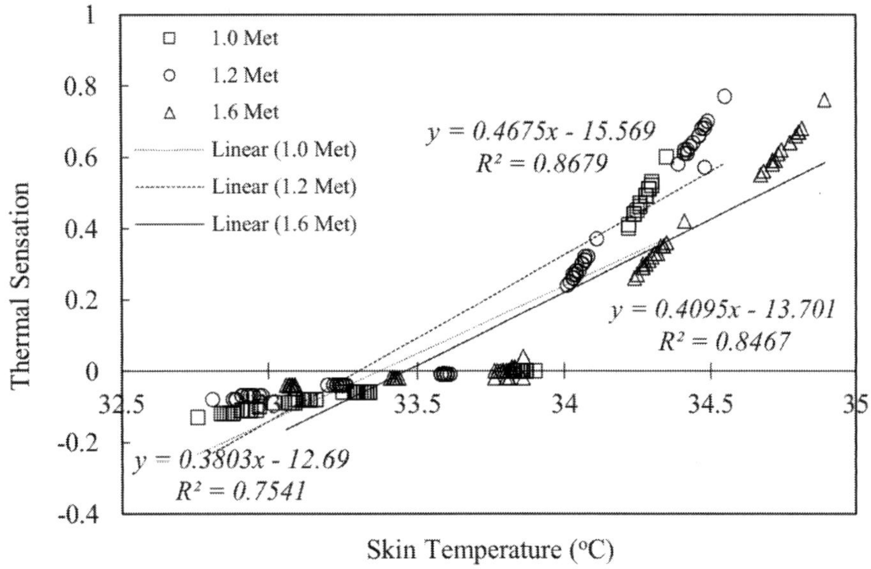

Fig. 4. The relationship between skin temperature and thermal sensation under three metabolic rate conditions.

4.2 Heat balance and heat loss of the human body

In this section, a heat balance analysis of the human body based on the specific test conditions was presented. The amount of heat stored inside the body was calculated through the increase in body temperature. The tympanic temperature measurements showed that the core body temperature is relatively stable throughout the experiment (see Table 3) and the difference between the measurements was statistically insignificant. Therefore, there was no heat storage in the body between the body temperature measurements and all the generated heat was dissipated to the environment through conduction, convection, radiation and evaporation.

Table 3. Body core temperature.

Mean body core temperature (°C)	Body core temperature Std. Dev.
36.85	0.029

The evaporative heat losses for males and females are similar were calculated as 5.01 W/m^2 respectively as illustrated in Table 4.

Table 4. Evaporative heat loss in males and females.

Gender	Average Met	Average surface area (m^2)	Duration of test (hour)	Evaporative heat loss (W/m^2)
Male	1.27	1.79	3.20	5.01
Female	1.27	1.60	3.20	5.01

An analysis was conducted to determine the relationship between the heat conduction through the body and the thermal comfort. The difference between the core temperature and skin temperature (ΔT_{cr}-T_{sk}) was normalized for body mass index (*BMI*) to take into account the insulation properties of the body. A significant correlation was detected

DOI: 10.1051/matecconf/20179001070

between the ΔT_{cr}-T_{sk} and the thermal comfort at $\alpha = 0.05$ (Pearson correlation: 0.58). Fig. 5 showed the relationship between these two variables. The correlation was also significant without *BMI* normalization, however, with a lower correlation coefficient (Pearson correlation: 0.21). This shows that thermal discomfort increases with the increased thermal strain on the body, which is the difference between the core body and the skin temperature. It can be seen that core body temperature a straight line fit data well (R^2=0.58).

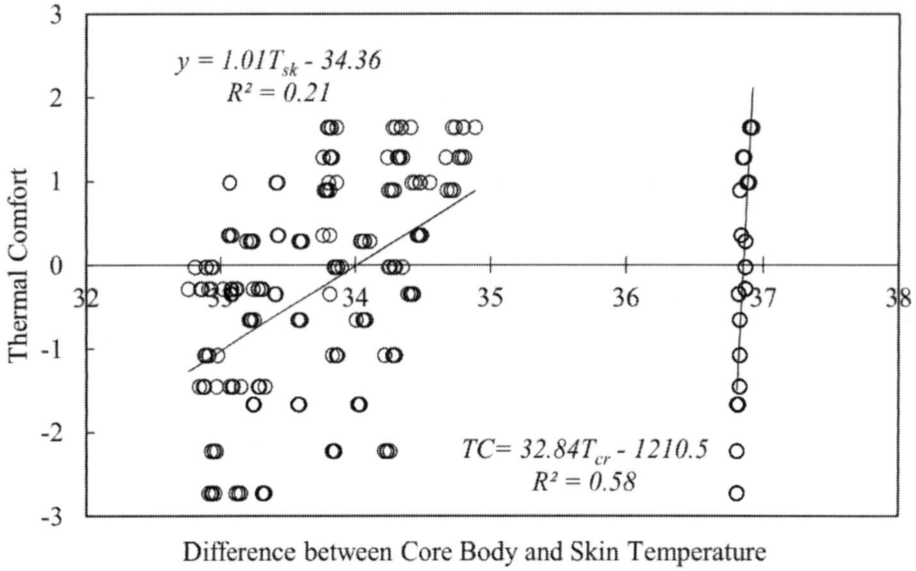

Fig. 5. Correlation between thermal comfort and the difference between core body and skin temperature.

5 Conclusions

In this paper, the effect of the increased activity level of thermal sensation was investigated. Moreover, theoretical analysis and experiment were combined to study the metabolic rate on skin temperature in different thermal conditions. The conclusions of this paper can be drawn as follows:

a) Air temperature is the one factor which affects the metabolic rate in different thermal conditions.
b) The mathematical model was established according to experiment and the thermal sensation range of skin temperature to the metabolic rate.

The conclusions of this study can provide some references to set a standard for air conditioning design and to set operating parameters of different functional architectures. Besides that, this study can achieve the goals of thermal comfort and energy conservation to some extent.

The authors would like to thank the Higher Education Ministry and University Malaysia Pahang for the financial support and assistance from a research grant.

References

1. S. Tsushima, S.I. Tanabe, K. Utsumi, Workers' awareness and indoor environment quality in electricity-saving office, Building and Environment, **88**(6), 10-19, (2015).

2. R.J. de Dear, T. Akimoto, E.A. Arens, G. Brager, C. Candido, K.W. Cheong, B. Li, N. Nishihara, S.C. Sekhar, S. Tanabe, J. Toftum, H. Zhang, Y. Zhu, Progress in thermal comfort research over the last twenty years, Indoor Air, **23**(6), 442-461, (2013).
3. L. Lan, P. Wargocki, Z. Lian, Quantitative measurement of productivity loss due to thermal discomfort, Energy and Building, **43**(5), 1057-1062, (2011).
4. O. Seppanen, W.J. Fisk, Some quantitative relations between indoor environment quality and work performance or health, HVAC & R Res, **12**(4), 957-973, (2006).
5. M. Luo, R. de Dear, W. ji, C. Bin, B. Lin, Q. Quyang, Y. Zhu, The dynamics of thermal comfort expectations: The problem, challenge and implication, Building and Environment, **95**, 322-329, (2016).
6. C. Dai, L. Lan, Z. Lian, Method for determination of optimal work environment in office buildings considering energy consumption and human performance, Energy and Building, **76**(11), 278-283, (2014).
7. J.K. Vanos, J.S. Warland, T.J. Gillespie, N.A. Kenny, Review of the physiology of human thermal comfort while exercising in urban landscapes and implications for bioclimatic design, International Journal Biometeorol, **54**(4), 319-334, (2010).
8. S.S. Ahmad, Thermal comfort and building performance of naturally ventilated apartment building in the Klang valley: A simulation study, *Proceeding of the energy in buildings (sustainable symbiosis) seminar*, 115-132, (2006).
9. A. Ahmad, *Case study: Thermal comfort study demonstration low energy office (LEO)*, M.Sc Thesis, UKM, Malaysia, (2004).
10. Malaysia Standard, *MS 1525 Code of practice on energy efficiency and use of renewable energy for non-residential buildings*, Department of Standards Malaysia, (2001).
11. ASHRAE, *ASHRAE handbook of fundamentals, comfort and health*, American Society of Heating, Refrigerating and Air-Conditioning Engineers, Inc, Atlanta, USA, (2009).
12. ISO 7730, *Moderate thermal environments: Determination of the PMV and PPD indices and specifications for thermal comfort*, International Organization for Standardization, (1994).
13. S. Abdul Rahman, K.S. Kannan, A study of thermal comfort in naturally ventilated classrooms: Towards new indoor temperature standards, *Proceeding of the Asia Pacific Conference on the Built Environment*, Kuala Lumpur, Malaysia, (1997).
14. M. Zainal, C.C. Keong, Thermal comfort and energy conservation in factory buildings, *Proceedings of the 7th International Conference on Indoor Air Quality and Climate*, **2**, 601-606, (1996).
15. R.F. Goldman, Environmental ergonomics: Whence what wither, *Proceeding of the 11th International Conference on Environmental Ergonomics*, 39-47, (2005).
16. Y. Yao, Z. Lian, W. Liu, Q. Shen, Experimental study on skin temperature and thermal comfort of the human body in a recumbent posture under uniform thermal environments. Indoor Built and Environment, **16**(6), 505-518, (2007).
17. A.P. Gagge, J.A.J. Stolwijk, Y. Nishi, The prediction of thermal comfort when thermal equilibrium is maintained by sweating, ASHARE Transaction, **75**, 108-125, (1969).
18. I.A. Strigo, F. Carli, C. Bushnell, Effect of ambient temperature on human pain and temperature perception, Anesthesiology, **92**, 699-707, (2000).
19. S. Tanabe, N. Nishihara, M. Haneda, Indoor temperature, productivity and fatigue in office tasks, HVAC & R. Research, **13**, 623-633, (2007).
20. P.O. Fanger, Calculation of thermal comfort: Introduction of a basic comfort equation, ASHRAE Trans, **73**(2), 1-20, (1967).
21. McIntyre, *Indoor climate*, Applied Science Publisher, London, (1980).

 DOI: 10.1051/matecconf/20179001070

22. L. Duanmu, L. Yu, H. Shu, Influence of human body heat storage and adaptation on thermal body, Refrig. Air Cond. Electr. Power Mach, **95**(01), 30-33, (2004).
23. C. Yang, T. Yin, M. Fu, Study on the allowable fluctuation ranges of human metabolic rate and thermal environment parameters under the condition of thermal comfort, Building and Environment, **103**, 155-164, (2016).
24. A.P. Gagge, J.A.J. Stolwijk, Y. Nishi, Comfort and thermal sensations and associated physiological responses at various ambient temperatures, Environmental Research, **1**, 1-20, (1967).
25. P.O. Fanger, *Thermal comfort: Analysis and applications in environmental engineering*, Copenhagen, Denmark, Danish Technical Press, (1970).

Numerical investigation on soot particles emission in compression ignition diesel engine by using particulate mimic soot model

Fadzli Ibrahim[1,2], *Wan Mohd Faizal* Wan Mahmood[1,*], *Shahrir* Abdullah[1], and *Mohd Radzi* Abu Mansor[1]

[1]Department of Mechanical and Materials Engineering, Faculty of Engineering and Built Environment, Universiti Kebangsaan Malaysia, 43600 UKM Bangi, Selangor, Malaysia.
[2]Mechanical and Aerospace Technology Division, Science and Technology Research Institute for Defence (STRIDE), 43000 Kajang, Selangor, Malaysia.

Abstract. Research via computational method, specifically by detailed-kinetic soot model offers much more advantages than the simple model as more detailed formation/oxidation process is taken into consideration, thus providing better soot mass concentration, soot size, soot number density as well as information regarding other related species. In the present computational study, investigation of in-cylinder soot concentration as well as other emissions in a single cylinder diesel engine has been conducted, using a commercial multidimensional CFD software, CONVERGE CFD. The simulation was carried out for a close-cycle combustion environment from inlet valve closing (IVC) to exhaust valve opening (EVO). In this case, detailed-kinetic Particulate Mimic (PM) soot model was implemented as to take benefit of the method of moment, instead of commonly implemented simple soot model. Analyses of the results are successfully plotted to demonstrate that the soot size and soot mass concentration are strongly dependent on the detailed soot formation and oxidation process rates. The calculated of soot mass concentration and average soot size at EVO provide the end value of 29.2 mg/m^3 and 2.04 x 10^{-8} m, respectively. Besides, post-processing using EnSight shows the qualitative results of soot concentration along simulation period in the combustion chamber.

1 Introduction

Fast evolution in diesel engine technology promises a good return in engine performance and efficiency. But stringent emission law in recent times forces diesel engine to be designed not only for superior performance but also with lower emission level. Therefore, particulate emission will still be a hot issue to be debated and solved, so that emissions emitted comply with current law.

One of a major part of particulate matters, soot, definitely gives bad impact not only to the environment but also to human well-being. This undesirable element was reported to cause global warming, acid rain, smog, odors, respiratory and other health hazards [1–4]. A

* Corresponding author: faizal.mahmood@ukm.edu.my

DOI: 10.1051/matecconf/20179001071

number of researchers successfully discovered the toxic effect of soot to human health [5–8] and the cause for a decrease in the overall engine performances [9–12].

Soot, normally in particulate form, was originally formed by the incomplete combustion process. Their details characteristics in terms of physical and chemical properties from their multiple sources – in-cylinder, in exhaust and in-oil, are still continuously explored and deliberated [13,14]. Since years ago, investigating soot emission by computational method was become as one of the topics of choice, apart from the experimental method. This method is become more demanding due to its undisputable advantages in terms of time and cost saving. Current development of computer programs with advanced precision of mathematical models brings simulation works closer to the real engine combustion environment.

In conducting soot emission analysis by computational technique, the soot model developed is the most important factor in determining the accuracy level of the analysis. Modeling of soot emissions from diesel engines generally falls into two different approaches: (a) simple empirical models which relate engine operating parameters to soot emissions and (b) complex multi-step phenomenological models that model the individual steps relating to soot formation and oxidation, often also involves complex reaction mechanisms [15]. There are numerous researchers who studied the emissions characteristics by simulation analysis but with various engine parameters using different computer software such as KIVA, Star-CD, OPEN FOAM CFD, AVL FIRE, CONVERGE CFD and also ANSYS FLUENT.

The aim of this study is to investigate the in-cylinder soot formation process within pre-set engine conditions using detailed soot model, namely Particulate Mimic (PM) model. To accomplish it, a commercial Computational Fluid Dynamic (CFD) software, CONVERGE CFD has been used for detailed computation and analysis purposes while Ensight software has been applied for post-processing technique.

2 Methodology

2.1 Pre-processing setup

All the engine parameters and combustion conditions are based on a single-cylinder diesel engine, Yanmar TF90M. For simulation purposes, a multidimensional CFD software, CONVERGE CFD which was developed by Convergent Science Inc has been used to simulate combustion environment but limited to the close-cycle combustion which is from inlet valve closing (IVC) at -168° Crank Angle (CA) After Top Dead Centre (ATDC) to exhaust valve opening (EVO) at 138° CA ATDC.

For accurate geometry model generation, a computer-aided design (CAD) software, Unigraphic NX 10.0, was used as shown in Figure 1. It was then exported to the pre-processing software, CONVERGE CFD for the geometry checking and clean-up [16] as shown in Figure 2.

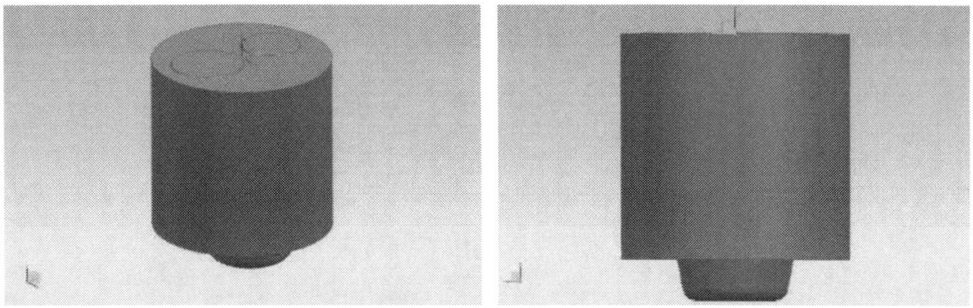

Fig. 1. Model of the combustion chamber created by Unigraphic NX 10.0.

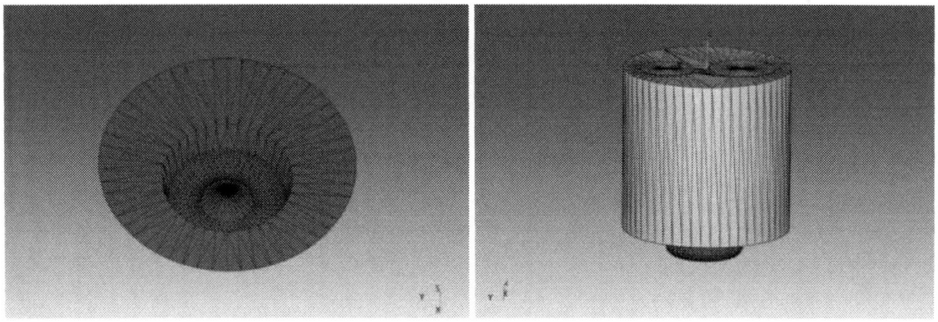

Fig. 2. Model of the combustion chamber cleaned-up by CONVERGE CFD.

Figure 3 shows the schematic diagram of the full combustion chamber complete with bowl-in-piston configuration. The details specification of the engine, operating conditions, initial conditions and fuel model used in the simulation are shown in Table 1, Table 2, Table 3 and Table 4. In the current test cases, no exhaust gas recirculation (EGR) was employed.

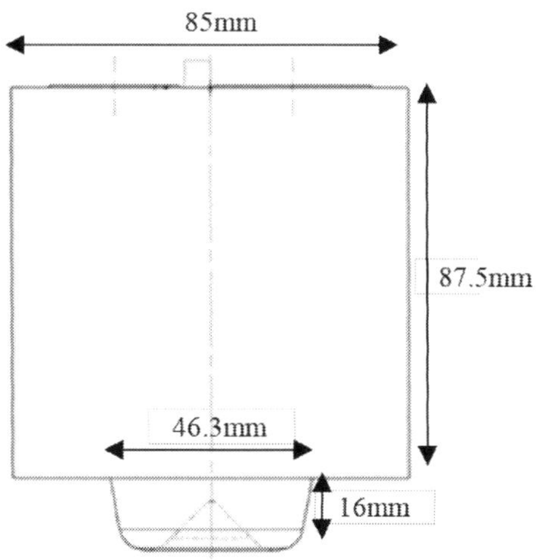

Fig. 3. Combustion chamber with piston bowl diagram.

Table 1. Test engine specification.

Parameters	Specifications
Engine make	Yanmar TF90M
Engine type	Water-cooled 4 cycle Diesel engine
Engine capacity (cc)	493
Number of cylinder	1
Bore × Stroke	85 × 87 mm
Injection system	Direct injection
Injection nozzle (n × Ø d)	4 × Ø 0.22 mm
Compression ratio	18.0:1

Table 2. Engine operating condition.

Parameters	Specifications
Engine speed	1600 rpm (max. torque)
Initial Swirl ratio (assumed)	1.8
Start of Injection (SOI)	-18° ATDC
Injection period (assumed)	15°
Injection pressure	196 bar
Injection quantity	19 mg
Injection profile	Single main injection

Table 3. Engine initial conditions (assumed).

Parameters	Specifications
In-cylinder temperature	315 K
In-cylinder pressure	105 KPa
Piston Head temperature	363 K
Cylinder Liner temperature	319 K
Piston temperature	403 K

Table 4. Fuel specification.

Parameters	Specifications
Fuel model	Diesel Euro5M
Density @ 15°C	832.2 Kg/m^3

2.2 Mesh setup

One of the aims in conducting this computational analysis is having full flexibility in mesh setup. By applying the extra feature embedded in CONVERGE CFD software so-called Adaptive Mesh Refinement (AMR), the computational time could be reduced significantly, instead of conventional setting with the similar mesh size for the whole simulation. This feature allows users to refine mesh size accordingly to certain spatial and temporal conditions. In this study, the overall mesh resolution of 1.0 mm was found to sufficiently produce grid independent results, as the comparison of different mesh size shown in Figure 4. This mesh size was resulted in 624,556 elements at the time when the intake valve closes. Injector region that was assumed as the most essential segment was set to 0.25 mm of mesh size. But it will only be activated during AMR feature which in this case starts from -18° CA onwards.

DOI: 10.1051/matecconf/20179001071

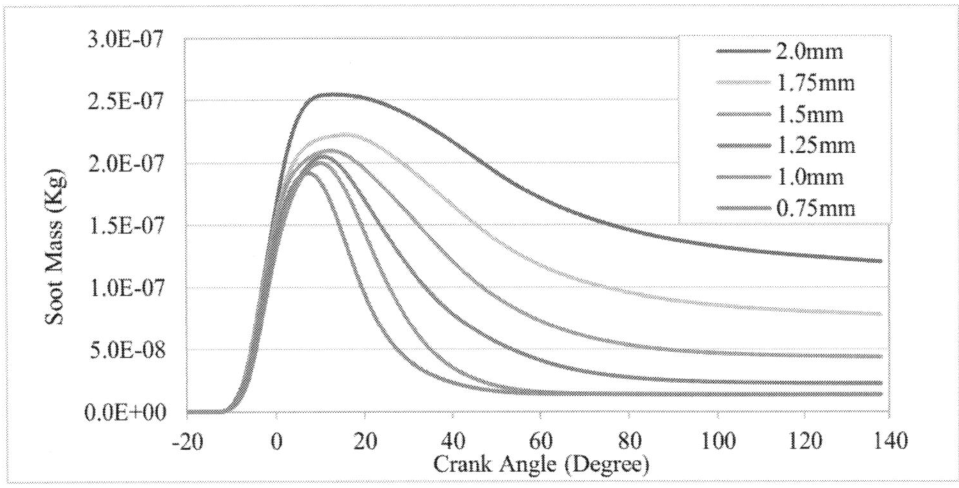

Fig. 4. Soot mass comparison for different mesh size.

Figure 5 shows the difference of mesh setting according to spatial and temporal condition with regard to AMR setup.

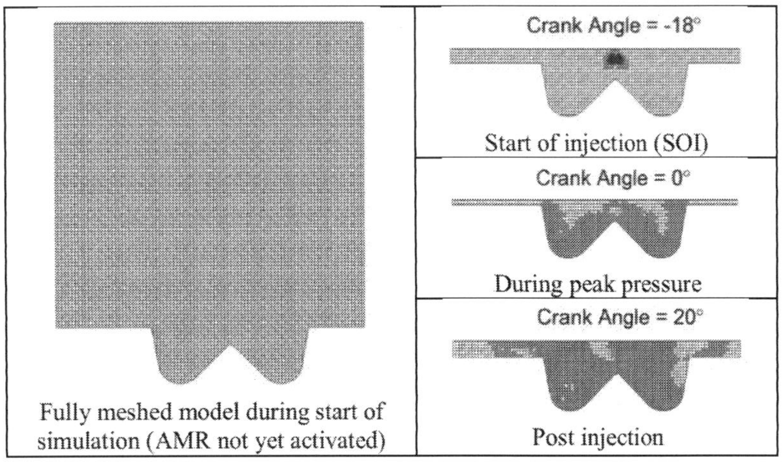

Fig. 5. The diagram of the meshed model during start of the simulation, follow by the difference of mesh condition with regard to AMR setup.

2.3 Processing setup

Despite of typically used two-steps soot model, a more detailed soot model, Particulate Mimic (PM) which based on the method of moments was applied. Via this model, all the basic physical and chemical processes for the formation of soot are measured which are particle inception, coagulation, condensation, surface growth and oxidation [17,18]. The dynamics of the soot particle characteristics can be described by following equations [16];

$$\frac{dM_r}{dt} = \dot{M}_{r,pi} + \dot{M}_{r,con} + \dot{M}_{r,coag} + \dot{M}_{r,sg} - \dot{M}_{r,ox} \qquad (1)$$

where $\dot{M}_{r,pi}$, $\dot{M}_{r,con}$, $\dot{M}_{r,coag}$, $\dot{M}_{r,sg}$ and $\dot{M}_{r,ox}$ are the rates of particle inception, condensation, coagulation, surface growth and oxidation for the rth moment of the particle size distribution function (PSDF), respectively. The moments can be defined as;

$$M_r = \sum_{i=1}^{\infty} i^r N_i \qquad (2)$$

where N_i is the number density of soot particles of size class i.

As part of diesel fuel, only n-heptane (C_7H_{16}) will be considered as a fuel species for further analysis, acetylene (C_2H_2) as a formation species, OH and O_2 as the oxidation species, and soot density is assumed to be 1.84701 g/cm^3 [19]. All these chemical species will be used to model diesel combustion chemistry based on Zeuch mechanism [20], producing a total of 588 reaction mechanism which contains of 5 elements and 121 species. Additional species of A_3R_{5-} as a soot precursor is set for the detailed soot model. In addition to the soot model, other dedicated sub-models are used to solve each sub-process, including spray breakup, turbulence and combustion. These sub-models are listed in Table 5.

Table 5. The computational sub-models applied.

Phenomenon	Model
Soot formation	PM model based on method of moment [17,18]
NOx formation	Extended Zeldovich [21]
Spray breakup	KH-RT model [22,23]
Drop drag	Dynamic model [24]
Collision and coalescence	NTC model [25]
Spray-wall interaction	Rebound/slide model [26,27]
Vaporization	Multi component fuel
Turbulence	RNG k-Ɛ model [28]
Combustion	SAGE [29]

All the computational models are then executed using CONVERGE CFD as a solver, and the results are visualized using post-processing software, EnSight 10.1. This post-processor software was used to animate or visualize the complex datasets obtained from computational analyses [30]. All the executed results of soot and other gas emissions characteristics are presented in the next section.

3 Results and discussion

3.1 Temperature and pressure effects

Reaction rates for both soot formation and oxidation process are strongly dependent on pressure and temperature effect. While these two factors are increased, the rate of oxidation increases more rapidly than the rate of formation [31]. As to show the relevance to the sooting process, changes of the pressure and temperature along the simulation time are plotted and shown in Figure 6.

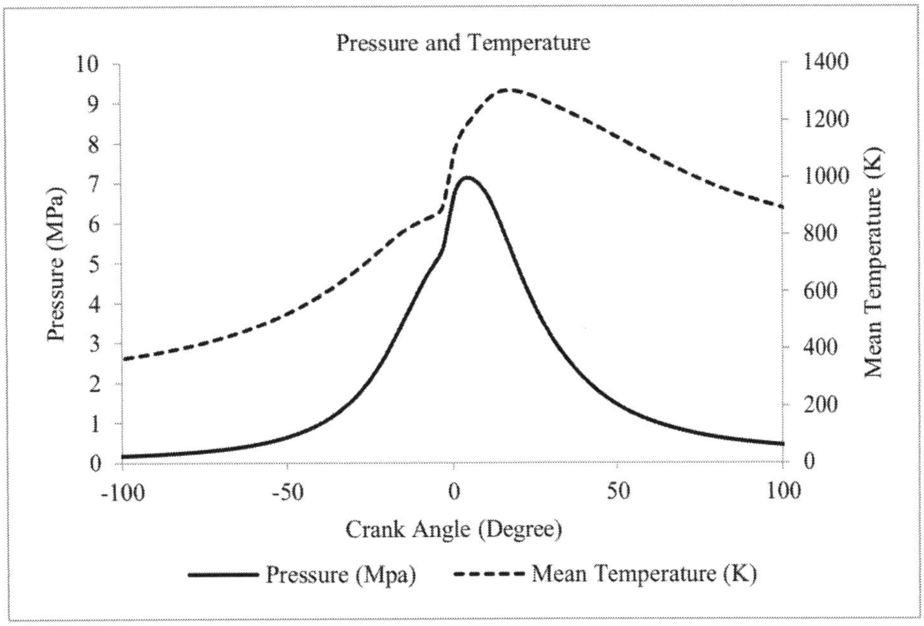

Fig. 6. Temperature and pressure change in the simulation.

The pressure and temperature value at IVC were set to 0.105 MPa and 315 K respectively, as an initial in-cylinder condition. As the piston moving upward during the compression stroke, these two values keep increasing until their maximum limit of 7.13 MPa at 4.5° CA for pressure and 1303.47 K at 17.3° CA for temperature. So the value of soot concentration which will be described in the subsequent section is strongly influenced by these parameters.

3.2 Soot concentration

Based on Eq. 1 and Eq. 2, soot mass concentration is plotted as shown in Figure 7. PM soot model calculates the soot formation since early of simulation time even with very tiny soot mass concentration value at 1.92 x 10^{-15} mg/m^3. The value from PM model however gradually increases to the peak value of 655.83 mg/m^3 at 16.3° CA° before decreases to the end value of 29.2 mg/m^3. As reported by Symonds et. al (2007) [32], the experimental values from Diesel Particulate Filter (DPF) and Diesel Mobility Spectrometer (DMS) fell within the range of 20.5 mg/m^3 to 67.7 mg/m^3, as shown at end of simulation in Figure 7. So the end computed value from simulation was well agreed with the experimental value.

PM as a detailed soot model considers all the important process in sooting formation, i.e. nucleation, fragmentation, surface growth, condensation, coagulation and oxidation, as compared to simpler soot model, i.e. Hiroyasu's soot model, which only considers soot formation and oxidation process only in general form. As to confirm whether this value of soot mass complies with the current emission regulation, the end value at EVO which assumed to be the mass that will be emitted to the atmosphere through the exhaust pipe should be converted into a comparable unit with emission regulation values.

 DOI: 10.1051/matecconf/20179001071

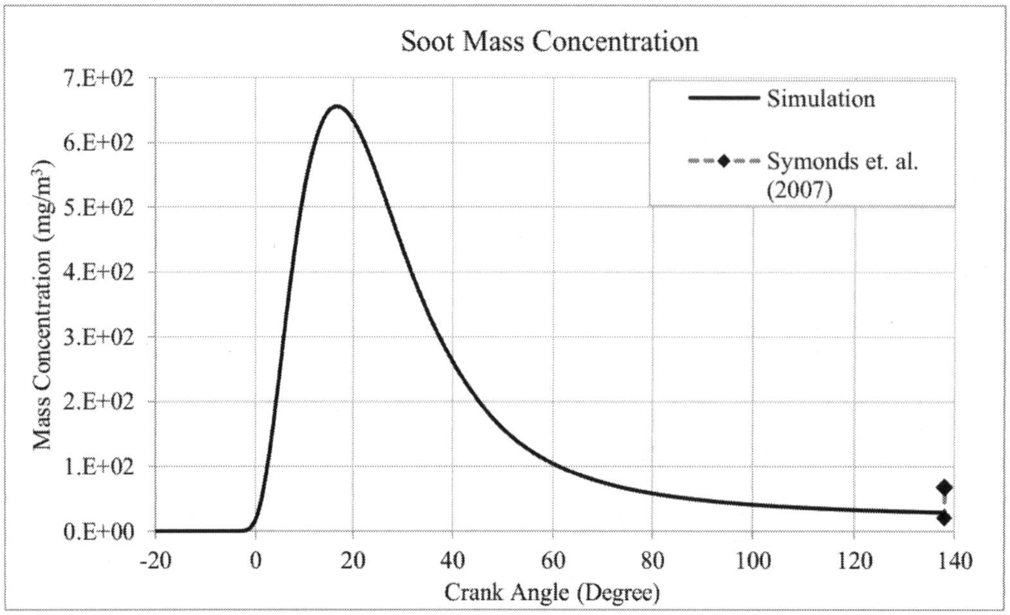

Fig. 7. Soot mass concentration of PM soot model.

Another extra parameter provided by PM soot model is soot average diameter, which is not calculated by the simple model. Initial soot size plotted was 8.66 x 10^{-10} m and keeps increasing until its final value of 2.04 x 10^{-8} m at EVO. This final value was perfectly agreed with the average soot size as suggested by Giechaskiel et. al. (2014) [33] which reported the values were fall within 10 nm to 30 nm. The average soot diameter is plotted and shown in Figure 8. The rapid increment of this value is principally due to relatively higher surface growth rate than oxidation rate. Similar to soot mass value, the information of soot size is also essential to comply with the current emission regulations.

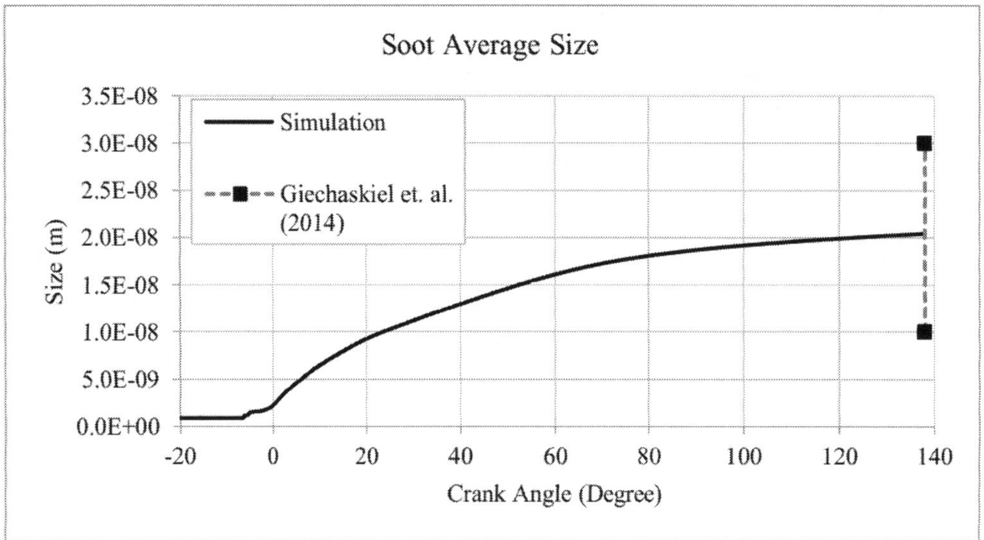

Fig. 8. Average soot diameter by PM soot model.

Using EnSight 10.1 as a post-processor, the net soot mass concentration on a cross-sectional plane of the combustion chamber can be visualized as presented in Figure 9. Soot mass concentration plotted from early injection period until end of the simulation are consistent with the net soot mass shown in Figure 7 during combustion process. Additionally, this qualitative result clearly presented that particles with higher mass value are more concentrated at the lower part of the cylinder (i.e. squish/bowl region) instead of the upper side of the cylinder. Towards the end of simulation, almost all of the particles with lower soot mass value occupies the upper side and near the cylinder wall, and these lighter soot particles are assumed to be emitted during exhaust stroke while a part of it will remain and deposited onto the cylinder wall.

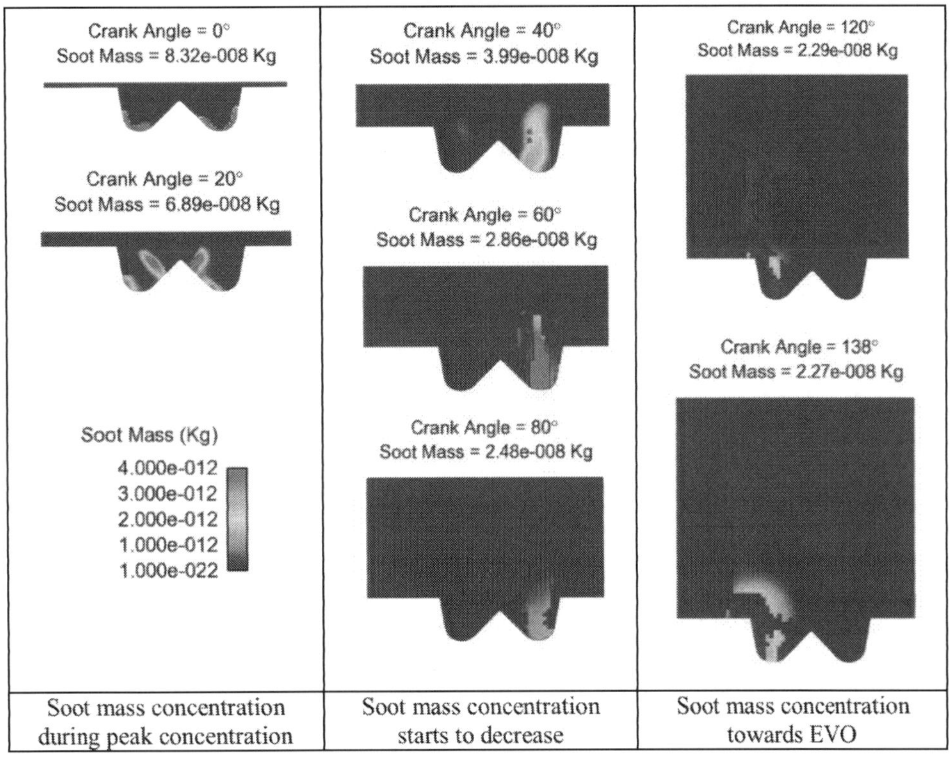

Fig. 9. Soot mass concentration in the combustion chamber.

4 Conclusions

In conducting soot emission analysis by computational technique, applying more complex soot model will give more advantages than simpler or most commonly applied two-step soot model. This study applied PM soot model, which is based on the method of moments as a detailed model. The calculated of soot mass concentration at EVO shows decreasing value from peak to the end value of 29.2 mg/m^3. Average soot size was recorded to keep increasing since early in the simulation, at the start of combustion, until its final value of 2.04 x 10^{-8} m at EVO. All the data obtained are strongly related to each of soot formation and oxidation process rates. All those values are well agreed with the suggested values as reported by some researchers through their experimental activities, thus suggesting that the PM soot model was well configured and the results are acceptable to almost similar actual combustion environment.

Further qualitative analysis by post-processing software demonstrates that soot mass concentration pattern is comparable with calculated value. It is demonstrated that heavier soot particles are more concentrated at the lower part of the cylinder (squish/bowl region) instead of the upper side of the cylinder. Until the end of simulation, lighter soot particles which occupy the upper side of the cylinder are expected to either move out through exhaust manifold or deposited onto the cylinder wall.

Nowadays, development of CFD software in simulating the engine combustion phenomenon with higher complexity has become more advanced, thus allowing more detailed soot model to be analyzed in more efficient ways. In order to endorse the results from this simulation method, validation by real combustion event through experimental work will remain an important criterion which is planned to be performed in the next future. Overall, CONVERGE-CFD software provides a good opportunity for soot particle study by various soot models.

The authors would like to acknowledge the support from Ministry of Higher Education of Malaysia through its research grant of FRGS/1/2013/TK01/UKM/02/2. All the funding is well translated by Department of Mechanical and Materials Engineering, Faculty of Engineering and Built Environment, Universiti Kebangsaan Malaysia (UKM) for its beneficial uses. The authors are pleased to acknowledge Public Service Department support for Fadzli Ibrahim through the HLP Programme. Special thanks to STRIDE for granting the use of its facilities regarding to this project.

References

1. V. Ganesan, *Internal Combustion Engines* (McGraw-Hill, 2004).
2. P. Eastwood, *Particulate Emissions from Vehicles* (John Wiley & Sons, 2008).
3. W.W. Pulkrabek, *Engineering Fundamentals of the Internal Combustion Engine* (Pearson Prentice Hall, 2004).
4. M.F.Gómez-Rico, I. Martín-Gullón, A. Fullana, J.A. Conesa, R. Font. J. Anal. Appl. Pyrolysis **68-69**, 527–546 (2003).
5. B. Brunekreef, S.T. Holgate. Lancet. **360,** 1233–1242 (2002).
6. J. Kagawa. Toxicology. **181-182**, 349–353 (2002).
7. N. Englert. Toxicol. Lett. **149**, 235–242 (2004).
8. K. Kim, E. Kabir, S. Kabir. Environ. Int. **74,** 136–143 (2015).
9. G. Woschni, K. Huber. SAE Tech. Pap. No. 910297 1–8 (1991).
10. R. Mainwaring. SAE Tech. Pap. No. 971631 47–64 (1997).
11. S. George, S. Balla, M. Gautam. Wear. **262,** 1113–1122 (2007).
12. S. George, S. Balla, V. Gautam, M. Gautam. Tribol. Int. **40,** 809–818 (2007).
13. F. Ibrahim, W.M.F. Wan Mahmood, S. Abdullah, M.R. Abu Mansor. Def. S&T Tech. Bull. **8**, 141–152 (2015).
14. F. Ibrahim, W.M.F. Wan Mahmood, S. Abdullah, M.R. Abu Mansor. *ReCAR 2015* (2015).
15. I.M. Kennedy. Prog. Energy Combust. Sci. **23**, 95–132 (1997).
16. K.J. Richards, P.K. Senecal, E. Pomraning. *CONVERGE (v2.2.0)*. (Convergent Science Inc. 2014).
17. A. Kazakov, M. Frenklach. Combust. Flame **114**, 484–501 (1998).
18. M. Frenklach, H. Wang. *Twenty-Third Symp. Combust.*, 1559-1566 (The Combustion Institute, 1990).
19. M.Y. Choi, A. Hamins, G.W. Mulholland, T. Kashiwagi. Combust. Flame. **99**, 174–186 (1994).
20. T. Zeuch, G. Moréac, S.S. Ahmed, F. Mauss. Combust. Flame. **155**, 651–674 (2008).
21. J. B. Heywood. *Internal Combustion Engine Fundamentals*. (USA, 1988).

22. R.D. Reitz, F.V. Bracco. *Encycl. Fluid Mech.*, (Gulf Pub. 1986).
23. L.M. Ricart, J. Xin, G.R. Bower, R.D. Reitz. SAE Tech. Pap. No. 971591. 13–31 (1997).
24. A.B. Liu, D. Mather, R.D. Reitz. SAE Tech. Pap. Ser. No. 930072. 1–13 (1993).
25. D.P. Schmidt, C.J. Rutland. J. Comput. Phys. **164**, 62–80 (2000).
26. J.D. Naber, R.D. Reitz. SAE Tech. Pap. Ser. No. 880107. 1–23 (1988).
27. M.A. González, Z.W. Lian, R.D. Reitz. SAE Tech. Pap. Ser. No. 920579. 1–13 (1992).
28. Z. Han, R.D. Reitz. Combust. Sci. Technol. **106**, 267–295 (1995).
29. P.K. Senecal, E. Pomraning, K.J. Richards, T.E. Briggs, C.Y. Choi, R.M. McDavid, M.A. Patterson. SAE Tech. Pap. 2003-01-1043 (2003).
30. Computational Engineering International Inc. *EnSight User Manual for version 10.1*. (Computational Engineering International, Inc. 2013).
31. D.R. Tree, K.I. Svensson. Prog. Energy Combust. Sci. **33**, 272–309 (2007).
32. J.P.R. Symonds, K.S. Reavell., J.S. Olfert, B.W. Campbell, S.J. Swift. J. Aerosol Sci. **38**, 52–68 (2007).
33. B. Giechaskiel, M. Maricq, L. Ntziachristos, C. Dardiotis, X. Wang, H. Axmann, A. Bergmann, W. Schindler. J. Aerosol Sci. **67**, 48–86 (2014).

Investigation of passenger car using Macpherson strut for suspension system pt.1: Vehicle behaviour variation of time response

Tanti Ali[1,*], *Rosli* Abu Bakar[1], *Gan* Leong Meng[1] and *Zarina* Amat Jafar[1]

[1]Faculty of Mechanical Engineering, Universiti Malaysia Pahang, 26600 Pekan, Pahang, Malaysia

Abstract. It has been decade, researchers has been conducting researches on the topics concerning vehicle behavior. Suspension system, driving maneuver and road profile are the particular parameters in order to achieve the aim in vehicle behavior understanding. This paper combined these three criteria by means of using a passenger car equipped with MacPherson strut front suspension undergoes different driving behavior. The objective of this paper is to study the effect of passenger car equipped with MacPherson strut front suspension system vehicle behavior based on different driving maneuvers. For this study, Proton Persona Sedan 1.6 Manual Transmission Base Line was used to investigate the MacPherson strut suspension system. Data were collected using DEWESoft Software. As the velocity and direction varies with time, the vehicle response subjected to stationary excitation, while it varies with different velocity and different type of road. Acceleration, deceleration and bumpy test the suspension mechanism support the weight of the vehicle yet to cushion bumps and holes in the road. It can be concluded that the MacPherson strut suspension system has an effect on not only vehicle behavior but also comfort ride. These findings provide the following insights for future research in suspension vibration in order to optimize the performance of the MacPherson strut suspension system.

1 Introduction

Road vehicle suspensions producing momentum and guidance forces are essential for driving safety. The driver is controlling these forces during accelerating, braking and steering the vehicles [1]. For a road vehicle, the suspension system isolates the occupants or cargo from severe levels of shock and vibration induced by the road surface. This isolation from road induced shock and vibration also improves the longevity and durability of the vehicle [2].

Just as importantly, the suspension system enables the wheels to maintain contact with the road, assuring stability and control of the vehicle [2]. The suspension system allows the springs and shock absorbers to absorb the energy of the bump, so drivers and passengers can have smooth ride. While doing these two jobs, the suspension system must not allow

* Corresponding author: tanty.ali@gmail.com

loose, uncontrolled movement of the tire and wheel and must keep the alignment of the tire as correct as possible [3,4]. Fig. 1 shows MAcPherson strut suspension diagram.

Fig. 1. MacPherson strut suspension.

Earle was the one whom patented MacPherson strut with the idea of locating the lower end of an inclined strut system tangentially with transverse link (or track control arm) and longitudinally by means of the lever arm of the anti-roll bar [2]. Thus, this paper will study the effect of passenger car equipped with MacPherson strut front suspension system vehicle behavior based on different driving maneuvers.

1.1 Quarter car model

In general, there are two kinds of approaches in dynamic analysis of a vehicle. The first approach relies on experimental analysis and field test; the second one utilizes computer simulation to conduct a numerical analysis. Fig. 2 shows quarter vibration model in suspension system analysis [5,6,7].

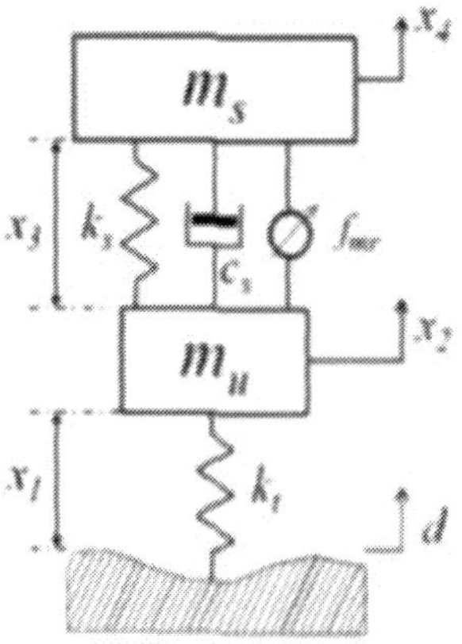

Fig. 2. Quarter car vibration model.

The general equation of motion for any vibration model assigned to vehicle can be defined as follow [5,6,7,8,9]:

$$[M]\{\ddot{X}\}+[C]\{\dot{X}\}+[K]\{X\}=\{P\} \tag{1}$$

Where: $[M]$ is mass matrix, $[C]$ is damping coefficient matrix, $[K]$ is stiffness matrix and $\{P\}$ is defined as imposed force on the vehicle by the road.

2 Design of experiment

In this study, the experiment is using Proton Persona Sedan 1.6 Manual Transmission Base Line. This test car is equipped with MacPherson strut for the front suspension and multi-link with stabilizer for the rear suspension. In addition, the test car is equipped with wire potentiometer, accelerometer, GPS, speed, tire slip angle, gyroscope, rotating wheel sensor that connected to the data acquisition system that has been installed with DEWESoft Software to collect and monitor the data. To check the performance on varying speed ranging from 20 to 120 km/h [10]. The gear shifting for the experiment need to follow the maximum speed of each gear (Table 1).

Table 1. Maximum speed of gear.

Gear	Maximum Speed
1	1-20 km/h
2	21-40 km/h
3	41-60 km/h
4	61-80 km/h
5	>81 km/h

2.1 Acceleration

The testing started recording at the static condition (velocity 0 km/h). The pedal is pressed based on the maximum velocity for each gear (Table 1). This test is run for five times to gain the consistent results. This procedure is repeated for each different speed which on paved road 0-60 km/h and 0-90 km/h whiles the unpaved road is at 0-40 km/h and 0-60 km/h.

2.2 Deceleration

The car is drove until reached 90 km/h and the speed is maintained for deceleration 90 km/h to 0 km/h. When the car reached the first cone reference spot (i) as in Fig. 3, the test is started to record and the clutch paddle is pressed. The velocity is cotrolled based on the maximum velocity for each gear (Table 1). As the front tire car reach the second cone reference spot (ii), the clutch paddle is released and the brake pedal is pressed steadily until the car has stopped within the stopping range. The recording is stopped when the final speed is at 0 km/h. The test is repeated five times. This procedure is repeated for each different speed which on paved road 60-0 km/h and 90-0 km/h whiles the unpaved road is at 45-0 km/h and 65-0 km/h.

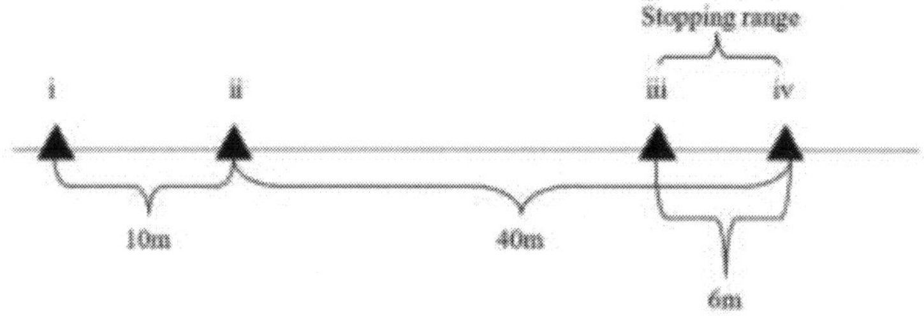

Fig. 3. Cone placement for deceleration testing.

2.3 Cornering

Test car is drove at the inner diameter of the road until reached the speed of 25 km/h. The driver needs to follow the maximum velocity for each gear (Table 1) that has been set. When the velocity speed reached 25 km/h the recording started. Based on Fig. 4, it ended as the car finishes a complete circle of the roundabout which this can be referred on the GPS reading. This test is repeated for five times to ensure the reliability of data gained. The test is repeated at the outer diameter for 40 km/h and a 50 km/h while at the roundabout with inner diameter is running the initial velocity at 25 km/h and 40 km/h. At the inner diameter, 50 km/h velocity is not conducted due to safety purpose which is already been considered.

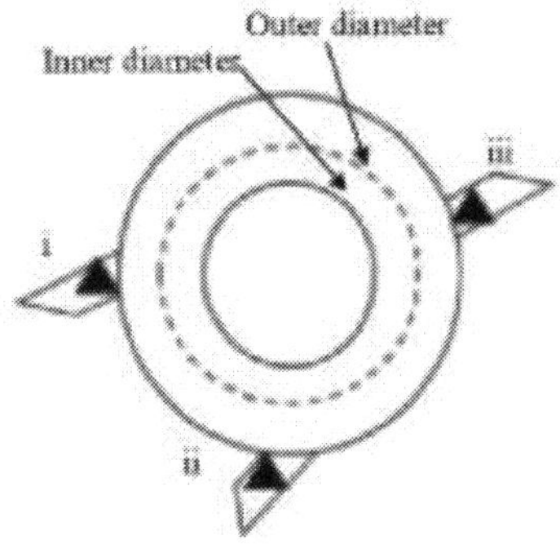

Fig. 4. Cone placement for cornering testing.

2.4 Bumpy

The test car is drove until velocity 30 km/h is reached and maintained. The driver has to press the pedal based on the maximum velocity for each gear (Table 1). The test is recorded when the front tire reached the first cone spot (i) as in Fig. 5 and ended when the rear tire passed by the second cone (ii). This test is repeated five times and then followed by different constant velocity at 40 km/h and 50 km/h.

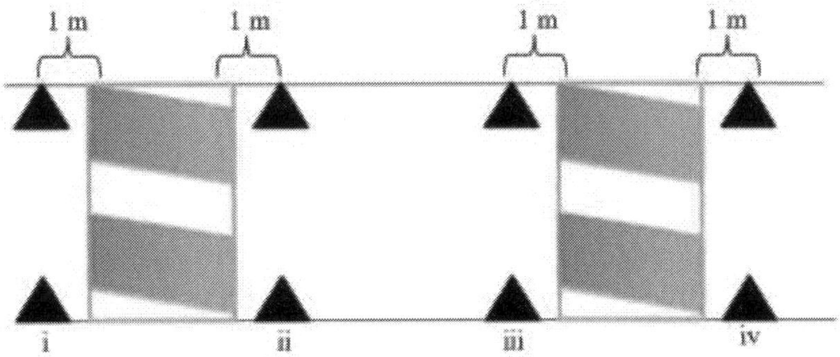

Fig. 5. Cone placement for bumpy testing.

3 Results and discussion

3.1 Variation of time response

In this part showed the variation of testing velocity in time response. The Fig. 6 until Fig. 9 below showed the relationship of the velocity and direction of the vehicle with the increment of time. These graph clarified the different of the driver driving behavior. The pattern of graph displayed the velocity is increased during acceleration (Fig. 6) and decreased during deceleration (Fig. 7).

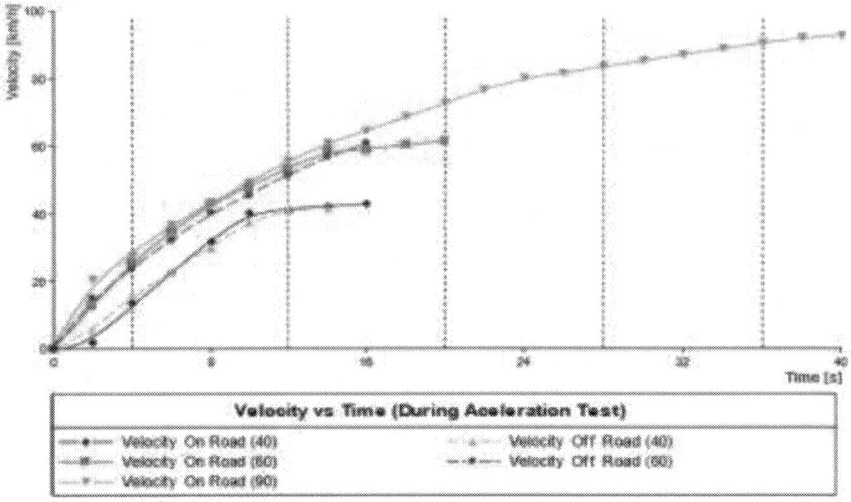

Fig. 6. Graph of Velocity versus Time during the acceleration test.

The time response for the acceleration test is longest during the test 90km/h on road that results in 40 seconds. As for the first acceleration test from 0 km/h to 40 km/h the testing is only at the 2nd gear. After the driver reached the speed vehicle at 40 km/h the driver need to maintain until 5 seconds. This is for storing data by DEWESoft whereby the results will be constant at final velocity at the end of the testing.

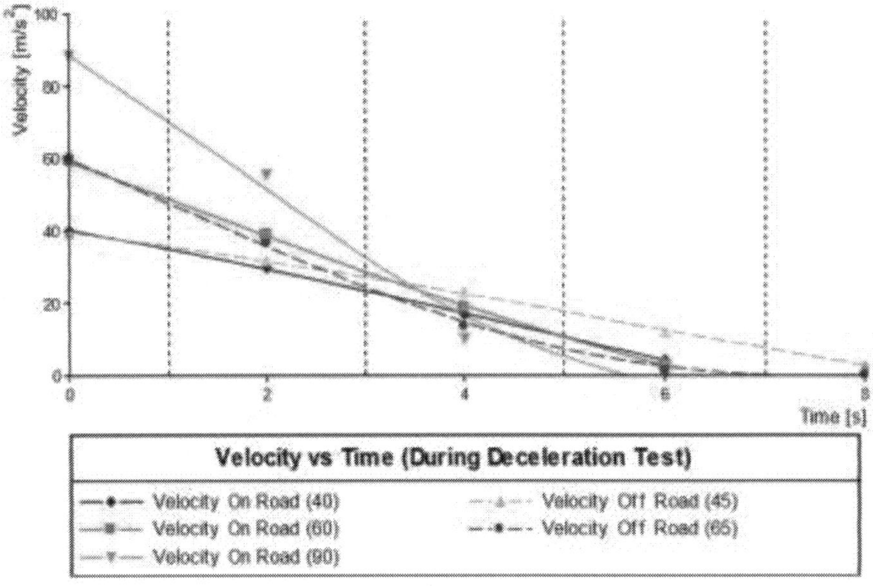

Fig. 7. Graph of Velocity versus Time during the deceleration test.

The longest time taken during the deceleration test on the off road at the speed of 45 km/h. Due to the off road condition, the tire undergo slip condition. In addition, during the deceleration test, the clutch paddle is pressed when the car is almost to stop in order to avoid engine end abruptly. So, this experiment demonstrated that the performance of the suspension system is typically rated as to provide passenger comfort and avoid hitting [11].

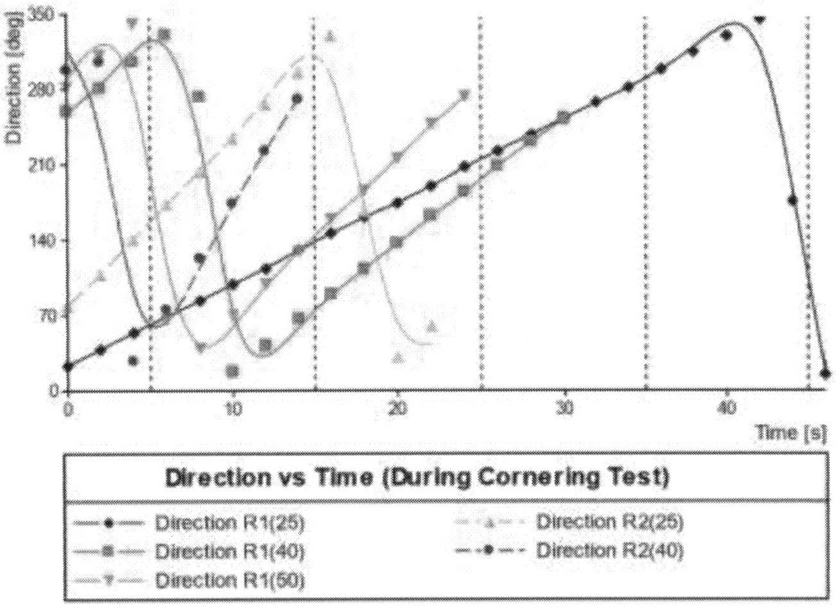

Fig. 8. Graph of Direction versus Time during the cornering test.

The direction of the test car is projectile in the cornering test as shown in Fig. 8, it can be observed that during the testing, test car is at constant speed. The test car results in constant after five second on the bumpy test. Besides, deflections of front suspensions at different speed are shown in Fig. 9. It is assumed that the deflection for right and left wheels is approximately same [7]. Nevertheless, these graphical representation varied concerning on comparison between speed, vehicle acceleration and road roughness effects [9].

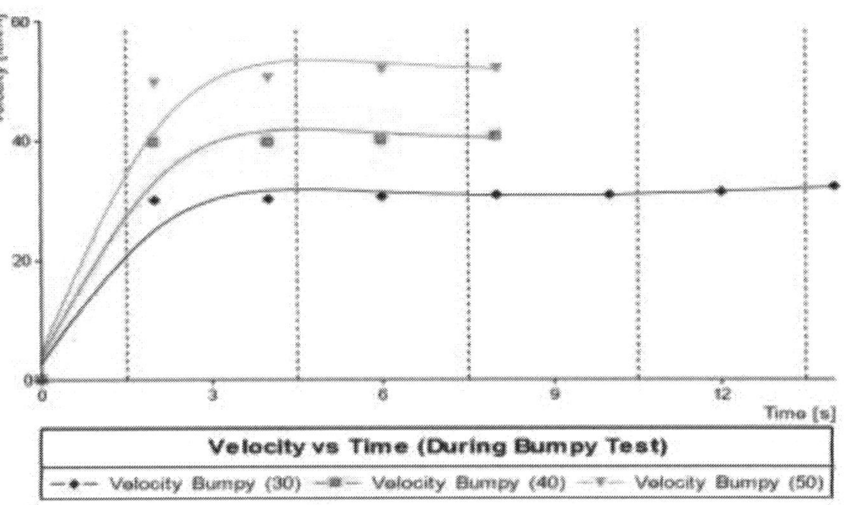

Fig. 9. Graph of Velocity versus Time during the bumpy test.

As the velocity and direction varies with time, the vehicle response subjected to stationary excitation, while it varies with different velocity and different type of road. Theoretically, throughout acceleration, deceleration and bumpy test the suspension mechanism not only support the weight of the vehicle but it cushioned the bumps and holes in the road. During the steady state cornering the suspension acted to sustain traction between the tires and the road in order to align the wheels [13].

The term "ride comfort" is related to absorption of the energy from road bumps and its dissipation without causing undue oscillation in vehicle. Undue oscillation, also known as vibration, affects the ride quality of the passenger and driver [11]. The term "road handling" relates with vehicle weight transfer from side to side and front to back during cornering and braking action. Thus, a vehicle with good road handling minimize the weight transfer and thus will add in increasing the traction force between vehicle and road surface [12,13].

4 Conclusions

Generally, road vehicle suspensions are producing vertical forces in the contact of the tires on the road that provide horizontal contact forces for vehicle propulsion and guidance. These static contact forces existed by dynamic loads due to suspension system behavior based on the unevenness surface. This paper has achieved the objective of showing the pattern of vehicle behavior with different driving maneuvers based on experimental results. It can be concluded that the MacPherson strut suspension system has an effect on not only vehicle behavior but also comfort ride. Conversely, both of ride comfort and vehicle handling requirements are conflicting while designing a vehicle suspension. Consequently, future study can be continued by analyzing the MacPherson strut suspension behavior in order to optimize the standard of ride comfort and vehicle handling under all driving conditions.

The authors would like to thank the Faculty of Mechanical Engineering Universiti Malaysia Pahang (FKM-UMP) final year students Year 2010 for the encouragement to carry this research work. Also the technical staff of FKM-UMP for the assistance during the experimental session using the UMP Test Car.

References

1. W. Schiehlen, I. Iroz, Uncertainties in road vehicle suspensions, *Procedia IUTAM*, **13**, 151-159, (2015).
2. D. Bastow, G. Howard, J.P. Whitehead, Car suspension and handling, *Wallendale: Society of Automotive Engineers Inetrnational*, (2004).
3. T.W. Birch, *Automotive suspension and steering system*, New York: Delmar Publisher, (1999).
4. W.H. Crouse, D.L. Anglin, *Automotives mechanics*, New York: McGraw-Hill, 658-710, (1993).
5. H.B. Dehkordi, *Vibration and force analysis of lower arm of suspension system*, Universite Du Quebec, (2014).
6. A. Kanjanavapastit, A. Thitinareumit, Estimation of a speed hump profile using quarter car model, Procedia Social and Behavioral Sciences, **88**, 265-273, (2013).
7. A. Gupta, V. Rastogi, Effects of various roadconditions on dynamic behavior of heavy road vehicle, Procedia Engineering, **144**, 1129-1137, (2016).
8. E. Alvarez-Sanchez, A quarter-car suspension system: car body mass estimator and sliding mode control, Procedia Technology, **7**, 208-214, (2013).

9. M. Hadpe, D.S. Deshmukh, P.M. Solanki, Vibration analysis of a two wheeler (Analytically), International Journal of Innovative Research in Science, Engineering and Technology, **3**, 415-421 , (2014).
10. M.P. Nagarkar, G.J.V. Patil, R.N.Z. Patil, Optimization of nonlinear quarter car suspension-seat-driver model, Journal of Advanced Research, (2016).
11. A. Mehmood, A.A. Khan, Vibration analysis of damping suspension using car models, International Journal of Innovation and Scientific Research, **9**, 202-211, (2014).
12. S.K. Sharma, V. Pare, M. Chouksey, B.R. Rawal, Numerical studies using full car model for combined primary and cabin suspension, Procedia Technology, **23**, 171-178, (2016).
13. T.Z.S. Ali. *Vibration investigation for passenger car front suspension system under various road condition and driving maneuvers.* Universiti Malaysia Pahang, (2010).

Effects of fractal grid on emissions in burner combustion by using fuel-water-air premix injector derived from biodiesel crude palm oil (CPO) base

Mirnah Suardi[1,*], *Amir* Khalid[1], *Mohd Azahari* Razali[1], *Azwan* Sapit[1] and *Shahrin Hisham* Amirnordin[1]

[1]Combustion Research Group (CRG), Centre for Energy and Industrial Environment Studies (CEIES), Faculty of Mechanical and manufacturing Engineering, Universiti Tun Hussein Onn Malaysia, 86400 Parit Raja, Batu Pahat, Johor

Abstract. The alternative fuel is attracted good attention from worldwide especially for renewable and prevention energy such as biodiesel. Biodiesel is one of the hydrocarbon fuels and it has potential for external combustion. As one of the different solutions to these problems, rapid mixing of biodiesel-water-air technique is one of the most significant approaches to improve the combustion and reduce the emissions. The gas emission can be reduced by two methods. First is by improving an injector with fractal and the other is by using a biodiesel-water mixture as an alternative fuel. Mixing of water with fuel in the combustion process is a low cost and effective way. This research used biodiesel Crude Palm Oil (CPO) as fuels in which blended with diesel. This study investigated the effects of water content and equivalence ratio on emissions with the rapid mixing injector. Fuels used are diesel, CPO5, CPO10 and CPO15 and the exhausts gaseous tested are CO, CO_2, HC and NO_X. The gas emissions processes are tested by using the gas analyzer. In this research, water premix of percentage up to 15vol% and blending biodiesel ratio was varied from 5vom% - 15vol%. The result shows that increasing of water content will effected decrement of CO, CO2 and HC emissions but increasing the NO_X emissions.

1 Introduction

Emission are main issue in the observation of external combustion characteristics. In combustion process, Nitrogen oxide (NO_X) reacts with water and other compounds to form various acidic compounds, fine particles and ozone. These pollutants can remain in the air for days or even years. NO_X are known to be immediately dangerous to human health and environment. Effects of NO_X are decreases in lung function, resulting in difficulty breathing, shortness of breath, and other symptoms [1-3]. Besides that, NO_X can

* Corresponding author: mirnahsuardi@yahoo.com

be reacted with other pollutants to form toxic chemical as well as contribute to the formation of acid rain. CO is highly toxic to human body even at lower levels [4].

In the future, the growth of population and economic require more energy and natural resources. Increasing in environmental issues and the cost of fossil fuel has resulted in research on alternative fuel such as biodiesel. Biodiesel as the renewable resource continue to be of interest research in order to achieve a sustainable energy economy thus reducing dependence on fossil fuels. Moreover, worldwide demand on the biodiesel as an alternative fuel is also increasing due to the energy sources crisis.

Advantage of biodiesel is an environmentally friendly fuel, which is one of the clean and renewable energy resources. The fuel is usually made from animal fat or vegetable oil revenue trans-esterification reaction. The oxygen content in biodiesel is 11% to 15%, which results in increasing the combustion process and reducing the emission from diesel engines [5-6]. The development of biodiesel is increase in particularly for Crude palm oil (CPO). CPO is readily available, safe to store and handle, and most importantly, totally renewable. Its negligible sulphur content compared with diesel's reduces the possibility of acid rain caused by sulphur dioxide emission [7].

CPO can be made into a biodiesel through the process of trans-esterification of triglycerides with methanol. The product of this process is known as palm oil methyl-ester (POME), or palm oil diesel. Studies conducted on POME [7-9] have shown promising results compared with diesel, in terms of combustion process and emission. However, POME is more expensive than CPO, due to the chemical and mechanical processing involved. The interest in CPO as a diesel substitute stems from its simple production process, which makes it less expensive than POME.

The variation of flame speed with mixture strength roughly follows that of flame temperature. In almost all cases, the maximum value occurs at an equivalence ratio of between 1.05 and 1.10. Notable exceptions to this general rule are hydrogen and carbon monoxide [10-12]. Equivalence ratio (ϕ) is an informative paranormal uses normally uses to show the composition of quality where the oxidize fuel is rich, lean or stoichiometry. The equivalence ratio for this research was 0.6 up to 1.4, which occur in the mixing.

Rapid premixing in the combustion process is to improve the combustion and reduce the gas emission especially NO_X emissions [13-14]. This study investigated the effect of water content and equivalence ratio on combustion and emissions with the rapid mixing injector. The fuels used are diesel, CPO5, CPO10 and CPO15. The gas emission processes are tested by using the gas analyzer. The exhausts gaseous tested are CO, CO_2, HC and NO_X. In this research, water premix of percentage up to 15vol% and blending biodiesel ratio was varied from 5vol% - 15vol%.

2 Experimental setup

The study used three types of biodiesel fuel blends and diesel fuel as a base line. Crude palm oil (CPO) was used to produce the biodiesel blending varied from 5vol %(CPO5) – 15vol %(CPO15). The biodiesel blends was prepared using biodiesel blending machine. Fuel properties are shown in Table 1. Figure 1 show the schematic diagram of the experiment and the schematic diagram of rapid mixing injector that being used. The experiment starts by injecting the fuel, air and water into the injector. The injector system uses fuel-air internal rapid mixing concept. The premixing injector was firstly designed by Y. Kidoguchi [10]. The air is pressurized at 0.25 bar. The fuel flow rate and water flow rate are controlled by flow meter and speed controller. Fuel-water-air is mixed at mixing chamber. Then, the mixtures are sprayed out from the nozzle. The mixtures will be ignited using blow torch. The emissions were measured using Autocheck 4/5gas. The Autocheck 4/5 analyser was used to measure the concentration of gas; HC, CO, CO_2, NO_X. In this study, the composition of the gas being measured is on the exhaust products available on burning furnace. These analyses were inserted at the end of exhaust tail pipe through exhaust probes and the sampling rates for all devices were set 45 seconds. The details of the experimental condition are in Table 2.

Table 1. Properties of fuels.

Fuel Type	Properties			
	Density (g/cm^3)	Kinematic Viscosity (cP)	Flashpoint (°C)	Water Content (ppm)
STD	0.833736	3	80	79.6
CPO5	0.837048	3	91.5	120.1
CPO10	0.837664	2.9	92	158.6
CPO15	0/840428	3	93.5	219
CPO20	0.841172	3.1	94.5	294.7
CPO25	0.841716	3	97	363.3
CPO30	0.845852	3.2	97.5	397.1
CPO35	0.844816	3.4	99.5	426.9
CPO40	0.848236	3.2	100	558

Table 2. Experimental condition.

Air Compressor	Model	QUASA HDC-D3050
	Capacity, L/min	200
	Pressure, kg/cm^2	8
Water Pump	Model	SFDP1-014-080-22-Seaflo
	Voltage, V	12
	Flow rate, L/min	5.1
Fuel Pump	Model	CNY-3805
	Pressure, bar	3
	Flow rate, L/Hr	100
DC Voltage Regulator	Model	Teletron TC-1206A
	Current, A	64 (max)
Operating condition	Air Pressure, bar	0.25
	Air Density, kg/m^3	1.16
	Ambient Temperature, K	300
	Water Percentage	0-15%
	Equivalence Ratio	0.6-1.4

3 Results and discussion

3.1 Effects of Diesel and CPO 5 vol% under variants water content on emissions

Figure 2 shows the concentration of CO, CO_2, HC and NO_X against equivalence ratio at variants water content. As shown in the figure, the NO_X emissions decrease when the amount of water content is increases between ϕ=1 and ϕ=1.4, where pressure of the water weakens the combustion flames and decreases the maximum temperature of the diesel. The CO and HC increase drastically for ϕ=0.6 to ϕ=0.8 despite little change observed in CO_2. As observed, increasing percentage of water content has decrease the HC as combustion mixture is rich. Highest content of HC occurs at diesel fuel which is 33 ppm at ϕ =1.4 followed by CPO5 (5vol%), CPO10 (10vol%) and CPO15 (15vol%).

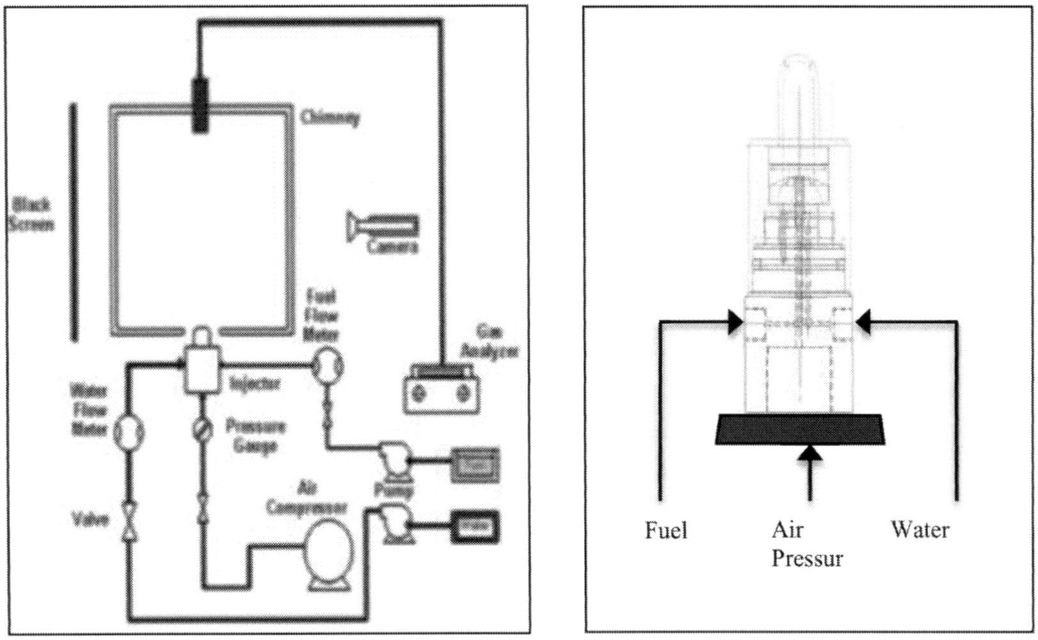

(a) (b)

Fig. 1. (a) Schematic diagram of experiment. ; (b)Schematic diagram of rapid mixing injector.

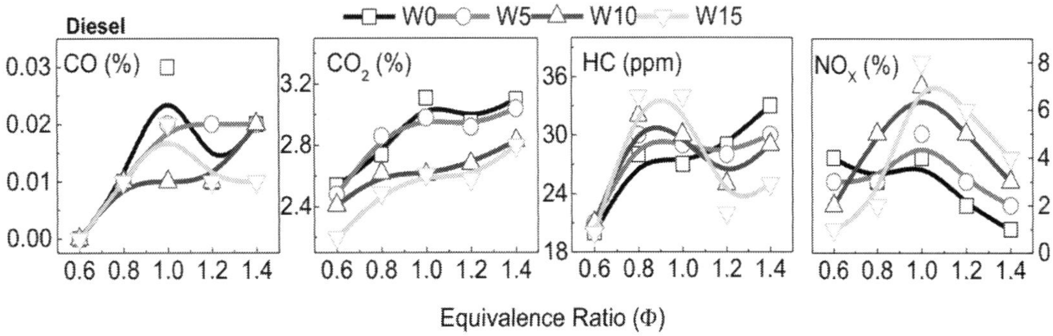

Fig. 2. Effect of water content on Diesel.

Based on the Figure 3, due to the properties of the biodiesel CPO the NO_X was increases which by the difference in physical and chemical properties of diesel with biodiesel. Biodiesel CPO base has significant oxygen content compared to the diesel and higher content of oxygen to fuel in combustion which can leads to better complete combustion to produce higher temperature and more NO_X formation. Biodiesel CPO also produces lower HC compare to the diesel. This because the homogeneous flame produces less HC and also helps to increases oxidation as well as reduces the amount of HC formation. HC is decreases when the increases of water content which shows similar trend as diesel. The lowest amount of HC is 21-28 ppm at all water content when ϕ= 0.6. The amount of CO_2 also decreases as the water content increases with lower value compared to diesel. Higher amount of oxygen in CPO reduces fuel carbon content which able to reduces to CO_2.

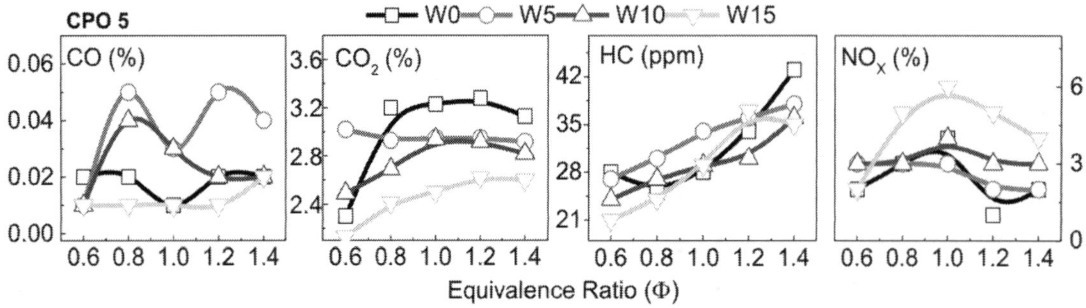

Fig. 3. Effect of water content on CPO5.

3.2 Effects of high blending ratio of CPO biodiesel and variants water content on emissions

The effect of water content on CPO 10 is shown in Figure 4. In the figure, increment of water content will affected decreasing of percentage of CO_2. Higher oxygen content at W15 because the more complete combustion occurs when more water is added. CO shows ununiformed trend which CPO10 decreases to lowest for W10 at equivalence ratio of 1.4 which is 0.01%. The amount of HC was lowest when increasing of water content. Apparently, increasing of water content will decrease the NO_X at equivalence ratio of ϕ=0.8 to 1.0.

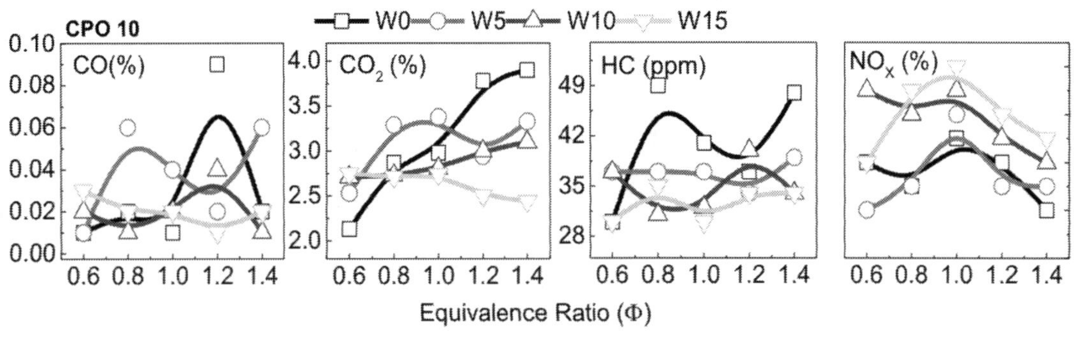

Fig. 4. Effect of water content on CPO10.

Shown in Figure 5 is the effect of water content on CPO15. CO, CO_2 shows similar trend also since CO is less generated for biodiesel CPO because the concentrated oxygen in biodiesel would enhance the oxygenated fuels like biodiesel and reduce to amount of CO. Biodiesel CPO also produce lowest NO_X at W0 which 2%. Considering to the results, increases of percentage of water content, the HC emission has directly proportional increasing and inversely proportional to the NO_X.

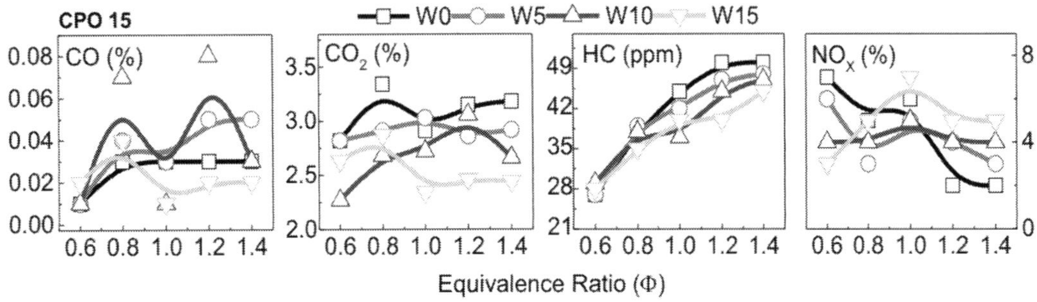

Fig. 5. Effect of water content on CPO15.

4 Conclusions

Rapid mixing of biodiesel-air-water technique is one of the most significant approaches to the reduction in emissions, such as nitrogen oxides (NO_X). The solution of this issue is by using biodiesel fuel as an alternative fuel from crude palm oil (CPO). In addition, the CPO is one of the most economical options for producing biodiesel due to the biodegradable properties and preserves energy. The emission of NO_X was significantly lower at high temperatures and low oxygen concentration as compared to low temperatures and high oxygen concentration.

For diesel fuel, the increases of percentage of water content, the NO_X emission has proportionally decreasing. CPO has significant oxygen content compared to the diesel and higher content of oxygen to fuel in combustion which can leads to better complete combustion to produce higher can temperature and more NO_X formation. CPO also produces lower HC compare to the diesel. This because the homogeneous flame produces less HC and also helps to increases oxidation as well as reduces the amount of HC formation. HC increases as the equivalence ratio increases. Observation on CO emission produces similar trend with HC and CO_2. Higher amount of oxygen in CPO reduces fuel carbon content which able to reduces to CO_2 where temperature show similar trend with diesel.

The authors also would like to thank the Ministry of Higher Education Malaysia for supporting this research under Exploratory Research Grant Scheme (ERGS) vot. E033 and partly sponsored by the Centre for Graduate Studies UTHM.

References

1. Payri R, Desantes JM, Salvador FJ, Manin J. "*Influence On Diesel Injection Characteristics And Behavior Using Biodiesel Fuels*". SAE paper 2009-01-0851; **200**.(2009)
2. Khalid, A., "*Effect of Ambient Temperature and Oxygen Concentration on Ignition and Combustion Process of Diesel Spray*", Asian Journal of Scientific Research, Volume **6**, Issue 3, 2013, Pages 434-444, Asian Network for Scientific Information, DOI: 10.3923/ajsr.2013.434-444 (2013)
3. A. Khalid, N. Jaat, A. Sapit, A. Razali, B. Manshoor, I. Zaman and A.A. Abdullah, "*Performance And Emissions Characteristics Of Crude Jatropha Oil Biodiesel Blends In A Diesel Engine*", International Journal of Automotive and Mechanical Engineering (IJAME) ISSN: 2229-8649 (Print); ISSN: 2180-1606 (Online); Volume **11**, pp. 2447-

2457, January-June 2015, DOI: http://dx.doi.org/10.15282/ijame.11.2015.25.0206. (2015)

4. Khalid, A., Tamaldin, N., Jaat, M., Ali, M.F.M., Manshoor, B., Zaman, I., "*Impacts of Biodiesel Storage Duration on Fuel Properties and Emissions*", Procedia Engineering, Volume **68**, 2013, Pages 225-230, Elsevier Ltd, DOI: 10.1016/j.proeng.2013.12.172 (2013)
5. Y. Kidoguchi, T. Yatsufusa, and D. Nakagawa, "*Improvement of Emissions and Burning Limits in Burner Combustion using an Injector on the Concept of Fuel-water Internally Rapid Mixing*." (2011)
6. T. Yatsufusa, T. Kumura, Y. Nakagawa, and Y. Kidoguchi, "*Advantage of Using Water-Emulsified Fuel on Combustion and Emission Characteristics*," pp. 2–7. (2009)
7. A. Hussein, A. Hamid, and R. Atan, "*Spray characteristics of jet – swirl nozzles for thrust chamber injector*," Aerosp. Sci. Technol., vol. **13**, no. 4–5, pp. 192–196, (2009)
8. X. Miao, Y. Ju, and X. Wang, "*Experimental Investigation on Premixed Combustion in a Diesel Engine with Ultra-Multihole Nozzle*," no. 1, (2009)
9. I. Zahmatkesh and M. Moghiman, "*From Turbulent Spray Flames*," vol. **30**, (2006)
10. ArthurH.Lefbvre and D. R. Ballal, *Gas Turbine Combustion Alternative Fuels and Emissions*. (Third Edition. 2010)
11. Y. Yoshimoto, M. Tsukahara, T. Murayama, "*Studies on the Microexplosion of Emulsified Fuels: 1st Report, Effects Of Fuel Properties, Water Contents and Particle Sizes on the Microexplosion on A Hot Surface*", J. JSME (B), 55 (519), pages 3538-3543 (1989)
12. Y. Xu, C.G. Zhu, Y.B. Shen, D. Poulikakos, "*Holographic Investigation of Microexplosion in An Emulsified Diesel Oil Spray, Atomization and Sprays*", 1993, 3 (2), pages 193-202 (1993)
13. Michele Battistoni, Carlo Nazareno Grimaldi. "*Numerical Analysis Of Injector Flow And Spray Characteristics From Diesel Injectors Using Fossil And Biodiesel Fuels*". Applied Energy 97, pages 656–666 (2012)

Predicting the performances of a CAMPRO engine retrofitted with liquefied petroleum gas (LPG) system using 1-dimensional software

M. Hazeem Kamaruddin[1,*], *Shahrul Azmir* Osman[1], *Mas* Fawzi[1], *Norrizal* Mustaffa[1] and *Mohd Mustaqim* Tukiman[1]

[1]Automotive Research Group (ARG), Center of Energy and Industrial Environment Studies (CEIES), Faculty of Mechanical and Manufacturing Engineering, Universiti Tun Hussein Onn Malaysia, 86400 Parit Raja, Johor Darul Takzim, Malaysia

Abstract. Recently, the depletion of petroleum resources and the impact of exhaust emission caused by combustion towards environmental has been forced to all researchers to come out with an alternative ways to prevent this situation become worse. Liquefied petroleum gas (LPG) is the most compatible and have a potential to become a source of energy for internal combustion engine. Unfortunately, the investigation of LPG in internal combustion engine among researcher still have a gap in research. Thus, in this study a 1-Dimensional simulation CAMPRO 1.6L engine model using GT-Power is developed to predict the performances of engines that using LPG as a fuel for internal combustion engine. The constructed model simulation will throughout the validation process with the experimental data to make sure the precision of this model. The validation process shows that the results have a good agreement between the simulation model and the experimental data. As a result, the performance of LPG simulation model shows that a Brake Torque (BT), Brake Power (BP) and Brake Mean Effective Pressure (BMEP) were significantly improved in average of 7% in comparison with gasoline model. In addition, Brake Specific Fuel Consumption (BSFC) also shows an improvement by 5%, which is become more economic. Therefore, the developed GT-Power model offer a successful fuel conversion to LPG systems via retrofit technology to provide comprehensive support for implementation of energy efficient and environmental friendly vehicles.

1 Introduction

The depletion in petroleum resources have been a popular issue among others country recently. Excessive use of fossil based fuels exhausts the reserves and also increases the air pollution [1]. The researchers and automotive manufacturer are forced to concentrate on finding alternatives to conventional petroleum fuels. An alternative fuel must be technically feasible, economically competitive, environmentally acceptable, and readily available [2]. Various potential alternative fuels have been as expected, such as, natural gas, biodiesel,

* Corresponding author: hazeemkamaruddin@gmail.com

methanol, ethanol, hydrogen, and liquefied petroleum gas (LPG) [3]. Among of alternative fuel, as an environmentally friendly and economic fuel, LPG can help keep planetary citizens moving while minimizing the impact of road transport on climate, weather, natural resources, and the human health [4].

LPG (also known as "Autogas") is a gas product of petroleum refining primarily consisting of Propane, some propylene, Butane and other light hydrocarbons [5]. Around 60% of total amount of LPG produced is recovered directly from oil and gas fields in which case no actual refining is needed. The remaining 40% is formed as a by-product in crude oil processing either in distillation phase or after treatment processes [6]. LPG is suitable and convenient as a fuel for internal combustion engine due to the higher Octane number but a low Cetane number which is close to 105 and also it has high calorific value compared to other gaseous fuels [7], [8]. LPG can be liquefied at low pressures, in the range 0.7–0.8 MPa, and low atmospheric temperatures. It has higher auto ignition temperature, greater flame velocity and wider flammability limits make LPG a better spark-ignition (S.I) engine fuel than gasoline [9].

Based on the established literatures, Z. Salhab et. al. [10] revealed that that the use of LPG instead of conventional gasoline will mean a reduction in low engine BP with loss of power 7%, also a reduction in BSFC about 20% - 30%, where the engine was converted to operate either on gasoline or on LPG using fuel injection in spark-ignition four stroke outboard engine.

According Tasic et al. [11], they found that the LPG fuel had potential reduced the emissions in comparison with gasoline; CO (-30%), CO_2 (-10%) HC (-30%) and NO_X (-41%) for urban test cycle (ECE 15) method. Meanwhile for the extra urban cycle (EUDC) method, the LPG fuel also the improvement in promoting a lower emission vehicles; CO (-10%), CO_2 (-11%) HC (-51%) and NO_X (-77%). The comparison were made by using Opel Zafira with four cylinder 1800cc Ecotec engine as the test vehicle which is equipped with Landi Renzo retrofit kit.

Mistry et. al. [12] have been conducted an experimental works and they found that the brake power developed is higher in case of LPG engine whereas heat carried by jacket water is covered by exhaust gases and unaccountable losses are higher in case of petrol engine. They also observed that, LPG has higher fuel consumption compared to petrol.
The works by Shankar K. S. et. al. [13], they was investigated the effect of variation in ignition timing on the performance of a four cylinder multipoint port fuel injection gasoline engine which is retrofitted to run with LPG injection. When using LPG as a fuel, better performance has been observed when ignition timing is set at 6° BTDC. Advancing the idle ignition timing has also resulted in reduced CO (-1%) and HC (-50ppm) emissions. But the advanced ignition timing shows an increase (+1400ppm) in NO_X emissions.

In terms of numerical analysis, Ali M. Pourkhesalian et. al. [14], the environment coding using Matlab has been developed and its results are validated with experimental data. The author's mentioned that this simulated engine model could be used as a powerful tool to investigate the performance and emission of a given S.I engine fuel by alternative fuels including hydrogen, propane, methane, ethanol and methanol. Eventually, it is concluded that gasoline produce more power than the all being tested alternative fuels. The BP produced by propane is less of gasoline by 10%, respectively. The BSFC of propane fueled engine is approximately 9% less than gasoline.

The works by Hakan Bayraktar et. al. [15], a mathematical cycle model has been arranged for both gasoline and propane, and a numerical analysis code for this model has been developed. Comparison shows that if LPG fueled S.I engines are operated at the same conditions with those of gasoline fueled S.I engines, significant improvements in exhaust emissions can be achieved. LPG decreased volumetric efficiency, thus reduces the engine effective efficiency and consequently increases specific fuel consumption.

In Malaysia scenario, the use of LPG as an alternative fuel via retrofit system are still questionable and tremendous potential of filling the gap in research. Therefore In this present paper, a 1-dimensional GT-Power engine simulation program is applied to investigate the engine performances such as BP, BT, BMEP and BSFC for S.I. engine.

2 Methodology

This study is focused on the investigation of engine performances by using LPG as an alternative fuel for S.I engine. This investigation also focused on constructing an engine simulation model based on a 1-dimensional GT-Power software. The input data that are required is based on the CAMPRO engine specifications; cylinder geometry, intake and exhaust system geometries, valve dimension, and engine operating conditions were measured accordingly. However, the undefined data and parameters were taken by using manual procedure of measurements and default data from the developer of GT-Power, Gamma Technologies LLC, since that information is not provided in the outsources database [16].

The experimental works was performed by using chassis dynamometer (Dynapack) to evaluate the engine performances such as BP, BT, BMEP and BSFC. The engine operating condition which is running at various engine speed 1500 rpm to 7000 rpm at wide open throttle (WOT) conditions. The engine system parameter that was used in the engine crank train is based on Table. 1. The details of model simulation was discussed in next section: engine modelling. In terms of data measurements, Kistler Pressure Sensor (6118B) coupled with Kistler Charge Amplifier (5018A) were used to capture the behavior of in-cylinder conditions. Meanwhile, Ono-Sokki flow meter (FM2500) and detector (FZ2100) were used to measure gasoline fuel consumption and LPG mass flow meter (M250SLPM).

Table 1. Engine system parameter.

	Gasoline engine	LPG engine
Engine type	4 stroke	
Number of cylinders	4 cylinders	
Bore	76 mm	76 mm
Stroke	88 mm	88 mm
Fuel injection	Multipoint fuel injection	Liquid sequential injection
Fuel type	Gasoline	LPG
Compression ratio	10:1	10:1

3 Engine modelling

3.1 Model simulation

The simulation model that consists of three basic parts that built begin from the intake system, engine system and until exhaust system. Features in GT-Power software are consist of object component from flow and mechanical library. **Fig. 1.** shows a constructed simulation model based on LPG and gasoline. There are several specific values of input parameters including the air-fuel ratio (AFR), engine speed, and injection timing were defined in the both models. The AFR for gasoline and LPG engine are based on experimental database, where normally, the experimental AFR is in the range of 14.7 and 15.7 respectively [17].

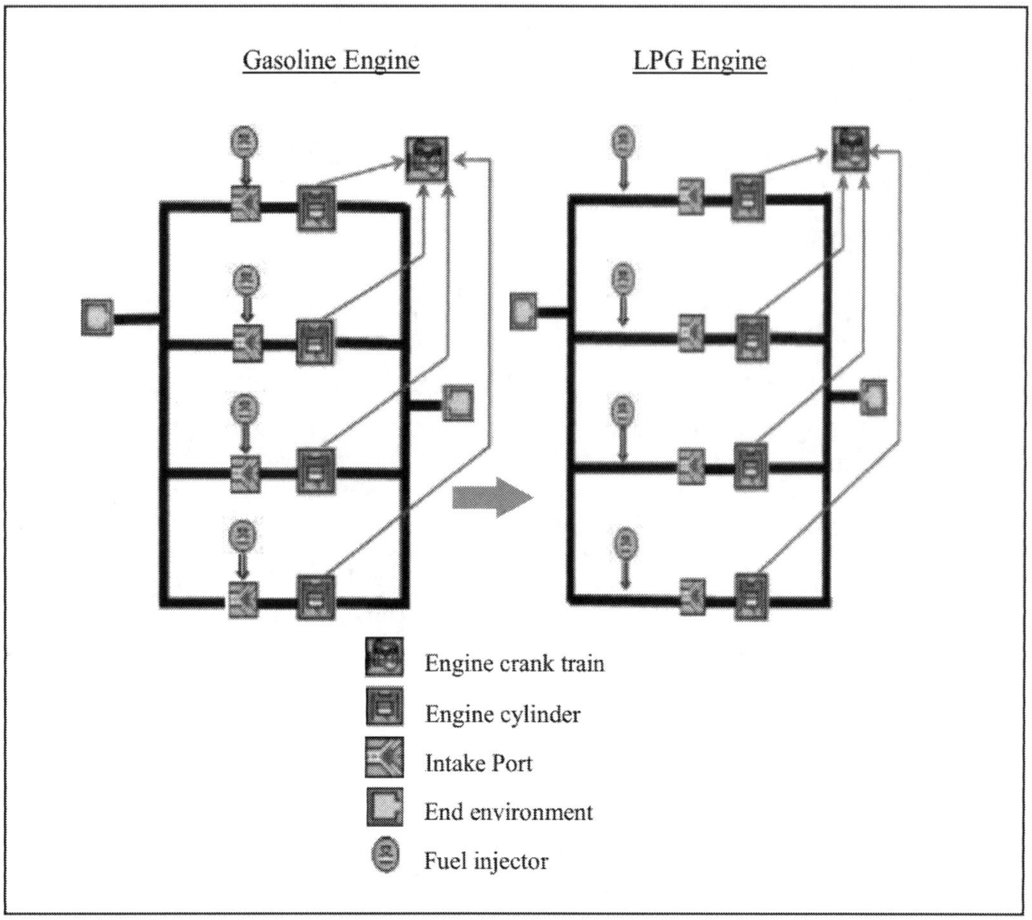

Fig. 1. Engine model in GT-Power simulation.

The intake system consists of air cleaner parts, throttle body, intake manifold and fuel injector. The intake manifold was created from a series of intake runner by using pipe-round object and flowsplitTRight object. The difference part between both models is a position of the fuel injector. The object used is InjAF-RatioConn which required AFR as input data. In gasoline model, the injector is located at the intake port of the manifold. Meanwhile, LPG model is on the intake runner pipe.

Second part of the engine model is the Powertrain where generally consist of engine geometry, cylinder geometry and valve dimension. Non-predictive combustion model known as EngCylCombSI-Wiebe is chosen as a combustion model in an engine cylinder object which is preferred according to the condition of (S.I) engine. Furthermore, this combustion model simply imposes a burn rate as a function of crank angle. The burn rate is represented by crank angle where 50% fuel burned and the duration of fuel burned. Here, the burn rate is calculated from measured in-cylinder combustion pressure at various engine speeds with full load conditions [18], [19], [20]. Each cylinder consists of 2 ValveCammConn object on intake and exhaust system.

The exhaust system is the last part of engine modelling which is exhaust manifold that used bend-pipe-round object, catalytic converter, muffler and tailpipe object. It is to consider that the simulation model will represent as actual engine condition. Lastly, once the model was completed, the engine operating conditions of simulation models are defined as engine speed 1500 to 7000 rpm and at wide open throttle (WOT). The results that obtained are used to match up the model to the experimental data.

3.2 Model validation

To achieve the accuracy of the simulation model, the model has been correlated to the experimental data based on fuel type. Refer to Fig. 2. The correlated model of gasoline model simulation by refer to in-cylinder condition (pressure) using gasoline as fuel for combustion, are very close to the experimental data, with the average differences of less than 4%. Moreover, the simulation results for LPG as Fig. 3. has also been verified, whereas the BT and BP predicts difference less than 7.42%. Last but not least, based on the validation process, the results between simulation models and experimental data has been well confirmed that represents as similar as actual engine condition.

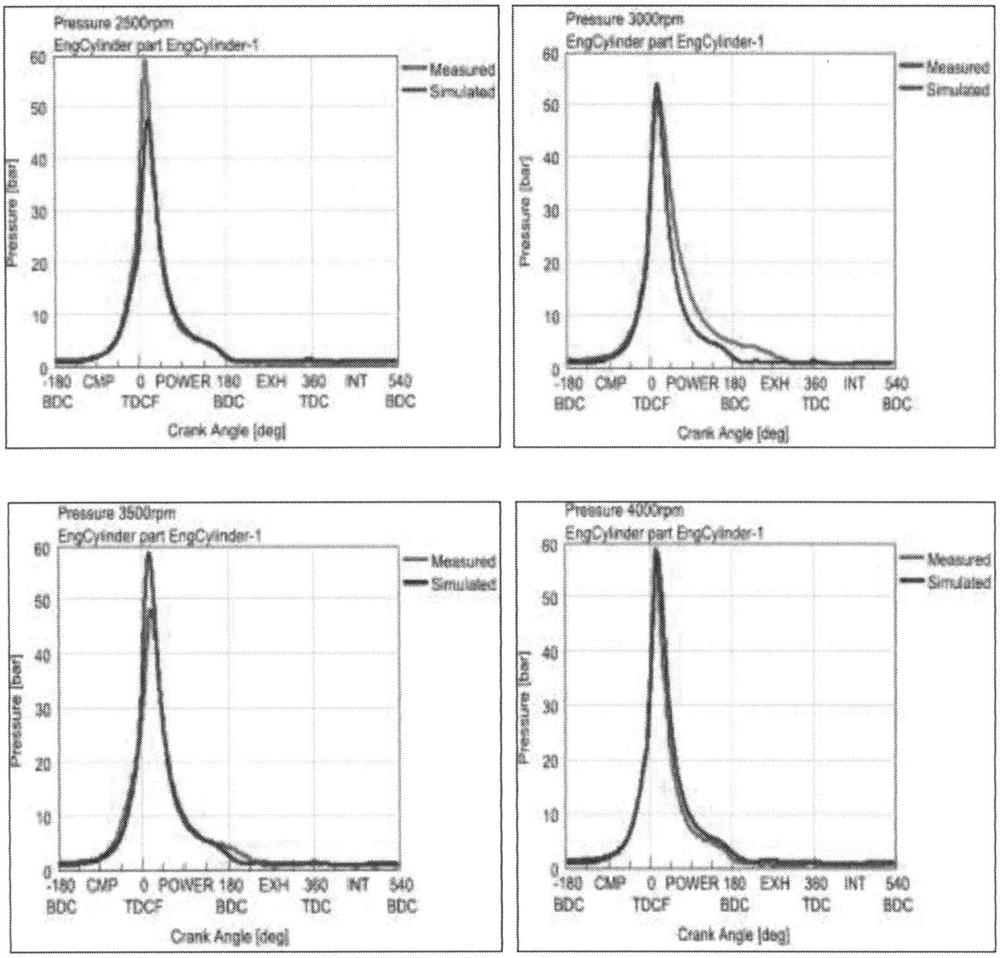

Fig. 2. Gasoline engine model correlation.

4 Results and discussion

Based on Fig. 4(a), shows that the BT at various engine speeds. The model simulation predicts that BT for LPG is higher compared to gasoline. At low engine speeds, LPG has been increased gradually from 1500 rpm until reach 2000 rpm at point maximum is 132.92 Nm. The engine experiences performance downfall due to a phenomenon known as torque loss at 2500 rpm until 3000 rpm. The torque dip phenomenon is caused by the pressure wave that is not coherent with the valve opening [21]. However, the BT a slowly recovered to the highest point of 144.33 Nm at engine speeds 3000 rpm until 4500 rpm. Meanwhile, at higher engine speed, the performance reduced as engine speed increase.

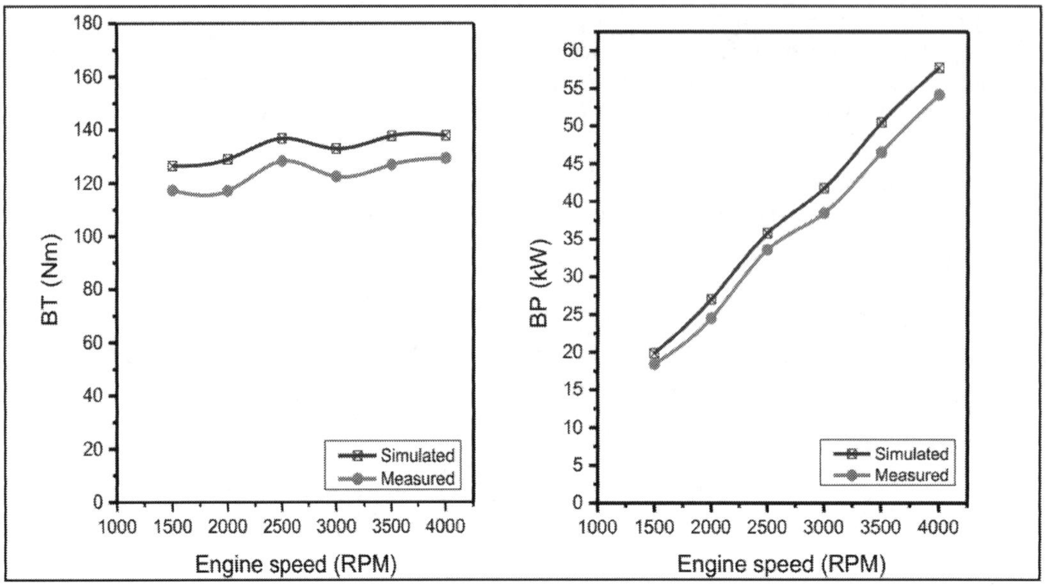

Fig. 3. LPG engine model correlation.

Refer to the Fig. 4(b), shows differences of BP at various engine speeds. It is predicted that LPG have an improvement as an average difference is 6.55%, where the highest noticed at 7000 rpm with 86.63 kW. This is because of LPG contain a higher heating value (46.4 MJ/kg) compared to gasoline (44.0 MJ/kg). Moreover, LPG contained a higher octane number than gasoline.

As depicted in Fig. 4(c), in term of BSFC results, the simulation model predicted around 4.47% saving improvement BSFC of LPG as engine speed increased from 1500 rpm until it reach 4000 rpm. For higher engine speed starting from 4000 rpm onwards, the percentage of fuel saving was predicted in an average of 6.24%. According to Fig. 4(d). In the perspective of BMEP, LPG can boosted a S.I. engine about 6% as engine speed increased, where the maximum BMEP is 11.35 bar was predicted at 4500 rpm.

5 Conclusions

As a conclusion, performances of LPG has been predicted by using 1-dimensional model simulation GT-Power software based on CAMPRO 1.6L engine. The model has been verified with the experimental results and it's shown a good agreement. Based on results, it shows that for those user who are concerning about the fuel economy and boosting the engine performances, a retrofitting S.I. engine via LPG fuelling system is the best option to be considered. The conclusion can be summarized as below:

i. LPG produced higher BP, BT and BMEP for entire engine speeds with an average difference is 7% than gasoline.
ii. BT increased around 7% average where the highest of BT was predicted at 144 Nm @ 4500 rpm
iii. BP for predicted an improvement on an average of 6.55%, where the maximum value noticed at 86.63kW @ 7000 rpm

iv. The variation of BSFC that predicted by simulation model was shown to be 5% lower than gasoline engine. Thus, the use of LPG as an alternative fuel will give benefit in fuel saving.

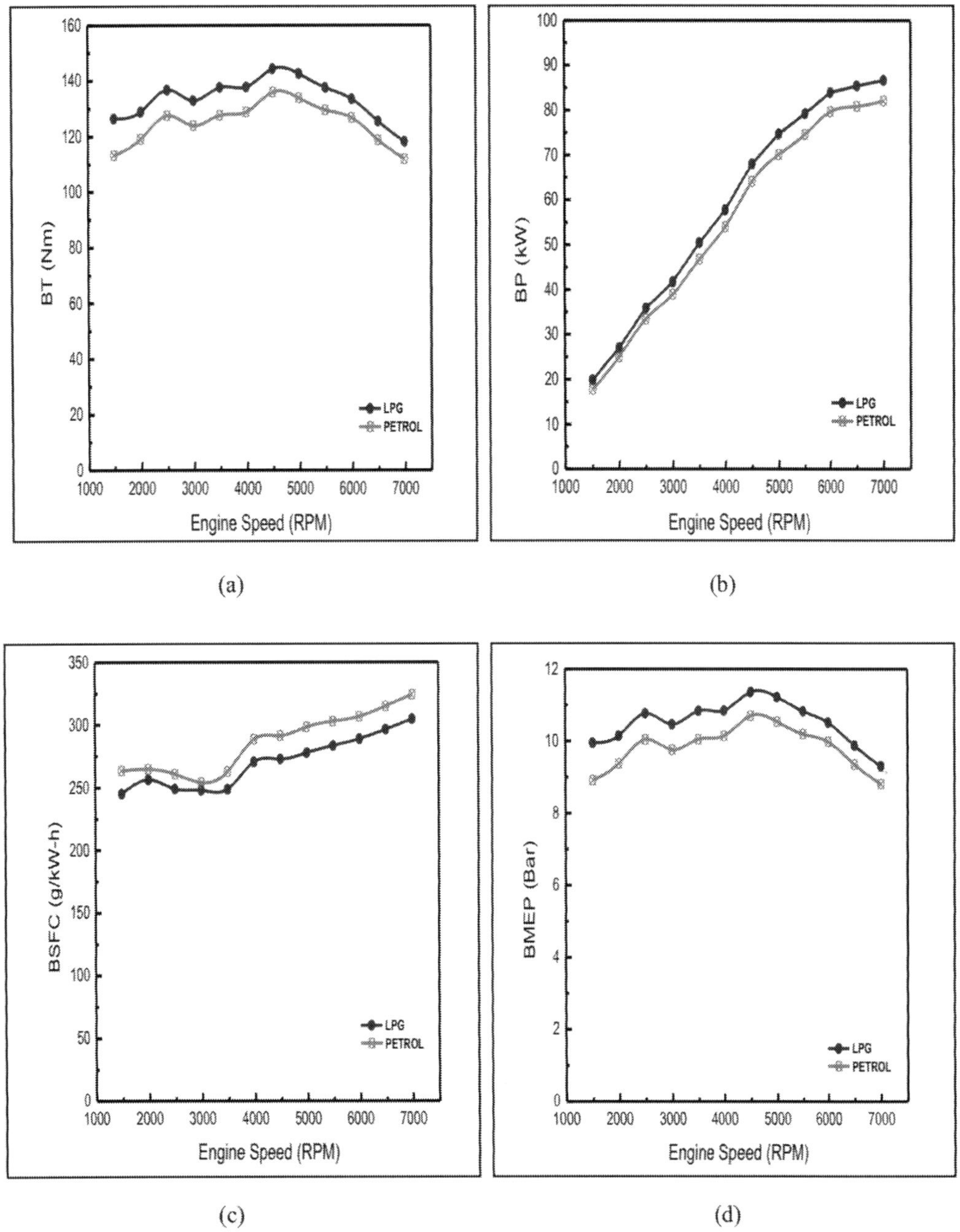

Fig. 4. Engine performances based on GT-Power.

Appreciation and acknowledgment to the Universiti Tun Hussein Onn Malaysia (UTHM) for providing the financial support under UTHM Contract Grant, VOT: U427. The author also would like to express highest gratitude to anyone who contributed to the research work.

References

1. B. Ashok, S. Denis Ashok, and C. Ramesh Kumar, "LPG diesel dual fuel engine – A critical review," *Alexandria Eng. J.*, vol. **54**, no. 2, pp. 105–126, (2015)
2. L. C. Meher, D. Vidya Sagar, and S. N. Naik, "Technical aspects of biodiesel production by transesterification - A review," *Renew. Sustain. Energy Rev.*, vol. **10**, no. 3, pp. 248–268, (2006)
3. M. Balat, "Potential importance of hydrogen as a future solution to environmental and transportation problems," *Int. J. Hydrogen Energy*, vol. **33**, no. 15, 4013–4029, (2008)
4. European LPG Association, "Autogas in Europe , The Sustainable Alternative: An LPG Industry Roadmap," (2013)
5. L. Raslavičius, A. Keršys, S. Mockus, N. Keršiene, and M. Starevičius, "Liquefied petroleum gas (LPG) as a medium-term option in the transition to sustainable fuels and transport," *Renew. Sustain. Energy Rev.*, vol. **32**, pp. 513–525, (2014)
6. "WLPGA - World LP Gas Association," (2006)
7. H. L. MacLean and L. B. Lave, "Evaluating automobile fuel/propulsions system technologies," *Prog. Energy Combust. Sci.*, vol. **29**, no. 1, pp. 1–69,(2003)
8. H. S. Tira, J. M. Herreros, A. Tsolakis, and M. L. Wyszynski, "Characteristics of LPG-diesel dual fuelled engine operated with rapeseed methyl ester and gas-to-liquid diesel fuels," *Energy*, vol. **47**, no. 1, pp. 620–629, (2012)
9. K. Lee and J. Ryu, "An experimental study of the flame propagation and combustion characteristics of LPG fuel," *Fuel*, vol. 84, no. **9**, pp. 1116–1127, (2005)
10. Z. Salhab, M. G. Qawasmi, H. Amro, M. Zalloum, M. S. Qawasmi, and N. Sharawi, "Comparative performance and emission properties of spark-ignition outboard engine powered by gasoline and LPG," *Jordan J. Mech. Ind. Eng.*, vol. **5**, no. February, pp. 47–52, (2011)
11. T. Tasic, P. Pogorevc, and T. Brajlih, "Gasoline and Exhaust Emission Comparison," *Adv. Prod. Eng. Manag.*, vol. **6**, no. 2, pp. 87–94, (2011)
12. C. S. Mistry and A. H. Gandhi, "Experimental Investigation on Multi-cylinder Engine Using Petrol and LPG as A Fuel," *SAE Tech.*, vol. 2004, no. 724, pp. 26–31, 2004.
13. M. P. Shankar K. S, "MPFI gasoline engine combustion, performance and emission characteristics with LPG injection," vol. 2, no. 4, pp. 761–770, 2011.
14. A. M. Pourkhesalian, A. H. Shamekhi, and F. Salimi, "Alternative fuel and gasoline in an SI engine: A comparative study of performance and emissions characteristics," *Fuel*, vol. 89, no. 5, pp. 1056–1063, 2010.
15. H. Bayraktar and O. Durgun, "Investigating the effects of LPG on spark ignition engine combustion and performance," *Energy Convers. Manag.*, vol. 46, no. 13–14, pp. 2317–2333, 2005.
16. Perusahaan Otomobil Nasional (PROTON), "Petrol/Diesel Engine Overhaul S4P," pp. 71–76, 2004.
17. J. B. Lheywood, *Internal Combustion Engine*, vol. 21.
18. A. J. Alimin, M. Y. Ismail, and S. A. Osman, "Predicting the Performance and Emissions Characteristics of a Medium Duty Engine Retrofitted with Compressed Natural Gas System Using 1-Dimensional Software," *Appl. Mech. Mater.*, vol. 660, pp. 468–473, 2014.
19. Gamma Technologies Inc, "GT-Power User's Manuals V7.0," no. September, 2009.
20. M. F. M. Said, Z. A. Latiff, A. Saat, M. Said, and S. F. Z. Abidin, "Analysis of Variable Intake Runner Lengths and Intake Valve Open Timings on Engine Performances," *Appl. Mech. Mater.*, vol. 663, pp. 336–341, 2014.
21. A. K. M. Mohiuddin, A. Rahman, A. Arfah, and M. A. Farraen, "Investigation and Probable Solution of Low End Torque Dip Problem of Campro 1 . 6L Engine," vol. 3, no. 2, pp. 176–186, 2008.

Investigating the influences of liquid LPG injection on spark ignition (SI) engine

Mohd Mustaqim Tukiman[1,2,*], *Norrizal* Mustaffa[1,2], *Mas* Fawzi[1,2], *Shahrul Azmir* Osman[1,2]

[1]Faculty of Engineering Technology, [2]Faculty of Mechanical and Manufacturing Engineering.
[2]Automotive Research Group (ARG), Center for Energy and Industrial Environment Studies (CEIES), Universiti Tun Hussein Onn Malaysia, 86400 Parit Raja, Batu Pahat, Johor, Malaysia.

Abstract. Liquefied petroleum gas (LPG) is one of the alternative fuels that becoming popular to be use in spark ignition engine (SI). This paper briefly presents the influence of energy content to the engine output of 1.6L SI engine of Proton Gen 2. The engine was coupled to a chassis dynamometer and few related apparatus were employed in determine the engine behavior. All data collected were illustrated in graph for further analysis. The engine shows comparable engine output, however, the engine requires some tuning in order to fully utilize the energy content of LPG.

1 Introduction

LPG is a clean fuel and also known as "Autogas". LPG consists of a mixture of propane (C_3H_8) and butane (C_4H_{10}). Presently, LPG is one of the alternative fuels that received much attention as a replacement of traditional fuels due to their economical costs, high octane numbers, high calorific values, availability in large volume and lower polluting exhaust emissions The usage of LPG especially in SI engine has broadly been used in a few countries such as Italy, Turkey, Russia and Korea [1].

The usage of LPG as alternative fuel especially in gasoline engine gives many advantages especially in reducing dangerous knocking phenomena while running the engine at the medium high load condition and offers better exhaust emissions production [2]. It's also provides better cold start performance in gasoline engine because its rate of vaporization is very fast during the intake stroke and the amount of homogeneous mixture in the combustion chamber is substantially increased [3-6]. Therefore, excessive enrichment of the air-fuel ratio at low coolant temperature is discarded.

Currently, the conversion technology of SI engine into the LPG system had been expanded since it was introduced. The technology continues to grow as the emissions regulation became stringent. Based on the LPG delivery and system control, LPG generation and technology in SI engine may be categorized into five generations [7,8]. The first four generations run on gaseous LPG and the fifth generation operate with liquid LPG. Table 1 shows the summary of the LPG generation in SI engine.

The present work aims to characterize the engine output with respect to the energy content of the gasoline and LPG at various throttle position (TP). The engine performance

* Corresponding author: mohdmustaqimakim@gmail.com

and exhaust emissions were examined. Mapping of the air-fuel mixing for each TP also was recorded for the tested engine.

Table 1. LPG generations [7,8].

Generation	Description
First	Mechanically controlled LPG carburetion system, open loop LPG system, suitable for carburetor engine. Also known as the Induction system. Simplest type of LPG kit. Introduces gas into the engine in a vapor form through the inlet manifold.
Second	Electronically controlled, close loop, single point injection LPG system, suitable for early Electronic Fuel Injection (EFI) gasoline engine.
Third	Multipoint injection LPG system.
Fourth	Multipoint sequential injection also known as the Sequential Gas Injection System (SGIS). Injects LPG as a vapor for each cylinder very close to the inlet valves.
Fifth	Multipoint sequential liquid LPG injection, Unlike the first to fourth generation, LPG is fed into the engine in its liquid form and without vaporizing.

1.1 Energy content

Each fuel has different energy content and this energy content will significantly affect the outcomes of combustion especially in internal combustion engine. In view of gasoline and LPG, the energy content per weight was shown in Table 2 and it's presented that energy content of LPG was slightly higher as compared to the gasoline fuel. Logically, the engine performance produced by LPG will be higher than gasoline but the benefit of LPG energy content only may be useful with the proper setting and adjustment. This is due to the LPG is a gaseous fuel meanwhile gasoline is a liquid fuel. In this study, the fifth generation of LPG was used.

Table 2. Energy content of gasoline and LPG [9-11].

Gasoline (kJ/kg)	LPG (kJ/kg)
43500	46500

2 Experimental details

2.1 Test engine

A 1.6 liter spark ignition engine naturally aspirated of Proton Gen-2 was selected as the test vehicle. The detailed specifications of the engine are as summarized in Table 3.

2.2 Experimental apparatus

In order to obtain the experimental data of the dedicated SI engine and LPG converted SI engine, series of experiments were carried out in analyzing the engine characteristics. The engine was coupled to the Dynapack Chassis Dynamometer that produced the performance data such as torque and brake power. The chassis dynamometer used is capable to run and simulate various types of engine testing including actual road operation. The exhaust emissions of the engine were captured using Autocheck Gas 4/5 emissions analyzer. The sampling time of the emissions analyzer was setup at 5 seconds for every measurement

 DOI: 10.1051/matecconf/20179001075

taken. The mass flow rate of fuel was measured using Ono Sokki fuel flow detector (FZ-2100) that employed coriolis principle. The injection duration and TP were monitored real time during experiments by scan tool Bosch KTS570. In fixing the TP at certain percentage for example at 25% TP, a special tool for pedal press was employed. For the LPG operation, dedicated software was used to determine the injection duration. Schematic diagram of the experimental setup is shown in Fig. 1.

Table 3. Proton Gen 2 engine specifications.

Engine model	S4PH – 1.6
Number of cylinders	4
Orientation	East-West
Valve train	DOHC 16V
Combustion chamber	Pentroof type
Bore x stroke	76.0mm x 88.0mm
Compression ratio	10.0 : 1
Fuel injection	Indirect injection (MPI)
Lubrication system	Pressure feed, full-flow filtration
Cooling system	Water-cooled
Maximum torque	148Nm (4000rpm)
Maximum power	110hp (6000rpm)

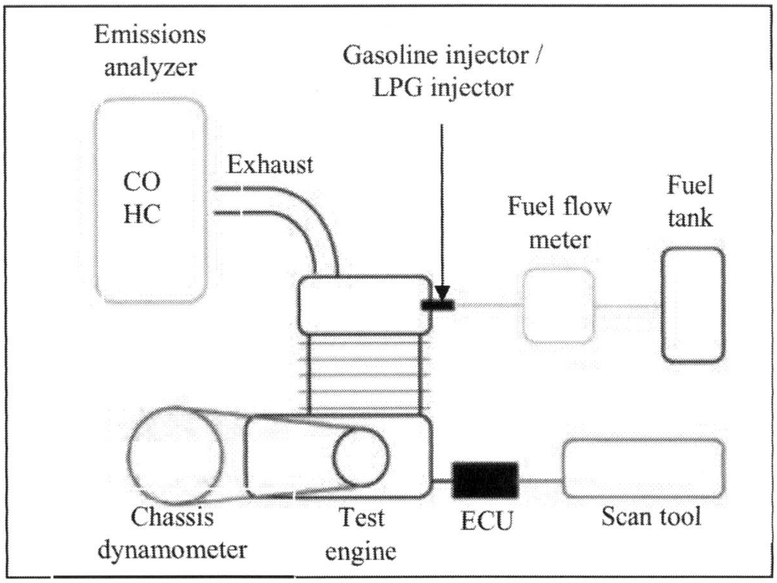

Fig. 1. Schematic diagram of experimental setup.

2.3 Experimental methodology

The steady state test was selected to evaluate complete this study. Steady state means the dedicated engine will be tested based on engine speed starts with the idling condition up to 4000rpm with the increment of 500rpm. The TP was varies at four different positions that were 25%, 50%, 75% and 100% for each engine speed tested. Table 4 described the details of the steady state test conditions conducted. All the experiments were repeated three times for each operating condition in reducing the error and obtain the trustable results.

Table 4. Proton Gen 2 engine specifications.

Engine speed (rpm)	Throttle position (%)
Idling	25 50 75 100
1500	
2000	
2500	
3000	
3500	
4000	

3 Results and discussion

Fig. 2 shows the brake thermal efficiency (BTE) of the tested engine for both gasoline and LPG fuels at four different TP. The BTE of gasoline at all TP was found increase as the engine speed increase meanwhile the BTE of LPG was found slightly decrease with the increment of engine speed. As the energy content is the function of BTE, it clearly described the effectiveness of the engine transform the energy content of both tested fuels into mechanical energy. Based on Fig. 2, LPG has better BTE at below 3000 engine speed and gasoline shows superior BTE after 3000 engine speed onwards. It clearly described that the engine running by gasoline has poor performance at low engine speed and only effective at higher engine speed. This may be attributed to the engine characteristic that has bad engine setting and tuning by manufacturer at its low engine speed. However, LPG shows almost consistent performance throughout the engine speed. Even though the flame speed of LPG is lower than gasoline, it presents reliable BTE and its may be improved by the adjustment of the ignition timing.

Since the energy content of LPG was a little bit higher than gasoline, quantity of gasoline injected was slightly higher than LPG. This is due to the energy compensation in order to produce the same engine output. This is clearly illustrated in Fig. 3 whereby the quantity of fuel injected was directly influenced by the injection duration of each fuel. Injection duration of LPG fuel at all TP was recorded lower than gasoline and the pattern of injection duration were found similar for both fuels. In addition, fuel quantity injected was found increase as the engine TP increase. The brake specific fuel consumption (BSFC) also shown in Fig. 3 where the BSFC also exhibits comparable pattern with BTE where the trade-off between these two fuels was at 3000 engine speed. This exhibits that the engine consumed higher gasoline at below 3000 engine speed but produced lower BTE and vice versa for LPG at all TP.

The brake power produced for both fuels were shown in Fig. 4 and the brake power was increased as the engine speed increase. The brake power produced for both fuels was almost identical at all TP. This is due to the LPG was injected in the liquid form rather than gaseous form. The liquid injection will leads to the benefits of its energy content by improved the volumetric efficiency due to the rapid vaporization that resulting in cooler and denser of the air-fuel mixture flowing into the combustion chamber. In other view, liquid injection of LPG would created the Joule-Thompson effects during the air-fuel mixing process that eventually produced less unburned fuel.

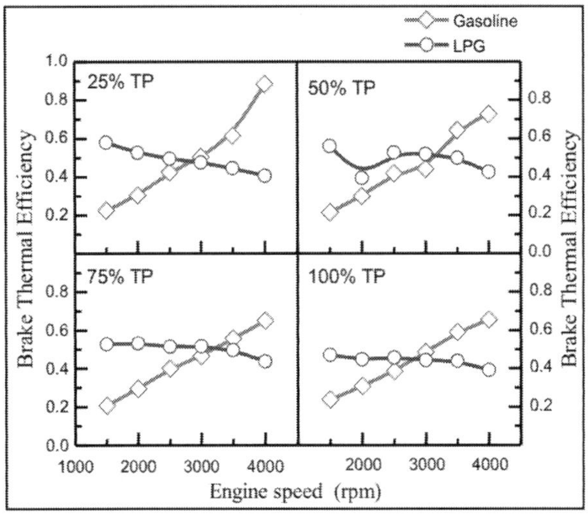

Fig. 2. Brake thermal efficiency of gasoline and LPG at various throttle position.

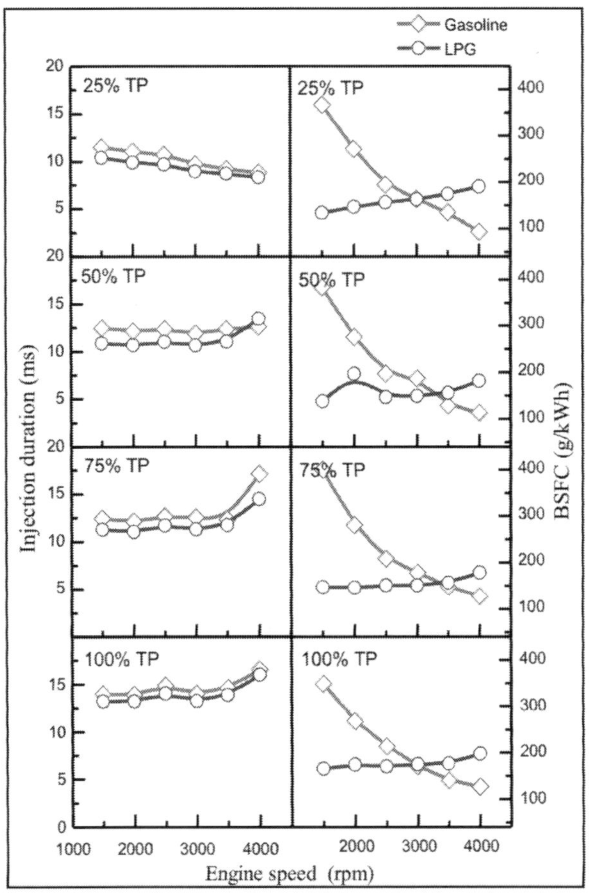

Fig. 3. Injection duration and BSFC of gasoline and LPG at various throttle position.

 DOI: 10.1051/matecconf/20179001075

For the emissions, carbon monoxide (CO) and hydrocarbon (HC) were analyzed and the data of both gases as per shown in Fig. 5. CO emission at 25%, 50% and 75% showed very lower value and this indicated that the combustion of both tested fuel occurred at lean combustion. However, at 100% TP, CO was recorded higher throughout the entire engine speed and only at the 4000 engine speed of 50% and 75%. This is obviously described that incomplete combustion occurred due to less oxygen during combustion.

In general, HC emission of LPG was higher than gasoline at all TP and this HC emission is in agreement with the CO emission. This depicted that lots of LPG was unburned as compared to gasoline especially at 100% TP. This may be due to the inaccurate spark timing and improper air-fuel mixing. Since the flame speed of LPG was lower than gasoline and energy content of LPG was higher than gasoline, the adjustment and proper setting of spark timing is required in order to fully utilize the LPG properties.

Other than that, the characteristics of air-fuel mixture of the tested engine were analyzed based on the data collected. In general, the air-fuel mixture mapping may be divided into four different conditions based on TP as shown in Table 5. The air-fuel mixture was mapped to the rich condition at 4000rpm of 50% and 75% TP and also at all engine speed of 100% TP. This is due to the engine preservation whereby the rich mixture will decrease the combustion temperature and cools the engine that reduced the tendency of engine faulty.

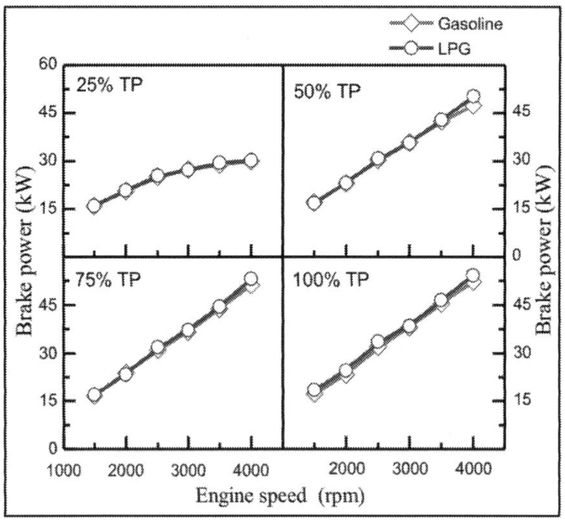

Fig. 4. Brake power of gasoline and LPG at various throttle position.

4 Conclusions

Based on the obtained results and some analysis conducted, LPG offers excellent promise as an alternative fuel for the spark ignition engine due to its higher energy content as compared to the gasoline fuel. However, the injection phase and spark timing shall be adjusted to the optimum setting.

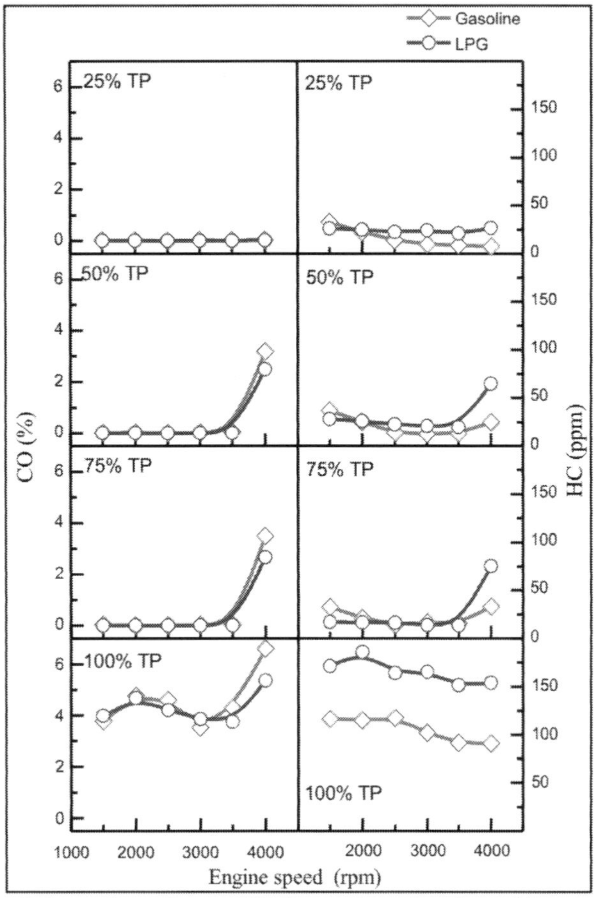

Fig. 5. CO and HC emissions of gasoline and LPG at various throttle position.

Table 5. Air-fuel mixture mapping of tested engine at four different TP.

	Engine speed (rpm)					
TP (%)	**1500**	**2000**	**2500**	**3000**	**3500**	**4000**
25	Lean	Lean	Lean	Lean	Lean	Lean
50	Lean	Lean	Lean	Lean	Lean	Rich
75	Lean	Lean	Lean	Lean	Lean	Rich
100	Rich	Rich	Rich	Rich	Rich	Rich

The authors would like to thank the Universiti Tun Hussein Onn Malaysia for supporting this research under the Short Term Grant VOT. U383.

References

1. Autogas Incentive Policies, A country-by-country analysis of why and how goverments promoto autogas and what works. World LP Gas Association. (2012)
2. Pipitone. E, Beccari. S, SAE Technical Paper. (2010)
3. Arcoumanis, C. Bae, C. Crookes, R. Kinoshita. E, Fuel **87**, (2008)
4. Kwak, H. Myung, Park, Fuel **86** (2007)
5. Myung, Choi, K. Kim, J. Lim, Y. Lee, Park, Energy **44**, 1(2012)

6. Myung, Kim, J. Choi, K. Hwang, Park, Fuel **94**, (2012)
7. Raslavičius, L. Keršys, A. Mockus, S. Keršienė, N. Starevičius, Renewable and Sustainable Energy Reviews **32**, (2014)
8. Biscoff, R. Akple, M. Turkson, R. Klomegah, Scenario of the emerging shift from gasoline to LPG fuelled cars in Ghana: A case study in Ho Municipality, Volta Region. Energy Policy **44**, (2012)
9. Saraf, Thipse, Saxena, SAE Tech. Pap. (2008)
10. Saraf, Thipse, Saxena, SAE Tech. Pap. (2009)
11. Park, Park, Y. Oh, S. Lee, Y. Kim, T. Y. Kim, Kang, SAE Tech. Pap. (2013).

Experimental study on the effect of alternator speed to the car charging system

Rozdman K. Mazlan[1,*], *Reduan* M. Dan[1,2], *Mohd* Z. Zakaria[1,2], *Abdul* H. A. Hamid[1]

[1]Faculty of Mechanical Engineering, Universiti Teknikal Malaysia Melaka, Hang Tuah Jaya, 76100 Durian Tunggal, Melaka, Malaysia.
[2]Centre for Advanced Research on Energy, Universiti Teknikal Malaysia Melaka, Hang Tuah Jaya, 76100 Durian Tunggal, Melaka, Malaysia.

Abstract. In this paper, we present our work, which is doing an energy audit on alternator's current output and battery's voltage based on alternator speed. Up until today, the demand for power in automobile is ever increasing. As technology advances, more and more electrical devices were produced and being installed in vehicles. To cope with the demand, alternator has been designed and modified so that it can produce enough power. This research is to study the effect of alternator speed to the charging system. The car used in this experiment is Proton Preve 1.6 Manual. In both ISO 8854 and SAE J 56, alternator testing and labelling standards indicate that the rated output an alternator is the amount of current that it is capable of producing at 6,000 RPM. Three different constant speed of engine which is 750 RPM as idle speed, 1500 RPM and 3000 RPM as cruise speed were taken as parameter. The speed of the alternator was measured using tachometer, digital multi-meter was used to measure battery's voltage, and AC/DC Clamp was used to measure alternator current output. The result shows that the faster the alternator spin, the more power it can produce. And when there is more power, the faster the charging rate of the battery.

1 Introduction

In our modern world, electricity plays a pivotal role in driving industry and powering the systems used to meet our agricultural, health care, educational, and commercial needs. Among the among impacted industry is the automobile industry. The electrical power demands or requirements in vehicle transportation have been rising rapidly for many years and are expected to continue to rise (Fig. 1). This flow of direction is driven by the replacement of engine-driven loads with electrically-powered versions, and by the introduction and establishment of a wide range of new functionality in vehicles [1]. The continuous increase in power requirements is pushing the limits of typical and traditional automotive power generation and control technology, and is motivating the development of both higher-power and higher-voltage electrical systems and components [2].

* Corresponding author: rozdman.edir@gmail.com

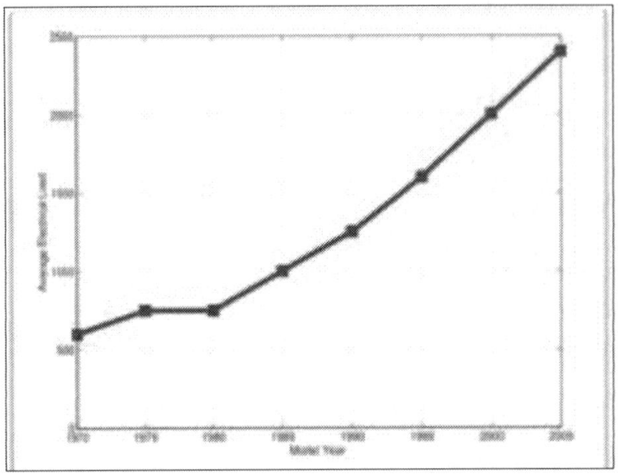

Fig. 1. Automotive electrical power requirements [1].

One of the machine that has been through many process of improvement in the automobile industry is the alternator. Alternators are all around us. It is almost impossible or hardly an engine which does not have an alternator, or maybe more, running alongside in cars, trucks, buses, recreational vehicles, boats, aircraft, farm machines, earth moving equipment, and all kinds of stationary engines. Clear as the daylight, the alternators are doing their job very well since we hear so little about them. Well, some of us may not know what an alternator is. And if something does go wrong, most of the more popular models have their replacement available almost instantly, waiting on dealer shelves, in alternator shops, often even in department stores [3].

Every automobile has a charging system. For a car, it consists of the alternator, regulator, and the interconnecting wiring. Charging system act as main source of electrical energy, and provide power to electrical components in the car while the engine is running. If there is no charging system, battery will be the source of power. But until the battery is weak, and there is no power source to charge the battery, then this will cause the engine to has less current to fire the spark plugs. In this case, the engine will be forced to stop running.

The use of an increasing number of electrically operated or actuated accessories is increasing the power demand from the alternator. The loads include electrically heated windows, seats, and catalysts; motors for cooling fans (and possibly coolant pumps); and electric actuators for power steering, seat adjustment, and possibly brakes. In 1990, a typical requirement was 500 W. Today, it is 1 to 1.5 kW, and by 2005, it is forecast to be approximately 3 kW [4]. With a 12 V system, 3 kW would require a current of 250 amps. Modern and latest car nowadays has been equipped with more electrical devices, or simply said, wide range of functionality. Devices such as DVD player, rear camera, USB port for phone charging, LED light, etc. All these devices require more power to be generated by the alternator. Conventional alternator may only produce power for about 500 Watt to 600 Watt. But nowadays, a single alternator can produce power up to 2500 Watt, depending on the power demand.

The power that can be produced by an alternator is determined by the alternator rating. The output of the alternator ranges from 40 Amp up to 120 Amp. For high-amp alternator, it can generate up to 300 Amp. Commercialized alternator is usually 12 V. For some systems, especially for high power system, the alternator varies from 24 V up to 48 V. The input of the alternator is speed dependents. It means that, the higher the speed of rotation of the alternator shaft, the more power the alternator produce. But vice versa, the output of the alternator is demand dependents. The alternator will only give out power depending on the demand from the system.

Most cars, while the engine is running, have a charging system that will generally produce a voltage between 13.5 and 14.4 volts. The charging system needs to produce more voltage than the battery rated voltage in order to overcome the internal resistance of the battery [5]. This is due to the fact that the current needed to recharge the battery would not flow at all if the charging system's output voltage was on par with the battery voltage. The charging rate will be faster if there is a greater difference of potential (voltage) between the battery's voltage and the alternator's output voltage [6].

For battery charging to occur, the alternator's voltage must exceed the battery's voltage. Alternator may not generate sufficient charging voltage until alternator speed is greater than about 2000 RPM. Typically, alternators have their full output rated at 6000 RPM but can continue to spin up to 12,000 RPM or more without any additional increase in output. The speed of an alternator is different for different type of car. The speed of an alternator depends on the speed of the engine. For a racing car, the ratio speed between engine and alternator is usually 1:1. For a drag car, the ratio is usually 1:2. And for street use, the ratio is usually 1:3.

When voltage causes current to flow, energy is converted. This is described as power. The unit of power is the watt [7]. As with Ohm's law, any one value can be calculated if the other two are known.

$$\text{Power} = \text{Voltage X Current} \tag{1}$$
$$P = VI \text{ or } I = P/V \text{ or } V = P/I$$

A vehicle charging system has been represented in Figure 2 as three blocks, i.e. the alternator, battery, and vehicle loads. When engine is not running, the alternator voltage is less than the battery voltage so current flows from the battery to the vehicle loads and the alternator diodes prevent current flowing into the alternator. When engine is running, the alternator output is greater than the battery voltage, so current flows from the alternator to the vehicle loads as well as the battery. This implies that alternator output voltage must always be above the battery voltage during operation of the engine. However, the actual voltage used is critical and depends on a number of factors [8].

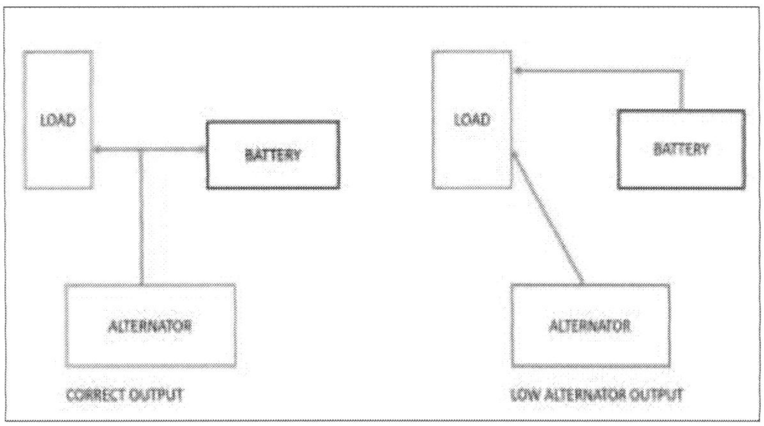

Fig. 2. Block diagram representing the vehicle charging system.

The relationship between alternator speed and alternator current is shown in Figure 3. The current output of the alternator is speed-dependent. The speed of the alternator depends

 DOI: 10.1051/matecconf/20179001076

on the turns ration ranging from 1:2 to 1:3 (crankshaft to alternator). The greater the speed, the greater the output. The rated current is output at an alternator speed of 6000 RPM [9].

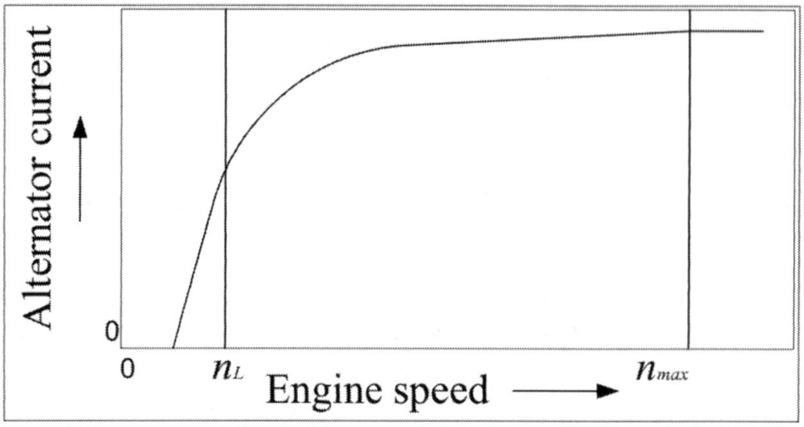

Fig. 3. Characteristic curve at maximum alternator current at constant voltage [9].

While the relationship between charging current, charging time, charging voltage, and alternator speed is shown in Figure 4 and Figure 5. The higher the current, the faster the charging rate of battery. When the current generated is greater than the current demand, then only the battery will start charging [10].

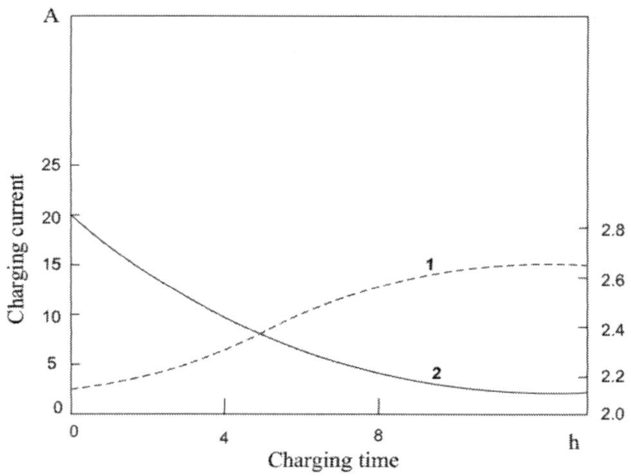

Fig. 4. Battery charging [10].

The purpose of this research is to find out the possibility of creating regenerating system to power added system to the car by channellings the unused current or excessive current from the alternator. In doing so, experiments needed to be done to find the relationship between alternator speed and the power it can produce, with the assists and knowledge gained from educational materials.

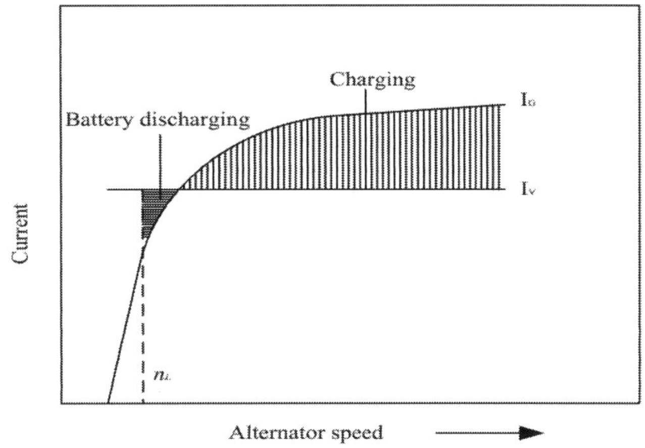

Fig. 5. Alternator output I_G as a function of alternator speed [10].

2 Experimental setup

2.1 Design of experiment

The car used as experimental study object is Proton Preve 1.6 Manual. The alternator of the car has a rating of 90 A 12 V. No modification is done to the car, especially the alternator. And no extra electrical devices were added. Only stock items were being used. The schematic diagram of the experimental study is shown in Figure 6. Car's battery and alternator are being tested first before continuing with the experiment to make sure they are in good condition and fully functioning.

A digital multi-meter was used to measure the voltage value of the battery. A tachometer was used to measure the speed of the alternator. Timing Light is a device which was used to measure the speed of the engine by clamping its clamp onto one of the cable at spark plug inlet. An AC/DC Clamp was used to measure the value of DC current at the alternator output.

The speed of the alternator and the speed of the engine was measured and compared to get the speed ration. This method is a lot easier compare to the method which one needs to measure the diameter of the alternator shaft and the diameter of the engine shaft to get the speed ration. The alternator is driven by the engine.

2.2 Experimental procedure

The parameters of the experimental study are alternator speed (RPM), DC current value form alternator output (Amp), and battery voltage value (V). Three different speed of engine has been set constant, which are 750 RPM as idle speed, 1500 RPM and 3000 RPM as cruise speed. Car's battery and alternator are being tested first before continuing with the experiment to make sure they are in good condition and fully functioning. A fully charged battery should read over 12.6 volts. The actual output voltage produced by the alternator will typically be about 1-1/2 to 2 volts higher than battery voltage. At idle, most charging systems will produce 13.8 to 14.3 volts with no lights or accessories on.

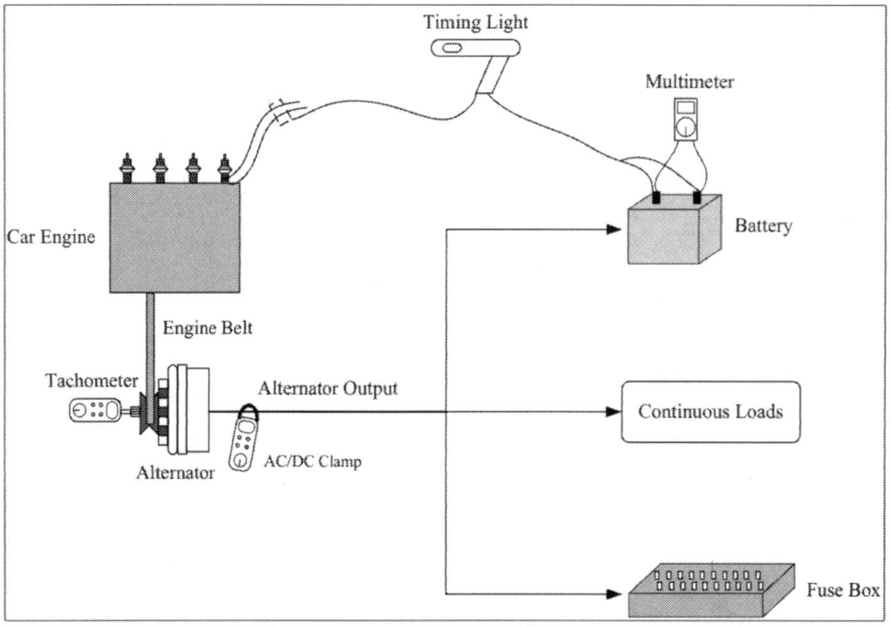

Fig. 6. Schematic diagram of experiment setup.

Table 1 shows the electrical components in the car. The electrical components acted as load to the alternator. The electrical components were turned on one by one in sequences. These steps applied to all three different speed. The value of current at alternator output and battery voltage was taken after each time the electrical component was turned on.

Table 1. Proton Preve electrical component.

No	Electrical components
1	Radio
2	Air-cond
3	Front lamp
4	Hi-beam
5	Fog light
6	Rear lamp (brake)
7	Hazard, wiper
8	Interior lamp
9	Radiator fan

After all the available electrical components had been turned on, and the measurement has taken, all of the electrical components were turned off, and leave the car idle for five minutes. The speed remained constant. After five minutes, the battery voltage was measured again.

Every time before switch to new alternator speed, the battery needed to be sure that it has fully charged.

3 Results and discussion

3.1 Experiment data

Table 1 shows the available electrical devices in Proton Preve 1.6 Manual. All the electrical devices were turned on and act as load to the alternator. The sequence has been set to ensure accuracy and consistency of the result. Table 2 is the measured speed of alternator and crankshaft. Table 3 shows the voltage reading of the battery after all the electrical components had been turned off for 5 minutes. The purpose is to study the relationship between current and charging rate of the battery.

Table 2. Speed difference between engine and alternator.

Engine speed (RPM)	Alternator speed (RPM)
750	1560
1500	3150
3000	6100

Table 3. Battery voltage value after five minutes after all the electrical components turned off.

Engine speed (RPM)	Battery voltage (V)
750	13.1
1500	13.7
3000	14.0

3.2 Analysis

The result of the experiment is as shown in Fig. 7.

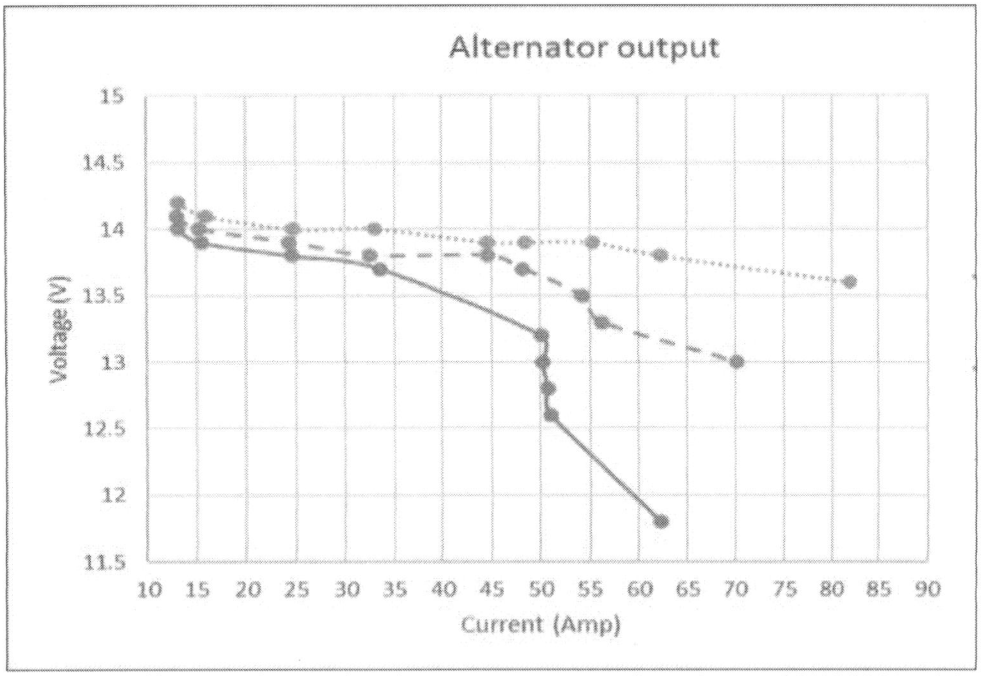

Fig. 7. Graph of battery voltage against alternator output current for engine speed 750 RPM (straight line), 1500 RPM (dashed line), and 3000 RPM (dotted line).

From the data shown in Table 2, it shows that the alternator has a speed ration of 2:1 to the engine. It means, the speed of the alternator is twice the speed of the engine. Different car has different alternator-engine speed ration, depending on the suitability of the system. Normally, the speed ration has been set so that the alternator can produce enough power according to the car demand. The current output of the alternator is speed-dependent. For an engine idling speed, the alternator can only supply some of its rated current if it has a conventional turns ration ranging from 1:2 to 1:3 (crankshaft to alternator). By definition, the rated current is output at an alternator speed of 6000 RPM.

Based on the graph shown in Figure 7, it tells that the higher the speed of the alternator, the more power it can produce as proved in Figure 3. But as the alternator output current increase, the battery voltage drops. This means that the battery is not receiving enough power for recharge. A fully charging battery should have a voltage reading above 13.7 V. When the battery is receiving low power, it will take more time to be fully charged. The slower the speed of the alternator, the more voltage drops significantly as the current increase. Power (P) is measured in watts (W), or for high power in kilowatts (kW). 1 kW is equal to 1000 W. Power used by an electrical device is easily calculated. Just take the voltage across the device and multiply that by the current through it. Power = voltage x current [7].

At current load 10 Amp to 35 Amp, the voltage drop for all three different speed are only slight different. The behaviour starts to change at current load more than 35 Amp. This can be said that the alternator can provide steady power for current below than 35 Amp even at the slowest speed. Speed of the alternator influence the power capability it can supply. At low speeds, the alternator can't satisfy the combined load and the shortfall would usually be supplied by the main battery. But this is not necessary true, as the alternator still can supply the demand, but not at full. Only if the demand is exceeding the alternator rating, then only would the battery start to supply power.

All the electrical components still functioning, and the engine still running and not lugging down. But in term of Power (watt), there was decrease in value. Alternator cannot provide maximum power required, but still can function at minimum power required. The car will still run, but it is noticeable that the headlights get progressively dimmer. This is a sign that the battery is not fully charge. This may cause difficulty to start the engine later as there will be not enough power to spark the engine.

Alternator supply all the current demand for the electrical devices, while at the same time charging the battery. Table 3 shows the battery voltage value after all the electrical devices being turned off and be left for five minutes. A full charged battery should has a reading of 14.2 V to 14.4 V. Here, based on Table 3, it shows that the higher the speed of alternator, the closest the voltage value of battery to fully charged value after five minutes. The faster the speed of alternator, the faster it is the rate of charging. And also, the faster the transfer or the moving of electron in the system. The faster the speed of the alternator, the faster the battery will be fully charged. As shown in Table 3, at idle speed 750 RPM, even after 5 minutes, the voltage reading was nowhere near the fully charged value. A battery voltage reading of 13.7 V and above while the engine is running is consider as good. But a battery voltage reading of 14.0 V and above is better.

When the engine is running, the alternator supplies electricity which, depending on the voltage level in the vehicle electrical system (determined by the alternator speed and the consumers drawing current), is normally enough to power the consumers and charge the battery as well. If the equipment current draw in the vehicle electrical system is greater than the alternator current (when engine is idling), the battery is discharged. The vehicle system voltage falls to the voltage level of the battery from which current is drawn. If the equipment current draw is less than the alternator current output, a proportion of the current flows to the battery and acts as a battery charging current. The vehicle system voltage increases to the set-point value specified by the voltage regulator.

The current output of the alternator is speed-dependent. For an engine idling speed, the alternator can only supply some of its rated current if it has a conventional turns ration ranging from 1:2 to 1:3 (crankshaft to alternator). By definition, the rated current is output at an alternator speed of 6000 RPM.

The electrical consumers have a variety of switch-on durations. A distinction is made between continuous loads, long-time loads, and short term loads. The electrical load requirements encountered during vehicle operations are not constant.

In vehicle electrical systems, the battery is charged using voltage limitation, where the battery charge current reduces automatically as the steady-state voltage rises. This prevents damage due to overcharging, and makes sure that the battery has a long service life. Battery charger, the alternator, on the other hand, still operates using constant current. Once the full state of charge has been reached, charging continues with only a slightly reduced, or possibly a constant current. Shortly after discharging begins, the voltage in the storage battery drops to a value which only changes relatively slowly if discharging continues. Battery discharge over a period of time, even if it is not subjected to loads. In other means, when no electrical equipment is connected.

The alternator must furnish the vehicle electrical system with a sufficient supply of current under all operating conditions in order to ensure the state of charge in the battery is always maintained at an adequate level. This is to ensure that the amount of current generated by the alternator under actual operating conditions is at least equal to the consumption of all electrical equipment within the same period.

Typical ratios in the passenger car sector lie between 1:2.4 and 1:3, in the commercial vehicle sector up to 1:5. In extreme cases like commuter traffic, the alternator runs for up to two-thirds of the total switch-on duration when the engine is idling, which is in the speed range with the lowest performance efficiency.

4 Conclusions

In this paper, the experiment was done to prove the theories, and to find the relationship between alternator speed and the charging system. The experiment was aimed to gain better understanding and better knowledge about charging system. The result shows that the faster the alternator, the more power it can produce. And when there is more power, the faster the charging of battery as the power to be supplied to other electrical consumer is more than enough. Current is inverse with voltage. When the alternator current is high at low speed, battery voltage will drop, and will need more time to get fully charged. But when alternator output current is low at low speed, the battery voltage is high as the battery get enough power for recharging. The Proton Preve used in the experiment has a crankshaft-alternator speed ratio of 1:2. At idling speed, the alternator still manages to supply the current demand from all the consumers, but has to reduce the current needed by the battery to get fully charge. As the speed of the alternator increase, it supplies more current to the consumers thus increase performance efficiency. Although Figure 7 shows that there is a decreases of battery voltage at 34 A and sudden drop of battery voltage at 50 A at idle speed, which cause by the increase of loads to the alternator and leads to low charge of battery, but in reality, this situation is rarely happen. In reality, most driver only turn on less equipment in the car such as radio, air-cond, and lights during night. All these only take up to 40 A of current, and the alternator can afford the demand and still can fully charge the battery even at idle speed.

The authors would like to be obliged to to Kementerian Pengajian Tinggi for providing financial assistance, and to Fakulti Mekanikal, Universiti Teknikal Malaysia Melaka for providing laboratory facilities under project no FRGS/1/2014/TK06/FKM/02/F00209.

References

1. J. M. Miller, "Multiple voltage electrical power distribution system for automotive applications," in Proc. 31st Intersociety Energy Conversion Engineering Conference (IECEC), Washington, DC, Aug. (1996).
2. J. M. Miller, D. Goel, D. Kaminski, H.-P. Schoner, and T. M. Jahns, "Making the case for a next generation automotive electrical system," in Proc. IEEE-SAE International Conference on Transportation Electronics (Convergence), Dearborn, MI, Oct. (1998).
3. Edgar J. Beyn, "The 12 Volt Doctor's Alternator Book", C. Plath North American Division, Spa Creek Inc, Annapolis, Maryland. (1986).
4. Smith, M. "It's Volts That Counts," in Automotive Engineering, **Vol. 26**, No. 9,2001, pp. 52-55. (2001).
5. Paul Brand, "Chapter 4 Alternator Charging System," in How To Repair Your Car, Motorbook International, pp. 61, (2006).
6. "Charging System Basic" Retrieved from http://www.bcae1.com/charging.htm [Last access 6 June 2016]
7. Stewart Robertson, *Chapter 3, Predicting Circuit Behaviour*, in Automotive Electrical Maintenance. Oxford, Elsevier, pp. 30, (1994)
8. Dr. N.K. Giri, "Automobile Mechanical," New Delhi, Khanna Publisher, pp. 376, (2009).
9. Bosch, *Automotive Handbook*, 7th Edition, Wiley, England, pp. 1009, (2007).
10. SAE International, *Automotive Handbook*, 8th Edition, Robert Bosch, Plochigen, pp. 952-968, (2011).

Omnidirectional configuration and control approach on mini heavy loaded forklift autonomous guided vehicle

Norsharimie Adam[1,*], *Mohd* Aiman[1], *Wan Mohd* Nafis[1], *Addie* Irawan[1], *Mohamad* Muaz[2], *Mohamad* Hafiz[2], *Akhtar Razul* Razali[2] and *Sheikh Norhasmadi* Sheikh Ali[3]

[1]Robotics and Unmanned Systems (RUS) research group, Faculty of Electrical and Electronics Engineering, Universiti Malaysia Pahang, 26600 Pekan, Pahang, Malaysia.
[2]Manufacturing Focus Group (MFG), Faculty of Mechanical Engineering, Universiti Malaysia Pahang, 26600 Pekan, Pahang, Malaysia.
[3]Vacuumschmelze (M) Sdn. Bhd, Lot 3465, Tanah Putih, 26600 Pekan, Pahang, Malaysia.

Abstract. This paper presents the omnidirectional configuration and control approach on Mini Heavy Loaded Forklift Autonomous Guided Vehicle (MHeLFAGV) for flexibility maneuverability in confine and narrow area. The issue in turning motion for nonholonomic vehicle in confine area becoming a motivation in MHeLFAGV design to provide holonomic vehicle with flexible movement. Therefore an omni-wheeled named Mecanum wheel has been configured in this vehicle design as well as omnidirectional control algorithm. MHeLFAGV system is developed with collaboration and inspired from Vacuumshmelze (M) Sdn. Bhd. Pekan, Pahang in order to have a customized mini forklift that able to work in a very confined warehouse (170cm x 270cm square) with heavy payload in a range of 20-200kg. In electronics control design, two stages of controller boards are developed namely as Board 1 and 2 that specifically for movement controller board and monitoring controller board respectively. In addition separate module of left, right, forward, backward, diagonal and zigzagging movement is developed as embedded modules for MHeLFAGV system's control architecture. A few experiments are done to verify the algorithm for each omnidirectional movement of MHeLFAGV system in the wide area. The waypoint of MHeLFAGV movement is plotted using Global Positioning System (GPS) as well as a digital compass by mapping the longitude and latitude of the vehicle. There are slightly different between the targeted movements with recorded data since Mecanum wheeled affected by the uneven surface of the landscape. The experiment is also further on moving in confine are on the actual targeted warehouse.

[*] Corresponding author: sharimie.adam@gmail.com

1 Introduction

Numerous research in nonholonomic systems has been widely studied especially on designing autonomous and guided vehicle. An autonomous vehicle can be in nonholonomic and holonomic mechanism depending on application requirement. However, most of the existed commercial industrial standard autonomous wheeled vehicles are nonholonomic with fixed driving wheels and focusing on wide area operations. There is an issue on dealing with confine area, especially for the heavy loaded task. Therefore omnidirectional configuration and control in the autonomous wheeled vehicle are essential. For the case of wheeled vehicle selection of omni-tyres are very important in order to realize omnidirectional movement. As for example mecanum wheel that consists of rollers around the surface body of the wheel depending on the size of mecanum, the bigger size the more roller is assembled[1]. As shown in Figure 1 (design is customized based on bought products from AndyMark, Inc.) the mecanum's roller assembled in 'axis-b' is slight skew to 'axis-a' wheel which usually 45 degrees of angle between 'a' and 'b' axis. In [2] an Omni-wheel is proposed by configuring three mecanum wheels with more controllable as well as steerable to use.

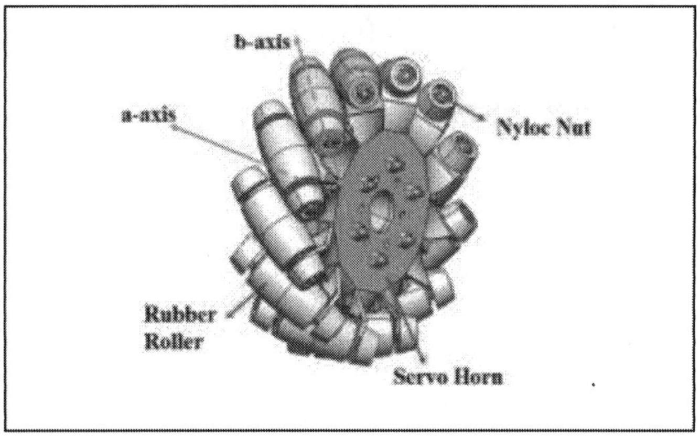

Fig. 1. Mecanum wheel design for MHeLFAGV.

On the other hand, model to control four wheel mecanum vehicle is proposed by [3-9] to develop the control model for the system to move more accurate and efficient by defining the velocity and the center of the wheel on the ground. The performance shows that the velocity of each wheel effect the smoothness of vehicle movement. Several wheel maneuver planning has been developed that incorporate kinodynamic constraint. Kinodynamic planning is aimed at solving kinematic constraints and dynamic constraints simultaneously, for designing the control input from the initial state and can be designed from the control input in one step[10]. Therefore, it has given the advantage that able to decide the control input simply, compared to existing planning [11, 12].

A survey of controllers' structure movements has been developed for vehicle systems that have an impact collision. For example, Zavala *et. al.* [13] introduced a family deadbeat feedback control laws to control a class of juggling-like systems. The contribution in their paper is a study of the intermediate control ability properties of the objects impact Poincare mapping [13, 14]. In previous work [15], the idea to collision free is by using the equations of motion. The difficulties with this two-stage planning approach are that the results of the kinematic planner may not be executable by the mobile robot due to limitations of the forces and torques available from the actuators. Other than that, the global optimality also cannot be guaranteed since only the optimal solution following the original kinematic path computed. Pan *et. al.*[16] in their research used a dynamic control planning method for a vehicle steered without Global Positioning Systems(GPS) and unmanned operation in unidentified circumstances. The proposed technique displayed effective in results, however, the vehicle is not stable when the velocity in a high state.

Therefore, this research and development have taken the initiative to involve on design and develop flexible heavy loaded vehicle named Mini Heavy Loaded Forklift Autonomous Guided Vehicle (MHeLFAGV) (patent applied) specifically for maneuvering in a confine area. The problems involve in the MHeLFAGV system are to transfer the items and work in confine space with 170cm x 270cm square with the range of 20-200kg payload. A holonomic mechanism is applied in MHeLFAGV design as well as Mecanum wheels for flexibility moves. Moreover, the wireless remotely control module is developed in this phase of development with two ways communication with Board 1 as well as Global Positioning System (GPS) and Digital Compass for monitoring in Board 2. On the other hand, embedded omnidirectional module is developed to realize the designed motion vectors of omnidirectional movement of MHeLFAGV regarding each Mecanum wheel. Several experiments are done for waypoint and longitude/altitude verification as well as moving in the targeted warehouse.

2 Overview on holonomic MHeLFAGV forklift system and configuration

2.1 MHeLFAGV system configuration

The unavailability of a commercial AGV system for the targeted confine area and for heavy loading used became the motivation of MHeLFAGV development. The system is currently developed with wireless remote control and sufficient enough to be used in targeted warehouse. As shown in Figure 2, mecanum wheels with overall size 996mm x 700mm is used for this vehicle and the forklift unit placed on the left side of its structure in order to balance the whole vehicle. Moreover, the system is expected to be able to loading and unloading as well as transferring from point to point a heavy copper spool with flexible movement. The systems consist of three main components; Direct Current (DC) motor controller, Radio Frequency (RF) remote controller, and four industrial grade DC motor.

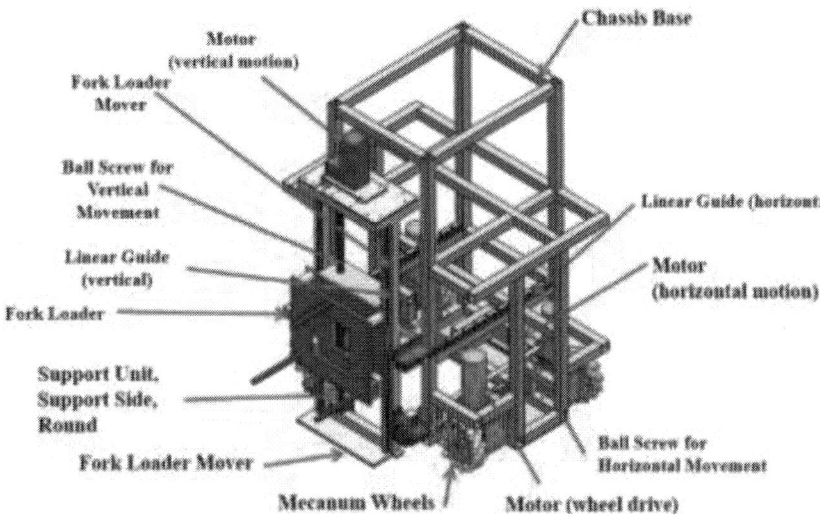

Fig. 2. Overview of design structure of MHeLFAGV system.

DOI: 10.1051/matecconf/20179001077

2.2 Control system architecture of MHeLFAGV system

The movement of the MHeLFAGV system with mecanum wheel is possible by controlling the four wheel independently that include speed and rotational. The direction of is navigated by rollers in mecanum wheel as depicted in Figure 3. These rollers are having a unique shape with an axis of rotation at 45° to the plane of the wheel and 45° to a line through the center of the roller parallel to the axis of rotation of the wheel[6] as discussed and labeled in Figure 1.

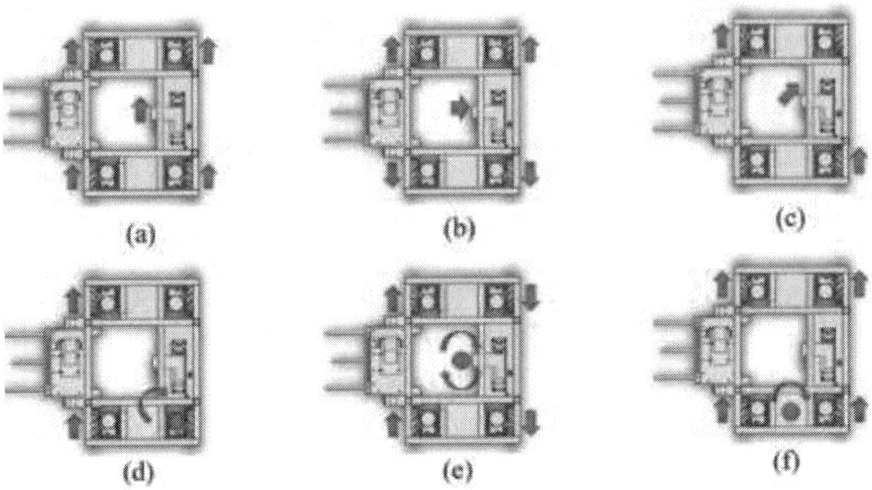

Fig. 3. Omnidirectional vector mapping for MHeLFAGV system; (a)Straight ahead, (b) Side Way, (c) Diagonal, (d)Turn Around, (e) Concerning, (f) Turn of Rear Axis.

Fig. 4. shows the dimension of MHeLFAGV base, from a top view the mecanum wheel is placed symmetrically. The roller of wheel acts as a reference point to place all four wheel performed in multidirectional movement system. The material used is a heavy duty which made from mild steel that can load over 150kg and the DC motor has the very high torque that could move the MHeLFAGV that weight about 200 kg.

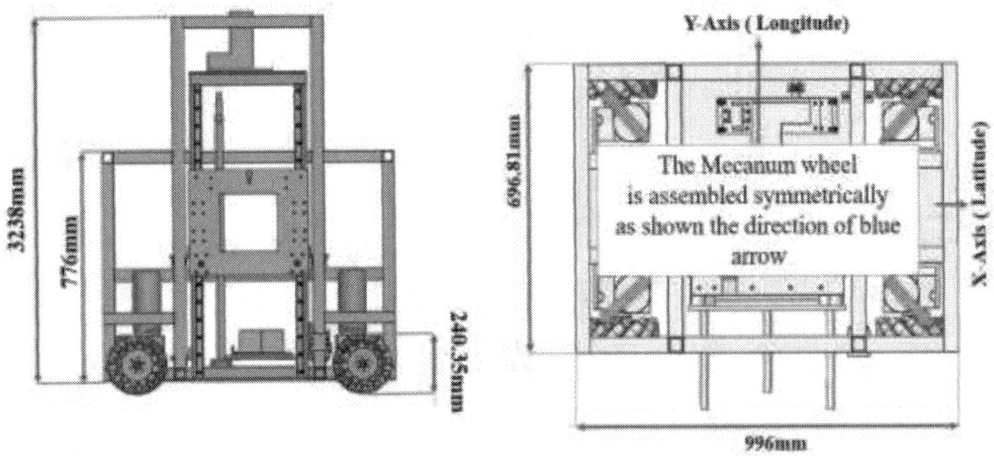

(a) (b)

Fig. 4. The dimension of MHeLFAGV base. (a) Side view; (b) Top view.

 DOI: 10.1051/matecconf/20179001077

In developing controller unit, two stages of controller boards are developed for this MHeLFAGV systems; movement controller board (Board 1) and monitoring controller board (Board 2). Moreover, several embedded modules are developed for each omnidirectional moving vectors including Zigzagging and Diagonal movement. A wireless remotely control modules are also prepared remotely controlled trajectory motion. On the other hand, Board 2 is developed as separated electronic system for MHeLFAGV motion vector mapping. As mentioned in the previous section Global Positioning System (GPS) unit is configured with Digital Compass is used and attached with MHeLFAGV system for longitude/latitude data acquisition.

The purpose of this separation approach is to guarantee a reliability of two ways real-time control and communication between user and MHeLFAGV system. As depicted in Figure 5, Board 1 is designed to consist of a microcontroller and dual channel motor controller that drive four wheels constantly. Three DC motor controllers connected to a microcontroller; Motor Controller 1 for front wheels, Motor Controller 3 for rear wheels and Motor Controller 2 for grasping/fork lifter. A potentiometer and Liquid Cristal Display (LCD) is configured for DC motor speed setup and calibration via its motor controller. SKPW-RX is an RF receiver that configured for the remote control connection between Board 1 and remote control as shown in Figure 6(a).

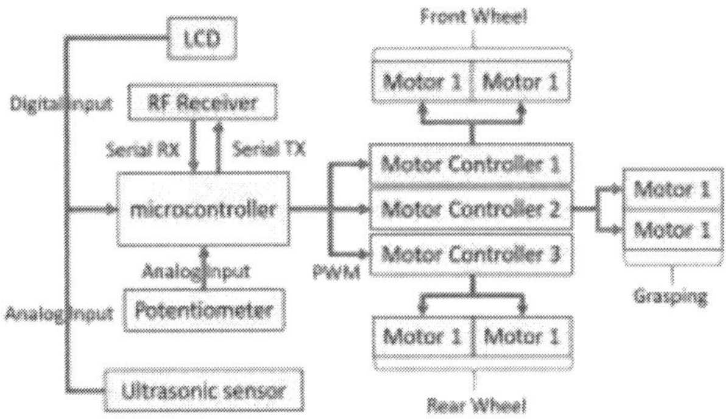

Fig. 5. Board 1 control schematic in MHeLFAGV system.

Each button on remote controller as shown in Figure 6(a) is programmed with reference to the omnidirectional vector as defined in Figure 3. The implementation The selected remote controller has function same as a potentiometer in terms of physical motion but producing pulses output as indication signal. Basically, this joystick can be slowly toggled to control the speed of movement of MHeLFAGV and producing a range of digital signals that useful for MHeLFAGV Cartesian inputs as shown in Figure 6(b). Separated board namely as Board 2 is developed with a combination of the GPS, Digital Compass sensor, and a microcontroller as shown in Figure 7. As mentioned earlier this second board is purposely developed to ensure a reliable real-time data logging for positioning and movement of MHeLFAGV. The GPS module transmitting the raw data of position and movement of MHeLFAGV to the microcontroller for longitude and latitude data encoding and saved into Micro Personal Computer (μPC).

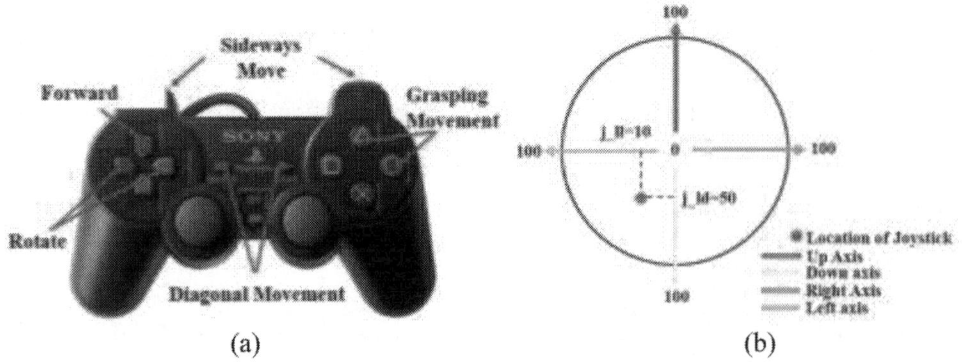

Fig. 6. MHeLFAGV Movement Controller (a) Physical feature of remote controller/joystick; (b) Remote controller/joystick button default mapping.

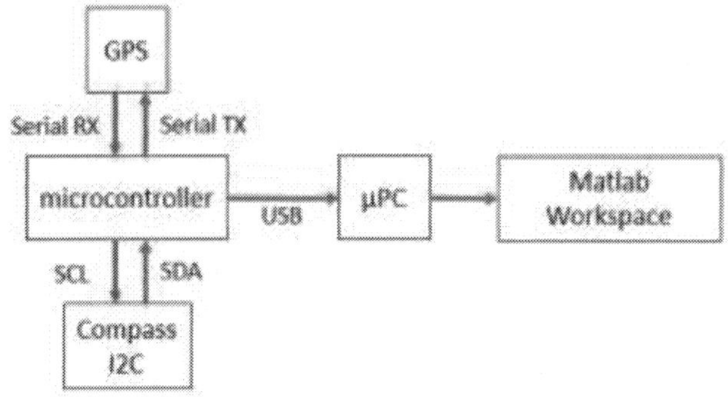

Fig. 7. Overall schematic diagram of Board 2.

3 Kinematic of MHeLFAGV system

For control planning, it is essential to model the kinematics of the system for desired input conversion to the driven elements. As shown in Figure 8, the MHeLFAGV system wheel consists of a fixed standard mecanum wheel with rollers attached to the wheel circumference. When viewing a MHeLFAGV from the top view, X_1 and Y_1 provide the final output coordinates of the system while *'a'* is the vehicle's width from wheel center to the MHeLFAGV center and *'b'* is likewise the MHeLFAGV length while *'l'* is the distance from the geometric center *'C'* to the wheel center *'D'*. Complete kinematics of an omnidirectional system with four wheels mecanum will be considered in this section.

Considering each of the mecanum wheels with mass (*mwi* = *mw*) and moment of inertia (*Iwi* = *Iw*) with the vehicle base has mass (*mb*) and moment of inertia (*Ib*). Figure 9 illustrated the system state variables $q=[x_I \quad y_I \quad \theta]^T$ in inertial laboratory frame *'I'* and input torques $\tau = [\tau_1 \quad \tau_2 \quad \tau_3 \quad \tau_4]^T$ acting on each wheel of the MHeLFAGV. In general, the vehicle base should be well balanced with the center of mass near the geometric center *'C'*. However, the model is generalized for bases where the center of mass *'C''* is offset from *'C'* by distances a_1 and a_2. As shown in Figure 9, the friction forces between each MHeLFAGV wheel are stated as $f = [f_1 \quad f_2 \quad f_3 \quad f_4]^T$.

 DOI: 10.1051/matecconf/20179001077

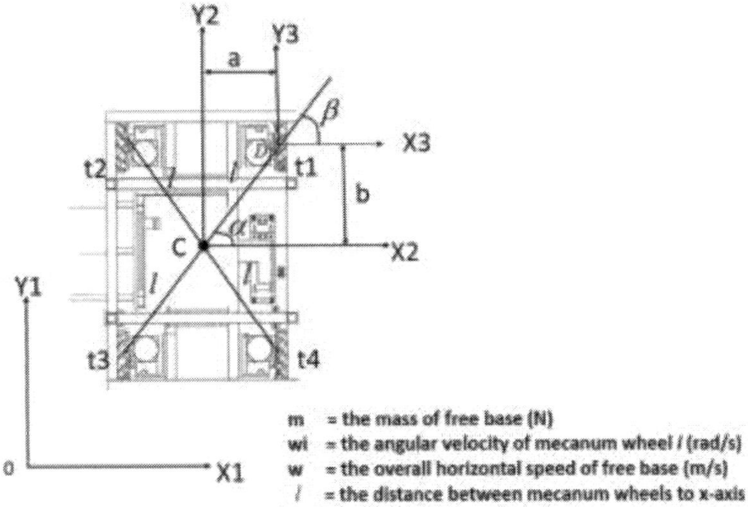

Fig. 8. Kinematic model of MHeLFAGV system.

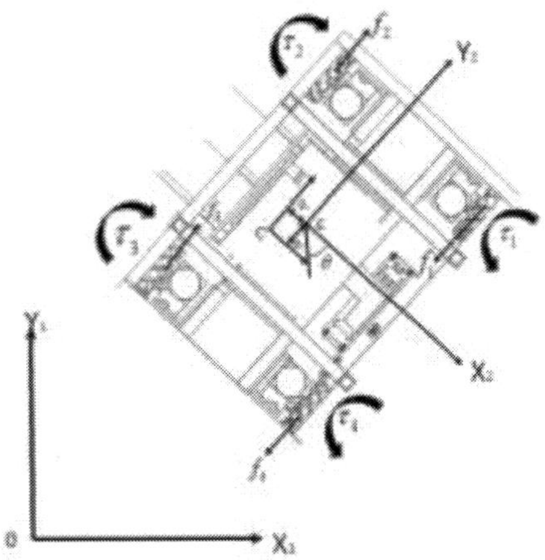

Fig. 9. System coordinates with input torques defined positively for Clockwise (CW) torques looking outward from the vehicle base.

The derivation of the system equations of this research shows that the kinematics of the MHeLFAGV wheel system rolling without slippage occurs. The components of inverse kinematic equations are tabulated in Table 1.

Table 1. Parameters of the each wheel.

Wheel	α_i	β_i	γ_i
t1	$\tan^{-1}(b/\alpha)$	$-\tan^{-1}(b/\alpha)$	$-(\pi/2+\pi/4)$
t2	$\pi-\tan^{-1}(b/\alpha)$	$\tan^{-1}(b/\alpha)$	$(\pi/2+\pi/4)$
t3	$\pi+\tan^{-1}(b/\alpha)$	$-\tan^{-1}(b/\alpha)$	$-(\pi/2+\pi/4)$
t4	$2\pi-\tan^{-1}(b/\alpha)$	$\tan^{-1}(b/\alpha)$	$(\pi/2+\pi/4)$

With reference to the equation (1) in order to find the direction of each wheel's rotation during the experiment, the equation of $\alpha = \tan^{-1}(b/\alpha)$ will be used. The equation is modified from the previous research works due to the positive or negative direction of γ_1.

$$\begin{bmatrix} t_1 \\ t_2 \\ t_3 \\ t_4 \end{bmatrix} = \left(\sqrt{2}/r\right) \begin{bmatrix} \sqrt{2/2} & \sqrt{2/2} & l\sin\left(\pi/4-\alpha\right) \\ \sqrt{2/2} & -\sqrt{2/2} & l\sin\left(\pi/4-\alpha\right) \\ -\sqrt{2/2} & -\sqrt{2/2} & l\sin\left(\pi/4-\alpha\right) \\ -\sqrt{2/2} & \sqrt{2/2} & l\sin\left(\pi/4-\alpha\right) \end{bmatrix} \begin{bmatrix} \cos\theta & \sin\theta & 0 \\ -\sin\theta & \cos\theta & 0 \\ 0 & 0 & 1 \end{bmatrix} \begin{bmatrix} x_1 \\ y_1 \\ \theta \end{bmatrix} \quad (1)$$

4 Experimental results

Several experiments are done on MHeLFAGV completed system on both wide areas and targeted confine warehouse. The movement topology of the mecanum principle in Figure 10 depicts all moving vectors setup for MHeLFAGV system as discussed in the previous section. The test is done in remotely controlled mode. Moreover, each point longitude and latitude are recorded for every 2 seconds.

Fig. 10. Movement planning for MHeLFAGV.

The first test is done in wide as shown in Figure 11 where programmed omnidirectional moving vector is verified by using customize wireless remote control as trajectory input in a range of 100meters. According to the MHeLFAGV physical movement, RF communication performed very established connection even the system. With reference to the Figure 12, the coordination of latitude-longitude shows slightly fluctuated if compared to the desired signal as a factor of friction and uneven floor surfaces. However, the pattern of the desired movement is still followed by the vehicle in all omnidirectional movement vectors as shown from Figure 12(a) to 12(d). Moreover, this experiment shows that wireless remote controller with RF is successfully established and controlling the MHeLFAGV to move all vectors of omnidirectional movement. On the other hand, the GPS system with digital compass able to localize MHeLFAGV way point with an additional uFL connector to ensure GPS system having fast connection recovery with satellite.

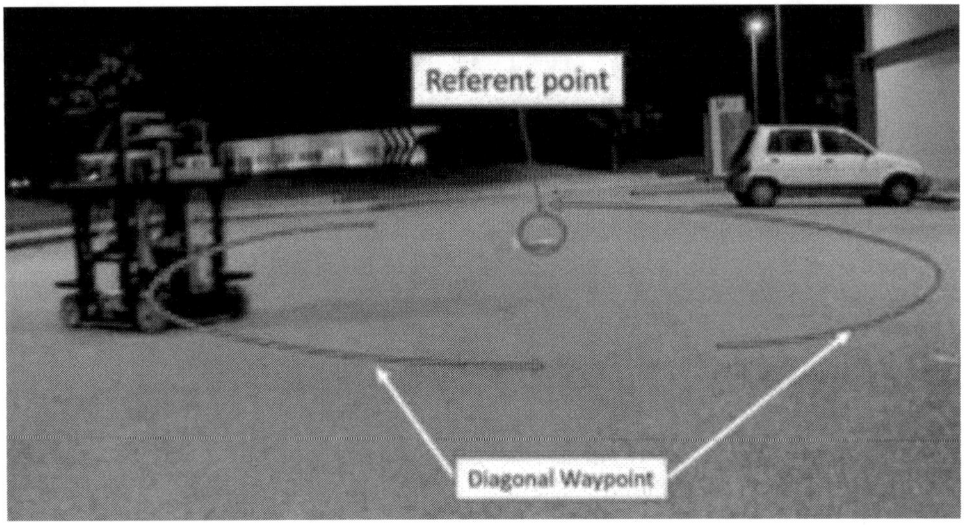

Fig. 11. Sample of MHeLFAGV experiment snap shot at wide area.

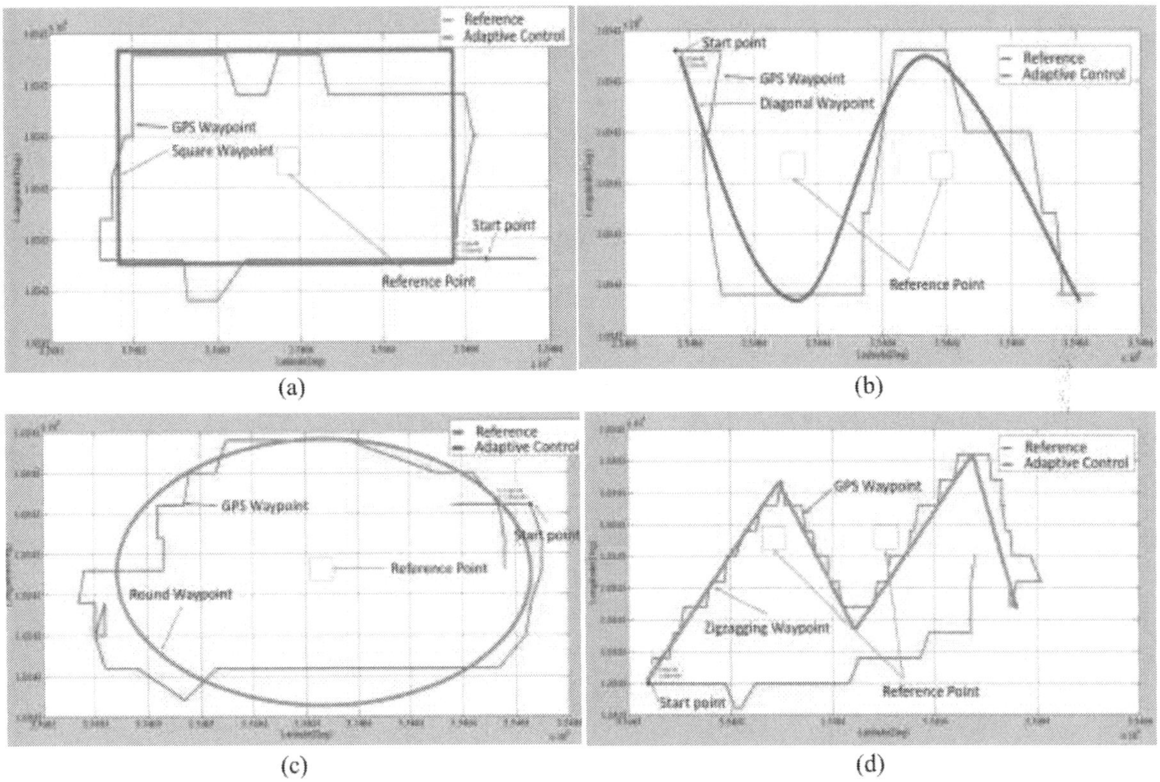

Fig. 12. The series testing of the movement of MHeLFAGV system.

The experiment is continued in the actual warehouse site at Vacuumshmelze (M) Sdn. Bhd. Pekan, Pahang, Malaysia. Through a series of movement tests, it shows that MHeLFAGV suitable to be deployed in the targeted line area as shown in Figure 13. The same omnidirectional movement was applied to move MHeLFAGV at this site as for the case of picking one of the heavy spools on the rack.

Fig. 13. Snapshot of MHeLFAGV on-site test.

5 Conclusions

This paper proposed a holonomic and omnidirectional design for a mini heavy loading AGV system named MHeLFAGV. Through a series of tests, it was shown that the programmed omnidirectional moving vectors were established and successfully deployed in various area including confine warehouses. It is remarked that experiment results show the durability of the designed AGV system to move almost accurately on the uneven surfaces with reference to the desired input. Improvement on the precision, dynamic motion, and autonomous movement are needed for MHeLFAGV system in order to comply with the industrial scale requirement and this will be the next step in this line of research.

This research and development are supported by Vacuumshmelze (M) Sdn. Bhd. Pekan, Pahang in collaboration with Universiti Malaysia Pahang (UMP) Research Grant (RDU150348).

References

1. Ramirez-Serrano, A. and R. Kuzyk. *Modified Mecanum Wheels for Traversing Rough Terrains*. in *2010 Sixth International Conference on Autonomic and Autonomous Systems*. 2010.
2. Jüttler, B., et al., *Classical Techniques for Applied Geometry and Kinematics of The Mecanum Wheel.* Computer Aided Geometric Design, 2008. **25**(9): p. 784-791.
3. Tlale, N. and M.d. Villiers. *Kinematics and Dynamics Modelling of a Mecanum Wheeled Mobile Platform*. in *Mechatronics and Machine Vision in Practice, 2008. M2VIP 2008. 15th International Conference on*. 2008.
4. Hoang, G., H.K. Kim, and S.B. Kim. *Control of Omini-Directional Mobile Vehicle for Obstacle Avoidance Using Potential Function Method*. in *Control Conference (ASCC), 2013 9th Asian*. 2013.
5. M. O, T., et al. *Design and Development of an Autonomous Omni-Directional Mobile Robot with Mecanum Wheels*. in *Automation, Quality and Testing, Robotics, 2014 IEEE International Conference on*. 2014.

DOI: 10.1051/matecconf/20179001077

6. Vlantis, P., et al. *Fault tolerant control for omni-directional mobile platforms with 4 mecanum wheels*. in *2016 IEEE International Conference on Robotics and Automation (ICRA)*. 2016.
7. Chang Lin Lih and Y.S. Hao, *Modeling and Adaptive Control of an Omni-Mecanum-Wheeled Robot.* 2013. **4**: p. 166-179
8. Jia, Y., X. Song, and S.S.D. Xu. *Modeling and Motion Analysis of Four-Mecanum Wheel Omni-Directional Mobile Platform*. in *Automatic Control Conference (CACS), 2013 CACS International*. 2013.
9. Qi-Ye, Z., et al. *A TSK Fuzzy Model and Adaptive Sliding-Mode Controller Design for Four-Mecanum-Wheel Omni-Directional Mobile Free-Bases*. in *Chinese Automation Congress (CAC), 2015*. 2015.
10. Jiwung, C., *Kinodynamic Motion Planning for Vehicle* International Journal of Advanced Robotic Systems, 2015. **11**(90).
11. Haustein, J.A., et al. *Kinodynamic Randomized Rearrangement Planning Via Dynamic Transitions between Statically Stable States*. in *Robotics and Automation (ICRA), 2015 IEEE International Conference on*. 2015.
12. Hongchuan, W. and S. Ferrari, *A Geometric Transversals Approach to Sensor Motion Planning for Tracking Maneuvering Targets.* Automatic Control, IEEE Transactions on, 2015. **60**(10): p. 2773-2778.
13. A. ZAVALA, B.B., *On the Control of a One Degree of Freedom Juggling Robot.* 1999.
14. Dranga, O. and I. Nagy. *Stability Analysis of Feedback Controlled Resonant DC-DC Converter Using Poincare Map Function*. in *Industrial Electronics, 2001. Proceedings. ISIE 2001. IEEE International Symposium on*. 2001.
15. Shiller, Z. and H.H. Lu. *Robust Computation of Path Constrained Time Optimal Motions*. in *Robotics and Automation, 1990. Proceedings., 1990 IEEE International Conference on*. 1990.
16. Pan, Z., et al. *Dynamic Motion Planning For Autonomous Vehicle In Unknown Environments*. in *Intelligent Vehicles Symposium (IV), 2011 IEEE*. 2011.

Transient modelling of heat loading of phase change material for energy storage

W.M. Asyraf[1], *Anusuiah* Vasu[1], *Ftwi Y.* Hagos[1,2,*], *M. M.* Noor[1,2], *Rizalman* Mamat[1,2]

[1]Faculty of Mechanical Engineering, Universiti Malaysia Pahang, 26600 Pekan, Pahang, Malaysia
[2]Automotive Engineering Centre, Universiti Malaysia Pahang, 26600 Pekan, Pahang, Malaysia

Abstract. As the development of solar energy is getting advance from time to time, the concentration solar technology also get the similar attention from the researchers all around the globe. This technology concentrate a large amount of energy into main spot. To collect all the available energy harvest from the solar panel, a thermal energy storage is required to convert the heat energy to one of the purpose such as electrical energy. With the idea of energy storage application that can be narrow down to commercial application such as cooking stove. Using latent heat type energy storage seem to be appropriate with the usage of phase change material (PCM) that can release and absorb heat energy at nearly constant temperature by changing its state. Sodium nitrate ($NaNO_3$) and potassium nitrate (KNO_3) was selected to use as PCM in this project. This paper focus on the heat loading process and the melting process of the PCM in the energy storage using a computer simulation. The model of the energy storage was created as solid three dimensional modelling using computer aided software and the geometry size of it depend on how much it can apply to boil 1 kg of water in cooking application. The materials used in the tank, heat exchanger and the heat transfer fluid are stainless steel, copper and XCELTHERM MK1, respectively. The analysis was performed using a commercial simulation software in a transient state. The simulation run on different value of velocity but kept controlled under laminar state only, then the relationship of velocity and heat distribution was studied and the melting process of the PCM also has been analyzed. On the effect of heat transfer fluid velocity, the higher the velocity resulted in higher the rate of heat transfer. The comparison between the melting percentages of the PCMs under test conditions show that $NaNO_3$ melts quite faster than KNO_3.

1 Introduction

From day to day, the dependence on fossil fuel energy is quite distressing. Fossil fuel energy will face depletion due to the drastic increase in world energy consumption. The non-renewable energy usage is also one of the leading reasons for the production of greenhouse gases. In recent decades, researchers have been studying the usage of

[*] Corresponding author: ftwi@ump.edu.my

environmental friendly and renewable energy that could be harvested around the globe with no cost [1, 2]. Solar energy is one of the clean and abundantly available renewable energy resources that can replace the non-renewable energy resources [3]. In the application of solar cooking using concentrated solar power, thermal energy storage (TES) medium is required to store the heat energy obtained from solar energy [4]. Phase change material (PCM) that can store and release heat energy according to necessity by changing its state from solid to liquid or vice versa is likely to be the most suitable type of material that can be used as TES. Heat loading process of the PCM during absorption of heat energy that leads to the phase change from solid to liquid is to be analyzed in this paper.

Lewis et al.[5] have reported that there are three major components needed for solar energy utilization [5]. These include solar capture, conversion, and storage. Solar capture and conversion can be achieved by photo-voltaic cells (PVs). The solar energy utilization and storage are expected to be economical and affordable to enhance its usage. Unfortunately, the cost of material needed to utilize this renewable energy is very expensive compared to fossil fuels usage. There should be an advancement in the approach to utilize this very important sustainable energy in a more economical way. Finding new material for thermal capture during its availability as well as new thermodynamic cycle to harvest the solar energy to produce a useful fuel is necessary. Chwieduk [6] stated that there are two methods of utilizing solar energy available. The first method is indirect utilization method. In this case, renewable energy is used as a secondary effect of solar energy. Hydro energy, ocean energy, and wind energy are the examples of secondary effect of solar energy. These energies are accomplished by photosynthetic processes and is mostly related to biomass and biofuels. The second method is direct utilization of solar energy. In this case, there are two fundamental methods of energy conversion available. These includes photo-thermal conversion of energy and solar - photoelectric conversion of solar energy radiation. The photo-thermal conversion is applied in active low temperature solar water and air heating system, passive low temperature solar water heating, high temperature solar heating system with solar concentrator and electric power system.

Edina Milisic [7] stated that, thermal energy storage must be able to store and release a large amount of energy. Using PCM as a storage medium will completely depend on the latent heat energy for the energy storage system. Unlike sensible heat energy, latent heat type shows less temperature change during energy storage because when the material changes its phase, the combination of pressure and temperature leads to the aggregate state. In the aggregate state, the latent heat will be stored and released almost at constant temperature, thus even a small difference in temperature can be used to store a large amount of latent heat energy. In general, energy storage can be distinguished as sensible heat, latent heat and thermochemical or any combination of it. Atul Sharma et al. [8] investigated about how thermal energy storage works with the change in internal energy of the material. This article includes the sensible heat, latent heat, and thermochemical or any combination of these for further discussion. Sensible heat storage (SHS) is a thermal energy stored by raising the temperature of liquid or solid material. SHS will develop the heat capacity and change in temperature during the loading and unloading processes of heat transfer. The specific heat and temperature change in the storage medium will affect the amount of heat stored in SHS. Another approach is latent heat storage (LHS). In this case, the energy is stored when a material undergoes a change in state that is liquid-to-solid or gas-to-liquid or vice versa. Regin et al. [9] stated that developing a suitable containment for PCM will lead to a better utilization of PCM and heat transfer of fluid. The PCM storage containers with several geometric designs have their own pro and cons. PCM storage container material should meet the requirement of strength, flexibility, thermal stability and corrosion resistance to the PCM material. PCM capsules made of container material of characteristics as above enhance the latent heat storage and at the same time acts as a barrier to protect the latent heat storage material from harmful interaction with environment. PCM capsules storage provides sufficient surface for heat transfer and structural stability for easy application.

Agyenim et al. [10] have reported that the most important factors need to be considered in PCM energy storage are the thermal and geometric parameters of the container material required for a fixed quantity of PCM. This criteria has a direct impact to heat transfer characteristic and has major influence to the melting time and performance of the PCM storage unit. The geometric parameters of the PCM container must correspond to the melting time and daily insolation process from a solar collector at specific location in order to ensure long term thermal performance of PCM storage systems. Cabeza et al. [11-15] highlighted a technique to encapsulate the PCM to avoid the liquid phase flow away from the specific location applied. The two principles of PCM encapsulation include micro-encapsulation and macro-encapsulation. Micro-encapsulation is a small, spherical, rod shaped particle that is covered in a thin, high molecular weight polymeric film. On the other hand, macro-encapsulation comprises the PCM in some form of tube, pouches, sphere or panel that serves directly as heat exchanger.

Arefeh Hesaraki [16] stated that PCM are widely used to store latent heat energy. This is due to PCM's ability to absorb a large amount of heat energy during its change in state at constant temperature. Hence, PCM changes its phase by absorbing or releasing heat energy without much change in temperature. The heat transfer enhancement and performance of the thermal energy system (TES) depends on the thermal conductivity of the PCM. To prevent heat loss, the PCM storage can be insulated but if the heat loss occurs through vaporization of materials, there is no clear method of preventing to avoid it. There are limited study on the high temperature TES employing PCM both experimental and numerical. The effect of heat transfer fluid velocity and fin spacing is conducted for medium temperature application (paraffin wax) by Ogoh and Groulx [17] in a cylindrical heat storage tank and the fins arranged horizontally. The same paraffin wax material is also utilized the study of heat transfer characteristics by Jesumathy et al. [18]. The general objective of this study is to develop a further understanding and modelling of loading process of phase change material (PCM). Therefore the current investigation is narrowed down to these specific objectives of analysis of the heat requirement for solar cooking applications, design and sizing of the thermal energy storage (TES) and execution of a transient simulation on heat energy distribution, heat loading process of thermal energy storage for solar cooking application using phase change material.

2 Methodology

Solar cooking model required a solar collector that will absorb the heat and radiation energy from the sun. Then the absorber will transfer the heat energy to the heat transfer fluid (HTF). The heat transfer from the HTF will heat up the PCM in the thermal energy storage (TES). To minimize the heat lost to surrounding the TES should be well insulated. The focus of the current study is only the TES system. The problem definition consists of a thermal heat storage tank with its volume filled by PCM and a heat transfer fluid channel enters at the bottom center of the tank and leaves at the top center of tank. The work is progressed in such a way that the effect of heat transfer fluid in the exchanger tube is studied with different PCMs.

2.1 Geometry design

The geometry of thermal energy storage is designed based on the application requirement to boil 1 kg of pure water. The heat requirement to boil 1 kg of pure water is calculated in order to obtain the minimal volume of PCM needed in the TES. The law of calorimetric to determine heat energy needed to boil 1 kg of water based on Eq. (1),

$$Q = c.m(100°\text{C} - T) + \Delta h_{vap}.m \quad (1)$$

where, c=specific heat of pure water, m=mass of water, T=initial temperature, and Δh_{vap}= specific enthalpy of vaporization of water. The heat energy calculated when the thermal efficiency of TES is assumed to be 70% is stated as total energy stored, E_{stored}. Using Eq. (2), the mass and volume of the PCM can be obtain.

$$m_{PCM,solid} = \frac{E_{stored}}{h_{PCM}}, \quad V_{PCM} = \frac{m_{PCM}}{\rho_{PCM}} \quad (2)$$

The value obtained from Eq. (2) is volume and mass of PCM in room temperature that is in solid state. PCM will change its state to liquid during heat loading process, thus the mass and volume of the PCM in liquid state also need to be considered and can be obtained using Eq. (3) [19].

$$\rho_{PCM,liquid} = 1000[2.112 - 0.0006891(Melting\ temp\ °\text{C})] \quad (3)$$

With the help of the above data, dimensional parameters of the TES design can be defined using CAD software. Based on the previous calculation from Eqs. (1) – (3), the volume of the TES should not be less than 0.017 m^3. Therefore, the cylindrical shape of TES can be designed using the volume equation given below. The heat transfer formula can yield the parameters such as TES length l = 0.25 m, diameter = 0.3 m and heat exchanger tube length; l = 0.35 m.

$$V_{cylinder} = \frac{\pi D^2 L}{4}, V_{pipe} = \frac{\pi D^2 L}{4} \quad (4)$$

$$Q_{conv} = hA(T_s - T_\infty), A = \pi D \quad (5)$$

where, h = $385 W/m^2.K$, T_s= Surface temp°C, T_∞= Surrounding temp (°C) , D= 0.015 m (diameter of copper tube).

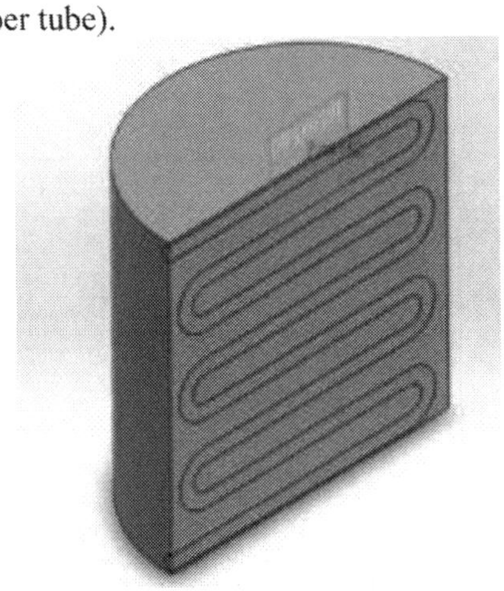

Fig. 1. Complete 3D model of TES with heat exchanger tube.

The geometrical structure considered for the present study is shown in Figure 1. It was created using Solid works modelling software. In this design, minimum dimensions and parameters are considered in such a way it can boil only 1 kg of pure water to make it more simple and economic.

2.2 Boundary conditions

The main boundary condition of this experiment excludes any heat flux from other sources except heat from the heat transfer fluid (HTF) in the heat exchanger tube (HET). It was assumed that heat was carried out through the HTF at temperature of 500 to 600°C. All the walls are treated as adiabatic walls to avoid heat loss. To avoid material transfer between PCM and air, the interface domain is considered as a wall. The initial temperature was set to room temperature. The side wall of HET provide heat flux for HTF. The domain is simplified by cutting it into half by using the symmetry assumption between all four directions. The surface of the tank walls are considered insulated. This can be achieved by applying the heat flux value equals to zero. Located at the bottom of the HET, the HTF flows upwards in the tube. The temperature of HTF is maintained about 350°C . This is the highest melting point of the PCMs to be tested. The gauge pressure is considered as zero bar for outlet boundary for the temperature output of various rate of flow of HTF in the tube. The initial part of the process is conducted at room temperature and state of the PCM is solid. Stainless steel and pure copper are the metals used for TES and HET respectively. The materials used for PCM are KNO_3 and $NaNO_3$. The mass, momentum and energy equation in the solution are monitored in such way the residuals are converged to a value not more than 10^{-6}.

2.3 Governing equation

In a specific range of temperature, problems in solidification or melting can be solved using FLUENT simulation. FLUENT has the option to track the liquid-solid front explicitly and enthalpy-porosity formulation. In this closed system, the liquid-solid mushy zone or porous zone was neglected due to the fluid material of PCM being fixed in a closed energy storage. Therefore, there is no additional momentum equation added into the simulation. The flow domain is separated into discrete cells and analysis is done by finite volume method. Solving the governing equations are controlled in the point of discrete cells. The partial differential equations were evaluated as algebraic equations using the finite volume method.

The list of differential and algebraic equations can be written as in Eqs. (6) – (14):

Continuity Equation:

$$\frac{\partial \rho}{\partial t} + \frac{\partial}{\partial x_j}\left[\rho u_j\right] = 0 \tag{6}$$

Momentum Equation:

$$\frac{\partial}{\partial t}(\rho u_i) + \frac{\partial}{\partial x_j}\left[\rho u_i u_j + p\delta_{ij} - \tau_{ji}\right] = 0 \tag{7}$$

Energy Equation:

$$\frac{\partial}{\partial t}(\rho e_0) + \frac{\partial}{\partial x_j}\left[\rho u_j e_0 + u_j p + q_j - u_i \tau_{ij}\right] = 0 \tag{8}$$

Stokes law was assumed for mono-atomic gases with Newtonian fluid. Hence, the viscous stress is as follows:

$$\tau_{ij} = 2\mu S_{ij} \tag{9}$$

The trace-less viscous strain-rate:

$$S_{ij} = \frac{1}{2}\left(\frac{\partial u_i}{\partial x_j} + \frac{\partial u_j}{\partial x_i}\right) - \frac{1}{3}\frac{\partial u_j}{\partial x_k}\delta_{ij} \tag{10}$$

Fourier's law heat-flux:

$$q_j = -\lambda\frac{\partial T}{\partial x_j} = -C_p\frac{\mu}{Pr}\frac{\partial T}{\partial x_j} \tag{11}$$

Laminar Prandtl number Pr is:

$$Pr = \frac{C_p\mu}{\lambda} \tag{12}$$

An equation of state is necessary to close these equations. An ideal gas was assumed so that the following equation to be valid:

$$\gamma = \frac{C_p}{C_v}, \quad P = \rho RT, \quad e = C_v T, \quad C_v - C_p = R \tag{13}$$

where, C_v, C_p and R are constant.

Hence the total energy:

$$e_0 = e + \frac{U_k U_k}{2} \tag{14}$$

2.4 Numerical simulation

The 3D geometry is then exported into the simulation commercial computer software. It is used to simulate the mechanism of heat exchange and the melting distribution of PCM. This commercial simulation software provides a step-by-step workspace, so that the analysis cannot be proceeded if the previous step is not complete.

By using a tetrahedral mesh with independent on solid interface, the model is discretized. Solid works file is exported in SAT format to make it easier to edit in Ansys. The setting of Ansys Fluent for numerical solutions are listed in this section. The geometry of the setup has undergone various checks on the mesh through Fluent and the report has appeared in the console. This is done to make sure that the minimum value of the mesh is always positive and the values are selected in the specified range. Once checking the grid of mesh model is done, the required units are selected. The general setting shows the solver settings and discretization details such as equation required for solving the problem and additional physics applications required regarding the solution of the simulation. The Table 1 provides the basic solver settings.

Table 1. Basic solver settings.

Model	Settings
Space	3D
Solver	Pressure based
Time	Unsteady, 1st-Order explicit
Viscous	Laminar
Heat transfer	Enable
Solidification and Melting	Enable

Solidification and melting is enabled in order to satisfy phase change condition. The value of mushy zone is neglected assuming that the PCM is already in the TES.

The properties of the materials under test are specified including the thermo-physical properties. Stainless steel and copper are the metals used for TES container and heat exchanger tube, respectively. KNO_3 and $NaNO_3$ are the PCM materials tested in this work and air is used as forced convection fluid. The settings provide for PCM are given in Table 2.

Table 2. Thermo-physical properties of major material in the simulation.

Property	Units	Method	KNO_3	$NaNO_3$
			Value(s)	
Density	kg/m^3	Piecewise continuous	$\rho = 2293.6 - 0.7497T$	
Specific Heat	J/kgK	Piecewise continuous	$C_P = 5806 - 10.833T + 7.2413 \times 10^{-3}T^2$	
Thermal Conductivity	w/mK	constant	0.5	
Viscosity	kg/ms	Piecewise continuous	$\mu = 0.4737 - 2.297 \times 10^{-3} + 3.731 \times 10^{-6}T^2 - 2.019 \times 10^{-9}T^3$	
Melting Heat	kJ/kg	constant	116	174
Solidifying Temperature	K	constant	600	573
Liquefying Temperature	K	constant	610	583

XCELTHERM MK1 is a commercial liquid by Radco used as HTF in the heat exchanger tube. The setting and properties of the HTF are given in Table 3.

Table 3. Thermo-physical properties of XCELTHERM MK1 as HTF [2].

Property	Units	Method	Value(s)
Density	kg/m^3	constant	695.3
C_P (Specific Heat)	J/kg-K	constant	2.62
Viscosity	cSt	constant	0.146
Thermal Conductivity	w/m-K	constant	0.076

3 Results and discussion

This section discusses the results of the design and simulation using Ansys Fluent software. Besides, It also discusses the trend of the observation and the reason behind it. Simulation results will focus on the heat transfer fluid velocity, type of phase change material that may affect the heat transfer rate of the thermal energy storage tank, and the melting phase distribution of phase change material.

3.1 Effect of heat transfer fluid velocity

The simulation of thermal energy storage is assumed to undergo a laminar fluid flow. The heat exchanger tube with diameter of 0.015 m is used in this work. Heat transfer fluid with density of 695.3 kg/m^3 (based on the online database) and dynamic viscosity of 6.88347 × 10-7 kg/m·s is considered for this experiment [20]. For fluid flow to be laminar, the Reynolds number should be less than 2300, the Reynolds number in this simulation is therefore considered to be 2200. The highest velocity of the heat transfer fluid at laminar state is about 0.15 m/s. This is calculated using Reynolds number formula, Eq. (15). The maximum and minimum temperatures of the PCM are calculated at specific time interval of 30 min for particular fluid velocity. The simulation is carried out for different values of fluid velocities as follows 0.01 m/s, 0.035 m/s, 0.07m/s, 0.1 m/s, 0.15 m/s and 0.2 m/s These values of velocities were inserted into the simulation to investigate the relationships.

$$Re = \frac{\rho V D}{\mu} \quad (15)$$

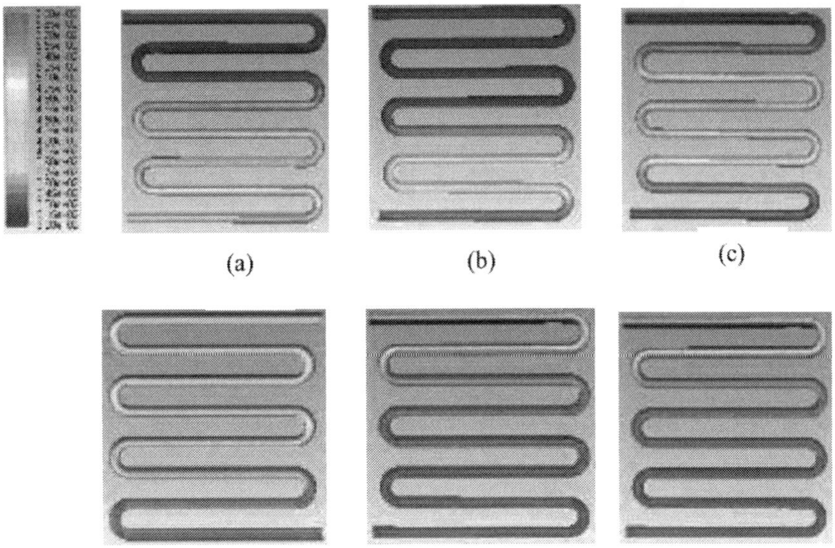

Fig. 2. Temperature distribution in K for different heat transfer fluid velocities (a) 0.01 m/s; (b) 0.035 m/s; (c) 0.07 m/s; (d) 0.1 m/s; (e) 0.15 m/s; (f) 0.2 m/s.

The minimum and maximum temperature yields at specific time of 30 min are 300 K and 600 K, respectively. The effect of the heat transfer fluid velocity on the temperature of both the heat transfer fluid pipe and on the PCM are depicted in Figures 2 and 3. As shown in Figure 2, the temperature distribution contour is roughly almost the same for both fluid velocities (e) 0.15 m/s and (f) 0.2 m/s. This is probably because, the velocity of heat transfer fluid turns into turbulence state above 0.15 m/s. As stated earlier, the FLUENT simulation software was set to be in laminar model only, therefore the simulation will not proceed if the heat transfer fluid model enter the turbulence state. The results appeared to be roughly the same because the software will only simulate within laminar stage. The rate of heat transfer is directly proportional to mass flow rate. It also states that mass flow rate is directly proportional to fluid velocity. Thus, it can be accepted that the higher the heat transfer fluid velocity, the higher the rate of heat transfer. This happened because, when the fluid in tube is in high velocity, more new and hot fluid from solar collector flows into the tube that replaces comparatively cooler fluid in the tube. At the end, the temperature of the PCM will be higher as the heat transfer fluid transfers more heat to the PCM. Therefore, the fluid velocity at 0.15 m/s should be the optimum velocity for the thermal energy storage tank design.

On the other hand, the temperature of the PCM is observed to be less effected with the heat transfer fluid velocity. The effect is only visible from 0.01-0.035 m/s. Then if become constant from 0.035- 0.2m/s. This is in contrary to the report by Ogoh and Groulx [17]. However, the report by Ogoh and Groulx is based on turbulent modeling and with various fin spacing and the PCM is medium temperature heat storage material. Therefore, the model should be extended to the turbulence so that full picture of the simulation could be seen.

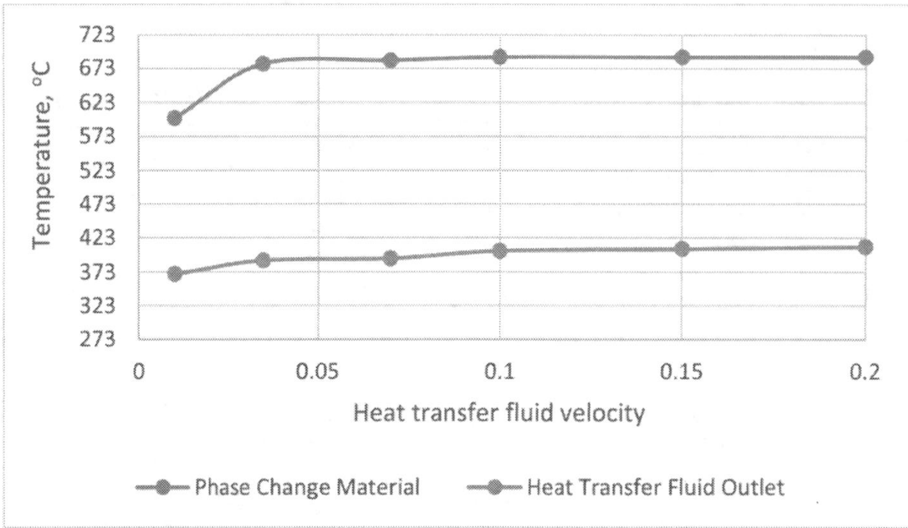

Fig. 3. Effect of heat transfer fluid velocity on the temperatures of the phase change material and the HTF outlet.

3.2 Effect of phase change material type

In the current study, the effect of PCM types on the melting rate during the heat charging of the thermal energy storage is investigated by taking sodium nitrate salt ($NaNO_3$) and potassium nitrate salt (KNO_3). These two PCMs are selected to study the effect of temperature for different fluid velocity in HTF. Different types of PCMs have their own physical and thermal properties (Table 2) and those properties decide the rate of temperature distribution during heating process.

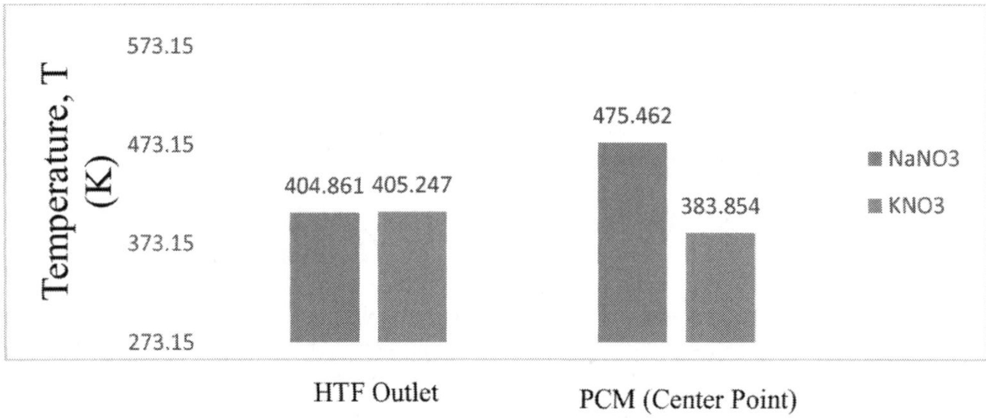

Fig. 4. Temperature of heat transfer fluid and PCM for different types of PCMs.

As shown in Figure 4, the temperature of heat transfer fluid outlet are about the same for both types of PCMs. However, if the analysis focuses on the center point of the PCM temperature, the temperature of the sodium nitrate salt was higher than potassium nitrate salt about 92°C. This result attributed to the latent heat of the sodium nitrate is higher than potassium nitrate salt. Thus, the sodium nitrate salt is better as compared to potassium nitrate salt to be used as phase change material in this thermal energy storage tank design.

 DOI: 10.1051/matecconf/20179001078

3.3 Melting phase rate of PCM

From the result of simulation, the melting rate of the PCM was investigated. At the room temperature the initial condition of the PCM was solid. Minimum HTF inlet velocity of the laminar flow that has been considered for modelling was the fluid velocity in this case. The simulations provided detailed phase distributions of PCM. The liquid fraction is labelled from the scale 0 (blue contour) to 1 (red contour). In which '0' indicates the solid state and '1'indicates fully liquid state.

The liquid fraction contour phase distribution is non-uniform in PCM due to very low thermal conductivity and poor heat distribution of HTF in the heat exchanger tube. Heat is distributed non-uniformly along the heat exchanger tube from the bottom (inlet) to the top (outlet). When the temperature at the inlet is higher than the outlet, then the PCM is likely to melt from the inlet level. That is the starting zone of melting process. The conduction of heat transfer is prevailed in the system from the beginning of the process until PCM starts to change its phase. After the PCM starts melting then both heat conduction from HTF and natural convection in the liquid PCM is act in the system. The more progress in melting process is caused by natural convection. So, as the convection process increases then rate of melting also increases.

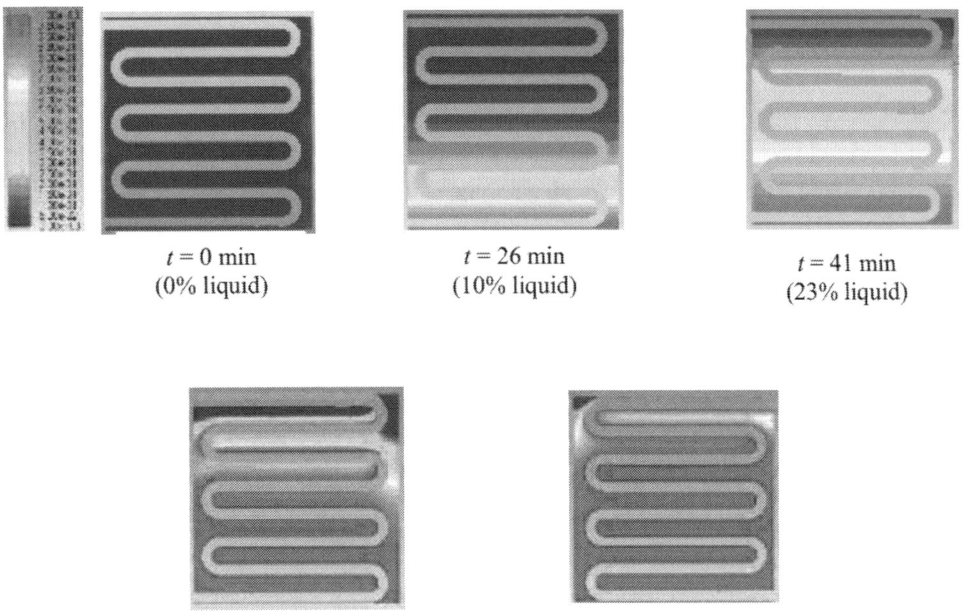

$t = 0$ min (0% liquid) $t = 26$ min (10% liquid) $t = 41$ min (23% liquid)

$t = 63$ min (50% liquid) $t = 75$ min (85% liquid)

Fig. 5. Liquid fraction of PCM melting phase distribution.

Figure 5 shows the variation of liquid fraction contour phase distribution with time for $NaNO_3$. The variation of liquid fraction indicates that the PCM is completely solid in the starting point of simulation that is t=0 min. Thermodynamic properties of the PCM showed variance with time in the simulation setup, hence the PCM starts phase change when having a high heat storage capacity. By comparing the melting phase of PCM with variance of time in Figure 5, it can be observed that the bottom part of PCM is melted easily and quickly but much time is required to melt the PCM that are close to the container wall. This may be attributed to the gravity effect. The contour is seems to be perpendicular to the heat exchanger tube wall due to heated moving fluid in it. At *t*=63 min, more than 50% of the PCM is melted. The PCM was melted about more than 85% when the time reached *t*=75 min, but unfortunately due to the limitation of this simulation, the complete melting phase of PCM seems to be difficult to achieve, Figure 5. Therefore, the final results obtained does not include complete melting state of PCM. Some results displayed in the simulation happens to differ and disturbed due to quality of mesh that can be set under limitation of the simulation software.

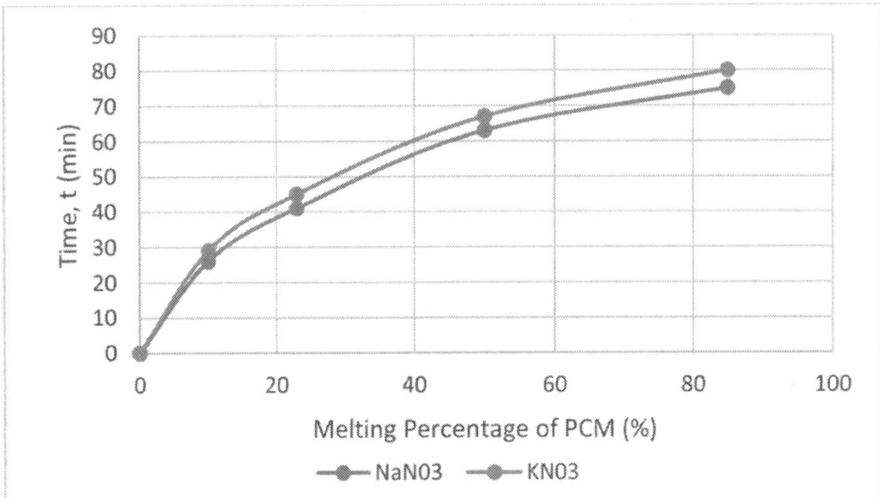

Fig. 6. Time verses PCM Melting rate for both $NaNO_3$ and KNO_3.

The comparison between the melting percentages of the PCMs under test show that sodium nitrate melts quite faster than potassium nitrate, Figure 6. This is because the melting temperature of potassium nitrate (336°C) is slightly higher than sodium nitrate (308°C). Sodium nitrate also has a higher latent heat energy compare to potassium nitrate that enable to absorb and store more heat energy. In fact, thermal conductivity of nitrate salt is very low, at the beginning of the simulation, however the melting process takes quite a long time but as the time increase, the PCM will be able to absorb and store more heat energy from the heat exchanger tube because of additional heat transfer from the melted layer of the PCM.

4 Conclusions

The PCM usage in medium range heat energy storage have a great potential in solar cooking applications. This work studied and analyzed the relationship between the HTF velocity, type of PCM and PCM heat loading process. Numerical analysis is done by using Ansys Fluent. The characteristics and applications of the PCMs are also included in this study. In this project PCMs are analyzed only during their melting process not in the solidification process (heat unloading process). This work concentrated in studying the time taken by the PCM to distribute temperature. Forced convection of air is included because the PCM is not completely filled up in the tank, thus the presence of air domain is considered in the application continuously. The maximum temperature reached, HTF velocity, melting fraction are computed in unsteady state transient analysis. PCMs of different range of temperatures can be tested in the same way for future usage in many solar applications like water heating etc.

References

1. M. Bayray, Mustefa, A., Yohannes, F., Kiros, H., Haileslasie, A., Gebray, P., Hayelom, M., Dagne, A.: Wind energy data analysis and resource mapping of geba catchment, north ethiopia. Wind Engineering **37**(4), 333-345 (2013).
2. A.M. Mahmud, Kahsay, M.B., Hailesilasie, A., Hagos, F.Y., Gebray, P., Kelele, H.K., Gebrehiwot, K., Bauer, H., Deckers, S., De Baerdemaeker, J.: Solar Energy Resource Assessment of the Geba Catchment, Energy Procedia **57**, 1266-1274 (2014).
3. D.A. Tesfamariam, Kahsay, M.B., Kahsay, M.T., Hagos, F.Y.: Modeling and Experiment of Solar Crop Dryer for Rural Application. Journal of Chemical and Pharmaceutical Sciences (2015).
4. A. Tesfay, Hagos, F., Yohannes, K., Nydal, O., Kahsay, M.: Preparing side charging of PCM storage: theoretical and experimental investigation. In: IOP Conference Series: Materials Science and Engineering 2015, vol. 1, p. 012021. IOP Publishing
5. S.L. Nathan, and Daniel G. Nocera: Powering the planet: Chemical challenges in solar energy utilization. **103**(43), 15729–15735 (2006).
6. D. Chwieduk: Solar energy utilisation. Opto-electronics Review **12**(1), 13-20 (2004).
7. J.W. Tester: Sustainable Energy: Choosing Among Options. The MIT Press (2005).
8. V.T. Atul Sharma, C. R. Chen, and D. Buddhi: Review on thermal energy storage with phase change material and applications. . Renewable and Sustainable Energy Reviews **13**, 318–345 (2009). doi:10.1016/j.rser.2007.10.005
9. S.C.S. A Felix Regin, and J S Saini: Heat transfer characteristics of thermal energy storage system using PCM capsules: A review. Renewable and Sustainable Energy Reviews **12**, 2438–2458 (2007).
10. N.H. Francis Agyenim , Philip Eames , and Mervyn Smyth A review of materials, heat transfer and phase change problem formulation for latent heat thermal energy storage systems (LHTESS). (2009).
11. L. Cabeza, Roca, J., Noguees, M., Mehling, H., Hiebler, S.: Long term immersion corrosion tests on metal-PCM pairs used for latent heat storage in the 24 to 29° C temperature range. Materials and corrosion **56**(1), 33-39 (2005).
12. L. Cabeza, Roca, J., Nogués, M., Mehling, H., Hiebler, S.: Immersion corrosion tests on metal-salt hydrate pairs used for latent heat storage in the 48 to 58°C temperature range. Materials and Corrosion **53**(12), 902-907 (2002).
13. L.F. Cabeza, Castell, A., Barreneche, C., Gracia, A.d., Fernández, A.I.: Materials used as PCM in thermal energy storage in buildings: A review. (2011).
14. L.F. Cabeza, Illa, J., Roca, J., Badia, F., Mehling, H., Hiebler, S., Ziegler, F.: Immersion corrosion tests on metal-salt hydrate pairs used for latent heat storage in the 32 to 36°C temperature range. Materials and Corrosion **52**, 140-146 (2001).
15. L.F. Cabeza, Svensson, G., Hiebler, S., Mehling, H.: Thermal performance of sodium acetate trihydrate thickened with different materials as phase change energy storage material. Applied Thermal Engineering **23**(13), 1697-1704 (2003).
16. A. Hesaraki: CFD modeling of heat charging process in a direct-contact container for mobilized thermal energy storage. MALARDALEN UNIVERSITY SWEDEN (2011)
17. W. Ogoh, Groulx, D.: Effects of the heat transfer fluid velocity on the storage characteristics of a cylindrical latent heat energy storage system: a numerical study. Heat Mass Transfer **48**(3), 439-449 (2012). doi:10.1007/s00231-011-0888-3
18. S.P. Jesumathy, Udayakumar, M., Suresh, S.: Heat transfer characteristics in latent heat storage system using paraffin wax. Journal of mechanical science and technology **26**(3), 959-965 (2012).
19. U. Ullmann, Tamme, R., Bauer, T., Hahne, E.: Heat storage media. (2013)
20. Radco: Solar Heat transfer Fluids: XCELTHERM MK1 SOLAR. In: Radco (ed.). (2014).

Study of fuel inlet temperature variations on palm biodiesel operating with a diesel engine

Mohd Hafizil Mat Yasin[1,*], *Rizalman* Mamat[2], *Ahmad Fitri* Yusop[2], *Mohd Hafiz* Ali[1] *and Mohamad Izuan* Izzudin[2]

[1]Department of Mechanical Engineering, Politeknik Kota Kinabalu, 88450 Kota Kinabalu, Sabah, Malaysia
[2]Faculty of Mechanical Engineering, Universiti Malaysia Pahang 26600 Pekan Pahang, Malaysia

Abstract. Biofuel includes biodiesel is introduced to overcome the air pollution problems. Biodiesel is the sustainable and alternative fuel that derived from edible and non-edible oil produced from bio-originated species, plants, and animals. This paper presents the simulation analysis on the effect of fuel temperature variations on diesel engine performance and emission using palm-biodiesel. A one-dimensional numerical analysis, Diesel RK software is used to simulate a single cylinder, four-stroke diesel engine on the performance and emission when operated at a full load condition using fuel temperature variations ranging 300 K to 333 K. Results showed that simulated results for brake power and brake torque were 0.7% each when compared to the highest and lowest fuel temperature ranging from 333 K to 300 K. Moreover, higher fuel temperature for palm biodiesel tends to produce lower exhaust gas temperature and brake specific fuel consumption at a constant engine speed of 2400 rpm. It can be concluded that from the study, fuel temperature variations of biodiesel could produce a significant effect regarding engine performance and emission that influence the driving economy of the engine.

1 Introduction

Biodiesel is a substitute fuel that derived from edible and non-edible oils of bio-originated feedstocks through the transesterification process. Each fuel possesses its physicochemical properties that can be used to identify the fuel type. In general, biodiesel has comprised all types of fatty acid methyl esters (FAMEs) originated from different sustainable bio-feedstocks, chemically produced using the transesterification process with methanol as a chemical catalyst [1]. The transesterification process is the most credible method in biodiesel production which requires an alkaline liquid catalyst, usually sodium or potassium methoxide for the chemical reaction between triglycerides and alcohol. Biodiesel and glycerol formation during transesterification process is the result of the chemical reaction between an alcohol and fatty acids [2]. In general, most of the edible and non-edible oils from various bio-feedstocks are qualified to be transesterified into biodiesel fuel [3]. Biodiesel apparently has similar composition and characteristics include Cetane

* Corresponding author: hafizil@polikk.edu.my

number, energy content, phase change, density and viscosity as compared to fossilized fuel [4 - 5]. Moreover, edible oils include vegetable oils are practical to be used in common diesel engines without any functional problems when several methods are adopted in reducing their high density and viscosity including petrol-diesel, pyrolysis, micro-emulsification (co-solvent blending) and transesterification.

Biodiesel feedstocks comprise edible and non-edible oils from sustainable bio-originated feedstocks of plants and animals. Among vegetable oils that been recognized as among most global influential vegetable oils is palm oil. This vegetable oil has been considered the suitable candidate for replacing fossilized fuel due to its higher oil yield and easily accessible. Palm oil is harvested from the oil palm tree that planted through the tropical rainforest climate regions include Indonesia, Malaysia, West Africa and other countries. Previous investigations were conducted to study the different proportion of biodiesel blends with diesel in reducing density and viscosity when operating with diesel engines [6 – 8]. Diesel engines can be fuelled directly or blended with various diesel proportions with biodiesel on some modifications or no modification at all [9 – 12]. However, the use of chemical solvents and additional processes in reducing biodiesel's density and viscosity require more cost and equipment to be involved which should need to be considered for the benefit-cost analysis. One simple method to be introduced for reducing the density and viscosity of palm biodiesel is applying the heating element approach on the biodiesel which consistently been heated at the specific temperature before entering the engine fuel pump. Many research works have been conducted previously and are still continuing on the research to determine the influence of fuel temperature in reducing density and viscosity [13].

Furthermore, higher fuel temperature contributes to the higher injection pressure, producing shorter ignition delay in the combustion process that improves cylinder pressure as well as resulting higher energy content. In consequence, there are many experimental works have been carried out investigating the influence of fuel temperature variations of a diesel engine regarding engine performance, combustion and emission characteristics operating with biodiesel. However, despite experimental works that are not feasible in terms cost, time and manpower, there are some suggested approaches including computational simulation and modelling approaches. In the present, there are three categories of diesel engine simulation models that consist, zero-dimensional models, quasi-dimensional models, and multi-dimensional models. Besides modelling approaches, computational simulation approach that is using specific mathematical models could be implemented economically and efficiently for evaluating and optimizing the engine design regarding the effects of fuel on diesel engines.

Diesel-RK is a multi-zone, full cycle, one-dimensional engine simulation software which has been developed by Razleytsev, Andrey Kuleshov, and others at Bauman Moscow State Technical University and is still developed until today [14 - 15]. This software is designed for simulating and optimizing thermodynamic working processes of two, and four-stroke engines consist all types of air boosting. Interestingly, this simulation is capable of determining engine performance parameters, combustion and emission prediction and optimization. It predicts fuel consumption, piston bowl shape and fuel injection profile multi parameters for advancement with enhancement in multiple fuel injection, fuel spray location and design in different models of engines.

This paper presents the simulated data of palm biodiesel when fuelling a single cylinder diesel engine in Diesel RK software. The objective of the present work is to investigate the fuel temperature variability on engine performance and emission operating with palm biodiesel. Investigated parameters in the study are included brake power, brake torque, brake specific fuel consumption (bsfc) and exhaust gas temperature regarding on the inlet fuel temperature variations.

2 Methodology

Fuel chemical compositions mainly attribute to the properties of any fuel that influence the performance and emission characteristics of the engine. Palm biodiesel is a sustainable and feasible alternate fuel (a major candidate fuel in replacing mineral diesel) for diesel engines with a chemical formula of CH_3-OOR_n. It is produced from the palm oil through a transesterification process with the presence of alcohol as a catalyst. Palm biodiesel is a clean burning fuel that been regulated firmly under ASTM and EN. It has a higher viscosity and density when compared to mineral diesel, which attributes to lower atomization process. However, palm biodiesel has higher oxygen content and lower sulphur content contribute to the emission reduction. Moreover, higher fuel consumption of palm biodiesel is mainly attributed to the lower heating value (LHV) of palm biodiesel that exhibited an average value of 37.08 MJ/kg that is much lower compared to that mineral diesel. Improvement in combustion process could be highly achieved with the presence of higher oxygen content, higher stoichiometric of fuel/air ratio and higher C/H fraction ratio that leads to the reduction of black smoke and soot emissions. The high stoichiometric fuel/air ratio, higher oxygen content, and higher C/H ratio enhance the combustion process thus the soot and smoke emission reduced [2]. The thermo-physical fuel properties of palm biodiesel compared to the mineral diesel are listed in Table 1.

Table 1. Vapour fuel properties of diesel and biodiesel.

Vapour Fuel Properties	Diesel	Palm Biodiesel
Carbon Atom per Molecule	13.5	18.82*
Hydrogen Atom per Molecule	23.6	34.39*
Oxygen Atom per Molecule	0	2*
Nitrogen Atom per Molecule	0	0*
Density (kg/m^3)	830	852**
Lower Heating Value (J/kg)	4.32 × 107	4.61 × 107 **
Min. Valid Temperature (K)	200	100
Max. Valid Temperature (K)	1200	1200

In the present work, intake system, powertrain model, exhaust system, engine cylinder and valve train are simulated in a one-dimensional (1D) engine simulation software, Diesel-RK. The effect of the tested fuels, conventional diesel and palm biodiesel on the engine performance and emission characteristics were simulated. The application of a four-stroke, one-cylinder diesel engine in a one-dimensional simulation is simulated and shown in this paper. Table 1 lists the specification of the diesel engine and Figure 1 presents the computational model of the specific diesel engine. The simulation model used standard atmospheric pressure at 1 bar and the initial atmospheric temperature of 298 K (25 °C). Initial intake air composition is assumed to be fresh air, with O_2 and N_2 as the primary gases. Furthermore, the second segment of the simulated model includes the engine cylinders, fuel injection system, intake, and exhaust system.

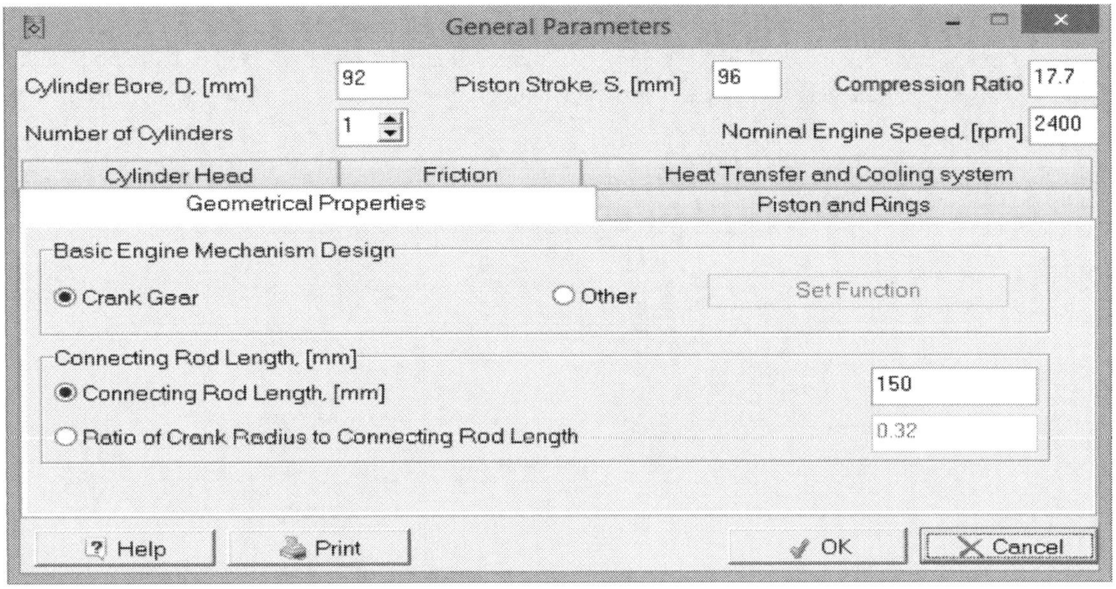

Fig. 1. A computational model of the engine [15].

Engine cylinder input parameters include the start of the cycle, cylinder geometry object, and initial state name. The start of the cycle of the engine can be defined as the crankshaft movement that represented by the crank angle and the intake valve closed. The crank angle for the engine crankshaft is 720 degree for four cycles. However, the crank angle value does not affected the simulation predictions but only specifies the starting and ending angle within one full cycle. As for the integrated and measured engine predictions, the dimensions of the bore, stroke, and connecting rod are referring to the actual engine which been mentioned in Table 2. Engine cylinder geometry data are included a bore, stroke, wrist pin to crank offset, compression ratio (CR), TDC clearance height and connecting rod length. The input for the engine simulation consists the cylinder quantity, type of cylinder arrangement and engine type. The details of the engine parameters used for the engine simulation are described in Table 2.

Table 2. Diesel engine specifications.

Parameter	Value
Type	Horizontal, water-cooled 4-cycle diesel
Combustion system	Direct injection
Number of Cylinder	1
Bore (mm)	92
Stroke (mm)	96
Compression ratio	17.7
Displacement (ℓ)	0.638
Connecting Rod Length (mm)	150
Rated continuous output, hp / rpm (kW)	10.5 / 2400 (7.8)
Maximum torque, kgf.m / rpm	4.42 /1800
Injection timing, deg	bTDC 17.0
Injection pressure, kg/cm^2	200
Injection hole diameter x number of holes	0.26 x 4
Injection angles	150°

Table 3 presents the engine test condition that described the overall condition for the engine operation. There were engine parameters that used to characterize engine operation which include engine speed, load, fuel temperature variations and ambient temperature. These parameters

Table 3. Test condition.

Test Parameter	Details
Engine speed, n	2400 rpm
Load	100% at 200 cycles
Fuel temperature, Tf	27°C (300 K)
	35°C (308 K)
	40°C (313 K)
	50°C (323 K)
	60°C (333 K)
Naturally aspirated air temperature, Tba	30°C (303 K)

3 Results and discussion

Engine parameters and operation were analysed to provide detailed information on engine characteristics when operated at different fuel temperature variations using palm biodiesel. The engine was simulated at different fuel temperature from 300 K to 333 K at a constant engine speed of 2400 rpm. Engine performance variations will be assessed based on the brake power, engine brake torque, exhaust gas temperature and brake specific fuel consumption (bsfc).

Figure 2 presents the effect of fuel temperature variations on engine brake power and torque variations operating with palm biodiesel and mineral diesel at a constant engine speed of 2400 rpm. Brake power is calculated when the power absorption device such as dynamometer is attached to the engine drive shaft. Higher fuel temperatures tend to produce higher injection pressure that causes lower ignition delay that resulted in an increase in brake power [16]. Shorter ignition delay causes a premature start of combustion (SOC). It is shown in the figure that palm biodiesel at higher fuel temperature (333 K) tends to produce higher brake power with 8.18 kW when compared to palm biodiesel with the lowest fuel temperature (300 K) which is 8.12 kW. It is found that the difference brake power for highest and lowest fuel temperature is 0.7% at a constant engine speed of 2400 rpm. It is well understood that fuel heating value affects the engine power [17], [18]. It can be described as when the fuel temperature is decreased; the energy level is also decreased.

Engine torque for fuel temperature variations with engine speed is presented in Figure 2(b). It is found from the figure that the maximum reduction in engine torque to be achieved was 0.7.% at the engine speed of 2400 rpm from highest to the lowest fuel temperature of palm biodiesel. Torque increases as the fuel temperature increase at a constant engine speed. This condition occurs due to the engine unavailability to consume a full charge of air at higher engine speed. Moreover, higher fuel temperature tends to produce higher fuel injection pressure at full atomization, as the fuel density is decreased.

As a result, similar engine torques tends to be achieved when operating with two different fuels with same injection pressure.

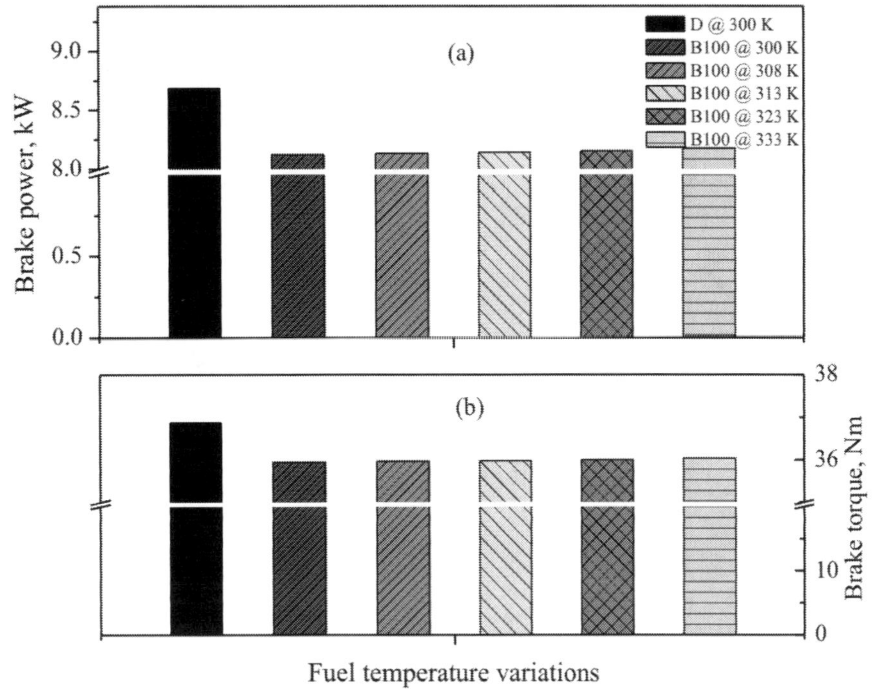

Fig. 2. (a) Engine brake power and (b) torque variations corresponding to the fuel temperature variations.

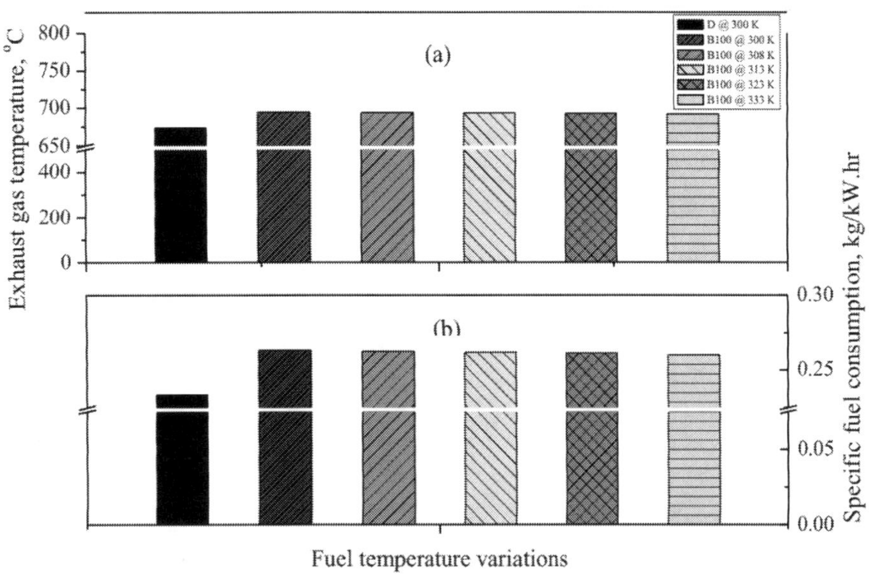

Fig. 3. Effect of fuel temperature variations on (a) exhaust gas temperature and (b) specific fuel consumption (sfc).

Figure 4 shows the effects of fuel temperature variations on exhaust gas temperature and specific fuel consumption at a constant engine speed of 2400 rpm. In this Figure 3(a), it depicts variations of exhaust gas temperature that attributed from different fuel temperatures of palm biodiesel with diesel as a reference fuel. It is found from the study that palm biodiesel operating with lower temperature (300 K) had produced higher exhaust gas temperature at 694.6 K than palm biodiesel with higher temperature (333 K) at 691.9 K. However, as for diesel, it produces lower exhaust gas temperature than palm biodiesel operating with different fuel temperatures. Nidal et. al. [10] and Selman et. al. [9] obtained similar results with this study.

While as for Figure 3(b), this figure shows the brake specific fuel consumption (bsfc) vary with the fuel temperature variations that corresponding to the constant engine speed. It is found that the resemblance trends for the fuel temperature variations occurred for palm biodiesel. Minimum bsfc (0.249 kg/kW-hr) for palm biodiesel was obtained at the maximum temperature of 333 K, while the maximum bsfc (0.251 g/kW-hr) was determined at the minimum temperature of 300 K. Lower energy content in the fuel contributes to the higher bsfc in the engine. Results clearly show that as the fuel temperature increases, the energy content as well increases, contributing lowest bsfc for the fuel temperature at 333 K when compared to the minimum temperature at 300 K. Lower specific fuel consumption with better performance in torque and power are obviously most desirable with lesser emission [13].

4 Conclusions

The study is conducted on the effect of fuel temperature variations corresponding to the constant engine speed of 2400 rpm operating with palm biodiesel on a single cylinder diesel engine. Following conclusions can be summarised as:

i. Higher fuel temperature contributes to the higher injection pressure, producing shorter ignition delay in the combustion process.
ii. Shorter ignition delay causes the earlier start of combustion (SOC), improving brake power and engine torque for the engine operation at a constant engine speed
iii. An increase in fuel temperature resulting higher energy content, contributing lower bsfc which is most obviously desired by any engine manufacturer

Authors are thankful to Universiti Malaysia Pahang for providing the financial support under Universiti Grant (RDU160309) for pursuing this research work at Automotive Engineering Center (AEC).

References

1. P. Adewale, M.-J. Dumont, and M. Ngadi, "Recent trends of biodiesel production from animal fat wastes and associated production techniques," *Renew. Sustain. Energy Rev.*, **45**, 574–588 (2015)
2. N. Pragya, K. K. Pandey, and P. K. Sahoo, "A review on harvesting, oil extraction and biofuels production technologies from microalgae," *Renew. Sustain. Energy Rev.*, **24**, 159–171 (2013)
3. M. M. K. Bhuiya, M. G. Rasul, M. M. K. Khan, N. Ashwath, and A. K. Azad, "Prospects of 2nd generation biodiesel as a sustainable fuel—Part: 1 selection of feedstocks, oil extraction techniques and conversion technologies," *Renew. Sustain. Energy Rev.*, **55**, 1109–1128 (2015)
4. M. H. Mat Yasin, T. Yusaf, R. Mamat, and A. Fitri Yusop, "Characterization of a diesel engine operating with a small proportion of methanol as a fuel additive in biodiesel blend," *Appl. Energy*, **114**, 865–873.(2014)

5. S. Moka, M. Pande, M. Rani, R. Gakhar, M. Sharma, J. Rani, and A. N. Bhaskarwar, "Alternative fuels: An overview of current trends and scope for future," *Renew. Sustain. Energy Rev.*, **32**, 697–712 (2014)
6. S. Lahane and K. A. Subramanian, "Effect of different percentages of biodiesel–diesel blends on injection, spray, combustion, performance, and emission characteristics of a diesel engine," *Fuel*, **139**, 537–545 (2015)
7. M. Gülüm and A. Bilgin, "Density, flash point and heating value variations of corn oil biodiesel–diesel fuel blends," *Fuel Process. Technol.*, **134**, 456–464 (2015)
8. A. Broatch, B. Tormos, P. Olmeda, and R. Novella, "Impact of biodiesel fuel on cold starting of automotive direct injection diesel engines," *Energy*, **73**, 653–660 (2014)
9. S. Aydin, C. Sayin, and H. Aydin, "Investigation of the usability of biodiesel obtained from residual frying oil in a diesel engine with thermal barrier coating," *Appl. Therm. Eng.*, **80**, 212–219 (2015)
10. N. H. Abu-Hamdeh and K. A. Alnefaie, "A comparative study of almond and palm oils as two bio-diesel fuels for diesel engine in terms of emissions and performance," *Fuel*, **150**, 318–324 (2015)
11. L. Yu, Y. Ge, J. Tan, C. He, X. Wang, H. Liu, W. Zhao, J. Guo, G. Fu, X. Feng, and X. Wang, "Experimental investigation of the impact of biodiesel on the combustion and emission characteristics of a heavy duty diesel engine at various altitudes," *Fuel*, **115**, 220–226 (2014)
12. C. C. Barrios, A. Domínguez-Sáez, C. Martín, and P. Álvarez, "Effects of animal fat based biodiesel on a TDI diesel engine performance, combustion characteristics and particle number and size distribution emissions," *Fuel*, **117**, 618–623 (2014)
13. R. Rahim, R. Mamat, M. Y. Taib, and A. A. Abdullah, "Influence of Fuel Temperature on Diesel Engine Performance operating with Biodiesel Blend," *J. Mech. Eng. Sci.*, **2**, June 2012, 226–236 (2012)
14. T. Janhunen, "HCCI-Combustion in the Z Engine," *SAE Tech. Pap. 2012-01-1132*, pp. 1–8 (2012)
15. A. Kuleshov, "Multi-Zone DI Diesel Spray Combustion Model for Thermodynamic Simulation of Engine with PCCI and High EGR Level," *SAE Int. J. Engines*, **2**, 1, 1811–1834 (2009)
16. A. M. Liaquat, H. H. Masjuki, M. a. Kalam, and I. M. Rizwanul Fattah, "Impact of biodiesel blend on injector deposit formation," *Energy*, **72**, 813–823 (2014)
17. I. K. Hong, J. R. Lee, and S. B. Lee, "Fuel properties of canola oil and lard biodiesel blends: Higher heating value, oxidative stability, and kinematic viscosity," *J. Ind. Eng. Chem.*, **22**, 335–340 (2015)
18. E. Sadeghinezhad, S. N. Kazi, A. Badarudin, C. S. Oon, M. N. M. Zubir, and M. Mehrali, "A comprehensive review of bio-diesel as alternative fuel for compression ignition engines," *Renew. Sustain. Energy Rev.*, **28**, 410–424 (2013)

Investigation of diesel-ethanol blended fuel properties with palm methyl ester as co-solvent and blends enhancer

Norhidayah Mat Taib[1], *Mohd Radzi* Abu Mansor[1,2,*], *Wan Mohd Faizal* Wan Mahmood[1,2], *Fais* Ahmad Shah[1], and *Nik Rosli* Nik Abdullah[3]

[1]Department of Mechanical and Materials Engineering, Faculty of Engineering and Built Environment, Universiti Kebangsaan Malaysia, 43600 UKM Bangi, Selangor, Malaysia
[2]Centre for Automotive Research, Universiti Kebangsaan Malaysia, 43600 UKM Bangi, Selangor, Malaysia
[3]Faculty of Mechanical Engineering, Universiti Teknologi MARA, 40450 Shah Alam, Selangor, Malaysia

Abstract. Diesel engine is known as the most efficient engine with high efficiency and power but always reported as high fuel emission. Malaysia National Automotive Policy (NAP) was targeting to improve competitive regional focusing on green technology development in reducing the emission of the engine. Therefore, ethanol was introduced to reduce the emission of the engine and while increasing its performance, Palm methyl ester was introduced as blend enhancer to improve engine performance and improve diesel-ethanol blends stability. This paper aimed to study the characteristics of the blends and to prove the ability of palm-methyl-ester as co-solvent in ethanol-diesel blends. Stability and thermophysical test were carried out for different fuel compositions. The stability of diesel-ethanol blended was proved to be improved with the addition of PME at the longer period and the stability of the blends changed depending on temperature and ethanol content. Density and viscosity of diesel-ethanol-PME blends also give higher result than diesel-ethanol blends and it's proved that PME is able to increase density and viscosity of blends. Besides, heating value of the blends also increases with the increasing PME in diesel-ethanol blends.

1 Introduction

The power of diesel engine in producing a better power output and efficiency especially to the heavy load transportation makes the diesel engine as one of the main focus of the researchers to innovate and develop the engine potential to a better performance and better emission. Diesel engine is a compression ignition engine that the power and efficiency of the engine are strongly depending on the compression ratio of the engine. However, diesel engine gives high carbon emission since diesel fuel was made up from crude oil that leads to high carbon emission and pollution if the incomplete combustion occurs during the

* Corresponding author: radzi@ukm.edu.my

combustion process. Along with the NAP (National Automotive policy) review in 2014, biofuels was selected as the main focus as new green technology initiative for the development of technology and human capital where the new goals of NAP 2014 are to develop a competitive and capable domestic automotive industry by pointed Malaysia as the regional automotive hub in Energy Efficient Vehicle (EEV) and to increase value-added activities in a sustainable way while continuously developing domestic capabilities [1].

Therefore, ethanol was chosen as effective biofuels to reduce the emission concentration from diesel combustion by contributing its potential due to its availability in large volume and second generation process in future [2]. Ethanol can be blended with diesel in the tank or injected into the cylinder to reduce the exhausted pollution [3]. Ethanol is easy to produce because it is produced from the fermentation of sugar such as sugar cane, grass, corn, barley sugar, etc [4]. Besides, it is capable of reducing the harmful exhaust emission such as CO, CO2, SOx, smoke opacity and particulate matter [5]. However, ethanol is not miscible in diesel fuel due to their physical properties difference and ethanol contains high oxygen content that leads to high NOx emission [6][7] and knocks problem. Besides, ethanol has extremely low cetane number [8] that increase the ignition time delay which leads to knock problem. All ethanol thermo-physical properties are explained in Table 1. Therefore, an emulsifier or co-solvent are needed to maintain the stability of diesel ethanol blend [2] and to act as fuel enhancer to improve engine performance and reducing NO emission. Many researchers have studied about the performance and emission of the diesel ethanol blend. They found out that the brake thermal efficiency and fuel consumption of blends may increase [5] and engine power will slightly reduce as well as the emission. Brake thermal efficiency are technically related to the heating value of the blends. Although the heating value of ethanol blends usually very low, the BTE of the engine may not always be increased depending on the brake power of the engine. Therefore, as the percentage of ethanol in diesel increase, thermal efficiency also may decrease [9].

Stability of diesel ethanol blends is strongly dependent on a few factors such as temperature [10], water content of ethanol, aromatic hydrocarbon, paraffin contents in diesel [3]. Stability problem of diesel ethanol blend may lead to the very dangerous situation to engine event to the most catastrophic consequences [11]. This is because ethanol is known as good water absorber and even a small amount of ethanol was added to diesel, the ethanol will pick up and absorb the water and when the saturation point of the blends are reached, two or three distinct layers are formed. If the pickup tube draws the fuel at water layer, the engine will experience thermal shock or hydro lock problem [10].

Palm Methyl Ester (PME) is an example of fatty acid methyl ester where the physical properties of fatty acid are closer to diesel fuel. PME are made of palm oil by esterification process. In esterification process, a glyceride reacts with an alcohol with the presence of a catalyst to form a fatty acid ester with alcohol. Malaysia was known as the world largest palm oil manufacturer and largest exporter which producing almost 15.8 million tons per year[12] and palm oil production process was the cheapest process among another biodiesel source. PME has the ability as fuel enhancer especially for diesel ethanol blends since it has very good properties as shown in table 1 with high cetane number that balancing the cetane number reduction when ethanol was added to diesel fuel. Cetane number is a very important parameter in the engine where the higher cetane number gives better engine efficiency with lower ignition delay in the engine. Besides, PME has low sulphur content that able to reduce Sox that lead to the poisonous environment. From the emission view, all biofuels such as PME and ethanol produces lower emission of CO, CO2, SOx, PM and HC compared to diesel. However, the oxygen content in biofuels gives some effect on NOx emission because oxygen contents in biofuels lead the combustion into lean combustion and at a very high-temperature, nitrogen gas may split to produce NOx. When discussing the diesel ethanol stability, PME has the special ability as co-solvent or amphiphile (surface reactant agent) [2] where PME has the most stable HLB (Hydrophile-Lipophile Balance) value for w/o emulsions. PME was used as co-solvent to stabilize the blends at a different temperature.

Table 1. Properties of diesel, ethanol and PME [13].

	Diesel euro 2M	Ethanol	PME
Chemical formula	C_3-C_{25}	C_2H_5OH	C19H35O2
Density (kg/L) at 15°C	0.81-0.87	0.796	0.8754
Kinematic viscosity, (mm^2/s)	1.9-4.1	1.4	4.55
Cetane number	49	8	70
Cloud point, °C	19		15
Flashpoint, °C	60	55	170
Sulphur contents, (mg/kg)	500		2
Heating value, (MJ/kg)	43.4	28.8	36.76
Autoignition temperature (°C)	315	422.8	342

The blend stability test is a method to determine the most stable blend for a further method for fuel properties. The stability of the blend is an important parameter to predict the highest ethanol percentage in avoiding the separation process of the fuel. If the separation of fuel happened after the addition of PME, the blend was categorized as a non-stable blend. The properties of the blends such as viscosity and heating value always related to engine efficiency. This fuel properties research was conducted to find the best blend by using the relation of viscosity, density, and heating value to the fuel consumption and engine performance. Viscosity played a very important role in combustion quality where it is strongly related to the injector in the engine. The viscosity of the fuel in the regular diesel engine must obey the injector regulation for a diesel engine with a standard viscosity of diesel fuel. The viscosity of fuel must be checked and have to maintain to the closest limit of diesel fuel viscosity which is around 1.7-4.1 cSt at 40°C [14]. Controlling the viscosity of fuel helps to provide a proper atomization of fuel during combustion [15] which leads to complete combustion and help the emission control due to incomplete combustion[16]. The density and viscosity of diesel-ethanol blends were said to decrease as the ethanol contents increase which proved by De Gang Li [17]as shown in Figure 1.

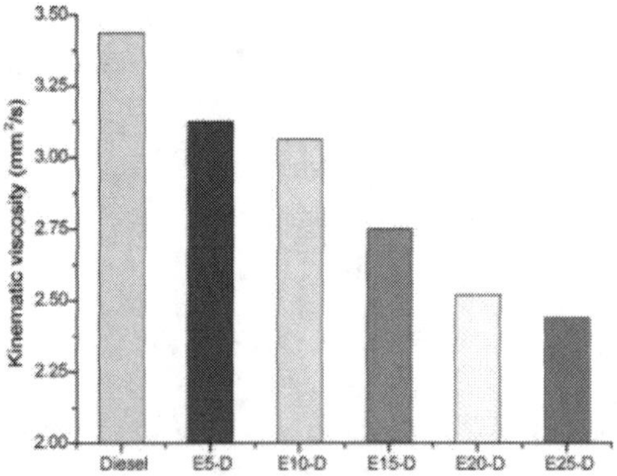

Fig. 1. Kinematic viscosity of blend fuels and diesel fuel [17].

Lower heating value is an another important properties in fuel to produce a better engine performance and engine emission. Fuel conversion efficiency or brake thermal efficiency of the engine is the ratio of brake power to the fuel power where the fuel power are related to the lower heating value [18]. The relation was explained in eq. (1) below where fuel power is inversely proportional to BTE.

$$\text{Brake thermal efficiency (BTE)} = \frac{\text{Brake power}}{\text{fuel power}} \quad (1)$$

Fuel power is related to the lower heating value which shown in eq. (2) where the product of mass flow rate of fuel, m_f and low heating value, LHV are indicated the fuel power and it is shown that the heating value of fuel is directly proportional to fuel power.

$$Fuel\,power = m_f * LHV \quad (2)$$

Ethanol and biodiesel release very low heating in combustion due to its high oxygen content [13]. From the relation shown in eq. 2, the lower heating value of the diesel ethanol blends may reduce the fuel power and increase the brake thermal efficiency. Fuel power is the power returned if the energy released from burning fuel in the engine that converted into mechanical power that produces through the shaft rotation. Thermal efficiency of the engine usually was increased when the engine speed increase up to the maximum value and decreased at some points due to momentum loss of the engine motion when the piston moves too fast [19]. Therefore, all the properties studies are very important in combustion engine for better emission and performance. In accordance with this importance, this research was focused on studying the properties of the diesel-ethanol-PME blends which including stability, density, viscosity and heating value. This research also aimed to proved that the stability of diesel-ethanol blends can be improved by using PME as co-solvent and blends enhancer.

2 Experimental setup

2.1 Materials

Diesel euro 2M was used to study the blend properties with hydrous alcohol (Ethanol 95 %) with a different percentage. The fuel percentage of diesel ethanol blends are shown in table 2. The blends were placed in the flasks with a maximum volume of 100 ml for each blend. 10 ml of PME was added into each diesel ethanol blend to study the stability of diesel ethanol PME blends

Table 2. Diesel-ethanol blends percentage for blend stability.

Composition	Diesel (%)	Ethanol (%)
D100	100	0
E10	90	10
E20	80	20
E30	70	30
E40	60	40
E50	50	50
E60	40	60
E70	30	70
E80	20	80
E90	10	90

Besides the blend stability test, a few blend properties were handled to obtain the thermal physical properties such as density, viscosity, and lower heating value. All these parameters are the requirement for the simulation analysis. The fuel percentage of the blends was explained in table 3.

Table 3. Diesel-ethanol blends percentage for density, viscosity and heating value test.

Composition	Diesel (%)	Ethanol (%)	PME (%)
E40B10	50	40	10
E35B15	50	35	15
E30B20	50	30	20
E25B25	50	25	25
E20B30	50	20	30
E15B35	50	15	35
E10B40	50	10	40
E5B45	50	5	45

2.2 Fuel properties test

2.2.1 Blend stability

Diesel-ethanol blend is well known as immiscible blends that may affect the engine condition. Diesel ethanol blend was mixed with different percentages to study the separation effects and time of separation of the blends. Besides, the stability of the blends was studied at a different temperature which is at 27°C and 16°C. Based on the theory, stability of the diesel ethanol blends is strongly dependent on temperature. The stability of blends was observed during the day up to 3 times and the time separation was recorded.

2.2.2 Density

The density of the blends was measured by using a hydrometer Kittiwake onboard test kit. The density was measured at two different temperature of 16°C and 27°C with the relation of density as explained in eq. (3). Density is the ratio of the mass to the volume of the fuel determined in ASTM D4052. After the fuel was mixed, the fuel was filled in a 100ml beaker and the mass of fuel was measured. The mass of a 100ml glass beaker was 19.8737g. The average of mass and volume were obtained after repeating the measurement for three times.

$$\rho = \frac{m}{V} \tag{3}$$

2.2.3 Viscosity

The viscosity is a measure of the resistance of the fuel to the gradual deformation by shear stress or tensile stress or uses the concept of the thickness of fuel. Viscosity also is strongly related to fuel consumption in the combustion process. The viscosity of fuel is measured by immersing a stainless steel ball into an onboard viscometer where the gravitational principle was applied. The ball will travel from top to bottom of glass under gravity influence. The specification of viscometer used in this experiment was described in table 4.

Table 4. Specification of the onboard viscometer.

Specification	Diesel (%)
Brand	Kittiwake
Density of stainless steel,ρ (g/ml)	8.02
Viscosity constant, k	≈0.34

All this specification are used in finding the kinematic viscosity based on the traveling time of stainless steel ball by using the equation (4). Travelling time of the steel ball was recorded from the top to the bottom of the glass which marked with a red line and was repeated for 3 times for better accuracy.

$$\mu = t \times k \times (\rho - \rho_f) \quad (4)$$

2.2.4 Heating value

Lower heating value (LHV) is an important consideration when selecting a fuel because engine efficiency usually quoted based on lower heating value. Heating value is an important parameter in determining the fuel power in combustion process within fuel and air in the combustion chamber. Heating value is determined by using IKA C 200 Oxygen bomb calorimeter based on the standard test method ASTM D240. A small amount of fuel sample placed in sample cup was weighed and placed in sealed bomb. A sufficient amount of oxygen supplied to the sealed bomb at 30 bar. Then, the sealed bomb was placed in the container filled with bath water at 15 ℃. The temperature rise measured in the bomb chamber resulting the lower heating value or calorific value at the specific mass of fuel. Calorific value is defined as the energy released when a unit mass of fuel are completely burned in the air. Therefore, to achieve a certain amount of fuel power, a sufficient amount of fuel needed to provide maximum engine power.

3 Results and discussion

Diesel-ethanol PME blends have very special properties which change the behavior of the diesel fuel in diesel engine. Ethanol gives a very big impact on the diesel fuel because it has very low density, low viscosity, low heating value and high water content. Due to the high water and oxygen content in ethanol, the blend of diesel and ethanol at normal condition resulting in a poor combination. This is observed when the separation layer or known as lipid bilayer is formed in between the solution of diesel and ethanol after some period. When ethanol was added into diesel fuel from 10% up to 90%, the formation of lipid bilayer were formed only for E40 (40%), E50 (50%), E60 (60%) and E70 (70%). Before the separation layer was formed, the blend was in crystalline single phase condition. While E90, E80, E10, E20 and E30 formed in single clear phase from the beginning. This is because the diesel contents in E90 and E80 were too small to see the formation of the layer formed as well as ethanol contents in E10, E20, and E30. However, the separation layer for E30 was formed at a lower temperature. This proves that the blending stability is change depending on the temperature and ethanol content in diesel fuel. The separation result at different temperature was explained in triangle phase diagram as shown in figure 2 (a) and (b). After the addition of 10% of biodiesel, the blends were stable up to more than diesel ethanol blends.

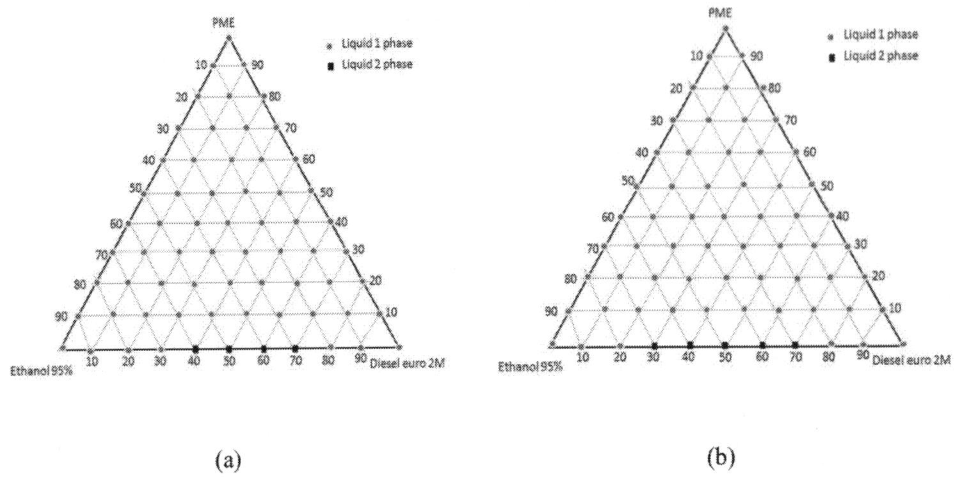

Fig. 2. Triangle phase diesel-ethanol PME blends at 27C (a) and 16C (b).

Since the density of ethanol is much lower than diesel fuel, the separation layer formed placed ethanol as an upper layer and diesel as a lower layer and separated by a lipid bilayer. The height of the lipid bilayer was measured and analyzed as in figure 3.

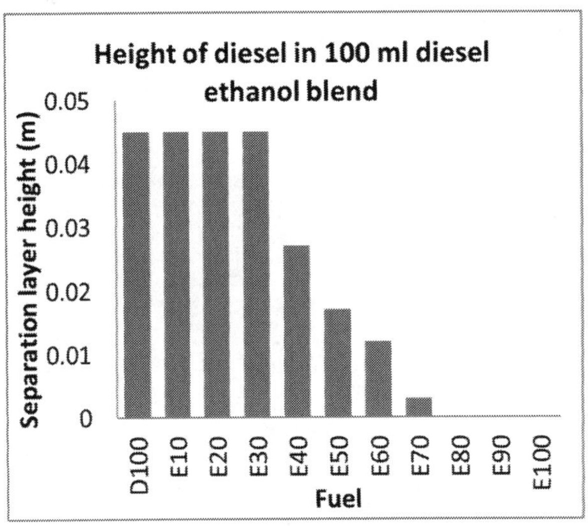

Fig. 3. Height of separation layer formed at 27 ℃.

The separation behavior is then explained with the separation time differences for the blends where the behavior of lipid bilayer may become the main reason of the change. Lipid bilayer is membrane layer that formed from two solution layer that contains hydrophobic head and hydrophilic tail and allows the molecule to pass through it and separate the fuel in meantime. From the graph shown in figure 4, the separation time of the blends are not uniform and always changed after a few experiments for separation were made. From the observation, E60 and E70 give a very unstable result where time separation of E60 and E70 are referred to the right axis and E340 and E50 are refer to left axis.The separation time behavior was observed for E30, E40, E50, E60 and E70 to study the separation layer behavior. This change happened due to the mechanical change applied to the blends. When a small force applied to the blends, the compact membrane layer molecule changes its structure and allow more diesel molecule to pass through the membrane caused the complete separation to occur. From this observation, the determination of separation time for optimization is not valid. Adding a co-solvent like PME gives the advantage for the blend to keep stable for a longer period where there is no separation occur after a longer period.

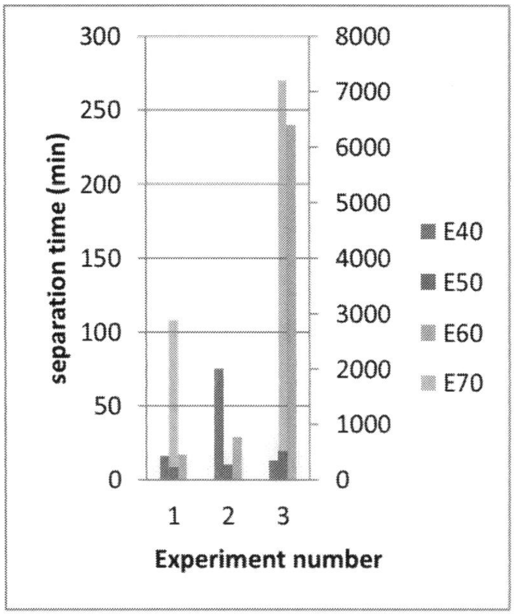

Fig. 4. Separation time of diesel-ethanol blends at 16℃.

Ethanol is a fuel that has a very low density and viscosity compare to diesel and biodiesel. From the graph shown in figure 6, the density of the blend decrease as the increasing of ethanol percentage and higher density will lead to higher viscosity as well [20]. This density of the blends are related to the viscosity which shown in figure 5 and from the equation of kinematic viscosity, the viscosity is strongly dependent on the density of the fuel where the density also decrease as the increasing of ethanol in blends. This proved that ethanol has lower density and viscosity. As the addition of PME in diesel ethanol blends, the density and the viscosity of the blends increased since the viscosity of PME is naturally higher than diesel and ethanol. In this blends, ethanol has the lowest density and viscosity. Therefore, the addition of PME in diesel was able to stabilize the density and viscosity of the blends to meet the standard regulation of blends' viscosity in a regular engine.

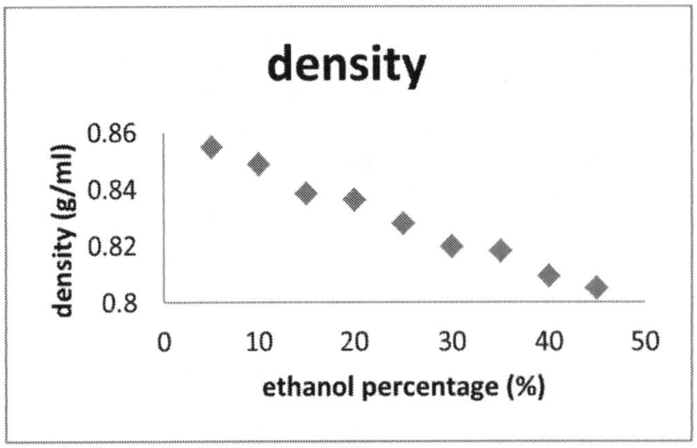

Fig. 5. Density of diesel-ethanol-PME blends at 27℃.

Viscosity plays an important role for engine performance and fuel consumption. In order to maintain the engine without any modification on engine cylinder and injector, the viscosity of the blends has to meet the diesel viscosity regulation which is around 1.7-4.1 cSt (1.7E-06 – 4.1E-06 m^2/s). From the graph in figure 6, the viscosity of the blends measured obeyed the viscosity regulation of diesel engine injector except for the blend with5%, 10% ethanol. This is because PME content in these blends is very high up to 40 and 45%. PME has a very high viscosity that affects the blends viscosity. The viscosity of diesel-ethanol-PME blends is increased around 10% when compared to the viscosity of diesel ethanol from [17] in figure1. The higher viscosity of fuel will increase the amount of fuel injected due to the delay of nozzle needle valve opening cause the reduction of energy conversion effectiveness. This is because the injector valve will need more fuel to be injected to meet the required heating value for the perfect piston motion.

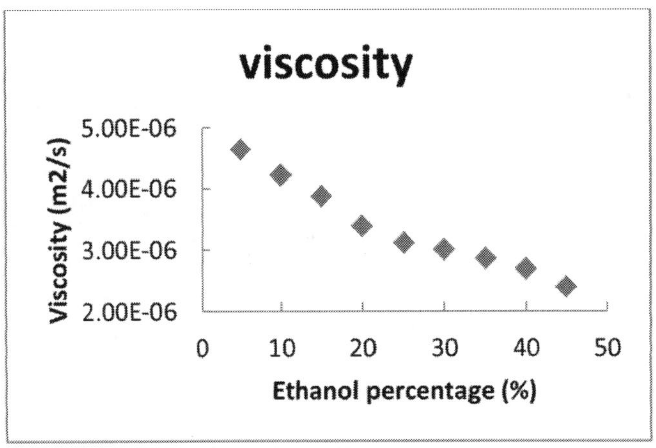

Fig. 6. Viscosity of diesel-ethanol-PME blends at 27℃.

Since biodiesel has a very low heating value, the heating value or calorific value of blends also decreased. Heating value is always related to engine performance and emission. From the graph shown in figure 7, the heating value of blends is decreased as the percentage of ethanol decrease and PME increase. This is because the water contents in biofuels are very high. The water content is vaporized during the combustion, taking up the heat produced from the combustion and lowering the heating value of fuel[6]. Since the heating values of the blends are decreased as the biofuels increased in blends, the brake thermal efficiency of the engine can be predicted to increase

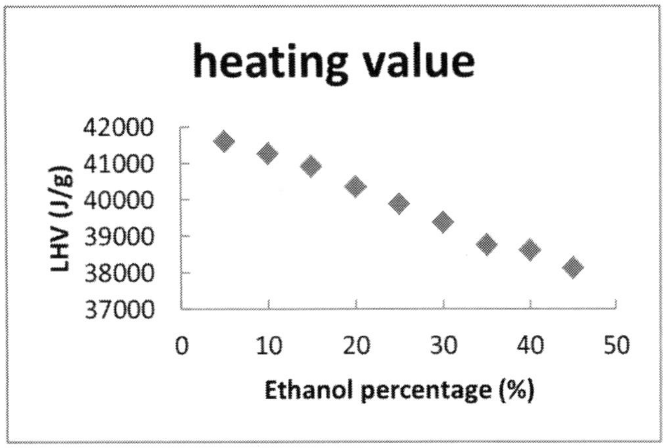

Fig. 7. Low heating value of diesel-ethanol-PME blends.

4 Conclusions

In this paper, the blends properties were observed and studied to find the best blends that regulate the standard engine requirement such as density, viscosity and low heating value of the blends. As the biofuels were used in diesel blends, the behaviors of the fuels are changed due to the biofuels properties. Diesel-ethanol blends are very unstable and formed a separation layer at some period and when PME was added into the blends, the stability of the blends are more stable up to more than two weeks. Based on the density and viscosity test, ethanol affects the density and viscosity of the blends since ethanol has lower density and viscosity. When more than 40% of PME added to diesel, the viscosity value was too high and beyond the standard viscosity regulation for standard engine injection. Therefore, the fuel blends with higher than 40% PME was not suitable to be used in the normal engine without any modification.

This research was supported by FRGS/2/2013/TK01/UKM/02/1 grant and special thanks to Universiti Teknologi Mara, UiTM Shah Alam in providing instruments for fuel properties experiment.

References

1. "National Automotive Policy (NAP) 2014," (2014).
2. L. Pidol, B. Lecointe, L. Starck, and N. Jeuland, "Ethanol-biodiesel-Diesel fuel blends: Performances and emissions in conventional Diesel and advanced Low Temperature Combustions," *Fuel*, **vol. 93**, no. x, pp. 329–338, (2012).
3. J. Lei, L. Shen, Y. Bi, and H. Chen, "A novel emulsifier for ethanol-diesel blends and its effect on performance and emissions of diesel engine," *Fuel*, **vol. 93**, no. 1, pp. 305–311, (2012).
4. S. A. Shahir, H. H. Masjuki, M. A. Kalam, A. Imran, I. M. R. Fattah, and A. Sanjid, "Feasibility of diesel–biodiesel–ethanol/bioethanol blend as existing CI engine fuel: An assessment of properties, material compatibility, safety and combustion," *Renew. Sustain. Energy Rev.*, **vol. 32,** pp. 379–395, (2014).
5. E. Torres-Jimenez, M. S. Jerman, A. Gregorc, I. Lisec, M. P. Dorado, and B. Kegl, "Physical and chemical properties of ethanol-diesel fuel blends," *Fuel*, **vol. 90**, no. 2, pp. 795–802, 2011.
6. S. S. Reham, H. H. Masjuki, M. A. Kalam, I. Shancita, I. M. Rizwanul Fattah, and A. M. Ruhul, "Study on stability, fuel properties, engine combustion, performance and

emission characteristics of biofuel emulsion," *Renew. Sustain. Energy Rev.*, **vol. 52**, pp. 1566–1579, (2015).
7. S. K. Hoekman and C. Robbins, "Review of the effects of biodiesel on NOx emissions," *Fuel Process. Technol.*, **vol. 96**, pp. 237–249, Apr. (2012).
8. A. Chotwichien, A. Luengnaruemitchai, and S. Jai-In, "Utilization of palm oil alkyl esters as an additive in ethanol–diesel and butanol–diesel blends," *Fuel*, **vol. 88**, no. 9, pp. 1618–1624, Sep. (2009).
9. C. Sayin, "Engine performance and exhaust gas emissions of methanol and ethanol-diesel blends," *Fuel*, **vol. 89**, no. 11, pp. 3410–3415, (2010).
10. "Phase separation in ethanol blended gasoline," *LCBA marketing group*, (2009). [Online]. Available: http://www.lcbamarketing.com/phase_separation_in_ethanol_blen.htm.
11. M. Lapuerta, O. Armas, and R. García-Contreras, "Stability of diesel–bioethanol blends for use in diesel engines," *Fuel*, **vol. 86**, no. 10–11, pp. 1351–1357, (2007).
12. D. Choo, Y. May, and P. C. Wei, "Outlook of Palm Biodiesel in Worldwide Biodiesel Development (1)," (2007).
13. S. D. B. S. Bhd, "Diesel biodiesel properties," (2016).
14. L. F. Ramírez-Verduzco, J. E. Rodríguez-Rodríguez, and A. D. R. Jaramillo-Jacob, "Predicting cetane number, kinematic viscosity, density and higher heating value of biodiesel from its fatty acid methyl ester composition," *Fuel*, **vol. 91**, no. 1, pp. 102–111, (2012).
15. W. A.J, *Diesel engine*, **3rd edition**. (1991).
16. B. B. Sahoo, N. Sahoo, and U. K. Saha, "Effect of engine parameters and type of gaseous fuel on the performance of dual-fuel gas diesel engines-A critical review," *Renew. Sustain. Energy Rev.*, **vol. 13**, no. 6–7, pp. 1151–1184, (2009).
17. G. Labeckas, S. Slavinskas, and M. Mažeika, "The effect of ethanol-diesel-biodiesel blends on combustion, performance and emissions of a direct injection diesel engine," *Energy Convers. Manag.*, **vol. 79**, pp. 698–720, (2014).
18. A. Alahmer, J. Yamin, A. Sakhrieh, and M. A. Hamdan, "Engine performance using emulsified diesel fuel," *Energy Convers. Manag.*, **vol. 51**, no. 8, pp. 1708–1713, Aug. (2010).
19. P. Kwanchareon, A. Luengnaruemitchai, and S. Jai-In, "Solubility of a diesel-biodiesel-ethanol blend, its fuel properties, and its emission characteristics from diesel engine," *Fuel*, **vol. 86**, no. 7–8, pp. 1053–1061, (2007).

AUTHOR INDEX

AUTHOR INDEX

AUTHOR INDEX